Brief Contents

EVOLUTION

SECOND EDITION

MARK RIDLEY
Departments of Anthropology and Biology
Emory University, Atlanta, Georgia

Blackwell
Science

Editorial offices:
238 Main Street, Cambridge, Massachusetts 02142 USA
Osney Mead, Oxford OX2 0El, England
25 John Street, London WC1N 2BL, England
23 Ainslie Place, Edinburgh EH3 6AJ, Scotland
54 University Street, Carlton, Victoria 3053, Australia
Arnette Blackwell SA, 224, Boulevard Saint Germain, 75007 Paris, France
Blackwell Wissenschafts-Verlag GmbH Kurfürstendamm 57, 10707 Berlin, Germany
Zehetnergasse 6, A-1140 Vienna, Austria

Distributors

USA

Blackwell Science, Inc.
238 Main Street
Cambridge, Massachusetts 02142
Telephone orders: 800-215-1000 or 617-876-7000
Fax orders: 617-492-5263

CANADA

Copp Clark, Ltd.
2775 Matheson Blvd. East
Mississauga, Ontario
Canada, L4W 4P7
Telephone orders: 800-263-4374 or 905-238-6074

AUSTRALIA

Blackwell Science Pty., Ltd.
54 University Street
Carlton, Victoria 3053
Telephone orders: 03-347-0300; Fax orders: 03-9349-3016

OUTSIDE NORTH AMERICA AND AUSTRALIA

Blackwell Science, Ltd.
c/o Marston Book Services, Ltd.
P.O. Box 2691
Abingdon
Oxon OX14 4YN
England
Telephone orders: 44-1865-791155

Acquisitions: Jane Humphreys
Development: Debra Lance and Jennifer Rosenblum
Production: Heather Garrison
Manufacturing: Lisa Flanagan
Cover: (design) Cia Boynton; (photo) Bruce Coleman Inc.
Typeset by Modern Graphics
Printed and bound by Quebecor

Printed in the United States of America
02 03 04 05 5 4

Library of Congress Cataloging-in-Publication Data
Ridley, Mark.
Evolution / Mark Ridley. —2nd ed.
p. cm.
Includes bibliographical references (p.) and index.
ISBN 0–86542–495–0 (case : alk. paper)
1. Evolution (Biology) I. Title.
QH366.2.R524 1996
575—dc20
96-18960

Contents

PART 4 Evolution and Diversity 369

Preface

The theory of evolution is by far the most important theory in biology, and it is always a pleasure to be a part of a class that is fortunate enough to study this subject. No other idea in biology is so scientifically powerful, or so intellectually stimulating. Evolution can add an extra dimension of interest to the most appealing sides of natural history — we will see, for example, how modern evolutionary biologists tend to argue that the existence of sex is the most profound puzzle of all, and quite possibly a mistake that half of all living creatures would be better off without.

Evolution gives meaning to the dryer facts of life as well. One of the delights of the subject involves recognizing ideas and facts within the disorienting technicalities of the genetics lab, and learning how deep theories about the history of life can hinge on measurements of the width of a region called "prodissoconch II" in the larval shell of a snail, or the number of ribs in a trilobite's tail. So great are the depth and range of evolutionary biology that students in every other classroom on campus must feel (as you can guess) trapped with more superficial and ephemeral materials.

The theory of evolution, as I have arranged it in *Evolution,* second edition, has four main components:

- ***Evolutionary genetics*** Population genetics provides the fundamental theory for evolution. If we know how any property of life is controlled genetically, population genetics can be directly applied to that property. We have such knowledge particularly for molecules (together with some, mainly morphological, properties of whole organisms); thus, molecular evolution and population genetics are therefore well integrated. Part 2 of the book considers the relationship that links these two subjects together.

- ***Adaptation and Natural Selection*** This second component of evolution is discussed in Part 3.

- ***Evolution and Diversity*** Evolution is also the key to understanding the diversity of life, and in Part 4 we consider such topics as the definition of species, the origin of new species, and classification and reconstruction of the history of life.

- ***Paleobiology and Macroevolution*** Finally, Part 5 considers evolution on the grand scale by examining the fossil record. This record is the main testing ground for these large evolutionary events, which occur over the geological time scale of tens or hundreds of millions of years.

Controversy is always tricky to deal with in any introductory text, and evolutionary biology has attracted more than its fair share of attention in this regard. When the textbook encounters a controversial topic, my first aim has been to explain the competing ideas in such a way that they can be understood on their own terms. In some cases (such as cladistic classification) I have then ventured to take sides; in other instances (such as the relative empirical importance of gradual and punctuated change in fossils) I have not taken a stand. Not everyone will agree with the positions I have taken, or indeed with my decisions in some cases not to take a position. Nevertheless, the book's success depends mainly on how well it enables a reader who has not studied evolution in depth to understand the various ideas and reach a sensible conclusion about them.

For the second edition of *Evolution*, I have added study and review questions at the end of the chapters. These questions are intended to enable readers working more or less by themselves to check whether they have understood the major points and acquired the main skills detailed in the chapter. A range of difficulties is presented in the exercises; I have indicated the few cases in which the question goes beyond the material presented in the chapter. The answers are found at the end of the book.

The book is intended as an introductory text, and I have subordinated all other aims to that end. I have attempted to explain concepts, wherever possible by example, and with a minimum of professional clutter. The principal interest, I believe, of evolutionary theory lies in regarding it as a set of ideas to ponder, and I have therefore tried in every case to introduce the ideas as soon as possible.

Evolution, second edition, is not a factual encyclopedia, nor (primarily) is it a reference work for research biologists. I have therefore sacrificed the usual convention, by which references are made through dates inserted in the text. Some references, it is true, can easily be made in this manner, but others, particularly the general references, could easily distract the reader from the most powerful discussions, such as those found at the end of a paragraph or a section. In this textbook, those textual positions are typically occupied by summary sentences and other matter more useful than directions to further reading.

Nonetheless, anyone who wishes to track down my sources will not have much difficulty. All the authors mentioned in the text are listed in the references at the end of the text, and in the handful of ambiguous cases I have given the date of the reference in the text. The Further Reading section at the end of each chapter is the main vehicle for references. In most cases, it should nearly always be possible to go directly to the appropriate source from the Further Reading section. I have referred to recent reviews when they exist, and the historical bibliography of each topic can be traced through those reviews.

I began writing the first edition of the book while I was a Research Fellow at St Catharine's College, Cambridge, and completed it after moving to Emory University; I have written the second edition while at Emory. I am most grateful to both institutions for the opportunities they provided. I am also grateful to Nick Barton, Michael Ghiselin, Alan Grafen, Eddie Holmes, Bruce MacFadden, Chris Sanford, Randy Skelton, Michael Turelli, John Turner, and Elizabeth Vrba for their error-reducing comments on various parts, large and small, of the book's first edition. My thanks, too, to the tolerant and eagle-eyed members of ANT 385, BIO 462, BIO 470, and ANT 503 at Emory, who cheerfully endured my chaotic manuscripts instead of a proper educational publication, and whose explicit (and implicit) experiences with the text have, more than anything else, inspired its final form.

For the second edition, I am further indebted to Michael Arnold, Diane Dodd, Richard Johnston, Alexey Kondrashov, Dolph Schluter, and Michael Turelli, all of whom kindly looked at bits, or more, of the revised manuscript; to the teachers who have pointed out confusing material, potentially misleading expositions, and simple arithmetic mistakes; and (again) to the students of ANT 362 and BIO 462 at Emory, who continue to inspire much of my writing.

Mark Ridley

CHAPTER 1

The Rise of Evolutionary Biology

THIS chapter first defines biological evolution, and contrasts it with some related but different concepts. It then discusses the rise of modern evolutionary biology from a historical perspective: we consider Darwin's main precursors; Darwin's own contribution; how Darwin's ideas were received; and the development of the modern "synthetic theory" of evolution.

1.1 Evolution means change in living things over long periods of time

Evolutionary biology encompasses a large body of science, one that grows ever larger. A list of its various subject areas could sound rather daunting. Evolutionary biologists now carry out research in some sciences, like molecular genetics, that are young and move rapidly, and in others like morphology and embryology that have accumulated their discoveries at a more stately speed over a much longer period. Evolutionary biologists work with materials as diverse as naked chemicals in test tubes, animal behavior in the jungle, and fossils collected from barren and inhospitable rocks.

A beautifully simple and easily understood idea—evolution by natural selection—can be scientifically tested in all of these fields. It is one of the most powerful ideas in all science, and is the only theory that can seriously claim to unify biology. It can give meaning to facts from the invisible world found in a drop of rainwater, or from the many colored delights of a botanic garden, or thundering herds of big game. The theory is also used to understand such topics as the geochemistry of life's origins and the gaseous proportions of the modern atmosphere. As Theodosius Dobzhansky, one of the twentieth century's most eminent evolutionary biologists, remarked in an often quoted but scarcely exaggerated phrase, "nothing in biology makes sense except in the light of evolution."

Evolution means change—change in the form and behavior of organisms between generations. The forms of organisms, at all levels from DNA sequences to macroscopic morphology and social behavior, can be modified from those of their ancestors during evolution. However, not all kinds of biological change are included in this definition (Figure 1.1). Developmental change within the life of an organism is not evolution in the strict sense, and the definition refers to evolution as "change between generations" to exclude developmental change. A change in the composition of an ecosystem, which is made up of a number of species, would also not normally be considered as evolution. Imagine, for example, an ecosystem containing 10 species. At time 1, the individuals of all 10 species are, on average, small in body size; the average member of the ecosystem is therefore "small." Several generations later, the ecosystem may still contain 10 species, but only five of the

Figure 1.1 Evolution refers to change within a lineage of populations between generations. (a) shows evolution in the strict sense of the word. Each line represents one individual organism, and the organisms in one generation are reproduced from the organisms in the previous generation. The composition of the population has changed, evolutionarily, through time. (b) Individual developmental change is not evolution in the strict sense. The composition of the population has not changed between generations and the developmental changes (from *a* to *a*') of each organism are not evolutionary. (c) Change in an ecosystem is not evolution in the strict sense. Each line represents one species. The average composition of the ecosystem changes through time: from 2 *a* : 1 *a*' at time 1 to 1 *a* : 2 *a*' at time 3. Within each species, however, no evolution occurs. The letters *a* and *a*' in all three figures stand for any pair of observable biological characters, such as small (*a*) and large (*a*') body size.

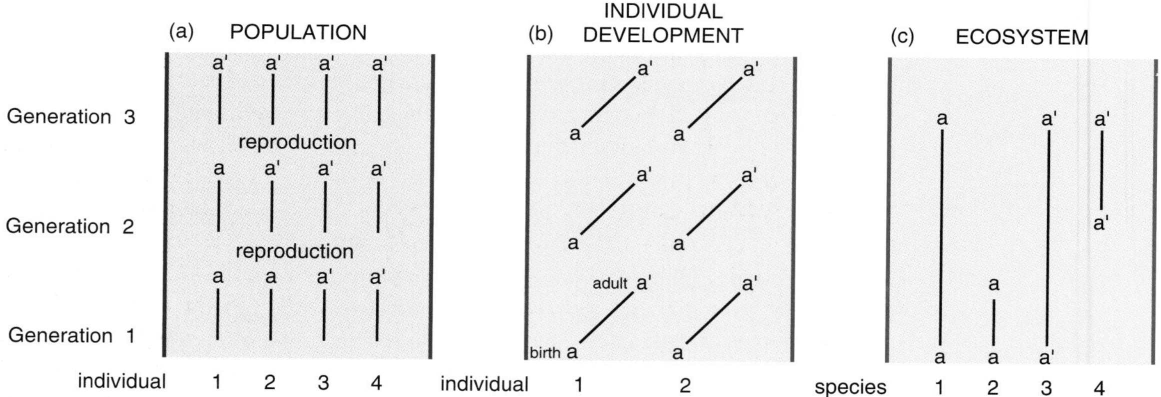

original small species remain; the other five have gone extinct and been replaced by five species, with large-size individuals, that have immigrated from elsewhere. The average size of an individual (or species) in the ecosystem has changed, even though no evolutionary change has occurred within any one species.

Most of the processes described in this book concern change between generations within a population of a species, and it is this kind of change we shall call evolution. When the members of a population breed and produce the next generation, we can imagine a *lineage* of populations, made up of a series of populations through time. Each population is ancestral to the descendant population in the next generation: a lineage is an "ancestor-descendant" series of populations. Evolution, then, consists of change between generations within a population lineage. Darwin defined evolution as "descent with modification," where the word "descent" refers to the way evolutionary modification takes place in a series of populations that are descended from one another.

Evolutionary modification in living things possesses some further distinctive properties. It depends on external environmental change and on random genetic innovation. At any moment, therefore, the form of future change is unpredictable, except conditionally. Moreover, the evolution of life has proceeded in a branching, tree-like pattern. The modern variety of species has been generated by the repeated splitting of lineages since the single common ancestor of all life.

The changes of politics, economics, human history, technology, and even scientific theories, are sometimes loosely described as "evolutionary." In this sense, "evolutionary" means mainly that they have changed through time, and perhaps not in a preordained direction. Human ideas and institutions can sometimes split during their history, but this division is not so characteristic and clear-cut a feature as it is in the history of life. Change and splitting provide two of the main themes of evolutionary theory.

1.2 *Living things show adaptations*

Adaptation is another of evolutionary theory's crucial concepts. Indeed, one of the main aims of modern evolutionary biology is to explain the forms of adaptation that we find in the living world. *Adaptation* refers to "design" in life—those properties of living things that enable them to survive and reproduce in nature.

The concept of adaptation becomes easier to understand with the help of an example. Many of the attributes of a living organism could be used to illustrate adaptation, because many details of the structure, metabolism, and behavior of an organism are well designed for life. Darwin's favorite example of this concept involved the woodpecker. Woodpeckers' most obvious adaptation is their powerful characteristically shaped beak. It enables woodpeckers to excavate holes in trees, allowing them to feed on the year-round food supply of insects that live under bark, insects that bore into the wood, and the sap of the tree itself. Tree-holes also make safe sites to build a nest. Woodpeckers have many other design features as well. Within the beak is a long, probing tongue, which is well suited to extracting insects from inside a tree-hole. They have a stiff tail that is used as a brace, short legs, and long curved toes for gripping onto the bark; woodpeckers even undergo a special type of molting in which the strong central pair of feathers (which are crucial in bracing) are saved and molted last. The beak and body design of the woodpecker are, therefore, adaptive. The woodpecker is more likely to survive, in its natural habitat, by possessing these attributes.

Camouflage is another, particularly clear, example of adaptation. Camouflaged species have color patterns and details of shape and behavior that make them less visible in their natural environment. Camouflage assists the organism to survive by making it less visible to its natural enemies and is, therefore, adaptive.

Adaptation is not an isolated concept referring to only a few special properties of living things. In humans, for example, it applies to almost any part of the body. Hands are adapted for grasping, eyes for seeing, the alimentary canal for digesting food, legs for movement; all these functions assist us to survive. Although most of the obvious things we notice are adaptive, not every detail of an organism's form and behavior may be adaptive (chapter 13). Adaptations are, however, so common that they must be explained. Darwin regarded adaptation as the key problem that any theory of evolution had to solve. In Darwin's theory—as in modern evolutionary biology—the problem is solved by natural selection.

Natural selection means that some kinds of individuals in a population tend to contribute more offspring to the next generation than do others. Provided that the offspring resemble their parents, any attribute of an organism causing it to leave more offspring than average will increase in frequency in the population over time; the composition of the population will then change automatically. Such is the simple, but immensely powerful, idea whose ramifying consequence we shall explore in this book.

1.3 *A short history of evolutionary biology*

We shall begin our discussion of evolution with a brief sketch of the historical rise of evolutionary biology, presented in four main stages:

1. Evolutionary, and non-evolutionary, ideas before Darwin
2. Darwin's theory (1859)
3. The eclipse of Darwin (c.1880–1920)
4. The modern synthesis (1920s–1950s)

1.3.1 *Evolution before Darwin*

The history of evolutionary biology really begins in 1859, with the publication of Charles Darwin's *On the Origin of Species,* although many of Darwin's ideas actually have an older pedigree. The most immediately controversial claim in Darwin's theory was that species have not been permanently fixed in form, but that one species has evolved into another. ("Fixed" here means unchanging.) Human ancestry, for instance, passes through a continuous series of forms leading back to a unicellular stage. Species fixity was the orthodox belief in Darwin's time, although others had questioned this idea before him. Naturalists and philosophers a century or two before Darwin had often speculated about the transformation of species. The French scientist Maupertuis discussed evolution, as did *encyclopédistes* such as Diderot; even Charles Darwin's grandfather Erasmus Darwin had questioned species fixity. However, none of these thinkers put forward anything we would recognize as a satisfactory theory to explain why species change. Instead, they were mainly interested in the factual possibility that one species might change into a second species.

The question was brought to an issue by the French naturalist Jean-Baptiste Lamarck (1744–1829). In his *Philosophie Zoologique* (1809), Lamarck argued that species change over time into new species. The way in which he envisioned the modification process was different from Darwin's and our modern idea of evolution in many respects, and historians prefer the contemporary word "transformism" to describe Lamarck's idea.[1]

Figure 1.2 illustrates Lamarck's conception of evolution, and how it differed from Darwin's and our modern concept. Lamarck supposed that lineages of species persisted indefinitely, changing from one form into another; lineages in his system did not branch and did not go extinct. Lamarck had a two-part explanation of why species change. The principal mechanism was an "internal force": some sort of unknown mechanism within an organism causing it to produce offspring slightly different from itself, such that when the changes had accumulated over many generations, the lineage would be visibly transformed, perhaps enough to be a new species.

1. The historical change in the meaning of the term "evolution" is a fascinating story in itself. Initially, it meant something like what we understand as "development" (e.g., growing up from an egg to an adult) rather than evolution: an unfolding of predictable forms in a preprogrammed order. The course of evolution, in the modern sense, is not preprogrammed; it is unpredictable in much the same way that human history is unpredictable. The change of meaning occurred around the time of Darwin; he did not use the word in *The Origin of Species* (1859), except in the form "evolved" which he used once as the last word in the book; but he did use it in *The Expression of the Emotions* (1872). It took a long time for the new meaning to gain widespread acceptance.

Figure 1.2 (a) Lamarckian "transformism," which differs in two crucial respects from (b) evolution as Darwin imagined it. Darwinian evolution is tree-like, as lineages split, and allows for extinction.

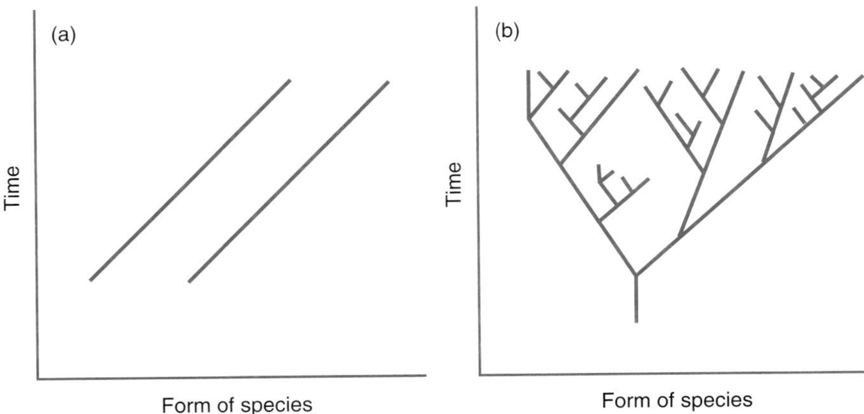

Lamarck's second (and possibly less important to him) mechanism is the one for which he is now remembered: the inheritance of acquired characters. Biologists use the word "character" as a shorthand for "characteristic": a *character* is any distinguishable property of an organism (not character in the sense of personality). As an organism develops, it acquires many individual characters, in this biological sense, due to its particular history of accidents, diseases, and muscular exercises. Lamarck suggested that a species could be transformed if these individually acquired modifications were inherited by the individual's offspring, and further modifications were added through time.

In his famous discussion of the giraffe's neck as evidence of inheritance of acquired characters, Lamarck argued that ancestral giraffes had stretched to reach leaves higher up in trees; the exertion caused their necks to grow slightly longer; their longer necks were inherited by their offspring, who thus started life with a propensity to grow even longer necks than their parents. Many generations of neck-stretching created the current form of the giraffe. Lamarck described the process as being driven by the "striving" of the giraffe, and he often described animals as "wishing" or "willing" to change themselves. His theory has sometimes been caricatured as suggesting that evolution happens by the will of the organism. However, the theory does not require any conscious striving on the part of the organism—only some flexibility in individual development and the inheritance of acquired characters.

Lamarck did not invent the idea of the inheritance of acquired characters. It is an ancient idea; Plato discussed it, for example. However, most modern thinking about the role of the process in evolution has been inspired by Lamarck, and the inheritance of acquired characters is now conventionally, if not entirely accurately, called Lamarckian inheritance.

Lamarck, as a person, lacked the genius for making friends, and his main rival, the anatomist Georges Cuvier (1769–1832), knew how to conduct a controversy. Lamarck had broad interests, in chemistry and meteorology as well as biology, but his contributions did not always receive the attention he felt they deserved. By 1809, Lamarck had already persuaded himself that there was a conspiracy of silence against his ideas. The meteorologists ignored his scheme of weather forecasts,[2] the chemists ignored his chemical system, and when the *Philosophie Zoologique* was finally published, Cuvier saw to it that this treatise was greeted with silence.

Actually, however, *Philosophie Zoologique* was an influential book. At least partly in reaction to Lamarck, Cuvier and his school made a belief in the fixity of species a virtual orthodoxy among professional biologists. Cuvier's school studied the anatomy of animals to discover the various fundamental, and unalterable, plans according to which the different types of organisms were designed, and Cuvier used these plans to establish that the animal kingdom had four main branches (called *embranchements* in French): vertebrates, articulates, molluscs, and radiates. Modern zoology recognizes a slightly different set of main groups, but the modern groupings do not radically contradict Cuvier's four-part system. Cuvier also established, contrary to Lamarck's belief, that species had gone extinct (section 23.1, p. 639).

Lamarck's ideas mainly became known in Britain through a critical discussion in Charles Lyell's *Principles of Geology* (1830–1833). Cuvier's influence was advanced through Richard Owen (1804–1892), who had studied with Cuvier in Paris before returning to England. Owen became generally regarded as Britain's leading anatomist, and by the first half of the nineteenth century most biologists and geologists had accepted Cuvier's view that each species had a separate origin and remained constant in form until it went extinct.

1.3.2 *Charles Darwin*

Meanwhile, Charles Darwin (Figure 1.3) was forming his own ideas. Darwin, after graduating from Cambridge, had traveled the world as a naturalist on board the *Beagle* (1832–1837). He then lived briefly in London before settling permanently in the country. His father was a successful doctor, and his father-in-law controlled the Wedgwood china business; Charles Darwin was a gentleman of independent means. The crucial period of his life, for our purposes, was the year or so after the *Beagle* voyage (1837–1838). As he worked over his collection of birds from the Galápagos Islands, he realized that he should have recorded which island from which each specimen came, because they varied from island to island. He had initially supposed that the Galápagos finches were all one species, but it became clear that each island had its own distinct species. How easy to imagine that they had evolved from a common ancestral finch! He was similarly struck by the way the ostrich-like birds called rheas varied among different places in South America; it was probably these observations of geographic variation that first led Darwin to accept the idea that species can change.

2. Lamarck's weather forecasts were notorious among his Parisian contemporaries. Each year, beginning in 1799, Lamarck published an *Annuaire Météorologique* that predicted (with scientific certainty) the weather for the next year. Unfortunately, some unforeseen accident always interfered, the weather turned out wrong, and it was necessary to modify the system in the next *Annuaire*. When Lamarck offered Napoleon a copy of the *Philosophie Zoologique*, the Emperor turned it down, supposing it was Lamarck's latest volume of weather forecasts.

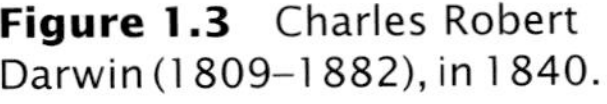
Figure 1.3 Charles Robert Darwin (1809–1882), in 1840.

The next important step was to invent a theory that explained why species change. The notebooks Darwin kept at the time still survive. They reveal how he struggled with several ideas, including Lamarckism, but rejected them all because they failed to explain a crucial fact: adaptation. His theory would have to explain not only why species change, but also why they are well designed for life. In Darwin's own words (in his autobiography):

> It was equally evident that neither the action of the surrounding conditions, nor the will of the organisms [an allusion to Lamarck], could account for the innumerable cases in which organisms of every kind are beautifully adapted to their habits of life— for instance, a woodpecker or tree-frog to climb trees, or a seed for dispersal by hooks or plumes. I had always been much struck by such adaptations, and until these could be explained it seemed to me almost useless to endeavour to prove by indirect evidence that species have been modified.

Darwin came upon the explanation while reading Malthus's *Essay on Population.* He continued:

> In October 1838, that is fifteen months after I had begun my systematic enquiry, I happened to read for amusement 'Malthus on population', and being well prepared to appreciate the struggle for existence which everywhere goes on from long-continued observation of the habits of animals and plants, it at once struck me that under these circumstances favorable variations would tend to be preserved and unfavorable ones to be destroyed. The result of this would be the formation of a new species.

Because of the struggle for existence, forms that are better adapted to survive will leave more offspring and automatically increase in frequency from one generation to the next. As the environment changes through time (for example, from humid to arid) different forms of a species will be better adapted to it than were the forms in the past; the better adapted forms will increase in frequency, and the now poorly adapted forms will decrease in frequency. As the process continues, eventually (in Darwin's words) "the result of this would be the formation of a new species." This process provided Darwin with what he called "a theory by which to work." And he started to work. He was still at work, fitting facts into his theoretical scheme, 20 years later when he received a letter from another traveling British naturalist, Alfred Russel Wallace (Figure 1.4). Wallace had independently arrived at a very similar idea to Darwin's natural selection. Darwin's friends, Charles Lyell and Joseph Hooker (Figure 1.5), arranged for a simultaneous announcement of Darwin and Wallace's idea at the Linnean Society in London in 1858. By then, however, Darwin was already writing an abstract of his full findings. That abstract is the scientific classic *On the Origin of Species.*

1.3.3 *Darwin's reception*

The reactions to Darwin's two connected theories—evolution and natural selection—differed. Many biologists almost immediately came to accept that evolution was true; species were not separately created and not fixed in form. A vigorous popular controversy sprung up about whether species evolved, or had separate origins, as the Bible seemed to suggest; in the United Kingdom, Thomas Henry Huxley, whom Darwin called "my general agent," particularly defended the new evolutionary view against religious attack.

Evolution was considerably less controversial among professional scientists. The new theory could, in some cases, make remarkably little difference to day-to-day biological research. The kind of comparative anatomy practiced by the followers of Cuvier, including Owen, lent itself equally well to a post-Darwinian search for pedigrees as to the pre-Darwinian search for "plans" of nature. The

Figure 1.4 Alfred Russel Wallace (1823–1913), photographed in 1848.

Figure 1.5 Joseph Dalton Hooker (1817–1911) on a botanical expedition in Sikkim in 1849 (after a sketch by William Tayler).

leading anatomists were by now mainly German; Carl Gegenbauer (1826–1903), one of the major figures, had soon reoriented his work to the tracing of evolutionary relationships between animal groups. The famous German biologist Ernst Haeckel (1834–1919) vigorously investigated the same problem, as he applied his "biogenetic law"—the theory of recapitulation (which we shall meet in section 21.3, p. 587)—to reveal phylogenetic pedigrees.

Although some kind of evolution was thus widely accepted, it is doubtful whether many biologists shared Darwin's own idea of it. In Darwin's theory, evolution is a branching tree, and local conditions largely determine what forms evolve at each stage. There is nothing automatically progressive about Darwinian evolution: if later species are better in some respect, it simply reflects the way things turned out; nothing guarantees the improvement. Most evolutionists of the late nineteenth and early twentieth centuries had a somewhat different conception of evolution, imagining evolution as one-dimensional and progressive. They often concerned themselves with describing mechanisms to explain why evolution should have an unfolding, predictable, progressive pattern (Figure 1.6).

While evolution—of a sort—was initially accepted, natural selection was just as surely rejected. People disliked the theory of natural selection for many reasons. We will not explain the arguments in any depth in this chapter; what follows represents only an introduction to the history of the ideas that we shall consider in the main body of the text.

One of the more sophisticated objections to Darwin's theory was that it lacked a satisfactory theory of heredity. All of the various theories of inheritance put forth at that time are now known to be wrong. Darwin preferred a "blending" theory of inheritance, in which the offspring blend their parental attributes; for example, if a red male mated with a white female, and inheritance "blended," the offspring would be pink. One of the deepest-hitting criticisms of the theory

Figure 1.6 (a) Darwin's theory suggests that evolution has proceeded as a branching tree. Note that *Homo* occupies an arbitrary position in the diagram—it does not have to be the right-hand extreme. The tree should be contrasted with the popular idea (b) that evolution is a one-dimensional progressive ascent of life. In Stephen Jay Gould's words, Darwinian evolution is a bush, not a ladder. (see also Figure 1.2)

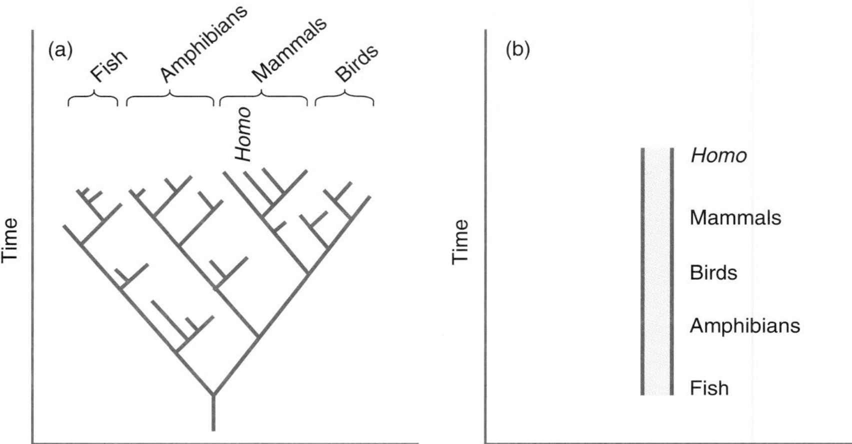

of natural selection pointed out that it could hardly operate at all if heredity blended (section 2.5, p. 29).

At a more popular level, two of the main objections to natural selection then (as now) were that it attributed evolution to chance and that some gaps between forms in nature could not be crossed if evolution was powered by natural selection alone. The anatomist St. George Jackson Mivart (1827–1900), for instance, in his book *The Genesis of Species* (1871), listed a number of organs that would not (he thought) be advantageous in their initial stages. In Darwin's theory, organs evolve gradually, and each successive stage must be advantageous to be favored by natural selection. Mivart retorted that, although a fully formed wing, for example, is advantageous for a bird, the first evolutionary stage of a tiny proto-wing might not prove any advantages.

Biologists who accepted the criticism sought to circumvent the difficulty by imagining processes other than selection that could work in the early stages of a new organ's evolution. Most of these processes belong to the class of theories of "directed mutation," or directed variation. They suggested that offspring, for some unspecified reason linked with the hereditary mechanism, consistently tend to differ from their parents in a certain direction. In the case of wings, the explanation by directed variation would say that the wingless ancestors of birds somehow tended to produce offspring with proto-wings, even though there was no advantage to it. (Chapter 13 deals with this general question; chapter 4 discusses variation.)

Lamarckian inheritance was the most popular theory of directed variation. Variation is "directed" in this theory because the offspring tend to differ from their parents in the direction of characteristics acquired by their parents. If the parental giraffes all have short necks and acquire longer necks by stretching, their offspring initially have longer necks, before any elongation by stretching.

Darwin accepted that acquired characters can be inherited. He even produced a theory of heredity ("my much abused hypothesis of pangenesis," as he called it) that incorporated the idea. In Darwin's time, the debate focused on the relative importance of natural selection and the inheritance of acquired characteristics; by the 1880s, however, the debate moved into a new stage. The German biologist August Weismann (1833–1914) produced strong evidence and theoretical arguments that acquired characteristics are not inherited, and the question became whether Lamarckian inheritance had any influence in evolution at all. Weismann initially suggested that practically all evolution was driven by natural selection, but he later retreated from this position.

Around the turn of the century, Weismann was a highly influential figure, but few biologists shared his belief in natural selection. Some, such as the British entomologist Edward Bagnall Poulton, were studying natural selection; the majority view, however, was that natural selection needed to be supplemented by other processes. An influential history of biology written by Erik Nordenskiöld in the 1920s could even take it for granted that Darwin's theory was wrong. About natural selection, he concluded "that it does not operate in the form imagined by Darwin must certainly be taken as proved." For Nordenskiöld, the only remaining question was "does it exist at all?"

By this time, Mendel's theory of heredity had been rediscovered. *Mendelism* (chapter 2) has been the generally accepted theory of heredity since the 1920s, and is the basis of all modern genetics. It was destined eventually to allow a revival of Darwin's theory, although its initial effect, in the first two decades of the century, was the exact opposite. The early Mendelians, such as Hugo de Vries and William Bateson, all opposed Darwin's theory of natural selection. They mainly researched the inheritance of large differences between organisms, and generalized their findings to evolution as a whole. They suggested that evolution proceeded in big jumps, by macromutations. A *macromutation* is a large and genetically inherited change between parent and offspring (Figure 1.7a). (Chapters 7, 13, and 19 discuss various perspectives on the question of whether evolution proceeds in small or large steps.)

Mendelism was not universally accepted in 1900–1920, however. Members of the other principal school, which rejected Mendelism, called themselves biometricians; Karl Pearson was one of the leading figures of this school. *Biometricians* studied small, rather than large, differences between individuals and developed statistical techniques to describe how frequency distributions of measurable characters (such as height) passed from parent to offspring population. They saw evolution more in terms of the steady shift of a whole population rather than the production of a new type from a macromutation (Figure 1.7b). Some biometricians were more sympathetic to Darwin's theory than were the Mendelians. Weldon, for instance, attempted to measure the amount of selection in crab populations on the seashore.

1.3.4 *The modern synthesis*

By the second decade of the century, research on Mendelian genetics had already become a major enterprise. It was concerned with many problems, most of which involved genetics rather than evolutionary biology. Within the theory of evolution, the main problem was reconciling the atomistic Mendelian theory

Figure 1.7 Early Mendelians and biometricians. (a) Early Mendelians studied large differences between organisms, and thought that evolution occurred when a new species evolved from a "macromutation" in its ancestor. (b) Biometricians studied small interindividual differences, and explained evolutionary change by the transition of whole populations. Mendelians were less interested in the reasons for small interindividual variations. The figure is a simplification: no historical debate between two groups of scientists and lasting for three decades can be fully represented in a single diagrammatic contrast.

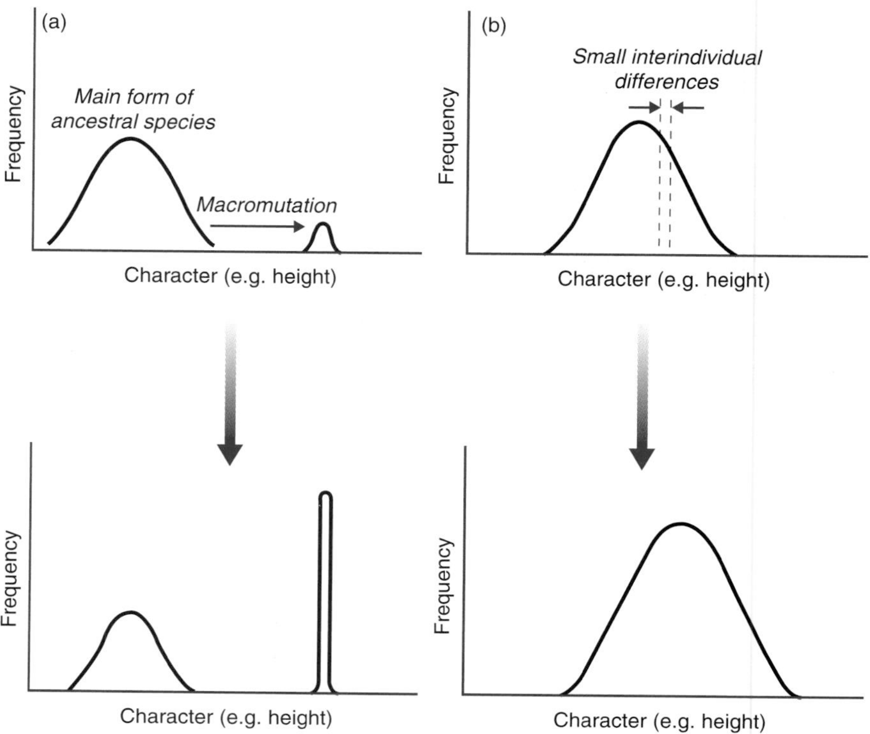

of genetics with the biometrician's description of continuous variation in real populations. This reconciliation was achieved by several authors in many stages, but a 1918 paper by R. A. Fisher is particularly important. In the paper, Fisher demonstrated that all the results known to the biometricians could be derived from Mendelian principles.

The next step was to show that natural selection could operate with Mendelian genetics. The theoretical work was mainly performed, independently, by R. A. Fisher, J. B. S. Haldane, and Sewall Wright (Figure 1.8). Their synthesis of Darwin's theory of natural selection with the Mendelian theory of heredity established what is known as *neo-Darwinism*, the *synthetic theory of evolution*, or the *modern synthesis*, after the title of a book by Julian Huxley, *Evolution: The Modern Synthesis* (1942). The old dispute between Mendelians and Darwinians was ended. Darwin's theory now possessed what it had lacked for half a century: a firm foundation in a well-tested theory of heredity.

The ideas of Fisher, Haldane, and Wright are known mainly from their great summary works, which were all written around 1930. Fisher published his book

Figure 1.8 (a) Ronald Aylmer Fisher (1890–1962), in 1912, as a Steward at the First International Eugenics Conference. (b) J. B. S. Haldane (1892–1964) in Oxford, United Kingdom, in 1914. (c) Sewall Wright (1889–1988), in 1928 at the University of Chicago.

The Genetical Theory of Natural Selection in 1930. Haldane published a more popular book, *The Causes of Evolution,* in 1932; it contained an appendix, "A mathematical theory of artificial and natural selection," summarizing a series of papers published from 1918 to 1932. Wright published a long paper on "Evolution in mendelian populations" in 1931; unlike Fisher and Haldane, Wright lived to publish a four-volume treatise (1968–1978) at the end of his career.

These classic works of theoretical population genetics demonstrated that natural selection could work with the kinds of variation observable in natural populations and the laws of Mendelian inheritance. No other processes are needed, nor are the inheritance of acquired characters, directed variation, or macromutations that produce extraordinary monsters necessary. This insight has been incorporated into all later evolutionary thinking, and the work of Fisher, Haldane, and Wright forms the basis for much of the material in chapters 5–9.

The reconciliation between Mendelism and Darwinism soon inspired new genetical research in the field and laboratory. Theodosius Dobzhansky (Figure 1.9), for example, began classic investigations of evolution in populations of fruitflies (*Drosophila*) after his move from Russia to the United States in 1927. Dobzhansky had been influenced by the leading Russian population geneticist Sergei Chetverikov (1880–1959), who had an important laboratory in Moscow until he was arrested in 1929. After his emigration, Dobzhansky both worked on his own ideas and collaborated with Sewall Wright. Dobzhansky's major book, *Genetics and the Origin of Species,* was first published in 1937 and its successive editions have been among the most influential works of the modern synthesis. We shall encounter several examples of Dobzhansky's work with fruitflies in later chapters.

In the 1920s, E. B. Ford (1901–1988) began a comparable, if more idiosyncratic, program of research in the United Kingdom. He studied selection in

Figure 1.9 Theodosius Dobzhansky (1900–1975) in a group photo in Kiev in 1924; he is seated second from the left at the front, in the great boots.

natural populations, mainly of moths, and called his subject "ecological genetics"; he published a summary of this work in a book called *Ecological Genetics* (1st edn, 1964). H. B. D. Kettlewell's (1901–1979) study of melanism in the peppered moth *Biston betularia* is the most famous piece of ecological genetic research (section 5.7, p. 103). Ford collaborated closely with Fisher; their best-known joint study was an attempt to show that the random processes emphasized by Wright could not account for observed evolutionary changes in the scarlet tiger moth *Panaxia dominula.*

Julian Huxley (Figure 1.10a) exerted his influence more through his skill in synthesizing work from many fields. His book *Evolution: The Modern Synthesis* (1942) introduced the theoretical concepts of Fisher, Haldane, and Wright to many biologists, by applying them to large evolutionary questions.

From population genetics, the modern synthesis spread into other areas of evolutionary biology. The question of how one species splits into two—the event is called *speciation*—was an early example. Before the development of the modern synthesis theory, speciation had often been explained by macromutations or the inheritance of acquired characters. A major book, *The Variation of Animals in Nature* (1936), by two systematists, G. C. Robson and O. W. Richards, accepted neither Mendelism nor Darwinism. Robson and Richards suggested that the differences between species are nonadaptive and unrelated to natural selection. Richard Goldschmidt (1878–1958), most famously in his book *The Material Basis of Evolution* (1940), argued that speciation was produced by macromutations, not the selection of small variants.

The question of how species originate is closely related to the question of population genetics, and has been discussed by Fisher, Haldane, and Wright. Dobzhansky and Huxley emphasized the problem even more. All of these scientists reasoned that the kinds of changes studied by population geneticists, if they took place in geographically separated populations, could cause the populations

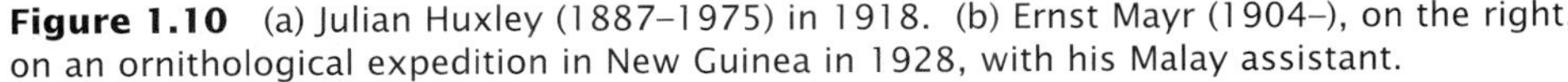

Figure 1.10 (a) Julian Huxley (1887–1975) in 1918. (b) Ernst Mayr (1904–), on the right, on an ornithological expedition in New Guinea in 1928, with his Malay assistant.

(a)

(b)

to diverge and eventually evolve into distinct species (chapter 16). The classic work addressing this issue was *Systematics and the Origin of Species* (1942) by Ernst Mayr. Like many classic books in science, it was written as a polemic against a particular viewpoint. Mayr's work was precipitated by Goldschmidt's *Material Basis* but criticized Goldschmidt from the viewpoint of a complete and differing theory—the modern synthesis—rather than narrowly refuting him; as a result, this book has a much broader importance. Both Goldschmidt and Mayr (Figure 1.10b) were born and educated in Germany and later emigrated to the United States. Mayr emigrated in 1930 as a young man; in contrast, Goldschmidt was 58 and had built a distinguished career when he left Nazi Germany in 1936.

A related development is often called "the new systematics," after the title of a book edited by Julian Huxley (1940). It refers to the overthrow of what Mayr called the "typological" species concept and its replacement by a species concept better suited to modern population genetics (chapter 15). The two concepts differ in how they address variation between individuals within a species.

A species, according to the typological concept, had been defined as a set of more or less similar-looking organisms, where similarity was measured relative to a standard (or "type") form for the species. A species then contains some individuals of the standard type, and other individuals who deviate from that type. The type individuals are conceptually privileged, whereas the deviants show some sort of error.

This concept of a species as type plus deviants was inappropriate in the theory of population genetics. The changes in gene frequencies analyzed by population geneticists take place within a "gene pool"—that is, a group of interbreeding organisms that exchange genes when they reproduce. The crucial unit becomes the set of interbreeding forms, regardless of how similar their appearances are. The idea of a "type" for a species is meaningless in a gene pool containing many genotypes: no one genotype is more of a standard form for the species than any other; some genotypes are not standard, and others deviate from it. No type form exists that could be used as a reference point for defining the species, so population geneticists came to define the members of a species by the ability to interbreed rather than by their morphological similarity to a type form. The modern synthesis had spread to systematics.

A similar treatment was given to paleontology by George Gaylord Simpson (Figure 1.11) in *Tempo and Mode in Evolution* (1944). Many paleontologists in the 1930s persisted in explaining evolution in fossils by *orthogenetic processes*—that is, some inherent (and unexplained) tendency of a species to evolve in a certain direction. Orthogenesis is an idea related to the pre-Mendelian concept of directed mutation, and the more mystical internal forces described by Lamarck. Simpson argued that no observations in the fossil record required these processes. All evidence was perfectly compatible with the population genetic mechanisms discussed by Fisher, Haldane, and Wright. He also demonstrated how such topics as rates of evolution and the origin of major new groups could be analyzed by techniques derived from the assumptions of the modern synthesis (see chapters 19–23).

By the mid-1940s, therefore, the modern synthesis had penetrated all areas of biology. The 30 members of a "committee on common problems of genetics, systematics, and paleontology" who met (with some other experts) at Princeton in 1947 represented all areas of biology. They shared a common viewpoint,

Figure 1.11 George Gaylord Simpson (1902–1984) with a baby guanaco in central Patagonia in 1930.

however—the viewpoint of Mendelism and neo-Darwinism. A similar unanimity of 30 leading figures in genetics, morphology, systematics, and paleontology would have been difficult to achieve before that date. The Princeton symposium was published as *Genetics, Paleontology, and Evolution* (Jepsen, Mayr, & Simpson, eds, 1949) and is now recognized as the symbolic point at which the synthesis had spread throughout biology. Of course, controversy remained within the synthesis proponents, and a counterculture emerged outside its boundaries. In 1959, two eminent evolutionary biologists—the geneticist Muller and the paleontologist Simpson—could still both celebrate the centenary of *The Origin of Species* with essays bearing (almost) the same memorable title: "One hundred years without Darwinism are enough."

In this book, we shall look in detail at the main ideas of the modern synthesis, and see how they are developing in recent research.

SUMMARY

1. Evolution means descent with modification, or the change in the form, physiology, and behavior of organisms over many generations of time. The evolutionary changes of living things occur in a diverging, tree-like pattern of lineages.
2. Living things possess adaptations: i.e., they are well designed in form, physiology, and behavior, for life in their natural environment.
3. Many thinkers before Darwin had discussed the possibility that species change over time into other species. Lamarck is the best known of these scientists. But in the mid-nineteenth century most biologists believed that species are fixed in form.
4. Darwin's theory of evolution by natural selection explains evolutionary change and adaptation.
5. Darwin's contemporaries mainly accepted his idea of evolution, but not his explanation of it by natural selection.
6. Darwin lacked a theory of heredity. When Mendel's ideas were rediscovered at the turn of the century, they were initially thought to argue against the theory of natural selection.
7. Fisher, Haldane, and Wright demonstrated that Mendelian heredity and natural selection are compatible; the synthesis of the two ideas is called neo-Darwinism or the synthetic theory of evolution.
8. During the 1930s and 1940s, neo-Darwinism gradually spread through all areas of biology and became widely accepted. It unified genetics, systematics, paleontology, and classical comparative morphology and embryology.

FURTHER READING

A popular essay about the adaptations of woodpeckers is by Diamond (1990a). On the history: Bowler (1989) and Mayr (1982b) provide general histories of the idea of evolution. On Lamarck and his context, see Burkhardt (1977), Barthélemy-Madaule (1982), Corsi (1988). On Darwin's ideas, see Ghiselin (1969). There are many biographies of Darwin; Browne (1995) is a recent one. Darwin's autobiography is an interesting source. A pleasant (if more demanding) way to follow Darwin's life is through his correspondence. Much was published in the life and letters (Darwin 1887) and *More Letters* (Darwin and Seward 1903). A modern scholarly edition is under way (Burkhardt and Smith 1985–). Bowler (cited above) discusses and gives references about the reception and fate of Darwin's ideas. On the modern synthesis, see also Provine (1971) and Mayr and Provine (1980). There are biographies of many of the key figures: Box (1978) for Fisher, Clark (1969) for Haldane (see also Haldane 1963 and Maynard Smith 1987a), Provine (1986) for Wright. Huxley (1970–1973) and Simpson (1978) wrote autobiographies. See Adams (1994) for Dobzhansky; see the papers in a dedicatory issue of *Evolution*, vol. 48 (1994), pp. 1–44, for Mayr. The two Darwin centennial essays by Muller and by Simpson alluded to in the text are Muller (1959) and Simpson (1961a).

Gould's popular essays, which first appear in *Natural History* magazine and are then anthologized (Gould 1977b, 1980a, 1983a, 1985, 1991, 1993, 1996), introduce many aspects of evolutionary biology; Gould (1989) discusses the popular misconception that evolution is a progressive ladder-climbing process. Jones (1993) is a popular book about human genetics and evolution. Keller and Lloyd (eds) (1992) contains pieces on many controversial "keywords" in evolutionary biology, including one by Dawkins on progress. The best single source in which to follow a range of modern evolutionary research is the monthly *Trends in Ecology and Evolution.*

STUDY AND REVIEW QUESTIONS

1. Review the ways in which biological evolution differs from individual development, changes in the species composition of ecosystems, and some other kinds of change that you can imagine.
2. What property of nature must any theory of evolution explain, if it is not (in Darwin's words) to be "almost useless"?
3. How did the main popular concept of evolution in the late nineteenth and early twentieth centuries differ from the conception of evolution in Darwin's theory?
4. What are the two theories that are combined in the synthetic theory of evolution?

CHAPTER 2

Molecular and Mendelian Genetics

THE chapter provides an introduction to the genetics that you need to understand the evolutionary biology contained in this book. It begins with the molecular mechanism of inheritance, and moves on to the Mendelian principles. It then considers how Darwin's theory almost required heredity to be Mendelian, because natural selection cannot operate with a blending mechanism of inheritance.

2.1 Inheritance is caused by DNA molecules, which are physically passed from parent to offspring

The molecule called DNA (deoxyribose nucleic acid) provides the physical mechanism of heredity in all living creatures. DNA carries the information used to build a new body, and to differentiate its various parts. DNA molecules exist inside almost all of the cells of a body, and in all the reproductive cells (or gametes). Its precise location in the cell depends on cell type.

There are two main types of cells: *eukaryotic* and *prokaryotic* (Figure 2.1). Eukaryotic cells have a complex internal structure, including internal organelles and a distinct region, surrounded by a membrane, called the nucleus; eukaryotic DNA exists within the nucleus. Prokaryotic cells are simpler and have no nucleus; prokaryotic DNA lies within the cell, but in no particular region. All complex multicellular organisms, including all plants and animals, possess eukaryotic cells, as do fungi (such as yeast) and protozoans (such as amebas). Prokaryotic cells are found in other unicellular organisms, particularly in bacteria and spirochetes.

Within a eukaryotic cell nucleus, the DNA physically is carried in structures called *chromosomes;* chromosomes can be seen through a light microscope at certain stages in the cell cycle. Individuals of different species characteristically have different numbers of chromosomes—each individual human has 46, for example, whereas a fruitfly *Drosophila melanogaster* has eight chromosomes, and other species have other numbers. The finer structure of the DNA is too small to be seen directly, but it can be inferred by the method of x-ray diffraction. The molecular structure of DNA was unraveled by Watson and Crick in 1953.

The DNA molecule consists of a sequence of units; each unit, called a *nucleotide*, consists of a phosphate and a sugar group with a *base* attached. The alternating sugar and phosphate groups of successive nucleotides form the backbone of the DNA molecule. The full DNA molecule consists of two, paired complementary strands, each consisting of sequences of nucleotides. The nucleotides of opposite strands are chemically bonded together. The two strands exist as a double helix (Figure 2.2).

Figure 2.1 The cells of a body have a fine structure (or "ultrastructure") made up of a number of organelles. Not all of the organelles illustrated here are found in all cells. Animal and fungal cells, for example, lack plastids; in contrast, all photosynthesizing organisms have them. Eukaryotes (i.e., all plants and animals) have complex cells with a separate nucleus. Within the nucleus, the DNA is here illustrated in the diffuse form called chromatin; when the cell divides, the chromatin coalesces into structures called chromosomes. Prokaryotes are simpler organisms, particularly bacteria, and they lack a distinct nucleus; their DNA lies naked within the cell.

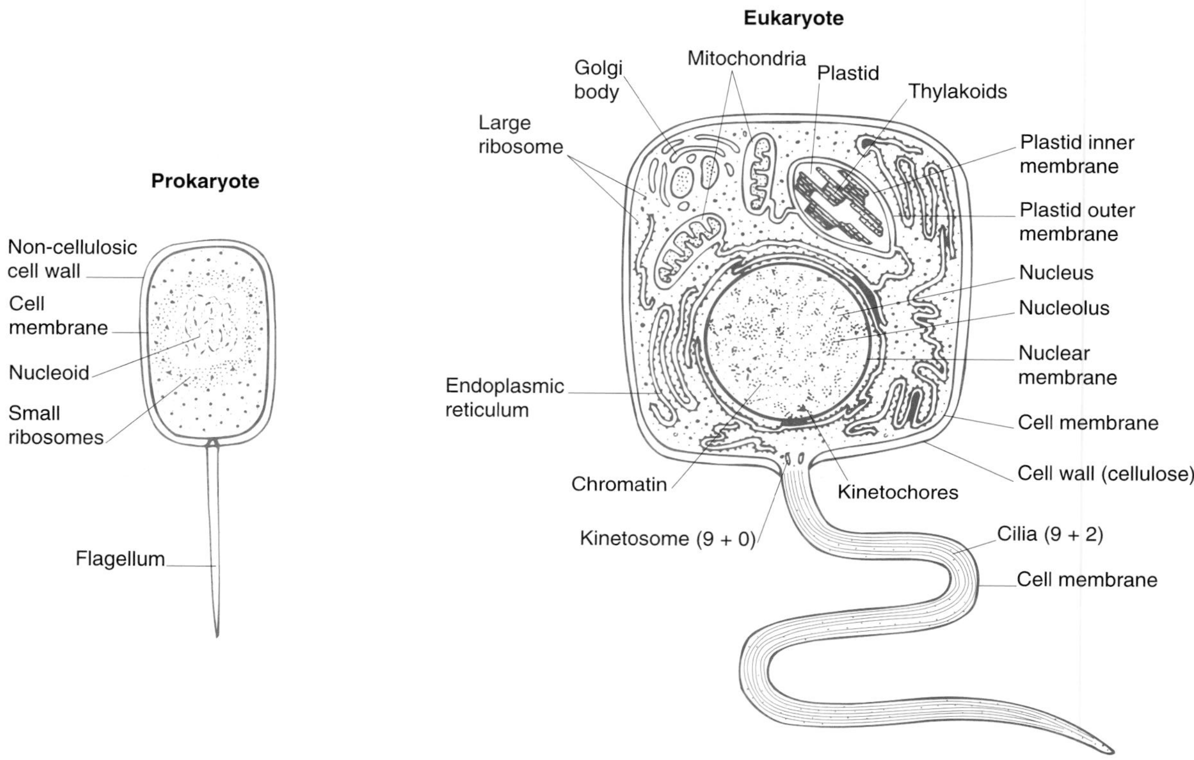

2.2 *The DNA structurally encodes the information used to build the body's proteins*

How does the DNA encode the information to build a body? The DNA in an individual human cell contains about 3×10^9 nucleotide units. This total length can be divided into *genes* and other kinds of DNA. Some genes lie immediately next to neigboring genes; others are separated by regions of varying length. Each gene contains the information that codes for a particular protein.

A crude but workable way to describe the role of proteins is to say that bodies are built from proteins. Different parts of the body develop their distinct characteristics because of the kinds of proteins of which they are made. Skin, for example, mainly consists of a protein called keratin; oxygen is carried in red blood cells by a protein called hemoglobin; eyes are made sensitive to light by

Figure 2.2 The structure of DNA. (a) Each strand of DNA is made up of a sequence of nucleotide units. Each nucleotide consists of a phosphate (P), a sugar, and a base (of which there are four types, here called G, C, T, and A). (b) The full DNA molecule has two complementary strands, arranged in a double helix. nm = nanometer = 1×10^9 meter.

(a) **Structure of single strand**

5' end of chain

3' end of chain

(b) **Structure of double strand**

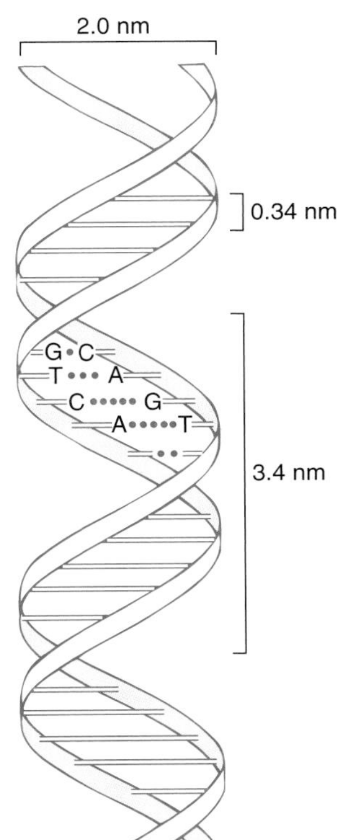

pigment proteins such as rhodopsin (rhodopsin itself consists of a protein called opsin combined with a derivative of vitamin A); and metabolic processes are catalyzed by a whole battery of proteins called enzymes (cytochrome c, for instance, is a respiratory enzyme, while alcohol dehydrogenase is a digestive enzyme). Coding for proteins by DNA ultimately enables the egg cell to develop into an adult body.

Proteins are made up of particular sequences of amino acids. Twenty different amino acids are found in most kinds of living things; each amino acid behaves chemically in distinct ways, such that different sequences of amino acids result in proteins with very different properties. A protein's exact sequence of amino acids determines its nature. Hemoglobin, for example, consists of two α-globin molecules, which are 141 amino acids long in humans, and two β-globins, which are 144 amino acids long; insulin is created by a sequence of 51 amino acids. Hemoglobin binds oxygen in the blood, whereas insulin stimulates cells (particularly muscle cells, and others) to absorb glucose from the blood. The different behaviors of hemoglobin and insulin result from the chemical properties of their different amino acids, arranged in their characteristic sequence.[1]

The idea that one gene encodes for one protein is a simplification. A more accurate statement would be that one cistron encodes for one polypeptide. A *cistron* is defined as the length of DNA that codes for a polypeptide. A polypeptide is a chain of amino acids—the polypeptides in a body are of various lengths. The description involving cistrons is more exact because some proteins, such as hemoglobin, are assembled from more than one polypeptide, and the different polypeptides are encoded in separate (rather than contiguous) regions of the genome. Hemoglobin is created from four polypeptides, and the cistrons for two of those polypeptides lie far from the cistrons for the other two. (Thus, all proteins are polypeptides, but not all polypeptides are proteins. Each of the four polypeptides that make up hemoglobin represents only a part of a protein.) No single "gene" encodes hemoglobin. However, for many purposes it is not grievously wrong to describe DNA as made up of genes (and noncoding regions), and genes as coding for proteins.

How exactly do genes in the DNA encode for proteins? The answer is that the sequence of nucleotides in a gene specifies the sequence of amino acids in the protein. DNA contains four types of nucleotides that differ only in the base part of the nucleotide unit; the sugar and phosphate groups remain the same in all four types. The four types of nucleotides are called adenine (A), cytosine (C), guanine (G), and thymine (T). Adenine and guanine belong to the chemical group called purines, while cytosine and thymine are pyrimidines. In the double helix, an A nucleotide in one strand always pairs with a T nucleotide in the other; and a C always pairs with a G (as in Figure 2.2b). If the nucleotide sequence in one strand was . . . AGGCTCCTA . . . , then the complementary strand would be . . . TCCGAGGAT. . . . Because the sugar and phosphate are constant, it is often more convenient to imagine a DNA strand as a sequence of bases, like the . . . AGGCTCCTA . . . sequence given above.

2.3 The information in the DNA is decoded by transcription and translation

There are four types of nucleotides, but 20 different amino acids. A one-to-one code of nucleotide encoding amino acid would therefore be impossible. In fact, a triplet of bases encodes one amino acid; the nucleotide triplet for an amino acid is called a *codon*. The four nucleotides can be arranged in 64 ($4 \times 4 \times 4$) different triplets, each of which codes for a single amino acid. The relation between triplet and amino acid is called the *genetic code*.

The mechanism by which the amino acid sequence is dictated by the nucleotide sequence of the DNA is understood in molecular detail. The full detail is unnecessary for our purposes, but we should distinguish two main stages. At

1. As chapters 4 and 7 discuss in more detail, the particular sequence of any one protein can vary within and between species. Thus, turkey hemoglobin differs from human hemoglobin, although hemoglobin still binds oxygen in both species. Other variants of hemoglobin are observed within a species, a condition called *protein polymorphism*. However, the sequences of all variants of hemoglobin from the same or different species are similar enough for them to be recognizably hemoglobins.

TABLE 2.1

The genetic code. The code is here expressed for mRNA. Each triplet encodes one amino acid, except for the three "stop" codons, which signal the end of a gene.

First Base in the Codon	Second Base in the Codon: U	C	A	G	Third Base in the Codon
U	Phenylalanine	Serine	Tyrosine	Cysteine	U
	Phenylalanine	Serine	Tyrosine	Cysteine	C
	Leucine	Serine	Stop	Stop	A
	Leucine	Serine	Stop	Tryptophan	G
C	Leucine	Proline	Histidine	Arginine	U
	Leucine	Proline	Histidine	Arginine	C
	Leucine	Proline	Glutamine	Arginine	A
	Leucine	Proline	Glutamine	Arginine	G
A	Isoleucine	Threonine	Asparagine	Serine	U
	Isoleucine	Threonine	Asparagine	Serine	C
	Isoleucine	Threonine	Lysine	Arginine	A
	Methionine	Threonine	Lysine	Arginine	G
G	Valine	Alanine	Aspartic acid	Glycine	U
	Valine	Alanine	Aspartic acid	Glycine	C
	Valine	Alanine	Glutamic acid	Glycine	A
	Valine	Alanine	Glutamic acid	Glycine	G

the first, an intermediate molecule called messenger RNA (mRNA) is transcribed from the DNA; the process is called *transcription.* mRNA has a structure similar to that of DNA, except that mRNA is single-stranded and uses a base called uracil (U) rather than thymine (T). The DNA sequence AGGCTCCTA would therefore have an mRNA with the sequence UCCGAGGAU transcribed from it. The genetic code is usually expressed in terms of the codons in the mRNA (Table 2.1). The mRNA sequence UCCGAGGAU, for example, codes for three amino acids: serine-glutamic acid-aspartic acid. The beginning and end of a gene are signaled by distinct base sequences, which (in a sense) punctuate the DNA message.

Transcription takes place in the nucleus. After the mRNA molecule has been assembled on the gene, it then leaves the nucleus and travels to one of the structures in the cytoplasm called ribosomes (see Figure 2.1); ribosomes are made of another kind of RNA called ribosomal RNA (rRNA). The ribosome is the site of the second main stage in protein production, in which the amino acid sequence is read off from the mRNA sequence and the protein is assembled. The process is called *translation.* The actual translation is achieved by yet another kind of RNA, called transfer RNA (tRNA).[2] tRNA has a base triplet recognition site, which binds to the complementary triplet in the mRNA, and has the appropriate amino acid attached at the other end (Figure 3.7, p. 55, shows the structure of tRNA). There are 61 possible types of tRNA, one for each coding triplet; all 61 are not found in any one cell because some tRNA molecules can bind to

2. There are three principal kinds of RNA: mRNA, rRNA, and tRNA. Both rRNA and tRNA molecules originate by transcription from genes in the DNA. Thus, it is not always true that genes code for proteins, or polypeptides; some genes code for RNA.

more than one triplet in the mRNA. (This phenomenon is called wobble, and the "wobble rules" govern which tRNA molecules bind which combinations of codon.) Thus, protein assembly consists of tRNA molecules lining up on the mRNA at a ribosome. Other molecules are also needed to supply energy and attach the RNAs correctly. Figure 2.3 summarizes the transfer of information in the cell.

Different proteins contain different numbers of amino acids, but a very approximate average figure is 100–300 amino acids. Insulin (51 amino acids) is a short protein; long proteins can include thousands of amino acids. The human genome may contain approximately 100,000 (10^5) genes (and therefore a human body makes use in its life of about 100,000 kinds of protein). A protein consisting of 300 amino acids needs roughly 1000 (10^3) nucleotides, which leads to the estimate of 10^8 coding bases in the human genome. Because the total length of an individual's DNA is about 3×10^9 bases, about 90% of the DNA in the genome is of uncertain function (see chapter 10).

In addition to the DNA on the chromosomes in the nucleus, certain organelles in the cytoplasm possess much smaller quantities of DNA (see Figure 2.1). Mitochondria—the organelles that control respiration—have some DNA, and in plants the organelles called chloroplasts that control photosynthesis also contain their own DNA. Mitochondrial DNA is inherited maternally: mitochondria are passed on through eggs but not through sperms.

2.4 *Mutational errors may occur during DNA replication and repair*

When a cell reproduces, its DNA and genes are physically replicated. Normally an exact copy of the parental DNA is produced, but sometimes replication accidentally introduces an error. Such errors are called *mutations*. The error may be immediately corrected by the repair machinery of the DNA, or it may not be. Moreover, the repair process itself can make a mistake and introduce a mutation. The new sequence of DNA that results from a mutation may code for a form of protein with different properties from the original. Mutations can happen in any cell, but the most important mutations, for the theory of evolution,

Figure 2.3 The transfer of information in a cell.

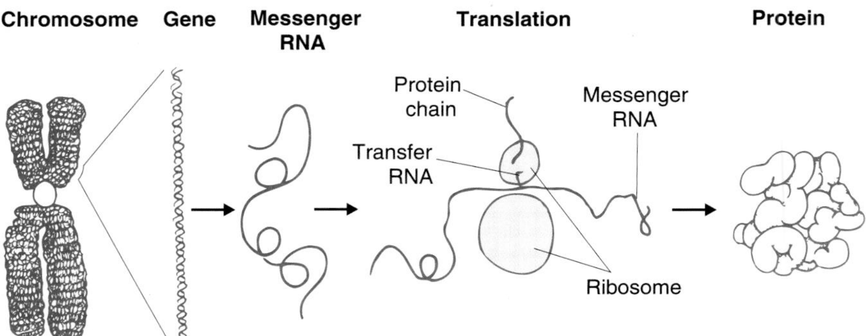

Figure 2.4 Different types of mutations. (a) Synonymous mutations—the base changes but the amino acid encoded does not. (b) Transition—change between purine types or between pyrimidine types. (c) Transversion—change from purine to pyrimidine, or vice versa. (d) Frame-shift mutation—a base is inserted. (e) Stop mutation—an amino acid encoding triplet mutates to a stop codon. The terms *transition* and *transversion* can apply to synonymous or amino-acid–changing mutations, but are only illustrated here for mutations that alter amino acids. The base sequence here is for the DNA. The genetic code is conventionally written for the mRNA sequence; thus, G must be transcribed to C, and so on, when comparing the figure with Table 2.1 (the genetic code). (The figure is stereochemically unconventional because the 3′ end has been put at the left and 5′ at right; but this detail is unimportant here.)

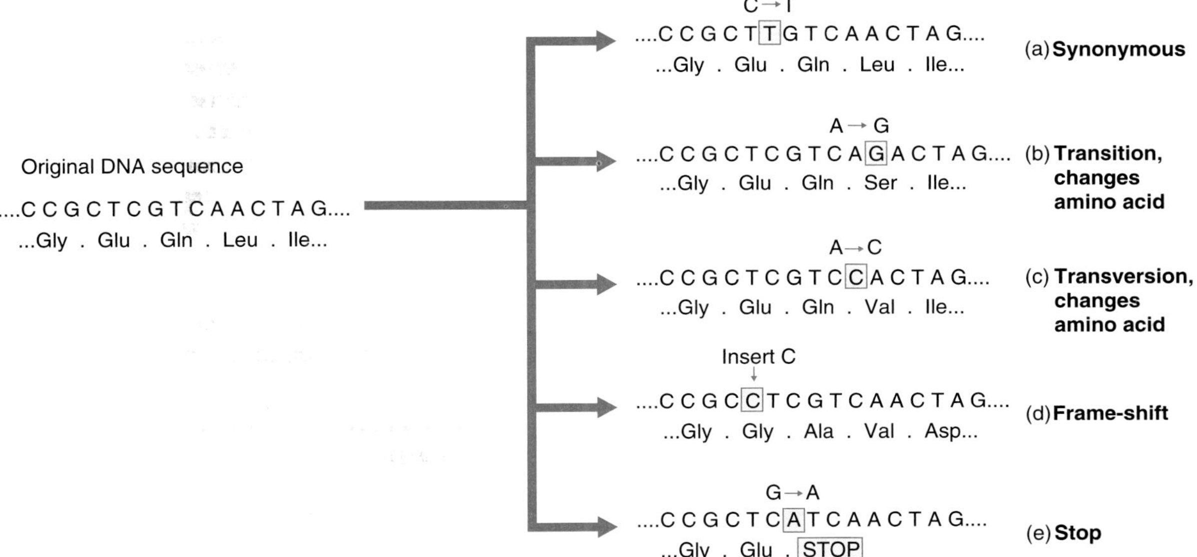

are those occurring in the production of the gametes. These mutations are passed on to the offspring, who may differ from their parents because of the mutation.

Various kinds of mutations can occur. In point mutation, a base in the DNA sequence changes to another base. The effect of a point mutation depends on the kind of base change (Figure 2.4a–c). Synonymous, or silent, mutations (Figure 2.4a) are mutations between two triplets that code for the same amino acid, and have no effect on the protein sequence; non-synonymous, or meaningful, point mutations do change the amino acid. Because of the structure of the genetic code (Table 2.1), most synonymous mutations occur in the third base position of the codon. Approximately 70% of changes in the third position are synonymous, whereas all changes in the second and most (96%) at the first position are meaningful. Another distinction for point mutations relates to transitions and transversions. Transitions are changes from one pyrimidine to the other, or from one purine to the other—that is, between A and G, and between C and T. Transversions replace a purine base by a pyrimidine, or vice versa—that is, from A or G to T or C (and from C or T to A or G). The distinction is interesting because transitional changes occur much more frequently in evolution than transversions.

Successive amino acids are read from consecutive base triplets. If, therefore, a mutation inserts a base pair into the DNA, it can alter the meaning of every base "downstream" from the mutation (Figure 2.4d). Such *frame-shift mutations* will usually produce a completely nonsensical, functionless protein. Another kind of mutation is for a previously coding triplet to mutate to a "stop" codon (Figure 2.4e); the resulting protein fragments will probably be functionless.

Mutations may also involve whole chunks of chromosomes rather than single bases (Figure 2.5). A length of chromosome may be translocated to another chromosome, or to another place on the same chromosome, or be inverted. Whole chromosomes may fuse, as has happened in human evolution; chimps and gorillas (our closest living relatives) have 24 pairs of chromosomes, whereas humans have 23. Some or all of the chromosomes may be duplicated.

It is more difficult to generalize about the phenotypic effects of these chromosomal mutations. If the break-points of the mutation divide a protein, that protein will be lost in the mutant organism. If the break occurs between two proteins, however, any effect will reflect whether the expression of a gene depends on its position in the genome. In theory, it might not matter whether a protein is transcribed from one chromosome or another; in practice, gene expression is probably at least partly regulated by relations between neighboring genes and a chromosomal mutation will then have phenotypic consequences.

Figure 2.5 Chromosomes can mutate by (a) deletion; (b) duplication of a part; (c) inversion; or (d) translocation. Translocation may be either "reciprocal" (in which the two chromosomes exchange equal lengths of DNA) or "nonreciprocal" (in which one chromosome gains more than the other). In addition, whole chromosomes may fuse, and whole chromosomes (or the whole genome) may duplicate.

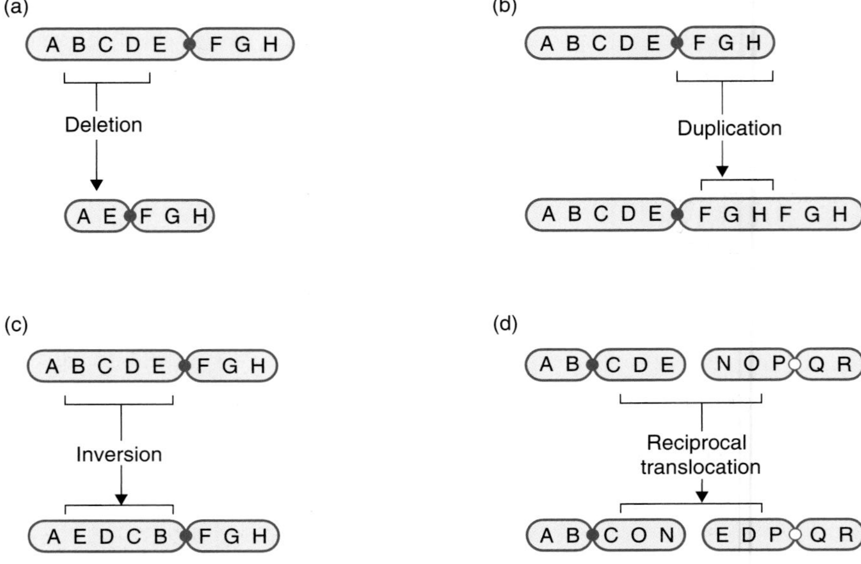

2.5 *The rates of mutation can be measured and estimated*

What is the rate of mutation? A direct method of finding out is to observe the rate at which a visible mutant phenotype arises in a laboratory population of a species. The method has been used with many laboratory species, such as bacteria, yeast, *Drosophila*, and mice. Table 2.2 summarizes some measurements, by direct observation, of rates of visible mutation. Many are mutations from a rare, often dysfunctional, form back to the normal form of the gene. A memorable figure from Table 2.2 is that mutation rates per gene are approximately 10^{-6}; however, these estimates, from direct observation, may be biased because the genetic changes are ones that are easily noticed. They may well have higher-than-average mutation rates, because changes with low mutation rates simply may not happen often enough to be noticed. Even if the rate is not simply biased, the sample remains biased because the changes are gross. Nonetheless, these observations form the basis of the classical genetical ball-park figure of 10^{-6} for the mutation rate per locus. When we examine molecular evolution, we will need the mutation rate per nucleotide, and we will learn how to estimate this rate in section 7.11.3 (p. 181). Section 9.12 (p. 247) describes how to estimate the mutation rate for the whole genome, and geneticists have many other methods of estimating mutation rates.

TABLE 2.2

Mutation rates for various genetic changes.

Organism	Character (gene → gene)	Rate
Bacteriophage (T_2)	Lysis inhibition, *rII* → *rII*$^+$	$1 \times 10^{-3} - 1 \times 10^{-9}$ or less
	Host range, *h*$^+$ → *h*	3×10^{-9}
Bacteria *(Escherichia coli)*	Lactose fermentation, *lac*$^-$ → *lac*$^+$	2×10^{-7}
	Phage T_1 sensitivity, T_1-*s* → T_1-*r*	2×10^{-8}
	Histidine requirement, *his*$^-$ → *his*$^+$	4×10^{-8}
	his$^+$ → *his*$^-$	2×10^{-6}
	Streptomycin sensitivity, *str-s* → *str-d*	1×10^{-9}
	str-d → *str-s*	1×10^{-8}
Algae *(Chlamydomonas reinhardi)*	Streptomycin sensitivity, *str-s* → *str-r*	1×10^{-6}
Fungi *(Neurospora crassa)*	Inositol requirement, *inos*$^-$ → *inos*$^+$	8×10^{-8}
	Adenine requirement, *ade*$^-$ → *ade*$^+$	4×10^{-8}
Corn *(Zea mays)*	Shrunken seeds, *Sh* → *sh*	1×10^{-5}
	Purple, *P* → *p*	1×10^{-6}
Fruitfly *(Drosophila melanogaster)*	Yellow body, *Y* → *y*, in males	1×10^{-4}
	Y → *y*, in females	1×10^{-5}
	White eye, *W* → *w*	4×10^{-5}
	Brown eye, *Bw* → *bw*	3×10^{-5}
Mouse *(Mus musculus)*	Piebald coat color, *S* → *s*	3×10^{-5}
	Dilute coat color, *D* → *d*	3×10^{-5}
Human *(Homo sapiens)*	Normal → hemophilic	3×10^{-5}
	Normal → albino	3×10^{-5}
Human bone marrow cells in tissue culture	Normal → 8-azoguanine resistant	7×10^{-4}
	Normal → 8-azoguanosine resistant	1×10^{-6}

As you may have noticed, the mutation rates in Table 2.2 show great variation. One reason behind this variation is that the mutations are of different kinds. When the mutant needed to restore the normal gene is something as easy as a transitional point mutation, it occurs at a high rate (10^{-3} or so); when a particular frame-shift, or deletion, is needed, the rates are much lower (10^{-9} or less).

We can also note that mutation rates can be increased by exposing organisms to certain chemicals, called *mutagens;* in bacteria, for example, caffeine is a mutagen. Many other chemicals act as mutagens, and, in some cases, the reason behind this role is understood at the chemical level. Nitrous acid, for example, directly converts bases in the DNA. Other mutagens are chemical analogues of bases and are incorporated into DNA: 5-bromouracil, for example, is a thymine analogue and increases the mutation rate when it is substituted for thymine bases in the DNA. X-rays also increase mutation rates, particularly the rates of breaks and rearrangements in chromosomes. These artificial influences on mutation rates are environmentally important, but may not have been of major importance in natural evolution.

2.6 *Diploid organisms inherit a double set of genes*

DNA is physically carried on chromosomes. Humans, as noted earlier, have 46 chromosomes, which consist of two sets of 23 distinct chromosomes. (More correctly, an individual has a pair of sex chromosomes—which are similar [XX] in females, but noticeably different [XY] in males—plus a double set of 22 non-sex chromosomes, called *autosomes.*) The condition of having two sets of chromosomes (and therefore two sets of the genes carried on them) is called *diploidy.* The estimate of 3×10^9 nucleotides in the human genome is for only one of the sets of 23 chromosomes; the total DNA library of a human cell has about 6×10^9 nucleotides.

Diploidy is important in reproduction. An adult individual possesses two sets of chromosomes. Its gametes (eggs in the female, sperm or pollen in the male) have only one set—a human egg, for example, has only 23 chromosomes before it is fertilized. Gametes are said to be *haploid.* They are formed by a special kind of cell division, called *meiosis;* in meiosis, the double set of chromosomes is reduced to result in a gamete with only one set. When male and female gametes fuse, at fertilization, the resulting *zygote* (the first cell of the new organism) has the double chromosome set restored, and it develops to produce a diploid adult. The cycle of genesis can then repeat itself. (In some species, organisms are permanently haploid. In this book, we will mainly be concerned with diploid species. Most familiar, non-microscopic species are diploid.)

Because each individual possesses a double set of chromosomes, it also has a double set of each of its genes. Any one gene is located at a particular place on a chromosome, called its *genetic locus;* an individual is therefore said to have two genes at each genetic locus in its DNA. One gene comes from the organism's father and the other from its mother. The combination of the two genes at a locus is called a *genotype.* The two copies of a gene in an individual may be the same, or slightly different (i.e., the amino acid sequences of the proteins encoded by the two copies may be identical or have one or two differences). If they are the same, the genotype is a *homozygote;* if they differ, it is a *heterozygote.* The different forms of the gene that can be present at a locus are called *alleles.*

Genes and genotypes are usually symbolized by alphabetic letters. For instance, if two alleles at the genetic locus under consideration are designated as *A* and *a,* an individual can then have one of three genotypes: *AA, Aa,* or *aa.*

The genotype at a locus should be distinguished from the *phenotype* it produces. If two alleles occur at a locus in a population, the two can combine into three possible genotypes: *AA, Aa,* and *aa.* (If there are more than two alleles, more than three genotypes are possible.) The genes will influence some property of the organism, and the property may or may not be easily visible. Suppose the genes influence color. The *A* gene might encode a black pigment and *AA* individuals would be black; *aa* individuals, lacking the pigment, might be white. The coloration is then determined by the phenotype controlled by the genotype at that locus: an individual's phenotype consists of its body and behavior as we observe them.

If we consider only the *AA* and *aa* genotypes and phenotypes in this example, then a one-to-one relationship exists between genotype and phenotype. However, a different relationship is possible, as can be illustrated by considering two possibilities for the phenotype of the *Aa* genotype. One possibility is that the color of *Aa* individuals is intermediate between the two homozygotes—that is, they are gray. In this case, three phenotypes exist for the three genotypes, in a one-to-one relationship. The second possibility is that the *Aa* heterozygotes resemble one of the homozygotes; they might be black, for instance. The *A* allele is then called *dominant* and the *a* allele *recessive.* (An allele is dominant if the phenotype of the heterozygote looks like the homozygote of that allele; the other allele in the heterozygote is then recessive.) If dominance occurs, only two phenotypes will be possible for the three genotypes, and there will no longer be a one-to-one relation between them. If you know only that an organism has a black phenotype, you do not know its genotype.

Any degree of dominance could appear at different genetic loci. Full dominance, in which the heterozygote resembles one of the homozygotes, and no dominance, in which it is intermediate between the homozygotes, are extreme cases. The phenotype of the heterozygote could lie anywhere between the two homozygotes. Instead of being either black or gray, it could exhibit any degree of grayness. Dominance is only one of a number of factors that complicate the relationship between genotype and phenotype. The most important such factor is the environment in which an individual grows up (see chapter 9).

2.7 *Genes are inherited in characteristic Mendelian ratios*

Mendelian ratios express the proportions of different genotypes in the offspring of parents of particular combinations of genotypes. The easiest case is a cross between an *AA* male and an *AA* female (Figure 2.6a). After meiosis, all males gametes and all female gametes contain the *A* gene. They combine to produce *AA* offspring. The Mendelian ratio is therefore 100% *AA* offspring.

Figure 2.6 Mendelian ratios. (a) For *AA* × *AA* cross. (b) For *AA* × *Aa* cross. (c) For *Aa* × *Aa* cross.

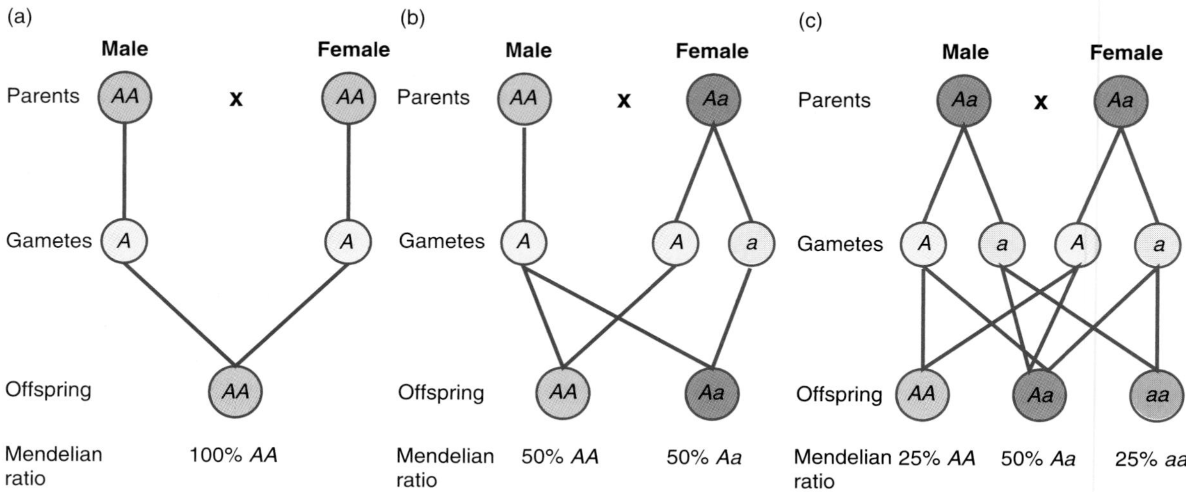

Now consider a mating between an *AA* homozygote and an *Aa* heterozygote (Figure 2.6b). Again, all of the *AA* individual's gametes contain a single *A* gene. When a heterozygote reproduces, half of its gametes contain an *A* gene, and half an *a* gene. The pair will produce *AA:Aa* offspring in a 50:50 ratio.

Finally, consider a cross between two heterozygotes (Figure 2.6c). Both male and female produce half *a* gametes and half *A* gametes. If we consider the female gametes (eggs, or ovules), half of them are *a;* half of these *a* female gametes will be fertilized by *a* sperm, and half by *A* sperm. The other half of the female gametes are *A*, of which half will be fertilized by *a* sperm and half by *A* sperm. The resulting ratio of offspring is ¼ *AA* : ½ *Aa* : ¼ *aa.*

The separation of an individual's two genes at a locus into its offspring is called *segregation.* The ratios of offspring types produced by different kinds of matings are examples of Mendelian ratios. They were discovered by Gregor Mendel in about 1856–1863. Mendel was a monk, later Abbé, in St. Thomas's Augustinian monastery in what was then Brünn in Austro-Hungary (now Brno in the Czech Republic).

Mendelian ratios can also exist for more than one genetic locus. If the alleles at one locus are *A* and *a*, and at a second *B* and *b*, then an individual will have a double genotype, such as *Ab/Ab* (double homozygote) or *Ab/ab* (single heterozygote). It has a double set of genes at each locus, one set from each parent. The segregation ratios now depend on whether the genetic loci lie on the same or different chromosomes. Recall that an individual human has a haploid number of 23 chromosomes and about 100,000 genes. Thus, there must be on average about 5000 genes per chromosome. Different genes on the same chromosome are described as being *linked.* Genes that are very close together are tightly linked, while those farther apart are loosely linked. Genes on separate chromosomes are unlinked.

The easy case involves two unlinked loci; the genes at the two loci then segregate independently. Imagine a cross in which only one of the loci is heterozygous, such as a cross between an *Ab/Ab* male and an *Ab/ab* female. All the genes at the B locus are the same, while at the A locus the male is *AA* and the female is *Aa*. The ratio of offspring will be ½ *AAbb* and ½ *Aabb*, a simple extension of the one-locus case.

Now suppose both loci are heterozygous. The ratios of B locus genotypes associated with each A locus genotype are those predicted by applying Mendel's principles independently to each locus. For instance, a cross between two *Bb* heterozygotes produces a ratio of offspring of ¼ *BB* : ½ *Bb* : ¼ *bb*, and this ratio will be the same within each A locus genotype. A cross between a male *AB/Ab* and a female *AB/ab* will generate ½ *AA* and ½ *Aa* offspring. Of the ½ that are *AA*, ¼ are *AB/AB*, ½ are *AB/Ab*, ¼ are *Ab/Ab*. The same concept applies to the ½ *Aa* genotypes. Add the two *A* genotypes and the total offspring ratios are:

AB/AB	*AB/Ab*	*Ab/Ab*	*AB/aB*	*AB/ab*	*Ab/ab*
⅛	¼	⅛	⅛	¼	⅛

The segregation of unlinked genotypes is called *independent segregation*.

When the loci are linked on the same chromosome, they do not segregate independently. At meiosis, when haploid gametes are formed from a diploid adult, an additional process called *recombination* occurs. The pair of chromosomes physically line up and, at certain places, their strands join together and recombine (Figure 2.7). Recombination shuffles the combinations of genes. If an individual inherited *AB* from its mother and *ab* from its father, and recombination occurs between the two loci, it will produce *Ab* and *aB* gene combinations in its gametes.

Recombination is a random process; it may or may not "hit" any point in the DNA. It occurs with a given probability, usually symbolized by r, between any two points on a chromosome. The probability r can be defined between nucleotide sites or genes. Thus, if the A locus and the B locus are linked, the chance of a recombination between them in an individual is r and the chance of no recombination is $(1 - r)$. In any one individual, recombination either does or does not happen, but the chance of recombination determines the frequencies of genotypes in the gametes produced by a population. If we consider a large number of *AB/ab* individuals, they will produce gametes in the following proportions:

gamete	*AB*	*Ab*	*aB*	*ab*
proportion	$\frac{1}{2}(1 - r)$	$\frac{1}{2}r$	$\frac{1}{2}r$	$\frac{1}{2}(1 - r)$

These fractions can be used to calculate the Mendelian ratio for a cross involving an *AB/ab* individual. The principle is logically easy to understand, but the ratios can be laborious to work out in practice. The case of independent segregation corresponds to $r = \frac{1}{2}$. That is, when the A and B loci lie on separate chromosomes, $r = (1 - r) = \frac{1}{2}$ and the *Ab/aB* parent produces *Ab, aB, AB*, and *ab* gametes in the ratio 1:1:1:1.

For any two genes, recombination can "hit" more than once between them in an individual (a process called "multiple hits"). If two hits occur between a pair of loci, they cancel one another and the chromosome has the same combination of genes as if no recombination occurred. Thus, it is more accurate to say that the

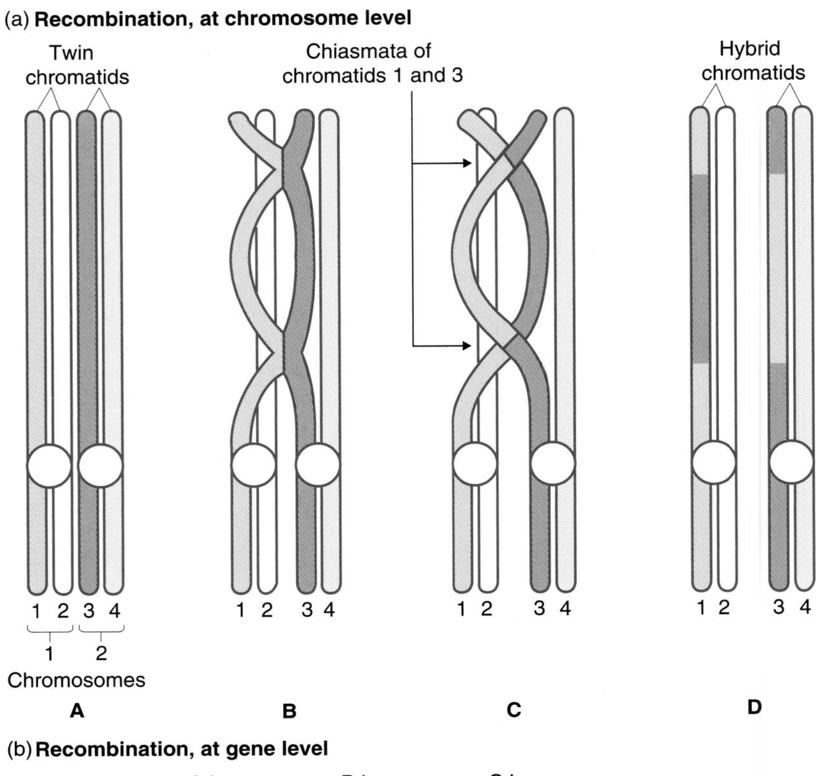

(b) **Recombination, at gene level**

A locus
Allele *A*
B locus
Allele *B*
C locus
Allele *C*
Chromosome 1
Recombination
Chromosome 2
Allele *a*
Allele *b*
Allele *c*

A *B* *c*

a *b* *C*

(c) **Recombination, at nucleotide level**

G G A G C T T G A T C A T C
Double helix, only one strand shown
Recom-bination
G G G G C T T G A A C A T C

G G A G C T T G A A C A T C

G G G G C T T G A T C A T C

Figure 2.7 Recombination, seen at the level of (a) chromosomes, (b) genes, and (c) nucleotides. At recombination, the strands of a pair of chromosomes break at the same point and the two recombine. The post-recombinational sequence of genes, or nucleotides, combines one strand from one side of the break-point with the other strand from the other side. In (b), the gene sequence in chromosome 1 changes from ABC to ABc; in (c), the nucleotide sequence of the chromosome with bases A and T (stippled nucleotides) changes to A and A. (For the nucleotide sequence, only one of the strands of the double helix is shown: each of the pairs of chromosomes has a full double helix with complementary base pairs, as in Figure 2.2.)

probability of recombination r equals the probability that an odd number of hits will occur, and the probability $(1 - r)$ equals the chance of no hits plus the chance of an even number of hits.

The Mendelian ratios, in which paired diploid genes segregate into haploid gametes and the gametes of different individuals then combine at random, are the basis of all the theory of population genetics that is discussed in chapters 5–9.

2.8 *Darwin's theory would probably not work if there was a non-Mendelian blending mechanism of heredity*

As chapter 1 described, Mendel's theory of heredity plugged a dangerous leak in Darwin's original theory, and the two theories together eventually came to form the synthetic theory of evolution, or neo-Darwinism. The problem was Darwin's lack of a sound theory of heredity; indeed, it had been shown even in Darwin's time that natural selection would not work if heredity was controlled in the way thought by most biologists before Mendel. Before Mendel, most theories of heredity were *blending* theories. We can see the distinction in much the same terms as have just been used for Mendelism (Figure 2.8). Suppose gene *A* causes its bearers to grow up dark blue in color, while gene *a* causes its bearers to grow up white. We can imagine that, as in the real world of Mendelism, individuals in our imaginary world of blending heredity are diploid and have

Figure 2.8 (a) Blending inheritance. The parental genes for dark blue (*A*) and white (*a*) color blend in their offspring, who produce a new type of gene (*A*′) coding for light blue color. (b) Mendelian inheritance. The parental genes are passed on unaltered by the offspring.

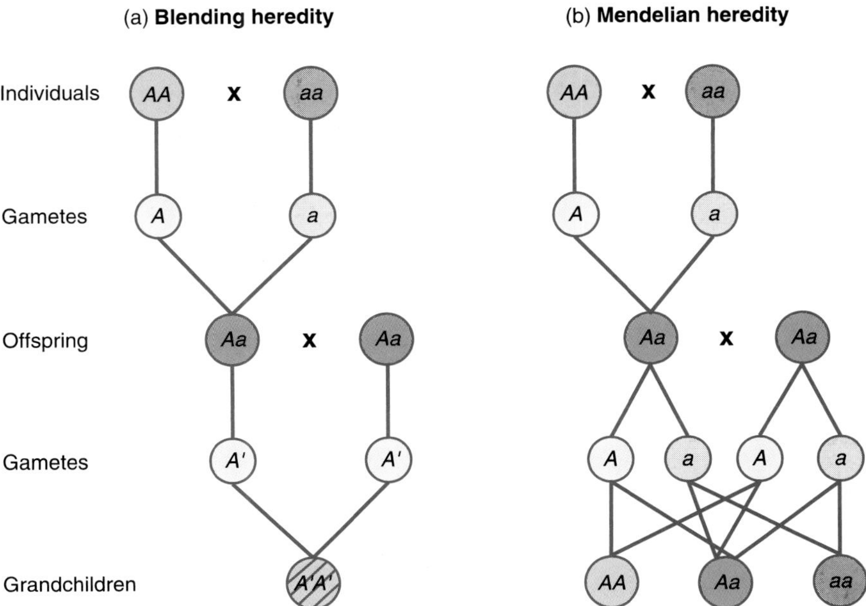

two copies of each "gene." An individual could then either inherit an *AA* genotype from its parents and have a dark blue phenotype, inherit an *aa* genotype and have a white phenotype, or inherit an *Aa* genotype and have a light blue phenotype. (Thus, in the Mendelian version of the system, we would say that no dominance exists between the *A* and *a* genes.)

The interesting individuals for this argument are the ones that have inherited an *Aa* genotype and have grown up to be light blue. They could have been produced in a cross of a dark blue and a white parent: the offspring will then be light blue whether inheritance is Mendelian (with no dominance) or blending. But consider the next generation. Under Mendelian heredity, the light blue *Aa* heterozygote passes on intact to its offspring the *A* and *a* genes it had inherited from its father and mother. Under blending heredity, an individual does not pass on the same genes as it inherited. If an individual inherited an *A* and an *a* gene, the two would physically blend in some way to form a new sort of gene (let us call it A') that causes light blue coloration. Instead of producing ½ *A* gametes and ½ *a*, the individual would then produce all A' gametes. This would affect the second generation. Whereas in Mendelian heredity the dark blue and white colors segregate out again in a cross between two heterozygotes, in the analogous cross with blending heredity they do not—all the grandchildren are light blue (see Figure 2.8).

Mendelism is an atomistic theory of heredity. Not only do discrete genes encode discrete proteins, but the genes are also preserved during development and passed on unaltered to the next generation. In a blending mechanism, the "genes" are not preserved. The genes that an individual inherits from its parents are physically lost, as the two parental sets are blended together. In Mendelism, it is perfectly possible for the *phenotypes* of the parents to be blended in the offspring (as they are in the initial $AA \times aa$ cross in Figure 2.8), but the genes do not blend. Indeed, the phenotypes of real mothers and fathers often do blend in their offspring, which led most students of heredity before Mendel to assume that inheritance must be controlled by some blending mechanism. However, the case of heterozygotes that are intermediate between the two homozygotes (i.e., no dominance) shows that the blending of phenotypes need not mean blending of genotypes. In fact, the underlying genes are preserved.

One way of expressing the importance of Mendelism for Darwin's theory is to say that it efficiently preserves genetic variation. In blending inheritance, variation is rapidly lost as extreme types mate together and their various "genes" are blended out of existence into some general mean form; in Mendelian inheritance, variation is preserved because the extreme genetic types (even if disguised in heterozygotes) are passed down from generation to generation.

Why does this preservation of genes matter for Darwinism? Our full discussion of natural selection comes in later chapters, and some readers may prefer to return to this point after they have read about natural selection in more detail. Armed with only the elementary account of natural selection in chapter 1, however, it is still possible to understand why Darwin "needed" Mendel. Figure 2.9 illustrates the argument.

Suppose that a population of individuals is white in color and has the *aa* genotype, and heredity blends (in the manner of Figure 2.8). For some reason, it is advantageous for individuals in this population to be dark blue in color:

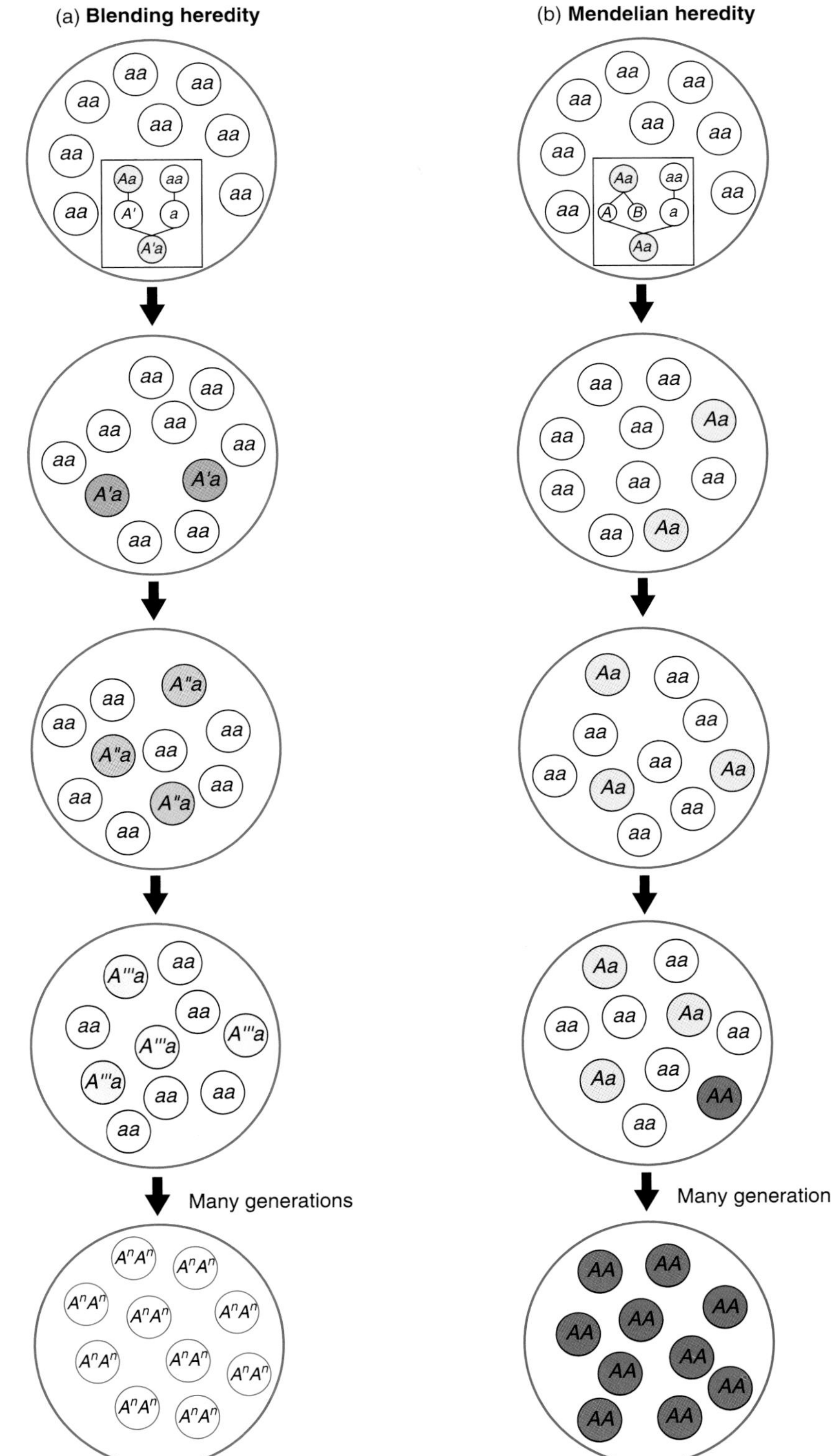

Figure 2.9 Two populations with 10 individuals each (real populations would have many more members), one with blending heredity and the other with Mendelian heredity. (a) Under blending heredity, a rare new advantageous gene is soon blended away. (b) Under Mendelian heredity, a rare new favorable gene can increase in frequency and eventually become established in the population. (See text for explanation.)

dark blue individuals would survive better and leave more offspring. Moreover, it is better to be somewhat blue (i.e., light blue) than to be white. Suppose now that a single new light blue individual somehow crops up by mutation, and it has an *Aa* genotype. This *Aa* individual will survive better than its *aa* fellow members of the population and produce more offspring. However, the advantageous gene cannot last long with blending. In the first generation, it produces A' gametes; these combine with *a* gametes (because all other members of the population are white) and produce $A'a$ offspring. We can suppose that these individuals are slightly lighter blue in color than the original *Aa* mutant; they still have an advantage, but it is less marked.

The $A'a$ individual's genes blend, such that all its gametes will have an A'' gene. When that unites with an *a* gamete (because still almost everyone else is white) an $A''a$ offspring results, which is even lighter blue in color. After a matter of time, the original favorable mutation will be blended almost out of existence. The best result possible would be a population that was very slightly less white than the initial population. A population of dark blue individuals could not possibly result from the mutation, because the original gene, which potentially could produce dark blue individuals, will cease to exist after only a single generation. The most famous objection was made in Darwin's time by a Scottish engineer, Fleeming Jenkin. Though Darwin had known about it since his early thinking in the 1840s, he never did find a wholly satisfactory way around it.

Mendelism was what Darwin needed. In the example just given, the original light blue mutation will be in an *Aa* heterozygote, and fully half its offspring will be light blue—because they are also *Aa* heterozygotes. Ample time exists for natural selection to increase the proportion of light blue individuals, and eventually enough will be produced so that two could potentially mate and produce some *AA* homozygotes among their offspring. It is now theoretically possible for a population of dark blue *AA* individuals to result (Figure 2.7b). Thus, natural selection is a powerful process when combined with Mendelian heredity, because Mendelian genes are preserved over time; in contrast, natural selection is a weak process when combined with blending inheritance, because potentially favorable genes are diluted before they can be established.

SUMMARY

1. Heredity is determined by a molecule called DNA. The structure and mechanisms of action of DNA are understood in detail.
2. The DNA molecule can be divided into regions called genes that encode for proteins. The code in the DNA is read to produce a protein in two stages: transcription and translation. The genetic code has been deciphered.
3. DNA is physically carried on structures called chromosomes. Each individual has a double set of chromosomes (one inherited from its father, the other from its mother), and therefore two sets of all its genes. An individual's particular combination of genes is called its genotype.
4. New genetic variation originates by mutational changes in the DNA. Rates of mutation can be estimated by direct observation.
5. When two individuals of given genotypes mate together, the proportions of genotypes in their offspring appear in predictable Mendelian ratios. The exact ratios depend on the genotypes in the cross.

6. Different genes are preserved over the generations under Mendelian heredity, which enables natural selection to operate. Before Darwin, it was generally (but wrongly) thought that the maternal and paternal hereditary materials blended in an individual rather than being preserved; if heredity did blend, natural selection would be much less powerful than under Mendelian heredity.

FURTHER READING

Any genetics text, such as Lewin (1994), Griffiths *et al.* (1993), or Weaver and Hedrick (1991), explains the subject in detail. See Lewontin's entry in Keller and Lloyd (1992) on the genotype-phenotype distinction. The classic statement of why Darwinism requires Mendelism, and does not work with blending heredity, is found in the first chapter of Fisher (1930).

STUDY AND REVIEW QUESTIONS

1. Review your understanding of the following genetic terms: DNA, chromosome, gene, protein, genetic code, transcription, translation, mRNA, tRNA, rRNA, mutation, synonymous, non-synonymous, frame-shift, inversion, genetic locus, meiosis, genotype, phenotype, homozygote, heterozygote, dominant, recessive, linked, unlinked, recombination.
2. What are the Mendelian ratios for the following crosses? (i) $AA \times AA$, (ii) $AA \times Aa$, (iii) $Aa \times Aa$, (iv) $AB/AB \times AB/AB$, when the A and B loci are unlinked, (v) $AB/AB \times AB/AB$, when the A and B loci are linked, (vi) $Ab/aB \times AB/ab$, when the A and B loci are unlinked, (vii) $Ab/aB \times AB/ab$, when the A and B loci are linked.

The Evidence for Evolution

How can it be shown that species change through time, and that modern species share a common ancestor? We begin with direct observations of change on a small scale and move out to more inferential evidence of larger-scale change. We then examine what is probably the most powerful general argument for evolution: the existence of certain kinds of similarities (called homologies) between species—similarities that would not be expected to exist if each species had originated independently. Homologies fall into hierarchically arranged clusters, as if they had evolved through a tree of life and not independently in each species. The order in which the main groups of animals appear in the fossil record makes sense if they arose by evolution, but would be highly improbable otherwise. Finally, the existence of adaptation in living things has no non-evolutionary explanation, although the exact way that adaptation can be used to suggest evolution depends on the alternative against which the argument is being made.

3.1 We distinguish three possible theories of the history of life

In this chapter, we will be asking whether, according to the scientific evidence, one species has evolved into another in the past, or whether each species had a separate origin and has remained fixed in form ever since that origin. For purposes of argument, it is useful to have some articulate alternatives. We can discuss three theories (Figure 3.1): (a) evolution, (b) transformism, in which species change, but the number of origins of species matches the number of species, and (c) separate creation, in which species originate separately and remain fixed. We will, therefore, consider the evidence for two evolutionary claims: (1) that species have changed in Darwin's sense of "descent with modification," and (2) that all species share a common ancestor, and that change occurs through a tree-like history.

Whether species have separate origins, and whether they change after their origin, are two distinct questions; some kinds of evidence, therefore, may bear upon one question but not the other. At this stage, we need not have any particular mechanism in mind to explain either how species spring into existence so easily in the theories of transformism and separate creation (Figure 3.1b, c), or how they change in form in the theories of evolution and transformism (Figure 3.1a, b). We merely suppose it could happen by some natural mechanism, and ask which of the three patterns is supported by the evidence.

In our examination of these issues, we will consider a number of lines of biological evidence. People differ in what they see as the main objection to the

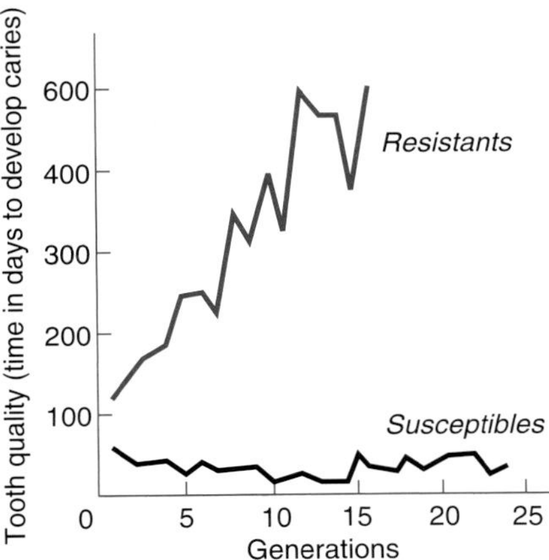

Figure 3.4 Selection for better and worse teeth in rats. Hunt *et al.* (1955) selectively bred each successive generation of rats from parental rats that developed caries later (resistants) or earlier (susceptibles) in life. The age (in days) at which their descendants developed caries was measured.

3.4 *Interbreeding and phenotypic similarity provide two concepts of species*

We are now close to the stage in the argument when we can consider evidence for the evolution of new species. Most of the evidence encountered so far has related to small-scale change within a species: house sparrows from all over North America are still classified as house sparrows, formally *Passer domesticus*, the same species as was introduced in the 1850s. The amounts of artificially selected change in pigeons and other domestic animals borders on the species level, but we must first develop a concept of a biological species before deciding whether the species barrier has been crossed.

All living creatures are classified into a Linnaean hierarchy. The species is the lowest important level in the hierarchy. Species, in turn, are grouped into genera, genera into families, and so on up through a series of levels. Figure 3.5 gives a fairly complete Linnaean classification of the wolf, as an example. If all life has descended from a single common ancestor, evolution must be capable of producing new groups at all levels in the hierarchy, from species to kingdom. Although we will look at the evidence for this development later in this chapter, for now we will concentrate on the species stage. What does it mean to say a new species has evolved?

This question unfortunately lacks a simple answer that would satisfy all biologists. We will discuss the topic fully in chapter 15, where several concepts of species will be discussed. In this chapter, however, we will take two of the most important species concepts and examine the evidence for the evolution of new species described by each. In arguing for evolution, we do not have to define a species. If someone asks, "What's the evidence that evolution can produce a new species?", we can reply, "Tell me what you mean by species, and I'll show you the evidence."

One important species concept is reproductive, defining a species as a set of organisms that interbreed among themselves but do not breed with members of

Figure 3.5 Each species in a biological classification is a member of a group at each of a succession of more inclusive hierarchical levels. The figure gives a fairly complete classification of the gray wolf *Canis lupus*. This way of classifying living things was invented by the eighteenth-century Swedish biologist who wrote under the latinized name Carolus Linnaeus.

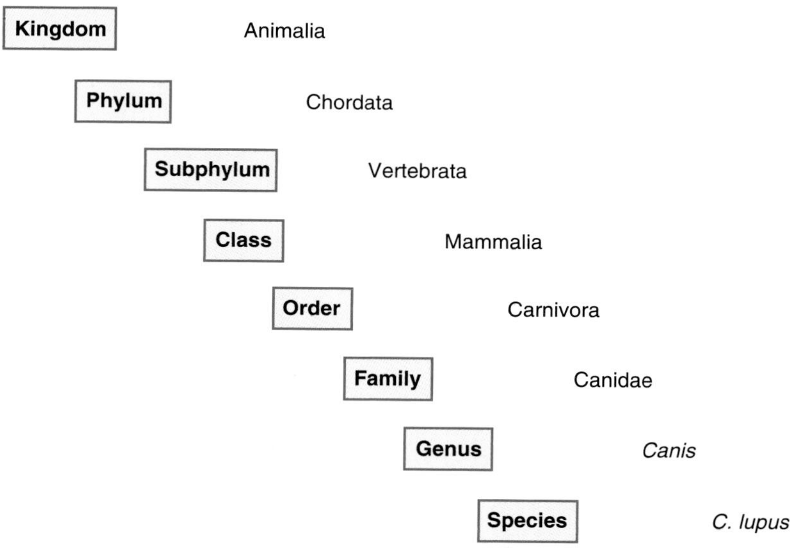

other species. Humans (*Homo sapiens*) are a separate reproductive species from the common chimpanzee (*Pan troglodytes*): any human can interbreed with any other human (of appropriate sex), but not with a chimp.

The second important concept uses phenotypic appearance: it defines a species as a set of organisms that are sufficiently similar to one another and sufficiently different from members of other species. This definition is less objective than the reproductive definition: it is clear whether the members of two population can interbreed, but it is less clear whether the two are sufficiently different to count as two phenotypic species. The final answer often lies with an expert who has studied the forms in question for years and acquired a good knowledge of the difference between species; formal methods of answering the question also exist. However, for relatively familiar animals we all have an intuitive phenotypic species concept. Again, humans and common chimpanzees belong to different species, and they are clearly distinct in phenotypic appearance. Common suburban birds, such as robins, mockingbirds, and starlings are separate species, and can be seen to have distinct coloration. Thus, without attempting a general and exact answer to the question of how great the differences between two species must be, we can see that phenotypic appearance might provide another species concept.

Because some biologists may reject one of these concepts, we should consider the evidence for the evolution of new species according to both concepts. As we move up the Linnaean hierarchy and reach categories above the species level, the members of a group become less and less similar. Two members of the same species, such as two wolves, are more similar than are two members of the same genus, but different species, such as a wolf and a silver-backed jackal (*Canis mesolemas*).

Members of the same class (Mammalia) can be as different as a bat, a dolphin, and a giraffe.

What degree of difference, in these taxonomic terms, has been produced by artificial selection in domestic animals? All domestic pigeons can interbreed, and are members of the same species in a reproductive sense. Such birds can be distinguished by their phenotypic appearance, however. Museum experts often have to classify birds from dead specimens, of unknown reproductive habits, and they make use of phenotypic characters of the bones, beak, and feathers. Darwin kept many varieties of pigeons, and in April 1856, when Lyell came for a visit, Darwin was able to show him how his 15 pigeon varieties differed enough to make "three good genera and about fifteen species according to the received mode of species and genera-making of the best ornithologists."

The variety of dogs (*Canis familiaris*) is comparable. To most human observers, the difference between extreme forms, such as a Pekinese and a St. Bernard, is much greater than that between two species in nature, such as a wolf and a jackal, or even two species in different genera, such as a wolf and an African hunting dog (*Lycaon pictus*). However, most domestic dogs are interfertile and belong to the same species in a reproductive sense. The evidence from domestic animals suggests that artificial selection can produce extensive change in phenotypic appearance—enough to produce new species and even new genera—but has not produced much evidence for new reproductive species. Evidence for the evolution of new reproductive species will be presented in a later section.

3.5 *Ring "species" show that the variation within a species can be extensive enough to produce a new species*

At any one time and place, an array of distinct species appears in nature. For example, a naturalist in southern California might have noticed two forms of the salamander *Ensatina*. One form, the species *Ensatina klauberi*, is strongly blotched in color, whereas the other, the species *E. eschscholtzii*, is more uniformly and lightly pigmented. Since the work of Stebbins in the 1940s, it had been suspected that they were two good species, meaning that they are distinct forms that do not interbreed where they coexist. For one site, 4600 feet up the Cuyamaca Mountains, San Diego County, Wake *et al.* (1986) confirmed that the two behaved as separate species. At that site, called Camp Wolahi, the two species coexist. No hybrid forms between them were found, however, and the genetic differences between the two species there suggested they had not interbred in the recent past. Salamander naturalists who visited Camp Wolahi would have no doubt they were looking at two ordinary species.

However, if those naturalists look for the two salamander species in other areas of southern California, the two species do not seem to be as distinct as at Camp Wolahi. Wake *et al.* sampled the salamanders from three nearby sites, and at all of them a small proportion (up to 8%) of individuals in the sample were hybrids between *E. eschscholtzii* and *E. klauberi*.

The picture becomes even clearer when the geographical scale is expanded. The salamanders can be traced westward from Camp Wolahi, to the coast, and northward, up the mountain range (Color Plate 1). However, in either direction, only one of the salamanders is present. The lightly pigmented, unblotched form *E. eschscholtzii* is found along the coast, while the blotched *E. klauberi* are located inland. Both forms can be traced to northern California, but they vary in form toward the north; the various forms have been given a series of taxonomic names, as can be seen in Color Plate 1. The forms meet again in northern California and Oregon, although only one form is found. In this location, the eastern and western forms have apparently merged completely.

Stebbins's interpretation of the salamanders' geographic pattern is as follows. Originally only one species lived in the northern part of the present range. The population then expanded southward, splitting down either side of the central San Joaquin Valley. The subpopulation on the Pacific side evolved the color pattern and genetic constitution characteristic of the coastal *E. eschscholtzii*, while the subpopulation inland evolved the blotches and the genetic constitution characteristic of *E. klauberi*. At various points down California, subpopulations leaked across and met the other form. At some of these meeting areas, the two forms interbred to some extent, and hybrids can be found; in such locations, the forms have not evolved apart enough to be separate reproductive species. When the forms reached the southern tip of California, however, the two lines of population have evolved far enough apart that when they meet, such as at Camp Wolahi, they do not interbreed because they are two normal species. Thus, the two species at Camp Wolahi are connected by a continuous set of intermediate populations, looped around the central valley.

The detailed picture is more complicated, but recent work supports essentially the same interpretation. One of the complications can be seen in Color Plate 1, which shows that the set of populations may not be perfectly continuous because of a gap in the southeastern part of the ring. Jackman and Wake (1994) showed that the salamander populations on either side of the gap are genetically no more different than are salamanders separated by an equivalent distance elsewhere in the ring. They suggest two interpretations. One is that salamanders lived in the gap until recently but are now extinct there; the other is that the blotched *Ensatina* are there and waiting to be found in the San Gabriel Mountains.

The salamander species *E. eschscholtzii* and *E. klauberi* in southern California are an example of a *ring species*. To imagine a ring species in the abstract, first imagine a species that is geographically distributed more or less in a straight line in space—for example, from east to west across America. The forms in the east and west might be so different that they could not interbreed, but this question is likely to remain unresolved because the two forms do not meet one another. Now imagine taking the line and bending it into a circle, such that the end points (formerly in the east and west) come to overlap in space. At this location, it will be possible to find out whether the two extremes do or do not interbreed. If they do interbreed, then the geographic distribution of the species takes the shape of a ring, but the animals will not be a "ring species" in the technical sense.

A proper ring species is one in which the extreme forms do not interbreed in the region of overlap. A ring species has an almost continuous set of intermediates between two distinct species, and these intermediates happen to be arranged in a ring. At most points in the ring, only one species exists. Where the the end points meet, two species may be found. (The statement that the extremes either do or do not interbreed is too categorical for real cases, which are typically more complicated. In the salamanders, for instance, hybridization occurs at some sites but not at others in southern California where the ring closes up. Thus, the real situation is not a simple ring, but can be understood as a ring species, with due allowance for "real world" complications.)

Ring species can provide important evidence for evolution, because they show that intra-specific differences can be large enough to produce an inter-species difference. The differences between species are, therefore, the same in kind (though not in degree) as the differences between individuals, and populations, within a species.

Natural variation comes in all degrees. At the smallest level, individuals demonstrate slight differences. Populations of a species show somewhat larger differences, and species are even more different. In a normal species, whose distribution of members may perhaps resemble the line we described earlier, the extreme forms may be very different from one another. We do not know, however, whether they are different enough to count as separate species in the reproductive sense. A supporter of the theory of separate creation might then argue that although individuals do vary within a species, that variation is too limited to give rise to a new species: the origin of new species is then not a magnified extension of the kind of variation we see within a species. In contrast, the extremes meet in ring species, and we can see that they form two species. Clearly, natural variation can, at least sometimes, be large enough to generate new species. Some species, therefore, have arisen without separate creation.

A slippery slope marks the path from interindividual variation to the difference between two species. Small individual differences, we know, arise by the ordinary processes of reproduction and development: we can *see* that each individual is not separately created. By extension, we can easily envision how the slightly larger differences between local populations arise without separate creation. In the case of the ring species of salamanders, this process can be seen to extend far enough to produce a new species. To deny it would require an arbitrary decision about where evolution stopped and separate creation started.

Suppose, for example, someone claimed that all salamanders found west of a point in northern California were separately created as a different species from all those located to the east of this point; at the same time, it is accepted that the variation within each species on either side of the point arose by ordinary natural evolutionary processes. The claim is clearly arbitrary and absurd. If evolution has produced the variation between salamanders in northern California and in mid-California on the coast, and between northern California and mid-California inland, it is nonsensical to suggest that the populations in the east and the west of northern California were separately created. The variation between any two points in the ring is of much the same kind, and the variation across the arbitrarily picked point will be just like the variation between two points to the left or right of it. Ring species, therefore, demonstrate the existence of a continuum from interindividual to interspecies variation. Natural variation is sufficient to break down the idea of a distinct species boundary.

The same argument, we shall see, can be applied to larger groups than species and, by extension, to all life. The idea that nature comes in discrete groups, with no variation between those groups, is a naive perception. If the full range of natural forms, in time and space, is studied, all the apparent boundaries become fluid.

3.6 *New reproductively distinct species can be produced experimentally*

The species barrier can also be broken by experiment. The varieties of artificially produced domestic animals and plants can differ in appearance at least as much as natural species, but they may be able to interbreed. Dog breeds that differ greatly in size probably interbreed little in practice, but it is still interesting to know whether we can make new species that unambiguously do not interbreed. It is possible to select directly for reduced interbreeding between two forms (section 16.4.2, p. 429).

More extreme, and more abundant, examples of new, reproductively isolated species come from plants. In a typical interbreeding procedure, we begin with two distinct, but related species. The pollen of one species is painted on the stigma of the other. If a hybrid offspring is generated, it is usually sterile: the two species are reproductively isolated. However, it may be possible to treat the hybrid in such a way as to make it fertile. For example, the chemical colchicine can often restore hybrid fertility by causing the hybrid to double its number of chromosomes (a condition called polyploidy). Hybrids so produced may be interfertile with other hybrids like themselves, but not fertile with the parental species. They represent a new reproductive species, and provide clear evidence that new species in the reproductive sense can be produced. If we add these plant examples to the examples of dogs and pigeons, we have now seen evidence for the evolution of new species according to both the reproductive and the phenotypic species concepts.

The first artificially created hybrid polyploid species was a primrose, *Primula kewensis*. It was formed by crossing *Primula verticillata* and *P. floribunda*. *Primula kewensis* is a distinct species: a *P. kewensis* individual will breed with another *P. kewensis* individual, but not with members of *P. verticillata* or *P. floribunda*. *P. verticillata* and *P. floribunda* have 18 pairs of chromosomes each, and simple hybrids between them also have 18 chromosomes. These hybrids are sterile. In contrast, *P. kewensis* has 36 chromosomes and is a fertile species. The chromosome doubling in this case was not induced artificially, by colchicine treatment, but occurred spontaneously in a hybrid plant.

Hybridization, followed by the artificial induction of polyploidy, is now a common method of producing new agricultural and horticultural varieties. Most garden varieties of irises, tulips, and dahlias, for example, are artificially created species. Their numbers are dwarfed by the huge numbers of artificial hybrid species of orchids, which are being formed at a rate estimated at approximately 300 new species per month.

Polyploid hybrids are not only a mechanism of artificial speciation. They are also a major source of new natural plant species (in this case, the polyploidy must occur spontaneously, by mutation, rather than being artificially induced). To decide how common polyploid hybrids have been in the origin of new plant species, we first need a method to recognize the hybrids.

One simple method is to examine the chromosomes in a group of related plant species; the form and number of chromosomes in the set of species may suggest which originated as hybrid species from two other species in the group. It is quite common for a genus, or a family, of flowering plants to consist of species with simple multiples of a basic number of chromosomes (N): the different species might have N, $2N$, $4N$, . . . chromosomes. The obvious interpretation is that many, or all, of the species with higher numbers of chromosomes originated by polyploidization from species with fewer chromosomes.

A more detailed method involves comparing the proteins (by gel electrophoresis) or DNA sequences of the species, again to see whether one species's set of genes appears to be a hybrid of the genes in two other species. By these and other methods, Grant estimated that roughly 50% of flowering plant species originated as polyploid hybrids. Other estimates, by Lewis and Stebbins, place this figure at 70%–80%.

The most powerful method of showing that a species originated as a hybrid is to re-create it from its ancestors, by hybridizing the conjectural parental species experimentally. This technique was first used with a common European herb, *Galeopsis tetrahit*, which Müntzing in 1930 successfully created by hybridizing *G. pubescens* and *G. speciosa*. The artificially generated *G. tetrahit* can successfully interbreed with naturally occurring members of the species. This method is more time-consuming than simple chromosome counts and has only been used with a small number of species. We return to this topic in section 16.7 (p. 446), when we discuss an example, from the *Tragopogon* of the Washington–Idaho region, of two new species that have originated by natural hybridization and polyploidy.

In conclusion, it is possible to make new reproductively isolated species, and by a method that has been highly important in the origin of new natural species.

3.7 *Small-scale observations can be extrapolated over the long term*

We have now seen that evolution can be observed directly on a small scale: the extreme forms within a species can be as different as two distinct species, and, in both nature and experiments, species will evolve into forms highly different from their starting point. It would be impossible, however, to observe in the same direct way the entire evolution of life from its common, single-celled ancestor a few billion years ago. Human experience is too brief.

As we extend the argument from small-scale observations, like those described in moths, dogs, and salamanders, to the history of all life we must shift from observation to inference. It is possible to imagine, by extrapolation, that if the small-scale processes we have seen were continued over a long enough period they could have produced the modern variety of life. The principle behind this theory is called *uniformitarianism*. In a modest sense, uniformitarianism means merely that processes seen by humans to operate could also have operated when humans were not watching; it also refers to the more controversial claim that processes operating in the present can account, by extrapolation over long periods, for the evolution of the earth and life. For instance, the long-term persistence of the processes we have seen in moths and salamanders could result in the evolution of life.

The principle of uniformitarianism is not peculiar to evolution, but is used in all historical geology. When the persistent action of river erosion is used to explain the excavation of deep canyons, the reasoning principle again is uniformitarianism.

Figure 3.6 All modern tetrapods have a basic pentadactyl (five-digit) limb structure. The forelimbs of a bird, human, whale, and bat are all constructed from the same bones even though they perform different functions. Adapted with permission from Strickberger (1990). © 1990 Boston: Jones and Bartlett Publishers.

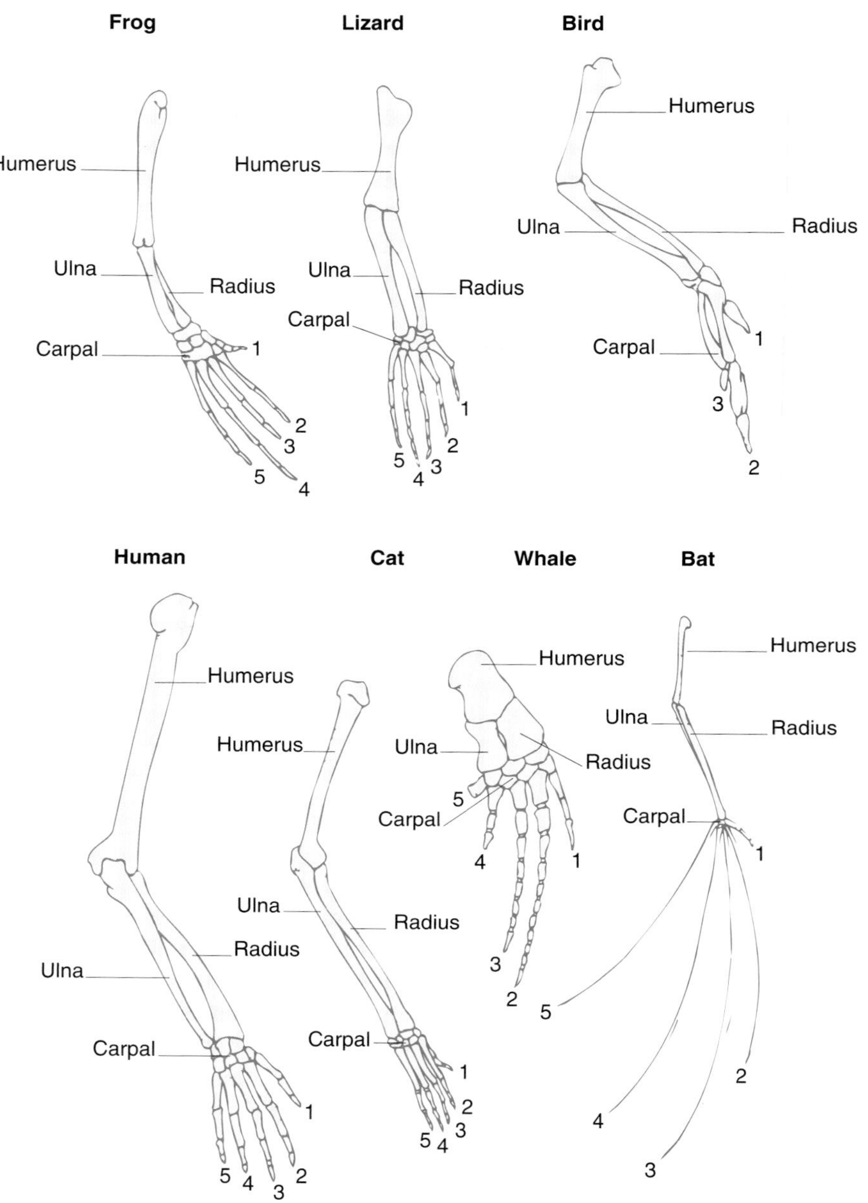

hemoglobin, but when injected with the mRNA they produce rabbit hemoglobin. The machinery for decoding the message must therefore be common to rabbits and *E. coli*. If the machinery is present in both of these organisms, it is a reasonable inference that all living things have the same code. (Recombinant DNA technology is built on the assumption of a universal code.) Minor variants of the code, which have been found in mitochondria and in the nuclear DNA of a few species, do not affect this argument.

Why should the code be universal? Two explanations are possible: that the universality results from chemical constraint, or that the code represents a historical accident.

According to the chemical theory, each particular triplet would have some chemical affinity with its amino acid. GGC, for example, would react with glycine in some way that matched the two together. Several lines of evidence suggest that this explanation is wrong. First, no such chemical relation has been found (and not for want of looking), and it is generally thought that one does not exist. Second, the triplet and amino acid do not physically interact in the translation of the code. They are both held on a tRNA molecule, but the amino acid is attached at one end of the molecule, while the site that recognizes the codon on the messenger RNA lies at the other end (Figure 3.7).

In addition, certain mutants can change the relation between triplet code and amino acid (Figure 3.8), suppressing the action of another class of mutants. Some of the triplets in the genetic code are "stop" codons: they signal that the protein has come to an end. If a triplet within a coding region mutates to a stop codon, the protein is not manufactured. Examples of these mutations are well known in bacterial genetics, and a mutation to the stop codon UAG, for example, is called an amber mutation. Once a bacterial culture with an amber mutation has been formed, it is sometimes possible to find other mutations that suppress the amber mutation. These mutants are normal, or near-normal, bacteria. The amber-suppressing mutants work by changing the coding triplet on a class of

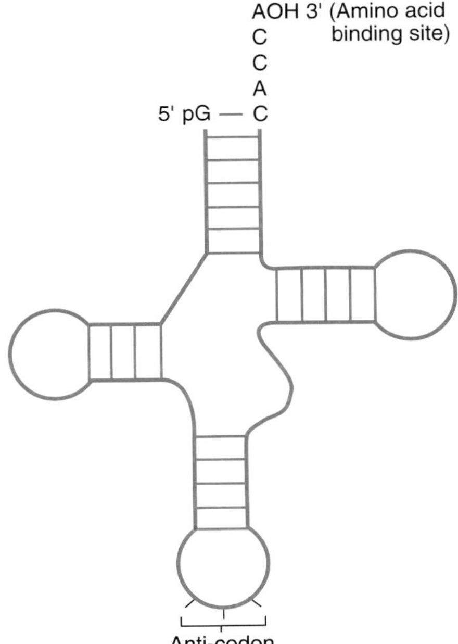

Figure 3.7 Transfer RNA molecule. The amino acid is held at the other end of the molecule from the anti-codon loop where the triplet code of the messenger RNA molecule is read.

Figure 3.8 Mutations that suppress amber mutations suggest that the genetic code is chemically alterable. For example, (a) the normal codon is UUG and encodes leucine. (b) The UUG mutates to the stop codon UAG (this is called an amber mutation). (c) A tRNA for tyrosine mutates from AUG to AUC (which recognizes UAG) and suppresses the amber mutation by inserting a tyrosine.

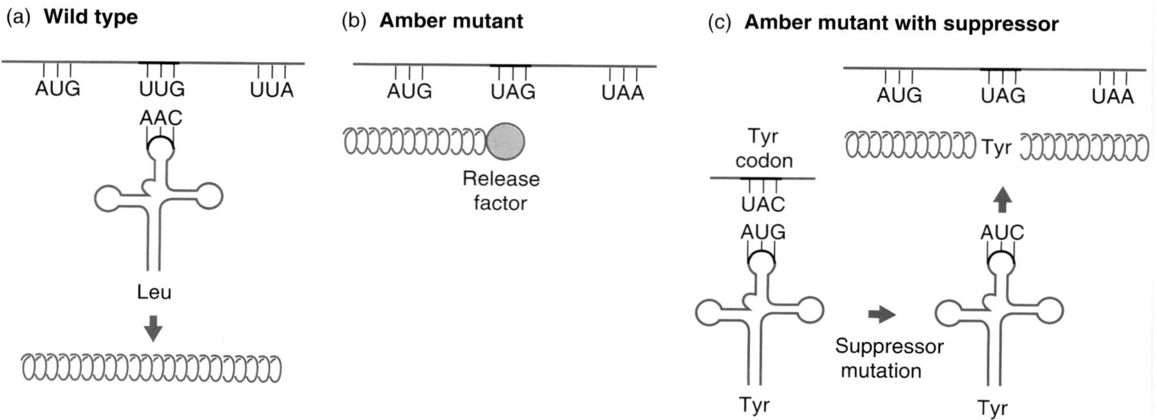

amino-acid–bearing tRNA to make it bind to UAG. The UAG codon then encodes an amino acid rather than halting transcription. The fact that the relation between amino acid and codon can be changed in this way shows that the same genetic code has not been forced on all species by some unalterable chemical constraint.

If the genetic code is not chemically determined, why is it the same in all species? The most popular theory is that it emerged as a historical accident. The code is arbitrary, in the same sense that human language is arbitrary. In English the word for a horse is *horse*, in Spanish it is *caballo*, in French it is *cheval*, and in ancient Rome it was *equus*. No reason exists why one particular sequence of letters rather than another should signify that familiar perissodactylic mammal. Therefore, if we find more than one person using the same word, it implies that these individuals have learned the word from a common source. That is, the discovery implies common ancestry. Thus, when the Starship *Enterprise* boldly descends on one of those extragalactic planets where the aliens speak English, the correct inference is that the locals share a common ancestry with one of the English-speaking peoples of the Earth. If they had evolved independently, they would not use English.

All living species use a common, but equally arbitrary, language in the genetic code. The code is thought to have evolved early on in the history of life, with one early form turning out to be the common ancestor of all later species. (Notice that saying all life shares a common ancestor is not the same as saying life evolved only once.) The code is then what Crick called a "frozen accident." The original choice of a code was an accident; once it had evolved, however, it would be strongly maintained. Any deviation from the code would be lethal. An individual that read GGC as phenylalanine instead of glycine, for example, would bungle all its proteins, and probably die at the egg stage.

The universality of the genetic code is important evidence that all life shares a single origin. In Darwin's time, morphological homologies like the pentadactyl limb were recognized, but these homologies are shared between fairly limited groups of species (like all the tetrapods). Cuvier (section 1.3.1, p. 6) had arranged all animals into four large groups according to their homologies. For this reason Darwin suggested a limited number (but perhaps more than one) of common ancestors for modern living species. Molecular homologies, such as the genetic code, now provide the best evidence that all life has a single common ancestor.

Homologous similarities between species provide the most widespread class of evidence that living and fossil species have evolved from a common ancestor. The anatomy, biochemistry, and embryonic development of each species contains innumerable characters like the pentadactyl limb and the genetic code: characters that are similar between species, but would likely be different if the species had independent origins. Homologies, however, are usually more persuasive for an educated biologist than for someone seeking immediately intelligible evidence for evolution.

The most obvious evidence for evolution is taken from direct observation of change; no one will have any difficulty in seeing how the examples of evolution in action, from moths and artificial selection, suggest that species are not fixed in form. The argument from homology is inferential, and more demanding. You must understand some functional morphology, or molecular biology, to appreciate that tetrapods would not share the pentadactyl limb, or all species share the genetic code, if they originated independently.

Nonetheless, some homologies are immediately persuasive—in particular, the homologies, such as vestigial organs, in which the shared form appears to be positively inefficient. If we consider the vertebrate limb, but move in from its extremities to the junction where it joins the spine, we find another set of bones—at the pectoral and pelvic articulations—that are recognizably homologous in all tetrapods. In most species, these bones are needed for the limb to be able to move; in a few species, such limbs have been lost. Modern whales, for instance, do not have hind limbs with bony supports. If we dissect a whale, we find at the appropriate place down the spine a set of bones that are clearly homologous with the pelvis of any other tetrapod (Figure 3.9). They are vestigial in the sense that they are no longer used to provide articulation for the hind limb; their retention suggests that whales evolved from tetrapods rather than being independently created.

Calling an organ vestigial does not necessarily mean it is functionless. Some vestigial organs may be truly functionless, but it is always difficult to confirm universal negative statements. Fossil whales called *Basilosaurus,* which lived 40 million years ago, had functional pelvic bones and may have used them when copulating; the vestigial pelvis of modern whales arguably is still needed to support the reproductive organs. However, that possibility does not count against the argument from homology. Why, if whales originated independently of other tetrapods, should they use bones that are adapted for limb articulation to support their reproductive organs? If they were truly independent, some other support would be used.

In homologies like the pentadactyl limb and the genetic code the similarity between species is not actively disadvantageous. One form of genetic code would probably be as good as almost any other, and no species suffers for using the actual genetic code found in nature. However, some homologies do appear positively disadvantageous (section 13.10, p. 359). One of the cranial nerves, as

Figure 3.9 Whales have a vestigial pelvic girdle, even though they do not have bony hind limbs. The pelvic bones are homologous with those of other tetrapods.

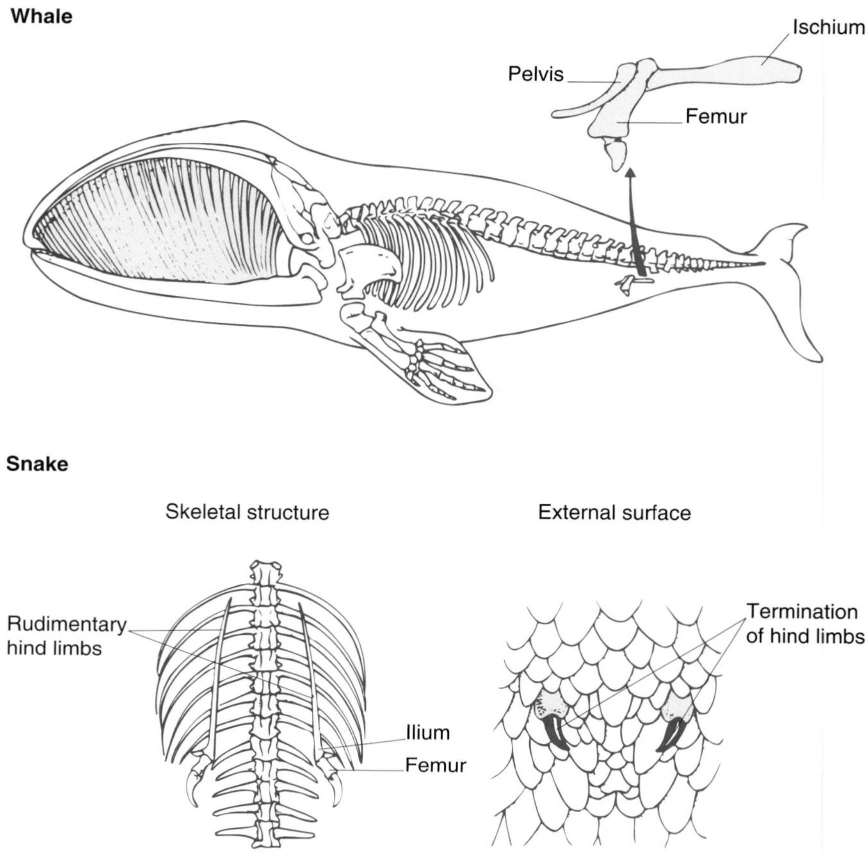

we shall see, goes from the brain to the larynx via a tube near the heart (see Figure 13.12). In fish, this path is a direct route. The same nerve in all species follows the same route, however. In the giraffe, it results in an absurd detour down and up the neck, so that the giraffe has to grow perhaps 10–15 feet more nerve than it would with a direct connection. The recurrent laryngeal nerve, as it is called, is surely inefficient. It is easy to explain such an efficiency if giraffes have evolved in small stages from a fish-like ancestor. Why giraffes should have such a nerve if they originated independently . . . well, we can leave that to others to try to explain.

3.9 *Different homologies are correlated, and can be hierarchically classified*

Different species share homologies, which suggests they are descended from a common ancestor, but the argument can be made both stronger and more revealing. Homologous similarities are the basis of biological classifications (chapter 14). Groups like "flowering plants," "primates," or "cats" are formally defined by homologies. The reason homologies are used to define groups is that they fall into a nested, or hierarchical, pattern of groups within groups; in addition, different homologies consistently fall into the same pattern.

A molecular study by Penny *et al.* (1982) illustrates the point, and shows how it argues for evolution. Different species can be more or less similar in the amino acid sequences of their protein, just as they can be more or less similar in their morphology. The pre-Darwinian distinction between analogy and homology is more difficult to apply to proteins. Our functional understanding of protein sequences is less well advanced than for morphology, and it can be difficult to specify an amino acid's function in the way we can for the pentadactyl limb. Actually, the functions of many protein sequences are understood, but the chemistry requires a great deal of explanation. For the argument here, it only needs to be accepted that *some* of the amino acid similarities between species are not functionally necessary, in the same way that all tetrapods do not have to have five-digit limbs. A protein contains a large number of amino acids, so this point need not be controversial. If we accept that some amino acids are homologous in the pre-Darwinian sense, we can see how their distribution among species suggests evolution.

Penny *et al.* examined protein sequences in a group of 11 species, using the pattern of amino acid similarities to work out the "tree" for the species. Some species have more similar protein sequences than others, and the more similar species are grouped more closely in the tree (chapter 17). The observation that suggests evolution is as follows. We start by describing the tree for one protein. We can then work the tree out for another protein, and compare the trees. Penny *et al.* worked out the 11-species tree for each of the five proteins, and observed that the trees for all five proteins are very similar (Figure 3.10). For 11 species, there are 34,459,425 different possible trees, but the five proteins suggest trees that form a small subclass from this large number of possible trees.

The similarities and differences in the amino acid sequences of the five proteins are correlated: if two species have more amino acid homologies for one of the proteins, they are also likely to have more such homologies for the other proteins. Thus, any two species are likely to be grouped together for any of the five proteins. If the 11 species had independent origins, there is no reason why their homologies should be correlated. In a group of 11 separately created species, some would no doubt show more similarities than others for any particular protein. But why should two species that are similar for cytochrome *c*, for example, also be similar for β-globin and fibrinopeptide-A? The problem is even more difficult because, as Figure 3.10 shows, all five proteins show a similar pattern of branching at all levels in the 11-species tree. It is easy to see how a set of independently created objects might show hierarchical patterns of similarity in any one respect. These 11 species have been classified hierarchically for five different proteins, however, and the hierarchy in all five cases is similar.

If the species are descended from a common ancestor, we would expect the observed pattern. All of the five proteins have been evolving in the same pattern of evolutionary branches, and we therefore expect them to show the same pattern of similarities. The hierarchical pattern of, and correlations among, homologies is evidence for evolution.

Figure 3.10 Penny *et al.* constructed the best estimate of the phylogenetic tree for 11 species using five different proteins. The "best estimate" of the phylogenetic tree is the tree that requires the smallest number of evolutionary changes in the protein. (a) For α-globin, six different trees were found, all requiring the same minimum number of changes. (b) For β-globin, six different trees were found, all requiring the same minimum number of changes. (c) For fibrinopeptide-A, there was one best tree. (d) For fibrinopeptide-B, there were eight equally good trees. (e) For cytochrome *c*, there were six equally good trees. The important point is how similar these trees are for all five proteins, given the large number of possible trees for 11 species. Reprinted with permission from Penny *et al.* (1982). Copyright 1982 MacMillan Magazines Limited.

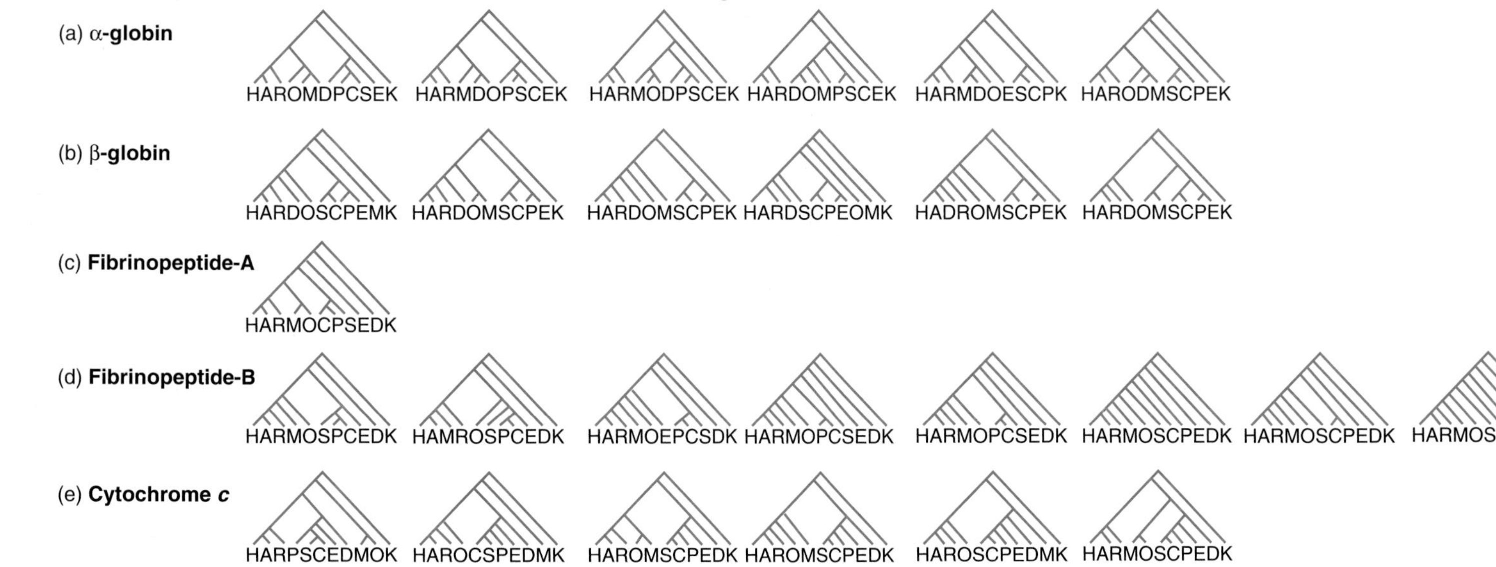

Consider an analogy involving a set of 11 buildings, each of which was independently designed and built. We could classify them into groups according to their similarities. Some might be built of stone, others of brick, others of wood; some might have vaults, others ceilings; some arched windows, others rectangular windows; and so on. It would be easy to classify the buildings hierarchically with one of these properties, such as building material. This classification would be analogous, in Penny *et al.*'s study, to making the tree of the 11 species for one protein. The same buildings could then be classified by another property, such as window shape; this strategy is analogous to classifying the species by a second protein. Some correlations between the two classifications of the buildings would probably emerge as a result of functional factors. Perhaps buildings with arched windows would be more likely to be built of brick or stone than of wood. However, other similarities would just be nonfunctional, chance associations in the particular 11 buildings in the sample. Perhaps the white-colored buildings in this 11-member set also happened to have garages, whereas the red buildings tend not to incorporate such structures. The argument for evolution concentrates on these inessential, rather than functional, patterns of similarity.

The analogy of Penny *et al.*'s result in the case of the buildings would be as follows. We should classify 11 buildings by five independent sets of characters (such as building material, window shape, and incidence and type of garage). We should then observe whether the five classifications all grouped the buildings in the same way. If the buildings were erected independently, no reason exists why they should show functionally unnecessary correlations. We would have no reason to expect that buildings that were similar for window shape, for example, would also be similar with respect to number of chimneys, angle of roof, or the arrangement of chairs indoors.

Of course, some innocent explanation might be found for any such correlations. (Indeed, if correlations *were* found in a real case, some explanation would be required.) Maybe they could all be explained by class of owner, or region, or common architects. Nevertheless, it would be clear that the buildings were not really independently created. If they were independently created, it would be very puzzling if they showed systematic, hierarchical similarity in functionally unrelated characteristics.

In the case of biological species, we do find this sort of correlation between characters. Figure 3.10 shows how similar the branching patterns are for five proteins, and the same conclusion could be drawn from any well-researched classification in biology. Biological classifications, therefore, provide an argument for evolution. If species had independent origins, we should not expect that, when several different (and functionally unrelated) characters were used to classify them, all the characters would produce strikingly similar classifications.

3.10 *There is some fossil evidence for the transformation of species*

Diatoms are single-celled, photosynthetic organisms that float in the plankton. Many species grow beautiful glasslike cell walls, and these can be preserved as fossils. Figure 3.11 illustrates the fossil record for the diatom *Rhizosolenia* between

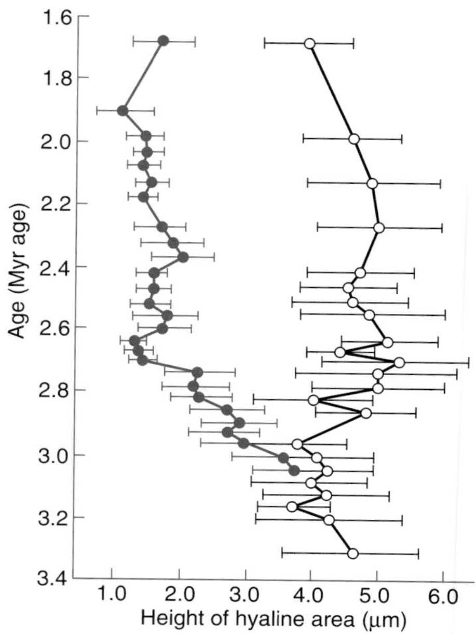

Figure 3.11 Evolution of the diatom *Rhizosolenia*. The form of the diatom is measured by the height of the hyaline (glasslike) area of the cell wall. ○—○ indicates forms classified as *R. praebergonii*, ●—● indicates *R. bergonii*. Bars indicate the range of forms at each time. Reprinted, by permission of the publisher, from Cronin and Schneider (1990).

3.3 and 1.6 million years ago. About 3 million years ago, a single ancestral species split into two. A comprehensive fossil record details the change at the time of the split.

The diatoms in Figure 3.11 show that the fossil record can be complete enough to reveal the origin of a new species, but such explicit examples are rare. In other cases, the fossil record is less complete and large gaps appear between successive samples (see chapter 19). In these instances, less direct evidence of smooth transitions between species exists. The gaps are usually long, however (perhaps 25,000 years in a good case, and millions of years in less complete records). One of the gaps allows enough time for large evolutionary changes, and no one need be surprised that fossil samples from either side of a gap in the record show large changes.

In other respects, as we saw at the beginning of the chapter (section 3.1), the fossil record provides important evidence for evolution. Against alternatives other than separate creation and transformism, the fossil record provides valuable evidence that the living world has not always been identical to its present form. The sheer existence of fossils proves some kind of change occurred, although it does not have to have been change in the sense of descent with modification.

3.11 *The order of the main groups in the fossil record suggests that they have evolutionary relationships*

The main subgroups of vertebrates, according to a conventional classification, are fish, amphibians, reptiles, birds, and mammals. It is possible to deduce that their order of evolution must have been fish, then amphibia, then reptiles, then mammals—not, for example, fish, then mammals, then reptiles, then amphibia

Figure 3.12 (a) Anatomical analysis of modern forms indicates that amphibians and reptiles are evolutionarily intermediate between fish and mammals. This order fits with (b) the geological succession of the major vertebrate groups. The width of each group indicates the diversity of the group at that time. Reprinted, by permission of the publisher, from Simpson (1949).

(a) **Anatomy of modern forms suggests order of evolution**

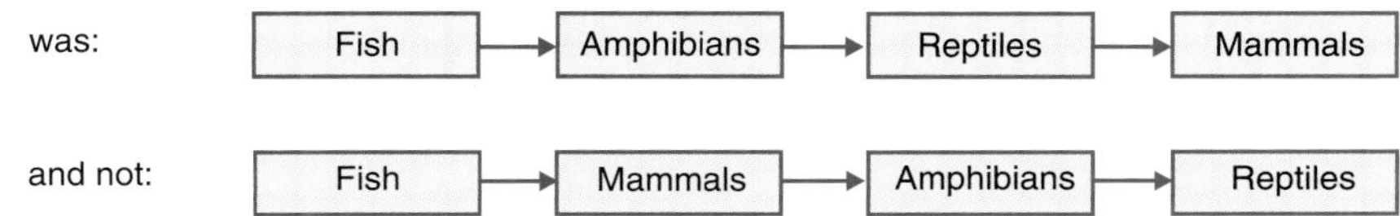

(b) **Order of main vertebrate groups in fossil records**

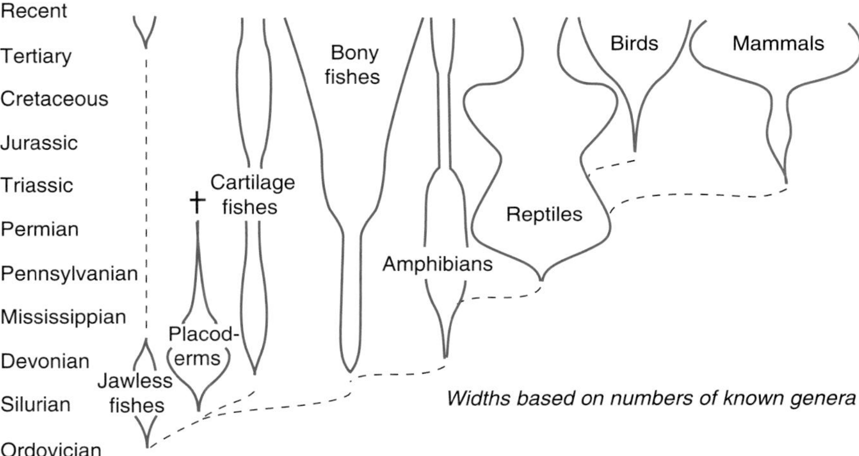

(Figure 3.12a). The deduction follows from the observation that an amphibian (such as a frog) or a reptile (such as an alligator) is intermediate in form between a fish and a mammal. Amphibians, for instance, have gills, like fish, but have four legs, like reptiles and mammals, rather than fins. If fish had evolved into mammals, and then mammals had evolved into amphibians, the gills would have been lost in the evolution of mammals and then regained in the evolution of amphibia—which is much less likely than that amphibia evolved from fish, retaining their gills, and the gills were then lost in the origin of mammals. (Chapter 17 discusses these arguments more fully.) Gills and legs are just two examples: the full list of characters placing amphibians (and reptiles, by analogous arguments) between fish and mammals would be long indeed. The forms of modern vertebrates alone, therefore, enable us to deduce the order in which they evolved.[3]

3. Strictly speaking, based on the argument given here, mammals could have come first and evolved into reptiles, the reptiles into amphibia, and the amphibia into fish. However, we can extend the argument by including more groups of animals, back to a single-celled stage. The fish would then be revealed, in turn, as an intermediate stage between amphibians and simpler animals.

The inference, from the modern forms, can be tested against the fossil record. The fossil record supports this conclusion: fish, amphibians, reptiles, and mammals, appear in the fossil record in the same order as they should have evolved (Figure 3.12b). The fit provides good evidence for evolution, because if fish, amphibians, reptiles, and mammals had been separately created, we should not expect them to appear in the fossil record in the exact order of their apparent evolution. Fish, frogs, lizards, and rats would probably appear as fossils in some order, if they did not appear at the same time. However, there is no reason to suppose they would appear in one order rather than another. Thus, it is a revealing coincidence when the fossil record verifies the evolutionary order. Similar analyses performed with other large and well-fossilized groups of animals, such as the echinoderms, have found the same result.

The argument can be stated another way. Haldane once said he would give up his belief in evolution if someone found a fossil rabbit in the Precambrian. His reasoning was that the rabbit, which is a fully formed mammal, must have evolved through reptilian, amphibian, and piscine stages and should not therefore appear in the fossil record a hundred million years or so before its fossil ancestors.

Creationists have appreciated the power of this argument. Various claims have been made for fossil human footprints contemporary with dinosaur tracks. Whenever one of these claims has been properly investigated, it has been exploded: some cases have turned out to have been carved fraudulently, others were carved as tourist exhibits, and still others are perfectly good dinosaur footprints. The principle of the argument is valid, however. If evolution is correct, humans could not have existed before the main radiation of mammals and primates, which took place after the dinosaurs had become extinct. The fact that no such human fossils have been found—that the order of appearance of the main fossil groups matches their evolutionary order—is one way in which the fossil record provides good evidence for evolution.

3.12 *Summary of the evidence for evolution*

At this point, we have discussed three main classes of evidence for evolution: direct observation, on the small scale; homology; and the order of the main groups in the fossil record. The small-scale observations work most powerfully against the idea of species fixity; by themselves, they are almost equally good evidence for evolution and for transformism (Figures 3.1a, b). They show, by uniformitarian extrapolation, that evolution at least could have produced the whole history of life. Stronger arguments for large-scale evolution come from classification and the fossil record. The geological succession of the major groups and most classical morphological homologies strongly suggest that these large groups have a common ancestor. The more recently discovered molecular homologies, such as the universal genetic code, extend the argument to the whole of life—and favor evolution (Figure 3.1a) over both transformism and creationism (Figures 3.1b, c).

The theory of evolution can also be used to make sense of, and to analyze, a large array of additional facts. As we study the different areas of evolutionary biology, it is worth keeping the issue of evidence in mind. How, for example, could we explain the molecular clock (section 7.9, p. 170) if species have independent origins? Or the difficulties of deciding whether closely related forms are different species (chapter 15)? Or the relative changes in prezygotic and postzygotic reproductive isolation through time (section 16.9, p. 452)? Or the unique branching pattern of chromosomal inversions in the Hawaiian fruitflies (section 17.14, p. 490)?

3.13 *Creationism offers no explanation for adaptation*

Another powerful reason why evolutionary biologists do not take creationism seriously is that creationism offers no explanation for adaptation. Living things are well designed, in innumerable respects, for life in their natural environments. They have sensory systems to find their way around, feeding systems to catch and digest food, and nervous systems to coordinate their actions. The theory of evolution offers a mechanical, scientific theory for adaptation: natural selection.

Creationism, by contrast, provides no explanation for adaptation. When each species originated, it must have already been equipped with adaptations for life, because the theory holds that species are fixed in form after their origin. An unabashedly religious version of creationism would attribute the adaptiveness of living things to the genius of God. Even this theory does not actually explain the origin of the adaptation, but simply pushes the problem back one stage (section 13.1, p. 338). In the scientific version of creationism (Figure 3.1c), supernatural events do not take place, and we are left with no theory of adaptation at all. Without a theory of adaptation, as Darwin realized (section 1.3.2), any theory of the origin of living things fails.

3.14 *Modern "scientific creationism" is scientifically untenable*

The evolution of life remains one of the great discoveries in scientific history, and the arguments in its favor are a fascinating topic. In modern evolutionary biology, the question of whether evolution happened is no longer a topic of research, because the question has been answered; the issue remains controversial outside science, however. Christian fundamentalists, some of them politically influential, in the United States have supported various forms of creationism and have been trying since the 1920s—sometimes successfully, sometimes unsuccessfully—to insert them into school biology curricula.

What relevance do the arguments of this chapter have for these forms of creationism? For a purely scientific form of creationism, the relevance is straightforward. As we have seen, the creationism of Figure 3.1c, which simply suggests that species had separate origins and been fixed since the development of the original species, is refuted by the evidence. The scientific creationism of Figure 3.1c did not address the mechanism by which species originated and therefore need not assert that the species were created by God: a supporter of Figure 3.1c might merely say that species originated by some natural mechanism, the details of which are not yet understood. However, it is unlikely that anyone would now seriously support the theory of Figure 3.1c without a belief that the species originated supernaturally. Such a case does not deal with a scientific theory.

This chapter has confined itself to the scientific resources of logical argument and public observation. Scientific arguments employ only observations that anyone can make, as distinct from private revelations, and consider only natural causes, as distinct from supernatural origins. Indeed, two good criteria to distinguish scientific from religious arguments are whether the theory invokes only natural causes, or needs supernatural causes as well, and whether the evidence is publicly observable or requires some sort of faith. Without these two conditions, no constraints are placed on the argument.

It is, in the end, impossible to show that species were not created by God and have remained fixed in form, because to God (as a supernatural agent) everything is permitted. Likewise, it cannot be shown that the building (or garden) you are in, and the chair you are sitting on, was not created supernaturally by God 10 seconds ago from nothing. At the moment of creation, God (a supernatural agent) could also have adjusted your memory and those of all other observers. That is why supernatural agents have no place in science.

Two final points are worth making. First, although modern "scientific creationism" closely resembles the theory of separate creation in Figure 3.1c, it also possesses the added feature of specifying the time at which all species were created. Theologians working after the Reformation were able to deduce, from some plausible astronomical theory and rather less plausible Biblical scholarship, that the events described in the first chapter of "Genesis" occurred about 6000 years ago. In our own time, fundamentalists have retained a belief in the recent origin of the world. A statement of creationism in the 1970s (and the one legally defended in court in Arkansas in 1981) included, as a creationist tenet, "a relatively recent inception of the earth and living kinds." Scientists, on the other hand, accept a great age for the earth because of radioactive dating and cosmological inferences from the background radiation. Cosmological and geological time are important scientific discoveries, but we have ignored them in this chapter because our subject has been the scientific case for evolution—religious fundamentalism is another matter.

Finally, it is worth stressing that there need be no conflict between the theory of evolution and religious belief. It is not an "either/or" controversy, in which accepting evolution means rejecting religion. No important religious beliefs are contradicted by the theory of evolution, and religion and evolution should be able to coexist peacefully in anyone's set of beliefs about life.

SUMMARY

1. A number of lines of evidence suggest that species have evolved from a common ancestor, rather than being fixed in form and created separately.
2. On a small scale, evolution can be seen taking place in nature, such as in the color patterns of moths, and in artificial selection experiments, such as those used in breeding agricultural varieties.
3. Natural variation can cross the species border (for example, in the ring species of salamanders), and new species can be made artificially (as in the process of hybridization and polyploidy by which many agricultural and horticultural varieties have been created).

4. Observation of evolution on the small scale, combined with the extrapolative principle of uniformitarianism, suggests that all life could have evolved from a single common ancestor.
5. Homologous similarities between species (understood as similarities that do not have to exist for any pressing functional reason) suggest that the species descended from a common ancestor. Universal homologies—such as the genetic code—found in all living things suggest that all species are descended from a single common ancestor.
6. The fossil record provides some direct evidence of the origin of new species.
7. The order of succession of major groups in the fossil record is predicted by evolution, and contradicts the separate origin of the groups.
8. The independent creation of species does not explain adaptation; evolution, by the theory of natural selection, offers a valid explanation.

FURTHER READING

Eldredge (1982), Ruse (1982), and Futuyma (1983) consider creationism and the case for evolution at book length. Chapters 10–14 of the *Origin of Species* (Darwin 1859) are the classic account of the evidence for evolution. Johnston and Selander (1964, 1971) are the sources for the house sparrow, and see Sheppard (1975), Ford (1975), and Endler (1986) for further examples of evolution in action. The ring species of the salamander were first described by Stebbins (1949) and have most recently been studied by Wake and Yanev (1986), Wake *et al.* (1986, 1989), and Jackman and Wake (1994). On polyploidy in plants, see the references in chapter 16. On the genetic code, see Bulmer (1988a). Gingerich *et al.* (1990) describe the Eocene whale with functional hind limbs, and Ahlberg and Milner (1994) describe Devonian tetrapods with non-pentadactyl limbs. Gould (1989) describes the animals of the Burgess Shale. Wellnhofer (1990) describes *Archaeopteryx*. On adaptation, see Dawkins (1986). For the broader context, see Numbers (1992) for the history and Nelkin (1982) on educational, and Larson (1989) on legal, business.

STUDY AND REVIEW QUESTIONS

1. The average difference between two individuals increases as they are sampled from the same local population, two separate populations, two species, two genera, and so on up to two kingdoms (such as plants and animals). Up to approximately what stage in this sequence can evolution be observed in a human lifetime?

2. In what sense is the range of forms of life on Earth (a) arranged, and (b) not arranged, in distinct "kinds"?

3. Which of the following are homologies and which analogies, in the pre-Darwinian sense of the terms? (a) A dolphin flipper and a fish fin. (b) The five-digit skeletal structure of the dolphin flipper and of a frog foot. (c) The white underside coloration of gulls, albatrosses, and ospreys (all of which are seabirds and catch fish by air raids from above). (d) The number of vertebrae in the necks of camels, mice, and humans (they all have seven vertebrae).

4. The genetic code has been called a "frozen accident." In what sense is it an accident, and why was it frozen?
5. Imagine a number of sets containing 10 objects each, such as 10 books, 10 dishes for dinner, 10 gems, 10 vehicles, 10 politicians, . . . or whatever. For each set, devise two or three different ways of classifying them in hierarchical groups. (For example, 10 politicians might be classified first into two groups as left of center/right of center; those groups could then be classified by such criteria as average length of sound bites, number of scandals per year, gender, region represented, or whatever.) Do the different hierarchical classifications recognize the same sets of groups, or similar sets of groups, or are they unrelated? Think about why the different classifications are similar for some sets of groups and for some classificatory criteria, whereas for others they differ.
6. Why would Haldane have given up his belief in evolution if someone discovered a fossil rabbit in the Precambrian?

Natural Selection and Variation

This chapter first establishes the conditions for natural selection to operate, and distinguishes directional, stabilizing, and disruptive forms of selection. We then consider how widely in nature the conditions are met, and review the evidence for variation within species; the review begins at the level of gross morphology and works down to molecular variation. Variation originates by recombination and mutation, and we finish by looking at the argument to show that new variation is not "directed" toward improved adaptation.

4.1 *In nature, there is a struggle for existence*

The Atlantic cod (*Gadus callarias*) is a large marine fish, and an important source of human food. It is also one of the most fecund of vertebrates. An average 10-year-old female cod lays about 2 million eggs in a breeding season, and large individuals may lay more than 5 million eggs (Figure 4.1a). Female cod ascend from deeper water to the surface to lay their eggs. As soon as they are discharged, however, a slaughter begins. The plankton is a dangerous place for eggs. The billions of eggs released by cod alone are devoured by innumerable planktonic invertebrates, by other fish, and by fish larvae. Approximately 99% of cod eggs die in their first month of life, and another 90% or so of the survivors die before reaching an age of one year (Figure 4.1b). A negligible proportion of the 5 million eggs laid by a female cod in her lifetime will survive and reproduce: an average female in the cod population will produce only two successful offspring.

This figure, that on average two eggs per female survive to reproduce successfully, is not the result of an observation—it comes from a logical calculation. Only two cod can survive, because any other number would be unsustainable over the long term. It takes a pair of individuals to reproduce. If an average pair in a population produce less than two offspring, the population will soon go extinct; if they produce more than two on average, the population will rapidly reach infinity—which is also unsustainable. Over a small number of generations, the average female in a population may produce more or less than two successful offspring, and the population will increase or decrease accordingly. Over the long term, the average must be two. Thus, we can infer that of the 5 million or so eggs laid by a female cod in her life, 4,999,998 die before reproducing.

A life table can be used to describe the mortality of a population (Table 4.1). A life table begins at the egg stage and traces the proportion of the original 100% of eggs that die off at the successive stages of life. In some species, mortality is

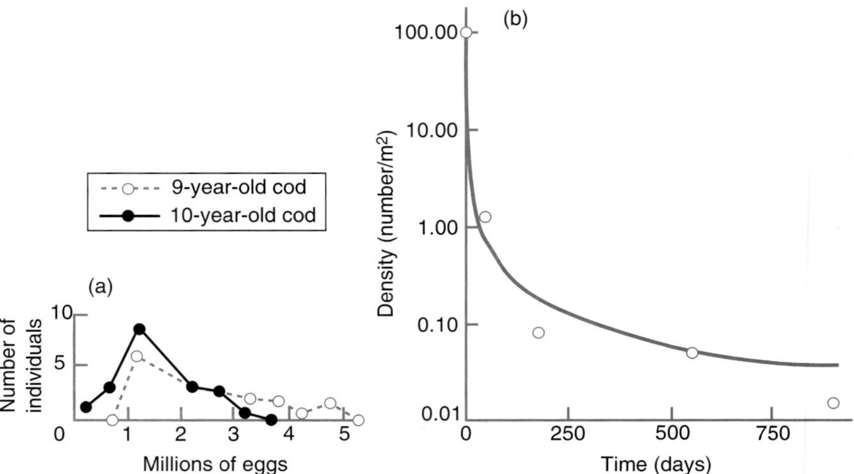

Figure 4.1 (a) Fecundity of cod. Notice both the large numbers, and that they are variable between individuals. The more fecund cod lay perhaps five times as many eggs as the less fecund; much of the variation is associated with size, because larger individuals lay more eggs. (b) Mortality of cod in their first two years of life. Reprinted, by permission of the publisher from May (1967) and Cushing (1975).

concentrated early in life; in other species, mortality has a more constant rate throughout life. All species experience mortality, however, which reduces the number of eggs and leads to a lower number of adults.

The condition of "excess" fecundity—in which females produce more offspring than survive—is universal in nature. In every species, more eggs are produced than can survive to the adult stage. The cod dramatizes the point in one way, because its fecundity and mortality are so high. Darwin dramatized the same point by considering the opposite kind of species—one that has an extremely low reproductive rate. The fecundity of elephants is low, but even they produce many more offspring than can survive. In Darwin's words:

TABLE 4.1

A life table for the annual plant *Phlox drummondii* in Nixon, Texas. The life table gives the proportion of an original sample (cohort) that survive to various ages. A full life table may also give the fecundity of individuals at each age. Reprinted with permission from Leverich and Levin (1978). Copyright 1978 University of Chicago Press.

Age Interval (days)	Number Surviving to Day *x*	Proportion of Original Cohort Surviving to Day *x*	Proportion of Original Cohort Dying during Interval	Mortality Rate per Day
0–63	996	1.000	0.329	0.005
63–124	668	0.671	0.375	0.009
124–184	295	0.296	0.105	0.006
184–215	190	0.191	0.014	0.002
215–264	176	0.177	0.004	0.001
264–278	172	0.173	0.005	0.002
278–292	167	0.168	0.008	0.003
292–306	159	0.160	0.005	0.002
306–320	154	0.155	0.007	0.003
320–334	147	0.148	0.043	0.021
334–348	105	0.105	0.083	0.057
348–362	22	0.022	0.022	1.000
362–	0	0	—	

> The elephant is reckoned the slowest breeder of all known animals, and I have taken some pains to estimate its probable minimum rate of natural increase; it will be safest to assume it begins breeding when thirty years old, and goes on breeding until ninety years old, bringing forth six young in the interval, and surviving till one hundred years old; if this be so, after a period of 740 to 750 years there would be nearly nineteen million elephants alive, descended from the first pair.

In elephants, then, as in cod, many individuals die between the egg and adult stages; they both have excess fecundity. This excess fecundity exists because nature provides inadequate resources to support all eggs that are laid and all young that are born. Only limited amounts of food and space are found in the world—thus, a population may expand to some extent, but logically there will come a point beyond which the food supply must limit its further expansion. As resources are depleted, the death rate in the population increases. When the death rate equals the birth rate, then the population will stop growing.

Organisms, therefore, in an ecological sense compete to survive and reproduce—both directly (for example, by defending territories) and indirectly (for example, by eating food that could otherwise be eaten by another individual). The actual competitive factors limiting the sizes of real populations constitute a major area of ecological study, and in different species different factors have been shown to operate. What matters here, however, is the general point that the members of a population, and members of different species, compete to survive, and this concept follows from the conditions of limited resources and excess fecundity. Darwin referred to this ecological competition as the "struggle for existence." The expression is metaphorical: it does not imply a physical fight to survive, although fights do sometimes occur.

The struggle for existence takes place within a web of ecological relations. Above an organism in the ecological food chain, there will be predators and parasites, seeking to feed off it; below the organism are the food resources it must consume to stay alive. At the same level in the chain are competitors, which may be vying for the same limited resources of food or space. An organism competes most closely with other members of its own species, because they have the most similar ecological needs to its own; other species, in decreasing order of ecological similarity, also compete and exert a negative influence on the organism's chance of survival.

In summary, organisms produce more offspring than—given the limited amounts of resources—can ever survive, and organisms therefore compete for survival. Only the successful competitors will reproduce themselves.

4.2 *Natural selection operates if some conditions are met*

The excess fecundity and consequent competition to survive in every species provide the preconditions for the process Darwin called natural selection. Natural selection is easiest to understand, in the abstract, as a logical argument, leading from premises to conclusion. The argument, in its most general form, requires four conditions:

Figure 4.2 Three kinds of selection. The top line shows the frequency distribution of the character (body size). For many characters in nature, this distribution has a peak in the middle, near the average, and is lower at the extremes. (The normal distribution, or "bell curve," is a particular example of this kind of distribution.) The second line shows the relation between body size and fitness, within one generation, and the third the expected change in the average for the character over many generations (if body size is inherited). (a) Directional selection. Smaller individuals have higher fitness, and the species will decrease in average body size through time. Figure 4.3 is an example. (b) Stabilizing selection. Intermediate-size individuals have higher fitness. Figure 4.4a is an example. (c) Disruptive selection. Both extremes are favored. If selection is strong enough, the population splits into two. Figure 4.5 is an example. (d) No selection. If no relation exists between the character and fitness, natural selection is not operating on it.

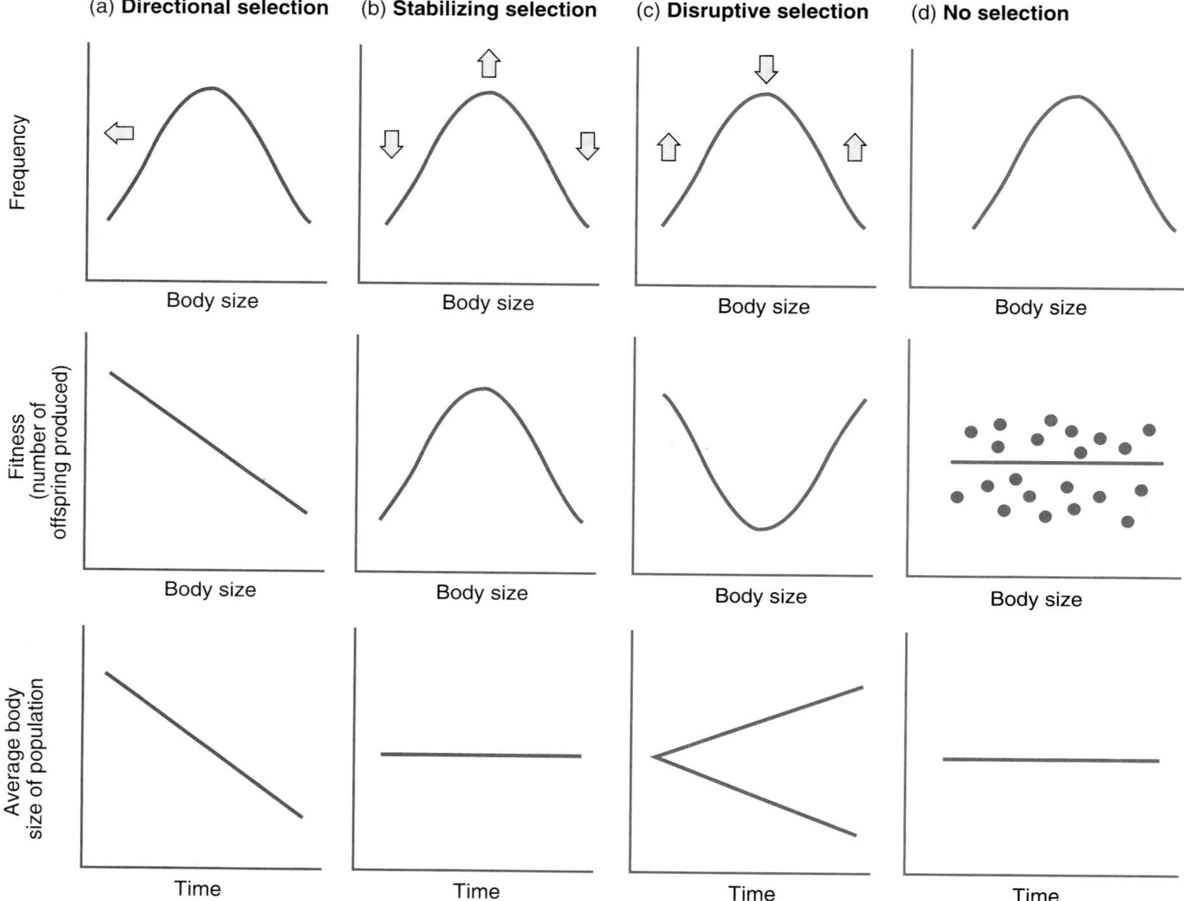

For example, pink salmon (*Onchorhynchus gorbuscha*) in the Pacific Northwest have been decreasing in size in recent years (Figure 4.3). In 1945, fishermen began being paid by the pound, rather than per individual, for the salmon they caught; fishermen responded by increasing their use of gill netting, which selectively takes larger fish. The selectivity of gill netting can be shown by compar-

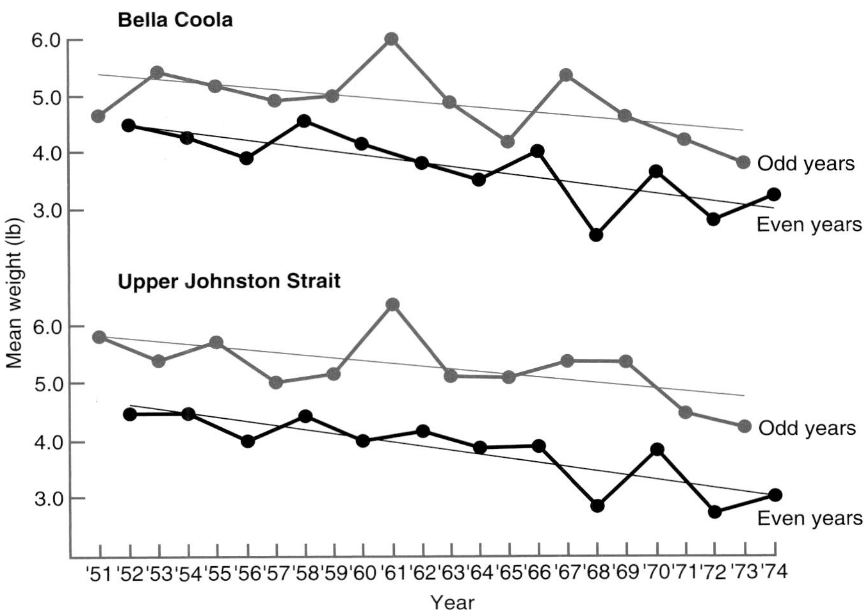

Figure 4.3 Directional selection created by fishing on pink salmon, *Onchorhynchus gorbuscha.* The graph shows the decrease in size of pink salmon caught in two rivers in British Columbia since 1950; it has been driven by selective fishing for the large individuals. Two lines are drawn for each river: one for the salmon caught in odd-numbered years, the other for even-numbered years. Salmon caught in odd-numbered years are consistently heavier, presumably because of the two-year life cycle of the salmon. From Ricker (1981). Reproduced with permission of the Minister of Supply and Services Canada, 1995.

ing the average size of salmon taken by gill netting with those taken by an unselective fishing technique: the difference ranged from 0.3 to 0.48 lb (0.14–0.22 kg). Therefore, after gill netting was introduced, smaller salmon had a higher chance of survival. Selection favoring small size in the salmon population was intense, because fishing is thorough—about 75–80% of the adult salmon swimming up the rivers under investigation were caught in these years. The average weight of salmon duly decreased, by about one-third, in the next 25 years.

A second (and in nature, more common) possibility is for natural selection to be *stabilizing* (Figure 4.2b). The average members of the population, with intermediate body sizes, have higher fitness than the extremes. Natural selection, therefore, acts against change in form, and keeps the population constant through time.

Studies of birth weight in humans have provided good examples of stabilizing selection. Figure 4.4a illustrates a classic result for a sample in London, in 1935–1946, and similar results have been found in New York, Italy, and Japan. Babies that were heavier or lighter than average did not survive as well as babies of average weight. Stabilizing selection has probably operated on birth weight in human populations from the time of the evolutionary expansion of our brains about 1–2 million years ago until the twentieth century, and probably continues to operate in most of the world. However, in the 50 years since Karn and Penrose's study, the force of stabilizing selection on birth weight has relaxed in wealthy countries (Figure 4.4b); in fact, by the late 1980s, it had almost disappeared. The pattern has approached that of Figure 4.2d: survival rates have become almost the same for all birth weights. Selection has relaxed because of improved care for premature deliveries (the main cause of lighter babies) and

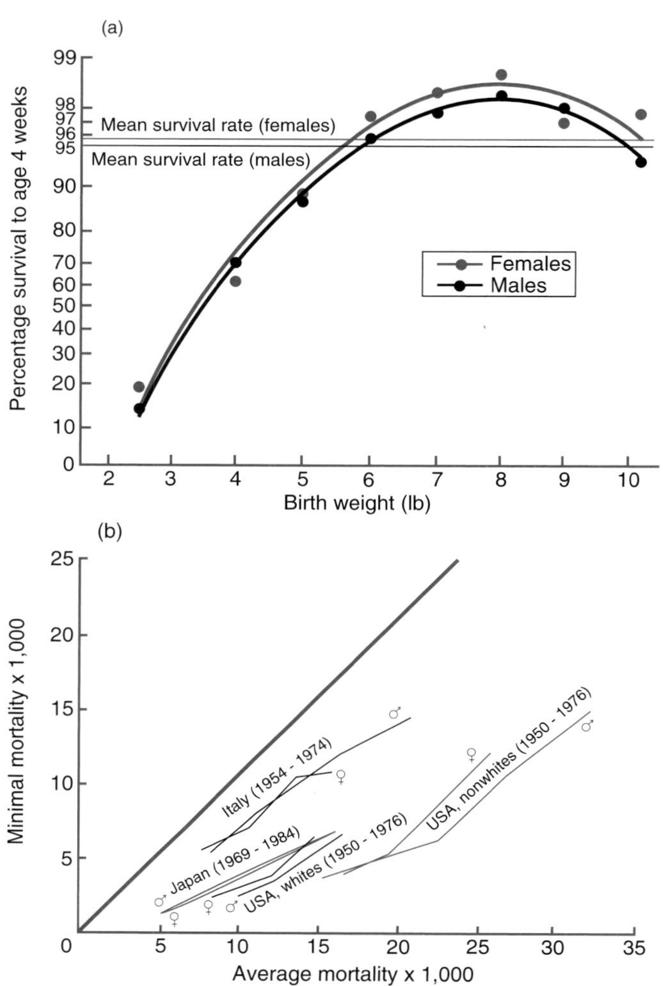

Figure 4.4 (a) The classic pattern of stabilizing selection on human birth weight. Infants weighing 8 lb (3.6 kg) at birth have a higher survival rate than heavier or lighter infants. The graph is based on 13,700 infants born in a hospital in London, England, during 1935–1946. (b) Relaxation of stabilizing selection in wealthy countries in the second half of the twentieth century. The *x*-axis is the average mortality in a population; the *y*-axis is the mortality of infants that have the optimal birth weight in the population (and so the minimum mortality achieved in that population). In (a), for example, females have a minimum mortality of about 1.5% and an average mortality of about 4%. When the average equals the minimum, selection has ceased, which corresponds to the 45° line (the "no selection" case in Figure 4.2d would produce a point on the 45° line). Note how the data for Italy, Japan, and the United States approach the 45° line through time. By the late 1980s, the Italian population had reached a point not significantly different from the absence of selection. From Karn and Penrose (1951) and Ulizzi and Manzotti (1988, for the Italian 1980s update see Ulizzi and Terrenato, 1992). Reprinted with the permission of Cambridge University Press.

increased frequencies of Caesarian deliveries for babies that are large relative to the mother (the lower survival of heavier babies was mainly due to injury to the baby or the mother during birth). In the 1990s, the stabilizing selection that had been operating on human birth weight for over a million years has all but disappeared in wealthy countries.

The third type of natural selection favors both extremes over the intermediate types—a condition called *disruptive* selection (Figure 4.2c). In nature, sexual dimorphism is probably a common example. In our discussion, we will use an experiment by Thoday and Gibson on fruitflies as an example of this selection mechanism. Thoday and Gibson bred from fruitflies with high, or low, numbers of bristles on a certain region of the body; individuals with intermediate numbers of bristles were prevented from breeding. After 12 generations of this disruptive

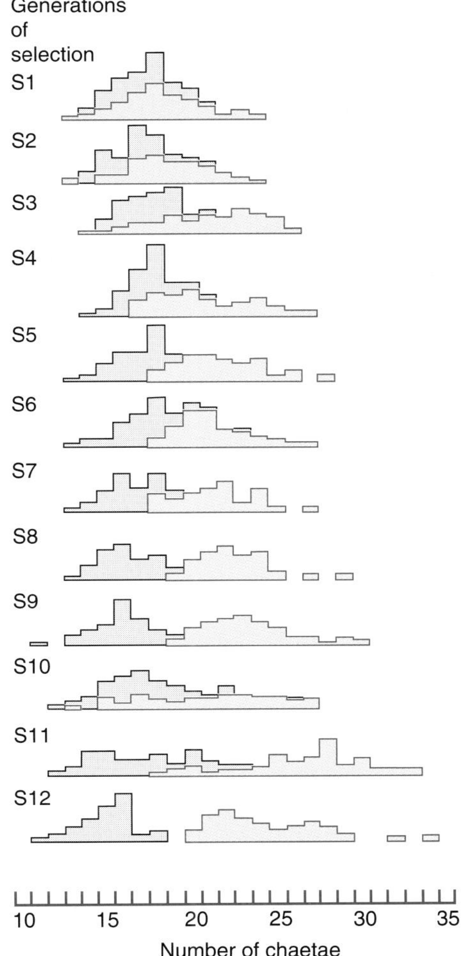

Figure 4.5 Experimental disruptive selection on sternopleural bristle number in the fruitfly *Drosophila melanogaster.* Individuals with many or few bristles were allowed to breed, while those with intermediate numbers were not. The population rapidly diverged. Adapted with permission from Thoday and Gibson (1962). Copyright 1962 Macmillan Magazines Limited.

selection, the population had noticeably diverged (Figure 4.5). Disruptive selection is of particular theoretical interest, because it can both increase the genetic diversity of a population (by frequency-dependent selection—section 5.13, p. 121) and promote speciation (chapter 16).

A final theoretical possibility is the absence of a relation between fitness and the character in question. In such a case, natural selection is not acting on the population (Figure 4.2d; Figure 4.4b also provides a near example of no selection).

4.5 *Variation in natural populations is widespread*

Natural selection will operate whenever the four conditions in section 4.2 are satisfied. The first two conditions are self-explanatory. It is well known that organisms reproduce themselves—this ability is often given as one of the defining properties of living things. It is also well known that organisms show inheritance.

Inheritance is produced by the Mendelian process, which is understood down to a molecular level. Not all the characters of organisms are inherited, and natural selection will not adjust the frequencies of non-inherited characters. Many characters are inherited, however, and natural selection can potentially work on them.

How much, and with respect to what characters, do natural populations show variation (the third condition) and, in particular, variation in fitness (the fourth condition)? Let us consider biological variation through a series of levels of organization, beginning with the organism's morphology, and working down to more microscopic levels. We will examine examples of variation, show how variation can be seen in almost all the properties of living things, and introduce some of the methods (particularly molecular methods) that are used to study variation.

At the morphological level, the individuals of a natural population will vary for almost any character we may measure. In some characters, like body size, every individual differs from every other individual; this condition is called continuous variation. Other morphological characters show discrete variation: they fall into a limited number of categories. Sex, or gender, is an obvious example; some individuals of a population are female, others male. This kind of categorical variation is found in other characters as well. Within a population of the peppered moth, for example, there are two main wing color forms (the dark, melanic form and the light, peppered form).

A population that contains more than one recognizable form is *polymorphic* (the condition is called polymorphism). There can be any number of forms in real cases, and they can have any set of relative frequencies. With sex, there are usually two forms; in the peppered moth, two main color forms are often distinguished, although real populations may contain three or more. The snail *Cepaea nemoralis*, which is polymorphic for its shell pattern (see section 11.1, p. 281), can exhibit many more forms. As the number of forms in the population increases, the polymorphic, categorical variation blurs into continuous variation.

Variation is not confined to morphological characters. If we descend to a cellular character, such as the number and structure of the chromosomes, we again find variation. In the fruitfly *Drosophila melanogaster*, the chromosomes exist in giant forms in the larval salivary glands and can be studied with a light microscope. They have characteristic banding patterns, and chromosomes from different individuals in a population show subtly varying banding patterns. One type of variant is called an *inversion* (Figure 4.6), in which the banding pattern—and therefore the order of genes—of a region of the chromosome is inverted. A population of fruitflies may be polymorphic for a number of different inversions.

Although chromosomal variation is more difficult to study in species that lack giant chromosomal forms, it is still known to exist. Populations of the Australian grasshopper *Keyacris scurra*, for example, may contain two (normal and inverted) forms for each of two chromosomes; thus, nine kinds of grasshopper are possible because an individual may be homozygous or heterozygous for any of the four chromosomal types. The nine differ in size and viability (Figure 4.7).

Chromosomes can vary in other respects, too. Individuals may vary in their number of chromosomes, for example. In many species, some individuals have one or more extra chromosomes, in addition to the normal number for the

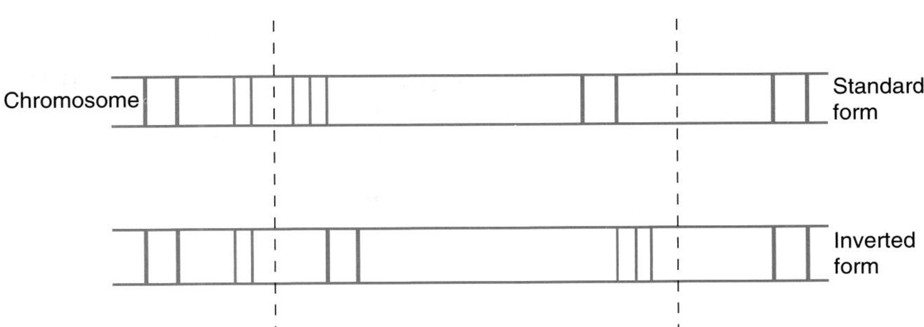

Figure 4.6 Chromosomes can exist in standard and inverted forms. The naming of the forms as "standard" and "inverted" is arbitrary. The inversion can be detected by comparing the fine structure of bands, as is diagrammatically illustrated here, or by the behavior of the chromosomes at meiosis.

species. These "supernumerary" chromosomes, which are often called B chromosomes, have been studied extensively in maize and in grasshoppers. In the grasshopper *Atractomorpha australis*, normal individuals have 18 autosomes; individuals have been found with from one to six supernumary chromosomes as well. Thus, the population is polymorphic with respect to chromosome number. Chromosomal variations other than inversions and B chromosomes are also

Figure 4.7 The Australian grasshopper *Keyacris scurra* is polymorphic for inversions for two chromosomes. The two chromosomes are called the CD and the EF chromosomes. The standard and inverted forms of the CD chromosome are called *St* and *Bl, respectively;* the standard and inverted forms of the EF chromosome are called *St'* and *Td, respectively.* v = relative viability at a site at Wombat, New South Wales, expressed relative to the viability of the *St/Bl St'/St'*, which is arbitrarily set as 1. x = mean live weight, and the picture illustrates relative sizes of the grasshoppers. From White (1973).

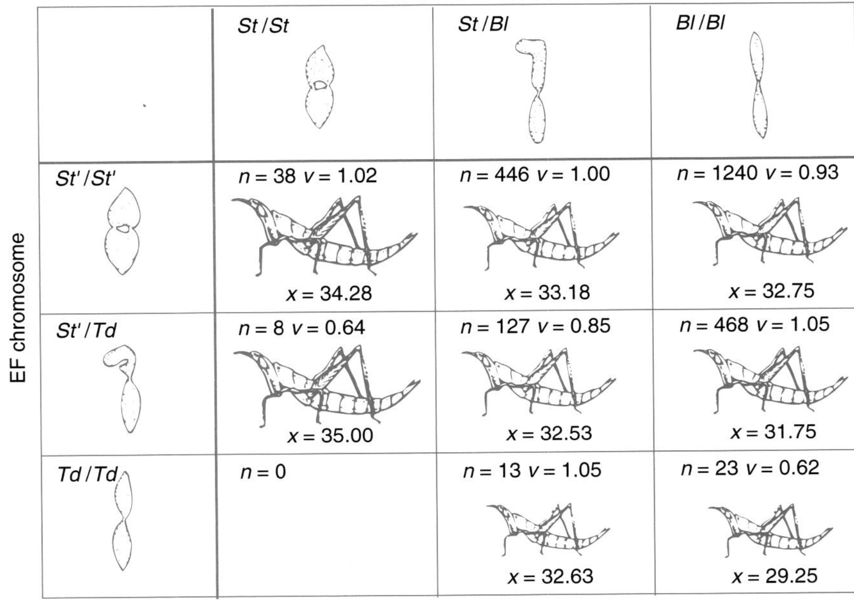

known, but these examples should suffice to make the point that individuals vary at the subcellular, as well as the morphological level.

The story is the same at the biochemical level, such as for proteins. Proteins are molecules made up of sequences of amino acid units. A particular protein, like human hemoglobin, has a particular characteristic sequence, which in turn determines the molecule's shape and properties. Do all humans have exactly the same sequence for hemoglobin, or any other protein? In theory, we could find out by taking the protein from several individuals and then working out the sequence in each of them, although this process would be excessively laborious. A much faster method, called *gel electrophoresis,* solves this problem by focusing on the electrical charges carried by amino acids. Different proteins—and different variants of the same protein—have different net electric charges because of their different amino acid compositions. If we place a sample of proteins (with the same molecular weight) in an electric field, those with the largest electric charges will move the most quickly. For the student of biological variation, the importance of the method is that it can reveal different variants of a particular type of protein. For a good example, we shall change to a less well-known protein than hemoglobin—an enzyme called alcohol dehydrogenase, which is found in the fruitfly.

Fruitflies, as their name suggests, lay their eggs in, and feed on, decaying fruit. They are attracted to rotting fruit because of the yeast it contains. Fruitflies can be collected almost anywhere in the world by leaving out rotting fruit as a lure; drowned fruitflies are usually found in a glass of wine left outside overnight after a garden party in the late summer. As fruit rots, it forms a number of chemicals, including alcohol, which is both a poison and a potential energy source. Fruitflies, and many other living species, deal with alcohol by a detoxifying enzyme called alcohol dehydrogenase. Gel electrophoresis reveals that in most populations of the fruitfly *Drosophila melanogaster*, alcohol dehydrogenase comes in two main forms. The two forms appear as different bands on the gel after the sample has been applied, an electric current put across the gel for a few hours, and the position of the enzyme exposed by a specific stain. The two variants are called slow (*Adh-s*) or fast (*Adh-f*) according to how far they have moved in the time. The multiple bands show that the protein is polymorphic; the enzyme called alcohol dehydrogenase is actually a class of two polypeptides with slightly different amino acid sequences. Gel electrophoresis has been applied to a large number of proteins in a large number of species; different proteins show different degrees of variability (see chapter 7). For now, you should focus on the fact that many of these proteins have been found to be variable: extensive variation occurs in proteins in natural populations.

If variation appears in every organ, at every level, among the individuals of a population, it is almost inevitable that variation will occur at the DNA level as well. The inversion polymorphisms of chromosomes that we encountered earlier, for example, result from inversions of the DNA sequence. However, the most direct method of studying DNA variation is to sequence the DNA itself. Let us consider alcohol dehydrogenase in the fruitfly once again. Kreitman isolated the DNA encoding alcohol dehydrogenase from 11 independent lines of *D. melanogaster* and individually sequenced them all. Some of the 11 had *Adh-f*, while others had *Adh-s;* the difference between *Adh-f* and *Adh-s* was always due to a single amino acid difference (Thr or Lys at codon 192).

Although the amino acid difference appears as a base difference in the DNA, it was not the only source of variation at the DNA level. The DNA is even more variable than the protein study suggests. At the protein level, only the two main variants were found in the sample of 11 genes, but at the DNA level there were 11 different sequences with 43 different variable sites. The amount of variation that we find is therefore highest at the DNA level. At the level of gross morphology, a *Drosophila* with two *Adh-f* genes is indistinguishable from one with two *Adh-s* genes, and gel electrophoresis resolves two classes of fly. At the DNA level, these two classes decompose into innumerable individual variants.

Restriction enzymes provide another method of studying DNA variation. Restriction enzymes exist naturally in bacteria, and a large number—more than 2300—of restriction enzymes are known. Any one restriction enzyme cuts a DNA strand at a particular sequence, usually consisting of 4–8 base pairs. For instance, the restriction enzyme called *EcoR1,* which is found in the bacterium *Escherichia coli,* recognizes the base sequence . . . GAATTC . . . . and cuts it between the initial G and the first A. In the bacterium, the enzymes help to protect against viral invasion, by cleaving foreign DNA; the enzymes can also be isolated in the lab and used to investigate DNA sequences. Suppose the DNA of two individuals differs, and that one has the sequence GAATTC at a certain site, whereas the other individual has some other sequence such as GTATT. If the DNA of each individual is put with *EcoR1,* only that of the first individual will be cleaved. The difference can be detected in the length of the DNA fragments: the pattern of fragment lengths will differ for the two individuals. The variation is called *restriction fragment length polymorphism* and has been found in all populations that have been studied.

In summary, natural populations show variation at all levels, from gross morphology to DNA sequences. When we examine natural selection in more detail, we can assume that in natural populations the requirement of variation is met, as well as the requirements of reproduction and heredity.

4.6 *Organisms in a population vary in reproductive success*

If natural selection is to operate, it is not enough that characters vary. The different forms of the character must also be associated with reproductive success (or fitness)—that is, the degree to which individuals contribute offspring to the next generation. Reproductive success is more difficult to measure than a phenotypic character like body size, and far fewer observations have been made of variation in reproduction than in phenotype. We have encountered some examples of reproductive variation in this chapter (section 4.5), and others will be presented later in the book. In our current discussion, we will look at an even more abundant sort of evidence, and at an abstract argument.

Whenever reproductive success in a biological population has been measured, it has been found that some individuals produce many more offspring than others. Figure 4.8 illustrates this variation in four species of orchids in the form

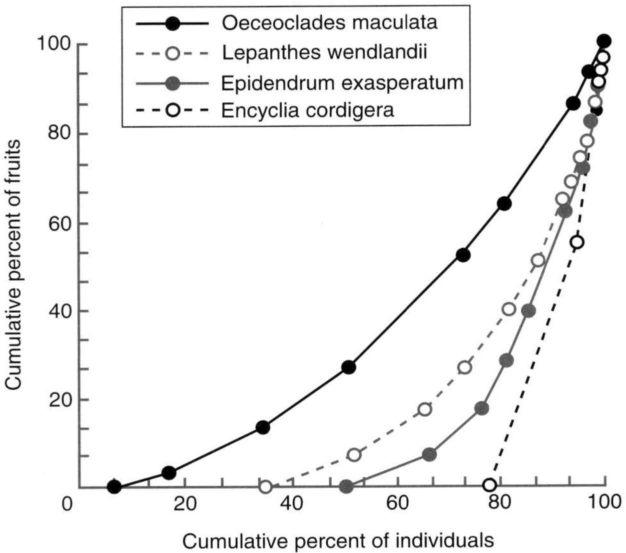

Figure 4.8 Variation in reproductive success within populations, illustrated by four species of orchids. The graphs plot the cumulative percentage of offspring produced by the plants, with the individual plants ranked from the least to the most successful. For instance, in *Epidendrum exasperatum*, the least successful 50% of individuals produce none of the offspring—that is, they fail to reproduce. The next 17% of individuals, moving up in the ranking of success, produce about 5% of the offspring in the population; and the next 10% produce about 13%; and so on. If every individual produced the same number of offspring, the cumulative percentage graph would be the 45° line. Graphs of this kind can be used to express inequality in a population generally. They were first used to express inequality in human wealth and are sometimes called Lorenz curves. Reprinted, by permission of the publisher, from Calvo (1990).

of a cumulative percentage graph. If every individual produced the same number of fruit (that is, the same number of offspring), the points would fall along the 45° line. In reality, the points usually start some way along the x-axis and fall below the 45° line, because some individuals fail to reproduce and a successful minority contribute a disproportionate number of offspring.

The differences between these four orchid species can be understood in terms of their relationships with insect pollinators. The reproductively egalitarian species *Oeceoclades maculata* reproduces by self-fertilization, and has no use for pollinators. The two intermediate species *Lepanthes wendlandii* and *Epidendrum exasperatum* are each capable of self-fertilization but can also be pollinated by insects. The highly inegalitarian *Encyclia cordigera*, in which 80% of the individuals fail to reproduce, requires insect pollination and is also unattractive to those insects.

Like some other orchids, *Encyclia cordigera* has evolved "deceptive" flowers that produce and receive pollen but do not supply nectar; such orchids "cheat" the insect and insects tend to avoid them (though not completely) in consequence. The amount of reproductive failure in orchids with such deceptive flowers can be remarkably high—higher than the 80% found with *Encyclia cordigera*. Gill, in another study, measured reproduction in a population of almost 900 individuals of the pink lady's-slipper orchid *Cypripedium acaule* in Rockingham County, Virginia, during 1977–1986. In that 10-year period, only 2% of the individuals managed to produce fruit. The rest had been avoided by pollinators and failed to breed. In four of the years, none of the orchids bred at all.

Clearly, the ecological factor determining variation in reproductive success in orchids is the availability of, and need for, pollinating insects. If they are unnecessary, all the orchids in a population produce a similar number of fruit; if the pollinating insects are necessary and scarce, because of the way the orchid "cheats" the pollinators, only a small minority of individuals may succeed in reproducing. Pollinators are a key factor in orchids, but other factors will operate in other species and ecological study can reveal why some individuals are more reproductively successful than others.

The results in Figure 4.8 show the amount of reproductive variation among the adults that exist in a population, but this variation is only applicable for the final component of the life cycle. Before this variation comes into play, individuals differ in survival because of other factors, and a life table (such as Table 4.1) can quantify that variation. A full description of the variation in lifetime success of a population would combine variation in survival from conception to adult and variation in adult reproductive success.

Examples like the peppered moth, or the pink salmon, show that natural selection can operate. Nevertheless, the question remains of how often natural selection operates in natural populations, and in what proportion of species. We could theoretically find out how widespread natural selection is by counting how frequently all four conditions apply in nature—a difficult task. The evidence of variation in phenotypic characters and of ecological competition has attracted interest because it suggests that the preconditions for natural selection to be operating are widespread—indeed, probably universal. Observers have always found variation in the phenotypic characters of populations, and ecological competition within them. Indeed, you do not need to be a professional biologist to recognize the presence of variation and the struggle for existence: they are almost obvious facts of nature.

It is logically possible that individual reproductive success varies in all populations in the manner of Figure 4.8, but that natural selection does not operate in any of them, because the variation in reproductive success is not associated with any inherited characters. Although this theory is logically possible, it is not ecologically probable. In almost every species, such a high proportion of individuals is doomed to die that any attribute increasing their chance of survival, in a way that might appear trivial to us, is likely to result in a higher than average fitness. In contrast, any tendency of individuals to make mistakes, slightly increasing their risk of death, will result in lowered fitness. Likewise, once an individual has survived to adulthood, its phenotypic attributes can influence its chance of reproductive success in many ways, and reproductive success is ultimately equivalent to fitness.

The struggle for existence and phenotypic variation are both universal conditions in nature. Variation in fitness associated with some of those phenotypic characters is, therefore, also likely to be very common. The argument is one of plausibility, rather than certainty: it is not logically inevitable that a population showing (inherited) variation in a phenotypic character will automatically support an association between the varying character and fitness. If such a relationship exists, natural selection will operate.

4.7 *New variation is generated by mutation and recombination*

The variation that exists in a population is the resource with which natural selection works. Imagine a population evolving increased body size. At first, there is variation and average size can increase. However, the population could evolve a limited amount if only the initial variation was possible, because it would soon reach the edge of available variation (Figure 4.9a). In existing human populations, for instance, height does not range much beyond about 8 feet, and evolution beyond that point would be impossible if existing variation were the only resources. Evolution from the origin, to the modern diversity, of life must have required more variation than existed in the original population. Where did the extra variation come from?

Recombination (in sexual populations) and mutation are the two main answers. As a population evolves toward individuals of larger body size, the genotypes encoding larger body size increase in frequency. At the initial stage, large body size was rare and only one or two individuals might have possessed genotypes for large body size. These individuals would probably interbreed with other individuals closer to the average size for the population and produce offspring of less extreme size. As the genotypes for large body size itself become the average, however, they are more likely to interbreed and produce new genotypes encoding even larger body size. As evolution proceeds, recombination among the existing genotypes generates a new range of variation (Figure 4.9b).

Mutation also introduces new variation, and in chapter 9 we will look in

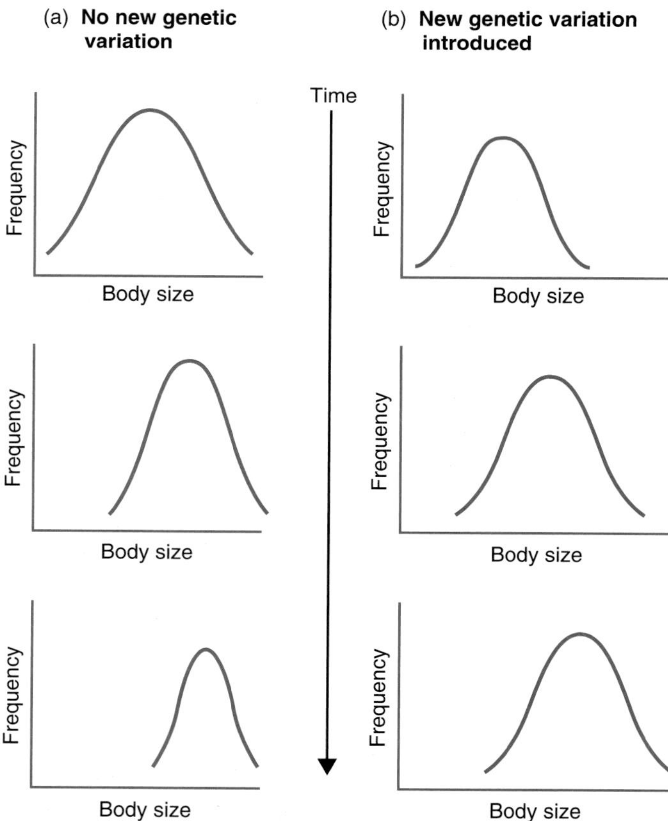

Figure 4.9 Natural selection produces evolution by working on the variation in a population. (a) In the absence of new variation, evolution soon reaches the limit of existing variation and comes to a stop. (b) Recombination generates new variation as the frequencies of the genotypes change during evolution. Evolution can then proceed further than the initial range of variation.

more detail at how selection and mutation balance one another to allow a certain amount of variation in a population. Here, we choose to notice only that new variation is continually created in natural populations.

4.8 *The variation created by recombination and mutation is random with respect to the direction of adaptation*

A fundamental property of Darwinism is that the direction of evolution, particularly of adaptive evolution, is uncoupled from the direction of variation. When a new recombinant or mutant genotype arises, there is no tendency for it to arise in the direction of improved adaptation. Natural selection imposes direction on evolution, using undirected variation. In this section, we define the alternative viewpoint (the theory of directed variation) and discuss why it is not accepted.

Consider the peppered moth example once again. When the environment changed, a new form of the species was favored. According to Darwin's theory, that environmental change does not itself cause mutations of the right form to appear: new mutations of all sorts arise constantly, but independently of favorable adaptations to the current environment. The alternative would be some kind of *directed mutation*, which would mean that when the environment changed to favor melanic moths, melanic mutations would be generated by some mutational mechanism.

Both factual and theoretical reasons exist to doubt that mutations are adaptively directed. The factual evidence comes from laboratory observations of spontaneous mutations: no evidence exists that these mutations take place toward novel adaptive needs of the organism. A classic experiment of Luria and Delbrück in 1943 on mutation in bacteria until recently served as the "textbook" demonstration that mutation is undirected. That particular experiment was effectively challenged by Cairns *et al.* (1988), however, and research on this topic is moving so quickly that any summary would be out of date before it was printed.

The strongest reason to doubt that mutations are adaptively directed is theoretical. In the case of the peppered moths, melanic moths probably already existed in the population at the time of the industrial revolution, and selection simply increased their frequency. Nevertheless, we can use the case as an example by assuming that no melanic moths were present when the environment changed. The industrial revolution imposed on the moths an environment that they had never encountered before. The environment (probably) was completely new. If the population did not contain any melanic moths, a mutation of melanic coloration was then needed for the moths to survive. Could it arise by directed mutation? At the DNA level, the mutation would have consisted of a set of particular changes in the base sequence of a gene. No genetic mechanism has been discovered that could direct the desired base changes to happen.

If we reflect on the kind of mechanism that would be needed, it becomes clear that an adaptively directed mutation would be practically impossible. The organism would have to recognize that the environment had changed, work out the change necessary to adapt to the new conditions, and then cause the correct base changes in the relevant parts of its DNA. These modifications would have been necessary to suit an environment that the species had never previously experienced. The situation would be analogous to humans describing subject-matter they had never encountered before in a language they did not understand: like a seventeenth-century American using Egyptian hieroglyphics to describe how to change a computer program. (Hieroglyphics were not deciphered until the discovery of the Rosetta Stone in 1799.) Even if it were possible to imagine, as an extreme theoretical possibility, directed mutations in the case of moth coloration, the changes in the evolution of a more complex organ (like the brain, circulatory system, or eye) would require a near miracle. Thus, mutation is thought not to be directed toward adaptation.

Although mutation is random and undirected with respect to the direction of improved adaptation, mutations are not necessarily non-random at the molecular level. The higher frequency of transitions than transversions is a case in point. It is not true that changes from the base *A*, for example, are equally probable to *G*, *T*, or *C*; a change to *G* may be 10 times more probable. Other molecular biases are known as well. For example, the two-base sequence CG demonstrates a tendency to mutate, when it has been methylated, to TG. (The DNA in a cell is sometimes methylated, for reasons that do not matter here.) After replication, a complementary pair of CG on the one strand and GC on the other will then have produced TG and AC. Species with high amounts of DNA methylation have (probably for this reason) low amounts of CG in their DNA.

These molecular mutational biases are not the same as changes toward improved adaptation, however. You cannot change a light-colored moth into a dark one just by causing transitional, rather than transversional, mutations in its DNA, or by converting a proportion of its CG dinucleotides into TG. Some critics of Darwinism have read that the Darwinian theory describes mutation as "random," and have then cited these sorts of molecular mutational biases as if they contradicted Darwinian theory. As we have seen, mutation can be non-random at the molecular level without contradicting the Darwinian theory. Darwinism simply rules out mutation directed toward new adaptation. Because of this confusion about the word "random," it is often better to describe mutation as "undirected" or "accidental" (the word Darwin used).

SUMMARY

1. Organisms produce many more offspring than can survive, which results in a "struggle for existence," or competition to survive.
2. Natural selection will operate among any entities that reproduce, show inheritance of their characteristics from one generation to the next, and vary in "fitness" (i.e., the relative number of offspring they produce) according to the characteristic they possess.
3. The increase in frequency of the melanic, relative to the light-colored, form of the peppered moth *Biston betularia* clearly illustrates how natural selection causes both evolutionary change and the evolution of adaptation.
4. Selection may be directional, stabilizing, or disruptive.
5. The members of natural populations vary with respect to characteristics at all levels. They differ in their morphology, their microscopic structure, their

chromosomes, the amino acid sequences of their proteins, and their DNA sequences.

6. The members of natural populations vary in their reproductive success. Some individuals leave no offspring, while others leave many more than average.
7. In Darwin's theory, the direction of evolution, particularly of adaptive evolution, is uncoupled from the direction of variation. The new variation that is created by recombination and mutation is accidental, and adaptively random in direction.
8. Two reasons suggest that neither recombination nor mutation alone can change a population in the direction of improved adaptation: no evidence exists that mutations occur particularly in the direction of novel adaptive requirements, and it is theoretically difficult to see how any genetic mechanism could have the foresight to direct mutations in this way.

FURTHER READING

An ecology text, such as Begon *et al.* (1990), will introduce life tables. For the theory of natural selection, see Darwin's original account (1859, chs. 3 and 4) and Endler (1986). Johnson (1976, 1979) introduces the kinds of selection. Law (1991) describes the selective effects of fishing. Travis (1989) reviews stabilizing selection. For a natural example of disruptive selection, see Smith (1993) on finch beaks.

Genetic variation is described in all the larger population genetics texts, such as Hartl (1988) and Hartl and Clark (1989), Hedrick (1983), Spiess (1989), and Wallace (1981). Lewontin (1982) describes variation in humans; White (1973) and Dobzhansky (1970) describe chromosomal variation. Variation in proteins and DNA will be discussed further in chapter 7, which gives references.

The authors in Clutton-Brock (1988) discuss natural variation in reproductive sucess; for plants, see Nilsson (1992) on orchids. References for the classic experiment on mutational directionality and the modern controversy are: Luria and Delbruck (1943), Cairns *et al.* (1988), Sniegowski and Lenski (1995), Thaler (1994), and Shapiro (1995). On biases in the direction of mutation at the DNA level, see Golding (1987).

STUDY AND REVIEW QUESTIONS

1. Use Figure 4.1b to construct a life table, like Table 4.1, for cod. (Use the densities/m^2 as numbers; you may also prefer to ignore the right-hand column, for daily mortality rates, which require logarithms.)
2. (a) Describe the four conditions needed for natural selection to operate. (b) What would happen in a population in which only conditions 1, 2, and 3 were satisfied? (c) What would happen in a population in which only conditions 1, 3, and 4 were satisfied?
3. Variation in reproductive success has been found in all populations in which it has been measured. Why is this observation alone insufficient to show that natural selection operates in all populations?

4. What sort of selection is taking place in populations a, b, and c in the graph?

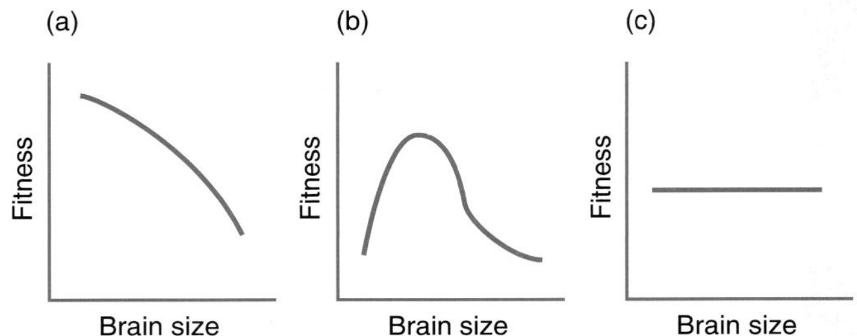

5. It is occasionally suggested that mutation is adaptively directed rather than random. Think through how a genetic mechanism of adaptively directed mutation would have to operate. For each component of the mechanism, how plausible is it that it could really exist?

6. For further thought: (a) On average only two offspring survive per parental pair. Why, therefore, does every pair in the population not produce exactly two offspring (rather than the more variable reproductive success we see in nature)? Such a reproductive strategy would lead to the same end consequence. (b) Why do some species produce a far greater "excess" far greater than others?

PART 2

Evolutionary Genetics

The theory of population genetics is the most important, most fundamental body of theory in evolutionary biology. It is the proving ground for almost all ideas in evolutionary biology: the coherence of an evolutionary hypothesis will usually remain in doubt until the hypothesis is expressed in the form of a population genetic model. We start with the simplest cases, and move on to more complex cases. The simplest case is when the population is large—large enough that we can ignore random effects; models of this kind are called deterministic. In chapter 5, we look at a simple deterministic model of natural selection. The model has only one genetic locus, and one allele of higher fitness is being substituted for an inferior allele. We also examine how natural selection can maintain variation at a single locus, in three circumstances. Examples from ecological genetics will illustrate the ideas. In small populations, the transfer of genes from one generation to the next cannot be understood purely deterministically, because random sampling may change the frequency of a gene.

Chapter 6 considers random effects in population genetics. In small populations, a new set of possibilities is opened up, in which evolution is not controlled by natural selection; the "neutral theory" suggests that molecular evolution is mainly driven by random processes. We will examine the evidence for and against the neutral theory and its selectionist alternative. We then move on to consider natural selection working simultaneously on more than one locus. Linkage between loci complicates the one locus model. With more than one locus, adaptive interactions between the genes become possible—which can make the existence of genetic recombination puzzling. The importance of higher-level interactions between gene loci remains controversial, as does the adequacy of the one-locus model when describing the real world.

As we move from two-loci evolution to multiple-loci evolution, we abandon Mendelian exactitude and use a quite different method: quantitative genetics. In quantitative genetics, the relations between individuals and between successive generations are described approximately and abstractly. This field is concerned with "continuous" characters, at the morphological level. As we saw in chapter 4, morphological characters show variation in natural populations, and we shall consider how to account for the observed level of variation. Finally, we consider some of the special evolutionary properties of multigene families and of the part of the DNA that does not appear to code for proteins and is made up of repeats of shorter unit sequences; some of these regions of the genome may be more or less parasitic "selfish DNA."

The Theory of Natural Selection

THE chapter introduces formal population genetic models. We first establish the variables with which the models are concerned, and the general structure of population genetic models. We look at the Hardy–Weinberg equilibrium, and see how to calculate whether a real population fits it. We then move on to models of natural selection, concentrating on the specific case of selection against a recessive homozygote. We apply the model to two examples: the peppered moth and resistance to pesticides. The second half of the chapter mainly concerns how natural selection can maintain genetic polymorphism. We examine selection-mutation balance, heterozygous advantage, and frequency-dependent selection, and we finish by looking at models that include migration in a geographically subdivided population. The theory in this chapter assumes that the population size is large; chapter 6 discusses how random effects operate in small populations.

5.1 Population genetics is concerned with genotype and gene frequencies

The human genome may contain some 100,000 gene loci. Let us focus on a locus at which more than one allele is present, because no evolutionary change can occur at a locus for which every individual in the population has two copies of the same gene. We shall be concerned in this chapter with models of evolution at a single genetic locus; these models are the simplest available in population genetics. In many real cases, evolutionary change probably occurs at many loci simultaneously; but we can keep the problem down to a manageable size by concentrating on only one of them.

The theory of population genetics at one locus is mainly concerned with understanding two closely connected variables: *gene frequency* and *genotype frequency*. Both of these variables are easy to measure. The simplest case is one genetic locus with two alleles (*A* and *a*) and three genotypes (*AA*, *Aa*, and *aa*). Each individual has a genotype made up of two genes at the locus and a population can be symbolized in the following manner:

Aa *AA* *aa* *aa* *AA* *Aa* *AA* *Aa*

This imaginary population includes only eight individuals. To find the genotype frequencies, we simply count the numbers of individuals with each genotype. Thus:

frequency of AA = 3/8 = 0.375
frequency of Aa = 3/8 = 0.375
frequency of aa = 2/8 = 0.25

In general, we can symbolize genotype frequencies algebraically, as follows:

genotype	AA	Aa	aa
frequency	P	Q	R

P, Q, and R are expressed as percentages or proportions, so in our population, $P = 0.375$, $Q = 0.375$, and $R = 0.25$ (they must add up to one, or to 100%). The genotype frequencies are measured by observing and counting the numbers of each type of organism in the population, and dividing by the total number of organisms in the population (the population size).

The gene frequency is likewise measured by counting the frequencies of each gene in the population. Each genotype contains two genes, and there are a total of 16 genes per locus in a population of eight individuals. In the population above,

frequency of A = 9/16 = 0.5625
frequency of a = 7/16 = 0.4375.

Algebraically, we can define p as the frequency of A and q as the frequency of a. Although p and q are always called "gene" frequencies, in a strict sense they are allele frequencies. That is, they are the frequencies of the different alleles at one genetic locus. The gene frequencies can be calculated from the genotype frequencies:

$$p = P + \tfrac{1}{2}Q \qquad (5.1)$$
$$q = R + \tfrac{1}{2}Q$$

(and $p + q = 1$). The calculation of gene frequencies from genotype frequencies is highly important, and we will refer back to these two simple equations many times in this chapter. Although the gene frequencies can be calculated from the genotype frequencies (P, Q, R), the opposite is not true: the genotype frequencies cannot be calculated from the gene frequencies (p, q).

Now that we have defined the key variables, we can see how population geneticists analyze changes in those variables through time.

5.2 *An elementary population genetic model has four main steps*

Population geneticists try to answer the following question: If we know the genotype (or gene) frequencies in one generation, what will they be in the next generation? It is worth looking at the general procedure for predicting these frequencies before examining particular models. The procedure is to break down the time from one generation to the next into a series of stages, and then to work out how genotype frequencies are affected at each stage. We can begin at any arbitrarily chosen starting point in generation n and then follow the genotype frequencies through to the same point in generation $n + 1$. Figure 5.1 shows the general outline of a population genetics model.

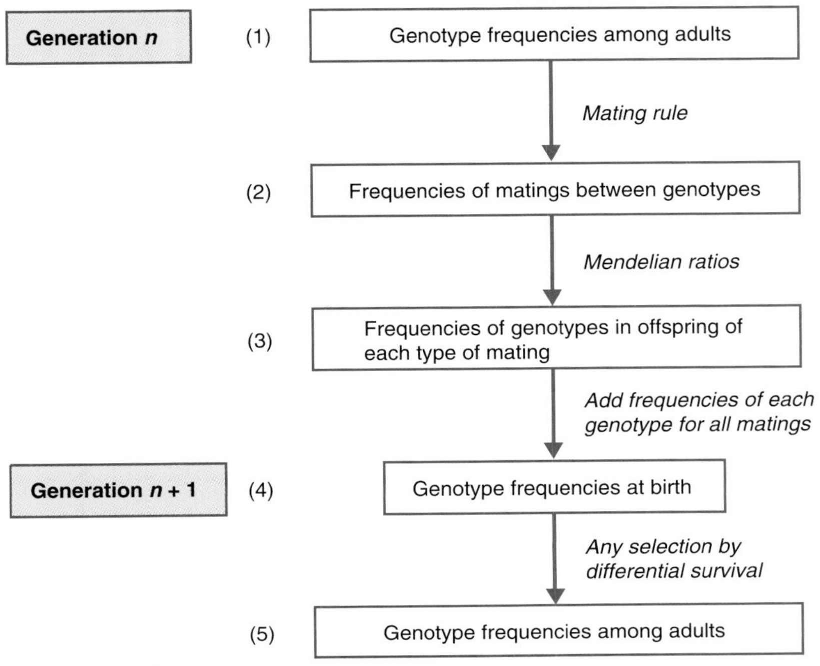

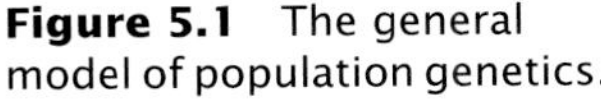
Figure 5.1 The general model of population genetics.

We start with the frequencies of genotypes among the adults in generation n. The first step is to specify how these genotypes combine to breed (called a mating rule). The second step is to apply the Mendelian ratios (chapter 2) for each type of mating. Third, we add the frequencies of each genotype generated from each type of mating to find the total frequency of the genotypes among the offspring, at birth, in the next generation. If the genotypes have different chances of survival from birth to adulthood, we multiply the frequency of each genotype at birth by its chance of survival to find the frequency among adults. When the calculation at each stage has been completed, the population geneticist's question has been answered.

Natural selection can operate in two ways—by differences in survival among genotypes or by differences in fertility. Two theoretical extremes are possible. At one extreme, the surviving individuals of all genotypes produce the same number of offspring, and selection operates only on survival. At the other extreme, individuals of all genotypes have the same survival, but differ in the number of offspring they produce (that is, their fertility). Both kinds of selection probably operate in many real cases, but the models we will consider in this chapter all express selection in terms of differences in chance of survival. This choice of models is not meant to suggest that selection always operates only on survival; rather, it is intended to keep the models simple and consistent.

The model, in the general form of Figure 5.1, may look rather complicated. However, we can cut it down to size by making some simplifying assumptions. The first two simplifying assumptions to consider are random mating and no selection (no differences in survival between genotypes from stage 4 to 5).

TABLE 5.1

Calculations needed to derive the Hardy–Weinberg ratio, for one locus and two alleles, *A* and *a*. (Frequency of *AA* = *P*, of *Aa* = *Q*, of *aa* = *R*.) The table shows the frequencies of different matings if the genotypes mate randomly, and the genotype proportions among the progeny of the different matings.

Mating Type	Frequency of Mating	Offspring Genotype Proportions
AA × *AA*	P^2	1*AA*
AA × *Aa*	*PQ*	½*AA* : ½*Aa*
AA × *aa*	*PR*	1*Aa*
Aa × *AA*	*QP*	½*AA* : ½*Aa*
Aa × *Aa*	Q^2	¼*AA* : ½*Aa* : ¼*aa*
Aa × *aa*	*QR*	½*Aa* : ½*aa*
aa × *AA*	*RP*	1*Aa*
aa × *Aa*	*RQ*	½*Aa* : ½*aa*
aa × *aa*	R^2	1*aa*

5.3 *Genotype frequencies in the absence of selection go to the Hardy–Weinberg equilibrium*

We will continue with the case of one genetic locus with two alleles (*A* and *a*). The frequencies of genotypes *AA*, *Aa*, and *aa* are *P*, *Q*, and *R*, respectively. If there is random mating and no selective difference among the genotypes, and we know the genotype frequencies in one generation, what will the genotype frequencies be in the next generation? The answer is found in the *Hardy–Weinberg equilibrium*.

Table 5.1 provides the calculation for this equilibrium. The mating frequencies follow from the fact that mating is random. To form a pair, we pick out at random two individuals from the population. What is the chance of an *AA* × *AA* pair? To produce this pair, the first individual we pick must be an *AA* and the second one also must be an *AA*. The chance that the first individual is an *AA* is simply *P*, the genotype's frequency in the population. In a large population, the chance that the second one is *AA* is also *P*.[1] The chance of drawing out two *AA* individuals in a row is, therefore, P^2. (The frequency of *Aa* × *Aa* and *aa* × *aa* matings are likewise Q^2 and R^2, respectively.) Similar reasoning applies for the frequencies of matings in which the two individuals have different genotypes. The chance of picking an *AA* and then an *Aa* (to produce an *AA* × *Aa* pair), for example, is *PQ*; the chance of picking an *AA* and then an *aa* is *PR*; and so on.

The genotypic proportions in the offspring of each type of mating are given by the Mendelian ratios for that cross. We can calculate the frequency of a

1. "Large" populations are not a separate category from "small" ones; populations come in all sizes. The random effects we consider in chapter 6 become increasingly important as a population becomes smaller. However, one rough definition of a large population is one in which the sampling of one individual to form a mating pair does not affect the genotype frequencies in the population: if one *AA* is removed, the frequency of *AA* in the population, and the chance of picking another *AA*, remains effectively *P*.

genotype in the next generation by addition. We look at which matings generate the genotype, and add the frequencies generated by all the matings. Let us consider the genotype *AA*. *AA* individuals, as Table 5.1 shows, come from *AA* × *AA*, *AA* × *Aa* (and *Aa* × *AA*), and *Aa* × *Aa* matings. We can ignore all the other types of mating. *AA* × *AA* matings have frequency P^2 and produce all *AA* offspring, *AA* × *Aa* and *Aa* × *AA* matings each have frequency *PQ* and produce ½ *AA* offspring, and *Aa* × *Aa* matings have frequency Q^2 and produce ¼ *AA* offspring. The frequency of *AA* in the next generation,[2] P', is then

$$P' = P^2 + \tfrac{1}{2}PQ + \tfrac{1}{2}PQ + 1/4Q^2 \tag{5.2}$$

This can be rearranged to give

$$P' = (P + \tfrac{1}{2}Q)(P + \tfrac{1}{2}Q)$$

$P + \tfrac{1}{2}Q$, we have seen, is simply the frequency of the gene *A*, *p*. Therefore

$$P' = p^2$$

The frequency of genotype *AA* after one generation of random mating is equal to the square of the frequency of the *A* gene. Analogous arguments show that the frequencies of *Aa* and *aa* are $2pq$ and q^2. The Hardy–Weinberg frequencies are then

genotype	*AA*	:	*Aa*	:	*aa*
frequency	p^2	:	$2pq$	:	q^2

Figure 5.2 shows the proportions of the three different genotypes at different frequencies of the gene *a*; heterozygotes are most frequent when the gene frequency is 0.5.

The Hardy–Weinberg genotype frequencies are reached after a single generation of random mating from any initial genotype frequencies. Imagine, for example, two populations with the same gene frequency but different genotype frequencies. One population has 750 *AA*, 0 *Aa*, and 250 *aa*; the other has 500 *AA*, 500 *Aa*, and 0 *aa*. In both instances, $p = 0.75$, and $q = 0.25$. After one generation of random mating, the genotype frequencies in both will become 558 *AA*, 375 *Aa*, and 67 *aa* if the population size remains 1000. (Fractions of an individual have been rounded to make the numbers add to 1000. The proportions are 9/16, 6/16, and 1/16.) After reaching those frequencies immediately, in one generation, the population remains at the Hardy–Weinberg equilibrium for as long as the population size is large, no selection occurs, and mating is random.

As we saw in section 5.1, in general it is not possible to calculate the genotype frequencies in a generation if you only know the gene frequencies. We can now see that it is possible to calculate, from gene frequencies alone, the genotype frequencies for the *next* generation, provided that mating is random, no selection occurs, and the population is large. If the gene frequencies in this generation

2. Population geneticists conventionally symbolize the frequency of variables one generation on by writing a prime. If P is the frequency of genotype *AA* in one generation, P' is its frequency in the next; if p is the frequency of an allele in one generation, p' is its frequency in the next generation. We will follow the same convention in this book.

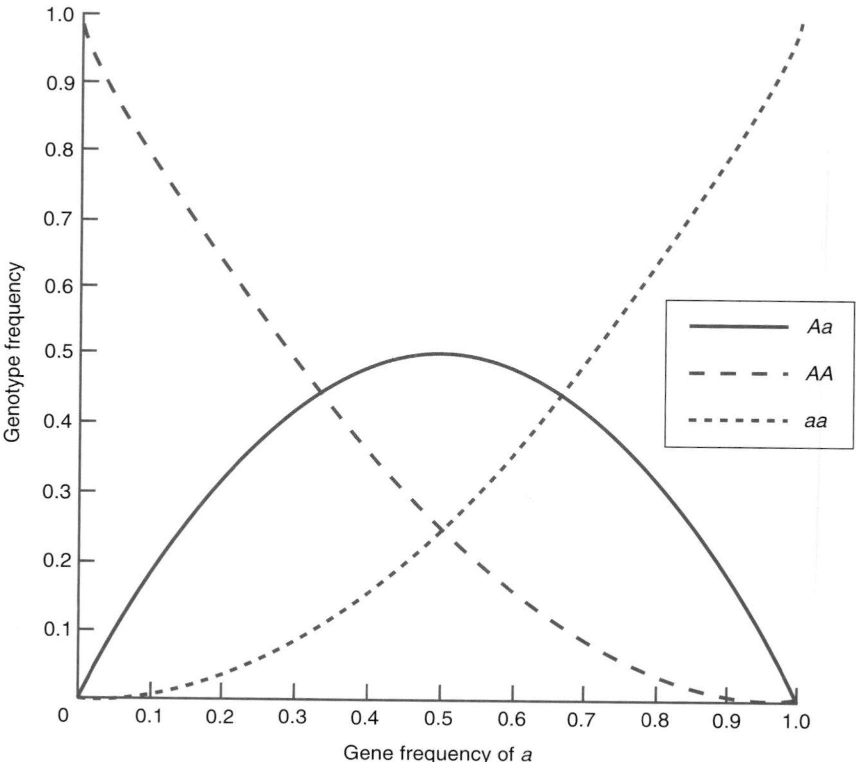

Figure 5.2 Hardy–Weinberg frequencies of genotypes *AA*, *Aa*, and *aa* in relation to the frequency of the gene *a* (q).

are p and q, in the next generation the genotype will have Hardy–Weinberg frequencies.

The proof of the Hardy–Weinberg theorem we have worked through was rather long-winded. We completed it to illustrate the general model of population genetics in its simplest case. However, for the particular case of the Hardy–Weinberg equilibrium, a more elegant proof can be given in terms of gametes.

Diploid organisms produce haploid gametes. We could imagine that the haploid gametes are all released into the sea, where they combine at random to form the next generation—a process called random union of gametes. In the "gamete pool" *A* gametes will have frequency p and *a* gametes will have frequency q. Because they are combining at random, an *a* gamete will meet an *A* gamete with chance p and an *a* gamete with chance q. From the *a* gametes, *Aa* zygotes will therefore be produced with frequency pq and *aa* gametes with frequency q^2. A similar argument applies for the *A* gametes (which have frequency p): they combine with *a* gametes with chance q, to produce *Aa* zygotes (frequency pq) and *A* gametes with chance p to form *AA* zygotes (frequency p^2). If we now add up the frequencies of the genotypes from the two types of gametes, the Hardy–Weinberg genotype frequencies emerge. We have now derived the Hardy–Weinberg theorem for the case of two alleles; the same argument easily extends to three or more alleles (Box 5.1).

You may be puzzled by the 2 in the frequency of the heterozygotes. It is a

BOX 5.1

The Hardy–Weinberg theorem for three alleles

We can call the three alleles A_1, A_2, and A_3, and define their gene frequencies as p, q, and r, respectively. We form new zygotes by sampling two successive gametes from a large pool of gametes. The first gamete we pick could be A_1, A_2, or A_3. If we first pick (with chance p) an A_1 allele from the gamete pool, the chance that the second allele is another A_1 allele is p, the chance that it is an A_2 allele is q, and the chance that it is an A_3 allele is r. From these three, the frequencies of A_1A_1, A_1A_2, and A_1A_3 zygotes are p^2, pq, and pr.

Now suppose that the first allele we chose had been an A_2 (which would happen with chance q). The chances that the second allele would again be A_1, A_2, or A_3 would be p, q, and r, respectively, giving A_1A_2, A_2A_2, and A_2A_3 zygotes in frequency pq, q^2, and qr. Finally, if we had picked (with chance r) an A_3 allele, we produce A_1A_3, A_2A_3, and A_3A_3 zygotes in frequency pr, qr, and r^2.

The only way to form the homozygotes A_1A_1, A_2A_2, and A_3A_3 is by picking two of the same kind of gamete and the frequencies are p^2, q^2, and r^2. The heterozygotes can be formed from more than one kind of first gamete and their frequencies are obtained by addition. The total chance of forming an A_1A_3 zygote is $pr + rp = 2pr$; of forming an A_1A_2 zygote is $pq + qp = 2pq$; and of an A_2A_3 zygote is $2qr$. The complete Hardy–Weinberg proportions are as follows:

$$\begin{matrix} A_1A_1 & : & A_1A_2 & : & A_1A_3 & : & A_2A_2 & : & A_2A_3 & : & A_3A_3 \\ p^2 & & 2pq & & 2pr & & q^2 & & 2qr & & r^2 \end{matrix}$$

simple combinatorial probability. Imagine flipping two coins and asking what the chances are of flipping two heads, or two tails, or one head and one tail. The chance of two heads is $(\frac{1}{2})^2$ and of two tails $(\frac{1}{2})^2$; the chance of a head and a tail is $2 \times (\frac{1}{2})^2$, because both a tail followed by a head, and a head followed by a tail, give one head and one tail. The head is analogous to allele *A*, and the tail to *a*; two heads to producing an *AA* genotype; and one head and one tail to a heterozygote *Aa*. The coin produces heads with probability ½, and is analogous to a gene frequency of $p = \frac{1}{2}$. The frequency $2pq$ for heterozygotes is analogous to the chance of one head and one tail, $2 \times (\frac{1}{2})^2$. The 2 arises because there are two ways of obtaining one head and one tail. Likewise, there are two ways of producing an *Aa* heterozygote: either the *A* gene can come from the father and the *a* from the mother, or the *a* gene from the father and the *A* from the mother. In either case, the offspring is *Aa*.

5.4 *We can test, by simple observation, whether genotypes in a population are at the Hardy–Weinberg equilibrium*

The Hardy–Weinberg theorem depends on three main assumptions: no selection, random mating, and a large population size. In a natural population, any of these conditions could be false; we cannot assume that natural populations will be at the Hardy–Weinberg equilibrium. In practice, we can ascertain whether a population is at the Hardy–Weinberg equilibrium for a locus simply by counting the genotype frequencies. From those frequencies, we first calculate the gene frequencies. Then, if the observed homozygote frequencies equal the square of

TABLE 5.2

The frequencies of the *MM, MN,* and *NN* blood groups in three American populations. Figures for expected proportions and *r* have been rounded.

Population		*MM*	*MN*	*NN*	Total	Frequency of *M*	Frequency of *N*
African Americans	Observed number	79	138	61	278		
	Expected proportion	0.283	0.499	0.219		0.532	0.468
	Expected number	78.8	138.7	60.8			
European Americans	Observed number	1787	3039	1303	6129		
	Expected proportion	0.292	0.497	0.211		0.54	0.46
	Expected number	1787.2	3044.9	1296.9			
Native Americans	Observed number	123	72	10	205		
	Expected proportion	0.602	0.348	0.05		0.776	0.224
	Expected number	123.3	71.4	10.3			

Specimen calculation for African Americans:

Frequency of *M* allele = (79 + (½ × 138))/278 = 0.532 = p
Frequency of *N* allele = (61 + (½ × 138))/278 = 0.468 = q
Expected proportion of *MM* = p^2 = $(0.532)^2$ = 0.283
Expected proportion of *MN* = $2pq$ = 2(0.532)(0.468) = 0.499
Expected proportion of *NN* = q^2 = $(0.468)^2$ = 0.219

Expected numbers = expected proportion x total number (n)

Expected number of *MM* = p^2n = 0.283 × 278 = 78.8
Expected number of *MN* = $2pqn$ = 0.499 × 278 = 138.7
Expected number of *NN* = q^2n = 0.219 × 278 = 60.8

their gene frequencies, the population is in Hardy–Weinberg equilibrium. If they do not, the population is not in Hardy–Weinberg equilibrium.

The *MN* blood group system in humans is a good example of using observation to test the Hardy–Weinberg equilibrium, because the three genotypes are distinct and the genes have reasonably high frequencies in human populations. Three phenotypes, *M, MN,* and *N* are produced by three genotypes (*MM, MN,* and *NN*) and two alleles at one locus. The phenotypes of the *MN* group, like the better known ABO group, are recognized by reactions with antisera. The antisera are made by injecting blood into a rabbit, which then makes an antiserum to the type of blood that was injected. If the rabbit has been injected with *M*-type human blood, it produces anti-*M* serum. Anti-*M* serum agglutinates blood from humans with 1 or 2 *M* alleles in their genotypes; likewise anti-*N* blood agglutinates the blood of humans with 1 or 2 *N* alleles. As a result, *MM* individuals are recognized as having blood that reacts only with anti-*M*, *NN* individuals react only with anti-*N*, and *MN* individuals react with both.

Table 5.2 gives some measurements of the frequencies of the *MN* blood group genotypes for three human populations. Are they at Hardy–Weinberg equilibrium? In Americans of European descent, the frequency of the *M* gene (calculated from the usual $p = P + \frac{1}{2}Q$ relation) is 0.54. If the population is at the Hardy–Weinberg equilibrium, the frequency of *MM* homozygotes (p^2) will be $0.54^2 = 0.2916$ (1787 in a sample of 6129 individuals); the frequency of *MN* heterozygotes ($2pq$) will be $2 \times 0.54 \times 0.46 = 0.497$ (3045 in a sample of 6129). As Table 5.2 shows, these values are close to the observed frequencies. In fact, all three populations are at Hardy–Weinberg equilibrium. We will see in section 5.6 that the same calculations do not correctly predict the genotype frequencies after selection has operated.

5.5 *The Hardy–Weinberg theorem is important conceptually and historically, and in practical research and the workings of theoretical models*

In the previous section, we saw how to determine whether a real population is in Hardy–Weinberg equilibrium. The importance of the Hardy–Weinberg theorem, however, is not mainly as an empirical prediction. We have no reason to think that genotypes in natural populations will generally have Hardy–Weinberg frequencies, because it would require both no selection and random mating. Real populations may often experience selection and non-random mating. Instead, the theorem creates interest on three fronts.

First, the theorem has historical and conceptual attraction. We saw in section 2.8 (p. 35) how blending inheritance rapidly blends the genetic variation in a population out of existence and the population becomes genetically uniform. With Mendelian genetics, variation is preserved and the Hardy–Weinberg theorem gives quantitative demonstration of that fact. The theorem was published in the first decade of the twentieth century, as Mendelism was becoming accepted, and it was historically influential in proving to people that Mendelian inheritance did allow variation to be preserved.

A second attraction of the theorem is its use as a springboard that launches us toward interesting empirical problems. If we compare genotype frequencies in a real population with the Hardy–Weinberg ratios, and they deviate, it suggests something interesting (such as selection or non-random mating) may be occurring that would merit further research.

A third interest is theoretical. The general model of population genetics (section 5.2 above) included five stages, joined by four calculations. The Hardy–Weinberg theorem simplifies the model wonderfully. If we assume random mating, we can go directly from the adult frequencies in generation n to the genotype frequencies at birth in generation $n + 1$, collapsing three calculations into one (Figure 5.3). If we know the adult genotype frequencies in generation n (stage 1), we only need to calculate the gene frequencies. The genotype frequencies at birth in the next generation (stage 2) must then have Hardy–Weinberg frequencies, because the gene frequencies do not change between the adults of

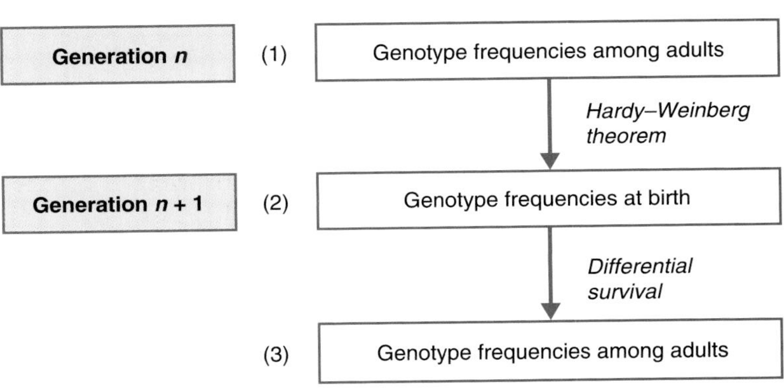

Figure 5.3 The general model of population genetics simplified by the Hardy–Weinberg theorem.

one generation and the new-born members of the next generation. A simple model of selection can concentrate on how the genotype frequencies are modified between birth and the adult reproductive stage (from stage 2 to stage 3 of Figure 5.3).

5.6 The simplest model of selection is for one favored allele at one locus

The simplest case involves natural selection operating on only one genetic locus, at which there are two alleles, one dominant to the other. Suppose that individuals with the three genotypes have the following relative chances of survival from birth to the adult stage:

Genotype	Chance of Survival
AA, *Aa*	1
aa	$1 - s$.

In this case, s is a number between zero and one; it is called the *selection coefficient*. Selection coefficients are expressed as reductions relative to the best genotype. If s is 0.1, then *aa* individuals have a 90% chance of survival, relative to 100% for *AA* and *Aa* individuals. These are relative values—in a real case the chance of survival from birth to reproduction of an individual with the best genotype might be 50% (and thus much less than 100%). If it was 50%, then an s of 0.1 would mean that *aa* individuals had a 45% chance of survival. (The convention of giving the best genotype a relative 100% chance of survival simplifies the algebra. If you are suspicious, check whether it makes any difference if the chances of survival are 50%, 50%, and 45%, respectively, for *AA*, *Aa*, and *aa*, rather than 100%, 100%, and 90%.) The chance of survival is the *fitness* of the genotype; we are assuming that all surviving individuals produce the same number of offspring. With the fitnesses given above, selection will act to eliminate the *a* allele and *fix* the *A* allele. (To "fix" a gene is genetic jargon for carrying its frequency up to one. When only one gene is present at a locus, it is said to be "fixed" or in a state of "fixation.") If s were zero, the model would lapse back to the Hardy–Weinberg case and the gene fequencies would be stable.

Notice, therefore, that alleles do not have any tendency to increase in frequency just because they are dominant, or to decrease because they are recessive. Dominance and recessivity describe only how the alleles at a locus interact to produce a phenotype. Changes in gene frequency are set by the fitnesses: if the recessive homozygote has higher fitness, the recessive allele will increase in frequency; if, as here, it has lower fitness the recessive allele decreases in frequency.

How rapidly will the population change through time? To find out, we seek an expression for the gene frequency of A (p') in one generation in terms of its frequency in the previous generation (p). The difference between the two, $\Delta p = p' - p$, is the change in gene frequency between two successive generations. The model has the form of Figure 5.3, and we shall work through both the general algebraic version and a numerical example (Table 5.3).

To begin with, at birth the three genotypes have Hardy–Weinberg frequencies, as they are produced by random mating among the adults of the previous

difference in color was controlled by one main locus. The original, peppered form (called *typica*) was one homozygote (*cc*) and the melanic form (called *carbonaria*) was another homozygote (*CC*), and the *C* allele is dominant. However, in other experiments the melanic allele was less dominant and the heterozygotes were intermediate; a number of different melanic alleles appear possible. Selection may have initially favored a melanic allele with no or weak dominance, and subsequently some other melanic alleles with stronger dominance. In any case, the degree of dominance of the melanic allele that was originally favored in the nineteenth century is uncertain, and it may have differed from the dominance shown by the melanic alleles that exist in modern populations.

The first estimates of fitnesses were made by Haldane. He resolved the problem of varying degrees of dominance by making two estimates of fitness. One estimate assumed that the *C* allele is dominant; the other assumed that the heterozygote is intermediate. The real average degree of dominance was probably between the two. Here we shall look only at the estimate for a dominant *C* gene.

5.7.2 *One estimate of the fitnesses is made using the rate of change in gene frequencies*

What were the relative fitnesses of the genes controlling the melanic and light coloration during this evolutionary change? For one method of calculating these fitnesses, we need measurements of the frequencies of the different color forms at at least two times. We can then estimate the gene frequencies from the genotype frequencies, and substitute them in equation 5.6 to solve for s, the selection coefficient.

The melanic form was first seen in 1848. but it was probably not a new mutation at that time. Although no earlier melanic specimens of the peppered moth are known, other moth species produced melanic forms in pre-industrial forests; the melanic form might therefore have been a rare form. Its frequency would probably be that of "mutation-selection balance." Mutation-selection balance means that the gene is disadvantageous and has a low frequency, determined by a balance between being formed by mutation and being lost by selection. A disadvantageous dominant gene in mutation-selection balance has a frequency of m/s, where m is its mutation rate and s its selective advantage (section 5.11). The values of m and s were unknown for the gene in the eighteenth and early nineteenth century, but Haldane guessed that $m = 10^{-6}$ and $s = 0.1$ for the melanic *C* gene. The gene would then have had a frequency of 10^{-5}. By 1898, the frequency of the light-colored (*typica*) genotype was 1–10% in polluted areas (it was not more than 5% near Manchester, for example, implying a gene frequency of about 0.2). Approximately 50 generations would have occurred between 1848 and 1898.

We now have all of the information that we need. What selective coefficient would generate an increase in its frequency from 10^{-5} to 0.8 in 50 generations? Equation 5.6 gives the selection coefficient in terms of gene frequencies in two successive generations, but between 1848 and 1898 there would have been 50 generations. The formula therefore has to be applied 50 times, which is most easily done by computer. A change from 10^{-5} to 0.8 in 50 generations, it turns out, requires $s \approx 0.33$: the peppered moths had two-thirds the survival rate of melanic moths (Table 5.5). Although these calculations are rough, they show how fitness can be inferred.

TABLE 5.5

Theoretical changes in gene frequencies in the evolution of melanism in the peppered moth, starting with an initial frequency of *C* of 0.00001 (rounded to zero in the table). *C* is dominant, and *c* is recessive. Genotypes *CC* and *Cc* are melanic, and *cc* is peppered in color. In the simulation, 1848 is generation 0.

Generation	Gene Frequency *C*	*c*
1848	0.00	1.00
1858	0.00	1.00
1868	0.03	0.97
1878	0.45	0.55
1888	0.76	0.24
1898	0.86	0.14
1908	0.90	0.10
1918	0.92	0.08
1928	0.94	0.06
1938	0.96	0.04
1948	0.96	0.04

5.7.3 *A second estimate of the fitnesses is made from the survivorship of the different genotypes in mark-recapture experiments*

The estimate of fitness can be checked against other estimates. The gene frequency change is believed to be produced by differential survival in nature, and we can try to measure the rate of survival of the two types and see whether the peppered moths survive only two-thirds as well as melanic moths. Kettlewell measured survival rates by performing mark-recapture experiments in the field. He released melanic and light-colored peppered moths in known proportions in polluted and unpolluted regions, and then later re-caught some of the moths (which are attracted to mercury-vapor lamps). Kettlewell then recorded the proportions of melanic and peppered moths in the moths recaptured from the two areas.

Table 5.6 gives some results for two sites, Birmingham (polluted) and Deanend Wood, an unpolluted forest in Dorset, England. The proportions in the recaptured moths are as we should expect: more light-colored *typica* in the Deanend Wood samples, more melanic *carbonaria* in the Birmingham samples. In Birmingham, melanic moths were recaptured at about twice the rate of light-colored ones, implying $s = 0.57$. This fitness difference is higher than the $s = 0.33$ implied by the change in gene frequency. Even if $s = 0.57$ is rounded down to $s = 0.5$, Haldane calculated that the change in gene frequency from 10^{-5} to 0.8 would have been accomplished in 27 generations (with complete dominance) or 37 generations (with heterozygotes intermediate), rather than the 50 or so generations actually taken.

The discrepancy is unsurprising because both estimates are uncertain, and it could be generated by a number of factors. The discrepancy may simply be a sampling error (the numbers in the mark-recapture experiment were small); the

TABLE 5.6

Frequencies of melanic and light peppered moths in samples recaptured at two sites in the United Kingdom, Birmingham (polluted) and Deanend Wood, Dorset (unpolluted). The observed numbers are the actual numbers recaught; the expected numbers are the numbers that would have been recaught if all morphs survived equally (= proportion in released moths × number of moths recaptured). The recaptured moths at Birmingham were taken over a period of about 1 week, at Deanend Wood, over about 3 weeks. Data from Kettlewell (1973).

Birmingham (Polluted)	Light	Melanic
Numbers recaptured:		
observed	18	140
expected	36	122
Relative survival rate	.5	1.15
Relative fitness	.5/1.15 = .43	1.15/1.15 = 1
Deanend wood (unpolluted)	**Light**	**Melanic**
Numbers recaptured:		
observed	67	32
expected	53	46
Relative survival rate	1.26	0.69
Relative fitness	1.26/1.26 = 1	0.69/1.26 = 0.55

frequency in 1848 may have less than 10^{-5}; and the genetics may have been different. In addition, light-colored moths would have migrated from unpolluted areas into polluted areas, and the fitness difference in the polluted areas alone would therefore exaggerate the advantage of the dark form in the moth population as a whole. The fitness differences in the 1950s probably also differed during 1848–1898. After clean-air laws were passed in the twentieth century, industrial pollution decreased, and the frequency of the light form increased again in formerly more polluted areas; this factor, however, may aggravate the discrepancy. Whatever the cause of the discrepancy, the two calculations do illustrate two important methods of estimating fitness.

5.7.4 *The details of the story are now known to be more complex*

What has recent research added to the story? One complication is that the genetic control involves at least three melanic alleles, not just one. An even more important discovery is that further selective forces are at work. Melanic forms in other species—including beetles, pigeons, and cats—have increased in frequency in polluted regions even though they do not suffer significant predation by birds. In ladybird beetles (*Adalia bipunctata*), the melanic form absorbs solar radiation more efficiently in smoky places and has an advantage for that reason; it may also have an advantage in breeding.

Even if we concentrate on the peppered moth *Biston betularia*, the geographic distribution of the types does not fit the simple story. The melanic form, for example, has a frequency of up to 80% in a region of the United Kingdom called East Anglia, where pollution is low (Figure 5.5). In polluted areas, the dark form does not seem to have a high enough frequency. It never exceeded about 95%,

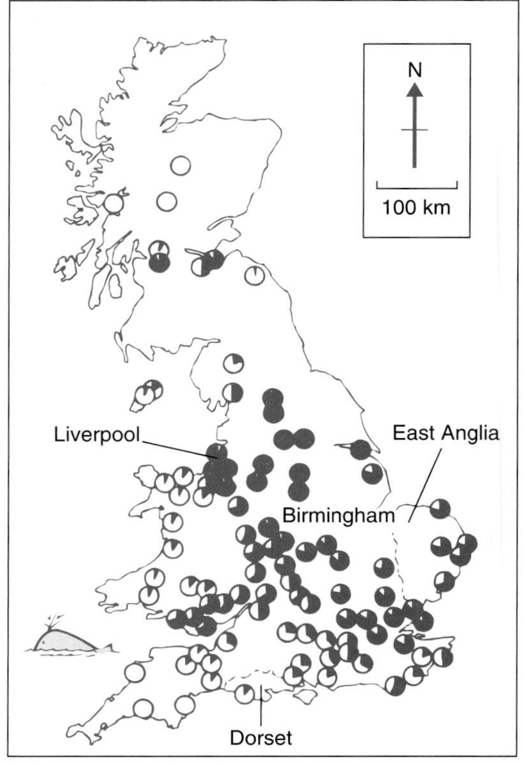

Figure 5.5 Frequency of melanic and peppered forms of the peppered moth in different parts of the United Kingdom. The filled-in part of each diagram represents the frequency of the melanic form in that area. Melanic moths are generally more prevalent in industrial areas, such as central England, but note the high proportion in East Anglia. Reprinted, by permission of the publisher, from Lees (1971).

despite its clearly better camouflage, which should have givent it a frequency of 100%.

Moreover, the decrease in frequency of melanic moths as pollution decreased seemed to outstrip what would be predicted from their camouflage alone. Between 1960 and 1975, the frequency of melanic moths near Liverpool decreased from 95% to 82%, even though they still appeared to be the better camouflaged form in 1975.

Evidence also suggests that the forms differ in fitness independently of bird predation. In 1980, Creed, Lees, and Bulmer collected all the measurements that had been made on survival to adulthood in the laboratory. They analyzed the results of 83 broods, containing 12,569 offspring; the original measurements had been made by many different geneticists in the previous 115 years. The viability of light-colored homozygotes, it turned out, was about 30% less on average than that of the melanic homozygote in the lab, where no bird predation occurs. While the reason behind this discrepancy is not known, it implies that the melanic genotype enjoys some "inherent" advantage. The fitness advantage detected in the lab, although it is probably only approximately similar to that in nature, implies that melanic moths would replace light ones even without bird predation in polluted areas. In unpolluted areas, light-colored moths remain only because birds eat more of the conspicuous melanic moths.

The final factor to be taken into account is migration. Male moths can fly long distances to find females, and a male peppered moth mates on average 2.5 km away from its birthplace. Migration may explain why melanic moths are found in some unpolluted areas like East Anglia and why light-colored moths persist in polluted areas where they are less well camouflaged.

The three factors—bird predation, inherent advantage to melanic genotypes, and migration—were incorporated in a simulation by Mani to try to explain the pattern of the gene in England. He could simulate the pattern most successfully with an "inherent" selective advantage (independent of bird predation) of 19% for the melanic homozygote (less than Creed *et al.*'s 30%), but the general success of Mani's simulation suggests that melanism in the peppered moth must be understood not only in terms of bird predation, but also in terms of migration and some other inherent selective advantage to the melanic moths.

In conclusion, the industrial melanism of the peppered moth is a classic example of natural selection, and illustrates the one-locus, two-allele model of selection. The model can be used to make a rough estimate of the difference in fitness between the two forms of moth using their frequencies at different times; the fitnesses can also be estimated from mark-recapture experiments. However, the one-locus, two-allele model is only an approximation to reality. In fact, several alleles are present (and their dominance relations are not simple); selection is not simply a matter of bird predation in relation to camouflage; and it seems that migration, as well as selection, is needed to explain the geographic pattern of gene frequencies.

5.8 *Pesticide resistance in insects is an example of natural selection*

Malaria is caused by a protozoan blood parasite (see section 5.12.2), and humans are infected with it by mosquitoes (family Culicidae—genera include *Aedes, Anopheles,* and *Culex*). It can therefore be prevented by killing the local mosquito population, and health workers have recurrently responded to malarial outbreaks by spraying insecticides such as DDT in affected areas. DDT, sprayed on a normal insect, is a lethal nerve-poison. When it first comes in contact with a local mosquito population, the population goes into abrupt decline. Subsequent events depend on whether DDT has been sprayed before.

On its first use, DDT is effective for several years. In India, for example, the pesticide remained effective for 10–11 years after its first widespread use in the late 1940s. DDT, on a global scale, was one reason why the number of cases of malaria declined to 75 million or so per year by the early 1960s. By then, however, DDT-resistant mosquitoes had already begun to appear. DDT-resistant mosquitoes were first detected in India in 1959, and they have increased so rapidly that when a local spray program is begun now, most mosquitoes become resistant in a matter of months rather than years (Figure 5.6). The malarial statistics reveal the consequence. The global incidence of the disease almost exploded, reaching an estimated 200 million victims per year by 1972 and 300 million per year now. Malaria currently kills about 3.5 million people per year, mainly children aged 1–4. Pesticide resistance was not the only reason for the increase, but it was important.

DDT becomes ineffective so quickly now because DDT-resistant mosquitoes exist at low frequency in the global mosquito population and, when a local population is sprayed, a strong force of selection in favor of the resistant mosquitoes is immediately created. It is only a matter of time before the resistant mosquitoes take over. A graph such as Figure 5.6 allows a rough estimate of the strength of selection. As with the peppered moth, we need to understand the genetics of the character, and to measure the genotype frequencies at two or more times. We can then use the formula for gene frequency change to estimate the fitness.

In order to use this formula, we must make a number of assumptions. One is that resistance is controlled by a single allele (we will discuss that assumption later). Another concerns the degree of dominance; the allele conferring resistance might be dominant, recessive, or intermediate, relative to the natural susceptibility allele. The case of dominant resistance is easiest to understand. (If resistance is recessive, we follow the same general method, but the exact result differs.) Let us call the resistance allele R and the susceptibility allele r. All mosquitoes that die in the mortality tests used in Figure 5.6 would then have been homozygous (rr) for susceptibility. Assuming (for simplicity rather than exact accuracy) Hardy–Weinberg ratios, we can estimate the frequency of the susceptibility gene as the square root of the proportion of mosquitoes that die in the tests. The selection coefficients are defined as follows, where fitness is measured as the chance of survival in the presence of DDT:

genotype	RR	Rr	rr
fitness	1	1	$1 - s$

If we define p as the frequency of R and q as the frequency of r, then equation 5.5 gives the change in gene frequency: selection is working against a recessive gene. Figure 5.6 shows the decline in the frequency of the susceptible mosquitoes, which are the recessive homozygotes. It is therefore easier to work with a formula for the change in q in one generation (Δq), rather than Δp (as on p. 101). The decrease in q is the mirror-image of the increase in p, and we simply need to put a minus sign in front of equation 5.5:

$$\Delta q = \frac{-spq^2}{1 - sq^2} \qquad (5.7)$$

The generation time is about one month. (The generations of mosquitoes overlap, rather than being discrete as the model assumes, but the exact procedure is similar in either case, and we can ignore the detailed correction for overlapping generations.) Table 5.7 shows how the genotype frequencies were read off Figure 5.6 in two stages, giving two estimates of fitness. Again, the formula for one generation is applied recurrently, for 8.25 and 4.5 generations in this case, to give an average fitness for the genotypes through the period. In Figure 5.6, it appears that the resistant mosquitoes had about twice the fitness of the susceptible ones—which is very strong selection.

The genetics of resistance in this case are not known, and the one-locus, two-allele model is an assumption only. The genetics are understood in some other cases, however. Resistance is often controlled by a single resistance allele.

TABLE 5.7

Estimated selection coefficients against DDT-susceptible *Anopheles culicifacies*, from Figure 5.6. Selection coefficients where relative fitness of susceptible type is $(1 - s)$. The estimate assumes the resistance allele is dominant. Simplified from Curtis *et al.* (1978).

Frequency of susceptible type			
Before	After	Time (months)	Selection coefficient
0.96	0.56	8.25	0.4
0.56	0.24	4.5	0.55

For example, Figure 5.7 shows that the resistance of the mosquito *Culex quinquifasciatus* to permethrin is due to a resistance (*R*) allele, which acts in a semi-dominant way, with heterozygotes intermediate between the two homozygotes. In houseflies, resistance to DDT results from an allele called *kdr*; *kdr* flies are resistant because they have fewer binding sites for DDT on their neurons. In other cases, resistance may be conferred not by a new point mutation, but by gene amplification. *Culex pipiens*, for instance, in one experiment became resistant to an organophosphate insecticide called temephos because individuals arose with increased numbers of copies of a gene for an esterase enzyme that detoxified the poison. In the absence of temephos, the resistance disappeared, which suggests

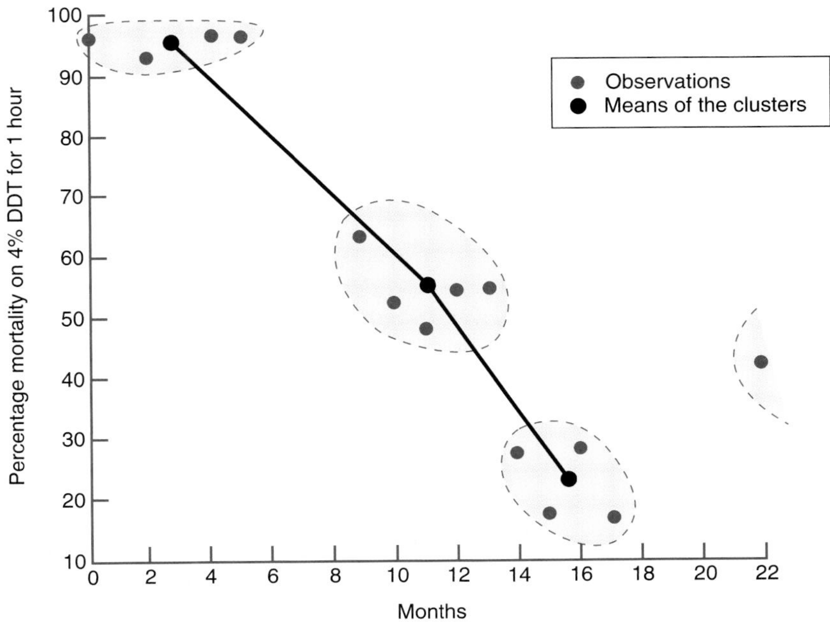

Figure 5.6 Increase in frequency of pesticide resistance in mosquitoes (*Anopheles culicifacies*) after spraying with DDT. A sample of mosquitoes was captured at each time indicated and the number that were killed by a standard dose of DDT (4% DDT for 1 hour) in the laboratory was measured. Reprinted, by permission of the publisher, from Curtis *et al.* (1978).

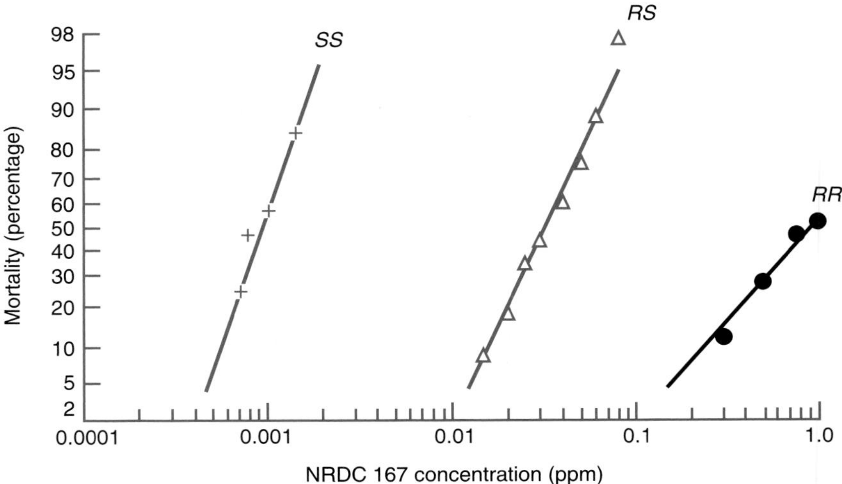

Figure 5.7 The mortality of mosquitos (*Culex quinquifasciatus*) of three genotypes at a locus when exposed to various concentrations of permethrin. The susceptible (*SS*) homozygote (+——+) dies at lower concentrations of the poison than the resistant homozygote (*RR*) (●——●). The heterozygote (*RS*) (Δ——Δ) has intermediate resistance. Reprinted, by permission of the publisher, from Taylor (1986).

that the amplified genotype must be maintained by selection. Gene amplification may be generally important in insecticide resistance; the semi-dominant behavior of resistance alleles in *C. quinquifasciatus* is, for example, compatible with this genetic mechanism. These examples suggest that insects can evolve resistance by a number of mechanisms, and Table 5.8 summarizes the main mechanisms that have been identified.

When an insect pest has become resistant to one insecticide, authorities often respond by spraying it with another insecticide. The evolutionary pattern we have seen here then usually repeats itself, and on a shorter time scale. On

TABLE 5.8

The main mechanisms of resistance to insecticides. Reprinted, by permission of the publisher, from Taylor (1986).

Mechanism	Insecticides affected
1. Behavioral	
Increased sensitivity to insecticide	DDT
Avoid treated microhabitats	many
2. Increased detoxification	
dehydrochlorinase	DDT
microsome oxidase	carbamates
	pyrethroids
	phosphorothioates
glutathione transferase	organophosphates (O-dimethyl)
hydrolases, esterases	organophosphates
3. Decreased sensitivity of target site	
acetylcholinesterase	organophosphates
	carbamates
nerve sensitivity	DDT
	pyrethroids
cyclodiene resistance genes	cyclodienes (organochlorines)
4. Decreased cuticular penetration	most insecticides

Long Island, New York, for example, the Colorado potato beetle (*Leptinotarsa septemlineata*) was first attacked with DDT. It evolved resistance to it in 7 years. The beetles were then sprayed with azinphosmethyl, and evolved resistance in 5 years; next came carbofuran (2 years), pyrethroids (another 2 years), and finally pyrethroids with synergist (one year). The decreasing time to evolve resistance is probably partly due to detoxification mechanisms that work against more than one pesticide. As more pests are sprayed with pesticides, more insects evolve resistance. According to a list developed in 1983, more than 400 insect species were resistant to one or more pesticide. The list grows longer every year.

Insecticide resistance matters not only in the prevention of disease, but also in farming. Insect pests currently destroy about 20% of world crop production, and it has been estimated that as much as 50% of crop production would be lost without the use of pesticides. Insect pests represent a major economic and health problem. The evolution of resistance to pesticides causes misery to millions of people, whether through disease or reduced food supply. The fact that insects can rapidly evolve resistance is not the only problem with using pesticides against pests—the pesticides themselves can cause ecological side-effects that range from irritating to dangerous. Nevertheless, pesticides were not present during the hundreds of millions of years in which insects had existed. The rapid evolution of resistance to pesticides since their introduction in the 1940s provides a marvellously clear example of evolution by natural selection (section 13.9, p. 353, extends the story).

5.9 *Fitnesses are important numbers in evolutionary theory and can be estimated by three main methods*

The fitness of a genotype, in the theory and examples we have met, consists of its relative probability of survival from birth to adulthood. The fitness also determines the change in gene frequencies between generations. These two properties of fitness allow two methods of measuring it.

The first method is to measure the relative survival of the genotypes within a generation. Kettlewell's mark-recapture experiment with the peppered moth provides an example of this approach. If we assume that the relative rate of recapture of the genotypes is equal to their relative chance of survival from egg to adulthood, we have an estimate of fitness. The assumption may be invalid, however. The genotypes may, for instance, differ in their chances of survival at some stage of life other than the time of the mark-recapture experiment; if the survival of adult moths is measured by mark-recapture, any differences among genotypes in survival at the egg and caterpillar stages will not be detected. Also, the genotypes may differ in fertility; fitnesses estimated by differences in survival are only accurate if all the genotypes have the same fertility. It is possible to test these assumptions by conducting further research. Survival can be measured at the other life stages, as can fertility. In a few cases, lifetime fitnesses have been measured comprehensively, by tracing survival and reproduction from birth to death.

The second method is to measure changes in gene frequencies between generations. We then substitute the measurements into the formula that expresses fitness in terms of gene frequencies in successive generations (equation 5.6).

Both fitness measurement methods have been used in many cases. The main problems are the obvious difficulties of accurately measuring survival (for the first method) and gene frequencies (for the second method). In addition, it may be difficult to understand the genetics of the characters: we need to know which phenotypes correspond to which genotypes to estimate genotype fitnesses.

A third method of estimating fitness will be discussed later, in the case of sickle-cell anemia (Table 5.9, p. 119). This method, which uses deviations from the Hardy–Weinberg ratios, can be used only when the gene frequencies in the population are constant between the stages of birth and adulthood, but the genotypes have different survival. Thus, it cannot be used in the examples of directional selection against a disadvantageous gene with which we have been concerned so far, because in those cases the gene frequency in the population changes between birth and adult stages.

We have discussed the inference of fitness in detail because the fitnesses of different genotypes are among the most important variables—perhaps *the* most important variables—in the theory of evolution. They determine, to a large extent, which genotypes we can expect to see in the world today. The examples we have examined, however, illustrate that fitnesses are not easy to measure. Making such estimates requires long time series and large sample sizes, and even then the estimates may be subject to "other things being equal" assumptions. Therefore, despite their importance, they have been measured in only a small number of the systems in which biologists are interested. (The absolute number of such studies is not small: a review of research on natural selection in the wild by Endler in 1986 contains a 24-page table listing all the work he had located. Fitnesses have been measured in only a minority—an unknown minority—of those studies of natural selection, but the number could still be non-trivial.) Many unsolved controversies in evolutionary biology implicitly concern values of fitnesses, but in systems in which it has not been possible to measure fitnesses directly with sufficient accuracy or in a sufficiently large number of cases. The causes of molecular evolution in chapter 7 is an example of one such controversy. In such cases, it is worth remembering the amount of work necessary to solve them by direct measurements of fitness.

5.10 *Natural selection operating on a favored allele at a single locus is not meant to be a general model of evolution*

Evolutionary change in which natural selection favors a rare mutation at a single locus, and carries it up to fixation, is one of the simplest forms of evolution. Sometimes evolution may happen that way, but nature is often more complicated. We have considered selection in terms of different chances of survival from birth to adulthood, but selection can also take place by differences in fertility, if individuals of different genotypes—after they have survived to adulthood—produce different numbers of offspring. The model assumed random mating among the genotypes, but mating may be non-random. Moreover, the fitness of a genotype may vary in time and space, and depend on the genotypes present at other loci (a subject covered in chapter 8). Much of evolutionary change probably consists of adjustments in the frequencies of alleles at polymorphic loci, as fitnesses fluctuate through time, rather than the fixation of new favorable mutations.

These complexities in the real world are important, but they do not invalidate—or trivialize—the one-locus model. The model should be used as an aid to understanding, not as a general theory of nature. In science, it is a good strategy to develop an understanding of nature's complexities by considering simple cases first and then building on them to understand the complex whole. Simple ideas rarely provide accurate, general theories, but they often provide powerful paradigms. The one-locus model is concrete and easy to understand. It represents a good starting point for the science of population genetics. Indeed, population geneticists have constructed models of all the complications listed above, and those models are all developments within the general method we have been studying.

5.11 *A recurrent disadvantageous mutation will evolve to a calculable equilibrial frequency*

The model of selection at one locus revealed how a favorable mutation will spread through a population. But what about unfavorable mutations? Natural selection will act to eliminate any allele that decreases the fitness of its bearers, and the allele's frequency will decrease at a rate specified by the equations given in section 5.6, but how does it handle a recurrent disadvantageous mutation that arises at a certain rate? Selection can never completely eliminate the gene, because it will contineually reappear by mutation. In this case, we can calculate the equilibrial frequency of the mutation: the equilibrium is between the mutant gene's creation, by recurrent mutation, and its elimination by natural selection.

To be specific, we can consider a single locus, at which there is initially one allele, *a*. The gene has a tendency to mutate to a dominant allele, *A*. We must specify the mutation rate and the selection coefficient (fitness) of the genotypes. Define m as the mutation rate from a to A per generation; we ignore back mutation (though actually this assumption does not matter). The frequency of $a = q$, and the frequency of $A = p$. Finally, we define the fitnesses as follows:

genotype	*aa*	*Aa*	*AA*
fitness	1	$1 - s$	$1 - s$

Evolution in this case will proceed to an equilibrial frequency of the gene *A* (we can write the stable equilibrium frequency as p^*). If the frequency of *A* is higher than the equilibrium, natural selection removes more *A* genes than mutation creates and the frequency decreases; the reverse is true if the frequency is lower than the equilibrium. At the equilibrium, the rate of loss of *A* genes by selection equals their rate of gain by mutation.

We can use that statement to calculate the equilibrial gene frequency p^*. What is the rate per generation of creation of *A* genes by mutation? Each new *A* gene originates by mutation from an *a* gene and the chance that any one *a* gene mutates to an A gene is the mutation rate m. A proportion $(1 - p)$ of the genes in the population are a genes. Therefore,

$$\text{total rate of creation of } A \text{ genes by mutation} = m(1 - p)$$

What is the rate at which A genes are eliminated? Each A gene has a $(1 - s)$ chance of surviving, or an s chance of dying. A proportion p of the genes in the population are A. Therefore,

$$\text{total rate of loss of } A \text{ genes by selection} = ps$$

At the equilibrium gene frequency (p^*)

$$\begin{aligned} \text{rate of gain of } A \text{ gene} &= \text{rate of loss of } A \text{ gene} \\ m(1 - p^*) &= p^*s \end{aligned} \tag{5.8}$$

This equation can be rearranged to give

$$\begin{aligned} m - mp^* &= p^*s \\ p^* &= \frac{m}{s + m} \end{aligned}$$

Of the two terms in the denominator, the mutation rate (perhaps 10^{-6}, as discussed in section 2.5, p. 29) will usually be much less than the selection coefficient (perhaps 10^{-1} or 10^{-2}). With these values $s + m \approx s$ and the expression is therefore usually given in the approximate forms

$$p^* = \frac{m}{s} \tag{5.9}$$

The simple result is that the equilibrium gene frequency of the mutation is equal to the ratio of its mutation rate to its selective disadvantage. The result is intuitive: the equilibrium is the balance between the rates of creation and elimination of the gene. To obtain the result, we used an argument about an equilibrium. We noticed that, at equilibrium, the rate of loss of the gene equals the rate of gain; this observation was used to calculate the exact result. This approach is a powerful method for deriving equilibria, and we shall use an analogous argument in the next section.

The expression $p = m/s$ can allow a rough estimate of the mutation rate of a harmful mutation just from a measurement of the mutant gene's frequency. If the mutation is rare, it will be present mainly in heterozygotes, which at birth will have frequency $2pq$. If p is small, $q \approx 1$ and $2pq \approx 2p$. Define N as the frequency of mutant bearers, which equals the frequency of heterozygotes: $N = 2p$. As $p = m/s$, $m = sp$; if we substitute $p = N/2$, $m = sN/2$. If the mutation is highly deleterious, $s \approx 1$ and $m = N/2$. The mutation rate can be estimated as half the birth rate of the mutant type. The estimate is clearly approximate, because it relies on a number of assumptions. In addition to the assumptions of high s and low p, mating is supposed to be random, although we usually cannot check the accuracy of this assumption.

Chondrosdystrophic dwarfism is a dominant deleterious mutation in humans. In one study, 10 births out of 94,075 had the gene, a frequency of 10.6×10^{-5}. The estimate of the mutation rate by the above method gives $m = 5.3 \times 10^{-5}$. However, it is possible to estimate the selection coefficient, enabling a more accurate estimate of the mutation rate. In another study, 108 chondrodystrophic dwarves produced 27 children, and their 457 normal siblings produced 582 children. The relative fitness of the dwarves was $(27 \times 108)/(457 \times 582) = 0.196$; the selection coefficient $s = 0.804$. Instead of assuming $s = 1$, we can use $s = 0.804$. Thus, the mutation rate $= sN/2 = 4.3 \times 10^{-5}$, a lower figure because lower selection allows the same gene frequency to be maintained by a lower mutation rate.

For many genes, we do not know the dominance relations of the alleles at the locus. A similar calculation can be done for a recessive gene, but the formula is different, and it differs yet again if the mutation has intermediate dominance. We can only estimate the mutation rate from $p = m/s$ if we know the mutation is dominant. The method is, therefore, unreliable unless its assumptions have been independently verified. However, the general idea of this section—that a balance between selection and mutation can exist and explain genetic variation—will be used in later chapters.

5.12 Heterozygous advantage

5.12.1 Selection can maintain a polymorphism when the heterozygote is fitter than either homozygote

We will now consider an influential theory as it applies to the case in which the heterozygote is fitter than both homozygotes. The fitnesses can be written:

genotype	*AA*	*Aa*	*aa*
fitness	$1 - s$	1	$1 - t$

Like *s*, *t* is a selection coefficient and has a value between 0 and 1. What happens here? Of the three possible equilibria, two are trivial. The cases $p = 1$ and $p = 0$ are stable equilibria, but only because no mutation is possible in the model. The third equilibrium—the interesting one—has both genes present, and we can calculate the equilibrial gene frequencies by a similar argument to the one in the previous section.

A genes and *a* genes are both removed by selection—the *A* genes because they appear in the inferior *AA* homozygotes and the *a* genes because they appear in *aa* homozygotes. At the equilibrium, both genes must have the same chance of being removed by selection. If an *A* gene has a higher chance of being removed than an *a* gene, the frequency of *a* is increasing, and vice versa. Only when the chance is the same for both will the gene frequencies be stable.

What is the chance that an *A* gene will die without reproducing? An *A* gene is either present (with chance *q*) in a heterozygote and survives, or present (with chance *p*) in an *AA* homozygote and has a chance *s* of dying. Its total chance of dying is therefore *ps*. An *a* gene similarly is either present (with chance *p*) in a heterozygote and survives, or present (with chance *q*) in an *aa* homozygote and has chance *t* of dying. Its chance of death is *qt*. At the equilibrium,

$$\text{chance of death of an } A \text{ gene} = \text{chance of death of an } a \text{ gene}$$

$$p^{*}s = q^{*}t \qquad (5.10)$$

Substitute

$$p^{*}s = (1 - p^{*})t$$

and rearrange

$$p^* = \frac{t}{s + t} \tag{5.11}$$

Similarly, if we substitute $q = (1 - p)$, then $q^* = s/(s + t)$. Now we have derived the equilibrial gene frequencies when both homozygotes have lower fitness than the heterozygote. The equilibrium has all three genotypes present, even though the homozygotes are inferior and selection works against them. They continue to exist because it is impossible to eliminate them. Matings among heterozygotes generate homozygotes. The exact gene frequency at equilibrium depends on the relative selection against the two homozygotes. If, for instance, *AA* and *aa* have equal fitness, then $s = t$ and $p = ½$ at equilibrium. If *AA* is relatively more unfit than *aa*, then $s > t$ and $p < ½$. Thus, fewer of the more strongly selected-against genotypes remain.

When heterozygotes are fitter than the homozygotes, therefore, natural selection will maintain a polymorphism. The result was first proved by Fisher in 1922 and independently by Haldane. Later we will consider in more detail why genetic variability exists in natural populations. In our discussion, we will test *heterozygous advantage,* which is one of several controversial explanations.

5.12.2 *Sickle-cell anemia is a polymorphism with heterozygous advantage*

Sickle-cell anemia is the classic example of a polymorphism maintained by heterozygous advantage. A nearly lethal condition in humans that is responsible for approximately 100,000 deaths each year, sickle-cell anemia is caused by a genetic variant of β-globin. If we symbolize the normal hemoglobin allele by *A* and the sickle-cell hemoglobin by *S*, then people who suffer from sickle cell anemia are *SS*. Hemoglobin *S* causes the red blood cells to become curved and distorted (sickle-shaped); they can then block capillaries and cause severe anemia if the blocked capillary is in the brain. About 80% of *SS* individuals die before reproducing. With such apparently strong selection against hemoglobin *S*, it was a puzzle why this condition persisted at quite high frequencies (10% or even more) in some human populations. The answer was first suggested by Haldane and confirmed by Allison.

Haldane actually discussed another anemic condition, but his argument applies to sickle-cell anemia as well. He compared a map of the incidence of malaria with a map of the gene frequency (Figure 5.8), and found that they were strikingly similar. Perhaps hemoglobin *S* provides some advantage in malarial zones. It turned out that, although *SS* is almost lethal, the heterozygote *AS* is more resistant to malaria than the homozygote *AA*. *AS* red blood cells do not normally sickle, but they do if the oxygen concentration falls. When the malarial parasite *Plasmodium falciparum* enters a red blood cell, it destroys (probably eats) the hemoglobin, which causes the oxygen concentration in the cell to decline; the cell then sickles and is destroyed, along with the parasite. The human survives because most of the red blood cells are uninfected and carry oxygen normally. Therefore, where the malarial parasite is common, *AS* humans survive better than *AA* individuals, who suffer from malaria.

Once the heterozygote had been shown physiologically to be at an advantage, the adult genotype frequencies can be used to estimate the relative fitnesses of the three genotypes. The fitnesses are as follows:

genotype	*AA*	*AS*	*SS*
fitness	$1 - s$	1	$1 - t$

If the frequency of the gene $A = p$ and the frequency of $S = q$, then the relative genotype frequencies among adults will be $p^2(1 - s) : 2pq : q^2(1 - t)$. If no selection occurs ($s = t = 0$), the three genotypes would have Hardy–Weinberg frequencies of $p^2 : 2pq : q^2$.

Selection causes deviations from the Hardy–Weinberg frequencies. Consider the genotype *AA*. The ratio of the observed frequency in adults to that predicted from the Hardy–Weinberg ratio will be $(1 - s)/1$. The frequency expected from the Hardy–Weinberg principle is found by the usual method: the expected frequency is p^2, where p is the observed proportion of $AA + 1/2$ the observed proportion of *AS*. Table 5.9 illustrates the method for a Nigerian population, where $s = 0.12$ ($1 - s = 0.88$) and $t = 0.86$ ($1 - t = 0.14$).

The method is only valid if the deviation from Hardy–Weinberg proportions relates to heterozygous advantage and the genotypes differ only in their chance of survival (not their fertility). If heterozygotes are found in excess frequency in a natural population, the heterozygote may have a higher fitness, but other reasons may come into play as well. Disassortative mating, for instance, can produce the same result (in this case, disassortative mating would mean that *aa* individuals preferentially mate with *AA* individuals). For sickle-cell anemia, the physiological observations showed that the heterozygote is fitter and the procedure is well justified. Indeed, in this case, although it has not been checked whether mating is random, the near lethality of *SS* means that disassortative mating will be unimportant; however, the assumption that the genotypes have equal fertility may well be false.

TABLE 5.9

Estimates of selection coefficients for sickle-cell anemia, using genotype frequencies in adults. The sickle-cell hemoglobin allele is *S*, and the normal hemoglobin (which actually consists of more than one allele) is *A*. The genotype frequencies are for the Yorubas of Ibadan, Nigeria. One small detail is not explained in the text. The observed/expected ratio for the heterozygote may not be equal to 1. In this instance, it is 1.12. All observed/expected ratios are, therefore, divided by 1.12 to make them fit the standard fitness regime for heterozygote advantage. From Bodmer and Cavalli-Sforza (1986). Copyright © 1976 by W.H. Freeman and Company. Used with permission.

Genotype	Observed Adult Frequency (O)	Expected Hardy–Weinberg Frequency (E)*	Ratio O/E	Fitness
SS	29	187.4	0.155	$0.155/1.12 = 0.14 = 1 - t$
SA	2993	2672.4	1.12	$1.12/1.12 = 1.00$
AA	9365	9527.2	0.983	$0.983/1.12 = 0.88 = 1 - s$
Total	12,387	12,387		

*Calculation of expected frequencies: gene frequency of S = frequency of SS + ½ (frequency of SA) = (29 + 2993/2)/12,387 = 0.123. Thus, the frequency of *A* allele = 1 − 0.123 = 0.877. From the Hardy–Weinberg theorem, the expected genotype frequencies are $(0.123)^2 \times 12{,}387, 2(0.877)(0.123) \times 12{,}387$, and $(0.877)^2 \times 12{,}387$, for *AA*, *AS*, and *SS*, respectively.

Figure 5.8 The global incidence of malaria coincides with that of the sickle-cell form of hemoglobin. Reprinted, by permission of the publisher, from Bodmer and Cavalli-Sforza (1976).

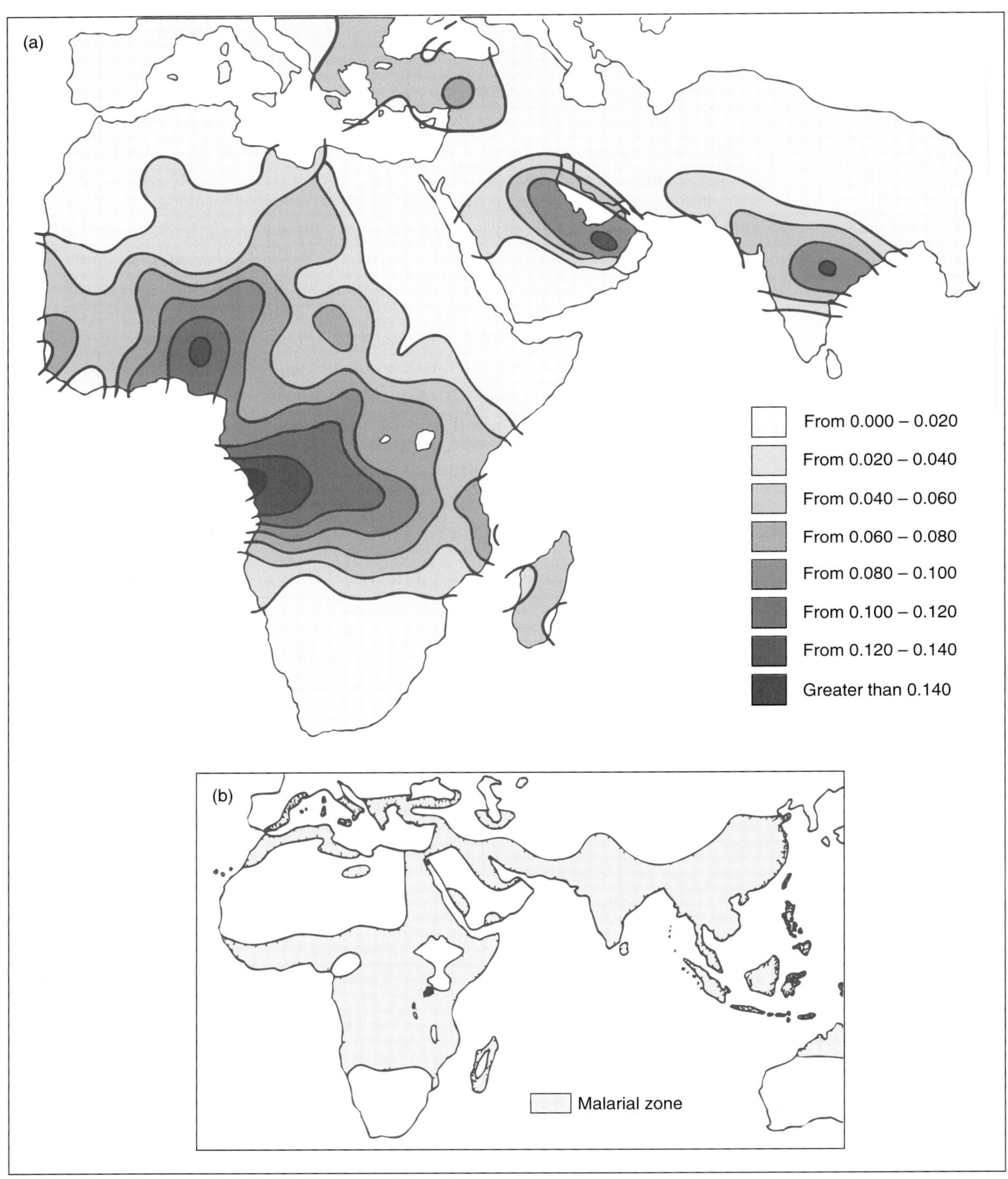

5.13 *The fitness of a genotype may depend on its frequency*

The next interesting complication is to consider selection when the fitness of a genotype depends on its frequency. In the models we have considered so far, the fitness of a genotype (1, $1 - s$, or whatever) was constant, regardless of whether the genotype was rare or common. It is also possible for the fitness of a genotype to increase (positively frequency-dependent) or decrease (negatively frequency-dependent) as the genotype frequency in the population increases.

Examples of both positive and negative frequency dependence can arise in systems of mimicry. Some poisonous species of animals, such as butterflies, have evolved recognizable color patterns, and predatory species (such as birds) learn to avoid feeding on these poisonous prey. Natural selection may then favor non-poisonous butterflies that have the same color pattern; they will be left alone by birds that have learned to avoid the poisonous types. This system is called Batesian mimicry. In many real cases, the mimic and model belong to separate species (and selection is, strictly speaking, number-dependent rather than frequency-dependent). Batesian mimicry can also exist within a species, and the fitness of each type then depends on its frequency.

Consider the non-poisonous mimics. When they are rare, birds will tend to avoid them, because they will have already have encountered a poisonous butterfly of the same appearance. When the non-poisonous type is common, the previous encounters of birds with butterflies of their appearance are more likely to have been rewarding. As a result, the birds will not avoid eating them, and their fitness will be lower. The fitness of the mimics is negatively frequency-dependent (Figure 5.9a).

In other butterflies, such as in central and south American *Heliconius* (section 8.3, p. 198), several morphs are present within a species, with each morph having a different color pattern. All of these morphs are poisonous. In this case the fitness of each morph increases as its frequency increases. When a morph is common, it will be more likely that birds will have learned to avoid them, whereas birds will not yet have learned to avoid a rare morph. An individual of a rare morph is, therefore, more likely to be the unlucky prey that educates the bird,

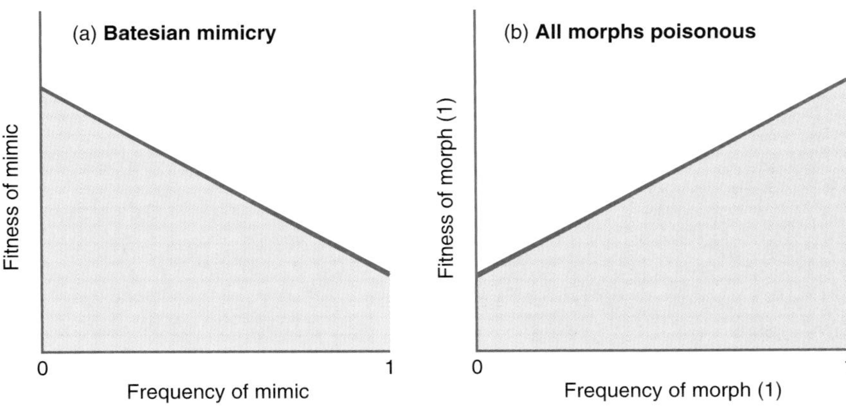

Figure 5.9 (a) With Batesian mimicry within a species, fitnesses are negatively frequency-dependent. (b) When all morphs of a species are poisonous, fitnesses become positively frequency-dependent. In a graph of fitness against frequency, any line other than a horizontal one is a case of frequency-dependent fitness.

and gets killed in the process. The fitness of each morph is positively frequency-dependent (Figure 5.9b). With positively frequency-dependent selection, a morph becomes more likely to be fixed as its frequency increases. Thus, we do not expect to find polymorphisms. Indeed, in *Heliconius*, only a single morph is present in any one area; the different morphs are found in different places.

With negatively frequency-dependent fitnesses (as in Batesian mimicry), it is possible for natural selection to maintain a polymorphism. When a genotype is rare, it is relatively favored by selection and will increase in frequency; as it becomes more common, its fitness decreases and at some point it may no longer be favored (Figure 5.10). At that point, the fitnesses of the different genotypes are equal and natural selection will not alter their frequencies—they are at equilibrium. The sex ratio is another case in which selection is frequency-dependent (section 11.5, p. 307).

In principle, it is a straightforward procedure to modify one of the genetic models we have considered for the cases of frequency dependence. We simply write the fitness of a genotype as a function of its frequency. One simple example involves a gene frequency of $A = p$, and a gene frequency of $a = q$:

genotype	*AA*	*Aa*	*aa*
fitness	$2(1 - p)$	1	$2(1 - q)$

In this case, fitness is negatively frequency-dependent. The fitnesses of the genotypes are 2, 1, and 0 when A is rare ($p \approx 0$), and 0, 1, and 2 when A is common. In more general models, the algebra becomes more difficult than in the frequency-independent case. Frequency-dependent selection is most easily investigated by means of graphs like Figure 5.9.

In theory, the graphical relation between fitness and frequency can have any shape, whether straight-lined or curved (Figure 5.11). The shape of the curve is theoretically important. It controls whether a polymorphism is possible, as well as the exact dynamics of gene frequency changes through time. The theoretical

Figure 5.10 (a) *AA* and *aa* have negatively frequency-dependent fitnesses; that of *Aa* is frequency-independent. To the left of the point (p^*) where the lines for *AA* and *aa* cross, *A* is favored and increases in frequency; the reverse is true to the left of the point. (b) Natural selection will take the population from any initial gene frequency to the gene frequency at which the fitnesses of *AA* and *aa* are equal.

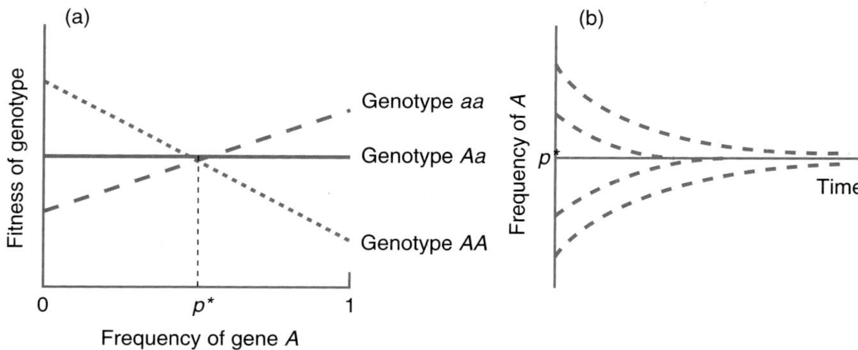

Figure 5.11 The relation between the fitness and the frequency of a genotype can theoretically have any shape.

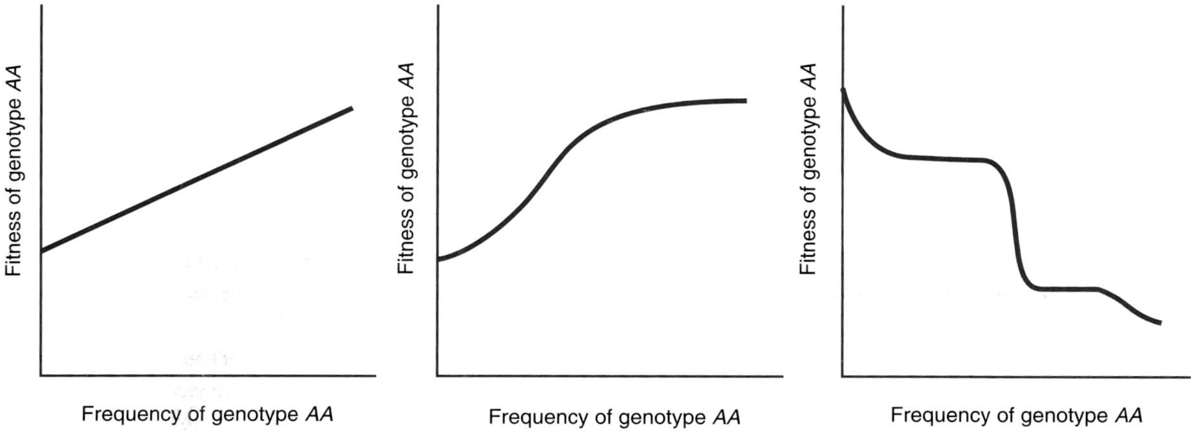

possibilities in models of this type are remarkably rich. They can, for instance, produce oscillations in gene frequency (section 11.3.6, p. 293). The main point here, however, is the possibility that, with negatively frequency-dependent fitnesses, natural selection can maintain a polymorphism.

5.14 *Multiple niche polymorphism can evolve in a heterogeneous environment*

Normal individual fruitflies (*Drosophila*) have orange-red colored eyes. In some mutant forms, however, the red pigment is lacking and the fruitflies' eyes are white. These eye color mutations affect the fruitfly's response to light. A normal *Drosophila*, with red eyes, is attracted to light, though repelled by very bright light; white-eyed mutants are more sensitive and (if given the choice) will settle in places of lower light intensity.

Jones and Probert performed the following experiment with normal and white-eyed *Drosophila simulans*. They first set up cages, with a mixture of the two genotypes, in either white or red light. In both cases the white-eyed mutants were at a disadvantage and selection reduced their frequency (Figure 5.12a). Jones and Probert also placed the two types of flies together in a cage that was illuminated with red light in one half and white light in the other half. In this case the two genotypes were maintained (Figure 5.12a). As you would expect, the white-eyed flies concentrated in the red light half of the cage, and the normal red-eyed flies in the half with white light—that is, the flies showed "habitat selection" (Figure 5.12b).

The polymorphism in the cage with both white and red light illustrates a *multiple niche polymorphism*, a topic first discussed by Levene in 1953. We can think of the parts of the cage with differing illumination as different "niches," analogous to distinct regions in the species' natural environment. When both

Figure 5.12 (a) Selection between normal and white-eyed genotypes of *Drosophila simulans* in three types of environment. (top) In white light, the white-eyed genotype is removed by selection, as it is in (middle) red light. But (bottom) when the flies are placed in cages red and white light, both genotypes persist. The experimental cages were divided in two by a barrier that the flies could walk through. In each case, the population was started with a mixture of the two genotypes, and the graphs plot the change in frequency of the white-eyed genotype for five replicate experiments. (b) In a cage with white light in one half and red light in the other (the condition of the lowest of the three experiments shown in (a)), the normal red-eyed fruitflies concentrate in the white light and the white-eyed mutants concentrate in the red light. Reprinted with permission from Jones and Probert (1980). Copyright 1980 Macmillan Magazines Limited.

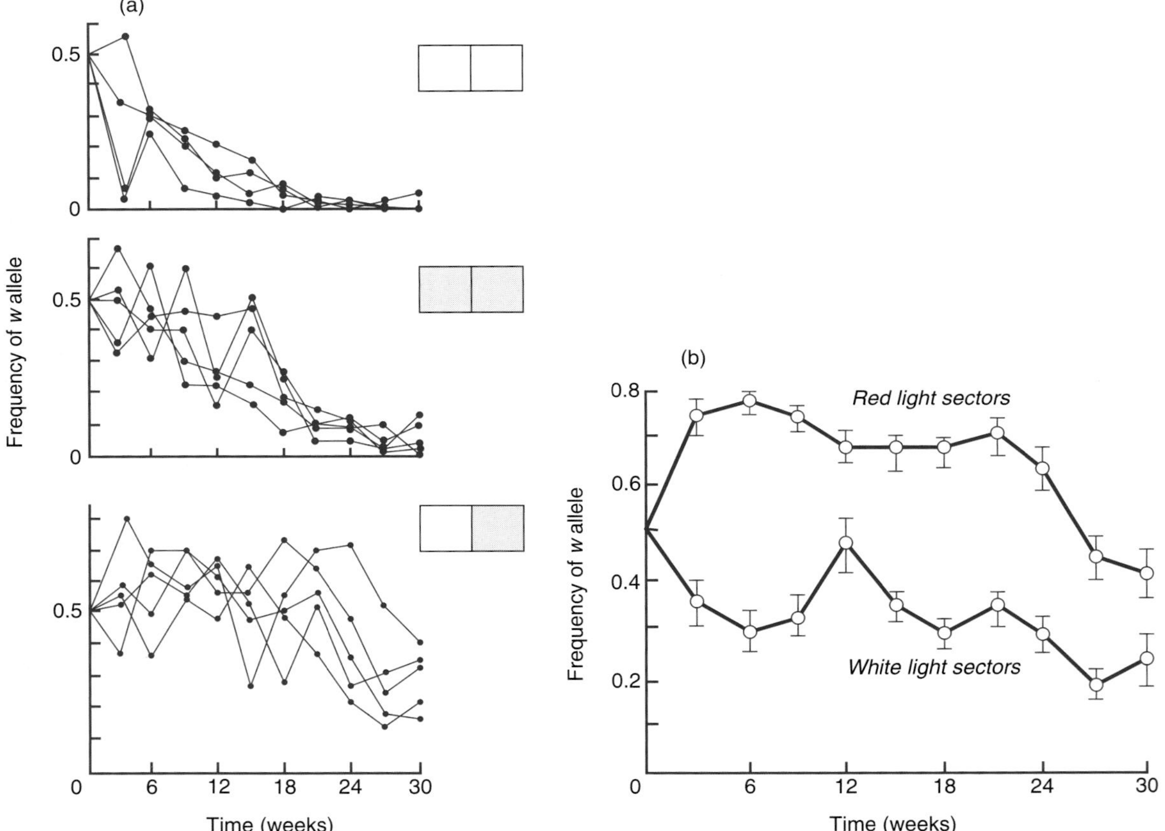

red and white niches are present, each type of fly chooses to live in the type of niche where it has higher fitness. Thus, normal flies go to the region with white light, leaving enough resources for the white-eyed mutants to survive where the light is red.

As many as two conditions are needed to create a multiple niche polymorphism. In nature, the members of a species occupy a variety of niches that vary in their physical conditions (whether they are dark, light, dry, damp, and so forth) and biological conditions (what predators and parasites are locally active, what food is available). The first (and crucial) condition is that different genotypes should have different fitnesses in the various niches. In theory, this condition alone could produce a polymorphism (though no such examples are known).

The simplest case would have two niches, A and B, with genotypes *AA* and *Aa* better adapted to A and *aa* is better adapted to B. The fitnesses are as follows:

genotypes	*AA*	*Aa*	*aa*
fitness in niche A	1	1	$1 - s$
fitness in niche B	$1 - t$	$1 - t$	1

In principle, it is easy to write a model in which the frequency of niche A is N and of niche B is $(1 - N)$. If the genotypes experience niche types in the frequencies, the fitness of *AA* will be $[N + (1 - N)(1 - t)]$. The same pattern applies to the other genotypes (the equations are fairly ugly). In this instance, we need notice only the possibility that a polymorphism can evolve. The relative frequencies of the genotypes at equilibrium will depend on the proportions of the two niches: if niche A is more common, then gene *A* has a higher frequency; the reverse is true if niche B is more common.

A stable polymorphism can more easily evolve if *habitat selection* occurs (as was the case in Jones and Probert's fruitflies), in which each genotype "chooses" the correct habitat in which to live. Individuals with genotype *AA* or *Aa* would live in niche type A and those with *aa* would pick niches of type B. A formal model would then have an even more complex expression for the fitness of a genotype, incorporating not only the frequency of each niche type but also the genotype's chance of living in it. However, the qualitative conclusion is, once again, that multiple niche polymorphisms can arise, and they are more plausible if habitat selection occurs and a relation exists between a genotype's fitness and the niche in which it lives.

Multiple niche polymorphism is a form of frequency-dependent polymorphism. When the genotype *AA* is rare, it experiences relatively little competition in the niches for which it is well adapted, and it increases in frequency. As *AA* becomes more common, it increasingly occupies all the niches of type A and has to survive in B type niches as well. Thus, its advantage decreases as its frequency increases. The fitness of each genotype is negatively frequency-dependent, like in Figure 5.9.

The purpose of sections 5.11–5.14 has been to illustrate the different mechanisms by which natural selection can maintain polymorphism. In chapter 6 we will study another mechanism that can maintain polymorphism and that does not rely on natural selection: genetic drift. In chapter 7, we will tackle the question of how important these mechanisms are in nature.

5.15 Subdivided populations require special population genetic principles

5.15.1 A subdivided set of populations has a higher proportion of homozygotes than an equivalent fused population: this is the Wahlund effect

So far we have considered population genetics within a single, uniform population. In practice, a species may consist of a number of separate populations, each more or less isolated from the others. The members of a species might, for example, inhabit a number of islands, with each island population being separated by the sea from the others. Individuals might migrate between islands from time to time, but each island population would evolve to some extent independently.

A species with a number of more or less independent subpopulations is said to have *population subdivision.*

Let us examine what effect population subdivision has on the Hardy–Weinberg principle. Consider a simple case in which there are two populations (called population 1 and population 2), and we concentrate on one genetic locus with two alleles, *A* and *a*. Suppose allele *A* has frequency 0.3 in population 1 and 0.7 in population 2. If the genotypes have Hardy–Weinberg ratios, then they will have the frequencies, and average frequencies, in the two populations shown in Table 5.10. The average genotype frequencies are 0.29 for *AA*, 0.42 for *Aa*, and 0.29 for *aa*. Now suppose that the two populations are fused together. The gene frequencies of *A* and *a* in the combined population are $(0.3 + 0.7)/2 = 0.5$, and the Hardy–Weinberg genotype frequencies are as follows:

genotype	*AA*	*Aa*	*aa*
frequency	0.25	0.5	0.25

The large, fused population contains fewer homozygotes than are present, on average, for the set of subdivided populations. This finding is a general, and mathematically automatic, result. The increased frequency of homozygotes in subdivided populations is called the *Wahlund effect.*

The Wahlund effect has a number of important consequences. First, we must know about the structure of a population when applying the Hardy–Weinberg principle to it. Suppose, for example, we had not known that populations 1 and 2 were independent. We might have sampled from both, pooled the samples indiscriminately, and then measured the genotype frequencies. We would find the frequency distribution for the average of the two populations (0.29, 0.42, 0.29), but the gene frequency would apparently be 0.5. Thus, the population would appear to contain more homozygotes than expected from the Hardy–Weinberg principle. We might suspect that selection, or some other factor, favored homozygotes. In fact, both subpopulations remain in Hardy–Weinberg equilibrium and the deviation is due to the unwitting pooling of the separate populations. We therefore need to search for population subdivision when interpreting deviations from Hardy–Weinberg ratios.

Second, when a number of previously subdivided populations merge together, the frequency of homozygotes will decrease. In humans, this merger can lead to a decrease in the incidence of rare recessive genetic diseases when a previously isolated population comes into contact with a larger population. The recessive

TABLE 5.10

The frequency of genotypes *AA*, *Aa*, and *aa* in two populations when *A* has frequency 0.3 in population 1 and 0.7 in population 2. The average genotypes are calculated assuming the two populations are of equal size.

Genotype	*AA*	*Aa*	*aa*
Frequency	$(0.3)^2 = 0.09$	$2(0.3)(0.7) = 0.42$	$(0.7)^2 = 0.49$ population 1
	$(0.7)^2 = 0.49$	$2(0.7)(0.3) = 0.42$	$(0.3)^2 = 0.09$ population 2
Average	$0.58/2 = 0.29$	$0.84/2 = 0.42$	$0.58/2 = 0.29$

disease is expressed only in the homozygous condition, and when the two populations start to interbreed, the frequency of those homozygotes declines.

5.15.2 *Migration acts to unify gene frequencies between populations*

When an individual migrates from one population to another, it carries genes that are representative of its own ancestral population into the recipient population. If it successfully establishes itself and breeds, it will transmit those genes between the populations. The transfer of genes is called *gene flow*. If the two populations originally had different gene frequencies and if selection does not operate, migration (or, to be exact, gene flow) alone will rapidly cause the gene frequencies of the different populations to converge. We can measure how rapidly this convergence occurs in a simple model.

Consider again the case of two populations and one locus with two alleles (A and a). Suppose this time that population 2 is much larger than population 1 (2 might be a continent and 1 a small island off it). In this case, practically all the migration is from population 2 to population 1. The frequency of allele a in population 1 in generation t is written $q_{1(t)}$; we can suppose that the frequency of a in the large population 2 is not changing between generations and write it as q_m. (We are interested in the effect of migration on the gene frequency in population 1 and can ignore all other effects, such as selection.) If we pick any one allele in population 1 in generation $(t + 1)$, it will either be descended from a native of the population or from an immigrant. Define m as the chance that it is a migrant gene. (Earlier in the chapter, m was used for the mutation rate: now it is the *migration* rate.) If our gene is not a migrant (chance $(1 - m)$) it will be an a gene with chance $q_{1(t)}$, whereas if it is a migrant (chance m) it will be an a gene with chance q_m. The total frequency of a in population 1 in generation $(t + 1)$ is

$$q_{1(t)} = (1 - m)q_{1(t)} + mq_m \tag{5.12}$$

This equation can be rearranged to show the effect of t generations of migration on the gene frequency in population 1. If $q_{1(0)}$ is the frequency in the 0th generation, the frequency in generation t will be

$$q_{1(t)} = q_m + (q_{1(0)} - q_m)(1 - m)^t \tag{5.13}$$

(From $t = 1$ it is easy to confirm that this equation is a rearrangement of the previous equation.) The equation says that the difference between the gene frequency in population 1 and population 2 decreases by a factor of $(1 - m)$ per generation. At equilibrium, $q1 = qm$ and the small population will have the same gene frequency as the large population (Figure 5.13). In Figure 5.13, the gene frequencies converge in about 30 generations with a migration rate of 10%. Similar arguments apply if, instead of one source and one recipient population, the source consists of a set of many subpopulations, and p_m is their average gene frequency, or if two populations both send migrants to, and receiving them from, one another.

Migration will generally unify gene frequencies among populations rapidly in evolutionary time. In the absence of selection, migration represents a strong force for equalizing the gene frequencies of subpopulations in a species. Provided that the migration rate is greater than zero, gene frequencies will eventually

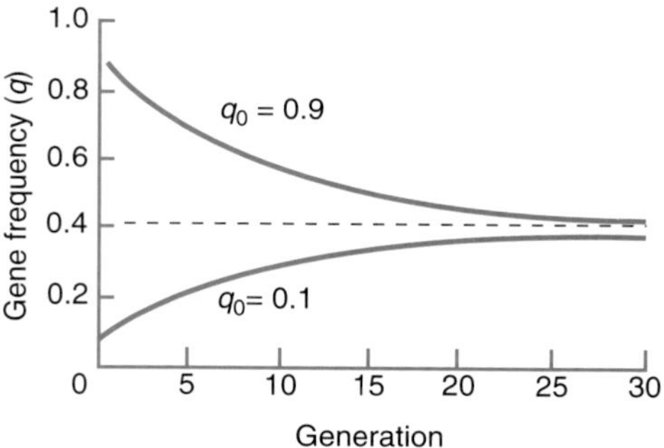

Figure 5.13 Migration causes the rapid convergence of gene frequencies in the populations exchanging migrants. Here a source population with gene frequency $q_m = 0.4$ sends migrants to two subpopulations, with initial gene frequencies of 0.9 and 0.1. They converge, with $m = 0.1$, onto the source population's gene frequency in about 30 generations.

equalize. Even if only one successful migrant appears per generation, gene flow inevitably draws the population's gene frequency to the species' average. Gene flow thus acts to bind the species together.

5.15.3 *The convergence of gene frequencies by gene flow is illustrated by the human population of the United States*

The *MN* blood group is controlled by one locus with two alleles (section 5.2). Frequencies of the *M* and *N* alleles have been measured—for example, in European and African Americans in Claxton, Georgia, and among West Africans (whom we can assume to be representative of the ancestral gene frequency of the African American population of Claxton). The *M* allele frequency is 0.474 in West Africans, 0.484 in African Americans in Claxton, and 0.507 in the European Americans in Claxton. (The frequency of the *N* allele = 1 − the frequency of *M* allele.) The gene frequency among African Americans is intermediate between the frequencies for European Americans and for the West African sample. Individuals of mixed parentage are usually categorized as African American and, if we ignore the possibility of selection favoring the *M* allele in the United States, we can treat the change in gene frequency in the African American population as due to "migration" of genes from the European American population. The measurements can then be used to estimate the rate of gene migration. In equation 5.13, q_m = the gene frequency in the European American population (the source of the "migrant" genes), $q_o = 0.474$ (the original frequency in the African American population), $q_t = 0.484$. As an approximate figure, we can suppose that the black population has been in the United States for 200–300 years, or about 10 generations. Then

$$0.484 = 0.507 + (0.474 - 0.507)(1 - m)^{10}$$

This equation can be solved to find $m = 0.035$. That is, for every generation, an average of about 3.5% of the genes at the *MN* locus have migrated from the white population to the black population of Claxton. (Other estimates by the same method but using different gene loci suggest slightly different figures, closer to 1%. The important point here is not the particular result; it is to illustrate how the population genetics of migration can be analyzed.) Notice again the rapid rate of genetic unification by migration: in only 10 generations, one-third of the gene frequency difference has been removed (after 10 generations the difference is 0.507 − 0.484, compared with the original difference of 0.507 − 0.474).

5.15.4 *A balance of selection and migration can maintain genetic differences between subpopulations*

If selection is working against an allele within one subpopulation, but the allele is continually being introduced by migration from other populations, it can be maintained by a balance of the two processes. We can analyze the balance between the two processes using the same arguments as put forth for selection-mutation balance and heterozygous advantage. The simplest case is again for one locus with two alleles.

Imagine selection in one subpopulation is working against a dominant A allele. The fitnesses of the genotypes are

genotype	*AA*	*Aa*	*aa*
frequency	$1 - s$	$1 - s$	1

The A allele has frequency p in the local population. Suppose that in other subpopulations, natural selection is more favorable to the gene A, and it has a higher frequency in those subpopulations (p_m on average). In this case, p_m will then be the frequency of A among immigrants to the local population. In the local population, A genes are lost at a rate ps per generation. They are gained at a rate $(p_m - p)m$ per generation, where m is the proportion of genes that are immigrants in a generation. Immigration increases the frequency in the local population by an amount $p_m - p$ because gene frequency is increased only in that the immigrating population has a higher frequency of A than the local population. If the immigrating gene frequency is the same as the local gene frequency, immigration has no effect.

Three outcomes are possible. If migration is powerful relative to selection, the rate of gain of A genes by immigration will exceed the rate of loss by selection. The local population will be swamped by immigrants. The frequency of the A gene will increase until it reaches p_m. If migration is weak relative to selection, the frequency of A will decrease until it is locally eliminated. The third possibility is an exact balance between migration and selection. An equilibrium (with local frequency of $A = p^*$) will occur if

$$\text{rate of gain of } A \text{ by migration} = \text{rate of loss of } A \text{ by selection}$$

$$(p_m - p^*)m = p^* s \tag{5.14}$$

$$p^* = p_m\left(\frac{m}{s - m}\right) \tag{5.15}$$

In the first case, migration unifies the gene frequencies in both populations, much in the same manner as described in section 5.15.2: migration is so strong relative to selection that it appears as if selection were not operating. In the second and third cases, migration is not strong enough to unify the gene frequencies and regional differences should appear in the gene frequency, so that gene frequency would be higher in some places than in others. In the third case, polymorphism occurs within the local population, where A is maintained by migration even though it is locally disadvantageous.

This section has made two main points. First, a balance of migration and selection is another process to add to the list of processes that can maintain polymorphism. Second, we have seen how migration can be strong enough to unify gene frequencies between subpopulations; alternatively, if migration is weaker the gene frequencies of different subpopulations can diverge under selection. This theory is also relevant in the question of the relative importance of gene flow and selection in maintaining biological species (section 15.2.6, p. 414).

SUMMARY

1. In the absence of natural selection, and with random mating in a large population, the genotype frequencies at a locus move in one generation to the Hardy–Weinberg ratio; the genotype frequencies are then stable.
2. It is easy to observe whether the genotypes at a locus are in the Hardy–Weinberg ratio. In nature, they will often not satisfy this ratio, because the fitnesses of the genotypes are not equal, mating is non-random, or the population is small.
3. A theoretical equation for natural selection at a single locus can be written by expressing the frequency of a gene in one generation as a function of its frequency in the previous generation. The relation is determined by the fitnesses of the genotypes.
4. The fitnesses of the genotypes can be inferred from the rate of change of gene frequency in real cases of natural selection.
5. From the rate at which the melanic form of the peppered moth replaced the light-colored form, the melanic form must have had a selective advantage of approximately 50%.
6. The geographic pattern of melanic and light-colored forms of the peppered moth cannot be explained simply by the selective advantage of the better camouflaged form. An inherent advantage to the melanic form, and migration, are also needed to explain the observations.
7. The evolution of pesticide resistance in insects is, in some cases, due to rapid selection for a gene at a single locus. The fitness of the resistant types can be inferred, from the rate of evolution, to be as much as twice that of the non-resistant insects.
8. If a mutation is selected against but continues to arise, the mutation settles at a low frequency in the population. This situation is called selection-mutation balance.
9. Selection can maintain a polymorphism when the heterozygote is fitter than the homozygote, when fitnesses of genotypes are negatively frequency-dependent, and when different genotypes are adapted to different niches (this third situation is a special case of the second).
10. Sickle-cell anemia is an example of a polymorphism maintained by heterozygous advantage.
11. Subdivided populations have a higher proportion of homozygotes than an equivalent large, fused population.
12. Migration, in the absence of selection, rapidly unifies gene frequencies in different subpopulations. It can also maintain an allele that is selected against in a local subpopulation.

FURTHER READING

There are a number of textbooks about population genetics. Bodmer and Cavalli-Sforza (1976), Crow (1986), Edwards (1977), Hartl (1988), and Maynard Smith (1968, 1989) are more introductory. More comprehensive works include Cavalli-Sforza and Bodmer (1971), Crow and Kimura (1970), Hartl and Clark (1989), Hedrick (1983), Li (1976), Spiess (1989), and Wallace (1981). Dobzhansky (1970) is a classic study, Lewontin *et al.* (eds, 1981) contains Dobzhansky's most famous series of papers.

Ford (1975) and Sheppard (1975) include introductory accounts of the peppered moth; Kettlewell (1973) is the standard account by the main authority. Haldane (1924, 1958) are the classic papers on the selection coefficients. For more recent work, see Brakefield (1987), and *Biological Journal of the Linnean Society*, vol. 39, 301–371 (1990). On pests and pesticides, see Taylor (1986), Mallett (1989), Devonshire and Field (1991), May (1993), McKenzie and Batterham (1994), and the further discussion (and further reading) in chapter 13. See Endler (1986) on measuring fitness in general; Primack and Kang (1989) for plants; and Clutton-Brock (1988) for research on lifetime fitness.

The various selective means of maintaining polymorphisms are explained in the general texts. In addition, on frequency-dependent selection, see Clarke (1979) and the special issue of *Philosophical Transactions of the Royal Society of London*, B 319, 457–640 (1988). On multiple niche polymorphism, the classic paper is by Levene (1953); Hedrick (1986) and Antonovics *et al.* (1988) are recent reviews. Models of migration are explained in the population genetics texts. For a probable example of balanced migration and selection, see Camin and Ehrlich's (1958) study of the water snakes of Lake Erie, and the recent work of R. B. King (1993).

STUDY AND REVIEW QUESTIONS

1. The following table gives genotype frequencies for five populations. Which are in Hardy–Weinberg equilibrium? For those that are not, suggest some hypotheses to explain why they are not at equilibrium.

	Genotype		
Population	*AA*	*Aa*	*aa*
1	25	50	25
2	10	80	10
3	40	20	40
4	0	150	100
5	2	16	32

2. Consider the following genotypes:

genotype	*AA*	*Aa*	*aa*
birth frequency	p^2	$2pq$	q^2
fitness	1	1	$1 - s$

(a) What is the frequency of *AA* individuals in the adult population?
(b) What is the mean fitness of the population?

3. What is the mean fitness of this population?

genotype	*AA*	*Aa*	*aa*
birth frequency	⅓	⅓	⅓
fitness	1	$1 - s$	1

4. Consider a locus with two alleles, *A* and *a*. *A* is dominant and selection works against the recessive homozygote. The frequency of *A* in two successive generations is 0.4875 and 0.5. What is the selection coefficient (s) against *aa*? (If you prefer to do this calculation in your head rather than with a calculator, round the frequency of *a* in the first generation to 0.5 rather than 0.5125.)

5. What main assumptions are made in estimating fitnesses by the mark-recapture method?

6. Listed below are some adult genotype frequencies for a locus with two alleles. The polymorphism is known to be maintained by heterozygous advantage, working on survival. What are the fitnesses (or selection coefficients) of the two homozygotes, relative to a fitness of 1 for the heterozygote?

genotype	*AA*	*Aa*	*aa*
frequency among adults	⅙	⅔	⅙

7. Consider two populations of a species, called population 1 and population 2. Migrants move from population 1 to 2, but not the reverse. Now consider a locus with two alleles, *A* and *a*. In generation n, the gene frequency of *A* is 0.5 in population 1 and 0.75 in population 2; in generation 2, it is 0.5 in population 1 and 0.625 in population 2. (a) What is the rate of migration, measured as the chance an individual in population 2 is a first-generation immigrant from population 1? (b) If the rate of migration is the same in the next generation, what will the frequency of *A* be in population 2 in generation 3?

[Questions 8–10 are questions for further thought. They do focus on topics explicitly covered in the chapter, but are slight extensions of such subjects.]

8. What is the general effect of assortative mating on genotype frequencies, relative to the Hardy–Weinberg equilibrium, for (a) a locus with two alleles, one dominant to the other and (b) a locus with two alleles, and no dominance (the heterozygote is a distinct phenotype intermediate between the two homozygotes)? What is the effect on genotype frequencies of a mating preference, in which females preferentially mate with males of (c) the dominant phenotype, and (d) the recessive phenotype?

9. Derive a recurrence relation, giving the frequency of the dominant gene *A* for one generation (p') in terms of the frequency in any generation (p) and selection coefficient (s) for selection against the dominant allele.

10. Derive the expression for the equilibrium gene frequency (p^*) for mutation-selection balance when the disadvantageous mutation is recessive.

Figure 6.2 The chance that a founder population will be homozygous depends on the number of founders and the gene frequencies. If less variation and fewer founders are present, the chance of homozygosity is higher. Here the chance of homozygosity is shown for three different gene frequencies at a two-allele locus.

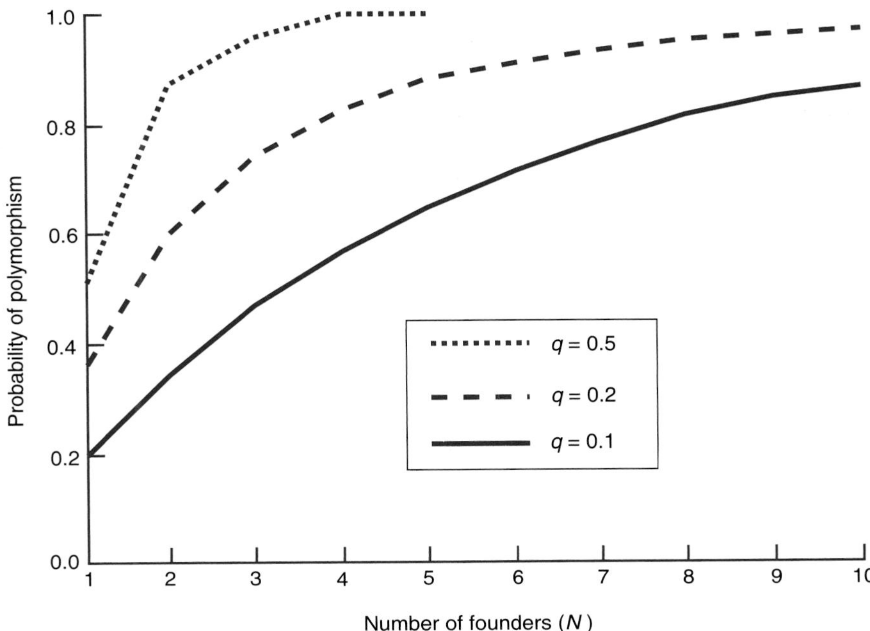

uniform. The interesting result is that founder events are not effective at producing a genetically monomorphic population. Even if the founder population is very small, with $N < 10$, it will usually possess both alleles. An analogous calculation could be performed for a population with three alleles, in which we asked the chance that one of the three would be lost by the founder effect. The resulting population would not then be monomorphic, but have two (instead of three) alleles. In general, then, founder events—whether by colonizations or population bottlenecks—are unlikely to reduce genetic variation unless the number of founders is tiny.

However, founder events can have other consequences. Although the sample of individuals forming a founder population is likely to include nearly all of the ancestral population's genes, the frequencies of the genes may differ from the parental population. Isolated populations often have exceptionally high frequencies of otherwise rare alleles, and the most likely explanation is that the founding population had a disproportionate number of those rare alleles. The clearest examples of this consequence involve human populations.

Consider the Afrikaner population of South Africa, which is mainly descended from one shipload of immigrants that landed in 1652, although later arrivals have added to the group. The population increased dramatically to its current level of 2,500,000. The influence of the early colonists is shown by the fact that almost 1,000,000 living Afrikaners bear the names of 20 original settlers.

The early colonists included individuals with a number of rare genes. The ship of 1652 contained a Dutch man carrying the gene for Huntington's disease, a lethal autosomal dominant disease. Most cases of the disease in the modern Afrikaner population can be traced back to that individual. A similar story can be told for the dominant autosomal gene causing porphyria variegata. Porphyria variegata results from a defective form of the enzyme protoporphyrinogen oxidase; carriers of the gene suffer a severe—even lethal—reaction to barbiturate anesthetics, and the gene was, therefore, not strongly disadvantageous before modern medicine. Approximately 30,000 carriers of the gene exist in the modern Afrikaner population, a far higher frequency than in the Netherlands. All of these carriers are descended from one couple, Gerrit Jansz and Ariaantje Jacobs, who emigrated from Holland in 1685 and 1688, respectively. Every human population has such "private" polymorphisms, which were caused by the genetic peculiarities of founder individuals in many cases.

Both of the examples involving the Afrikaner population involve medical conditions. The individual carriers of the genes have lower fitness than average, and selection will, therefore, act to reduce the frequency of the gene to zero. For most of history, the porphyria variegata gene may have had a similar fitness to its allele; it may have been a neutral polymorphism until its "environment" came to contain (in selected cases) barbiturates.

In contrast, the gene for Huntington's disease will have been consistently selected against. Thus, its present high frequency suggests that the founder population had an even higher frequency, because selection would have probably decreased its frequency in subsequent generations. Any particular founder sample would not be expected to show a higher than average frequency of the Huntington's disease gene. If enough colonizing groups set out, however, some of them are bound to have peculiar, or even very peculiar, gene frequencies. In the case of Huntington's disease, founders with more copies of the gene than average are known in instances other than the Afrikaner population. For example, 432 carriers of Huntington's disease in Australia are descended from a Miss Cundick who left England with her 13 children. In addition, a French nobleman's grandson, Pierre Dagnet d'Assigne de Bourbon, bequeathed all of the known cases of Huntington's disease on the island of Mauritius.

6.4 *One gene can be substituted for another by random drift*

The frequency of a gene is equally likely to decrease as to increase by random drift. On average, the frequencies of neutral alleles remain unchanged from one generation to the next. In practice, their frequencies drift up and down, and it is possible for a gene to enjoy a run of luck and be carried up to a much higher frequency—in the extreme case, its frequency could, after many generations, increase to 1 (become fixed) by random drift.

During each generation, the frequency of a neutral allele has a chance of increasing, a chance of decreasing, and a chance of staying constant. If it increases in one generation, it again has the same chance of increasing, decreasing, or staying constant in the next generation. Thus, the frequency has a small chance of increasing for two generations in a row (equal to the square of the chance of increasing in any one generation), a still smaller chance of increasing through three generations, and so on. For any one allele, fixation by random drift is very improbable. The probability is finite, however, and if enough neutral alleles, at enough loci, and over enough generations, are randomly drifting in frequency, one of them will eventually be fixed. The same process can occur, regardless of the initial frequency of the allele. It is less likely that a rare allele will reach fixation by random drift than a common allele, because it would take a longer run of good luck. Nevertheless, that scenario is still possible. Indeed, a unique neutral mutation has some chance of eventual fixation. Any one mutation is most likely to be lost, but if enough mutations arise, one will be bound to be fixed eventually.

As a result, a gene can be substituted by random drift. What is the rate of this kind of neutral evolution? We might expect it would occur more rapidly in smaller populations, because most random effects are more powerful in smaller populations. However, an elegant argument demonstrates that the neutral evolution rate exactly equals the neutral mutation rate, and is independent of population size. In a population of size N, there are a total of $2N$ genes at each locus. On average, each gene contributes one copy of itself to the next generation. Because of random sampling, however, some genes will contribute more than one copy and others will contribute none. When we look two generations ahead, those genes that contributed no copies to the first generation cannot contribute copies to the second, third, fourth, or any subsequent generation—once a gene fails to be copied, it is lost forever. In the next generation, some more genes will likewise "drop out" and be unable to contribute to future generations. Each generation, some of the $2N$ original genes are lost in this way (Figure 6.3).

If we look far enough ahead, we eventually come to a time when all $2N$ genes are descended from just one of the $2N$ genes now. This evolution occurs because some genes will fail to reproduce in every generation. Thus, we must eventually come to a time when all but one of the original genes have dropped out. That one gene will have experienced a long enough run of lucky increases and will have spread through the whole population. That is, it will have been fixed by genetic drift. Because the process is based entirely on luck, each of the $2N$ genes in the original population has an equal chance of being the remaining gene. Any one gene in the population, therefore, has a $1/(2N)$ chance of eventual fixation by random drift (and a $(2N - 1)/(2N)$ chance of being lost by it).

Because the same argument applies to any gene in the population, it also applies to a new, unique, neutral mutation. When the mutation arises, it will be one gene in a population of $2N$ genes at its locus (i.e., its frequency will be $1/(2N)$), and it has the same $1/(2N)$ chance of eventual fixation as does every other gene in the population. The mutation will, therefore, most likely be lost (probability of being lost $= (2N - 1)/(2N) \approx 1$ if N is large), but it does have a small ($1/(2N)$) chance of success. Thus, according to the first stage of the argument, the probability that a neutral mutation will eventually be fixed is $1/(2N)$.

The rate of evolution equals the probability that a mutation is fixed, multiplied by the rate at which mutations appear. We define the rate at which neutral mutations arise as u per gene per generation. (Note that u is the rate at which new selectively neutral mutations arise, not the total mutation rate. The total

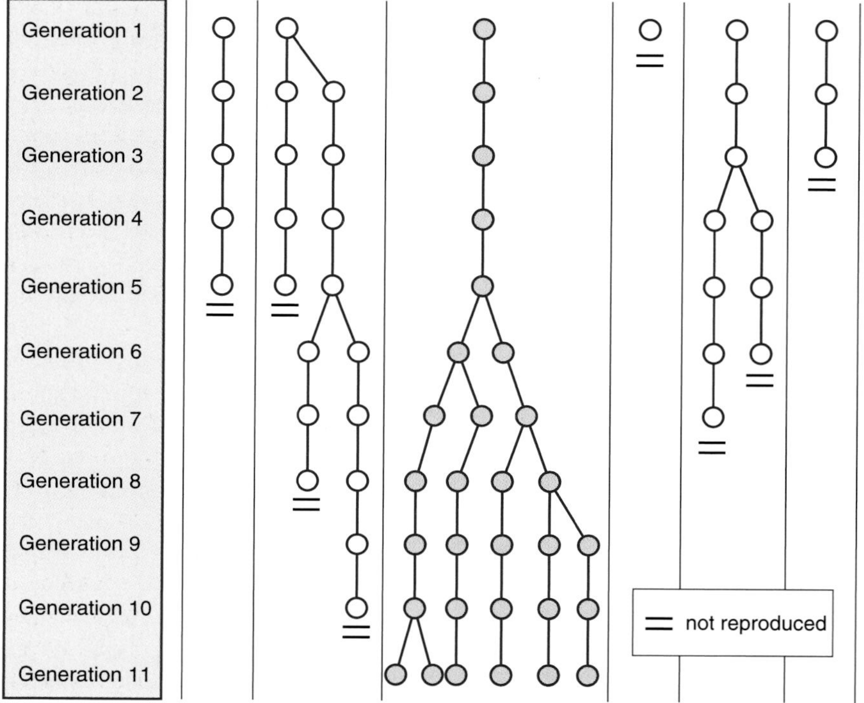

Figure 6.3 The drift to homozygosity. The figure traces the evolutionary fate of six genes; in a diploid species these genes would be combined each generation in three individuals. Every generation, some genes may by chance fail to reproduce and others by chance may leave more than one copy. Because a gene that fails to reproduce its line is lost forever, over time the population must drift to become made up of descendants of only one gene from an ancestral population. In this example, the population after 11 generations consists of descendants of gene number 3 (◯) in generation 1.

mutation rate includes selectively favorable and unfavorable mutations as well as neutral mutations.) At each locus, there are $2N$ genes in the population. The total number of neutral mutations arising in the population will be $2Nu$ per generation. The rate of neutral evolution is then $\frac{1}{2}N \times 2Nu = u$. The population size cancels out and the rate of neutral evolution is equal to the neutral mutation rate.

6.5 *The Hardy–Weinberg "equilibrium" is not an equilibrium in a small population*

Let us stay with the case of a single locus, with two selectively neutral alleles *A* and *a*. If genetic drift does not occur—if the population is large—the gene frequencies will remain constant from generation to generation and the genotype frequencies will also be constant, in Hardy–Weinberg proportions (section 5.3, p. 94). In a smaller population, however, the gene frequencies can drift around. The average gene frequencies in one generation will be the same as in the previous generation, and it might be thought that the long-term average gene and genotype frequencies will simply be those of the Hardy–Weinberg equilibrium, but with a bit of "noise" around them. That, however, turns out not to be correct. The long-term result of genetic drift is that one of the alleles will be fixed; the polymorphic Hardy–Weinberg equilibrium is unstable in a small population.

Suppose that a population consists of five individuals, containing five *A* alleles and five *a* alleles (obviously a tiny population, but the same point would apply if there were 500 copies of each allele). The genes are randomly sampled to produce the next generation. Imagine that six *A* alleles and four *a* alleles are sampled. This sample represents the starting point to produce the next generation. The most likely ratio in the next generation is six *A* and four *a*, because no "compensating" process pushes the ratio back toward five and five. Perhaps six *A* and four *a* are drawn in the next generation. The fourth generation might be seven *A* and three *a*, the fifth generation six *A* and four *a*, the sixth generation seven *A* and three *a*, then seven *A* and three *a*, eight *A* and two *a*, eight *A* and two *a*, nine *A* and one *a*, and then 10 *A*. The same process could have moved in the opposite direction, or begun by favoring *A* and then reversing to fix *a*—random drift is directionless. However, when one of the genes is fixed, the population is homozygous and will remain homozygous (Figures 6.3 and 6.4).

The Hardy–Weinberg equilibrium is a good approximation for all but small populations, and retains its importance in evolutionary biology. It is also true that, once we allow for random drift, the Hardy–Weinberg ratios are not an equilibrium. The Hardy–Weinberg ratios apply to neutral alleles at a locus and the Hardy–Weinberg result suggests that the genotype (and gene) ratios are stable over time. However, in a small population, gene frequencies drift about, and one of the genes will eventually be fixed. Only at that point will the system be stable. The true equilibrium of the Hardy–Weinberg system in a small population appears at homozygosity.

Some geneticists would say that this conclusion is overdramatized. The Hardy–Weinberg theorem has two features. One feature specifies a certain ratio of genotype frequencies, given the gene frequencies; this ratio results from simple algebra and is not affected by drift. The other feature is the claim that the genotypic ratios, once reached, are stable. This is true only in the absence of drift, because drift alters the gene frequencies. However, some population specialists would deny that this second claim is substantially affected by drift. They would prefer to say that the Hardy–Weinberg equilibrium refers only to a large population, and events in small populations are unrelated to the Hardy–Weinberg equilibrium. The important thing to remember is small populations do not support Hardy–Weinberg equilibrium.

6.6 *Neutral drift over time produces a march to homozygosity*

Over the long term, pure random drift causes the population to "march" to homozygosity at a locus. The process by which this march occurs has already been considered (section 6.4) and illustrated (Figure 6.3). All loci at which several selectively neutral alleles appear will tend to become fixed for only one gene.

It is not difficult to derive an expression for the rate at which the population becomes homozygous. First, we define the degree of homozygosity. Individuals in the population are either homozygotes or heterozygotes. Let $f =$ the proportion of homozygotes, and $H = 1 - f$ represent the proportion of heterozygotes (f comes from "fixation"). Homozygotes here include all types of homozygotes at

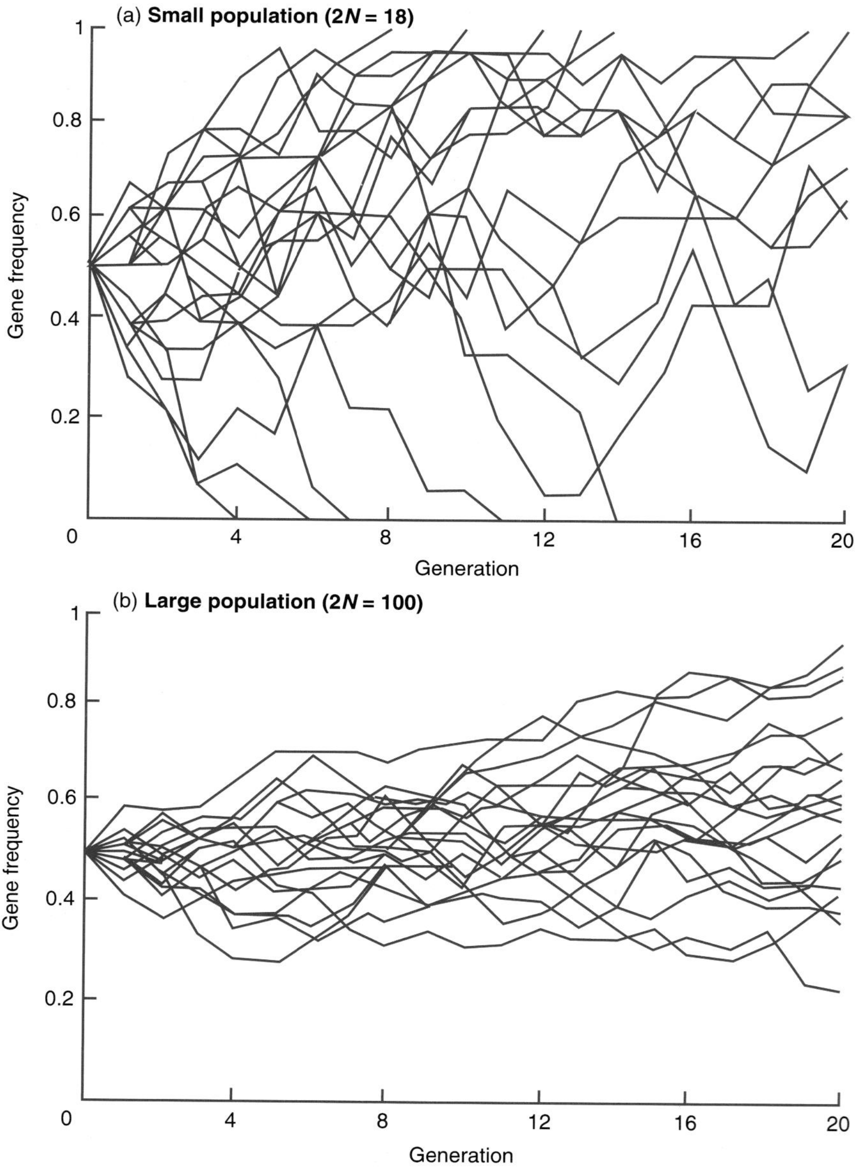

Figure 6.4 Twenty repeat simulations of genetic drift for a two-allele locus with initial gene frequency 0.5. (a) Small population ($2N = 18$); (b) larger population ($2N = 100$). Eventually one of the alleles drifts to a frequency of one. The other alleles are then lost. The drift toward homozygosity occurs more rapidly in a smaller population, but homozygosity is the final result in any small population without mutation.

a locus. If, for example, there are three alleles A_1, A_2, and A_3 then f is the number of A_1A_1, A_2A_2, and A_3A_3 individuals divided by the population size; H is the sum of all heterozygote types. N will again stand for population size.

How will f change over time? We will derive the result in terms of a special case: a species of hermaphrodite in which an individual can fertilize itself. Individuals in the population discharge their gametes into the water and each gamete has a chance of combining with any other gamete. New individuals are formed by sampling two gametes from the gamete pool. The gamete pool contains $2N$ gamete types, where "gamete types" should be understood as follows. There are $2N$ genes in a population made up of N diploid individuals. A gamete type consists of all gametes containing a copy of any one of these genes. Thus, if an individual, with two genes, produces 200,000 gametes, there will be, on average, 100,000 copies of each gamete type in the gamete pool.

To calculate how f, the degree of homozygosity, changes through time, we derive an expression for the number of homozygotes in one generation in terms of the number of homozygotes in the previous generation. We must first distinguish between *a* gene-bearing gametes in the gamete pool that are copies of the same parental *a* gene, and those that are derived from different parents. A homozygote may be produced in two ways: when two *a* genes from the same gametic type meet or when two *a* genes from different gametic types meet (Figure 6.5); the frequency of homozygotes in the next generation will be the sum of these two homozygote offspring.

The first way of making a homozygote is called "self-fertilization." There are $2N$ gamete types but, because each individual produces many more than two gametes, there is a chance $1/(2N)$ that a gamete will combine with another gamete of the same gamete type as itself. If such a combination occurs, the offspring will be homozygous. (If, as above, each individual makes 200,000 gametes, the gamete pool would contain $200{,}000N$ gametes. We first sample one gamete from the pool. Of the remaining gametes, practically 100,000 of them—99,999 in

Figure 6.5 Inbreeding in a small population produces homozygosity. A homozygote can be produced either by combining copies of the same gene from different individuals, or by combining two copies of the same physical gene. Here we imagine that the population contains six adults, which are potentially self-fertilizing hermaphrodites, and each adult produces four gametes. Homozygotes can then be produced by the kind of cross-mating assumed in the Hardy–Weinberg theorem (e.g., offspring number 2) or by self-fertilization (offspring number 1). Self-fertilization necessarily produces a homozygote only if its parent is homozygous (compare offspring 1 and 4).

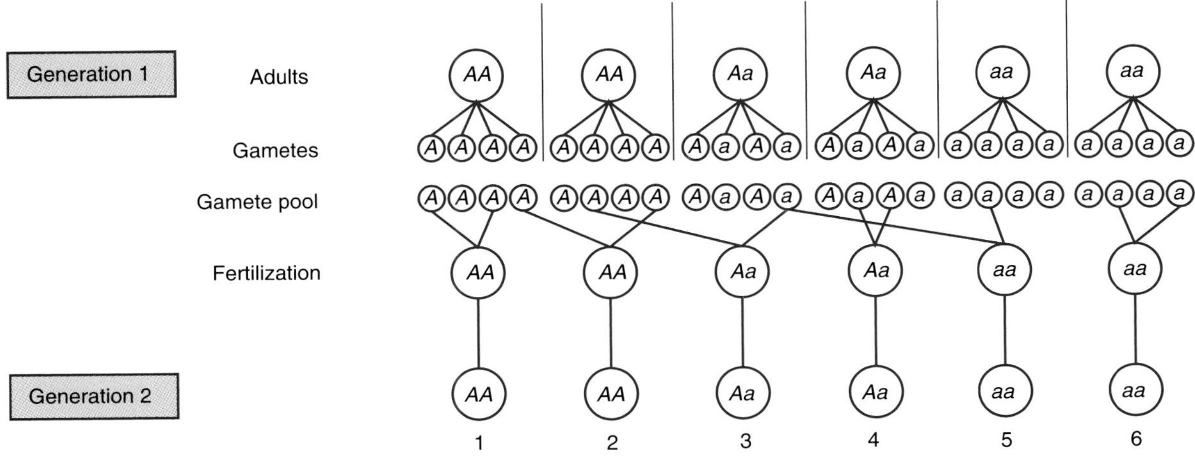

fact—are copies of the same gene. The proportion of gametes left in the pool that contains copies of the same gene as the gamete we sampled is 99,999/200,000N, or $1/(2N)$.)

The second way to produce a homozygote is by combining two identical genes that were not copied from the same gene in the parental generation. If the gamete does not combine with another copy of the same gamete type—chance $1 - (1/(2N))$—it will still form a homozygote if it combines with a copy made from the same gene but from another parent. For a gamete with an *a* gene, if the frequency of *a* in the population is p, the chance that two *a* genes meet is simply p^2, which is the frequency of *aa* homozygotes in the parental generation. For two types of homozygote, *AA* and *aa*, the chance of forming a homozygote will be $p^2 + q^2 = f$. In general, the chance that two independent genes will combine to form a homozygote is equal to the frequency of homozygotes in the previous generation. The total chance of forming a homozygote by this second method is the chance that a gamete does not combine with another copy of the same parental gene, $1 - (1/(2N))$, multiplied by the chance that two independent genes combine to form a homozygote (f). That is, $f(1 - (1/(2N)))$.

Now we can write the frequency of homozygotes in the next generation in terms of the frequency of homozygotes in the parental generation. This frequency is the sum of the two ways of forming a homozygote. Following the normal notation for f' and f (f' is the frequency of homozygotes one generation later),

$$f' = \frac{1}{2N} + \left(1 - \frac{1}{2N}\right)f \tag{6.2}$$

We can follow the same march to increasing homozygosity in terms of the decreasing heterozygosity in the population. *Heterozygosity* is a general measure of the genetic variation per locus in a population. A population may contain 1, 2, 3 . . . n alleles at a locus, and we can write the frequency of those alleles as $p_1, p_2, p_3 \ldots p_n$. In general, the frequency of the ith allele (where i can have any value from 1 to n, depending on the allele to which it refers) is p_i. Heterozygosity is then defined as:

$$H = 1 - \Sigma\, p_i^2$$

This definition can be understood intuitively. In a population of size N, there are $2N$ genes per locus. Imagine drawing out two of those genes at random. The heterozygosity is the chance that the two will be different alleles. If only one allele exists, its frequency is one, and $H = 1 - 1 = 0$. When there are n alleles, for any one of them p_i^2 is the chance of drawing two copies of that allele—p_i^2 is the sum of all chances of drawing two copies of one allele. $1 - p_i^2$ is one minus the total chance of drawing two identical alleles for all alleles, which equals the chance of drawing two different alleles.

If the population is in Hardy–Weinberg equilibrium, then the heterozygosity equals the proportion of heterozygous individuals. H is a more general definition of genetic diversity than the proportion of heterozygotes, however. The chance that two random genes differ measures genetic variation in all populations, regardless of whether they are in Hardy–Weinberg equilibrium. $H = 50\%$, for example, in a population consisting of half *AA* and half *aa* individuals (with no heterozygotes).

Heterozygosity can be shown, by rearrangement of the equation, to decrease at the following rate (the rearrangement involves substituting $H = 1 - f$ in the equation):

$$H' = \left(1 - \frac{1}{2N}\right)H \qquad (6.3)$$

That is, heterozygosity decreases at a rate of $1/(2N)$ per generation until it reaches zero. The population size N is again important in governing the influence of genetic drift. If N is small, the march to homozygosity is rapid. At the other extreme, we reencounter the Hardy–Weinberg result. If N is infinitely large, the degree of heterozygosity is stable, and there is no march to homozygosity.

Although this derivation might seem to apply for only a particular, hermaphroditic breeding system, the result can, in fact, be extended to all cases (a small correction is needed for the case of two sexes). The march to homozygosity in a small population proceeds because two copies of the same gene may combine in a single individual. In the hermaphrodite, this combination happens obviously with self-fertilization. If there are two sexes, a gene in the grandparental generation can appear as a homozygote, in two copies, in the grandchild generation. The process by which a gene in a single copy in one individual combines in two copies in an offspring is called *inbreeding*. Inbreeding can occur in any breeding system with a small population, and its likelihood increases as the population size decreases. With inbreeding, homozygosity is increased over the Hardy–Weinberg level generated by random mating because it is impossible for two copies of the same gene to combine to form a heterozygote. The important point is that, in a small population, homozygosity is increased relative to an infinite population. In an infinite population, two copies of the same gene never fertilize each other; in a small population, there is some chance they will cross-fertilize.

6.7 A calculable amount of polymorphism will exist in a population because of neutral mutation

It might appear that the theory of neutral drift predicts that populations should be completely homozygous. However, new variation will be contributed by mutation, and the equilibrial level of polymorphism (or heterozygosity) will actually be a balance between its elimination by drift and its creation by mutation.

We can now work out what that equilibrium is. The *neutral* mutation rate $= u$ per gene per generation. (As before, u is the rate at which selectively neutral mutations arise, not the total mutation rate.) To find out the equilibrial heterozygosity under drift and mutation, we modify equation 6.2 to account for mutation. If an individual was born a homozygote, and if neither gene has mutated, it remains a homozygote and all of its gametes will have the same gene. (We ignore the possibility that mutation produces a homozygote—for example, by a heterozygote *Aa* mutating to a homozygous *AA*; we are assuming that mutations produce new genes.) For a homozygote to produce all of its gametes with the same gene, *neither* of its genes must have mutated. If either of them has mutated, the frequency of homozygotes will decrease. The chance that a gene has not mutated = $(1 - u)$ and the chance that neither of an individual's genes has mutated = $(1 - u)^2$.

Now we can simply modify the recurrence relation derived earlier. The frequency of homozygotes will be as before, but multiplied by the probability that they have not mutated to heterozygotes:

$$f' = \left[\frac{1}{2N} + \left(1 - \frac{1}{2N}\right)f\right](1 - u)^2 \qquad (6.4)$$

Homozygosity (f) will not increase to 1, but will converge to an equilibrial value. The equilibrium lies between the increase in homozygosity by drift, and its decrease by mutation. We can find the equilibrium value of f from $f^* = f = f'$. In this equation, f^* indicates a value of f that is stable in successive generations ($f' = f$). Substituting $f^* = f' = f$ in the equation gives (after a minor manipulation):

$$f^* = \frac{(1 - u)^2}{2N - (2N - 1)(1 - u)^2} \qquad (6.5)$$

The equation simplifies if we ignore terms in u^2, which will be relatively unimportant because the neutral mutation rate is low. Then

$$f^* = \frac{1}{4Nu + 1} \qquad (6.6)$$

The equilibrial heterozygosity ($H^* = 1 - f^*$) is:

$$H^* = \frac{4Nu}{4Nu + 1} \qquad (6.7)$$

This important result gives the degree of heterozygosity that should exist for a balance between the drift to homozygosity and new neutral mutation. The expected heterozygosity depends on the neutral mutation rate and the population size (Figure 6.6). Because the march toward homozygosity is more rapid if the population size is smaller, it makes sense that the expected heterozygosity is lower if N is smaller. Heterozygosity is also lower if the mutation rate is lower, as we should expect. In sum, the population will be less genetically variable for neutral alleles when population sizes are smaller and the mutation rate lower.

6.8 *Population size and effective population size*

What is "population size"? We have seen that N determines the effect of genetic drift on gene frequencies. But what exactly is N? In an ecological sense, N can be measured by counting items, such as the number of adults in a locality. However, for the theory of population genetics with small populations, the estimate obtained by ecological counting is only a crude approximation to the "population size," N, implied by the equations. What matters is the chance that two copies of a gene will be sampled as the next generation is produced, which is affected by the breeding structure of the population. A population of size N contains $2N$ genes. The correct interpretation of N for the theoretical equations

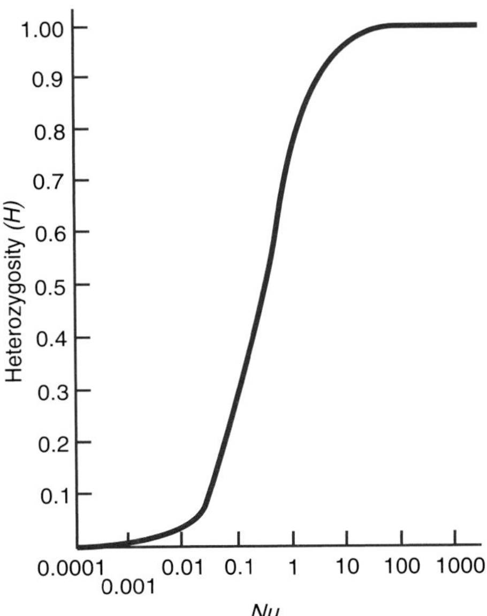

Figure 6.6 The theoretical relation between the degree of heterozygosity and the parameter *Nu* (the product of the population size and neutral mutation rate).

is that N has been correctly measured when the chance of drawing two copies of the same gene is $(1/2N)^2$.

If we draw two genes from a population at a locality, we may be more likely for various reasons to get two copies of the same gene than would be implied by the naive ecological measure of population size. Population geneticists therefore often write N_e (for "effective" population size) in the equations, rather than N. In practice, effective population sizes are usually lower than ecologically observed population sizes. The relation between N_e, the effective population size implied by the equations, and the observed population size N can be complex. A number of factors are known to influence effective population size.

1. Sex ratio. If one sex is rarer, the population size of the rarer sex will dominate the changes in gene frequencies. It is much more likely that identical genes will be drawn from the rarer sex, because fewer individuals are contributing genes to the next generation. Sewall Wright proved in 1932 that in this case

$$Ne = \frac{4N_m \cdot N_f}{N_m + N_f} \tag{6.8}$$

where N_m = number of males, N_f = number of females in the population.

2. Population fluctuations. If population size fluctuates, homozygosity will increase more rapidly while the population goes through a "bottleneck" of small size. N_e is disproportionately influenced by N during the bottleneck, and a formula can be derived for N_e in terms of the harmonic mean of N.
3. Small breeding groups. If most breeding takes place within small groups, then the effective population size will be much smaller than if there is population-wide random mating (called "panmixis"). The degree of homozygosity will

exceed that predicted from the size of the breeding groups drawn from the total population. Breeding groups can be small either because of geographic subdivision of the population or because of non-random mating. If certain types of individuals consistently choose a particular kind of mate, then the mating will be conducted through only a part of the whole population even without geographic subdivision. In either case, the chance that copies of the same gene will be "drawn" together is increased relative to the case with a large population. Several ways of creating population subdivisions are possible to derive expressions for N_e. Effective population size is mainly influenced by the degree of subdivision and the migration rate between the subgroups: the more migration occurs, the more N_e tends to N.

4. Variable fertility. If the number of successful gametes varies between individuals (as it often does among males when sexual selection operates, see chapter 11), the more fertile individuals will accelerate the march toward homozygosity. Again, the chance that copies of the same gene will combine in the same individual in the production of the next generation is increased and the effective population size is decreased relative to the total number of adults. Wright showed that if k is the average number of gametes produced by a member of the population and σ_k^2 is the variance of k (see Box 9.1, p. 233, for the definition of variance), then

$$N_e = \frac{4N - 2}{\sigma_k^2 + 2} \tag{6.9}$$

For $N_e < N$, the variance of k must be greater than random. If k varies randomly, as a Poisson process, $\sigma_k^2 = k = 2$ and $N_e = N$.

These are all quite technical points. The N_e found in the equations for neutral evolution is an exactly defined quantity, but it is difficult to measure in practice. It is usually less than the observed number of adults, N. $N_e = N$ when the population mates randomly, is constant in size, has an equal sex ratio, and experiences approximately Poisson variance in fertility. Natural deviations from these conditions produce $N_e < N$. How much smaller N_e is than N is difficult to measure, though it is possible to make estimates by the formulas we have previously used. Other things being equal, species with more subdivided and inbred population structures have lower N_e than more panmictic species.

SUMMARY

1. In a small population, random sampling of gametes to produce the next generation can change the gene frequency. These random changes are called genetic drift.
2. Genetic drift has a larger effect on gene frequencies if the population size is small than if the population size is large.
3. If a small population colonizes a new area, it is likely to carry all the ancestral population's genes, but the gene frequencies may be unrepresentative.
4. One gene can be substituted for another by random drift. The rate of neutral substitution is equal to the rate at which neutral mutations arise.
5. In a small population, in the absence of mutation, one allele will eventually be fixed at a locus. The population will eventually become homozygous. The

Hardy–Weinberg equilibrium does not apply to small populations. The effect of drift is to reduce the amount of variability in the population.

6. The amount of neutral genetic variability in a population will be a balance between its loss by drift and its creation by new mutation.
7. The "effective" size of a population, which is the population size assumed in the theory of population genetics for small populations, should be distinguished from the size of a population that an ecologist might measure in nature. Effective population sizes are usually smaller than observed population sizes.

FURTHER READING

Population genetics texts, such as those of Crow and Kimura (1970), Crow (1986), Ewens (1979), Gale (1990), Hartl and Clark (1989), Hedrick (1983), Spiess (1989), and Wallace (1981) explain the theory of population genetics for small populations. Lewontin (1974), Kimura (1983), and Nei (1987) also explain much of the material. Wright (1969) is more advanced. Beatty's entry in Keller and Lloyd (1992) explains the history of ideas, including Wright's theories, about random drift. Li (1977) is an anthology of many of the classic papers of stochastic population genetics. Kimura (1983) also contains a clear account of the parts of the theory most relevant to his neutral theory and discusses the meaning of effective population size. For the medical examples of founder events in humans, see Dean (1972) and Hayden (1981).

STUDY AND REVIEW QUESTIONS

1. A population of 100 individuals contains 100 *A* genes and 100 *a* genes. If no mutation occurs and the three genotypes are selectively neutral, what would you expect the genotype and gene frequencies to be over the long term, such as 10,000 generations?

2. Explain (a) the meaning of "random" in random sampling, and the reason why drift is more powerful in smaller populations, and (b) the argument as to why all genes at any locus (such as the insulin locus) in the human population are descended from one gene in an ancestral population.

3. What is the heterozygosity (*H*) of the following populations?

Genotype numbers

AA	*Aa*	*aa*	*H*
25	50	25	
50	0	50	
0	50	50	
0	0	100	

4. If the neutral mutation rate is 10^{-8} at a locus, what is the rate of neutral evolution at that locus if the population size is (a) 100 individuals? (b) 1000 individuals?

5. What is the probability in a population of size N that a gene will combine (a) with another copy of itself to produce a new individual? (b) with a copy of another gene?
6. Show how equation 6.2 can be manipulated to produce equation 6.3, and how equation 6.6 can be manipulated to produce equation 6.7.

CHAPTER 7

Molecular Evolution and the Neutral Theory

THE chapter discusses the relative importance of two processes in driving molecular evolution: neutral drift and natural selection. We begin by defining what it means for these two processes, and particularly neutral drift, to be a general theory of molecular evolution. We then look at four classes of observation that have been used to test these theories: (1) the rate of evolution and magnitude of polymorphism; (2) the constancy of molecular evolution (or the "molecular clock"); (3) the relation between functional constraint on molecules and their rates of evolution; and (4) the relation between polymorphism and evolutionary rate in different molecules (or parts of molecules). We introduce the general concept of genetic load during test 1. We see how test 1 has proved inconclusive, and how the availability of DNA sequences has enriched tests 2, 3, and 4. DNA sequences contain both silent and replacement nucleotide changes, which enable the construction of some subtle tests. The controversy between neutral and selective theories remains unsettled.

7.1 Neutral drift and natural selection can both hypothetically explain molecular evolution

Evolution, at the molecular level, is observable as nucleotide (or base) changes in the DNA and amino acid changes in proteins. Both can be studied by examining the differences between species—if, for example, one species has nucleotide *A* at a certain site and another species has nucleotide *G*—and in polymorphisms within a species—if, for example, some individuals of a species have nucleotide *A* at a certain site while other individuals have *G*. Both polymorphism and evolutionary change between species can be explained by two processes we have already encountered: natural selection and neutral drift.

The main factors controlling the relative importance of neutral drift and natural selection are (as we saw in chapters 5 and 6) population size and the selection coefficients of the different genotypes. If the population is small and the selection coefficients low, genetic drift dominates, whereas natural selection dominates if the population and the selection coefficients are large. In this chapter we will mainly, though not completely, ignore the effects of population size. We will adopt this tactic not because population size is unimportant, but rather to simplify the discussion, and because reliable measurements are rarely available.

Whenever we consider neutral drift in this chapter, we will assume that population sizes are small enough for the process to operate.

We will also employ a slightly different notation for selection coefficients. In chapter 5, the genotype with the highest fitness was given a fitness of 1 and the other genotypes were given fitnesses like $(1 - s)$. Here we will be interested in whether one form of a molecule has a higher, lower, or equal fitness with another form. It will therefore be more convenient to talk about selection coefficients that are +, 0, or −: a +ve selection coefficient means natural selection favors the variant; −ve means it is selected against; 0 means it is neutral.

The evolution of modern species has incorporated millions of molecular changes. Let us think about them all, in the abstract. Natural selection and neutral drift could logically have produced any proportions of the changes (though the proportions of the two must add to one). The subject is normally discussed in terms of a controversy between two extreme positions, however. In 1968 and 1969, Kimura, and King and Jukes, respectively, suggested that most evolutionary change at the molecular level is driven by random drift rather than natural selection. Kimura called this view *the neutral theory of molecular evolution.* The neutral theory does not suggest that random drift explains all evolutionary change; natural selection is still needed to explain adaptation. It is possible that the adaptations we observe in organisms required only a minority of all evolutionary changes that have taken place in the DNA. The neutral theory states that evolution at the level of the DNA and proteins, but not at the level of adaptation, is dominated by random processes; most evolution at the molecular level would then be nonadaptive. We can contrast the neutral theory with its converse: the idea that almost all molecular evolution has been driven by natural selection.

The difference between the two ideas can be understood in terms of the frequency distribution for the selection coefficients of mutations, or genetic variants. (It does not matter here whether we talk about new mutations or the set of genetic variants existing in a population at a genetic locus. "Genetic variant" could be substituted for "mutation" throughout this paragraph.) Given a mutation of a certain selection coefficient, the theory of neutral drift or selection (as described in chapters 5 and 6) applies in a mathematically automatic way: if the selection coefficient is positive, the mutation increases in frequency; if it is negative, it is eliminated; if it is zero, the gene frequencies drift. The dispute is about how likely it is that mutations will arise with a certain selection coefficient.

Consider the nucleotide sequence of a gene in a living organism. The gene codes for a reasonably well-adapted protein. After all, the protein is unlikely to be a complete failure if the organism containing it is alive. Now consider all of the mutations that can be made in the gene. You could work down the gene, altering one nucleotide at a time, and ask, for each change, whether the new version was better, worse, or equally as good as the original gene. In a population of organisms in nature, mutations will occur and cause those kinds of change, in certain frequencies.

Almost all biologists agree that most of the changes will be for the worse; most will have negative selection coefficients. Adaptation is an unlikely physical state, and if a random change is made in an adapted protein it is likely to be for the worse.

On the other hand, biologists disagree about the relative frequency of the other two classes of mutations: neutral and selectively advantageous mutations.

If natural selection has produced most evolutionary change, many advantageous mutations must have occurred, whereas neutral mutations must have been rare. If neutral drift has produced most evolutionary change, the relative frequencies are the reverse. Figures 7.1a and 7.1b illustrate the two views, with the difference lying in the relative heights of the graph in the 0 and the + region. The high frequency of mutations in the − region is common to the two.

Imagine a gene of about 1000 nucleotides (corresponding to a protein of about 300 amino acids). There are 4^{1000} (or about 10^{600}) possible sequences of the gene. The protein encoded by the gene will have some function—for example, carrying oxygen in the blood (actually done by hemoglobin, which is made up of four polypeptides of slightly less than 150 amino acids each). The neutralist view of evolution suggests that, of the 10^{600} possible molecules, the great majority would fail to carry oxygen at all, and many would do so poorly. A minority of perhaps several hundred different sequences, all very similar to one another, would then code for proteins that carried oxygen equally well. What we observe as evolution would consist of shuffling around within this limited set of equivalent sequences. The selectionist alternative is that the few hundred variants are not equivalent, but that one works better in one environment, another in another environment, and so on; evolution consists of the substitution of one variant for another when the environment changes. The difference in viewpoint exists between people who think that a protein could have a fairly large number of equivalent forms, and those who think that any change in the amino acid sequence, no matter how small, is likely to make a difference in the protein's function (and its bearer's fitness) under some conditions.

Graphs like Figures 7.1a and 7.1b can be drawn because each genetic variant has a selection coefficient, which determines its evolutionary fate. However, the selection coefficient of a genetic variant will often depend on the environment in which it finds itself. Two genetic variants of an enzyme, for example, may

Figure 7.1 The neutral and selectionist theories postulate different frequency distributions for the rates of mutation with various selection coefficients. (a) According to the selectionists, exactly neutral mutations are rare and enough favorable mutations arise to account for all molecular evolution. (b) Neutralists believe more neutral, and hardly any selectively favored, mutations arise. (c) The theory of pan-neutralism, according to which all mutations are selectively neutral.

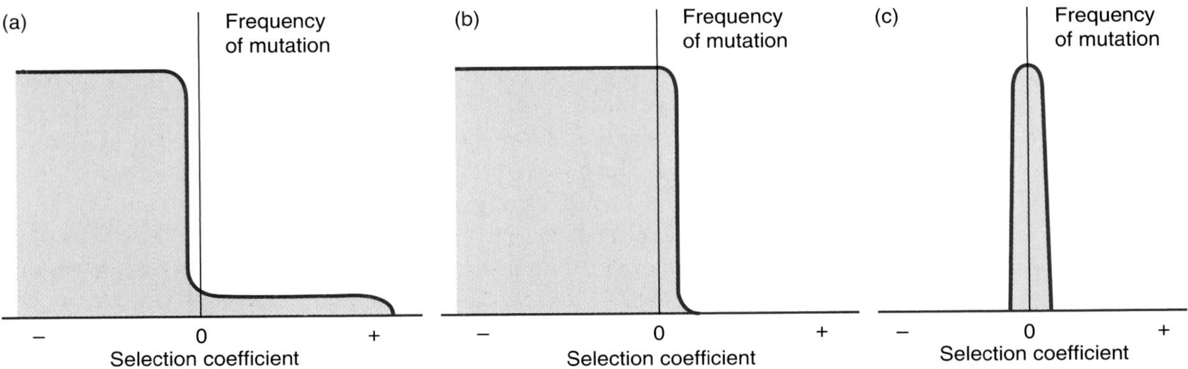

have identical fitnesses in one environment, but different fitnesses in another environment. In nature, during the life of an organism, a genetic variant has a fitness given by the average of all the environments it experiences, and the formulation in Figure 7.1 (with one fitness value for each genetic variant) is therefore theoretically valid.

When measuring fitnesses, or considering the biological reason why some variant exists, it may be necessary to think more subtly, however. It has been argued, for instance, that many mutants are "conditionally neutral" or "conditionally advantageous." That is, the mutants are neutral in some environments, but make a difference in other (perhaps rare) environments. In this chapter, we will continue to talk about "the" fitness, or selection coefficient, of a genetic variant, but the use of that term should not be taken to mean that a genetic variant always unconditionally has the same fitness.

Figure 7.1c shows *pan-neutralism,* another idea with which the neutral theory is often confused. Pan-neutrality means that all forms of a protein (or, at least, the majority of them) have equal fitness—that all mutations are neutral. How does this idea differ from the neutral theory? First, we need to distinguish mutation from evolutionary change. In the neutral theory, all evolutionary change is neutral, which does not mean that all mutations are neutral. In Figure 7.1b, all mutations that may contribute to evolutionary change are neutral, but the majority of mutations are deleterious and will be selected against. Deleterious mutations disappear from the population before they have any chance to show up as evolution. The neutral theory, therefore, does not rule out natural selection. It simply has a different use for it than the selectionist theory of molecular evolution. Selectionists use natural selection to explain why mutations are lost (when they are deleterious) or fixed (when they are advantageous). Neutralists, on the other hand, use selection to explain only why deleterious mutations are lost; they use drift to explain how new mutations are fixed.

Pan-neutralism is almost certainly false. Good empirical reasons exist to reject this theory. For example, pan-neutralism cannot explain why different genes evolve at different rates, or (see section 7.11.4) why different species have different biases in codon usage. Nor is it theoretically plausible. It is absurd to suggest that hardly any mutations are disadvantageous. Organisms, including their molecules, are adapted to their environments; we need reflect on only the efficiency of digestive enzymes—or any other biological molecule—for supporting life to realize that fact. If molecules are affected by adaptation, many (or most) changes in them will be for the worse. (Note that these objections apply only to pan-neutralism, not to Kimura's neutral theory.)

If the crucial difference between the selective and neutral theories involves the relative frequencies of neutral and selectively advantageous mutations, the direct way to test between them would simply be to measure the fitnesses of many genetic variants at a locus, and count the numbers with negative, neutral, or positive selection coefficients under certain environmental conditions. It has proved impossible to settle the controversy in this way. Measuring the fitness of even one common genetic variant represents a major research program; it would be completely impractical to measure the fitnesses of many rare variants.

Testing between the theories has been attempted by less direct means, however. We shall look at four of the main kinds of tests (although these four by no means exhaust the list of evidence that has been brought into this controversy). The four tests include the following:

1. The absolute rate of molecular evolution and degree of polymorphism, both of which have been argued to be too high to be explained by natural selection.
2. The constancy of molecular evolution, which has been argued to be inconsistent with natural selection.
3. The observation that functionally less constrained parts of molecules evolve at a higher rate, which has been argued to be the opposite of what the theory of natural selection would predict.
4. The observation that the degree of polymorphism and the rate of evolution of a molecule are not correlated in exactly the way the neutral theory predicts.

7.2 *The rates of molecular evolution and amounts of genetic variation can be measured*

In the 1960s, when the sequences of particular proteins became known for several species, biologists could work out the number of amino acid changes that occurred during the evolutionary divergence of a pair of species. The same calculation has been made with many DNA sequences, as they have become increasingly available since the 1980s. For any two species, the approximate age of their common ancestor can be estimated from the fossil record. The rate of protein evolution can then be calculated as the number of amino acid differences between the protein of the two species divided by 2 × the time to their common ancestor (Figure 7.2). If the species are humans and mice, their common ancestor probably lived about 80 million years ago. If we look at the sequence of a 100 amino acid protein in the two species and it differs at 16 sites, then the rate of evolution is

Figure 7.2 Imagine that some region of a protein has the illustrated sequences in two species. The evolutionary change has happened somewhere within the lineage connecting the two species via their common ancestor. An alanine has been substituted for a glycine in the lineage leading to species 2, or a glycine for an alanine in the lineage to species 1. Either way, the amount of evolution is one change, and it has taken place in twice the time from the species back to their common ancestor. That is, one change has occurred in $2t$ years.

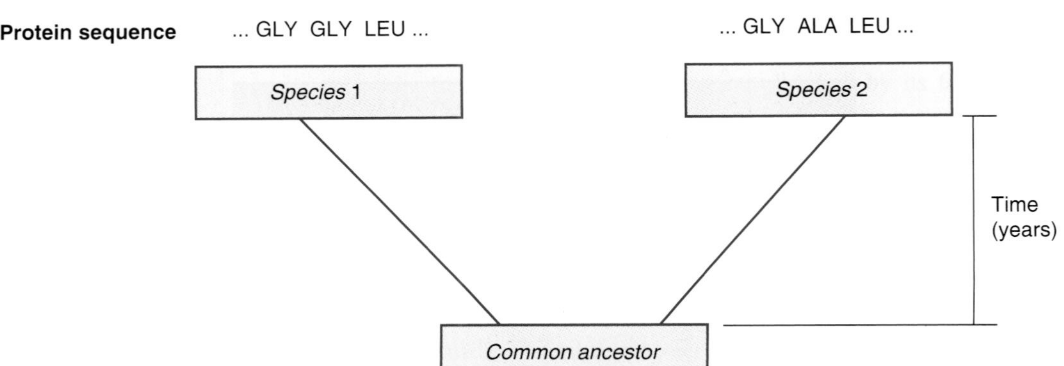

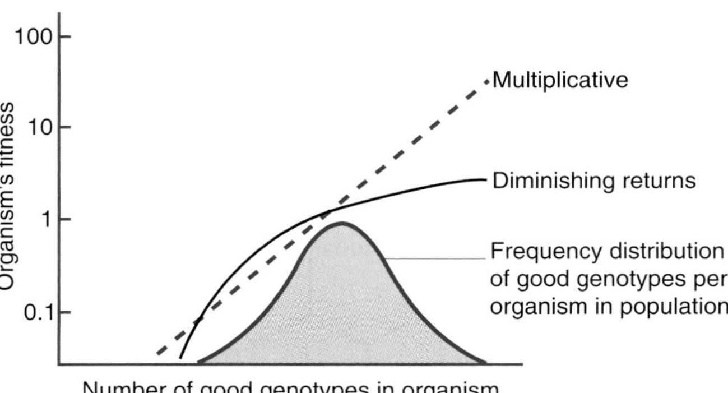

Figure 7.6 Relation between the number of good (high fitness) genotypes as opposed to bad (low fitness) genotypes in an organism and its fitness. Fitness on the *y*-axis is plotted on a logarithmic scale. If genotypes at different loci have independent effects on fitness, the graph is a straight line. There also could be a decreasing returns relation, for reasons discussed in the text. See also Figure 11.3 (p. 290).

having good genotypes at many loci. If we assume a heterozygous advantage, the good genotypes would be heterozygotes (the argument that follows applies to all kinds of selection). Now imagine working down the genome, converting good genotypes into inferior genotypes. How will the fitness of the organism decline? The question is clearly an empirical one, but it cannot be answered in general with factual evidence. We can, however, reason about it in the abstract.

The idea that all, or even most, loci independently affect an organism's fitness is arguably biologically naive. Suppose the development of the right eye is controlled by one locus and the left eye by another: an organism with good genotypes at both loci has two eyes, and a bad genotype at either locus causes the loss of the appropriate eye. What will the fitnesses be for organisms with two good genotypes, one bad and one good genotype, or two bad genotypes? The organisms with two eyes clearly have the highest fitness. Organisms with one bad eye will have lower fitness, but perhaps not by a dramatic amount. A one-eyed organism can compensate for its defect by various means: it can turn its head, scan the landscape repeatedly, and spend slightly longer in looking around than would a two-eyed organism. Most of the time it will manage as well as a two-eyed organism, and its fitness on average will not be much reduced. On the other hand, the organism with two bad genotypes is blind and its fitness will crash. In this case, the fitnesses are not multiplicative. The fitness of a doubly defective organism is much lower than predicted by combining the fitnesses of organisms with the two kinds of single defect.

In fact, the development of right and left eyes is presumably not controlled by two genetic loci, but the argument illustrates what may be a general point. The relation between the number of good genotypes in an organism and the log of its fitness may not be a straight line but may instead demonstrate the decreasing returns pattern (Figure 7.6). The reason to think that the argument may apply in general is that organisms possess back-up systems and compensation mechanisms. If we introduce a small number of errors or defects into the development of an organism, it will be able to make up for them. As the number of defects increases, however, they eventually overwhelm the back-up mechanisms and fitness declines more precipitously.

If the fitness relation has the decreasing returns shape, it is easier for natural selection to maintain variation at a large number of loci (by heterozygous advantage or any other selective means). The individuals with multiple homozygotes have very low fitness and are disproportionately likely to die, as compared with the multiplicative fitness case. When a multiple homozygote dies, it takes all of its homozygotes with it; it may have enough homozygotes to kill it many times over, but it only needs to die once to eliminate them all. Thus, one selective death kills more than one inferior genotype. Natural selection is more efficient at purging homozygotes when fitness decreases more rapidly as the number of inferior genotypes (homozygotes, in this case) in an organism increases.

Whether the fitness relation actually has one of the patterns shown in Figure 7.6, or some other pattern, is not known. The pattern you expect will depend partly on how "atomistic" or "holistic" your view of biological organisms is. If organisms are atomistically organized into many self-contained subunits, the pattern would be more likely to be the straight line; if they are holistically organized, with multiple overlapping subsystems that can compensate for one another, the pattern would be more likely to be a curve. It is probable that extra good genotypes provide some "decreasing return" and therefore some curve in the fitness pattern of Figure 7.6, but whether this level is enough for selection to explain the observed levels of molecular variation remains an open question.

The main points of this section are that the absurd figures for genetic loads assumed that the fitness effects of different loci are independent. Other kinds of fitness relation make it easier for selection to maintain variation, as they require much lower genetic loads. An analogous argument can be made to explain how selection could drive evolutionary change, rather than maintain variation, at multiple loci, without running up an impossible cost or substitutional load.

A final point to note is that natural selection can, in theory, maintain polymorphism by frequency dependence of fitnesses as well as by heterozygous advantage (section 5.13, p.121). A genetic load is unavoidable if heterozygous advantage exists, but it is possible for a polymorphism to be maintained by frequency dependence without any load at all. (This statement is true for a sex ratio, for example; no selective death is needed at equilibrium to maintain the 50:50 sex ratio. See section 11.5, p. 307.)

We have now completed the three selectionist answers to the problem of genetic load. The paradoxical genetic loads were apparently implied by the discovery, in the 1960s, of high levels of molecular variation and high rates of molecular evolution. As we have seen, the problem exists only if selection is hard, fitnesses are multiplicative, and heterozygous advantage is the only selective regime—none of which need be true.

7.8 *The first test, using observations of the absolute rates of evolution and levels of polymorphism, is indecisive*

Kimura's argument was that neutral drift can explain the high rate of molecular evolution and high levels of polymorphism, whereas natural selection cannot. His case against natural selection made assumptions about genetic loads that are unrealistic, however, and it is therefore unconvincing. Although the negative case against an explanation by natural selection is unconvincing, the positive case for it is no more powerful than for neutral drift. Neutral drift, we saw, can explain the observed evolutionary rates and heterozygosities by inserting more or less conjectural values for N and u into the equations for genetic drift. The actual values of N and u are unknown, and a neutralist could "explain" almost any evolutionary rate or heterozygosity by positing suitable values for N and u.

When natural selection is considered positively, as a possible explanation for molecular evolution, it turns out to occupy a similar position. Selectionists also do not generally know the value of the variables that control the rate of evolution. The probability of fixation of a selectively advantageous mutation with selective advantage s is approximately $2s$, which makes the rate of evolution $2sm$ where m is the rate of selectively advantageous mutations. However, m remains just as much of an unknown as u, and not much can be said about s apart from that it is probably low (1% or less). Like the neutralist, the selectionist can explain almost any observation by inventing suitable values for unknown numbers.

The objections to selective explanation for molecular variation and evolution were successfully countered, as we saw in the previous section, but it does not mean that selection *is* the explanation for the facts. Indeed, few biologists now think that heterozygous advantage explains more than a few polymorphisms. The main reason is empirical: examples have accumulated too slowly. The main example, sickle cell anemia, was one of the first polymorphisms to be understood in selective terms and it was natural to assume (tentatively) that the process might be general. Not much evidence has since been found of heterozygous advantage at other polymorphic genes; in fact, 30 years of research between the discovery of heterozygous advantage in sickle cell anemia and a review by Endler in 1986 revealed only another six examples, which represents a tiny percentage of the polymorphic genes that have been investigated.

Frequency dependence is another possible selective explanation. Some population geneticists, such as Clarke, have argued that it is a general cause of polymorphism. The possibility has not been widely accepted, but it cannot be ruled out either. No general arguments have been set forth against this theory, except perhaps a vague intuition that not enough biological interactions have the correct characteristics to be negatively frequency-dependent and to maintain 3000 polymorphisms in a fruitfly population. Some widespread biological processes such as the interaction of predators and prey and of parasites and their hosts (section 11.3.6, p. 294) produce frequency-dependent relations, although not necessarily stable polymorphisms. The importance of frequency-dependent selection in nature remains an open question. The same question plagues other kinds of selection, in which polymorphism is maintained by changes in the fitness of genotypes in time and space. In short, the observations about rates of molecular evolution and levels of polymorphism do not allow us to draw decisive conclusions.

7.9 Test 2: the rates of molecular evolution are arguably too constant for a process controlled by natural selection

The rate of molecular evolution can be measured for any pair of species by the method shown in Figure 7.2. Each pair of species has its own number of molecular differences and time to their common ancestor. We can plot the point defined by these two numbers for many pairs of species on a graph, as Figure 7.7 does

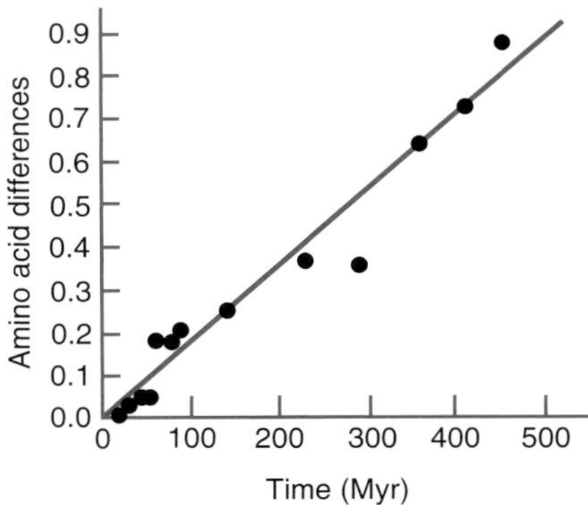

Figure 7.7 The rate of evolution of hemoglobin. Each point on the graph is for a pair of species, or groups of species, and the value for that pair was obtained by the method shown in Figure 7.2. Some of the points are for α-hemoglobin, others for β-hemoglobin. From Kimura (1983). Reprinted with the permission of Cambridge University Press, © 1983.

for hemoglobin. The striking property of the graph is that the points for the different species pairs fall on a straight line. Molecular evolution appears to have an approximately constant rate per unit time—that is, it shows a *molecular clock*. Evolutionary change at the molecular level ticks over at a roughly constant rate, and the amount of molecular change between two species measures how long ago they shared a common ancestor. (Molecular differences between species are used to infer phylogenies; see chapter 17.)

A graph like Figure 7.7 requires a knowledge of the time between the common ancestor and the evolution of each species pair. These times are estimated from the fossil record and are uncertain (see chapter 19); the results are not universally trusted. However, we can also test the constancy of molecular evolution by another method, which does not require absolute dates, and this test also suggests that molecular evolution is fairly clock-like (Box 7.1). An empirical controversy has arisen as to the constancy of the molecular clock; the statistical details of this issue are complex, and we will not discuss them here. It can reasonably be concluded at present that the rate of molecular evolution is constant enough to require explanation.

What does a constant rate imply about the roles of natural selection and neutral drift in molecular evolution? Kimura reasoned that constant rates are more easily explained by neutral drift than selection. Neutral drift has the property of a random process, and its rate will show the variability characteristic of a random process. Neutral mutations crop up at random intervals, but if they are observed over a sufficiently long time period the rate of change will appear to be approximately constant. Thus, neutral drift will drive evolution at a fairly constant rate. Natural selection, Kimura argued, does not produce such constant change. Under selection, the rate of evolution is influenced by environmental change as well as the mutation rate. It would require a surprisingly steady rate of environmental change, occurring over hundreds of millions of years, in organisms as diverse as snails, mice, sharks, and trees to produce the constant rate of change seen in Figure 7.7.

Moreover, if we examine characters (e.g., any adaptive morphological characters) that have undoubtedly evolved by natural selection, they do not appear to evolve at constant rates. Kimura's discussion of the evolution of the vertebrate

BOX 7.1

The Relative Rate Test

The relative rate test is a method of testing whether a molecule (or, in principle, any other character) evolves at a constant rate in two independent lineages. It was first used by Sarich and Wilson in 1973. Suppose we know the sequence of a protein in three species, *a*, *b*, and *c*, and we also know the order of phylogenetic branching of the three species (Figure B7.1). We can now infer the amounts of change in the two lines from the common ancestor of *a* and *b* to the modern species (x and y in Figure B7.1). If the protein evolved at the same rate in the two lineages, the number of amino acid changes between the common ancestor and *a* (x) should equal the number of changes between the common ancestor and *b* (y); that is, $x = y$. Using simple simultaneous equations, we can infer x and y. We know the differences between the protein sequences in *a* and *b* (k), *b* and *c* (l), and *a* and *c* (m). Thus,

$$k = x + y$$
$$l = y + z$$
$$m = x + z$$

We have three equations with three unknowns and can solve for x, y, and z. We then test whether the rates were the same by seeing whether $x = y$. Notice that we do not need to know the absolute date (or the identity) of the common ancestors.

The relative rate test can show only that a molecule evolved at the same rate in the two lineages connecting the two modern species with their common ancestor. It does not prove that the molecule always has a constant rate; it does not, in other words, confirm the molecular clock. If the identity of the relative rate is shown for many pairs of species, with common ancestors of very different antiquities, it is suggestive of (and consistent with) a molecular clock, but it is not conclusive evidence.

We can see why in a counterexample (Figure B7.2). Suppose that a molecule evolves at the same rate in

Figure B7.1 Phylogeny of three species: *a*, *b*, and *c*. In the figure, *k*, *l*, and *m* are the observed numbers of amino acid differences between the three species. The amounts of evolution (*x*, *y*, *z*) in the three parts of the tree can be simply inferred, as the text explains.

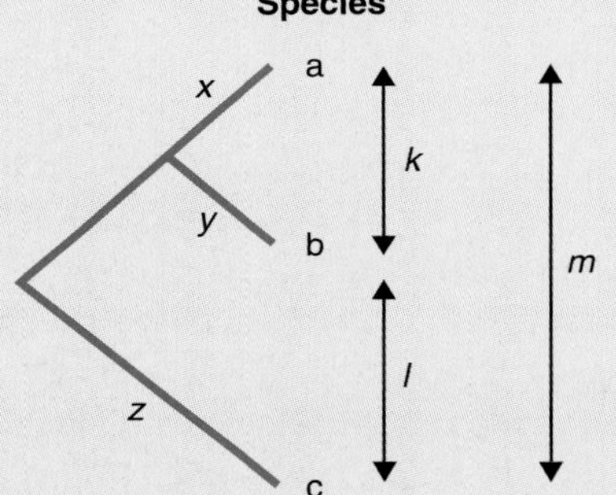

wing illustrates the argument. A long period preceded the evolution of the wing, during which the vertebrate limb remained relatively constant (in the form of the tetrapod limb of amphibians and reptiles); then came a shorter period when the wing originated and evolved; finally, a more or less finished wing form was fine-tuned over a long period. The wing undoubtedly evolved under the influence of natural selection, and the rate of change probably underwent large fluctuations between fast and slow evolution. The argument suggests that rates of molecular evolution are strikingly constant; it is also Kimura's reason for confining the neutral theory to molecules, and not applying it to the gross phenotypes of organisms. Molecular evolution does appear to proceed at a fairly constant rate, as would be expected for a random process. However, the rate of morphological all lineages at any one time, but that it has been gradually slowing down through evolutionary history. A pair of species with a common ancestor 100 million years ago will then show rate constancy according to the relative rate test (because the molecule evolves at the same rate in all lineages at any one time). Any other species pair—for instance, with a common ancestor 50 million years ago—will also show relative rate constancy. However, no molecular clock exists because the rate slows down through time. The relative rate test will not detect that the more recent species pair have a smaller absolute number of changes—absolute dates would be needed for that.

The same point would apply in the case of a trend in evolutionary rate with time, which does not have to be directional. The molecule could speed up and slow down many times in evolution. As long as the speeding up and slowing down apply to all lineages, the relative rate test will show equal rates of evolution in the two lineages. The relative rate test, therefore, cannot conclusively test the molecular clock hypothesis.

Further reading: Fitch (1976).

Figure B7.2 (a) The rate of evolution of a molecule has slowed down gradually through time, but the rate of evolution is always the same in all lineages at any one time. The molecule does not evolve like a clock (which would show up as a flat graph of rate against time). (b) For any pair of species, with common ancestors at any time, the amount of change will be the same in both lineages.

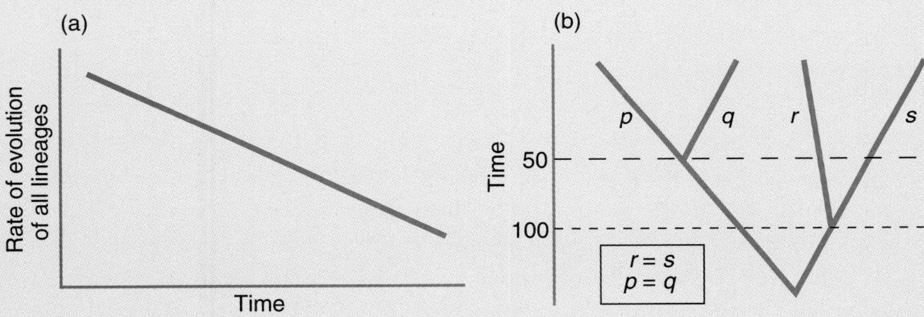

evolution has a different pattern, and was probably produced by the nonrandom process of selection.

Molecular evolution in "living fossils" provides a striking example both of the constant rate of molecular evolution and of the independence between molecular and morphological evolution. The Port Jackson shark *Heterodontus portusjacksoni* is a living fossil—a species that closely resembles its fossil ancestors (fossils over 300 million years old are very similar to the Port Jackson shark). Its molecules have been evolving very differently from its morphology. Hemoglobin duplicated into α and β forms before the ancestor of mammals and sharks, at the beginning of the chordate radiation. We can count the amino acid differences between α- and β-globin as a measure of the rate of molecular evolution in the lineages leading to the modern species. Table 7.4 reveals that changes have accumulated in the Port Jackson shark lineage at the same rate as the human lineage. Thus, the rates of molecular evolution in the two lineages are roughly equal. In contrast, the two lineages show large differences in their rates of morphological evolution. The Port Jackson shark lineage has hardly changed at all, but humans have evolved from fish-like ancestors, and passed through amphibian, reptilian, and several mammalian stages.

Moreover, as Table 7.4 shows, human β-globin is as different from human α-globin as it is from carp α-globin. This disparity has evolved despite the fact that human α- and β-globin have shared much more similar external selective pressures, being locked in the same kind of organisms throughout evolution, than have human β-globin and carp α-globin. The result suggests that the α- and β-globin molecules have been accumulating changes independently, at roughly constant rates, regardless of the external selective circumstances of the molecule. This evidence, in turn, suggests that most of the evolutionary changes in the globin molecule have been neutral shifts among equivalent forms, of equal adaptive utility. While the rates of morphological change vary greatly among the various evolutionary lineages of vertebrates, the rates of molecular evolution all seem to have been more similar.

7.10 *The generation time effect in the molecular clock*

7.10.1 *The protein molecular clock runs relative to absolute time, not generation time*

Wilson *et al.* (1977) provided the first powerful evidence that the rate of protein evolution is constant relative to absolute time, rather than the number of generations. Their arguments are based on comparisons between pairs of species. In each pair, one of the species had a long generation time (elephant or sperm whale, for example) and the other had a short generation time (mouse or rabbit, for example). Wilson *et al.* found 12 proteins that had been sequenced in one or more pairs of a long and a short generation time species, and they were able to make 25 species-pair comparisons in all. Each comparison used the calculations of the relative rate test (Box 7.1), and the rate of evolution was estimated for the lineage leading to the short and to the long generation time species (Figure

TABLE 7.4

Amino acid differences between the α- and β-globins, for three species pairs. Adapted, by permission of the publisher, from Kimura (1983).

	Number of Amino Acid Differences
Human α versus human β	147
Carp α versus human β	149
Shark α versus shark β	150

Figure 7.8 Wilson *et al.*'s method to test for a generation time effect on the rate of protein evolution. In the figure, *a*, *b*, and *c* are the numbers of evolutionary changes in the three segments of the tree; they are estimated from the pairwise molecular differences between the species using the method of Box 7.1. The "outgroup" can be any species known to have a more distant common ancestor with the pair of species being compared. The evidence suggests that $a \approx b$ for many molecules and species pairs, whereas *a* would be less than *b* if generation time influenced evolutionary rate.

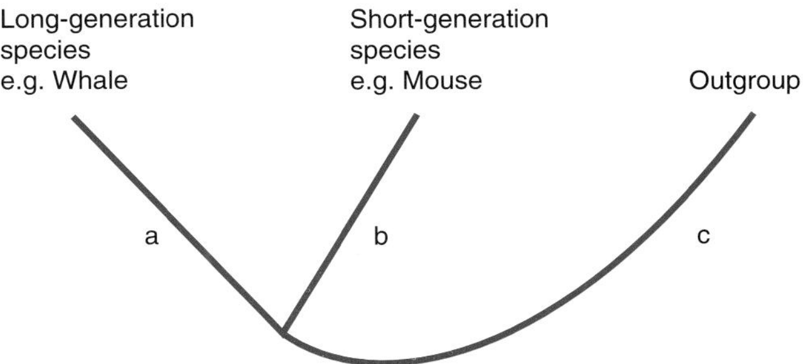

7.8). It is a reasonable assumption that the generation times of the stream of species leading up to the modern sperm whale were, on average, longer than those of the species leading to the modern mouse.

If generation times influence the rate of evolution, the amount of change in the long generation time lineage should be less (Wilson *et al.* reckoned it would be about ⅕ to ⅛ of the amount in the short generation time lineage, if the molecular clock keeps time relative to numbers of generations). In fact, the evidence suggested that the rate of evolution was about the same for the two lineages. Thus, the rate of evolution (and molecular clock) ticks at a rate proportional to absolute time, in years, rather than number of generations.

Notice that this result applies to the amino acid sequence of proteins. We will see later that silent changes in the DNA—changes that have no effect on the amino acid—may tell a different story. Some subsequent work hints at small generation time influences on the rate of protein evolution, but the influences, if they exist at all, are always much smaller for protein evolution than for silent changes in the DNA. For proteins, Wilson *et al.*'s conclusion that the molecular clock runs relative to years and not numbers of generations is reasonably well substantiated.

According to the neutral theory, the rate of evolution is equal to the neutral mutation rate. The rate will, therefore, be proportional to the time scale of the process-generating mutations. This relation is normally thought to imply generation time. Most mutations probably arise as copy errors during mitotic or meiotic replication. The mutation rate should therefore be proportional to the number of DNA replications in the germ line. Fewer such replications are produced per unit of absolute time in species with longer generations, because the number of mitotic divisions in the germ line is much less than proportional to longevity. For visible mutation, mutation rates per generation are not notice-

BOX 7.2

Slightly Deleterious Mutations in the Neutral Theory

A neutral mutation has the same fitness as the existing allele, or alleles, at its locus. Its evolution is controlled by neutral drift. However, random drift influences the evolution not only of neutral genes, but also of all genes to a greater or lesser extent. If one allele has a very different fitness from another, the two alleles' fate will mainly be governed by natural selection. (Even here, the fate of a new mutation is strongly influenced by chance; the chance that a mutation with selective advantage s will be lost by random drift is $1 - 2s$. A mutation with as high a selective advantage as 5% still has an approximately 90% chance of random loss.) As the selective difference between the two alleles decreases, neutral drift becomes more important. The other factor determining the relative importance of selection and drift is population size; random drift is more important in smaller populations.

The neutral theory, therefore, is not only relevant for neutral genes. To a rough approximation, genes with a selective advantage (or disadvantage) of $1/(2N)$ or less behave as effectively neutral genes. Thus, their evolutionary fate is influenced more by chance than selection.

Ohta realized that a number of difficulties in the neutral theory could be overcome if many mutations are slightly deleterious rather than exactly neutral. Selection does work, feebly, against slightly deleterious mutations (SDM), but their dynamics are dominated by drift. Ohta tackled three difficulties in the neutral theory: the level of heterozygosity, which is often too low for the neutral theory; the absence of a generation time effect in the molecular clock; and the possibility that the molecular clock is too variable for one driven by neutral drift (we will not discuss this factor because the evidence is too controversial). Neutralism, as we saw, was criticized for predicting levels of heterozygosity that are too high. However, the equilibrial heterozygosity is lower if most mutations are slightly deleterious, rather than exactly neutral. The theory can then account for the facts.

The argument for evolutionary rates is more complex. Under pure neutrality, the rate of evolution equals the mutation rate; it is independent of population size. But for slightly deleterious mutations, the evolutionary rate can depend on population size, because slightly deleterious mutations are more likely to behave as effectively neutral in a smaller population. As population size decreases, some of the class of slightly deleterious mutations move into the neutral zone; a smaller population has a higher effectively neutral mutation rate. When we allow for a class of slightly deleterious mutations, neutral evolution should proceed faster in smaller populations. (The effect is partly off-set by a higher total mutation rate in larger populations, because there are more individuals producing mutations; but in Ohta's model the net effect is for smaller populations to evolve faster.)

What kind of species have smaller population sizes? Ohta reasoned that species in which generation time is shorter tend to have larger populations. (Chao and Carr (1993) have recently demonstrated this relation.) She can then explain the pattern of evolutionary rates. For silent sites in DNA, most mutations are neutral and the observed higher rate of evolution in species with short generation times simply reflects their higher rate of mutation in absolute time. For amino acids in proteins, more of the mutations are deleterious; some will only be slightly deleterious. Then, where generation times are shorter (and populations larger), the slightly deleterious mutations will be influenced more by natural selection, and selected against. Where generation times are longer (and population smaller), more of the slightly deleterious mutations will behave as neutral and may drift to fixation. There is a balance between the total mutation rate in absolute time and the proportion of mutations that

ably longer in species with longer generation times (Table 2.2, p. 29). If (as seems reasonable) neutral mutations are like other mutations, neutral evolution should occur more slowly in species with longer generation times. Selectionists can, therefore, retort that neutralism, far from predicting the molecular clock, actually does not predict it. The neutralist's clock would be quite different from the clock behave as neutral. The balance, however, needs to be exact for the evolutionary rate in all species to be independent of generation time.

Certain difficulties in the neutral theory can be removed, or at least reduced, by postulating a large class of slightly deleterious mutations. The theory can still reasonably be called the neutral theory, because chance determines the fate of genes in evolution, but it is a modified version of the original theory (Figure B7.3). This modified theory can, in turn, be criticized. Gillespie, in particular, looked in detail at the population sizes required for Ohta's theory to explain the observations. Ohta qualitatively observed that the dependence of evolutionary rates on population size will shift the predictions closer to the observations. It is possible, however, to calculate exactly what values of *N* are needed. Gillespie's original papers should be consulted for the arguments, but his conclusion doubted whether neutral theory and genetic fact could be reconciled by realistic population size fluctuations.

Even if Ohta's theory does fit the facts, it still postulates a large class of slightly deleterious mutations for which no independent evidence exists. As a result, we do not know how important slightly deleterious mutations are in evolution.

In summary, Ohta suggested the neutral theory could explain the facts of protein polymorphism and the molecular clock better if a large number of slightly deleterious mutations arise. The idea is theoretically important, but its ability to explain the facts remains controversial.

Further reading: Ohta (1992) and Gillespie (1991).

Figure B7.3 (a) According to the theory of slightly deleterious mutations, there is a large class of mutations with selection coefficients that are negative, but so slightly negative that they behave as if they were neutral. (b) The theory is a slight modification of the original neutral theory, which postulated a large class of exactly neutral mutations (see Figure 7.1). The exact shape of the graphs to the left of the slightly deleterious region is unimportant. Both theories suggest that very few advantageous mutations, many neutral ones, and many deleterious ones occur; they differ in the frequencies of slightly deleterious mutations postulated.

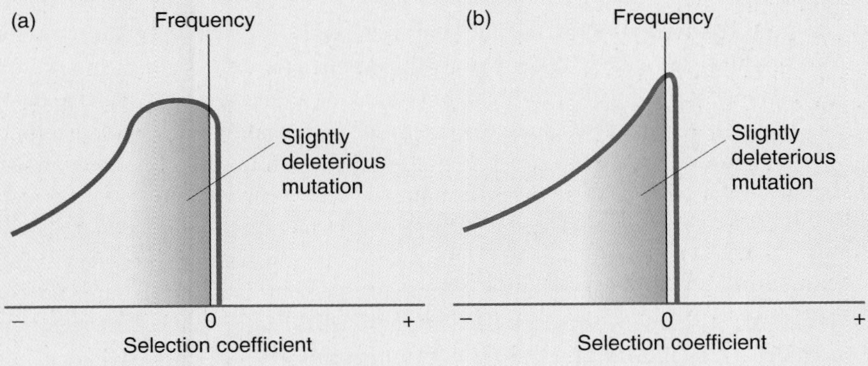

that has been found, and would tick in proportion to numbers of generations, not of years. (Box 7.2, however, considers a modified neutralist theory that can account for the absence of a generation time effect.)

Selection probably provides a better explanation than neutral drift for why the clock depends on absolute time. Under neutral evolution, the rate of evolution exactly equals the mutation rate; as a result, neutralism strongly predicts a relation between number of generations and evolutionary changes. Under selection, the rate of evolution can be controlled by the pace of environmental change, which could depend on absolute time.

Suppose, for instance, that two kinds of mammal, such as a mouse (generation time approximately 0.33 year) and an elephant (generation time approximately 33 years), were both separately evolving in relation to changes in parasites (e.g., bacteria) with short generation times. After 0.33 year, the bacterial population will have accumulated 0.33 year's worth of change, which acts as an agent of selection on the mouse population. On the other hand, after 33 years, the bacteria that parasitize elephants will have accumulated 100 times as much change, which will select 100 times as strongly on the elephants. The larger elephant generation time will, to some extent, be compensated by the stronger selection it will experience. The resulting rates of evolution in mice and elephants might be quite similar. This argument has been simplified, of course, but it suggests that natural selection can explain why the molecular clock keeps absolute, rather than generational, time.

7.10.2 *DNA sequences can now be used to test for a generation time effect in the molecular clock*

As an increasing number of DNA sequences have been analyzed, it has become possible to test for the constancy of molecular evolution in the DNA sequences. The genetic code contains 64 codons, of which 61 code for 20 amino acids (see chapter 2); the three-fold redundancy in the code means that not all base changes in the DNA cause amino acid changes in the protein. As an almost inevitable theoretical consequence, there cannot be an evolutionary molecular clock for both base changes and amino acid changes. We have seen that the clock for proteins keeps reasonably good time, although it ticks over according to absolute, not generational, time. What happens in the DNA?

The evidence suggests the presence of a generation time effect. We can distinguish the results for "silent" and "replacement" base substitutions. Silent (sometimes also called "synonymous") base changes do not result in an amino acid change in the protein; the protein is the same after a silent mutation as before. In contrast, replacement (sometimes also called "meaningful") base changes do alter the amino acid. It seems likely that silent mutations will more often be neutral than will replacement mutations.

Rodents, such as mice and rats, have shorter generation times than primates and artiodactyls (such as cows). For both silent substitutions and replacement substitutions, evolution is faster in rodents than in artiodactyls, and faster in artiodactyls than in primates. The effect is particularly marked in silent sites (Table 7.5). Silent substitutions occur more rapidly in species with shorter generation times, as the neutral theory predicted. Perhaps the silent parts of the DNA do, in fact, evolve neutrally. (We will have more to say about the rate of evolution in DNA sequences in later sections; here we are concerned only with whether they show a generation time effect.)

In conclusion, the absence of a generation time effect in the molecular clock for proteins suggests that neutral drift is not as important in their evolution as Kimura originally thought. For silent base changes in DNA, however, the clock depends on generation time and the neutral explanation is much more plausible.

TABLE 7.5

Rates of evolution in silent base sites are faster in groups with shorter generation times. Estimates have been made for various pairs of species, with each estimate being an average for a number of proteins; the number of sites is the total number of base sites (for all proteins) that have been used to estimate the rate. The divergence times, which are in millions of years, are uncertain; a range of estimates (in parentheses) have been made. Simplified from Li, Tanimura, and Sharp (1987).

Comparison	Number of Proteins	Number of Sites	Divergence	Rate ($\times 10^{-9}$ yr)	Generation Time
Primates					
Human versus chimpanzee	7	921	7 (5–10)	1.3 (0.9–1.9)	Long
Human versus orangutan	4	616	12 (10–16)	2 (1.5–2.4)	
Human versus Old World monkey	8	998	25 (20–30)	2.2 (1.8–2.8)	
Artiodactyls					
Cow versus goat	3	297	17 (12–25)	4.2 (2.9–6)	Medium
Cow/sheep versus goat	3	1027	55 (45–65)	3.5 (3–4.3)	
Rodents					
Mouse versus rat	24	3886	15 (10–30)	7.9 (3.9–11.8)	Short

7.11 *Evolutionary rate and functional constraint*

7.11.1 *Test 3: the more functionally constrained parts of proteins evolve at slower rates*

A protein contains functionally more important regions (such as the active site of an enzyme) and less important regions. It has consistently been found that the rate of evolution in the functionally more important parts of proteins is slower. Insulin, for example, is formed from a proinsulin molecule by excising a central region (Figure 7.9). The central region is then discarded, so its sequence is probably less crucial than that of the outlying parts that form the final insulin protein. The central part evolves six times more rapidly than the outlying parts.

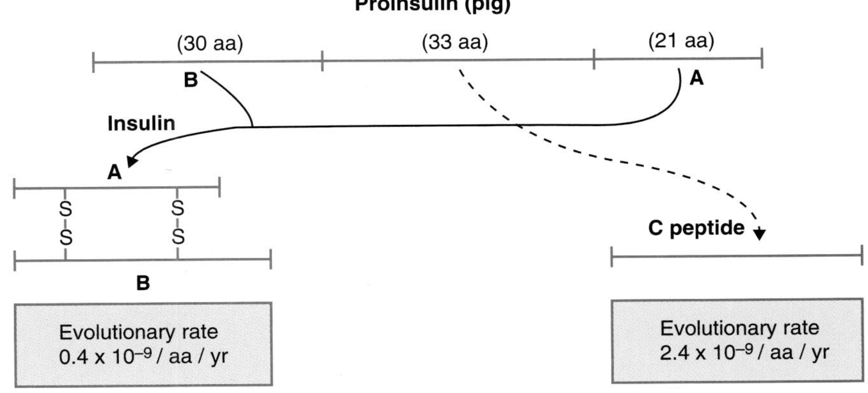

Figure 7.9 The insulin molecule is made by snipping the center out of a larger proinsulin molecule. The rate of evolution in the central part, which is discarded, is higher than that of the functional extremities. From Kimura (1983). Reprinted with the permission of Cambridge University Press, © 1983.

The same result has been found by comparing evolutionary rates in the active sites and in other regions of enzymes (or of proteins generally). The surface of a hemoglobin molecule, for example, may be functionally less important than the heme pocket, which contains the active site. The evolutionary rate is about 10 times faster in the surface region (Table 7.6).

A similar tendency may underlie differences in the rates of evolution of whole genes, or proteins. In Table 7.1 we saw that some proteins evolve faster than others. One generalization is that "housekeeping" genes, which control the basic metabolic processes of the cell, evolve slowly. In contrast, genes like the globins, fibrinopeptides, and immunoglobulins, which have more specialized functions and only operate in specific cell types, evolve more rapidly. This pattern is less clear-cut than the pattern seen within a gene for insulin and for hemoglobin, but it is possible that differing degrees of functional constraint are related to rates of evolution for different genes as a whole. A basic housekeeping gene may be more difficult to change during evolution than a gene with a more localized function.

7.11.2 *Both natural selection and neutral drift can explain the trend*

The neutral explanation for the relation between evolutionary rate and functional constraint follows. In the active site of an enzyme, an amino acid change will probably modify the enzyme's activity. Because the enzyme is relatively well adapted, the change is likely to be for the worse and may well spoil the enzyme's function. In the other parts of the molecule, the identity of an amino acid occupying a site may matter less, and a change is more likely to be neutral. The proportion of mutations that are neutral will be lower for the functionally constrained regions. As a result, if the total mutation rate is similar throughout the enzyme, the number of neutral mutations will be lower in the active site. The evolutionary rate will be lower as well. Geneticists have found a good fit between facts and prediction, and the relationship has been confirmed repeatedly. We will examine further examples involving DNA later.

What is the selective explanation for this trend? It has been argued that, if all evolution occurs by natural selection, there should either be no relation between functional constraint and evolutionary rate, or the opposite relation from what is found. Natural selection would arguably be more important in the functionally more important parts of proteins, and evolutionary rates for those regions should, therefore, be higher. The facts show that the opposite is true,

TABLE 7.6

Rates of evolution in the surface and heme pocket parts of the hemoglobin molecules. Rates are expressed as number of amino acid changes per 10^9 years. From Kimura (1983). Reprinted with the permission of Cambridge University Press, © 1983.

Region	α-globin	β-globin
Surface	1.35	2.73
Heme pocket	0.165	0.236

although, as Clarke pointed out, natural selection can explain those facts. The argument is often put in terms of a model that Fisher presented for adaptive evolution; the model illustrates why natural selection favors small changes more often than large changes (Box 7.3).

Mutations in a protein's active site will tend to have large effects; mutations in the outlying regions will have smaller effects. A change in amino acid in the active site is a virtual macromutation that will almost always make things worse. Natural selection will only rarely favor amino acid changes. However, a similar change in the less functionally constrained parts may have more chance of being a small fine-tuning improvement that would be favored by natural selection. Selection will then more often favor changes in the less constrained regions of molecules, because those parts offer more opportunities for fine-tuning. An apt analogy might be a radio, in which the more and less constrained regions of a protein are like two tuning knobs—one for fine adjustments and the other for larger movements. Once a radio station has been located, we make more use of the fine adjustment knob, to keep track with the signal. Likewise, changes outside the active site are more frequent in evolution.

Kimura (1983) rejected this argument as "completely false." It ignores the probability that a mutation, even if it is advantageous, eventually becomes fixed. The chance of fixation of a favorable mutation is, as we have seen, approximately $2s$ and, as a consequence, is lower for a mutation of smaller s. The fine-tuning mutants will, Kimura reasons, have lower selective advantage, and the larger proportion of selectively advantageous small mutations will be canceled by their higher chance of random loss.

Kimura's reasoning does not work for all cases. His argument works for a unique mutation. If a favorable mutation occurs recurrently, however, it will eventually be fixed even if its chance of survival is low each time. In that instance, the recurrent, advantageous mutations in unconstrained regions will eventually be fixed, while the recurrent disruptive changes in active sites will be rejected. The rate of evolution will be higher in the functionally less constrained parts of molecules, as is observed. Thus, the selective explanation is quite plausible, and both theories can explain the well-documented relation between functional constraint and evolutionary rate.

7.11.3 *Silent sites in the DNA evolve more rapidly than replacement sites*

The same relationships between functional constraint and evolutionary rate have been found, and the same explanations defended, for DNA as for proteins. Two properties of DNA sequences are particularly interesting: the relation between silent and replacement changes in the third codon position, and the evolutionary rate of *pseudogenes*.

A pseudogene is a region of a DNA molecule that clearly resembles the sequence of a known gene, but differs from it in some crucial respect and probably has no function. Some pseudogenes, for example, cannot be transcribed, because they lack promotors and introns.[2] Pseudogenes, once formed, are probably under

2. Promotors and introns are sequences that are needed for transcription, but are removed from the mRNA before it is translated into a protein. Some pseudogenes may have originated in mistaken reverse transcription of processed mRNA into the DNA.

BOX 7.3

Fisher's Model of Adaptive Evolution

Fisher's model has general importance, but it is convenient to introduce it at this point in the text. For almost any character, a relation exists between the form of the character and its fitness as shown in Figure B7.4. For the following discussion, it does not matter whether the graph has multiple peaks, or the detailed shape of the slope. Rather, all that matters is the presence of at least one fitness peak, with some "hills" leading up to it from all sides. At the peak, the organism is better adapted than any slightly different form. The abstract idea can apply to any kind of character—a morphological character like bone diameter, or a molecular one, such as a chemical property of an enzyme. We can assume that, in nature, the members of a species will lie somewhere near the peak, because organisms, if not perfectly adapted, are at least fairly well adapted.

Mutations will arise at random with respect to the direction of improved adaptation. A mutation is equally likely in any direction from where the species's location is; one is equally likely "up" hill as "down" hill. The important relation exists between

Figure B7.4 A general model of adaptation. For some trait (x), the fitness of an individual has an optimum at a certain value of x, and declines away from that point, creating a hill of fitness values. A mutation that changes the value of x also changes its bearer's fitness.

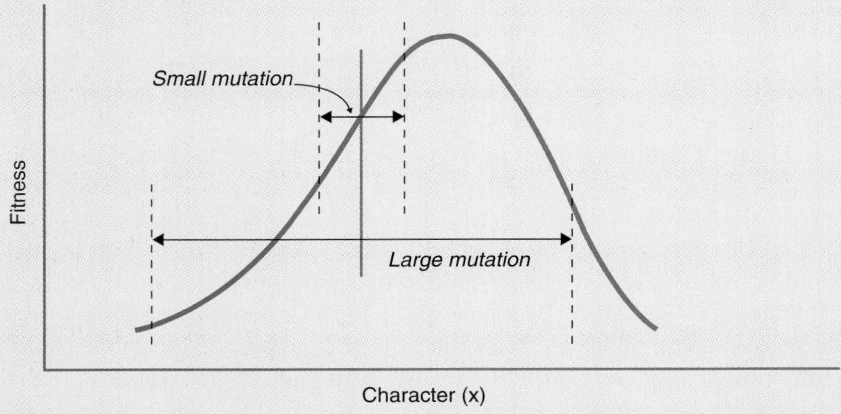

little or no constraint, and mutations will accumulate by neutral drift at the rate at which they arise. They will show pure neutral evolution in the pan-neutralist sense that all mutations are neutral.

The neutral theory predicts that pseudogenes should evolve rapidly—and they do (Table 7.7). It is tempting to use these rates of pseudogene evolution as direct estimates of the mutation rate. For instance, taking an approximate figure of one change per site per 10^9 years, we could multiply the rates for the whole genome, with a haploid content of about 3×10^9 bases in humans, to suggest that about three mutations occur on average per year (or more like 60 per generation in humans). Other pseudogenes evolve at other rates, and the evidence as a whole suggests that about 100 mutations on average appear between a parent and its offspring. However, as we will see below, some evidence also exists for the size of a mutation and the chance that it is an improvement.

Consider first a large mutation. If it is directed downhill, the mutation will make its bearer worse adapted. More interestingly, if it occurs in the uphill direction, a large enough mutation will overshoot the peak and make the organism worse adapted by taking it farther down the hill on the other side. This well-known argument shows that adaptations will not usually evolve by macromutations: if you make a large enough random change in a well-adjusted machine, the change will be for the worse.

Now consider a small mutation. If it is downhill, the mutation will again make its bearer worse adapted. If it is uphill, the mutation will probably be an improvement. The probability depends on the nearness of the species to the peak. If the species is far from the peak, quite a large mutation uphill can be favored; if it is close to the peak, the mutation must be very small.

During evolution, a species will wander up and down the hills around a peak because the peak will move as the environment changes. Fisher concluded that, in the limit of mutations of indefinitely small size, there is a ½ chance that they will be favorable. As the magnitude of the mutation increases, the chance it is favorable decreases to zero (Figure B7.5). If a character is related to its bearer's fitness as in Figure B7.4, smaller mutations are more likely to be advantageous.

Further reading: Fisher (1930), Kimura (1983), Orr and Coyne (1992).

Figure B7.5 Smaller mutations have a higher chance of being selectively advantageous. Macromutations are almost never advantageous. Most evolutionary change, therefore, is achieved by substituting mutations of small effect.

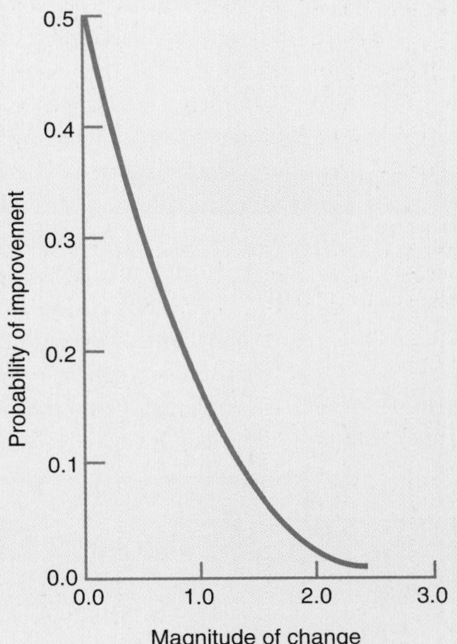

TABLE 7.7

Pseudogenes evolve at about the same rate as silent base changes. Rates are expressed in numbers of base changes per 10^9 years. The comparisons are for various genes and pseudogenes in the globin gene family. Simplified from Li, Tanimura, and Sharp (1987).

Species Pair	Divergence Time (million yr)	Evolutionary Rate	
		Pseudogenes	Silent Sites
Human versus chimpanzee	7	1.2	1.3
Human versus orangutan	15	1.0	2
Human versus rhesus monkey	25	1.5	2.2
Human versus owl monkey	35	1.6	—
Rhesus versus owl monkey	35	1.9	—
Cow versus goat	17	2.7	4.2

constraints on pseudogene evolution; if some mutations are selected against, then the total mutation rate must be higher than the evolutionary rate.

A selectionist could argue that the fast evolution of pseudogenes is due to particularly frequent fine-tuning in these molecules, but to date no one has advanced this theory. Selectionists will be more tempted to concede pseudogenes as an exception, but an exception that tells us nothing about the reason for faster evolution in other cases of reduced functional constraint. As a result, it is widely accepted that the rapid evolution of pseudogenes is due to neutral drift.

Silent base changes, which do not alter the amino acid, should be less constrained than replacement sites. Kimura had predicted, before DNA sequences were available, that silent changes would evolve more rapidly. It has now been confirmed that Kimura was correct; evolution in silent sites runs roughly five times as rapidly as evolution in replacement sites (Table 7.8). A few exceptions are known, including the antigen recognition site of the HLA genes (Figure 7.10). The debate over whether the usual pattern of rapid change in silent sites involves solely neutral drift or frequent selected fine-tuning has not been conclusively settled, but most people accept that neutral drift is the main force, if only because it is difficult to imagine why silent changes should be selected for so often.

If silent codon positions were selectively unconstrained, they should show pan-neutral evolution. They should then evolve in a different manner from proteins. Different proteins, as we have seen, evolve at different rates, which neutralists explain by different degrees of constraint (and selectionists by different intensities of selection). If the silent positions in the DNA encoding those same proteins are unconstrained, they should not show different evolutionary rates. Instead, they should evolve at the rate of a "universal" molecular clock (i.e., at their total mutation rate). This idea can be tested by comparing the rates of change in the silent sites of different genes to see whether evolutionary rates remain constant in silent sites. Initially, Miyata and others suggested that the

TABLE 7.8

Rates of evolution for meaningful (i.e., amino acid changing) and silent base changes in various genes. Rates are expressed as inferred number of base changes per 10^9 years. Simplified from Li, Wu, and Luo (1985).

Gene	Meaningful Rate	Silent Rate
β2-Microglobulin	1.21	11.77
Albumin	0.92	6.72
Histone H4	0.027	6.13
Immunoglobulin V_H	1.07	5.67
α-globin	0.56	3.94
β-globin	0.87	2.96
Parathyroid hormone	0.44	1.73
Average (38 proteins)	0.88	4.65

Note: The absolute numbers are imperfect because of difficulties in the fossil dating used in the rate calculations; comparisons between genes are meaningful, however.

Figure 7.10 Evolutionary rates of silent and amino acid replacing base changes in the human (HLA) and mice (H-2) major histocompatibility loci. The *y*-axis is the replacement minus the silent rate. The rate of silent changes is higher for most genes, which would give a negative value on the *y*-axis. This graph for the major histocompatibility loci breaks the regions of the molecules down into two parts: the antigen-binding site (light blue) and the rest of the molecule (dark blue). The regions outside the antigen-binding site show the normal pattern of evolutionary rates. But in the antigen-binding site, the rate of amino acid changing substitutions is higher than the rate of silent changes. Reprinted with permission from *Nature* from Potts and Wakeland (1990), using the data of Hughes and Nei (1988, 1989). Copyright 1988 Macmillan Magazines Limited.

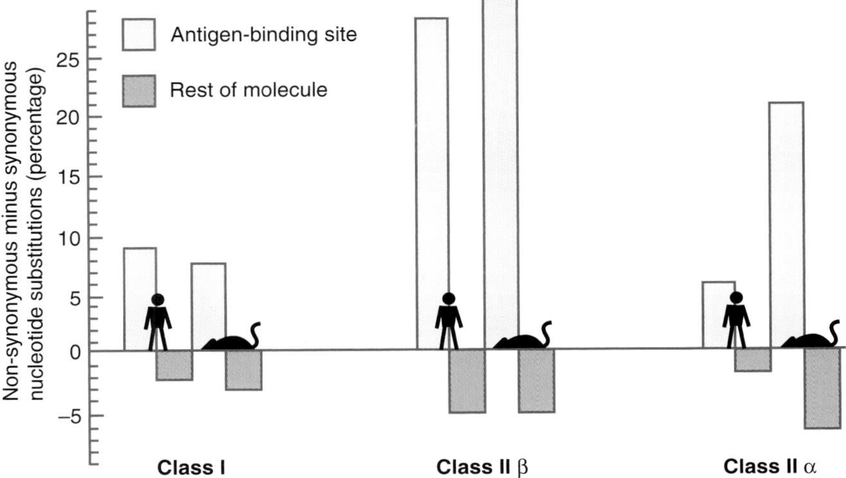

silent rates were constant, but more recent evidence has not supported them (see right-hand column of Table 7.8, where the rates vary almost 10-fold between different genes). Evolutionary rates are less variable in silent sites than in replacement sites, suggesting a lower degree of constraint, but they vary nonetheless. Also, evidence showing that evolutionary rates for a single gene can vary between different lineages (such as that provided by Bulmer *et al.* [1991]), also contradicts the idea of pan-neutral evolution.

Other, more direct, evidence gathered about selective constraints in silent base sites involves two kinds. The first concerns the relative rates of evolution in pseudogenes and silent positions. It was initially thought that pseudogenes evolve twice as fast as silent codon positions, but more recent evidence (Table 7.7) suggests that evolution occurs just as rapidly in silent codon positions as it does in pseudogenes. We can provisionally conclude that pseudogenes and silent sites evolve at about the same rate, although future evidence may suggest otherwise.

The question of these evolution rates is interesting, because when it was thought that pseudogenes evolve faster than any other DNA, it was tempting to argue that their evolution was completely unconstrained. If true, their rate of evolution would then be a direct estimate of the total mutation rate. We now think that pseudogenes evolve at the same rate as silent base changes, and it is argued that evolution is probably constrained even in pseudogenes to some extent. In that case, they should not be used to estimate the total mutation rate. The argument works because we have independent evidence that evolution is constrained in silent sites. This evidence is the topic of the next section.

7.11.4 *Codon usages are biased*

Biases in codon usage provide the second kind of evidence for constraints on silent sites. If all of the silent alternative codons were functionally equivalent, we should expect only random variation in the frequency of those codons in a species. In fact, consistent biases appear. Table 7.9 gives evidence for the six arginine codons in *Homo sapiens*, *Drosophila*, and *Escherichia coli*, from the work of Grantham. AGG is the most common in humans, but the rarest in *Drosophila* and *E. coli*; CGC is the most common in *Drosophila*, and CGU in *E. coli*. The relative rarity of, for example, AGA compared with CGU in *E. coli* suggests that mutations from CGU to AGA tend to be selected against. The changes are not all neutral. Pan-neutralism is not valid in this case. (Notice again that we are concerned only with the possibility that some codons, out of a synonymous set, are deleterious, and selected against; the nonrandom frequency distributions could then be explained. This explanation does not argue against Kimura's neutral theory. Logically, the nonrandom frequency distributions could arise because of positive selection for certain codons. The neutral theory would then be wrong, but that is another matter.)

How could selection discriminate among silent codons? The mechanism is not known, although two suggestions have been put forth. Both possibilities are theoretical only—it has not been shown that natural selection does operate on silent codon changes.

The first suggestion is that the nucleotide sequence controls the secondary structure of the DNA molecule; changes in nucleotides might then influence the molecular shape, which could make a difference to the organism's fitness. Silent substitutions would be as likely to influence structure as replacement substitutions, and selection would work on both for the same reason.

TABLE 7.9

Frequencies of six arginine codons in the DNA of three species. The table gives the percentages of arginine amino acids that are encoded by each of the six codons in various numbers of genes in the species. Simplified from Grantham, Perrin, and Mouchiroud (1986).

Codon	Human	*Drosophila*	*E. coli*
Arginine			
AGA	22%	10%	1%
AGG	23%	6%	1%
CGA	10%	8%	4%
CGC	22%	49%	39%
CGG	14%	9%	4%
CGU	9%	18%	49%
Total number of arginine codons	2403	506	149
Total number of genes	195	46	149

The second factor is transfer RNA. The different silent codons use, to some extent, different tRNA molecules. It has been observed, in yeast and in *E. coli*, that the frequency of use of the codons in a set of synonyms is correlated with the tRNA abundances in the cell (Figure 7.11). We should expect the relation if the frequencies of codons were set by some other factor; the tRNA abundances would then adjust (by natural selection) to the quantity needed. The relation could also arise if the tRNA abundances were fixed by some other constraining factor. In that case, the mutations among the silent codons would not be neutral, as mutations to codons whose tRNA was in short supply would be selected against relative to those with abundant tRNA.

Directional mutation pressure is another possible explanation for biases in codon usage. A tendency for A and T bases to mutate more often to G or C than do G and C to A or T would result in a build-up of GC bases. Codon biases would result automatically. It has, in fact, been suggested that GC is more stable than AT, and favored by mutation pressure. The task is then to apply these chemical mechanisms to explain the differences between species in their codon biases (Table 7.9).

In summary, the functionally more important parts of proteins (such as the active sites of enzymes) and DNA (such as replacement rather than silent sites) evolve more slowly than the less important parts. Two explanations have been suggested for the trend. The first theory is that it may arise because a higher proportion of mutations are neutral in the less important parts; this theory is the generally accepted explanation for the rapid rate of evolution for silent base changes. However, even these sites are not completely unconstrained—different taxonomic groups show different biases in codon usage and have different rates

Figure 7.11 Relative frequencies of codons match tRNA abundances. (a) The light blue columns (above) are the relative frequencies of six leucine codons in *E. coli;* the dark blue columns (below) are the relative frequencies of the corresponding tRNA molecules in the cell. The two sets of codons joined by a + sign are recognized by a single tRNA molecule. (b) Same relation, but in yeast. Notice the different bias in codon usage in the two species, which illustrates the point of Table 7.9. From Kimura (1983). Reprinted with the permission of Cambridge University Press, © 1983.

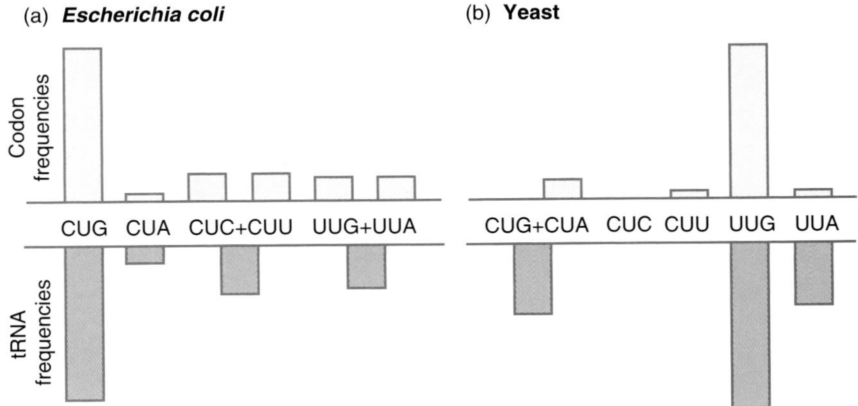

of evolution in silent sites. The second explanation for the trend is that natural selection works more readily with small changes, and a change by one amino acid is more likely to be a small change if it is found in a less important part of a molecule than if it is located in a more important part. Biologists do not agree as to whether selection or drift is the reason why the functionally important parts of proteins evolve slowly.

7.12 *DNA sequences provide strong evidence for natural selection on protein structure*

In chapter 4, when we considered the evidence for biological variation, we noticed that many DNA sequence variants can be uncovered if a single gel electrophoretic class of a protein is sequenced at the DNA level (section 4.5, p.80). This observation has important implications for molecular evolution. At the alcohol dehydrogenase (*Adh*) locus in the fruitfly, two electrophoretic classes (fast and slow) are present. When Kreitman sequenced proteins from the two classes, he found that the amino acid sequence of each was uniform, but many variants were observed in silent third codon positions. The combination of a fixed amino acid sequence and variable silent sites provides, as Lewontin has emphasized, evidence that natural selection operates to maintain the enzyme structure.

Two reasons have been suggested as to why the enzyme sequence, at the amino acid level, should be fixed within each gel electrophoretic class. In the "identity by descent" hypothesis, all copies of the Adh_f allele, for example, may be descended from an ancestral mutation, which had that sequence and has been passively passed from generation to generation. Eventually another mutation may arise and the Adh_f allele will then become two alleles; the constant sequence within a population merely indicates that not enough time has passed for such a mutation to occur. Alternatively, the Adh_f copies may all have the same sequence because the sequence is maintained by natural selection; when a mutation arises, selection removes it.

The observed variability distinguishes between these two hypotheses. The variability in the silent base sites means that there has been time for mutations to arise in the molecule. If mutations have arisen in silent sites, they will surely have occurred in replacement sites as well. Therefore, we can reason that the identity in amino acid sequence is unlikely to be identity by descent. Mutations in replacement sites have presumably not been retained because natural selection eliminated them.

If the Adh_f allele has been found to be fixed for one DNA sequence at all sites—both replacement and silent—we could not know whether the uniformity was due to selection or identity by descent. We would be in the same position as when we had only the gel electrophoretic evidence. Perhaps the fixity meant simply that no mutations had occurred. The DNA sequences thus provide evidence for selection that could not have been obtained with amino acid sequences alone.

The absence of amino acid sequence variation within the Adh_f (and Adh_s) allelic class is particularly striking because 30% of the enzyme consists of isoleucine and valine, which are biochemically very similar (and electrophoretically indistinguishable). A neutralist might have predicted that some of the valines could be changed to isoleucines, or vice versa. The only amino acid sequence variant is the one that causes the AdH_t/AdH_s polymorphism. Because abundant evidence exists that natural selection operates on that change, Lewontin concluded that there are actually no neutral amino acid sites in the 255 amino acid alcohol dehydrogenase enzyme of the fruitfly. Interestingly, that conclusion means that we could almost construct Figure 7.1 for alcohol dehydrogenase at the amino acid level; the resulting graph would have the appearance of Figure 7.1a. Natural selection is powerfully maintaining the amino acid sequence, while the silent base changes are probably neutral.

7.13 *Test 4: are the rates of evolutionary change of molecules correlated with their heterozygosities?*

For the fourth kind of test we return to the two phenomena we discussed in the first test: the degree of polymorphism (or heterozygosity) within species and the rate of evolutionary change between species. In this test, we will look not at their absolute rates, but at the relation between the two. The neutral theory predicts the two rates will be correlated. The same factor—the neutral mutation rate—explains both polymorphism and the rate of change, in the neutral theory. All polymorphism is "transient," and both polymorphism and the substitution of one gene for another are manifestations of the same process—the neutral drift of genes up or down in frequency. If we look at different genes (or different parts of one gene), we find some have higher heterozygosities than others, which the neutral theory explains by their having a higher neutral mutation rate. The same factor also produces a higher rate of change. We should, therefore, find a positive correlation between the rate of evolutionary change and the heterozygosity of different regions of the DNA.

If natural selection dominates molecular evolution, a positive correlation would not necessarily be expected. Natural selection can produce transient polymorphism. It takes time for selection to substitute one allele for another (although not as long as it takes with drift) and the population will be polymorphic while the substitution proceeds. The process has been seen in action in the peppered moth and some other examples, but it is generally supposed to be rare. If approximately 10,000 gene loci in a population are polymorphic, it is unlikely that substitutional events of the peppered moth kind are occurring at more than a minority of them in an average population.

This intuition could be wrong, but it seems more likely that most selected polymorphisms are not transient but stable, being due to frequency dependence, heterozygous advantage, or some other selective process. If selected polymorphisms are not transient, then different processes control evolutionary change and polymorphism. Occasional rounds of directional selection cause change, but frequency dependence (or whatever) causes polymorphism. If we look at different genes (or regions of one gene), the two can be correlated only by coincidence. Genes with high evolutionary rates will be those that experience strong and persistent directional selection; we have no particular reason to suppose they will also be the ones at which selection favors a high heterozygosity. Thus, there is now less reason to expect a positive correlation between the evolutionary rate and polymorphism.

What do the facts show? The relation was first examined in gel electrophoretic data for proteins, and broadly supported the neutral theory. That test was relatively coarse, however, and an expanding program of research is now making better controlled comparisons in the DNA.

The first important study of this kind was performed by McDonald and Kreitman in 1991. They did not compare different genes, but silent and replacement events within one gene, alcohol dehydrogenase (*Adh*) in two closely related species of fruitfly (*Drosophila*). McDonald and Kreitman developed a more exact version of the argument given above. For the two species, they measured the number of (fixed) differences for replacement and for silent nucleotide changes; they divided one by the other to give a ratio. They also measured the polymorphism within each species for replacement and silent sites and expressed them as a ratio. The two ratios should be equal if all the changes are neutral, for the following reason. The rate of neutral evolution equals the mutation rate (section 6.4, p.138), and the ratio of species differences for replacement sites/species differences for silent sites will, therefore, be the ratio of the neutral replacement change mutation rate/neutral silent change mutation rate. (The neutral replacement change mutation rate will arguably be the lower of the two, because it is less likely that any one replacement mutation will be neutral than any one silent mutation.) The amount of polymorphism under neutral drift is given by a more complicated formula (section 6.6, p. 146), but if the formula for the heterozygosity for neutral replacement changes is divided by that for the heterozygosity for silent changes, everything cancels except the respective mutation rates. The predicted replacement:silent ratio for the heterozygosities within species is then the same as for the number of differences between species.

The virtue of the test carried out by McDonald and Kreitman is that it is so well controlled. Many factors can influence both polymorphism and evolutionary rates, but all those factors should operate in the same way if we concentrate on a comparison for a single gene. The factors cancel out in the comparison and can be ignored. We can, therefore, test the neutral theory by seeing whether, in one gene in a pair of species, the ratio of evolutionary rates for replacement and silent sites equals the ratio of polymorphisms for replacement and silent sites.

The first study, for *Adh*, contradicted the neutral prediction. Too many replacement changes occurred between the species relative to the polymorphisms within the species. McDonald and Kreitman's explanation was that natural selection had favored certain amino acid changes in the evolution of the two species. Since their work, the same ratios have been measured in a number of other genes, but the results do not point to any simple conclusion (Figure 7.12). In three of the seven genes in Figure 7.12, replacement changes are more frequent between the species; in another three genes, the ratios are about the same (and in one, the gene *zeste*, the result is ambiguous). The results suggest that some of the genes may be evolving more in the neutral manner and others showing more of a deviation from it. It would take further work to establish this possibility more definitively. If we did find differences in the evolutionary modes between genes, it would raise further questions. For now, the main point is that the relation between evolutionary rates and levels of polymorphism offers a promising

Figure 7.12 Ratio of replacement to silent changes, for differences among species and for polymorphisms within a species, in seven genes in fruitflies. Each column shows the percentage of changes that cause an amino acid replacement (dark blue part) or that are silent (light blue part). There are two columns for each gene, one labeled P for sites that are polymorphic within a species, one labeled F for sites that are fixed for a different nucleotide in two different species. The neutral theory predicts that both ratios should be the same. In *boss*, *period*, and *yolk protein 2*, the ratios are equal; in *alcohol dehydrogenase*, *glucose 6 phosphate dehydrogenase*, and *jingwei*, the ratios are different; *zeste* is ambiguous. Slightly adapted from Brookfield and Sharp (1994), compiled from seven sources.

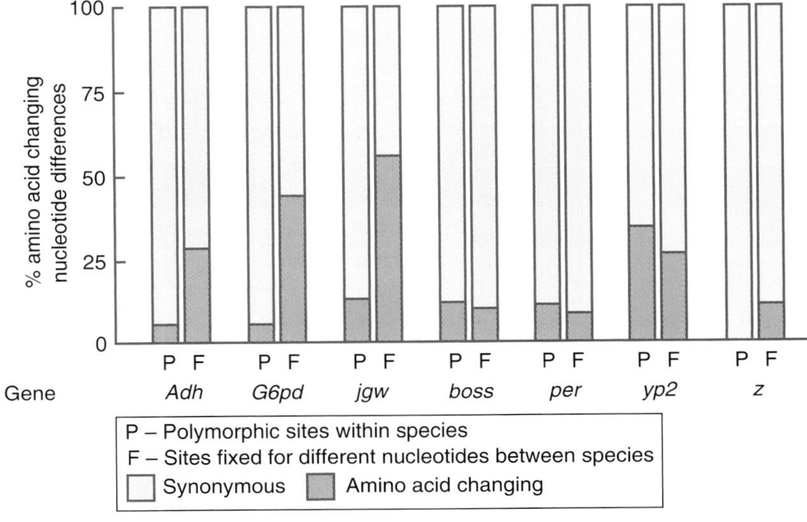

line of research. At this time, the evidence does not point to any simple conclusion—but that may be because no simple conclusion exists.

7.14 *The analysis of DNA sequences is only the most recent stage in a long controversy*

The controversy between the selective and the neutral theories has moved through several stages. The first stage was concerned with the amino acid sequences of proteins. Kimura suggested that the high rate, and constancy, of protein evolution, and high levels of heterozygosity, could not be explained by natural selection. It seems likely, however, that natural selection could explain these observations. Moreover, further evidence suggested that the levels of heterozygosity are too low for the neutral theory; evidence has also suggested that the molecular clock may be more erratic than the neutral theory predicts for proteins and that it ticks according to generational, rather than absolute, time.

More recently, the availability of base sequences for DNA has added another dimension to these tests. DNA contains both silent and replacement base positions. It seemed likely, or at least possible, that silent base changes—which do not alter the amino acid—would be neutral. Silent changes should have a higher rate of evolution than changes at replacement sites, and it has been confirmed that they do (with a few interesting exceptions, such as the HLA genes). Moreover, the evolutionary rate for silent sites should be proportional to generation time; the evidence suggests it is. Silent sites are not pan-neutral in their evolution, however. Constraints have also been suggested by the different patterns silent sites show in different taxonomic groups. The DNA sequence has also provided strong evidence that selection maintains the amino acid sequence of enzymes. New, more subtle tests have recently been developed for the pattern of heterozygosities and evolutionary rates in different regions of the same gene.

Thus, the controversy is not settled. However, it is unlikely that neutral drift is as important, and selection as unimportant, in protein evolution as Kimura originally suggested. His case against selection has been rebutted, and the facts do not comfortably fit the neutral predictions. As evidence for DNA sequences has accumulated, the controversy has increasingly moved away from simple generalizations about all molecular evolution, and toward understanding which kinds, or regions, of molecules evolve in which mode. For silent sites in the DNA, the neutral theory is widely accepted; the theory made clear predictions, many years before the facts were discovered, and the predictions have turned out correct. Ample room remains for disagreement about which molecular evolutionary changes were driven by selection, and which by neutral drift.

SUMMARY

1. The neutral theory of molecular evolution suggests that molecular evolution is mainly due to neutral drift. The mutations that have been substituted in evolution were selectively neutral with respect to the genes they replaced. Alternatively, molecular evolution may be mainly driven by natural selection.
2. Four main observations were originally interpreted in favor of the neutral theory: molecular evolution has a rapid rate, its rate has a clock-like constancy, it is more rapid in functionally less constrained parts of molecules, and natural populations are highly polymorphic.
3. Kimura argued that the high rate of evolution, and the high degree of variability of proteins, would, if caused by natural selection, impose a high genetic load.
4. Neutral drift can drive high rates of evolution and maintain high levels of variability, without imposing a genetic load.
5. The constant rate of molecular evolution gives rise to a "molecular clock."
6. Neutral drift should drive evolution at a stochastically constant rate. Kimura pointed to the contrast between uneven rates of morphological evolution and the constant rate of molecular evolution and argued that natural selection would not drive molecular evolution at a constant rate.
7. The molecular clock for proteins ticks over according to absolute time rather than generational time. For silent changes in DNA, lineages with shorter generation times probably evolve faster. Neutral drift should cause the molecular clock to run according to generational, not absolute, time.
8. Selection can operate without producing impossible genetic loads, and Kimura's original case for the neutral theory is no longer convincing.

9. The neutral theory explains the higher evolutionary rate of functionally less constrained regions of proteins by the greater chance that a mutation there will be neutral.
10. Selectionists explain the higher evolutionary rate of functionally less constrained regions of proteins by the greater chance that a mutation there will be a small, rather than a large, change.
11. The four properties observed in the evolution of proteins (rates of evolution, constancy of rates of evolution, levels of variability, and relation between functional constraint and evolutionary rate) have also been seen in DNA.
12. Pseudogenes and silent changes in third codon positions may be relatively functionally unconstrained. These parts of the DNA evolve faster than do the first two positions in codons and meaningful third base changes. Neutralists attribute this high rate of evolution to enhanced neutral drift.
13. For amino acids encoded by more than one codon, consistent biases appear in the frequencies of the codons. Changes between the silent codons are, therefore, not completely unconstrained.
14. The neutral theory predicts a positive relation between the degree of variability of a molecule and its rate of evolution. Genes, and parts of genes, that are more polymorphic than others are predicted to evolve at a higher rate. A comparison of the ratio of silent to meaningful nucleotide changes between and within a species contradicts the neutral prediction for some genes such as *Adh* but supports it for others.

FURTHER READING

Kimura (1983) is the principal reference on the neutral theory. He wrote shorter introductions as well (e.g., 1979, 1991). Kreitman and Akashi (1995) and Gillespie (1991) are modern, and Lewontin (1974) classical, reviews. Li and Grauer (1991, ch. 4) introduce some of the material. The volumes edited by Golding (1994) and Selander *et al.* (1991) contain relevant papers. See also Nei (1987). Population sizes and direct measures of fitnesses in arguably representative samples of genes were two topics touched upon briefly in section 7.1. Fisher and Ford (1947) is a classic attempt to show that population sizes were too large for drift to account for one observed case of evolution. See Wright (vol. 4 of 1969–1978). The work of Dykhuizen measures fitnesses of genetic variants directly under lab conditions; see Dykhuizen and Dean (1990) and the introduction by Lenski (1993), which contains further references.

Test 1. For the general idea of genetic loads, and the cost of selection, see Crow's entry in Keller and Lloyd (1992), Crow (1993), and the population genetics texts, such as Crow and Kimura (1970) or Wallace (1981). Kimura (1983) and Lewontin (1974) deal with these topics in relation to neutralism. Haldane (1957) is the classic paper on the cost of selection. Nevo (1988) reviews the natural levels of polymorphism detected by gel electrophoresis. On the controversy about the cause of polymorphism, see Wright (vol. 4 of 1969–1978), Endler (1986), and Koehn and Hiblish (1987). Many variants may be neutral in some environments but selected in others; see Hartl *et al.* (1985) and Watts (1991).

Test 2. This is a potentially important test, but many earlier attempts were unsettled by Bulmer (1988b). See Gillespie (1991) for recent estimates. Scherer (1990) reviews the molecular clock. Nichol *et al.*'s (1993) study of evolution in the vesicular stomatitis virus seems to show highly variable, non-neutral rates of evolution; see also Gillespie's (1993) commentary on it. The presence or absence of generation time effects is not finally settled in the literature. The effect in Table 7.5 employed geological dates, but the same trend has been found using the relative rate test (Bulmer *et al.* 1991, Ohta 1993). Ohta (1993) also considers another case in which the clock is often wrong—changes in duplicated genetic descendants from their common unduplicate ancestor.

Test 3. The male determining gene on the mammalian Y-chromosome (Whitfield *et al.* 1993, Tucker and Lundrigan 1993) is another recent example of excess replacement substitutions, like those shown in Figure 7.12. On the inference about alcohol dehydrogenase in fruitflies, see Lewontin (1985a,b, 1986) and Kreitman (1983). On biases in codon usage, see Bulmer (1988a), Sharp et al. (1995), and the general references given above.

Test 4. On the relation between polymorphism and evolutionary rate, see Brookfield and Sharp (1994) for a summary of, and references to, recent work, and Maynard Smith (1994) for a caveat. Hudson *et al.* (1987) describe a related test. Hudson (1994) reviews work on why levels of variation are lower in regions of low recombination rate.

STUDY AND REVIEW QUESTIONS

1. Draw the frequency distribution for the selection coefficients of the genetic variants at a locus, according to the neutral and the selectionist theories of molecular evolution.
2. Why is the neutral theory confined to *molecular* evolution, rather than being applied to all evolution?
3. Calculate the genetic load of the following population.

genotype	*AA*	*Aa*	*aa*
fitness	1	0.9	0.8
frequency	0.4	0.4	0.2

4. Review the arguments for and against the idea that molecular evolution occurs too rapidly to have been driven by selection.
5. Explain the relation between the degree of functional constraint on a molecule (or region of a molecule) and its rate of evolution by (a) the neutral theory, and (b) natural selection.
6. Do silent sites show pan-neutral evolution? What evidence bears upon the answer?
7. What is the formula for the ratio of

$$\frac{\text{average heterozygosity for silent sites}}{\text{average heterozygosity for replacement sites}}$$

if all evolution occurs by neutral drift? What is the ratio for rates of neutral evolution at silent and replacement sites? How can these formulas be used to distinguish whether a molecule has evolved purely by drift, or whether selection has also operated?

Two-Locus and Multi-Locus Population Genetics

We begin with an example of a character that is controlled by a multi-locus genotype. The set of genes that an individual inherits from one of its parents form a "haplotype," and the theory of population genetics for multi-locus systems traces haplotype frequencies through generations. This chapter has two main purposes. First, it introduces the population genetic theory for multi-locus systems, and the distinctive concepts that apply to it but not to one-locus models; we examine the multiple-locus concepts of linkage disequilibrium (together with its causes), recombination, and multiple-peaked fitness surfaces. Second, it describes the conditions under which single-locus models of evolution are inadequate and multi-locus models are necessary—the main condition is the existence of linkage disequilibrium.

8.1 *Mimicry in* Papilio *is controlled by more than one genetic locus*

The characters with which we dealt in earlier chapters were controlled by single genetic loci. Enzymes, such as alcohol dehydrogenase, are encoded by a single gene, and it is not much of a simplification to treat the polymorphism in the peppered moth as a set of genotypes at one locus. We will now move on to consider evolutionary changes at more than one locus.

The first example concerns a multi-locus polymorphism, and we can lead into it via a similar polymorphism that is controlled by a single locus. Both examples come from the same attractive group of butterflies called swallowtails. The swallowtails have a global distribution, and *Papilio* is the largest genus of them; their most striking characteristic is a "tail" on the hindwing. They come in many colors—gorgeous greens, subtle shades of reds and orange, and marbled patterns in white and gray—but the most common type has stripes of black and yellow. The North American tiger swallowtail *Papilio glaucus* is easy to recognize by its tiger stripes, as it flutters through woodland lanes or humid valleys. More accurately, *most* tiger swallowtails are easy to recognize in this way. In part of the species' range (roughly, to the southeast of a line from Massachusetts to south Minnesota and from east Colorado to the Gulf Coast), the standard form of the tiger swallowtail lives alongside another form of the same species. This second form is black, with red spots on its hindwings, and is called *nigra*; it is only found in females. The *nigra* form is not poisonous, but mimics another species, the pipevine swallowtail *Battus philenor,* which is poisonous. The *nigra*'s geographic distribution fits that of the pipevine swallowtail, and *nigra* is well protected there from predatory birds that have learned by stomach-churning experience not to eat butterflies that resemble pipevine swallowtails. The tiger swallowtail, therefore, has a *mimetic polymorphism.* It has both the standard non-mimetic tiger morph of yellow and black stripes, and a black mimetic morph.

The polymorphism of the tiger swallowtail *P. glaucus* looks almost simple when compared with the amazing array of female forms in the species *Papilio memnon* (Color Plate 2). *P. memnon* lives in the Malay Archipelago and Indonesia. Its male is non-mimetic, although its color is deep blue rather than yellow and black stripes. However, instead of one mimetic female form, *P. memnon* females come in a seemingly infinite variety. Their forewings show different geometric patterns of black and white; their hindwings, as well as varying in shape, can be colored in yellow, orange, or blood-red, and may or may not have a bright white spot; some have tails, others do not; the abdomen varies in color; and a spot at the butterfly's "shoulder" (i.e., at the base of the forewing near the head) called the epaulette, may be present in various shades of red.

Clarke and Sheppard, who have studied the species, suggest that each female form mimics a different model (Color Plate 2 shows six examples; notice that three of them have tails and three do not). Their evidence is not strong, as it comes only from the geographic ranges of mimic and model, and from superficial similarity of appearance (which is not exact in all cases). Good evidence for mimicry requires experimentally showing that birds that have learned to avoid the model will also then avoid the mimic; this research has been performed for the *nigra* form of *P. glaucus*, but not for *P. memnon.* However, we can accept as a working hypothesis that the apparently mimetic morphs of *P. memnon* are, indeed, mimetic. (*P. memnon* has other, non-mimetic forms as well, but they are not essential to the discussion here.)

Clarke and Sheppard were interested in the genetic control of this complex mimetic polymorphism. Crosses between the various morphs initially suggested the existence of a single genetic locus with many alleles. When two forms are crossed, the offspring usually either all resemble one of the parents, or contain a mixture of the two parental types, as will happen with one locus and simple dominance relations among alleles. For instance, if one morph has genotype A_1A_1, another A_1A_2, and A_2 is dominant to A_1, then an $A_1A_1 \times A_1A_2$ cross produces the same two classes of offspring (A_1A_1 and A_1A_2) as were present in the parents.

Soon the genetic story grew even more complicated. In addition to the mimetic and non-mimetic morphs of *Papilio memnon*, all of which exist in reasonable frequencies in nature, some much rarer morphs have been found. An example, in Java, is the rare morph called *anura* (Color Plate 2m). A specimen found in Borneo was sent to Clarke and Sheppard in Liverpool. When it was crossed with a known *P. memnon* morph, it behaved like another allelic form of the mimicry "locus." A closer look at *anura* suggests a different interpretation. *Anura*'s morphology mixes patterns from two of the common morphs—it has the wing color pattern of the morph *achates* (Color Plate 2g–i), but it lacks *achates*' tail.

Clarke and Sheppard's interpretation is that *anura* is not actually an allelic variant, but a recombinant; they also decided that the mimetic patterns of *P.*

memnon are not controlled by one locus but by a whole set of loci. If *anura* is a recombinant, then there must be at least one locus (call it T) controlling the presence (allele T_+) or absence (T_-) of a tail and at least one other locus (C) controlling the color patterns (C_1 for *achates*, and C_2, C_3, . . . alleles for other color morphs). In this case, *achates* would have a genotype consisting of one or two sets (depending on whether the alleles are dominant) of the two-locus genotype T_+C_1, and *anura* would have T_-C_1, after recombination between a tailless morph and *achates*. The loci in question are so tightly linked that these recombinants practically never arise in the lab—which is why the different multi-locus genotypes appear, when crossed, to segregate like single-locus genotypes. We can predict that if more than one locus really is involved, a sufficiently large number of crosses should be able to break one of the "alleles" (such as the *anura* "allele") into several real combinations of alleles at several loci. Darlington and Mather defined a set of genes that are so tightly linked that they behave like a single locus in a breeding experiment as a *supergene*.

From *anura* alone, it seemed that there must be at least two loci controlling the mimetic polymorphism of *P. memnon*; but other rare types have also been found. Some, for example, combine the forewing color of one morph and the hindwing pattern of another, suggesting that separate loci control the color of forewings and hindwings. When all the inferred recombinants are considered together, there appear to be at least five loci in the mimicry supergene: T, W, F, E, and B. These loci control, respectively, presence or absence of tail, hindwing patterns, forewing patterns, epaulette color, and body color. Thus, *anura* is a recombinant between the T locus and the other four loci. The common morphs, which mimic natural models, should each consist of a particular set of alleles at the five loci. The morph mimicking model species number 1, for example, might have genotype $T_+\,W_1F_1E_1B_1/T_+\,W_1F_1E_1B_1$, and another morph (mimicking a second model) might have $T_-\,W_2F_2E_2B_2/T_-\,W_2F_2E_2B_2$ or $T_-\,W_2F_2E_2B_2/T_+\,W_1F_1E_1B_1$. The recombinant genotypes such as $T_+\,W_1F_1E_2B_2$ do not exist naturally, except as very rare forms like *anura*.

The point to remember is that each of the morphs of *Papilio memnon* is thought to be controlled by a multi-locus genotype. This mechanism does not operate like the camouflage polymorphism in the peppered moth (section 5.7, p. 103), in which the different morphs are controlled by genotypes at one locus. A whole set of one-locus genotypes is needed to produce each of the swallowtail butterfly morphs. The genetics are not definitively confirmed, but we can use the idea for purposes of discussion.

8.2 *The genotypes at different loci in* Papilio memnon *are coadapted*

How will natural selection act on a rare recombinant morph of *Papilio memnon*, such as *anura* in Java? Successful mimicry requires as complete a resemblance as possible between a mimic and its model. A potential mimic that mixes the patterns needed to resemble two species will mimic neither as successfully as an organism that resembles one model in all respects. In fact, it will probably be selected against. The *anura* butterfly has the color pattern of *achates*, but will genotype contains two genes from the two parents. If the A and B loci are on the same chromosome, each haplotype represents a gene combination on a chromosome. Haplotypes can also be specified for loci on different chromosomes. The frequency of a haplotype in a population can be counted as the number of gametes bearing a particular combination of genes.

A haplotype can be specified for any number of loci. We will mainly discuss two-locus haplotypes, but the haplotypes in the *Papilio memnon* example had five loci, and the mimetic patterns of *Heliconius* are controlled by 12 or 15 loci. As we will see, to understand the evolution of haplotype frequencies, we need some concepts that do not exist for gene frequencies. Two-locus population genetics is, therefore, not simply a doubled-up version of single-locus population genetics.

8.5 *The frequencies of haplotypes may or may not be in linkage equilibrium*

We can begin by asking a question like the one that led to the Hardy–Weinberg theorem for one locus. In the absence of selection, and in an infinite population with random mating, what will be the equilibrium frequencies of haplotypes? The question for multiple loci will lead us to another important concept, called *linkage equilibrium.*

The simplest case is for two loci with two alleles each. The crucial trick is to write the haplotype frequencies in terms of the gene frequencies at each locus, plus or minus a correction factor, called D. Let the gene frequency in the population of $A_1 = p_1$, $A_2 = p_2$, $B_1 = q_1$, and $B_2 = q_2$. Then

Haplotype	Frequency in Population		
A_1B_1	a	=	$p_1q_1 + D$
A_1B_2	b	=	$p_1q_2 - D$
A_2B_1	c	=	$p_2q_1 - D$
A_2B_2	d	=	$p_2q_2 + D$

The total frequencies add up to one (that is, $p_1q_1 + p_1q_2 + p_2q_1 + p_2q_2 = 1$), and the sum of the two $+D$ and two $-D$ factors is zero. The important term to understand is D, which measures "linkage disequilibrium." Linkage equilibrium occurs when $D = 0$ and means that the alleles at the two loci are combined independently. The two B alleles would then be found with any one A allele (such as A_1) in the same frequencies as they are found in the whole population. If we examine all of the A_1 genes, q_1 of them are linked with B_1 genes and q_2 with B_2 genes; likewise, q_1 of the A_1 genes are linked with B_1 genes and q_2 with B_2 genes. At linkage equilibrium, the frequency of the A_1B_1 haplotype is p_1q_1. D measures the deviation from linkage equilibrium. If $D > 0$, an excess of A_1B_1 (and A_2B_2) haplotypes have arisen: A_1 is more often found with B_1 (and less often with B_2) than would be expected if they combined at random.

D is conventionally defined by adding the frequencies of A_1B_1 and A_2B_2 and subtracting that total from the sum of the frequencies of A_1B_2 and A_2B_1. It could also have been defined the other way around. Although other measures of the non-random combination of genes are possible, the points of the linkage equilibrium principle can be made with D alone.

Papilio memnon is an example of high linkage disequilibrium. If Clarke and Sheppard are correct, the allele T_+ is almost always combined with the other alleles W_1, F_1, E_1, and B_1 rather than with W_2, W_3, or W_4 (and equivalent alleles at the other loci). There is a large excess of the haplotypes $T_+ W_1F_1E_1B_1$, $T_- W_2F_2E_2B_2$, $T_- W_3F_3E_3B_3$, and so forth, while haplotypes such as $T_+ W_2F_2E_2B_2$, $T_+ W_1F_2E_2B_2$, or $T_+ W_1F_1E_2B_2$ are almost absent. The linkage disequilibrium in *P. memnon*, as we have seen, is caused by selection. In this section, however, we are asking how a set of haplotype frequencies should change through time in the absence of selection.

Let us return again to the haplotype A_1B_1. Its frequency has been defined as a in one generation. What will its frequency be in the next generation? (We can again use the notation a' as the frequency of A_1B_1 one generation on.) In the absence of selection, the frequencies of each gene will be constant, but the frequencies of the haplotypes can be altered by recombination. The frequency of A_1B_1 cannot be altered by recombination in double or single homozygotes. The number of A_1B_1 haplotypes coming out of an A_1B_1/A_1B_1 individual or an A_1B_1/A_1B_2 individual must equal the number going in, whether or not recombination occurs. The frequency can be altered only by recombination in the double heterozygotes A_1B_1/A_2B_2 and A_1B_2/A_2B_1. When recombination takes place in an A_1B_1/A_2B_2 individual, the number of A_1B_1 haplotypes is decreased; when it takes place in an A_1B_2/A_2B_1 individual, the number of A_1B_1 haplotypes is increased. To be exact, half of the genes of an A_1B_1/A_2B_2 double heterozygote are A_1B_1. When recombination hits between the loci the frequency of A_1B_1 decreases by an amount $-\frac{1}{2}$. Similarly, recombination in an A_1B_2/A_2B_1 individual increases the frequency of A_1B_1 by an amount $+\frac{1}{2}$.

The frequency of A_1B_1/A_2B_2 heterozygotes in the population is $2ad$ and of A_1B_2/A_2B_1 is $2bc$. The frequency of recombination between the two loci is defined as r. (Theoretically, r can have any value up to a maximum of 0.5, if the loci are on different chromosomes; its value lies between 0 and 0.5 for loci on the same chromosome depending on how tightly linked they are. See chapter 2.) Thus

$$a' = a - \frac{1}{2}r2ad + \frac{1}{2}r2bc$$

$$a' = a - r(ad - bc)$$

The expression $(ad - bc)$ is simply equal to the linkage disequilibrium, D. (We can confirm this equivalence by multiplying out $ad - bc$ from the definitions of a, b, c, and d.) If $D = 0$ (i.e., if the genes are randomly associated), the haplotype frequencies are constant, and $a' = a$. If there is an excess of A_1B_1 haplotypes, the excess decreases by an amount rD per generation. The same relation holds true for any successive pair of generations. We can see what is happening graphically if we substitute for a in the equation:

$$a' = p_1q_1 + D - rD$$

$$a' - p_1q_1 = (1 - r)D$$

The difference between a and p_1q_1 is the amount of "excess" of the A_1B_1 haplotype (i.e., the amount by which the frequency exceeds the random frequency). It is also equal to the linkage disequilibrium ($D = a - p_1q_1$). Therefore

$$D' = (1 - r)D$$

In the absence of selection and in an infinite random mating population, the amount of linkage disequilibrium undergoes exponential decay at a rate equal to the recombination rate between the two loci (Figure 8.2). In other words, the difference between the actual frequency of a haplotype such as A_1B_1 (a) and the random proportion (p_1q_1) decreases each generation by a factor equal to the recombination rate between the loci.

Over time, any non-random genic associations will disappear, as recombination will destroy the association. The higher the rate of recombination, the more rapid the destruction. The highest possible value of r is ½, which is true when the two loci are found on different chromosomes. Genic associations persist longer for tightly linked loci on the same chromosome, as we should intuitively expect.

The equilibrial haplotype proportions have $D = 0$. At equilibrium, the following frequencies apply:

Haplotype	Equilibrial Frequency
A_1B_1	$a = p_1q_1$
A_1B_2	$b = p_1q_2$
A_2B_1	$c = p_2q_1$
A_2B_2	$d = p_2q_2$

These haplotype frequencies, which we have encountered before, are called linkage equilibrium. It is called an "equilibrium" because, in the absence of selection, the action of recombination will drive the haplotypes to these frequencies and then keep them there.

Recombination randomizes genic associations over time. If an excess of one haplotype, such as A_1B_1 exists, recombination will tend to break it down, and A_1 will end up with B_1 and B_2 in their population proportions (q_1 and q_2) and B_1 will be found with A_1 and A_2 in their population proportions (p_1 and p_2). At linkage equilibrium, each of the two alleles at the A locus, A_1 and A_2, are then associated with B_1 in the same proportion.

In a way, linkage equilibrium is, for a two-locus system, an analogy to the Hardy–Weinberg equilibrium for the one-locus system. It describes the equilibrium that is reached in the absence of selection, and in an infinite, randomly mating population. Linkage equilibrium, however, is a property of haplotypes, not genotypes. A diploid individual has two haplotypes, and at the equilibrium

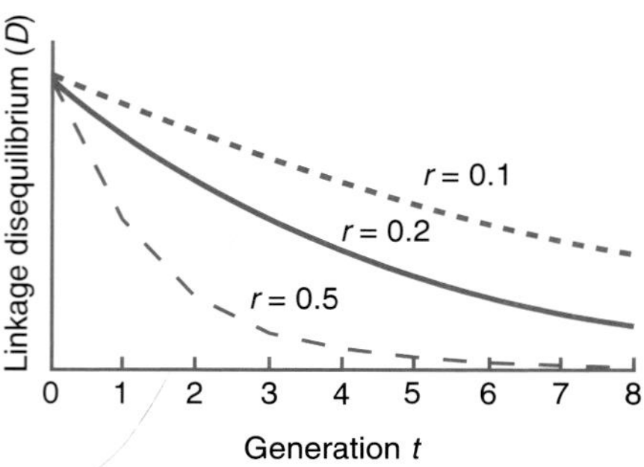

Figure 8.2 Non-random associations between genes at different loci are measured by the degree of linkage disequilibrium (D). Recombination between the loci breaks down the linkage disequilibrium, which decays at an exponential rate equal to the recombination rate between the loci.

the genotypes at each locus will be in Hardy–Weinberg proportions while the haplotypes are at linkage equilibrium. Notice also that whereas the Hardy–Weinberg equilibrium for one locus is reached instantly in one generation (section 5.3, p. 94), it takes several generations for linkage equilibrium to be reached.[2]

Linkage equilibrium has three important aspects for our purposes. The Hardy–Weinberg theorem for one locus was the simplest model in single-locus population genetics; it illustrated how to construct a model with recurrence relations for gene frequencies. The model of linkage equilibrium is, likewise, the simplest model for two loci and shows us how to construct a recurrence relation for haplotype frequencies.

Also like the Hardy–Weinberg theorem, linkage equilibrium provides a theoretical baseline telling us whether anything interesting is occurring in a population. Deviations from Hardy–Weinberg proportions in a natural population suggest that selection, non-random mating, or sampling effects may be operating. Likewise, if two loci are in linkage disequilibrium, we may also suspect that one or more of these variables are at work. If the first thing we had discovered about *Papilio memnon* had been its high linkage disequilibrium, we would have been led on to study how selection was operating on the loci, and perhaps ended up discovering the mimetic polymorphism. In fact, the direction of research in *P. memnon* took the opposite tack, but the general point—that linkage disequilibrium indicates something interesting—holds true.

Linkage equilibrium can also tell us whether the more complex two-locus theory is needed in a real case. To a rough approximation, the theory of population genetics for a single locus is satisfactory for populations in linkage equilibrium. It is when genes become non-randomly associated that a two-locus model is needed. The case we have been discussing can show why. At linkage equilibrium, A_1 and A_2 are equally associated with B_1. To understand evolution at the A locus we can then ignore the relative fitnesses of B_1 and B_2, because if B_1 is fitter than B_2, the association with B_1 will benefit A_1 and A_2 equally (and B_2 will equally detract from them). If A_1, for instance, is more associated with B_1 than is A_2 (i.e., linkage disequilibrium exists), then any advantage of B_1 over B_2 will passively give rise to an advantage to A_1. To understand frequency changes of A_1 we then need to know the relative fitnesses of B_1 and B_2, and the degree of association A_1 has with them. We need a two-locus model.

8.6 *The human HLA genes are a multi-locus gene system*

The HLA system in humans consists of a set of linked genes on human chromosome 6 that control "histocompatibility" reactions. When an organ is transplanted from one individual to another, it is immunologically rejected by the recipient

2. The terms "linkage equilibrium" and "linkage disequilibrium" are not very satisfactory. They were first used by Lewontin and Kojima in 1960. "Linkage disequilibrium" can exist without linkage—among genes on different chromosomes—and it can also exist at equilibrium. It is, however, an equilibrium under certain specifiable conditions (like the Hardy–Weinberg equilibrium). The word linkage is avoided in certain other terms, such as "gametic phase equilibrium." Nevertheless, linkage disequilibrium is the most commonly used term.

in a matter of days—skin grafts last only 2–15 days, for instance. The rejection implies that the immune system can distinguish between "self" and "foreign" cells, and the distinction is generally believed to be achieved by the products of the HLA genes. The best evidence for the HLA genes' role comes from the time course of kidney transplant rejection among siblings that either have or have not been matched for their HLA genes. For kidney transplants between HLA-matched siblings, more than 90% of transplants will survive after 48 months; among HLA-unmatched siblings, 90% survive for 4 months and only about 40% for 48 months.

The HLA system contains a number of genes (Figure 8.3); we will concentrate on two of them, called HLA-*A* and HLA-*B*. Each HLA locus, in a human population, is highly polymorphic. The *B* locus alone may contain 16 alleles with frequencies of 1–10% and many more rare alleles. For example, a sample of 874 people in France contained 31 different alleles at the *B* locus and another 17 alleles at the *A* locus. These degrees of variability are exceptionally high. More typical loci (outside the HLA) might have 1–5 alleles, many less than the number found in the HLA. The reason for the high variability is uncertain, but it would allow the HLA genotype of an individual even in a large population to be unique, which is presumably important in the distinction of self from foreign cell types.

Particular HLA alleles are associated with specific diseases. The strongest association found so far is between ankylosing spondylitis and the allele *B27*; 90% of people with the disease have the *B27* allele, compared with only 7% in the population at large.

The diversity of HLA types may reflect a history of coevolution between humans and disease agents. Over evolutionary time, disease agents may have tried to fool the immune system into treating the agent as part of the body, with the human population then responding by evolving new HLA alleles as new, reliable indicators of "self." This type of evolution would provide a further advantage to variability in the HLA loci. A heterozygous individual with two HLA proteins can compare itself with a possible invader in two ways. The invader must match a homozygote in only one respect, but a heterozygote must be matched in two independent respects. The HLA loci, therefore, probably show heterozygous advantage (section 5.11, p.115), and the same process may have

Figure 8.3 Genetic map of the human HLA loci on chromosome 6. The text concentrates on the *A* and *B* loci, but many more genes are found in the histocompatibility system. The map is not drawn exactly to scale.

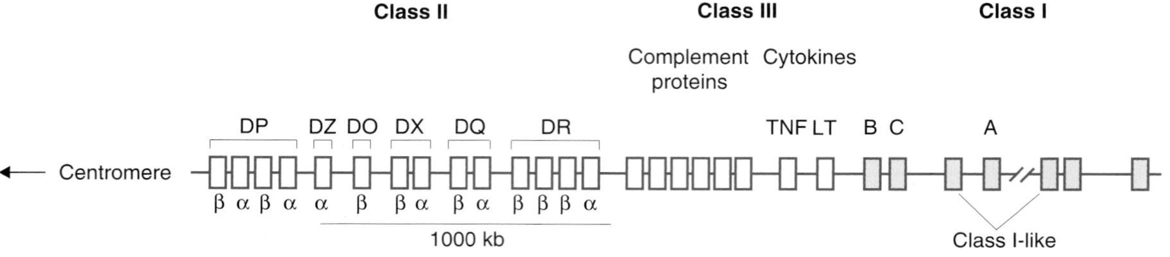

caused the exceptional pattern of evolution in silent and amino acid changing bases within codon triplets (Figure 7.10, p.185).

The HLA system also provides examples of linkage disequilibrium. Particular combinations of genes are found in greater than random proportions. North European populations, for example, characteristically demonstrate an excess of the *A1B8* haplotype. Figure 8.4a provides a more general picture of the linkage disequilibrium values for all combinations of *B* alleles and the allele *A1*. An analogous graph could be constructed for each *A* allele. In Figure 8.4a, D = 0.07 for *A1B8*. If *A1* and *B8* combined in their population proportions, *A1B8* would have a frequency of about 0.023 (2.3%). In reality, this haplotype is found in about 9.3% of individuals (0.093 - 0.023 = D = 0.07). In all, the HLA-*A* and HLA-*B* loci are associated with about six clear cases of linkage disequilibrium; *A1B8* and *A3B7* are the most striking of these cases. The reason why these haplotypes are found in greater than random proportions is unknown, but it is generally believed to be due to selection in favor of the gene combinations. Selection is not the only possible reason for linkage disequilibrium, however, as the next section will reveal.

Figure 8.4 (a) One example of the pattern of degrees of linkage disequilibrium (D) in the HLA: the linkage disequilibrium between 21 *B* alleles and the *A1* allele. An analogous graph can be drawn for every *A* allele. The *A1B8* haplotype occurs at a much higher than random frequency (note the gap in the *x*-axis). The *y*-axis shows the expected frequency of the haplotype if the alleles were associated at random. Thus, the observed frequency of a haplotype is the *y*-axis value + (or −) its *x*-axis value. (b) The linkage disequilibrium between 8 HLA loci: more closely linked loci show higher linkage disequilibrium. The *y*-axis is a kind of average linkage disequilibrium for the multiple alleles; see Hedrick *et al.* (1981) for the exact measure. Figure 8.3 is a map of the loci. Compare this graph with Figure 8.2 to see the effect of recombination. The *A* and *B* loci are indicated as examples; other points indicate other pairs of loci. Reprinted, by permission of the publisher, from Hedrick *et al.* (1991).

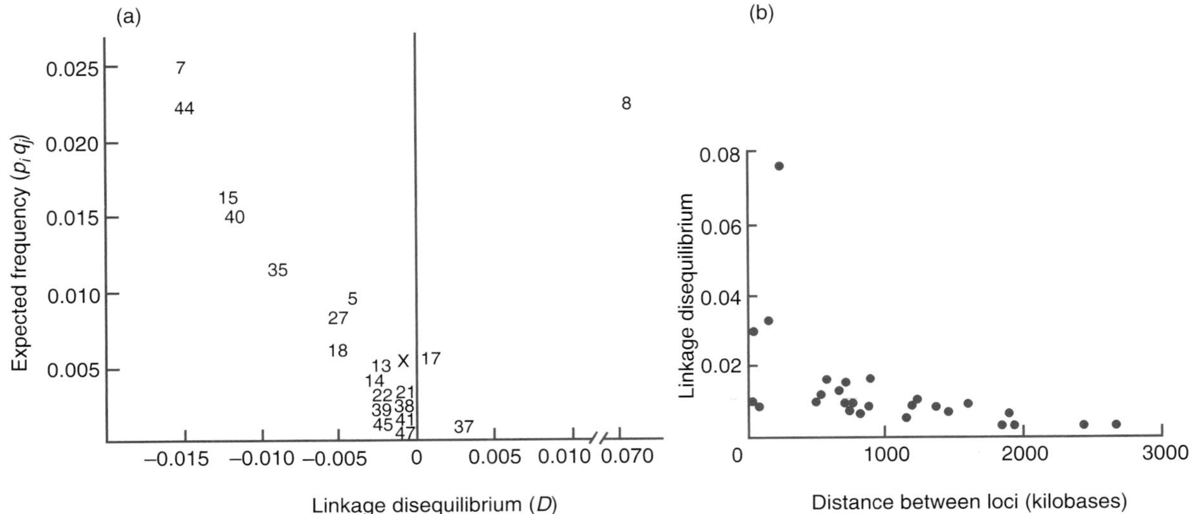

8.7 Linkage disequilibrium can exist for several reasons

Recombination breaks down non-random genic associations. In some cases, like *Papilio memnon* and the HLA genes, non-random associations continue to exist. What causes the linkage disequilibrium? In *Papilio* and in at least some of the HLA associations, this condition probably results from selection. If selection favors individuals with particular combinations of alleles, then it produces linkage disequilibrium. Selection is not the only possible cause for linkage disequilibrium, and a full study of a real case must examine three other main possibilities.

The first factor is that the equilibrium has not yet been reached. A number of generations are required for recombination to do its randomizing work and, particularly for tightly linked genes (both the *Papilio* supergene and the HLA loci are tightly linked), linkage disequilibrium can persist for some time. In the human HLA system, the average linkage disequilibrium between more closely linked loci is larger (Figure 8.4b).

A second process that may affect linkage disequilibrium is random drift in a finite population. Random processes have the interesting property that they cause persistent—not just transitory—linkage disequilibrium. If random sampling produces by chance an excess of a haplotype in a generation, linkage disequilibrium will have arisen. This statement is true for all four haplotypes; random sampling that produces an excess of any of them will disturb the state of linkage equilibrium. Any haplotype could be "favored" by chance, so the disequilibrium is equally likely to have $D > 0$ or $D < 0$. As a population approaches linkage equilibrium, all random fluctuations in haplotype frequencies will tend to occur in a direction away from the linkage equilibrium values. If a population is located far away from the point of linkage equilibrium, random sampling is equally likely to move it toward or away from the equilibrium. Most natural populations are probably near linkage equilibrium (see Figure 8.5), and the balance between the random creation of linkage disequilibrium and its destruction by recombination, in small enough populations, is such that linkage disequilibrium will persist.

The third possibility is non-random mating. If individuals with gene A1 tend to mate with B_1 types rather than B_2 types, A_1B_1 haplotypes will have excess frequency over that seen with random mating. (The exact effect depends on whether it is homozygous A_1/A_1 individuals that mate non-randomly, or the homozygotes and the A_1/A_2 heterozygotes, and on whether they mate preferentially only with B_1/B_1 homozygotes or with B_1/B_2 heterozygotes as well. The general effect of non-random mating on linkage disequilibrium is not complicated.)

The three processes other than selection probably account for some cases of linkage disequilibrium in nature. The process that has most interested evolutionary biologists, however, is natural selection. Let us now consider how we can model the effect of selection on haplotype frequencies.

8.8 Two-locus models of natural selection can be built

The effect of natural selection on haplotype frequencies in two-locus models, like its effect on gene frequencies in single-locus models, depends on the fitnesses

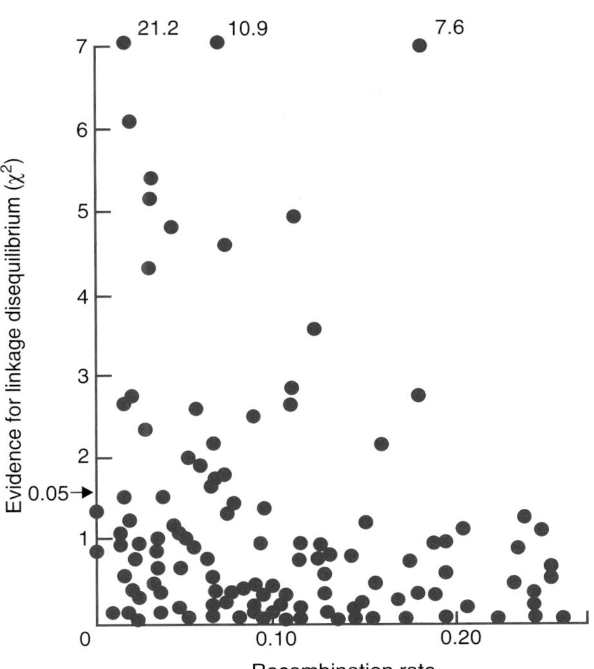

Figure 8.5 Linkage disequilibrium (on the *y*-axis) among gel electrophoretic samples of pairs of genes in *Drosophila*. It is plotted as a χ^2 value. The χ^2–value indicates the strength of the evidence for linkage disequilibrium—the higher the χ^2, the more linkage disequilibrium between that pair of genes. (The χ^2–value corresponding to a statistical significance of 0.05 is indicated.) Most of the gene pairs have insignificant or low linkage disequilibrium (i.e., low χ^2). The *x*-axis shows the rate of recombination between the pair of genes. Reprinted, by permission of the publisher, from Langley (1977).

of the genotypes. We have many possible ways to write down the fitness of each genotype. In one of the simplest two-locus models, the fitness of a two-locus genotype is the product of the fitnesses of its two single-locus genotypes; the model is realistic if the fitness effect of one locus operates independently of the genotype at the other. Suppose, for example, that the *A* locus influences survival from age 1–6 months, such that

genotype	A_1/A_1	A_1/A_2	A_2/A_2
chance of survival to age 6 months	w_{11}	w_{12}	w_{22}

The other locus influences survival from ages 6–12 months

genotype	B_1/B_1	B_1/B_2	B_2/B_2
chance of survival from 6–12 months	x_{11}	x_{12}	x_{22}

The total chance of surviving from age 1–12 months would then be the product of the two genotypes that an individual possessed because selection at age 1–6 is independent of selection at age 6–12 months:

	A_1/A_1	A_1/A_2	A_2/A_2
B_1/B_1	$w_{11} \times x_{11}$	$w_{12} \times x_{11}$	$w_{22} \times x_{11}$
B_1/B_2	$w_{11} \times x_{12}$	$w_{12} \times x_{12}$	$w_{22} \times x_{12}$
B_2/B_2	$w_{11} \times x_{22}$	$w_{12} \times x_{22}$	$w_{22} \times x_{22}$.

These fitnesses are called *multiplicative*. An individual's fitness for its two genotypes is found by multiplying the fitnesses of each of its one-locus genotypes. The genotypes are independent, in the sense that the effect of one genotype on survival is independent of the other locus. An individual with the genotype $A_1/$

A_1 has a chance of surviving from age 1–6 months of w_{11} whether its genotype at the other locus is B_1/B_1, B_1/B_2, or B_2/B_2.

Additive fitness interactions are another possibility. Imagine that the two loci encode two digestive enzymes. The enzyme produced by the *A* locus digests one kind of food, while the enzyme produced by the *B* locus digests another kind of food. The genotypes might enable a certain number of joules of energy to be obtained by feeding in a particular environment:

genotype	A_1/A_1	A_1/A_2	A_2/A_2	B_1/B_1	B_1/B_2	B_2/B_2
energy obtained	w_{11}	w_{12}	w_{22}	x_{11}	x_{12}	x_{22}

The fitness of an individual might be proportional to the total energy it obtains by feeding. Its fitness for the two loci would then be found by adding the fitnesses of its two genotypes together.

	A_1/A_1	A_1/A_2	A_2/A_2
B_1/B_1	$w_{11} + x_{11}$	$w_{12} + x_{11}$	$w_{22} + x_{11}$
B_1/B_2	$w_{11} + x_{12}$	$w_{12} + x_{12}$	$w_{22} + x_{12}$
B_2/B_2	$w_{11} + x_{22}$	$w_{12} + x_{22}$	$w_{22} + x_{22}$.

The next step is to derive a recurrence relation between the frequency of a haplotype in one generation and in the next generation. However, the algebra is tiresome and we do not need to work through it here. The procedure follows the same outline as for the single-locus case, with the additional factor of recombination. The recurrence relation for haplotype frequency takes account of the frequency and fitness of all genotypes in which a haplotype is found. It also adds and subtracts the number of copies gained and lost by recombination; we multiply by $(1 - r)$ the frequency of the double heterozygotes containing the haplotype and by r the frequency of the double heterozygote that can generate it if recombination occurs. The Mendelian rules are then applied, and the frequency of the haplotype in the next generation results.

Which kinds of selection cause linkage disequilibrium? The question is important because, as we have seen, two-locus models are particularly needed when linkage disequilibrium exists. The two types of fitness (multiplicative and additive) that we have just considered both result in equilibrial haplotype frequencies with zero linkage disequilibrium. (The possibility arises only if the equilibrium has both loci polymorphic. If one gene is fixed at either locus, $D = 0$ trivially. The fitnesses as written above (w_{11}, w_{12}, and so on) were frequency-independent. A doubly heterozygous equilibrium then requires heterozygous advantage at both loci: $w_{11} < w_{12} > w_{22}$, $x_{11} < x_{12} > x_{22}$. See chapter 5, section 5.12.1, p.117.) In conclusion, if linkage disequilibrium exists at any time that fitnesses are independent at the two loci, it will decay to zero as the generations pass.

The more interesting case is when the fitnesses of the two loci interact *epistatically*. Epistatic interaction means that the fitness effects of a genotype depend on the genotype with which it is associated at the other locus. The selection in the mimetic polymorphism of *Papilio memnon* is epistatic, for example.

We can reduce the situation in *P. memnon* to the most elemental theoretical form if we imagine that one locus controls whether the butterfly has a tail on its hindwing and a second controls coloration. Let T_+ (presence of tail) be dominant to T_- (absence). At the other locus, C_1 is dominant. C_1/C_1 and C_1/C_2 individuals have a color pattern that mimics a model species with a tail, whereas C_2/C_2 individuals are colored like a model species that has no tail. The relative fitness of each genotype depends on the genotype found at the other locus. For example, a T_+/T_+ genotype in the same butterfly as a C_2/C_2 will be less fit than a T_-/T_- with C_1/C_1. The fitnesses can be written as follows (the simplification relative to the earlier fitness matrixes arises because of dominance, and because one term exists for the fitness of both loci together rather than one term for each locus):

	T_+/T_+	T_+/T_-	T_-/T_-
C_1/C_1	w_{11}	w_{11}	w_{21}
C_1/C_2	w_{11}	w_{11}	w_{21}
C_2/C_2	w_{12}	w_{12}	w_{22}

In the case we discussed, w_{12} is the fitness of a butterfly with a tail and the color pattern of a model lacking a tail. Therefore, $w_{12} < w_{11}$ and w_{22}. The term w_{21} is the fitness of a butterfly without a tail, but the color pattern of a tailed model. Therefore, $w_{21} < w_{11}$ and w_{22}. Selection now favors the $T_+/-$ genotypes when they are with $C_1/-$, but not when with C_2/C_2; likewise, selection favors T_-/T_- when it is with C_2/C_2 but not when with $C_1/-$ (the dash implies it does not matter which gene is present, because of dominance). The fitness relations are epistatic, and a doubly polymorphic equilibrium can now occur. All four alleles will be present, and the haplotypes T_+C_1 and T_-C_2 will have disproportionately high frequencies. T_+C_2 and T_-C_1 haplotypes are selected against, because they often find themselves in poorly mimetic butterflies. Linkage disequilibrium ($D > 0$ in this case) exists at the equilibrium.

In general, selection can only produce linkage disequilibrium at equilibrium when the fitnesses of the genotypes at different loci interact epistatically. Not all epistatic fitness interactions generate doubly polymorphic equilibria with linkage disequilibrium, but all such equilibria do have epistatic fitnesses.

We have been discussing the different sorts of fitness interactions—multiplicative, epistatic, and so on—as properties of formal models. Real genes in real organisms will have fitness interactions as well, and the more important question deals with the characteristics of these interactions. In cases like *Papilio,* epistasis is present and powerful; these cases may be isolated examples rather than representing a general condition, however. Evolutionary biologists are interested in whether fitness interactions between loci are generally epistatic and generate strong linkage disequilibrium, or whether they are generally independent and generate linkage equilibrium. These two extremes roughly correspond to a more "holistic" and a more "atomistic" (or "reductionist") school of thought (which is not to say that they correspond to two clearly demarcated camps of biologists).

No general answer has been found for this question, but it is possible to make some observations. Different loci will tend to interact multiplicatively when they have independent effects on an individual's survival and reproduction. This interaction could could occur with loci and influence events at different times in an organism's life —though it is perfectly possible for such events to interact. Epistatic interactions are more likely for loci controlling closely interdependent parts of an organism. The extent to which we expect loci to interact epistatically will, therefore, loosely depend on how atomistic or holistic a view we have of the organism (see also section 7.7.2, p.167).

Notice that epistatic fitness interaction is not the same as mere physiological or embryological interaction. Fitness epistasis requires heterozygosity at two loci. Imagine a case in which the *A* locus controls, for example, muscle strength and the *B* locus controls metabolic rate. Muscles and metabolism interact in a physiological sense: when muscles are put to work, the metabolic rate increases. However, if the population is fixed for homozygotes at both loci (all individuals are A_1B_1/A_1B_1), then fitness epistasis cannot be present. The special condition of epistatic fitness requires heterozygosity at both loci, and the kind of fitness relations that we saw in the *Papilio memnon* example. Although this situation is often called fitness "interaction," the term interaction is used here in a technical, not a colloquial, sense.

Another, empirical method has been applied to answering the question of how commonly epistatic fitness interactions occur in nature. Linkage disequilibrium is produced by epistatic selection, and the degree of linkage disequilibrium in a population can be measured. If it is high, then epistatic selection may be common. The argument works in one direction but not the other. That is, because linkage disequilibrium may be caused by several factors (section 8.7), its existence does not demonstrate epistatic selection. If linkage disequilibrium is absent or low, however, we can infer that epistatic selection is unimportant in nature.

A few general surveys of the extent of linkage disequilibrium in natural populations have been made. An investigation by Maynard Smith *et al.* (1993) for bacteria found high levels of linkage disequilibrium in some species, such as *E. coli* (which lives in our, and other mammals', guts) but low levels in other species, such as *Neisseria gonorrhoeae.* The reason why many bacteria show linkage disequilibrium is that they reproduce asexually, and no recombination happens to decay the linkage disequilibrium. *N. gonorrhoeae* presumably has enough sexual exchange of genes to produce linkage equilibrium.

In eukaryotic organisms that are known to reproduce sexually, the evidence suggests that little deviation from linkage equilibrium appears in nature. The main evidence comes from gel electrophoretic surveys of protein polymorphisms, to see directly whether genes at different loci are associated.

Figure 8.5 illustrates some comprehensive results for the fruitfly *Drosophila.* Some evidence of linkage disequilibrium is found, but the results suggest that the level is low and most loci are in linkage equilibrium. One conclusion from this survey would be that, although epistatic interactions are important in particular cases (e.g., *Papilio*), they may not be of general importance at least in the evolution of sexually reproducing species.

Other biologists would disagree. They might be unconvinced by the evidence of Figure 8.5, perhaps calling it "limited" or "for a single species." The amount of interaction between loci that must go on during the development of a complex, organic body is so high that they would expect epistatic fitness interactions to be common. Such is the assumption of the school of thought that follows Wright, whose ideas we will discuss at the end of the chapter.

8.9 *Hitchhiking occurs in two-locus selection models*

When a gene is changing frequency at one locus over time, it can cause related changes at linked loci; conversely, events at linked loci can interfere with one another. Suppose, for instance, that directional selection is substituting one allele A' for another (A) at one locus, and a neutral polymorphism (B, B') appears at a linked locus. The frequency of B or B' (whichever was linked with A' when it arose as a mutant) will then be increased. If the new mutant A' arose on a B-bearing chromosome, then B will eventually be fixed together with the selected allele A' unless recombination splits them before A has been eliminated. The increase in the B allele frequency is due to *hitchhiking*.

Another possibility is for the polymorphism at the B locus to be a selectively "balanced" polymorphism, due to heterozygous advantage. Suppose that a selectively favored mutation A' arises at a linked locus, and that it happens to arise on the same chromosome as a B allele. The polymorphism at the B locus will now interfere with the progress of A'. As A' increases in frequency by directional selection, it will also increase the frequency of B. Because A' is linked to B, it will be more likely to appear in a body with a B/B homozygote than will its allele A, and less likely to be in a B/b heterozygote. B/b has higher fitness than B/B, and the selection against B/B individuals will also work against the A' gene. Depending on the selection coefficients at the two loci, and the rate of recombination between them, the heterozygous advantage at the B locus can slow the rate at which A' is fixed. The A' gene will then have to wait for a recombinational "hit" between the two loci before it can progress to fixation.

Hitchhiking has a further interest in relation to the neutral theory of molecular evolution. It alters the neutral theory's predictions for the level of heterozygosity. We saw in the last chapter (Figure 7.3, p.164) that observed heterozygosities are lower than those predicted by the neutral theory for plausible estimates of population size and mutation rate. Heterozygosities are not only too low, but also too uniform, given the variation in natural population sizes (see Figure 6.6, p. 147). Maynard Smith and Haigh pointed out that hitchhiking could bring the neutral theory more closely into line with the observations. As a favored allele is fixed, the heterozygosity declines at any linked locus that has a neutral polymorphism; the more closely linked the polymorphic locus, the greater the reduction in its heterozygosity. We do not know to what extent hitchhiking causes heterozygosities to deviate below the neutral theory's prediction; however, the hitchhiking effect of a favorable gene must undoubtedly act to reduce heterozygosities at linked neutral loci.

8.10 *Linkage disequilibrium can be advantageous, neutral, or disadvantageous*

The linkage disequilibrium in *Papilio memnon*'s mimetic polymorphism is advantageous. Natural selection favors individuals with the genic associations like $T_- W_2F_2E_2B_2$, whereas it works against recombinants like $T_+ W_2F_2E_2B_2$. An individual benefits from having the haplotypes that are in excess frequency in the population. Whole populations of *P. memnon* survive better than they would if the five loci existed in linkage equilibrium.

In other cases, the opposite is true. As seen in the previous section, as the favored allele A' increased in frequency, and the frequency of an allele, B, at a linked polymorphic locus was also increased by hitchhiking. The polymorphism at the B locus was caused by heterozygous advantage, and the linkage disequilibrium built up by selection on the A locus (creating an excess of the $A'B$ haplotype) was, therefore, disadvantageous. The individuals on average have lower fitness than if linkage equilibrium existed between the A and B loci, because the increase in the $A'B$ haplotype reduces the proportion of B/b heterozygotes. Natural selection will favor recombinant individuals that do not have the $A'B$ haplotype.

A third possibility is for linkage disequilibrium to be selectively neutral. An example of this case was provided by the hitchhiking of an allele at a neutral polymorphic locus with a selectively advantageous mutant at a linked locus. While the mutant is being fixed, linkage disequilibrium temporarily builds up between it and the alleles at nearby loci to which it is linked. It disappears when the mutant reaches a frequency of one.

The distinction between advantageous and disadvantageous linkage disequilibrium is crucial to understanding one of the major problems of evolutionary biology—why recombination exists.

8.11 *Why does the genome not congeal?*[3]

Genetic recombination undoubtedly exists, but it is not obvious why it occurs. The problem can be seen by considering two kinds of populations near equilibrium—one at linkage equilibrium and the other with linkage disequilibrium due to epistatic selection (or coadaptation between genetic loci, like in *Papilio memnon*). We will assume that mating is random and the population is large. What effect does recombination have in the first case, which includes linkage equilibrium? None at all—the frequencies of A_1B_1, A_1B_2, A_2B_1, and A_2B_2 are repeated from generation to generation, regardless of whether recombination happens. If there is no recombination, the haplotype frequencies are necessarily constant. If recombination is present, it creates and destroys each haplotype at the same rate, which is what is meant by linkage equilibrium. In the absence of linkage disequilibrium, therefore, recombination has no effect.

Suppose that linkage disequilibrium occurs, and it is due to coadaptation between loci. Let us imagine that A_1B_1 and A_2B_2 are coadapted pairs of alleles, like the tail genes and color genes in *P. memnon*, that are present in greater than random frequencies ($D > 0$). The actual amount of linkage disequilibrium will be a balance between the increase due to selection and the decrease due to recombination. In this case, recombination is disadvantageous. The lower the recombination rate, the stronger the association between the coadapted alleles that can be built up, and the fitter on average the individuals will be. Selection, therefore, acts to reduce the recombination rate between the loci.

3. In 1967, J. R. G. Turner published a strikingly titled paper, "Why does the genotype not congeal?" The paper is often mistakenly cited in the form of this section heading.

It might seem, then, that recombination either makes no difference or is disadvantageous. If there is linkage equilibrium, the genotype frequencies are unaltered by recombination; if beneficial associations between genes have built up, recombination tends to destroy them. So why does recombination exist? Why are recombination rates not reduced to zero? The answer to this question has not found universal agreement. The vague answer is probably that recombination is favored because of environmental change. This answer does not eliminate problems in specifying the kind of environmental change needed for recombination to be advantageous, and in showing that natural environments do change in the necessary way. Before we discuss environmental change in depth, we should first rule out one other possible explanation.

Natural selection cannot work on an organism if it shows no genetic variation. If no genetic variation influenced recombination rates, then recombination rates could not be reduced to zero. Is the recombination rate genetically variable? The most convincing test is to carry out an artificial selection experiment. If a trait varies genetically, an attempt to change it by artificial selection should be successful; if it does not vary genetically, the attempt should fail. Several experiments have selected successfully for both higher and lower recombination rates in fruitflies (Figure 8.6). Thus, absence of genetic variation is not the reason why recombination exists.

Moreover, in the case of linked groups of coadapted genes, recombination can be effectively eliminated by a chromosomal inversion. We do not know whether the *P. memnon* mimicry genes are contained within an inversion, but let us consider what would happen if they were. If a single recombination event occurs within an inverted chromosomal region, the resulting chromosomes lack certain genes and are, therefore, likely to be lethal (Figure 8.7). Only when recombination (or, more precisely, an odd number of recombination events) has not occurred in the inversion will the offspring be viable. Thus, the set of genes within the inversion are practically protected from recombination.

This point is not merely theoretical, because chromosomal inversions are known to exist. They are particularly easy to study in fruitflies (section 4.5, p. 78; section 17.14, p. 249), and Dobzhansky did much research on them. He emphasized the idea that chromosomal inversions protect sets of epistatically

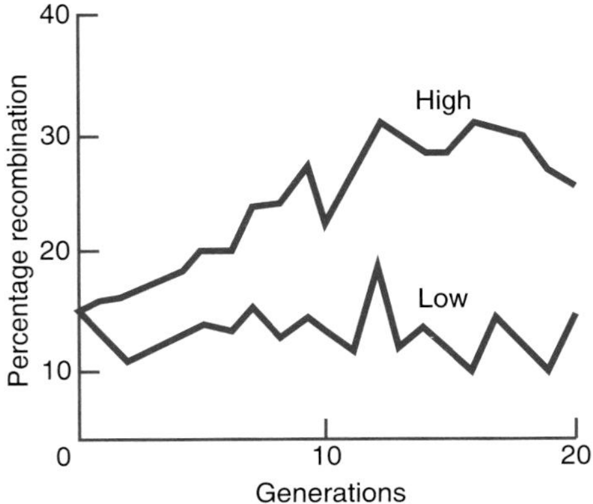

Figure 8.6 Artificial selection can increase or decrease the rate of recombination. These results are for the rate of recombination between two loci (called *Gl* and *Sb*) in lab stocks of the fruitfly *Drosophila melanogaster*. Reprinted, by permission of the publisher, from Kidwell (1972).

Figure 8.7 (a) The letters (*a*, *b*, *c*, . . .) indicate different genes along a chromosome. The genes *c-d-e* are an inversion polymorphism. (b) If recombination takes place within the inversion, the resulting chromosomes (c), which will become gametes, contain double copies of some genes but lack copies of other genes. Recombination within an inversion is, therefore, usually lethal. The only offspring of such an individual will be those in which recombination did not "hit" the inverted region.

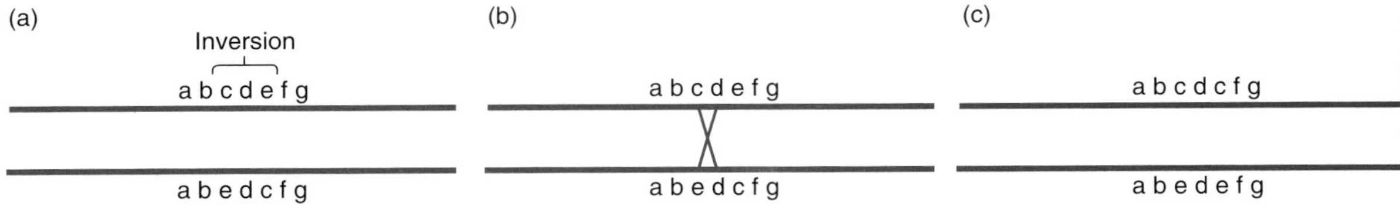

interacting genes, which is an important principle of chromosomal evolution. Notice, by the way, that the genes in an inversion will tend to behave as a "supergene"—a set of linked genes that segregate as if they were a single Mendelian locus. In a supergene, the recombinants will often die and therefore not be seen.

In summary, the recombination rate can be reduced. It is, therefore, unlikely that recombination persists in natural populations simply because natural selection cannot eliminate it for want of genetic variation.

Recombination, as we saw, is disadvantageous only when it disrupts coadapted sets of genes in linkage disequilibrium. Its disadvantage will depend on how much of this sort of linkage disequilibrium exists in nature. If the amount is low, the disadvantage will be small. As we have seen (Figure 8.5), data collected to date suggest quite low amounts of linkage disequilibrium. Coadapted gene complexes, in the sense of haplotypes occurring in greater than random frequencies, may be exceptional in nature. To this extent, the problem of recombination is reduced.

But what positive advantage might recombination provide? The general condition for recombination to be favored is that disadvantageous genic associations should exist, in disproportionately high frequency, in the population. Recombination is selected against when it breaks up favorable gene combinations. It is selected for when it breaks up unfavorable gene combinations. Hitchhiking is a case in point. When a selected polymorphic locus interferes with selection at a linked locus, recombination between the loci is advantageous. Strobeck *et al.* have formally demonstrated this advantage, although the argument may, in fact, may be a special circumstance. The process would be more common in a rapidly changing environment, because the fitnesses of different alleles would then be changing and selection attempting to adjust the frequencies of genes at many (potentially interfering) loci.

The problem of why recombination exists is closely related to the problem of why sex exists. We will look at a general theory that may solve both problems in section 11.3 (p. 284). That theory depends on the rate of deleterious mutation, and the discussion of selection-mutation balance in section 9.11 (p. 247) provides a background to it. In this chapter, however, the main point is that recombination is a puzzle. It can have clear disadvantages, and it does not exist simply because natural selection cannot remove it.

8.12 *Wright invented the influential concept of an adaptive topography*

Wright's idea of an *adaptive topography* is particularly useful for thinking about complex genetic systems, but it is easier to begin with the simplest case of a single genetic locus. The topography is a graph of mean population fitness ($\bar{w}$) plotted against gene frequency (Figure 8.8). (Adaptive topographies can also be drawn for fitness in relation to genotype frequencies. They can even be drawn with phenotypic variables on the x-axis; see, for example, Raup's analysis of shell shape, discussed in section 13.9 (p. 353). The graph in Box 7.3, p. 182, used in Fisher's theory of adaptation, is also similar.) We have repeatedly encountered the concept of mean fitness; it is equal to the sum of the fitnesses of each genotype in the population, each multiplied by its proportion in the population. In a case in which the genotypes containing one of the alleles have higher fitness than those of the alternative, the mean fitness of the population simply increases as the frequency of the superior allele increases and reaches a maximum when the gene is fixed (Figure 8.8a). When heterozygous advantage occurs, mean fitness is highest at the equilibrium gene frequency given by the standard equation (section 5.12.1, p. 117); it declines on either side, where more of the unfavorable homozygotes will be dying each generation than at the equilibrium (Figure 8.8b). The graph of this relation is also called a *fitness surface.*

In these two cases, natural selection carries the population to the gene frequency where mean fitness is at a maximum. With one favorable allele, the maximum mean fitness appears where the allele is fixed—and natural selection will act to fix the allele. With heterozygous advantage, the maximum mean fitness

Figure 8.8 A fitness surface, or adaptive topography, is a graph of the mean fitness of the population as a function of gene, or in some cases genotype, frequency. (a) If allele A (frequency p) has higher fitness than a (frequency $= 1 - p$), mean population fitness simply increases as the frequency of A increases. (b) With heterozygous advantage (fitnesses of genotypes AA:Aa:aa are $1 - s$:1:$1 - t$), mean population fitness increases to a peak at the intermediate frequency of A at which the proportion of heterozygotes is the maximum possible.

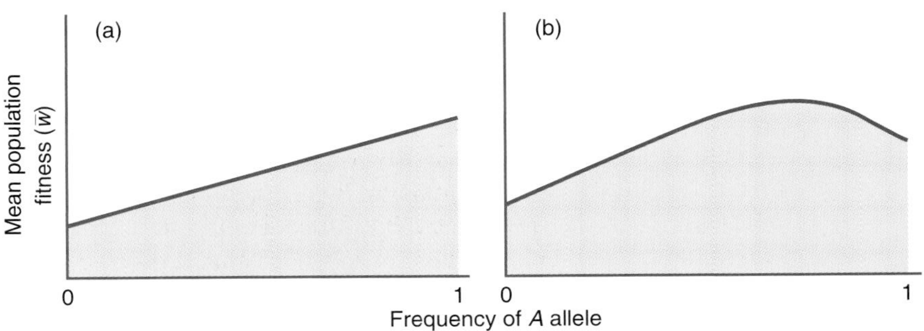

appears where the smallest number of homozygotes are dying each generation—and natural selection drives the population to an equilibrium where the amount of homozygote death is minimized. An interesting question in theoretical population genetics is whether natural selection always drives the population to the state at which the mean fitness is the maximum possible. In some cases, it does not, such as in frequency dependence (section 5.13, p. 121). When frequency dependence maintains a polymorphism, the fitness of a genotype is highest when it is rare. When a genotype is rare, natural selection acts to increase its frequency, making it more common. It is possible to devise theoretical cases in which the mean fitness is maximized at genotype frequencies that are not at equilibrium. In these instances, natural selection acts to reduce mean fitness by moving the genotype frequencies away from those points. If natural selection does not always maximize mean fitness, a further—and still unanswered—theoretical question is whether natural selection always maximizes some other function; we will not pursue the question here. Whatever the answer to this question, there are many cases in which natural selection maximizes simple mean fitness—a useful assumption for many purposes. We can then think of natural selection as a hill-climbing process, analogous to the hills in the adaptive topography (Figure 8.8).

Now consider a second locus. Selection can take place here as well, and the fitness surface for the two loci might resemble Figure 8.9a, a simple case in which one locus has heterozygous advantage and the other possesses a single favored allele. The idea of an adaptive topography can be extended to as many loci as interact to determine an organism's fitness, but the additional loci must be imagined, rather than drawn on two-dimensional paper.

Wright believed that, because the genes at different loci interact, a real multidimensional fitness surface would often contain multiple peaks, with valleys between them (Figure 8.9b). The kind of reasoning behind this theory is abstract rather than concrete. Imagine a large number of loci, many with more than one allele, and the alleles at the different loci interacting epistatically in their effects

Figure 8.9 Fitness surface for two loci. (a) Combination of patterns in Figures 8.8a and 8.8b: heterozygous advantage at locus *A* and one allele has higher fitness than the allele at locus *B*. (b) Two-locus fitness surface with two peaks.

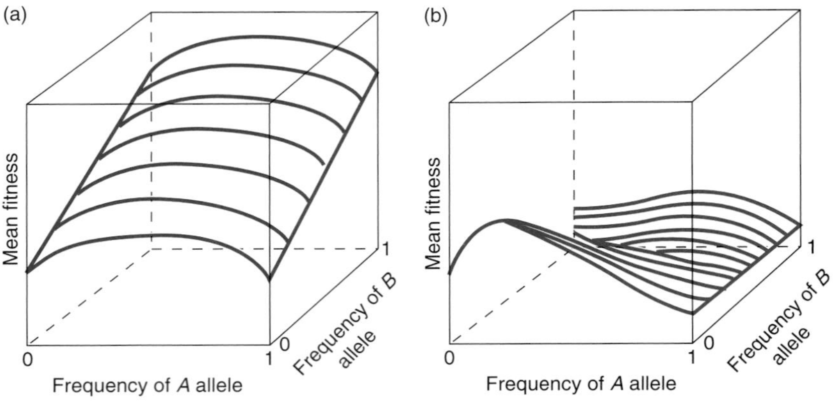

on fitness. Epistatic interactions, we now suppose, are common because organisms are highly integrated entities compared with the atomistic chromosomal row of Mendelian genes from which they develop. The genes will have to interact to produce an organism. As we saw above, developmental interactions among genes do not automatically generate epistatic fitness interactions among loci. The extent to which undoubted developmental interaction will produce a multiply peaked fitness surface is, therefore, open to question, but the possibility is plausible. (Wright called the genes that interact favorably to produce an adaptive peak an "interaction system.")

In the coadapted supergenes of *Papilio memnon*, the mimetic genotypes occupy fitness peaks and the recombinants occupy various fitness valleys. The actual shape of the adaptive topography in nature is, however, a more advanced question than can be tackled here. The point of this section is to define an adaptive topography, and to point out that its visual simplicity can be useful in thinking about evolution when many gene loci interact.

8.13 *The shifting balance theory of evolution*

Wright used his idea of adaptive topographies to create his general evolutionary theory. He imagined that real topographies would have multiple peaks, separated by valleys, and that some peaks would be higher than others. When the environment changed and competing species evolved new forms, the shape of the adaptive topography for a population would change as well. The surface would also change shape when a new mutation arose. A new allele at a locus could interact with genes at other loci differently from the existing alleles, and the fitnesses of the genes at the other loci would then be altered. Thus, genetic changes would take place at other loci to adjust to the new mutant. At the same time, natural selection would be a hill-climbing process, directing the population up toward the currently nearest peak. When the surface changed, the direction to the nearest peak might change, and selection would then send the population off in the new upward direction.

Natural selection, even in so far as it is a hill-climbing process (i.e., maximizing mean fitness), is only a *locally* hill-climbing process. In theory, the local fitness peak could lie in the opposite direction from a higher, or the global, peak (Figure 8.10).

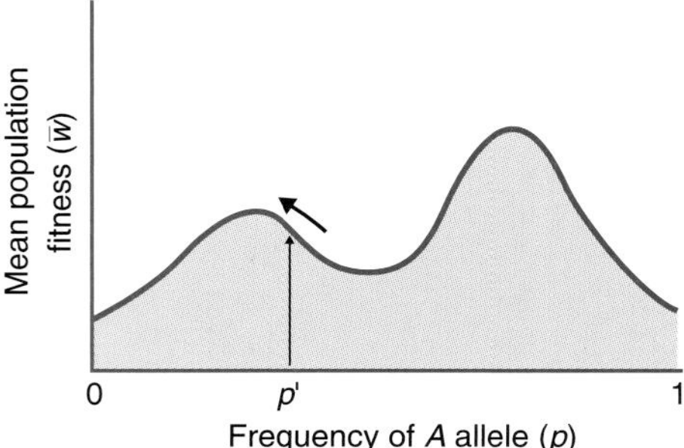

Figure 8.10 A two-peaked fitness surface with local and global maxima. Natural selection will take a population with gene frequency p' toward the local peak, and away from the peak with highest average fitness.

Natural selection, however, will direct the population to the local peak. Now suppose that the mean fitness of a population measures the quality of its adaptations, such that a population with a higher mean fitness has better adaptations than a population with a lower mean fitness. Because natural selection seeks out only local peaks, natural selection may not always allow a population to evolve the best possible adaptations. A population could be stuck on a merely locally adaptive peak. However, adaptation and mean fitness cannot always be equated in this way. In the simple case in which one allele is superior to another (Figure 8.8a), the organisms with the better genotype will also be better adapted. When frequency-dependence of fitness develops or when conflict emerges between group and individual adaptations (chapter 12), it is no longer true that the maximum of mean population fitness automatically corresponds to the best adaptation.

Wright was interested in how evolution could overcome the tendency of natural selection to become stuck at local fitness peaks. When fitness peaks correspond to optimal adaptations, the question is relevant to the evolution of adaptation; when they do not correspond, the question still has a technical interest in population genetics. Wright suggested that random drift could play a creative role. Drift will tend to make the population gene frequencies "explore" the terrain around their present position. By drift, the population could move from a local peak to explore the valleys of the fitness surface. Once it had explored to the foot of another hill, natural selection could start climbing uphill on the other side. If this process of drift and selection were repeated with different valleys and hills on the adaptive topography, a population would be more likely to reach the global peak than if it remained under the exclusive control of the locally maximizing process of natural selection.

Wright's full shifting balance theory includes ideas other than those of selection and drift within a local population. He also suggested that populations would be subdivided into many small local populations, and drift and selection would occur in each. The large number of subpopulations would multiply the chance that one subpopulation would find the global peak. If the members of a subpopulation at the highest peak were better adapted, they could produce more offspring; they would also produce more emigrants to the other subpopulations. Those other subpopulations would then be taken over by the superior immigrant genotypes. Thus, the whole species would evolve to the higher peak. As we can see, Wright's theory is an attempt at a comprehensive, realistic model of evolution. It includes everything: multiple loci, fitness interactions, selection within and between populations, drift, and migration. (The theory of adaptive peaks is also relevant to speciation; see section 16.4.4, p. 434.)

As always in science, complex realistic models are useful for some purposes and simple, idealized ones for others. Realistic models may describe nature more accurately, but they are less illuminating when explaining principles or studying the consequences of particular factors. For this reason, when we consider particular factors in isolation in this book, we limit ourselves to simple and idealized models that concentrate on the factor of interest. The encumbrance of all of the other factors that may influence gene frequencies in nature is needed when we are trying to describe the real world, but not when we are trying to understand arguments.

Fisher generally disagreed with Wright over the importance of random drift in evolution. Fisher maintained that natural populations are generally too large for drift to play an important role; Wright retorted that Fisher had misunderstood the role of drift in the shifting balance theory, which was to assist, rather than counteract natural selection. Much of their argument seems to have been at cross-purposes, but many of the points are interesting.

Fisher, for instance, doubted whether natural selection actually would confine populations to local peaks. Fisher was preeminently a geometric thinker and he pointed out that, as the number of dimensions in an adaptive topography increases, local peaks in one dimension tend to become points on hills in other dimensions (Figure 8.11). In the extreme case of an infinity of dimensions, natural selection will certainly be able to hill-climb all the way to the global peak without any need for drift. Each one-dimensional (Figure 8.8) or two-dimensional (Figure 8.9) peak will be crossed at its top by an infinity of other dimensions, and it is highly implausible that the fitness surface will turn downhill in all of them at that point. This argument is highly interesting, although it is, of course, purely theoretical. It refutes Wright's *theoretical* claim that natural selection will get stuck at local peaks, but leaves open the empirical question of how important selection and drift have been in exploring the fitness surfaces of nature.

Figure 8.11 As extra gene loci are considered (extra axes in the adaptive topography), it becomes increasingly likely that what appeared to be a local maximum in fewer dimensions will turn out to be a hillside or saddle point in more dimensions. In this case, the fitness surface for the *A* locus at a gene frequency of zero for the *B* allele is the same as in Figure 8.10. At other *B* gene frequencies, however, the local peak in the *A*-locus fitness surface disappears. If the population started in the valley at *B* gene frequency = 0 and *A* gene frequency = p', natural selection would initially move gene frequencies to the local peak, but it would eventually reach the global peak by continuous hill-climbing.

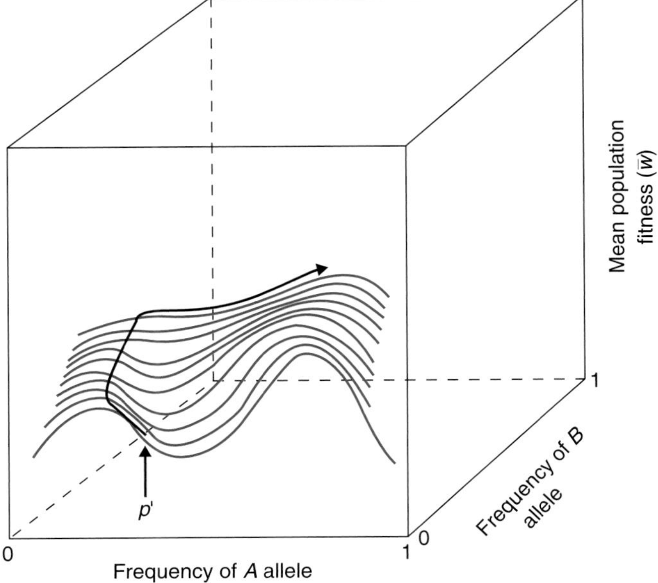

SUMMARY

1. Population genetics for two or more loci is concerned with changes in the frequencies of haplotypes, which are the multi-locus equivalent of alleles.
2. Recombination tends, in the absence of other factors, to make the alleles of different loci appear in random (or independent) proportions in haplotypes. An allele A_1 at one locus will then be found with alleles B_1 and B_2 at another locus in the same proportions as B_1 and B_2 are found in the population as a whole. This condition is called linkage equilibrium.
3. A deviation from the random combinatorial proportions of haplotypes is called linkage disequilibrium.
4. The theory of population genetics for a single locus works well for populations in linkage equilibrium.
5. Linkage disequilibrium can arise because of non-random mating, random sampling, and natural selection.
6. For selection to generate linkage disequilibrium, the fitness interactions must be epistatic. That is, the effect on fitness of a genotype (such as A_1/A_2) must vary according to the genotype with which it is associated at other loci.
7. Pairs of alleles at different loci that cooperate in their effects on fitness are called coadapted. Selection acts to reduce the amount of recombination between coadapted genotypes.
8. When selection works on one locus, it will influence gene frequencies at linked loci. The effect is called hitchhiking.
9. Recombination is selectively disadvantageous in so far as it breaks down favorable gene combinations.
10. The mean fitness of a population can be drawn graphically for two loci; the graph is called a fitness surface or an adaptive topography.
11. Wright suggested that real adaptive topographies will have many separate "hills" with "valleys" between them. Natural selection enables populations to climb the hills in the adaptive topography, but not to cross valleys. A population could become trapped at a local optimum.
12. Random drift could supplement natural selection by enabling populations to explore the valley bottoms of adaptive topographies.
13. It is questionable whether real adaptive topographies have multiple peaks and valleys. They might have a single peak, with a continuous hill leading up to it. Natural selection could then take the population to the peak without any random drift.
14. Two-locus population genetics uses a number of concepts not found in single-locus genetics. The most important of these ideas are haplotype frequency, recombination, linkage disequilibrium, epistatic fitness interaction, hitchhiking, and multiple-peaked fitness surfaces.

FURTHER READING

Ford (1975) and Sheppard (1975) both introduce mimicry in swallowtail butterflies. The multilocus genetics of mimicry in *P. memnon* and in *Heliconius* is explained by Turner (1977, 1984). Population genetics for two loci, as for one locus, is introduced in such standard textbooks as Crow and Kimura (1970), Hartl and Clark (1989), and Hedrick (1983). Hedrick *et al.* (1978) is about multilocus population genetics in particular. The HLA loci are introduced by Bodmer and Cavalli-Sforza (1976); Potts and Wakeland (1990) and Hedrick (1994) are more recent evolutionary reviews. Dobzhansky (1970) and Wright (1977) discuss chromosomal inversions and coadaptation. On hitchhiking, see Maynard Smith and Haigh (1974) and Kaplan *et al.* (1989). On recombination, see Maynard Smith (1978a), the references in chapter 11, Haig and Grafen (1991), and Rice (1994). On Wright's shifting balance theory, see Wallace (1968, ch. 16), Wright's four-volume treatise, particularly volumes 3–4 (1969–1978, and 1986), Lewontin (1974, final chapter), Provine (1986, ch. 9), Kondrashov (1992), and Wade (1992). See Provine (1986) and Turner (1987) for the question of whether fitness surfaces will be multiply peaked.

STUDY AND REVIEW QUESTIONS

1. The haplotype frequencies in four populations are shown below.

Genotype: A_1B_1/A_1B_1	A_1B_1/A_1B_2	A_1B_2/A_1B_2	A_1B_1/A_2B_1	A_1B_1/A_2B_2	A_1B_2/A_2B_2	A_2B_1/A_2B_1	A_2B_1/A_2B_2	A_2B_2/A_2B_2
3/16	1/16	0	1/16	3/8	1/16	0	1/16	3/16
1/16	1/8	1/16	1/8	1/4	1/8	1/16	1/8	1/16
1/81	4/81	4/81	4/81	16/81	16/81	4/81	16/81	16/81
0	1/81	8/81	2/81	8/81	26/81	7/81	27/81	2/81

Calculate the linkage disequilibrium in each.

	Frequency of: A_1B_1	A_1	B_1	Value of: D
(a)				
(b)				
(c)				
(d)				

2. What kinds of selection can be hypothesized to be operating in (a)–(d) of question 1?

3. In a large population, with random mating, and in the absence of selection, what haplotype frequencies would you expect in populations (a)–(d) of question 1 after a few hundred generations?

4. How can the theory of hitchhiking help to reconcile the neutral theory of molecular evolution with the levels of heterozygosity observed in natural populations?

5. Draw a fitness surface for a single locus with two alleles (A and a) when heterozygous advantage operates. Where is the equilibrium on the graph?

Quantitative Genetics

THE Mendelian genetics of beak size in finches is unknown, but the character shows evolutionary changes as the food supply changes through time. We begin by looking at beak size in Darwin's finches as an example of the kind of character studied by quantitative genetics. We then move on to the theoretical apparatus used to analyze characters that are controlled by large numbers of unidentified genes. The influences on these characters are divided into environmental and genetic factors, and the genetic influences are divided into those that are inherited and influence the form of the offspring and those that are not. A number called heritability expresses the extent to which parental attributes are inherited by offspring. With the theoretical apparatus in place, we can then apply it to a number of evolutionary questions, including directional selection, in both artificial and natural examples, and stabilizing selection. We investigate whether a balance between selection and mutation can explain the amount of genetic variation seen in nature, and consider an experiment in which the effect of selection is minimized to reveal the power of deleterious mutation.

9.1 Climatic changes have driven the evolution of beak size in one of Darwin's finches

Fourteen species of Darwin's finches live in the Galápagos archipelago, and many of them differ most obviously in the sizes and shapes of their beaks. A finch's beak shape, in turn, influences how efficiently it can feed on different types of food. Grant has been studying these finches since 1973, and his best evidence to demonstrate that beak size influences feeding efficiency comes from a comparison of two species (Color Plate 3a,b), the large-beaked *Geospiza magnirostris* and the smaller *G. fortis*, feeding on the same kind of hard fruit.

The large-beaked *G. magnirostris* can crack the fruit (called the mericarp) of caltrop (*Tribulus cistoides*) transversely, taking on average only 2 seconds and exerting an average force of 26 kgf. It can then easily eat all 4–6 seeds of the smashed fruit, taking only 7 seconds. The smaller *G. fortis* is not strong enough to crack *Tribulus* mericarps and instead twists open the lower surface, applying a force of only 6 kgf and taking 7 seconds on average to reach the seeds inside. Only one or two of the seeds can be obtained in this way, and it takes an average of 15 seconds to extract them. Thus, *Geospiza magnirostris* usually has an advantage with these large, hard types of food.

Smaller finches are probably more efficient with smaller types of food, although this capability is more difficult to show. Both large and small finches on the Galápagos eat small seeds, although there is an indirect reason (as we will see) to believe that smaller finches do so more efficiently. From the evidence we have examined so far, we should predict that natural selection would favor larger finches when large fruits and seeds are abundant. The prediction should apply both within and between species. A *G. magnirostris* looks like an enlarged *G. fortis*, and a larger individual *G. fortis* can probably deal with a large food item more efficiently than can a smaller conspecific, much as an average specimen of *G. magnirostris* is more efficient than an average *G. fortis*. When large seeds are common, we might expect the average size of a population of *G. fortis* to increase between generations, and vice versa when large seeds are rare—if beak size is inherited.

If beak size is inherited . . . but is it? Beak size is inherited if parental finches with larger than average beaks produce offspring with larger than average beaks. Grant measured the sizes of parental and offspring finches in several families and plotted the latter sizes against the former (Figure 9.1). Large-beaked parental finches do, indeed, produce large-beaked offspring: thus, beak size is inherited. It therefore makes sense to test the prediction that changes in beak size should follow changes in the size distribution of food items. The test was carried out by Grant for the species *Geospiza fortis*, on one of the Galápagos Islands, Daphne Major. Since the study began, this species has undergone two major, but contrasting, evolutionary events.

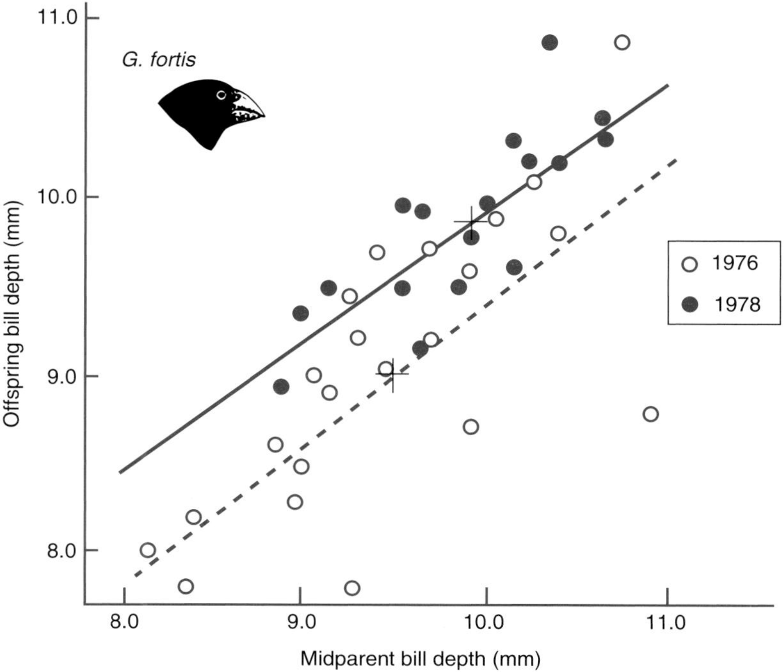

Figure 9.1 Parents with larger than average beaks produce offspring with larger than average beaks in *Geospiza fortis* on Daphne Major. Beak size is inherited. Reprinted, by permission of the publisher, from Grant (1986).

In the Galápagos, the normal pattern of seasons calls for a hot, wet season from about January to May to be followed by a cooler, dryer season through the rest of the year. In early 1977, for some reason, the rain did not fall. Instead of the normal progression of a wet season followed by a dry season, then another wet season and dry season, the dry season that began in mid-1976 continued until early 1978—one whole wet season was missed. The finch population of Daphne Major collapsed from about 1200 to about 180 individuals, with females being particularly hard hit; at the end of 1977, the sex ratio was about five males per female. As the sex difference shows, not all finches suffered equally. Smaller birds died at a higher rate. The reason, again, relates to the food supply. At the beginning of the drought, the various types of seeds were present in their normal proportions. *G. fortis* of all sizes take small seeds, and, as the drought persisted, these smaller seeds were relatively reduced in numbers. With time, average available seed size became larger (Figure 9.2). Now the larger finches were favored, because they could eat the larger, harder seeds more efficiently; and the average finch size increased as the smaller birds died off. (Females died at a higher rate than males because females are on average smaller.) Size, as we have seen, is inherited. The differential mortality in the drought, therefore, caused an increase in the average size of finches born in the next generation. *G. fortis* born in 1978 were about 4% larger on average than those born before the drought.

Four years later, in November 1982, the weather reversed. The rainfall of 1983 was exceptionally heavy and the dry volcanic landscape became covered with foliage in the periodic disturbance called El Niño (Color Plate 3c,d). Seed production was enormous. The theory developed for 1976–1978 could now be tested. The conditions had reversed: the direction of evolution should go into reverse as well. In the year after the 1983 El Niño event, an excess of small seeds appeared. If the smaller finches could, in fact, exploit them more efficiently, the smaller finches should survive relatively better. Grant again measured the sizes of *G. fortis* on Daphne Major in 1984–1985 and found that the smaller birds were indeed favored; finches born in 1985 had beaks about 2.5% smaller than those born before the El Niño downpours. His theory, that seed sizes control beak size in these finches, was confirmed.

The fluctuations in the direction of selection on beak shape—with smaller beaks favored in some years and larger ones in others—probably results in a kind of "stabilizing" selection over a long period of time such that the average size of beak in the population is the size favored by long-term average weather. Later in the chapter, we will see how the degree of selection can be expressed more exactly (Figure 9.9 will show the results for 1976–1977 and 1984–1985).

9.2 *Quantitative genetics is concerned with characters controlled by large numbers of genes*

The beak size of Galápagos finches illustrates a large class of characters that shows *continuous variation*. Simple Mendelian characters, like blood groups or the mimetic variation of *Papilio*, often have discrete variation. Many of the characters of species are like beak size in these finches, varying continuously, such that every individual in the population differs slightly from every other

Figure 9.2 During a drought in 1976–1977, (a) the population of *Geospiza fortis* decreased on the island of Daphne Major in the Galápagos archipelago, due to (b) the decline of the food supply. (c) The average size of the seeds available as food increased during the drought. Reprinted, by permission of the publisher, from Grant (1986).

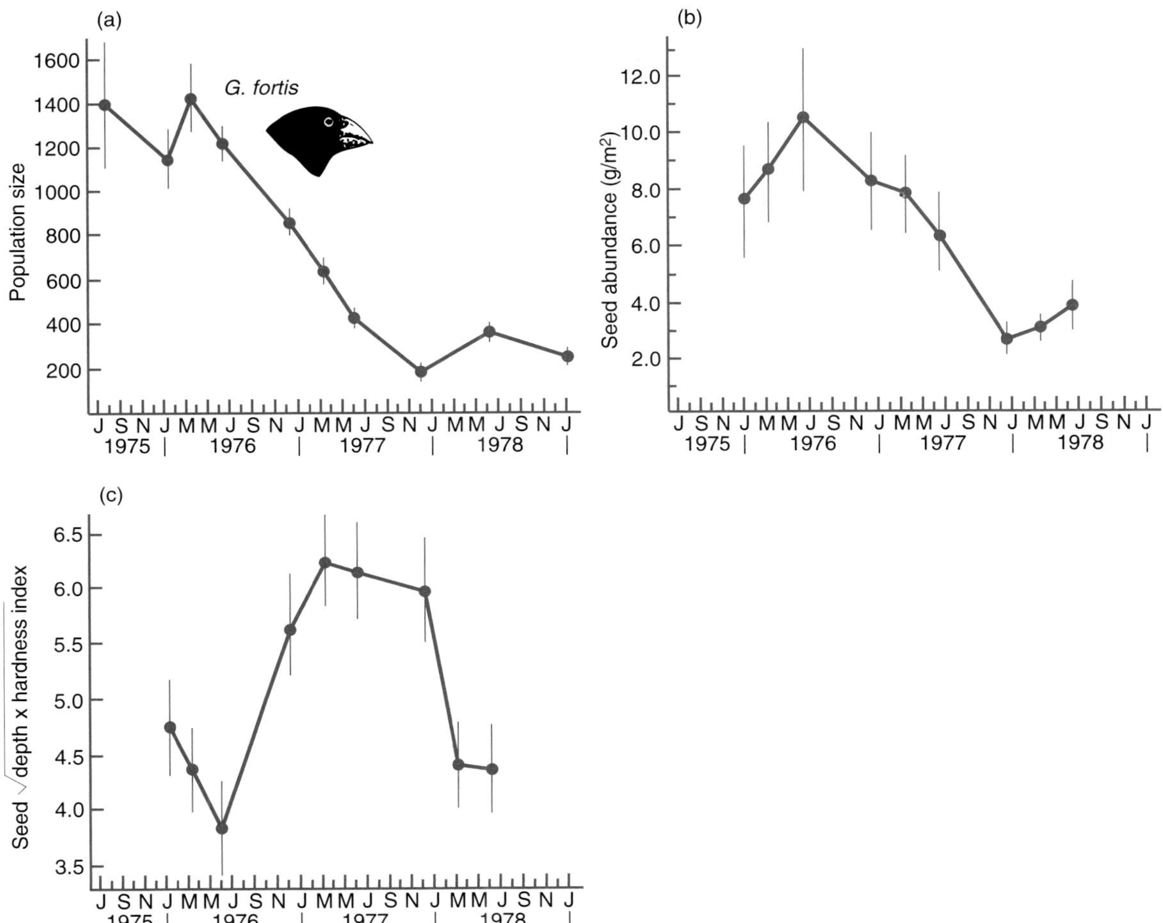

individual. There are no discrete categories of beak size in *G. fortis* or in most other species of birds.

The other important point about beak size is that we do not know the exact genotype that produces any given beak size. We can, however, say something about the general sort of genetic control it may have. Characters like beak size, which has an approximately normal frequency distribution (i.e., a bell curve), are probably controlled by a large number of genes, each of small effect (Figure 9.3). Imagine that beak size was controlled by a single pair of Mendelian alleles at one locus, with one dominant to the other, so that *AA* and *Aa* is long and *aa* is short. In this case, the population would contain two categories of individuals (Figure 9.3a). Now suppose that beak size was controlled by two loci with two alleles each. Beak size might now have a background value (e.g., 1 cm) plus the

Figure 9.3 (a) The phenotypic character—beak size, for example—is controlled by one locus with two alleles (*A* and *a*). *A* is dominant to *a*. There are two discrete phenotypes in the population. (b) The character is controlled by two loci with two alleles each (*A* and *a, B* and *b*); *A* and *B* are dominant to *a* and *b*. There are three discrete phenotypes. (c) Control by six loci, with two alleles each. (d) Control by many loci, with two alleles each. As the number of loci increases, the phenotypic frequency distribution becomes ever more continuous.

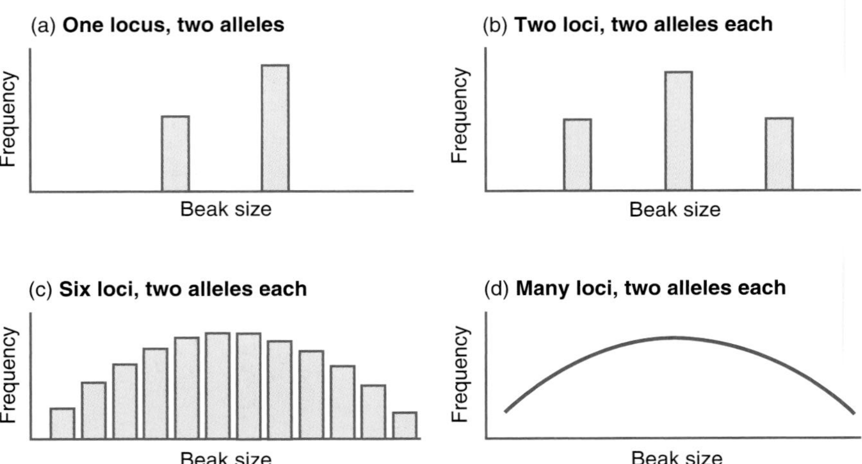

contribution of the two loci; an *A* adds 0.1 cm and a *B* subtracts 0.1 cm. *A* and *B* are dominant to *a* and *b*. Then *aaBb* and *aaBB* individuals have 0.9 cm beaks, *aabb, AaBB, AaBb, AABB,* and *AABb* have 1 cm beaks, and *AAbb* and *Aabb* have 1.1 cm beaks. Figure 9.3b is the frequency distribution if all alleles have frequency one-half and the two loci are in linkage equilibrium. The distribution now has three categories and has become more spread out. It becomes even more spread out if beak size is influenced by six loci (Figure 9.3c) and it begins to look normal for 12 loci (Figure 9.3d).

Thus, when a large enough number of genes influence a character, it will have a continuous, normal frequency distribution. The normal distribution can result if a large number of alleles are present at each of a small number of loci influencing the characters, or if fewer alleles appear at a larger number of loci. In this chapter, we will mainly discuss the theory of quantitative genetics as if there were many loci, each with a small number of alleles. Although this genetic system may well underlie many continuously varying characters, the theory applies equally well with a few (even one) loci containing many alleles at each.

Mendel had noticed in his original paper in 1865 that multifactorial inheritance (i.e., the character is influenced by many genes) can generate a continuous frequency distribution. This theory was not well confirmed until later work was carried out, particularly by East, Nilsson-Ehle, and others, circa 1910. Quantitative genetics is concerned with characters influenced by many genes, called *polygenic characters.* For a quantitative geneticist, five to 20 genes is a small number of genes to be influencing a character; whereas many quantitative characters may be influenced by more than a hundred, or even several hundred, genes. For characters influenced by a large number of loci, it ceases to be useful to follow the transmission of individual genes or haplotypes (even if they have been identified) from one generation to the next. The pattern of inheritance, at the genetic level, is too complex.

There is an additional complication. So far we have considered only the effect of genes. The value of a character, like beak size, will usually be influenced by the environment in which the individual grows up. Beak size is probably related to general body size, and all characters related to bodily stature will be influenced by the amount of food an organism happens to find during its life. If we take a set of organisms with identical genotypes and allow some to grow up with abundant food and others with limited food, the former will typically become larger. In nature, each character will be influenced by many environmental variables, some tending to increase it, others to decrease it. Thus, if we consider a class of genotypes with the same value of a character before the influence of the environment and add the effect of the environment, some of the individuals of each genotype will be made larger and others smaller in various degrees. This variation will produce a further "spreading out" of the frequency distribution. Any pattern of discrete variation in the genotype frequency distribution is likely to be obscured by environmental effects and the discrete categories converted into a smooth curve (Figure 9.4).

The small effects of many genes and environmental variables are two separate influences that tend to convert the discrete phenotypic distribution of characters controlled by single genes into continuous distributions. A continuous distribution for a character could, in principle, result from either process; quantitative genetics is mainly concerned with characters influenced by both factors. Quantitative genetics employs higher-level genetic concepts that are genetically less exact than those of one- or two-locus population genetics, but which are more useful for understanding evolution in polygenic characters. Instead of following changes

Figure 9.4 Environmental effects can produce continuous variation. (a) Twenty-five individuals in the absence of environmental variation all have the same phenotype, with a value for a character of 20. (b) Influence of one environmental variable. The variable has five states, and, based on the state in which state it grows up, an organism's character becomes larger or smaller or is not changed. The five states change the character by +10, +5, 0, −5, and −10, respectively, and an organism has equal chance of experiencing any one of them. (c) Influence of a second environmental variable. This variable also has five equally probable states, and they change the character by +10, +5, 0, −5, and −10, respectively. Of the five individuals in (b) with character value 10, one will get another −10, giving a value of 0, a second will get −5, giving 5, and so on. After the influence of both variables, the frequency distribution ranges from 0 to 40 and begins to resemble a bell curve. With many environmental influences, each of small effect, a normal distribution would result.

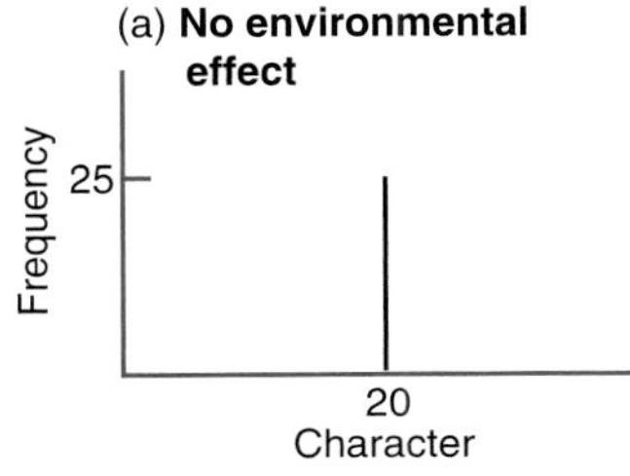

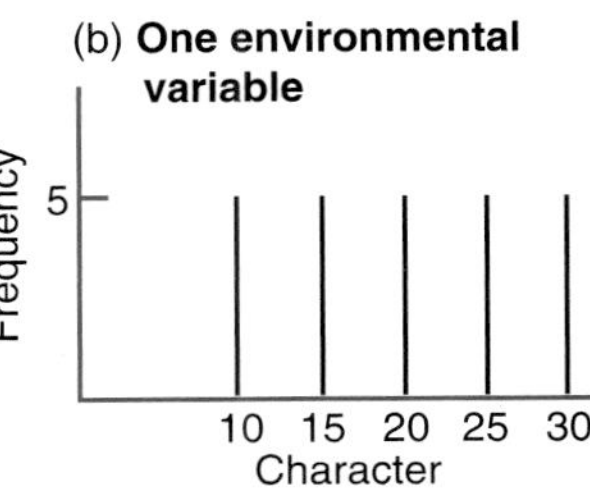

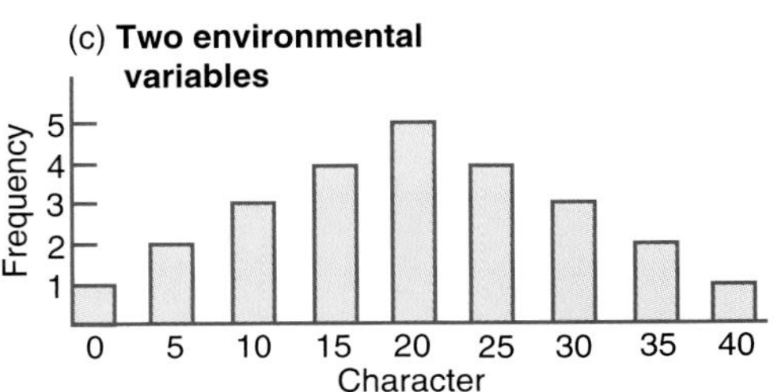

in the frequency of genes or haplotypes, we now follow changes in the frequency distribution of a phenotypic character. Quantitative genetics is important because so many characters have continuous variation and multi-locus control.

9.3 *Variation is first divided into genetic and environmental effects*

Quantitative genetics contains an unavoidable minimum of formal concepts that we need to understand before we can put it to use. Those formalities are the topic of this section and the next.

To understand how a quantitative character like beak size will evolve, we must "dissect" its variation. We tease apart the different factors that cause some birds to have larger beaks than others. Suppose, for example, that all the variation in beak size was caused by environmental factors—that is, all birds have the same genotype and they differ in their beak sizes only because of the different environments in which they grew up. Beak size could not then change during evolution (except for non-genetic evolution due to environmental change). For the character to evolve, it must be at least partly genetically controlled. We need to know how much beak size varies for environmental reasons, and how much for genetic reasons. However, even if different finches vary in their beak size for genetic reasons, it does not necessarily mean it can evolve by natural selection. As we will see, genetic influence must be classified even further, into components that allow evolutionary change and those that do not.

In quantitative genetics, the value of a character in an individual is always expressed as a deviation from the population mean. Thus, beak size will have a certain mean value in a population, and we can talk about environmental and genetic influences on an individual as deviations from that mean. The procedure is easy to understand if the population mean is viewed as a "background" value, and the influences leading to a particular individual phenotype are then expressed as increases or decreases from that value. Suppose there is one locus with two alleles influencing beak depth. AA and Aa individuals' beaks are 1 cm from top to bottom, aa individuals 0.5 cm. In this system, the environment has no effect. If the population average was 0.875 cm (as it would be for gene frequency of $a = \frac{1}{2}$), then the beak phenotype of AA and Aa individuals would be written as $+0.125$ cm and that of aa as -0.375 cm. In general, we symbolize the phenotype by P. In this case, $P = +0.125$ for AA and Aa individuals and $P = -0.375$ for aa.

Clearly, the value of P for a genotype depends on the gene frequencies. When the frequency of A is $\frac{1}{2}$, the genotypic effects are those just given. When the frequency of A is $\frac{1}{4}$, however, the population mean would be 0.71875 cm. For aa, P is now -0.21875; for AA and Aa, $P = +0.28125$.

In this example the phenotype is controlled only by a genotype, whose effect can be symbolized by G. G, like P, is expressed as a deviation from the population mean. In this case, for an individual with a particular genotype (because the environment has no effect)

$$\text{mean of population} + P = \text{mean of population} + G$$

The background population mean cancels from the equation, and can be ignored. We are then left (in this case in which the environment has no effect) with $P = G$.

The value of a real character will usually be influenced by the individual's environment as well as its genotype. If the character under study is related in some way to size, for example, it will probably be influenced by how much food the individual found during its development, and how many diseases it has suffered. These environmental effects are measured in the same way as for the genotype, as a deviation from the population mean. If an individual grew up in an environment that caused it to develop a larger beak than average, its environmental effect will be positive. In contrast, the environmental effect would be negative if it grew up in an environment giving it a smaller than average beak. The phenotype can then be expressed as the sum of environmental (E) and genotypic influences:

$$P = G + E$$

This simple relationship is the fundamental model of quantitative genetics. For any phenotypic character, the individual's value for that character (expressed as a deviation from the population mean) results from the effect of its genes and environment.

We must look further into the genotypic effect. We need to consider how to subdivide the genotypic effect, and why the subdivision is necessary. The main point can be seen in the one-locus example we have already used. The A gene is dominant, and both AA and Aa birds have 1 cm beaks ($P = +0.125$). (Because we are investigating the genotypic effect, we will ignore environmental effects, so that $P = G$.) Suppose we take an AA individual and mate it to another bird drawn at random from the population. The gene frequency $= 1/2$ and the random bird is AA with chance $1/4$, Aa with chance $1/2$, and aa with chance $1/4$. Regardless of the mate's genotype, all offspring will have beak phenotype $P = +0.125$, because A is dominant. Now suppose we take an Aa individual and mate it to a random member of the population. The average beak of their offspring is $(1/4 \times +0.125) + (1/2 \times 0) + (1/4 \times -0.125) = 0$. That is, the average offspring beak size is the same as the population average. Thus, for two genotypes with the same beak size ($P = +0.125$), one produces offspring with beaks like their parents, while the other produces offspring with beaks closer to the population average.

As we can see, some genotypic effects are inherited by the offspring and some are not. We can divide the genotypic effect into a component that is passed on and a component that is not. The component that is passed on is called the *additive effect* (A) and the component that is not is called (in this case) the *dominance effect* (D). The full genotypic effect in an individual is the sum of the two:

$$G = A + D$$

The additive effect has the greatest impact. The parent deviates from the population mean by a certain amount, and its additive genotypic effect is the part of that deviation potentially passed on to its offspring. However, when an individual reproduces, only half of its genes are inherited by its offspring. The offspring inherit only half the additive effect of each parent. Thus, the additive effect A for an individual is equal to twice the amount by which its offspring deviate from the population mean, if mating is random. For the AA parent, therefore, the additive effect is $+0.25$. (The full quantitative genetics of the AA individuals are: $G = +0.125$, $A = +0.25$, and $D = -0.125$.) The offspring of *Aa* birds deviate by zero, so their additive effect is twice zero (or simply zero). Thus, the amount by which *Aa* heterozygotes deviate from the population mean is entirely due to dominance, and is not inherited by their offspring. (For *Aa* individuals, $G = +0.125$, $A = 0$, $D = +0.125$.)

The division of the genotypic effect into additive and dominance components tells us what proportion of the parent's deviation from the mean is inherited, and reveals how the non-inheritance of the *Aa* individuals' genotypic effect is due to dominance. In practice, quantitative geneticists do not know the genotypes underlying the characters they study; they know only the phenotype. They might, for instance, focus on the class of birds with 1 cm ($P = +0.125$) beaks. In the example given above, the additive component of their phenotypic value depends on the frequencies of the *AA* and *Aa* genotypes. If all birds with 1 cm beaks are *Aa* heterozygotes, then none of the offspring will inherit their parents' deviation; if they are all *AA*, then half of the offspring will inherit the deviation.

Why is the additive effect of a phenotype so important? Once the additive effect for a character has been estimated, that estimate plays much the same role in quantitative genetics as the exact knowledge of Mendelian genetics in a one- or two-locus case (chapters 5 and 8). We use this estimate to predict the frequency distribution of a character in the offspring, given a knowledge of the parents. In a one-locus genetic model, we know the genotypes corresponding to each phenotype, and can predict the phenotypes of offspring from the genotypes of their parents. In the case of selection, the gene frequency in the next generation is easy to predict if we know selection allows only *AA* individuals to breed. In two-locus genetics, the procedure is the same. If the next generation is formed from a certain mixture of *Ab*/*AB* and *AB*/*AB* individuals, we can calculate its haplotype frequencies if we know the exact mixture of parental genotypes.

In quantitative genetics, we do not know the genotypes. We have only measurements of phenotypes, like beak size. If we can estimate the additive genetic component of the phenotype, then we can predict the offspring in a manner analogous to the procedure when the real genetics are known. When we know the genetics, Mendel's laws of inheritance tell us how the parental genes are passed on to the offspring; when we do not know the genetics, the additive effect tells us which component of the parental phenotype is passed on. Thus, estimating the additive effect is the key to understanding the evolution of quantitative characters. The estimates are practically made by breeding experiments. In the case of finches with 1 cm beaks in a population of average beak size 0.875 cm, the additive effect can be measured by mating finches with1 cm beaks to random members of the population. The additive effect is then two times the offspring's deviation from the population mean.

The genetic partitioning that we have undertaken so far applies reasonably well for one locus. In that case, dominance is the main reason why the genotypic effect of a parent is not exactly inherited in its offspring. When many loci influence a character, *epistatic interactions* between alleles at different loci can produce additional effects (see section 8.8, p. 208). Epistatic interactions, like dominance effects, are not passed on to offspring—they are not additive. They depend on particular combinations of genes and the effect disappears when the combinations are broken up (by genetic recombination). An example of an epistatic interaction would be for individuals with the haplotype A_1B_2 to show a higher deviation from the population mean than the combined average deviations of A_1 and B_2; the extra deviation is epistatic. Other non-additive effects can arise because of gene–environment interaction (when the same gene produces different phenotypes in different environments) and gene–environment correlation (when particular genes are found more often than random in particular environments).

A full analysis can take all of these effects into account. In a full analysis, such as the simple case given here, the aim is to isolate the additive effect of a phenotype. The additive effect is the part of the parental phenotype that is inherited by its offspring.

9.4 *The variance of a character is divided into genetic and environmental effects*

We return now to the frequency distribution of a character. We will continue to express effects as deviations from the population mean. If we consider an individual some distance from the mean, some of its deviation will be environmental, some genetic. In the genetic component, some parts will be additive, some dominant, some epistatic. These terms have been defined so that they add up to give the exact deviation of the individual from the mean. Any individual has its particular phenotypic value (P) because of its particular combination of environmental experiences and the dominance, additive, and interaction effects in its genotype (Figure 9.5). The different combinations of E, D, and A in different individuals explain why the character shows a continuous frequency distribution in the population.

The variation seen for the character in any one population could exist because of variation in any one of, or any combination of, these effects. Thus, the individual differences could all be due to different environmental effects, with every individual having the same value for G. Alternatively, it could be 25% due to the environment, 20% due to additive effects, 30% due to dominance effects,

Figure 9.5 The x-axis of the continuous distribution for a character can be scaled to have a mean of 0. Consider the individuals (called x) with phenotype $+P$. Their overall phenotypic value is the sum of their individual combinations of $E + A + D$; the individuals with character x can have any combination of E, A, and D such that $E + A + D = x$. Any individual's deviation from the mean is due to its individual combination of environmental, additive, and dominance (and other, such as epistatic) effects.

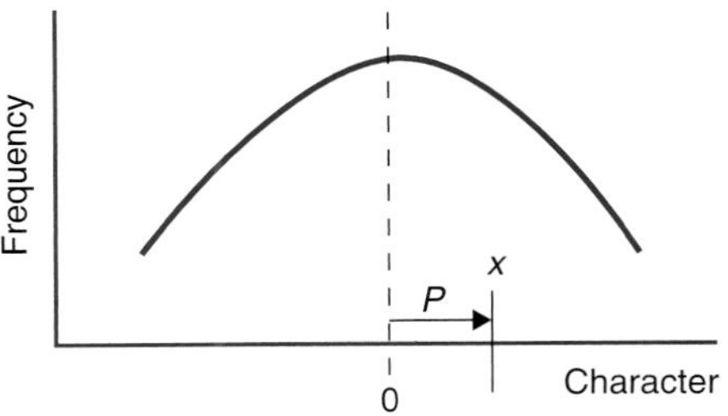

and 25% due to interaction effects. The proportion of variation related to the different effects matters when we consider how a population will respond to selection. If the variation exists because different individuals have different values of E, selection will not generate a response. On the other hand, if the variation is mainly additive genetic variation, the response will be large. The proportion of the variation resulting from different values of A in different individuals tells us whether the population can respond to selection.

The variability in the population related to any particular factor, such as the environment, is measured by the statistic called the *variance.* (Box 9.1 explains some statistical terms used in quantitative genetics.) Variance is the sum of squared deviations from the mean divided by the sample size minus one. For all values x of a character like beak size in a population, the variance of x is

$$V_X = \frac{1}{n-1} \Sigma (x_i - \bar{x})^2$$

We have seen how the total phenotype, genetic effect, environmental effect, and so on, can be measured for an individual. The measurements (P for phenotypic effect, and so on) are expressed as deviations from the mean. We can easily calculate what their variances are for a population.

$$\text{phenotypic variance} = V_P = \frac{1}{n-1} \Sigma P^2$$

$$\text{environmental variance} = V_E = \frac{1}{n-1} \Sigma E^2$$

$$\text{genetic variance} = V_G = \frac{1}{n-1} \Sigma G^2$$

$$\text{dominance variance} = V_D = \frac{1}{n-1} \Sigma D^2$$

$$\text{additive variance} = V_A = \frac{1}{n-1} \Sigma A^2$$

The phenotypic variance for a population, for example, expresses how spread out the frequency distribution for the character is. If it is spread out, with different individuals having very different values of the character, the phenotypic variance will be high. If the distribution is a narrow spike, with most individuals having a similar value for the character, phenotypic variance will be low.

9.5 ***Relatives have similar genotypes, producing the correlation between relatives***

As we saw in Figure 9.1 (p. 223), Grant measured the depth of the beak in many parent-offspring pairs in *Geospiza fortis*, and plotted the results on a graph. Each point indicates one offspring and the average of its parents. Beak depth in *G. fortis*, like many characters in most species, shows a correlation between parent and offspring. This correlation exists between relatives as well, for just as parents and offspring are similar to one another, so are siblings and more distant relatives similar to some extent.

Similarity among relatives may be either environmental or genetic, and the

BOX 9.1

Some Statistical Terms Used in Quantitative Genetics

The text mentions three main statistical terms. This box explains variance, but serves mainly as a reminder of the exact definition of covariance and regression. Reference should be made to a statistical text for fuller explanation.

Variance. The variability of a set of numbers can be expressed as a variance. Take a set of numbers, such as 4, 3, 7, 2, and 9. To calculate their variance, follow these steps.

1. Calculate the mean:

$$\text{mean} = \frac{4 + 3 + 7 + 2 + 9}{5} = 5$$

2. For each number, calculate the square of its deviation from the mean. For the first number, 4, it is $(5 - 4)^2 = 1$. Do likewise for all five numbers.
3. Calculate the sum of the squared deviations from the mean; for the five numbers, it is $1 + 4 + 4 + 9 + 16 = 34$.
4. Divide the sum by $n - 1$; $n = 5$ in this case.

$$\text{variance} = 34/4 = 8.5$$

The general formula for the variance of a character X is

$$V_X = \frac{1}{n-1}\Sigma(x_i - \bar{x})^2$$

where $\bar{x}$ is the mean and x_i is a standard notation for a set of numbers.

Here we have five numbers. In terms of the notation, that means that i can have any value from 1 to 5 and is the value of the character for each i. Thus $x_1 = 4$, $x_2 = 3$, and $x_5 = 9$. The summation in the general formula covers all values of i. If there had been 50 numbers, i would have varied from 1 to 50 and we should proceed as in the example for all 50 numbers. The variance describes how variable the set of numbers is. The higher the variance, the greater the differences among the numbers. If all numbers were the same (all $x_i = \bar{x}$), then their variance is zero.

Covariance. Imagine the individuals have been measured for two characters, X and Y. The covariance between the two is defined as

$$\text{cov}_{XY} = \frac{1}{n-1}\Sigma(x_i - \bar{x})(y_i - \bar{y})$$

Covariance measures whether an individual having a large value of X also has a large value for Y. If the x_i and y_i of an individual are both large, the product $x_i y_i$ will also be large; if y_i is not large when x_i is large, the product will be smaller. Generally, if X and Y covary, the product (and so the covariance) is large; if they do not, the sum of the products will come to zero.

Regression. The regression, symbolized by b_{XY}, between characters X and Y is their covariance divided by the variance of X.

$$b_{XY} = \frac{\text{cov}_{XY}}{V_X}$$

Regressions are used to describe the slopes of graphs and are, therefore, useful in describing the resemblance between classes of relatives. If X and Y are unrelated, $\text{cov}_{XY} = 0$ and $b_{XY} = 0$; if they are related, the covariance and regression can be positive or negative (Figure B9.1).

Figure B9.1 Relation between two variables (x and y). (a) Negative regression coefficient ($b < 0$). (b) No relation ($b = 0$). (c) Positive regression coefficient ($b > 0$).

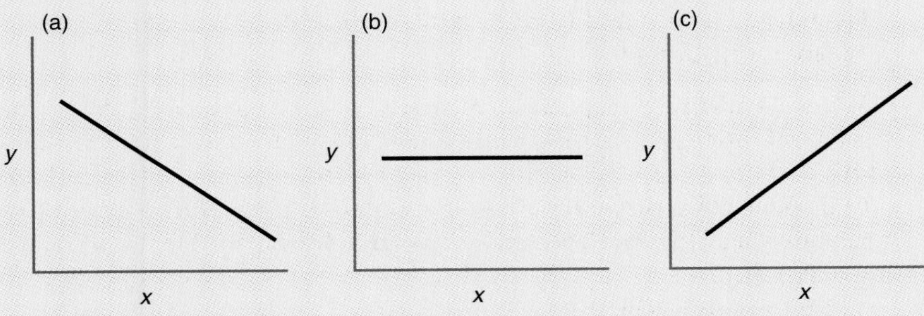

two effects may have varying importance in different kinds of species. In humans, in which parents and offspring live together in social groups, much of the similarity results from the family's common environment (i.e., gene–environment correlation). At the other extreme, in a species like a bivalve mollusc in which eggs are released into the sea at an early stage, relatives will not necessarily grow up in similar environments, and shared environments will not exert such a strong influence on similarity among relatives. Darwin's finches probably lie somewhere between these two extremes. It is easy to understand similarity among relatives that is caused by similar environments. In so far as relatives grow up in correlated environments, and environmental variation exists in a character, relatives will be more similar than non-relatives.

The similarity between relatives due to their shared genes is evolutionarily more important. It is possible to deduce this type of correlation among any two classes of relatives from the variance terms we have already defined. We will consider only one case—the correlation between parents and offspring—to see how it occurs. To simplify matters, we will assume that the environments of parent and offspring are not correlated. This assumption allows us to ignore environmental effects (because any environmental effect in the parent will not appear in the offspring—if the parent is larger than average because it benefited from a good food supply, that does not mean its offspring will share those benefits). Any correlation between parent and offspring will then be due to their genetic effects.

The genetic value of the character in the parent, we have seen, consists of several components of which only the additive component is inherited by the offspring. When mating is random, half of that additive component of the individual parent is diluted. At a locus, a parent has an additive deviation from the population mean in both its genes. When an offspring is formed, one of the parental genes goes into the offspring together with another gene drawn at random from the population (we are assuming random mating). The average value of the character in the offspring is, as we saw earlier, half of the additive value of the parent ($\frac{1}{2}A$); the average genetic value in the parent is $A + D$. The correlation between parent and offspring is the covariance between the two (Box 9.1):

$$\text{cov}_{\text{OP}} = \Sigma\ \tfrac{1}{2}A(A + D)$$

where all offspring–parent pairs are summed. The covariance can also be expressed as:

$$\text{cov}_{\text{OP}} = \tfrac{1}{2}\Sigma\ A^2 + \tfrac{1}{2}\Sigma AD$$

$\frac{1}{2}\Sigma AD = 0$ because A and D have been defined to be uncorrelated. That is, if an individual has a large value of A, we know nothing about whether its value of D will be large or small. Thus, the equation can be written as

$$\text{cov}_{\text{OP}} = \tfrac{1}{2}\Sigma A^2 = \tfrac{1}{2}V_A$$

To describe this relation in words, the covariance of an offspring and one of its parents is equal to half the additive genetic variance of the character in the population.

The expression for the covariance between one parent and its offspring is true for each parent. It requires only a small step (though we will not go into

TABLE 9.1

The correlations between several different classes of relatives

Relatives	Correlation
One parent and offspring	$\frac{1}{2}V_A$
Mid-parent and offspring	$\frac{1}{2}V_A$
Half siblings	$\frac{1}{4}V_A$
Full siblings	$\frac{1}{2}V_A + \frac{1}{4}V_D$

it here) to show that the same expression ($\frac{1}{2}V_A$) also gives the covariance between offspring and the midparental value. Other expressions can be deduced, by similar arguments, for the covariance between other classes of relatives (Table 9.1). The formulae are useful for estimating the additive variances of real characters. However, the estimates become most interesting, for the evolutionary biologist, when expressed in terms of the statistic called heritability.

9.6 *Heritability is the proportion of phenotypic variance that is additive*

The similarity between relatives in general, and between parents and offspring in particular, is governed by the additive genetic variance of the character. If a character has no additive genetic variance in a population, it will not be passed on from parent to offspring. For instance, many of the properties of an individual phenotype are accidentally acquired characters, such as cuts, scrapes, and wounds; if we measure these in parent and offspring they will show no correlation. Thus, $V_A = 0$ for such characters. Moreover, some characters, such as the number of legs per individual in a natural population of, for example, zebra, show practically no variation of any sort; for those characters, V_A trivially is zero. Additive variance is, therefore, often discussed as a fraction of total phenotypic variance. This fraction is called the *heritability* (h^2) of a character.

$$h^2 = \frac{V_A}{V_P}$$

Heritability is a number between zero and one. If heritability is one, all variance of the character is genetic and additive. Given that $V_P = V_E + V_A + V_D$, all terms appearing on the right other than *VA* must then be zero. In so far as the factors other than additive variance account for the variance of a character, heritability is less than one.

Heritability has an intuitive meaning. Consider a parent that differs from the population by a certain amount. If its offspring also deviate by the same amount, heritability is one; if the offspring have the same mean as the population, heritability is zero; if the offspring deviate from the mean in the same direction as their parents but to a lesser extent, heritability is between zero and one. Heritability, therefore, is the quantitative extent to which offspring resemble their parents, relative to the population mean.

How can we estimate the heritability of a real character? One method is to cross two pure lines. This approach is mainly of interest in applied genetics, where the problem might be to breed a new variety of crops; it holds little interest for the evolutionary biologist. The two other main methods involve measuring the correlation between relatives and the response to artificial selection, respectively.

Figure 9.1 is an example that uses the correlation between relatives to determine heritability. The slope of the graph, which shows the beak size in offspring finches in relation to the average beak size of the two parents, is equal to the heritability of beak size in that population. The reason is as follows. The slope of the line is the regression of offspring beak size on mid-parental beak size. The regression of any variable y on another variable x = $\mathrm{cov}_{xy}/\mathrm{var}_x$ (Box 9.1). The covariance of offspring and mid-parental value = $\frac{1}{2}V_A$ (Table 9.1) and the variance of the mid-parental beak size is equal to $\frac{1}{2}V_P$. (The variance is half of the total population variance because *two* parents have been drawn from the population and their values averaged; if Figure 9.1 had the value for one parent on the x-axis, its variance would be V_P). The regression slope simply equals V_A/V_P, which is the character's heritability. For beak depth in *Geospiza fortis* on Daphne Major, the regression and therefore heritability is 0.79.

9.7 A character's heritability determines its response to artificial selection

How can quantitative genetics be applied to understand evolution? We will consider two of the many ways possible here: directional selection and stabilizing selection. A third kind of selection—disruptive selection (section 4.4, p.73) —will not be discussed here. This section focuses on *directional selection*, which has been studied extensively through artificial selection experiments. Artificial selection is important in applied genetics, as it provides the means of improving agricultural stock and crops.

To increase the value of a character by artificial selection, we can use any of a variety of selection regimes. One simple form is truncation selection, in which the selector picks out all individuals whose value of the character under selection is greater than a threshold value, and uses them to breed the next generation (Figure 9.6). What will be the value of the character in the offspring generation? Let us define S as the mean deviation of the selected parents from the mean for the parental population; S is also called the selection differential. The response to selection (R) is the difference between the offspring population mean and the parental population mean. In this case, calculating the response to selection is found by regressing the character value in the offspring on that in the parents, where the parents are the individuals that were selected to breed. We plot the offspring's deviation against the parental deviation from the population average to produce a graph like that shown in Figure 9.1. The slope of the graph for parents and offspring is symbolized by b_{OP} and, as we saw in the previous section, for any character $b_{OP} = h^2$; that is, the parent-offspring regressional slope equals the heritability. Therefore

$$R = b_{OP}S$$

or

$$R = h^2S$$

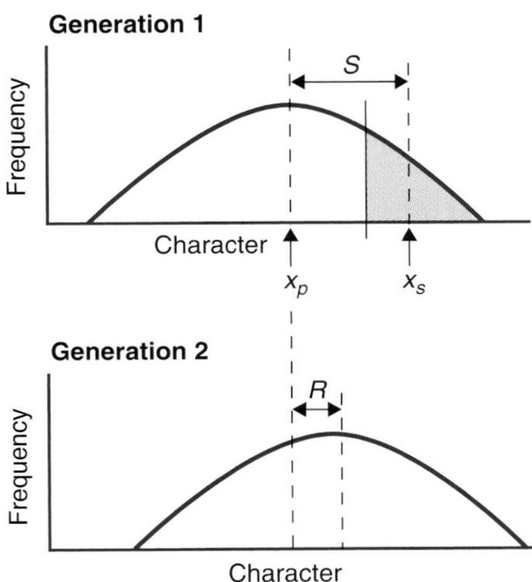

Figure 9.6 Truncation selection. The next generation is bred from those individuals (shaded area) with a character value exceeding a threshold value. The selection differential (s) is the difference between the whole population mean (x_p) and the selected subpopulation's mean (x_s). $s = x_s - x_p$. Because R is about 0.4 of S we could deduce in this case that $h^2 \approx 0.4$.

This result is important because it shows that the response to selection is equal to the amount by which the parents of the offspring generation deviate from the mean for their population multiplied by the character's heritability. (The response to selection or the parent-offspring regression can be used to estimate the heritability of a character; for a selected population, they are two ways of looking at the same set of measurements.)

A real example of directional selection may not take the form of truncation selection. In truncation selection, all individuals above a certain value for the character breed and all individuals below do not breed. All selected individuals contribute equally to the next generation. It is possible that there could be no sharp cut off, and that individuals with higher values of the character contribute increasing numbers of offspring to the next generation. However, the same formula for evolutionary response works for all forms of directional selection. The difference between the mean character value in the whole population and in those individuals that actually contribute to the next generation (if necessary, weighted by the number of offspring they contribute) is the "selection differential." This value can be plugged into the formula to find the expected value of the character in the next generation.

A population can respond to artificial selection only for as long as the directional selection lasts. Consider, for example, the longest-running controlled artificial selection experiment. Since 1896, corn has been selected, at the State Agricultural Laboratory in Illinois, for (among other things) either high or low oil content. As Figure 9.7 shows, even after 76 generations the response to selection for high oil content was not exhausted. However, it will eventually stop. As the corn is selected for increased oil content, the genotypes encoding for high oil content will increase in frequency and be substituted for genotypes for lower oil content. This process can proceed only until all the individuals in the population have the same genotype for oil content. At the loci controlling oil content, no additive genetic variance will then be left; heritability will have been reduced to zero and the response to artificial selection will terminate. In the Illinois corn

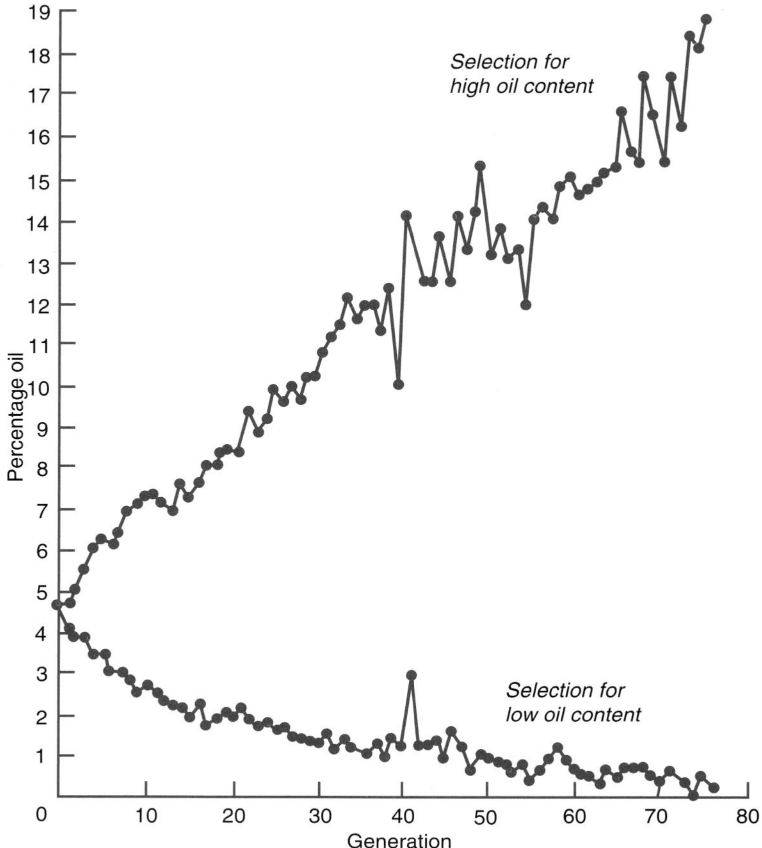

Figure 9.7 Response of corn (*Zea mays*) artificially selected for high or low oil content. The experiment began in 1896 when, from a population of 163 corn ears, the high line was formed from the 24 ears highest in oil content and the low line from the 12 ears with lowest oil content. Reprinted, by permission of the publisher, from Dudley (1977).

experiment, the process has not yet run its full course. The heritability of oil content in both the high and low selected lines has decreased through the experiment (Table 9.2), but it remains above zero—which is why the population continues to respond to selection.

In other artificial selection experiments, the complete process has been recorded. Figure 9.8 shows the response of a population of fruitflies to consistent directional selection for increased numbers of scutellar chaetae (i.e., bristles on a dorsal region of the thorax). Initially the population responded. Then, as the additive genetic variation was used up (or, as its heritability declined), the rate of change slowed to a stop in generations 4–14. It also appeared that, if selection was continued after the response stopped, the population suddenly began to respond again after an interval (in generations 14–17). The renewed bout of change is attributed to a rare recombinant or mutation that reinjected new genetic variation into the population.

The relation between response to selection (R), heritability, and selection differential (S) enables us to calculate any one of the three variables if the other two have been measured. For example, we saw in chapter 4 that fishing has selected for small size in salmon, because larger fish are selectively taken in the

TABLE 9.2

The heritability of oil content in corn populations after different numbers of generations of artificial selection for high or low oil content, in the Illinois corn experiment (Figure 9.7). The heritability declines as selection is applied. From Dudley (1977).

	Heritability of Oil Content	
Generation	High Line	Low Line
1-9	0.32	0.5
10-25	0.34	0.23
26-52	0.11	0.1
53-76	0.12	0.15

nets. The selection differential S can be estimated from three measurements: the average size of salmon caught in the nets, the average size of salmon in the population at the mouth of the river (before it is fished), and the proportion of the population that is removed by fishing. All three have been measured and lead to the estimates that the salmon who survive to spawn are roughly 0.4 lb smaller than the population average. Figure 4.3 (p.75)shows the response (R) for this population—the average size of the salmon decreased by about 0.1 lb between each two-year generation. We can, therefore, estimate the heritability, $h^2 = 0.1/0.4 = 0.25$.

In the finches, Grant measured the response to selection (R) and heritability for several characters related to body size, and used these values to estimate selection differentials. We saw that in *Geospiza fortis* that heritability of beak size is about 80%. After the bout of selection for large size in 1976–1977, the finches were approximately 4% larger. We can estimate the selection differential as $S = 0.04/0.8 = +5\%$. The results for several characters in three periods are shown

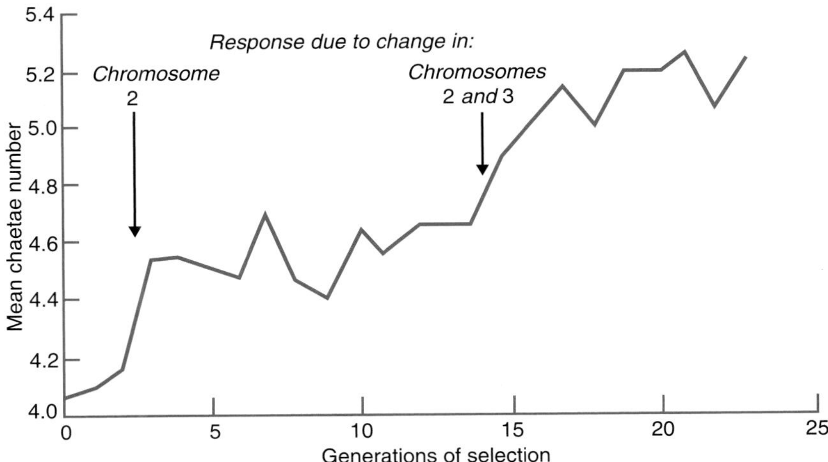

Figure 9.8 Artificial selection for increased numbers of scutellar chaetae in *Drosophila melanogaster*. The response took place in two rapid steps; the steps coincided with observable changes in the form of chromosomes 2, and 2 and 3, respectively. The changes are thought to have been recombinational events. Reprinted, by permission of the publisher, from Mather (1943).

Figure 9.9 (a) In the drought of 1976–1977, *Geospiza fortis* individuals with larger beaks survived better and the average size of the finch population increased. (b) In the normal years of 1981–1982, having a larger beak provided a slight advantage, but much smaller than during the drought. (c) After the 1983 El Niño, in 1984–1985, finches with smaller beaks survived better and the average size of the finch population decreased. The *x*-axis expresses the selection differential (*S*) in standardized form—that is, the mean for the survivors after selection minus the population mean before selection all divided by the standard deviation of the character. (The standard deviation is the square root of the variance; see Box 9.1 for the meaning of variance.) The value of *S* of about 5% for beak depth in the text corresponds here to a standardized *S* of about 0.6. Reprinted, with permission from *Nature,* from Gibbs and Grant (1987). Copyright 1987 Macmillan Magazines Unlimited.

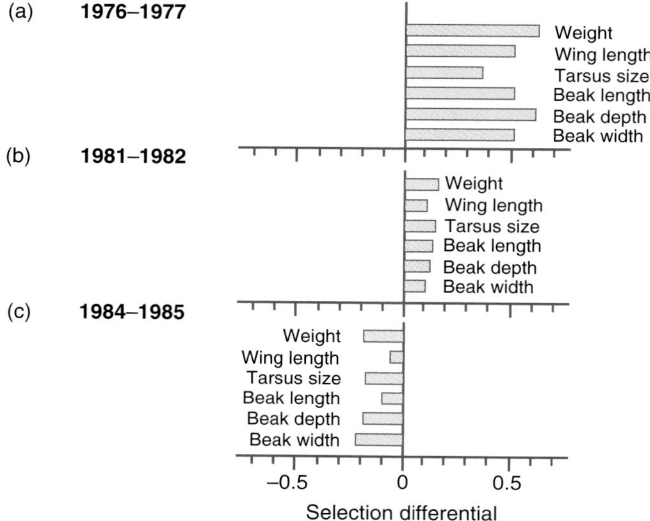

in Figure 9.9. As the direction of selection reversed from favoring larger beaks in 1976–1978 and smaller beaks in 1983–1985, the selection differentials reversed from positive values in 1976–1977 to negative values in 1984–1985.

In summary, the response to directional selection is controlled by the heritability of a character. The relation between the selection differential, response, and heritability can be used to estimate one of the variables if the other two are known. Sustained selection will exhaust the heritable variation, and the response will then stop.

9.8. *The relation between genotype and phenotype may be non-linear, producing remarkable responses to selection*

Figure 9.10 illustrates a remarkable artificial selection experiment. Scharloo selected a population of fruitflies for increased relative length of the fourth wing vein (Figure 9.10a). The figure shows the frequency distribution of vein lengths in the population for 10 generations. A length of 60–80 has been reached by about generation 5. At this stage the frequency distribution (amid the scatter often encountered in real experiments) starts to show a consistent bimodality.

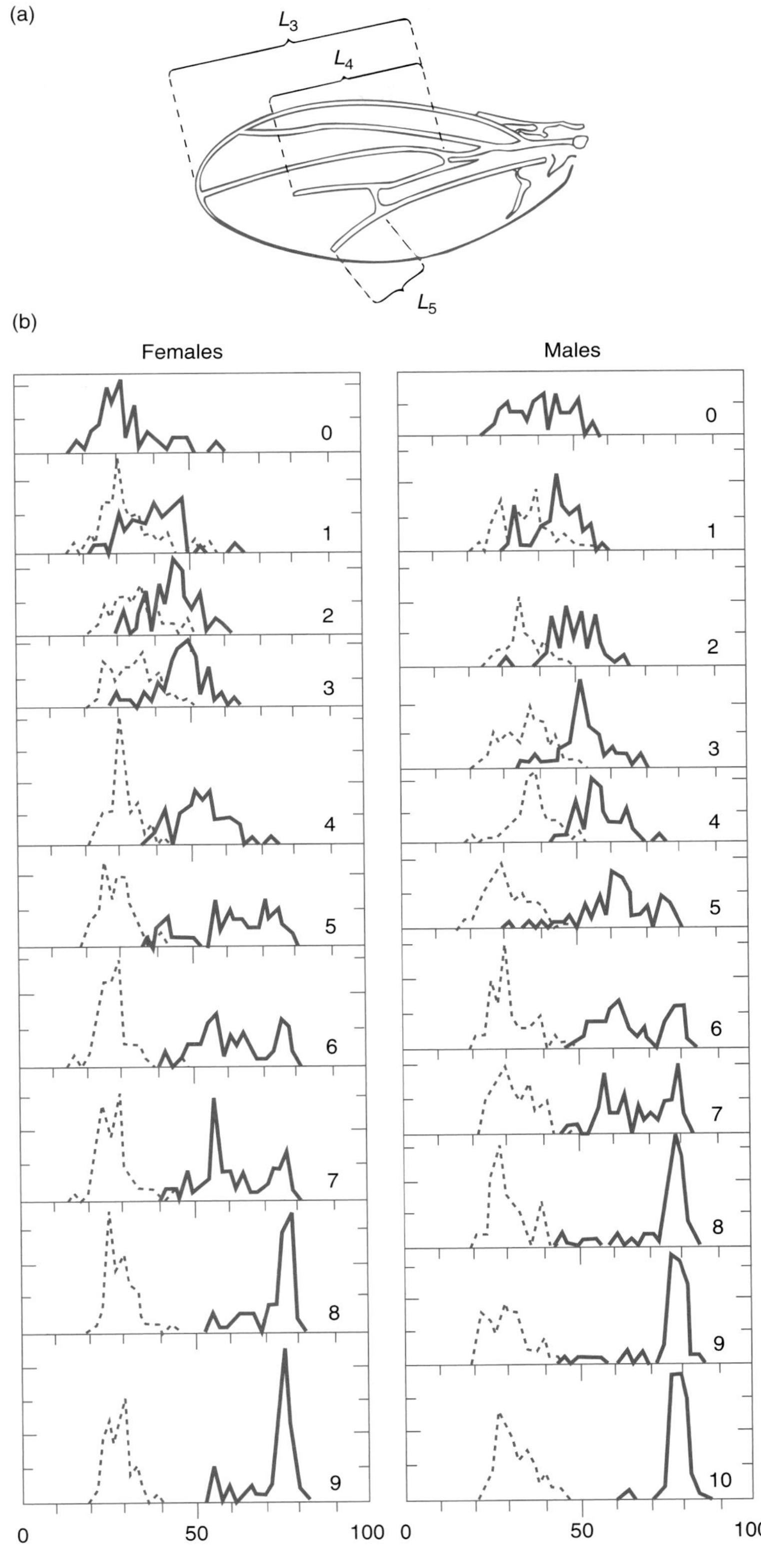

Figure 9.10 (a) The main veins in the wing of *Drosophila*. Relative length of the fourth vein was measured by the ratio L_4/L_3. (b) Artificial selection for relative length of fourth vein. The series on the left is for females, the series on the right for males. Each graph is a frequency distribution for the selected population. In both males and females, a bimodal distribution appears at vein lengths of about 60–80. Reprinted, by permission of the publisher, from Scharloo (1987). Copyright 1987 by Springer-Verlag.

This condition is clearest in generations 5–7, with only the high peak being maintained in generations 8–10. The experiment suggests that more complicated events can occur in artificial selection experiments than we have seen so far. What is going on?

The key to understanding the shape of the response is the relation between genotype and phenotype. A simple response, such as that for oil content (Figure 9.7) or bristle number (Figure 9.8), results when an approximately linear relation exists between genotype and phenotype (Figure 9.11a). Genotype here is expressed as a metrical variable. The easiest way to think of this relation is to imagine that the character is controlled by many loci; at each, some alleles (+) cause the phenotypic character to increase, and others (−) cause it to decrease. The more + genes an individual possesses, the higher its genotypic value (Figure 9.12a). When we select for an increase in the character, we pick the individuals with more + genes, enabling the value of the character to increase smoothly between generations in the manner of the Illinois corn oil experiment.

The approximately linear form of Figure 9.11a is not the only possible relation between genotype and phenotype (compare Figures 9.11b,c). The bimodal response in Figure 9.10 is thought to result from a threshold relation between genotype and phenotype (Figures 9.11c and 9.12). In Figure 9.12, the graph has been rotated 180° relative to the form in Figure 9.11; the *x*-axis (genotype) is drawn down the page on the left. The genotype is believed to control the amount of some vein-inducing substance. Vein length appears across the top of the graph. The relation between substance and vein length is hypothesized to contain a jump at vein length 60–80 (where the artificial selection response becomes bimodal).

Imagine the course of selection for longer wing veins with this threshold relation between genotype and phenotype. The population starts at the top left of the graph with relatively short veins. Initially, some variation appears in the population for genotype, with an associated normal distribution of vein lengths. Artificial selection produces flies with a higher concentration of the vein-inducing substance and with

Figure 9.11 Relation between genotype and phenotype. The genotypic value is for a polygenic character, in which higher values might be produced by more + alleles. The black frequency distribution for the phenotype at the bottom is what would be observed, given the shape of the graph, if the genotypic frequency distribution occurred as illustrated. (a) Linear relation. (b) "Canalized" relation. (c) Threshold relation. In the canalized relation, the phenotypic variance of a population is reduced relative to the variance in the causal genotypic factor; the reverse is true in the threshold relation.

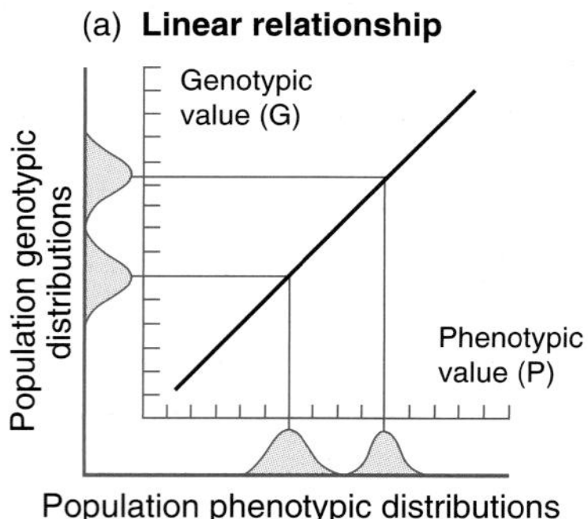

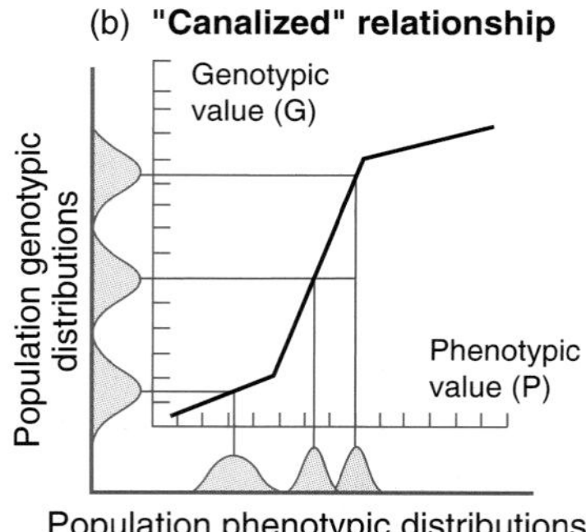

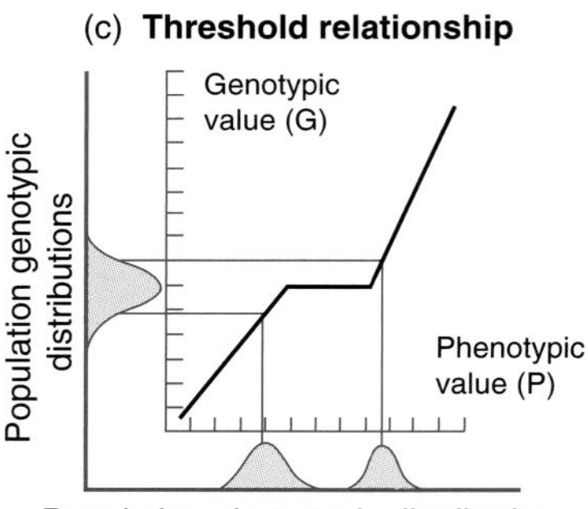

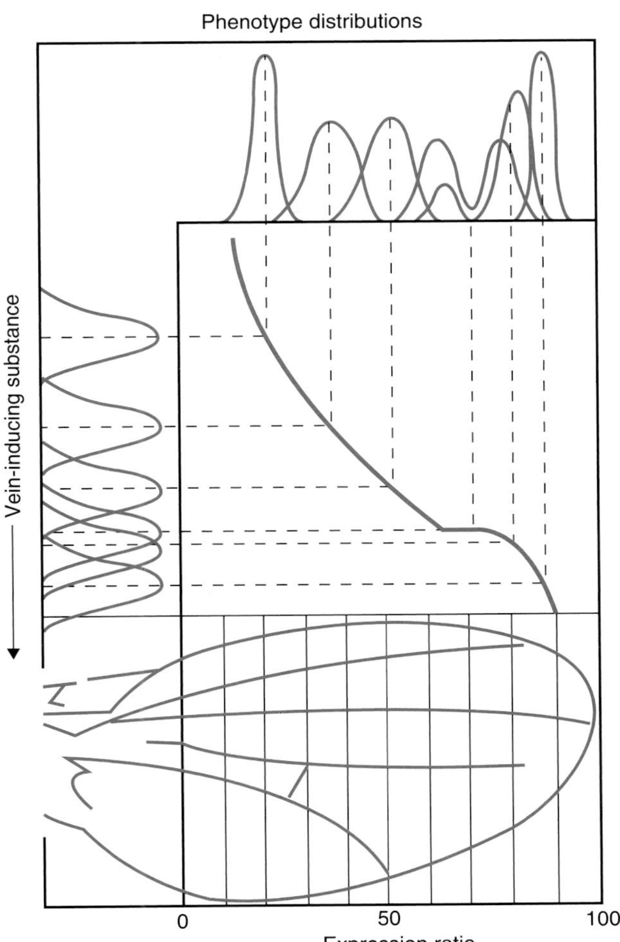

Figure 9.12 Model of relation between genotype and phenotype for *Drosophila* fourth wing vein. The genotype produces an amount of vein inducing substance. The phenotype (vein length) appears at the top; six different population frequency distributions are shown. The graph shows the relation between genotype and phenotype, and between the genotypic and phenotypic frequency distributions of a population. The phenotype in the expression ratio 60–80 range matches a threshold jump in the genotype-phenotype relation; a bimodal phenotypic frequency distribution arises there. Reprinted, by permission of the publisher, from Scharloo (1987). Copyright 1987 by Springer-Verlag.

correspondingly longer wing veins. This process continues until the population reaches a vein length of about 60–80. At that point, the normal (unimodal) variation for the amount of vein-inducing substance generates a bimodal distribution of vein lengths. The bimodal distribution will later disappear, as the population passes beyond the jump in the genotype-phenotype mapping function. Hence, we have the observed response to selection. The relation of genotype and phenotype for vein length in Figure 9.12 is only a hypothesis, but it does show how, in theory, a bimodal response to selection could arise.

The main points to take away from this discussison are that when the genotype-phenotype relation has the linear form of Figure 9.11a, a simple response to artificial selection occurs. The population changes until the genetic variation is exhausted. However, we have no reason to think that this process reflects the typical genotype-phenotype relation. When the relation is more complex, the response to artificial selection can be interestingly different, as the bimodal response to selection on wing vein length in fruitflies illustrates.

9.9 *Selection reduces the genetic variability of a character*

Natural populations have been undergoing the natural analogy of artificial selection for many generations. They can be thought of as being at the end of, rather than the beginning or anywhere else in, the selection experiment. For characters on which directional selection has been working, the heritability will have been reduced to a low level. We saw in Table 9.2 how the heritability of corn oil content has declined as the population has been selected. The more strongly selection works for a character, the more its heritability should decline, and we might expect that characters closely related to the reproductive fitness of individuals will show low heritability. (Figure 9.13, which is discussed in depth later, illustrates this point, as the heritabilities for life history characters appear lower than those for morphological characters.) In any case, directional selection acts to reduce genetic variation for the character.

For many characters, an intermediate rather than an extreme form is probably optimal; the development of this form is called *stabilizing selection.* (See section 4.4, p. 73, where Figure 4.4 illustrates how birth weight in humans demonstrates stabilizing selection.) How much genetic variation should we expect here? Before answering this question, we must distinguish between stabilizing selection and heterozygous advantage. At one locus, heterozygous advantage causes the maintenance of additive genetic variation (section 5.12, p.117); if it operated at a number of loci, it would also result in genetic variation in a polygenic system.

However, heterozygous advantage is not synonymous with stabilizing selection. The two concepts differ even in a one-locus system. If, at one locus, *Aa* is intermediate in phenotype between *AA* and *aa*, then any stabilizing selection would coexist with heterozygous advantage. However, if the heterozygote is not intermediate, the identity breaks down. If *AA* is intermediate in phenotype between *Aa* and *aa*, then any stabilizing selection would favor *AA*, not the heterozygote.

The difference between heterozygous advantage and stabilizing selection is more important for polygenic characters. Suppose the value of a character in an organism is influenced by a large number of loci. At each locus two alleles are possible. One increases (+) the value of the character, while the other decreases (−) it. An individual's haplotypes will then each consist of a series of alleles, and might, for example, be symbolized by − + + − + − − − + + (for 10 loci). An intermediate phenotype could be brought about by any genotype made up of half + genes and half-genes, such as any of the following:

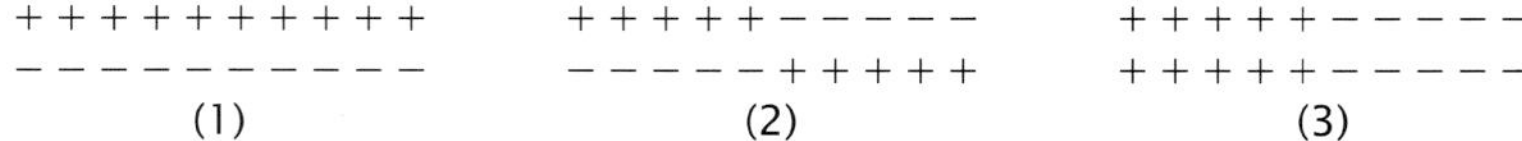

Cases (1) and (2) would produce heterozygous advantage. Selection could also favor the intermediate genotype (3), however, and in a population containing only this haplotype there is no heterozygous advantage.

How much genetic variability do we expect for a character subject to stabilizing selection? In a population containing the four haplotypes, the three genotypes (1), (2), and (3) will all have the same fitness, and we might expect a population to retain considerable genetic variation as these genotypes interbreed and produce a variety of offspring types. However, over a long period of time, genotypes like

Figure 9.13 Heritabilities of quantitative characters in *Drosophila*. Roff and Mousseau (1987) compiled heritability estimates from 130 studies and divided them into morphological, behavioral, and life history characters. The 130 studies are placed arbitrarily on the *y*-axis: morphological character 43, for instance, is "sternopleural bristle 2 sides" and morphological character 61 is "abdominal bristles 1 sternite." Each line consists of all heritability estimates in that study. Some characters are the subject of more than one study and will have more than one line in the figure. Morphological characters have heritabilities in the range 0.1–0.5; behavioral and life history characters have lower heritabilities, approximately 0.1. Reprinted, by permission of the publisher, from Roff and Mousseau (1987).

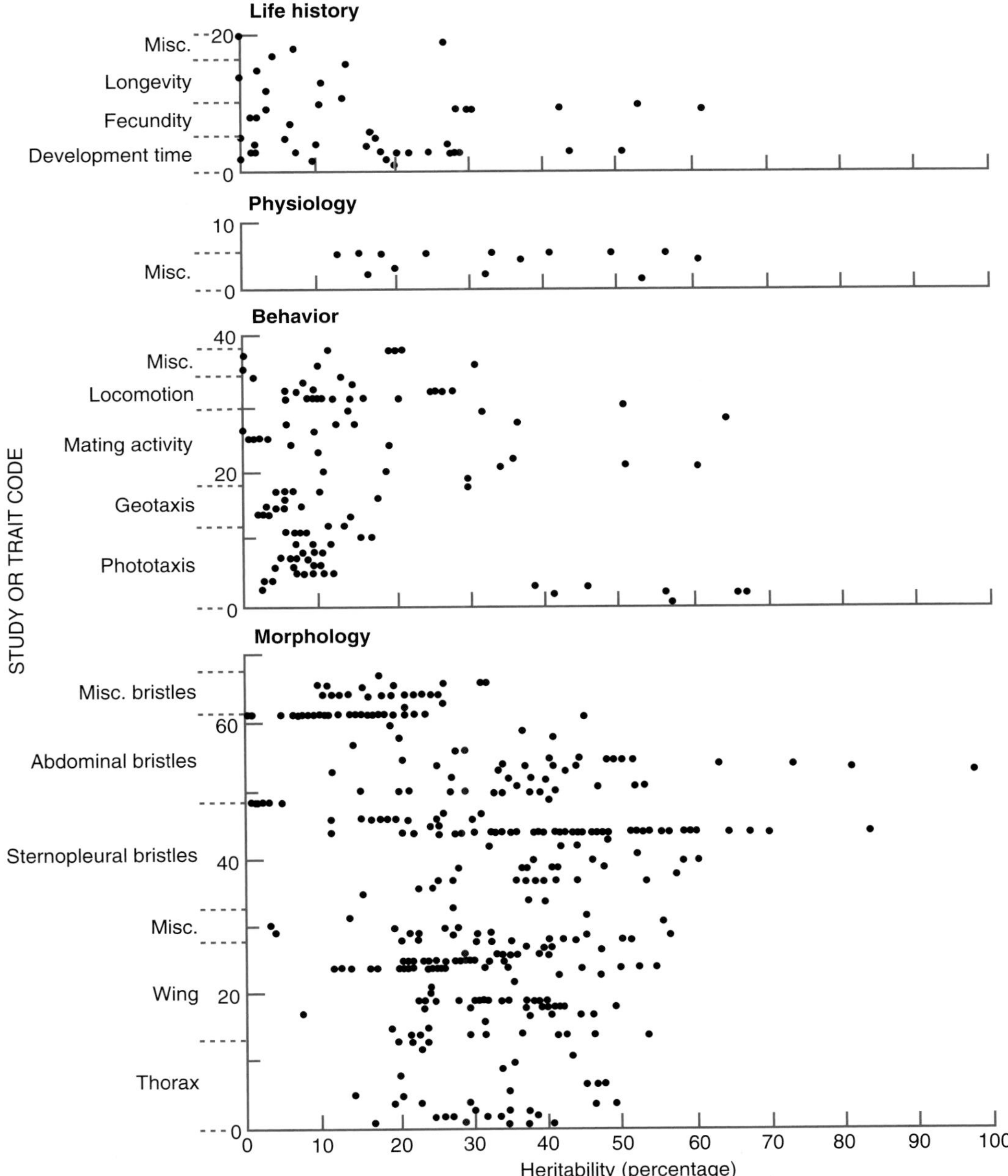

(3) that breed true will show an advantage, because all of this genotype's offspring have the optimal phenotype, whereas some of the offspring of genotypes (1) and (2) do not. In a population consisting of these three genotypes, selection slightly favors genotype (3). If the environment were constant for a long time, always favoring the same phenotype, selection should eventually produce a uniform population with a genotype like (3).

We can take the argument a stage further. Genotype (3) is not the only true-breeding homozygote that can produce the intermediate optimal form. All of the following also have this characteristic:

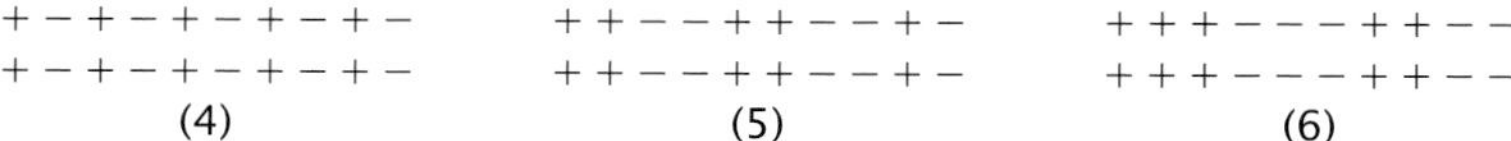

Suppose we have a population made up of genotypes (3)–(6), and selection favors the intermediate phenotype. What will happen now? Evolution will again tend toward a population with only one genotype, and that genotype should be a multiple homozygote. The reason behind this result is that any one of the homozygotes that has a slightly higher frequency than the others will have an advantage. Suppose, for example, that genotype (4) had a higher frequency than (3), (5), and (6). All genotypes will now be most likely to mate with a genotype (4). When genotype (4) mates with genotype (4), their offspring will all have the favored phenotype, identical to their parents. On the other hand, when genotypes (3), (5), and (6) mate with genotype (4), the offspring contain potentially disadvantageous genotypes. The offspring of a mating between a genotype (3) and a genotype (4), for example, will be + − + − + − + − + − / + + − − + + − − + − and has the favored intermediate phenotype. However, *its* offspring will contain disadvantageous recombinants. The end result is for selection to produce a uniform population, in which minority genotypes are selected against because they do not fit with the majority form. Selection eventually reduces the genetic variability to zero, even with stabilizing selection.

The argument leads to the following conclusion. If we start with a set of genotypes that influence a character subject to stabilizing selection, and if the environment remains constant for a long enough period, then selection ought eventually to produce a population containing only one, multiply homozygous, genotype.

9.10 *Characters in natural populations subject to stabilizing selection show genetic variation*

The conclusion of the previous section is contradicted by observable facts. Heritabilities can be measured for real characters, and many show significant genetic variation. Figure 9.13 summarizes some measurements for *Drosophila*. It suggests that typical values for heritability occur in the range 0.1–0.5. These estimates are for lab populations, and most reliable heritability estimates are unfortunately artificial. Some measurements have been taken from nature as well, such as Grant's measurements for the Galápagos finch, and these natural observations agree with the conclusion from Figure 9.13—real characters often have heritabilities higher than zero.

If stabilizing selection tends to eliminate heritable variation, why does it exist? The question is currently a matter of theoretical controversy, and we will look at two possible explanations: mutation–selection balance, and selective processes (like heterozygous advantage) that maintain variation.

9.11 *Selection–mutation balance is one possible explanation, but there are two models for it*

Some genetic variation exists in a population as mutation creates new deleterious genes and selection eliminates them. For any one locus, the amount of variation maintained by this selection–mutation balance is low because mutation rates are low (section 5.11, p. 115). For a polygenic character, however, mutation rates should be approximately multiplied by the number of loci influencing the character. A character controlled by 500 loci will have 500× the mutation rate of a one-locus character. The amount of variability that can be maintained is proportionally increased.

How much genetic variation will exist? This question has been approached from two directions. One approach, which was revived and developed by Lande in the 1970s, considers stabilizing selection on a continuous character (such as body size) controlled by many loci. Mutations at any of the loci can influence the character; because a genotype may be above or below the optimal value for the trait, a small random mutation has a 50% chance of being an improvement. The other approach, which was revived and developed by Kondrashov and Turelli in 1992, does not consider stabilizing selection on a phenotypic character, but supposes mutations occur at many loci and the great majority (many more than 50%) are deleterious. The result is a balance between selection and deleterious mutation at many loci.

We do not have space to go far into either theoretical system, or their relative merits. However, Box 9.2 gives an outline of Lande's theory. The conclusion of the box is tentative but suggests that mutation alone cannot explain the facts of Figure 9.13. We will not discuss the other theoretical system at all, but we can look at the experimental evidence measuring mutation rates of deleterious mutation at many loci. The results of this examination have general evolutionary importance.

9.12 *The rate of slightly deleterious mutations can be observed in experiments in which selection against them is minimized*

The first large-scale experiments to illustrate the destructive power of mutation when selection is minimized were performed by the Japanese geneticist Terumi Mukai. (They were indeed large-scale—Mukai's first experiment in 1964 made use of about 1.7 million flies.) The aim of the experiment was to reduce the effect of selection to a minimum for a number of generations, and measure the decline in fitness. Fitness declines because mutations accumulate—mutations that would normally be eliminated by selection. Mutations can be of varying degrees of destructiveness, from lethal to neutral. The rate of lethal mutations had been measured before Mukai's work, as had the rate of visible mutations

BOX 9.2

Can a Balance of Mutation and Stabilizing Selection Explain Genetic Variation in Quantitative Characters?

Lande tackled this question by developing formulae derived by Kimura (1965) and by Latter. We need a formula for the rate at which variation is removed by stabilizing selection and another for the rate at which it is added by mutation. The balance of these two rates then gives the amount of genetic variation we should see if only these two processes are operating.

The full argument is too long to present here, but we can follow the gist of it by simplifying the work of Turelli. The strength of stabilizing selection can be expressed by the ratio V_S/V_E. V_E is defined as the environmental variance of a character. V_S is the phenotypic variance of the part of the population that has been selected to produce the next generation (Figure B9.2). V_E acts as an upper limit on the strength of selection. The strongest possible selection happens when all the selected individuals that will produce the next generation have the same genotype; they would still show phenotypic variability equal to the environmental variance, V_E. If selection is less extreme, V_S will exceed V_E, because the selected individuals will contain more than one genotype. Turelli collected evidence to suggest that, in real cases of stabilizing selection, the ratio V_S/V_E ranged from 5 to 100; he used 20 as an approximate number.

Mutation will generate variability in the population. Clayton and Robertson defined V_M as the variance due to mutation. This variance consists of mutations of various effects on the character. If the organism contains n loci, numbered from 1. . .n, then mutation rate can be defined as μ_i and the average effect on the phenotype of a mutation as m_i at the ith locus (the effect, as usual, is measured as a deviation from the mean). The total variance in the character introduced by mutation, for a population of diploid individuals, is then calculated as follows:

$$V_M = 2\ \Sigma \mu_i m_i^2$$

The summation occurs over the n loci. (If we think of μ and m as averages for all n loci, we can write the formula as $V_M = 2n\mu m^2$.)

Lande collected several estimates of V_M. These estimates suggested that

$$V_M \approx 10^{-3} V_E$$

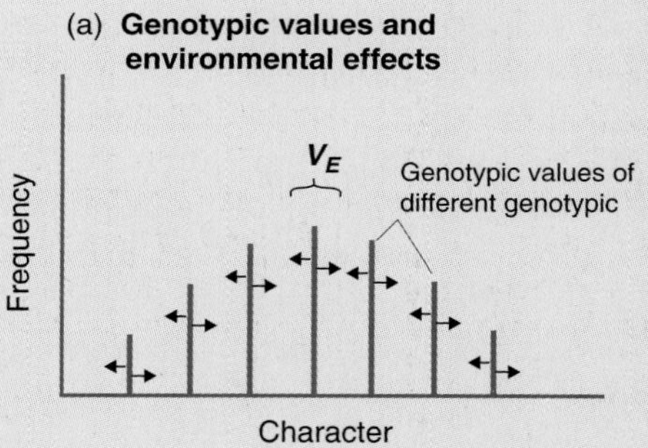

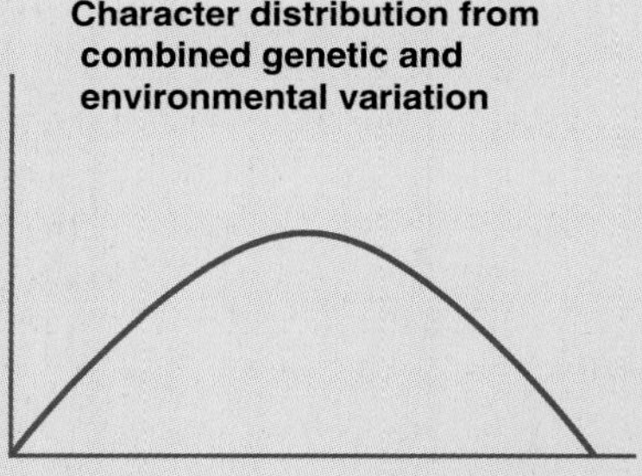

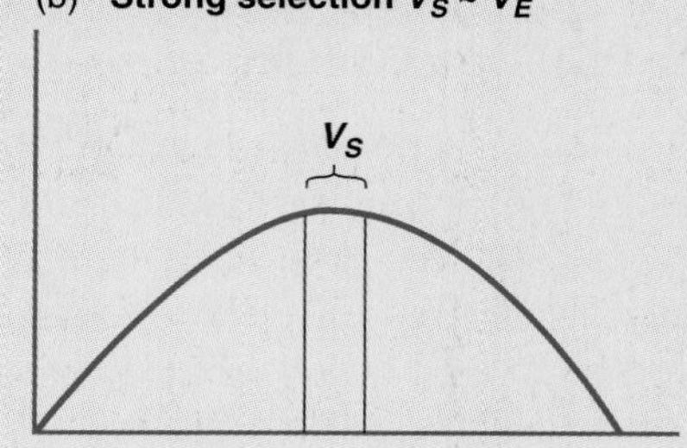

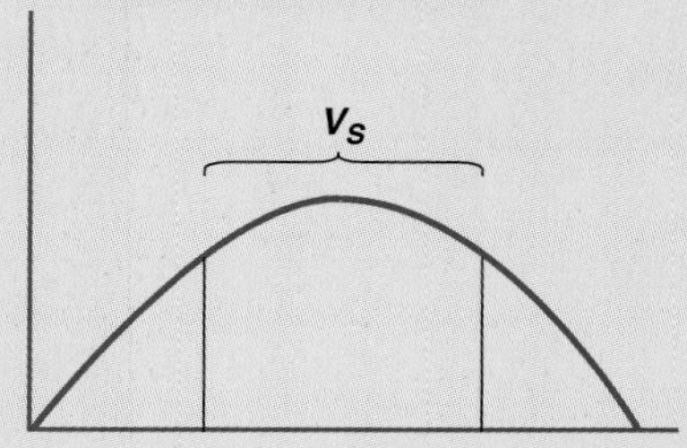

Figure B9.2 The degree of stabilizing selection can be described by the ratio V_S/V_E. (a) The frequency distribution of a character exists because there are several different genotypes (seven in this case) with different genotypic values, and the environment causes individuals with the same genotype to grow up with some deviation from the genotypic value. (b) If extremely strong selection operates, such that all selected individuals have the same genotype, they will still have variance of V_E. A V_S/V_E ratio of 1 is, therefore, the strongest possible selection. Under weaker selection, the selected individuals will have some genotypic as well as the environmental variation: $V_S/V_E > 1$.

If the influence of mutation and stabilizing selection are put together, the amount of genetic variability is given by a formula that Kimura first derived. We will not work through the derivation here, but the formula is

$$V_G \approx \sqrt{2nV_M V_S}$$

The reader can confirm that the observed values of $V_M \approx 10^{-3} V_E$ and $V_S \approx 20 V_E$ do, indeed, fit in with heritabilities of about 0.1–0.5. (Heritability = $V_G/(V_G + V_E)$. Therefore $V_G = V_E$ when heritability is ½. Try $n = 25$ in the formula.) We have now reached the conclusion of Lande's argument. Realistic values of mutation and stabilizing selection in polygenic systems can alone account for the observed levels of heritability.

Turelli, however, noticed that Kimura's formula assumes mutations of very small effect. More exactly, it requires $m_i^2 << V_{G(i)}$: the effect of the mutations at the *i*th locus is much smaller than the genetic variance at the locus. The condition can be tested crudely using Lande's own estimate of $V_M \approx 10^{-3} V_E$. The plausibility of the condition ($m_i^2 << V_{G(i)}$) depends on the mutation rate. It can be met if the genetic variance is due to many mutations of small effect, but if only a small number of mutations occurs, each mutation must have a larger effect. A conventional estimate of mutation rate of about 10^{-6} implies a small number of mutations, and the condition is violated. Thus, the assumption Kimura made to derive the formula above is wrong. Mukai's experiment (section 9.12) suggested mutation rates about 50 times higher than the conventional figure. In Mukai's experiment, however, the mutations were all deleterious, whereas not all the mutations implied in Lande's theory are deleterious. In any case, even if we include all of Mukai's mutation, the total mutation rate is still not high enough to meet the condition. Even a mutation rate 100 times higher, at 10^{-4}, Turelli calculated to imply $m_i^2 \approx 10 V_{G(i)}$.

It is possible to derive Kimura's formula again for $m_i^2 > V_{G(i)}$. Turelli performed this derivation, and concludes that Lande's argument cannot be saved in this way. To explain the observed heritability values, very high mutation rates and numbers of loci are needed, such as 1000 loci with mutation rates of 10^{-5} each. More realistic mutation rates (about 10^{-6}) and numbers of loci (<100) imply much lower levels of genetic variability than are observed. The conclusion, however, depends on the mutation rates used in the argument, and they are subject to uncertainty.

Lande's general idea was to use estimates for mutation rates, the degree of stabilizing selection, and genetic variability, to see whether a balance between stabilizing selection and mutation could explain the amounts of genetic variability. It has proved a fruitful line of research. The estimates used in the formulae are quite crude—they were made for artificial populations—and the conclusions may have to be modified. At present, it appears that the mutation-selection balance alone may not explain all of the levels of genetic variation that are observed in polygenic characters, as in Figure 9.13. Some further factor is probably needed—most likely natural selection in a form that actively favors variation.

Heterozygous advantage is an example of this question. In section 9.9 we assumed that an intermediate phenotype could be equally well produced by a + + + − − −/+ + + − − − homozygote as by a + + + + + +/− − − − − − heterozygote. As we saw, selection will fix the homozygotes in that case. If the heterozygotes were fitter for some reason, genetic variation would persist because of the segregation of recombinant genotypes from the heterozygotes. Why heterozygotes should be fitter is another question, but if they are, it would explain the maintenance of genetic variation. A second possibility is that selective equilibrium is not reached because the phenotype favored by selection changes rapidly. Darwin's finches may be an example of this case; the best beak size may change every few years, and genetic variation for beak size will persist because selection never favors one genotype long enough to fix it in the population. Frequency-dependent selection can also act to maintain variation. Any of these processes may operate in particular cases in addition to mutation, and they may account for the observed amounts of genetic variation in a character subject to stabilizing selection.

Further reading: For summaries: Barton and Turelli (1989), Bulmer (1989), and Turelli (1986, 1988). The original papers were Lande (1976) and Turelli (1984, 1985). See also Lynch (1988), Kondrashov and Turelli (1992), and Houle *et al.* (1994).

(if the effect of a mutation is visible, it is likely to be a large change). These measurements are of only limited interest because lethal and other large mutations probably constitute only a minority of all deleterious mutations. Geneticists suspected the generation of a large class of "slightly deleterious" mutations, which cannot be individually detected at the phenotypic level, and with fitness effects appearing in between the lethal and the neutral mutations. Mukai's experiment revealed the importance of those mutations empirically.

Mukai designed his experiment to transfer a chromosome through a number of generations without recombination or selection, to reveal the way mutations accumulate on it. The experiment used fruitflies (*Drosophila melanogaster*). Fruitflies have four pairs of chromosomes; they are distinguishable and referred to as chromosomes 1–4. The second chromosome, which was the subject of Mukai's study, makes up about 30% of the total genome. Mukai carried out genetic crosses to create an individual male fly that was homozygous for its entire second chromosome (symbolized +/+): the particular second chromosome came from a fly caught in nature and was not peculiar in any way—the fitness of the homozygotes was in the normal range for homozygotes of natural chromosomes. After creating 104 lines from the original chromosome, Mukai bred the flies through a number of generations, preventing recombination and minimizing selection.

Recombination needed to be eliminated so that the changes in the chromosome would be only due to mutation: if the chromosome could also recombine, some changes would be due to exchange with other versions of the 2nd chromosome. Mukai prevented recombination in the females by means of genetic tricks; no recombination occurs in male fruitflies.

Ideally, Mukai would have prevented all selection as well, but that goal is impossible. Instead he minimized it, by two procedures. First, he picked only one male, at random, to father the next generation in each line. Second, the flies were reared in optimal conditions for fly growth, so that as many as possible eggs should have been able to grow up as adults. Selection would not have been reduced to zero, because some of the flies probably had such poor genotypes that they failed to develop; this kind of selection in not excluded in the experiment. Nevertheless, a high proportion of suboptimal flies should have grown up and had a fair chance of being picked to father the next generation. Thus, very little selection operated against slightly deleterious mutations.

As the generations passed, we could have expected mutation to cause the average quality of the flies to decline. Mukai measured the viability (as one component of fitness) of the flies every 10 generations. He sampled individuals from each line and used them to generate flies that were homozygous (+/+) for the chromosome that had accumulated the mutations, or heterozygous for it and a standard lab chromosome (+/standard). The relative survival of the two types of fly were then measured. Any lower survival in the homozygotes relative to the heterozygotes should be due to the accumulated mutations, which appear in double dose in the homozygote but only single dose in the heterozygote (allowance has to be made for dominance). In fact, the relative viability of +/+ flies steadily declined (Figure 9.14).

The result of Mukai's experiment is important for two reasons. First, it reveals

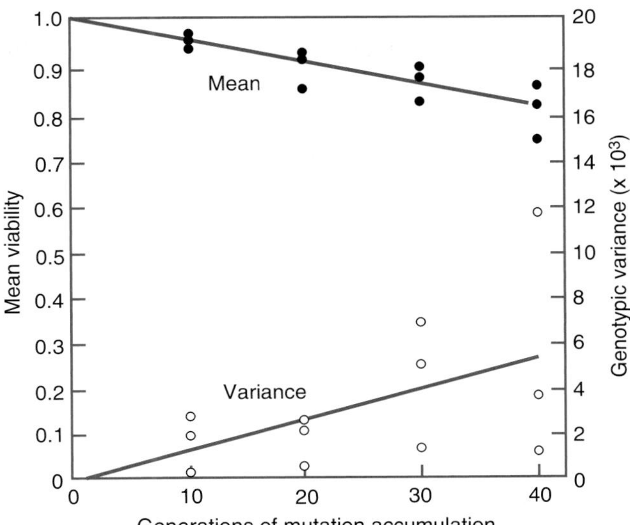

Figure 9.14 The mutational meltdown in fruitflies protected from selection. Note the decrease in viability. Viability is measured in flies that are homozygous for a chromosome that has experimentally accumulated mutations relative to flies that are heterozygous for the same chromosome. The decline is due to the accumulated deleterious mutations. There were 104 lines and the variance in viability among lines increases through time (see Box 9.1 for definition of variance). Reprinted, by permission of the publisher, from Mukai *et al.* (1972).

the destructive power of mutation. After 40 generations, enough mutations have accumulated in the + chromosome to reduce by 15% the viability of a homozygous +/+ fly relative to a heterozygote. When selection on a population is stopped (or at any rate minimized), the quality of the individuals in it deteriorates over the generations. This fact can be seen even more dramatically, if in less scientific form, by anyone who looks at the flies after a number of generations of the Mukai experience. Mukai himself did not describe his experimental flies, but after 50 generations or so in a similar experiment, the flies are noticeably less vigorous than normal flies.

Second, the quantitative rate of decline in relative fitness can be used to estimate the actual rate of deleterious mutation. We will not work through the full estimation procedure here (Crow 1993 describes it), except to say that it makes use of the variance in fitness (which is also shown in Figure 9.14). If the decline in fitness were caused by many mutations of small effect, the variance among lines would be small, whereas the variance would be large if it resulted from a smaller number of mutations of large effect. The actual variance can be used to derive the magnitude, and thus the implied rate, of the mutations responsible for the observed decline in fitness. Several estimates can be made, depending on the treatment of the numbers, but a low estimate is that the second chromosome in Mukai's experiment had a total mutation rate of 0.15 per generation (corresponding to 0.5 per whole genome per generation). That rate applies to all slightly deleterious mutations; thus, each mutation will have a particular effect on fitness, but we can estimate the average effect on fitness of all mutations. The total reduction in fitness, calculated by multiplying the mutation rate and the average fitness effect per mutation, was −0.003 per generation. Figure 9.14 shows that the number is approximately correct, because after 10 generations the relative fitness of a +/+ fly is about 3–4% lower than at the start of the experiment.

We can conclude that mutation introduces a large amount of genetic variation each generation, and for polygenic characters it is not implausible that selection–mutation balance could maintain quite large amounts of genetic variation at equilibrium. We can also draw one other incidental moral from Mukai's experiment concerning the role of natural selection in the living world. We have met natural selection as a force of change, in peppered moths, pestiferous insects, and Darwin's finches; we have encountered it as a force for constancy in human birth weight; and we have seen how it leads to the evolution of adaptation. Natural selection is all of these things, but now we have another way of looking at the process. We have just seen what happens to a population in which the effect of selection is minimized. We can imagine by extension how much worse things would be if no selection operated at all. Natural selection in a sense acts to support living things against a ferocious headwind of deleterious mutations that, in the absence of selection, would soon destroy them.

9.13 *Conclusion*

One- and two-locus population genetics is used for characters controlled by one or two loci and whose genetics is known. Quantitative genetics provides the techniques to understand evolution in characters that are influenced by a large number of genes, and for which the exact genotype (or genotypes) producing any given phenotype are unknown. It is possible that the majority of characters have this kind of genetics, in which case quantitative genetics would be appropriate for understanding the majority of evolution. At any rate, it is a highly important set of techniques.

In this chapter, we have seen how quantitative genetics divides up the variation in a character to recognize the component that controls how offspring resemble their parents; the component is called the additive genetic effect. The additive genetic effect plays the same role in quantitative genetics as a knowledge of Mendelian genetics in one- and two-locus population genetics.

The response to selection can be analyzed by means of the heritability of a character, which is the fraction of its variation due to additive genetic effects. However, even with simple directional selection, the exact response depends on the underlying genetic control. For example, the possible threshold relation between the genotype and phenotype for the wing veins of the fruitfly generates an interesting bimodal response to selection. Here, the heritability of the character would show strange changes as the character evolved. Directional selection unambiguously should continue to alter a character until its heritability is reduced to zero. With stabilizing selection, it might be thought that many genotypes could be maintained if they all produce the same intermediate phenotype. However, even here it can be argued that all but one of the genotypes should eventually be eliminated by selection. The argument appears to be contradicted by the facts, and it remains unsettled how mutation and the exact form of stabilizing selection stop populations from becoming genetically uniform.

SUMMARY

1. Quantitative genetics, which is concerned with characters controlled by many genes, considers the changes in phenotypic and genotypic frequency distributions between generations, rather than following the fate of individual genes.
2. The phenotypic variance of a character in a population can be divided into components due to genetic differences and to environmental differences between individuals.
3. Some of the genetic effects on an individual's phenotype are inherited by its offspring; others are not. The former are called additive genetic effects; the latter are due to such factors as dominance and epistatic interaction between genes.
4. The heritability of a character comprises the proportion of its total phenotypic variance in a population that is additive.
5. The heritability of a character determines its evolutionary response to selection.
6. The additive genetic variance can be measured by the correlation between relatives, or by artificial selection experiments.
7. The response of a population to artificial selection depends on the amount of additive genetic variability and on the relation between genotype and phenotype. If the relation is non-linear, strange bimodal responses can arise.
8. Stabilizing selection acts to reduce the amount of genetic variability in a population. However, polygenic characters show non-trivial values for heritability.
9. The level of genetic variation may be a balance between an input of new deleterious mutations and their removal by selection. Experiments in which the effect of selection is held to a minimum suggest the power of deleterious mutation.
10. The rate of deleterious mutation may not be high enough to explain the observed levels of genetic variation, and some other factors, such as a form of selection that maintains variation, may be needed to explain the observations.

FURTHER READING

Moving from the introductory to the advanced, the following are accounts of quantitative genetics: Falconer (1989), Lewontin's chapters on the subject in Griffiths *et al.* (1993), and vol. 2 of Wright (1968–1978). Charlesworth (1994) shows how quantitative genetics can be applied to age-structured populations—as is necessary for such problems as life histories and senescence. Hill (1984) is an anthology of classic papers, and Weir *et al.* (1988) is a more recent conference. On Darwin's finches, see Grant (1986, 1991), Grant and Grant (1995), and the popular book by Weiner (1994). Mitchell-Olds and Rutledge (1986) review the application to natural plant populations.

On selection, see also the references in chapter 4 for introductions. On the limits to the response, see Lee and Parsons (1968). Waddington (1957) and Rendel (1967) are classic discussions of the relation between genotype and phenotype for polygenic characters. Scharloo (1987, 1991) reviews non-linear selection responses and canalization.

On mutation–selection balance, Crow (1993) discusses the work of Mukai and others who followed in Mukai's footsteps; Mukai (1964) is the original. Houle *et al.* (1994) is a more recent experiment, applying the technique to the evolution of senescence; also see Rice (1994). Johnston and Schoen (1995) estimate the total mutation rate by another method, in self-fertilizing plants. Langley *et al.* (1981) is another study, focusing on mutation-selection balance for null alleles in natural populations of fruitflies, that leads to an estimate of the average selection coefficient in heterozygotes of 0.0014; see also Gillespie (1991, p. 60). For the theory, see the references in Box 9.2, together with Loeschcke (1987).

STUDY AND REVIEW QUESTIONS

1. Review the reasons why characters that are influenced by a large number of genetic and/or environmental effects will show a normal distribution.
2. Suppose the average spine length in a population of porcupines is 10 inches, and we take a number of individuals with 12-inch spines and mate each with a random member of the population. The offspring grow up with spines of average length 10½ inches. What is the additive effect in those 12-inch porcupines?
3. Given below are measurements, or estimates, of the phenotypic values and their additive genetic components in a population of nine porcupines.

1	2	3	4	5	6	7	8	9	V_P	V_A	h^2
−1	−5	−5	0	0	0	+5	+5	10			
−4	−2	−2	0	0	0	+2	+2	+5			

(a) Calculate V_P, V_A, and h^2. (b) What would you predict as the average spine length for the next generation if porcupines 8 and 9 were used to produce it?

4. Imagine we are selecting porcupines to make them pricklier. The heritability of prickliness is 0.75. Average prickliness before selection is 100 (and the standard deviation is also 100). The average prickliness of the porcupines that produce the next generation is 108. What will be the prickliness of the next generation?
5. If a character is subject to stabilizing selection, how might the genotype-phenotype relationship for it evolve?
6. Suppose stabilizing selection acts on a character controlled by eight gene loci, and the optimum individual should have six + genes and 10 − genes. What will happen over time to a population that initially contains individuals of two sorts: those that are + + + − − − − −/+ + + − − − − − and those that are + + + + + + − −/− − − − − − − −?
7. In section 2.5 you were told that an approximate figure for a per locus mutation rate is 10^{-6} per generation; in section 7.6 you learned that an approximate figure for the number of loci in a fruitfly is about 10,000; in section 9.12 you can read that the second chromosome of a fruitfly comprises about 30% of its genome and has a total mutation rate of about 0.15 per generation. Are these numbers consistent? If not, suggest some reasons for the inconsistency.

Genome Evolution

THE organization of genes into genomes is a large question, and this chapter looks at only two aspects of this issue. Both involve non-Mendelian inheritance, in which genes are copied laterally through the DNA rather than vertically from parent to offspring. We first consider concerted evolution by looking at the definition of a gene family, and how such families originate by duplication. We then see that gene families often show an evolutionary pattern that is difficult or impossible to explain if mutations are independent in each gene of the family; it can be explained if mutations move laterally, however. Next, we move on to the evolution of selfish DNA—that is, functionless sequences of DNA that are passively copied from generation to generation. We see that large quantities of non-coding DNA are present in eukaryotic species, and that (at least in some species) much of it is repetitive, being made up of duplications of certain unit sequences. We look at the kinds of repetitive DNA and the genetic processes that operate in them, such as slipping and unequal crossing over among tandem repeats, and transposition between scattered repeats. If non-coding repetitive DNA is selfish DNA, it would explain the apparently unnecessary excess of DNA in eukaryotes.

10.1 Non-Mendelian processes must be added to classical population genetics to explain the evolution of the whole genome

The theory of population genetics, as we have discussed it in chapters 5–9, has been concerned with discrete genetic loci that are inherited according to Mendel's laws. The theory shows how mutation, selection, drift, migration, and linkage determine changes in gene frequencies. It was first developed in the 1920s and 1930s, before the dawn of molecular genetics. Since the 1960s, however, it has become increasingly clear that only a part—and probably a small part, at that—of the DNA in an organism consists of discrete genes, each coding for a protein or having some regulatory function. Large amounts of non-coding DNA are present as well, which are often arranged in the form of repeating unit sequences. The genes themselves, moreover, can have distinctive arrangements on the chromosomes; related genes, for instance, may be distributed in clusters in the DNA. These discoveries do not contradict the theory of population genetics. Indeed, the theory of population genetics is the only theory we have to make sense of them. However, additional concepts are needed to explain this system. In this chapter we will concentrate on two such concepts: concerted evolution and selfish DNA. They are related in that both depend on non-Mendelian heredity.

Any theory of how a genetic (or any other) phenomenon evolves must specify how the phenomenon originated, and then how it spread through the population. In the standard theory of evolution, new forms originate as mutations at a locus and then spread by either natural selection or random drift. The ideas in this chapter describe mechanisms for the origin of variants, rather than for their spread through a population. Once the new genetic variant of the types discussed in this chapter has originated, it spreads or does not spread by the same processes of drift and selection as any other genes.

10.2 *Genes are arranged in gene clusters*

How are the many genes of an organism arranged in its DNA? The theory of population genetics would make good sense if single genes were scattered arbitrarily (although we saw in chapter 8 how selection can bring coadapted genes together into a supergene). Some genes do exist by themselves as single copies, but many genes are arranged along the chromosomes in groups of related genes. The groups are called *gene clusters.* Related genes may be arranged in more than one physical cluster, and a whole family of related genes is called a *gene family.* Gene clusters and gene families may vary in importance in different taxonomic groups. They seem to be much rarer, for example, in insects than in mammals, so most of this chapter will be concerned with mammalian and vertebrate genes. Gene clusters, where they exist, have peculiar evolutionary properties, which we can introduce by means of two well-studied cases: the ribosomal RNA genes and the globin gene family.

Eukaryotic ribosomes are made up of three classes of ribosomal RNA: 5S, 18S, and 28S. The 18S and 28S parts are formed from a large 45S precursor molecule that is transcribed from the DNA as a whole. Along the DNA, the 45S ribosomal genes are arranged in rows of multiple copies of a recognizable unit (Figure 10.1a); hundreds of these units are strung out in a row of *tandem repeats.* In the African clawed toad *Xenopus laevis,* the 400–600 tandem repeats all appear on one chromosome; in humans, approximately 300 repeats are scattered through five chromosomes in tandem repetitive blocks of various sizes. The organism presumably needs multiple copies of the gene to synthesize large quantities of the gene product. Other functional genes also exist in tandem repeats. The five histone genes, at least in *Drosophila* and sea urchins, have this structure. Although the histone genes of vertebrates are also a multiple gene family, they are scattered about, rather than appearing side by side.

The globin gene family has a slightly different arrangement. The hemoglobin molecule that transports oxygen in the blood comprises a number of components. Human hemoglobin is a tetramer. In an adult, it is made up of two α-globins and two β-globins; in the fetus, of two α- and two γ-globins; and in the embryo, two ϵ- and two ζ-globins. The sequences of these five globin types are similar and the genes that code for them occur in clusters on the chromosomes. In humans, two main clusters of globin genes exist: the α-globin cluster on chromosome 16 and the β-globin cluster on chromosome 11; the myoglobin gene on chromosome 22 is also part of the same family (Figure 10.1b). Unlike the ribosomal RNA genes, the globins are not simple tandem repeats of an identical

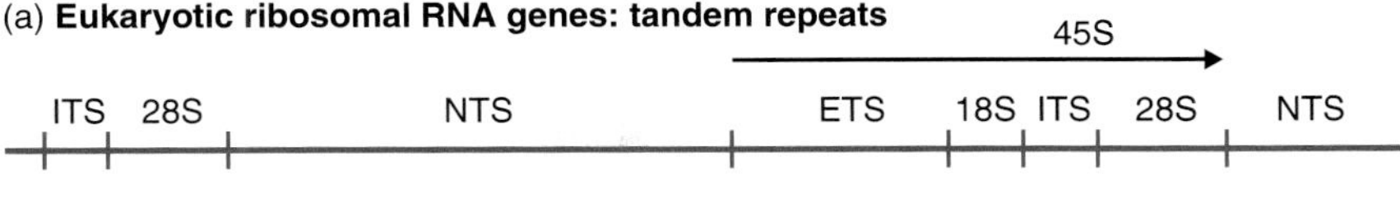

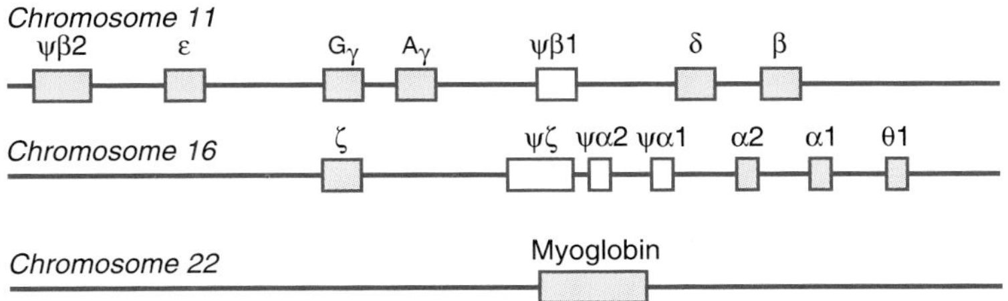

Figure 10.1 (a) Eukaryotic ribosomal RNA genes consist of long rows of tandem repeats of a unit genic organization, illustrated here. The transcribed part consists of the 18S and 28S genes together with an internal (ITS) and external transcribed spacer (ETS) region. The NTS part is a non-transcribed spacer. (b) The globin gene family consists of a number of clusters of related genes. In humans, illustrated here, there are three clusters. In other species, the globin gene in each cluster and the distribution of the clusters through the chromosomes differ from humans.

set of genes. Instead, the different globin genes represent distinct genes with distinct sequences. The ribosomal RNA genes and the globin genes, however, share the common property that the members of the gene family are linked in clusters. Many other examples of gene clusters are known, including the homeobox genes that are important in development and the HLA genes (Figure 8.3, p. 204).

10.3 *Gene clusters probably originated by gene duplication*

How could a cluster of genes, like the globin gene family, originate? For related genes strung out along a chromosome, duplication seems the most likely mechanism. Gene clusters are so common that duplication must have been a highly important mechanism during evolutionary history. Duplications probably arise by "mistakes" in recombination; the process is called *unequal crossing over.* A misalignment occurs, such that if two copies of a gene initially appeared on each strand, then after crossing over three copies appear on one strand and one copy appears on the other strand (Figure 10.2a). It is particularly easy to envisage a misalignment if a number of copies of a gene already occur in a row. The sequences of the two strands are matched up when they align for recombination and a gene could easily align with a non-complementary copy in the series of repeats (Figure 10.2b).

In the human globin genes, duplication may have been facilitated by a non-coding repeat sequence between the globin genes. For example, the two γ-globin genes on chromosome 11 (Figure 10.1b) each have a short repeat sequence on either side. If this repeated sequence existed before the duplication, it could have been the target for the mismatching that produced the duplication (Figure 10.2c). Certain variants of the human α-globin genes provide even more evidence for the importance of unequal crossing over among the globins. A normal human

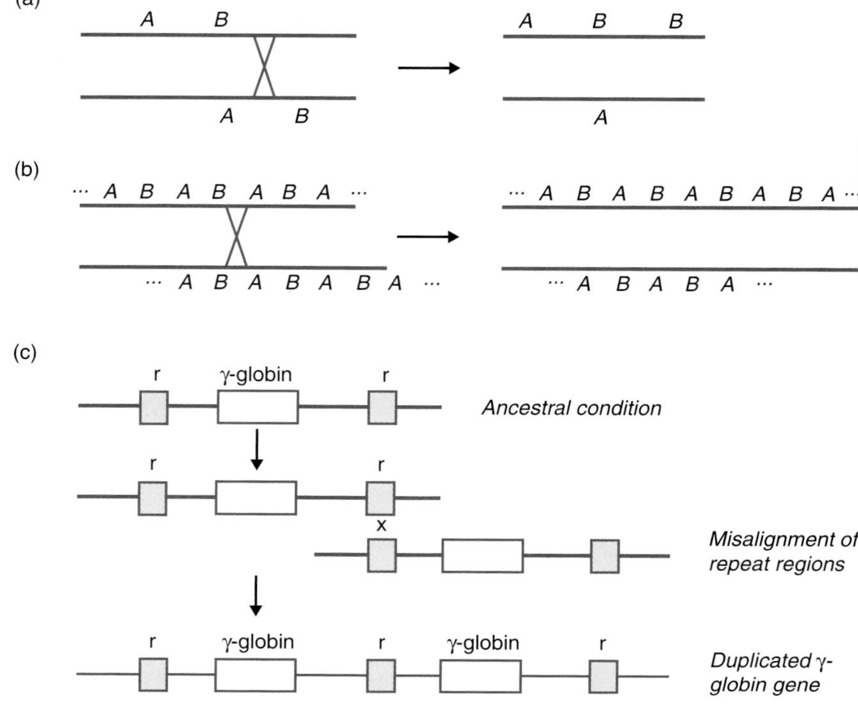

Figure 10.2 Unequal crossing over occurs when the sequences of the two chromosomes are misaligned at recombination. (a) In a simple case, chromosomes with three copies and one copy of a gene could be generated from two chromosomes with two genes each. (b) In practice, misalignment is more likely if a long series of copies of similar sequences is present. (c) The human γ-globin genes include a repeat sequence (solid box) on either side of the gene itself (open box); this repeat could have been mismatched, causing duplication.

has two α-globin genes, but individuals can have three, or a single, α-globin gene. (Homozygotes for these chromosomes are one genetic cause of thalassaemia.) The obvious explanation is that these copies were generated by unequal crossing over.

Once a duplication has originated in a single copy, it could spread through the population by drift or selection. After the duplication became fixed, the two genes could retain the same sequence (like in ribosomal RNA genes) if selection favored multiple copies of the gene. Alternatively, they could diverge to form a gene family (like the globins). A third possibility is that if only one of the copies of the gene is needed, the other might be switched off or decay by mutation and drift.

Heterozygous advantage (section 5.12, p. 117) at a locus may set up selection for gene duplication. Suppose that selection at a locus favors heterozygotes because an individual is better off with two versions of a protein—perhaps because each protein is adapted to slightly different conditions. While these two versions are supplied by a single locus, disadvantageous homozygotes are generated when two heterozygotes breed together. If the gene were duplicated, the two versions could be supplied by homozygotes for each at the two loci, and the disadvantageous segregational load (section 7.6, p. 161) disappears.

The human globin gene family, as we saw, is arranged in two (or three, including myoglobin) clusters on two (or three) chromosomes. The kind of duplication illustrated in Figure 10.2 may well explain the origin of many genes within each cluster, but is less likely to explain how more than one such cluster originates. Multiple clusters are more likely to have originated by translocation, by the duplication of part or all of a chromosome, or by polyploidy (i.e., the doubling, or multiplying, of the whole genome). For instance, suppose that one pair of chromosomes did not segregate equally at meiosis. One of the daughter cells would acquire a double set of the chromosome and the other would acquire none. The latter offspring would probably die, because it is missing the genes of the chromosome. Thus, the result is ultimately a mutation to a double set of the chromosome. As with gene duplication, after a chromosome has duplicated, its new set of genes could be maintained, diverge, or be suppressed, according to the outcome of drift or selective circumstances.

The globin gene family in two species of *Xenopus* supports this interpretation. In *Xenopus tropicalis*, unlike mammals, the α- and β-globin clusters are on the same chromosome. *X. laevis* (a tetraploid species), however, has two sets of the linked α- and β-globin gene cluster. If the α set were suppressed on one chromosome and the β set on the other, the mammalian arrangement would have evolved.

The sequences of the human globin genes can be arranged in a phylogeny (chapter 17 discusses phylogenetic inference). The phylogeny suggests that at least seven duplications have occurred in the history of the molecule. Once we have the tree, it is easy to use other molecular clock evidence to calibrate this scheme and estimate the times of the duplications (Figure 7.7, p. 170, shows the kind of evidence used—it gives the rate of evolution of globin genes, in various species pairs). The results are given in Figure 10.3. The difference in the sequence of the α- and β-globins, for instance, suggests they are derived from a duplication that happened approximately 500 million years ago.

These inferred dates pose an interesting paradox. If we compare the sequences of the human α_1- and α_2-globins, the split apparently happened very recently. In fact, it may post-date the split between humans and the great apes and it certainly post-dates the split between the great apes and the rest of the primates. In this case, the primates outside the great apes should have only one α-globin.

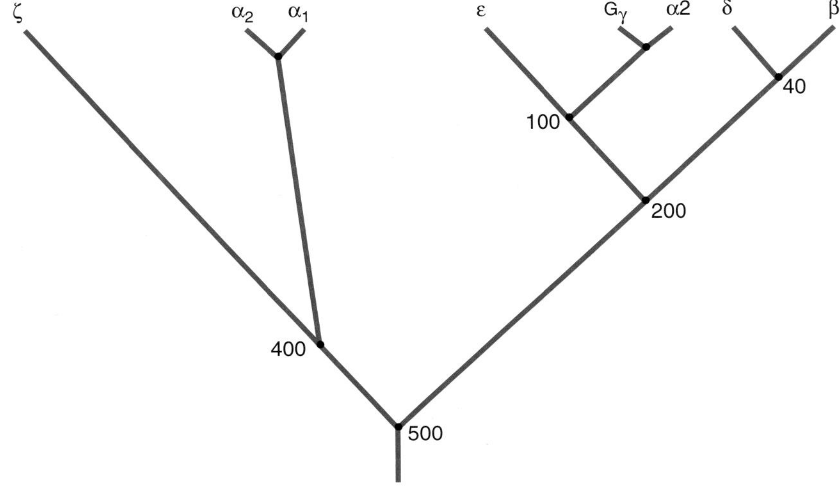

Figure 10.3 A phylogeny of the human globin genes. The genes multiplied by gene duplications, and the dates on the figure, are inferred from the molecular clock. In fact, as the text explains, the inferred dates are untrustworthy. Reprinted, by permission of the publisher, from Jeffreys *et al.* (1983).

The prediction, however, is false: all the primates have two α-globin molecules. The full taxonomic distribution of the α-globins suggests that the gene duplicated at least before the origin of the mammals (about 85 million years ago), and probably before the split between mammals and birds (about 300 million years ago). Yet the similarity of the α_1- and α_2-globins in humans suggests the genes duplicated about 1 million years ago. Which of the figures is correct? This paradox, which is a common characteristic of evolution in gene families, is called *concerted evolution.*

10.4 *The genes in a gene family often evolve in concert*

If each tandem repeat of the ribosomal RNA gene evolved independently by mutation, drift, and selection, the different genes might gradually diverge from one another over time. Figure 10.4 illustrates how this divergence might take place. In the figure, different mutations (numbered 2, 3, 4, 5, 6) arise in different genes occasionally in the lineages of two species and become fixed. Because the evolutionary changes accumulate independently both in the different genes and in the different species, we might expect that the similarity between the genes in a gene family within a species would be approximately the same as the similarity between copies of the same gene in different species. In the modern species *A* and *B* at the top of Figure 10.4, if you examine one gene in the gene family,

Figure 10.4 Evolution of a gene family in two evolutionary lineages (in separate species). If mutations arose independently at each locus, then to evolve by selection or drift, the different genes of a species should diverge at the same rate as the same gene in different species. Reprinted, by permission of the publisher, from Arnheim (1983).

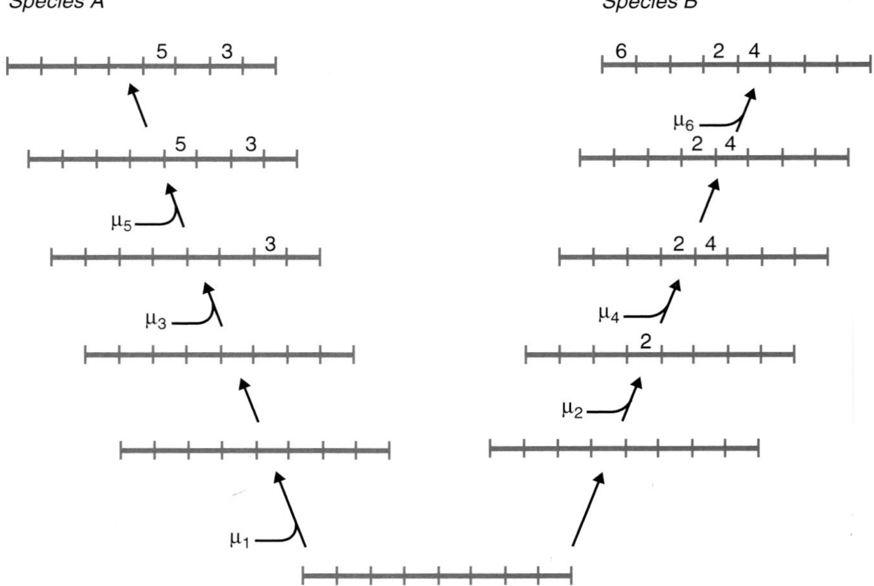

such as the gene at the extreme left, reveals one difference (due to mutation 6) between it in the two species; there is also one difference between this gene and the fifth and seventh gene within species *A*.

This structure provides only a rough prediction of how evolution could proceed. The different genes in a gene family within a species have been separate since the duplication, whereas copies of the same gene in different species have been separate since the species split apart. If the duplication occurs in both species, then the speciation event will be more recent than the duplication. Other things being equal, the copies of the same gene in different species should then be more similar than the duplicates within a species. On the other hand, correlated selection pressures may operate within a species, tending to make the duplicated copies within a species more similar.

And what about the facts? The first study, by Brown *et al.* (1972), was made for the genes that code for ribosomal RNA in *Xenopus*. The 18S and 28S regions (Figure 10.1a) both fit the expected pattern. The sequences were almost identical in all tandem repeats of both *X. borealis* and *X. laevis*, suggesting strong stabilizing selection. For the non-transcribed spacer region (NTS), however, the pattern was different. Again, within a species the sequence was similar between tandem repeats, but this sequence differed in the two species. The NTS regions of *X. borealis* and *X. laevis* showed no more similarity than would two quite different genes—whereas within a species the NTS regions all had the same sequence. Thus, the NTS region showed concerted evolution (Figure 10.5).

The α-globin genes of primates described earlier can be used to illustrate the same principle. All primates, we saw, have two α-globins; we can therefore assume that the common ancestor of primates had two α-globin genes. The sequence of each α-globin gene differs between primate species. In the great apes, for example, any two species differ by about 2.5 amino acid substitutions in each gene. However, within a species, the α_1- and α_2-globins differ by only about one-tenth of that amount (this difference has been shown to be true not only for humans (Figure 10.3), but also for chimpanzees, gorillas, and rhesus monkeys). If one gene accumulates about 2.5 amino acid changes in the time between two species, then two different genes (α_1 and α_2) that have been separated for 300 million years should have accumulated many more changes—if they have been evolving independently. The conclusion is that they have not evolved independently; they have evolved in concert.

Why do gene families show concerted evolution? The concept is very difficult to explain if the genes evolve by mutations that occur independently at each gene locus. Indeed, independent mutations combined with neutral drift can be ruled out as an explanation. If the α_1- and α_2-genes drifted at random independently, they should be much more different than observation shows them to be within each species. Independent mutations and selection can, in principle, explain concerted evolution (Figure 10.5a). The same mutations would have to arise independently in both α_1- and α_2-globins in each species and then be fixed by selection, perhaps because one variant of the two genes was favored in the conditions under which humans evolved, whereas another variant of both was favored in the conditions under which chimpanzees evolved.

For the two globins in a number of primate species, this theory is a possibility. However, as the gene family showing concerted evolution grows larger, the

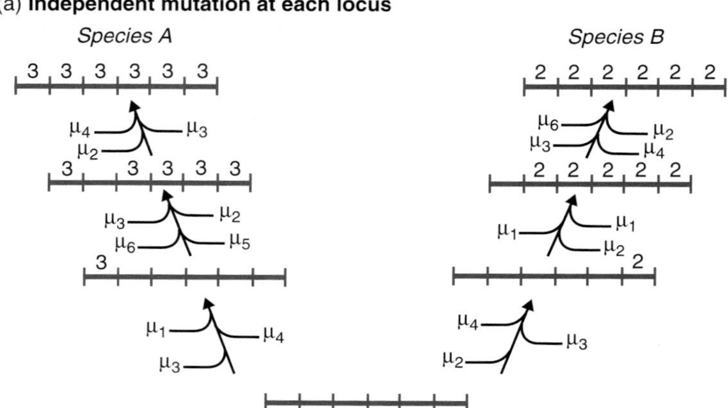

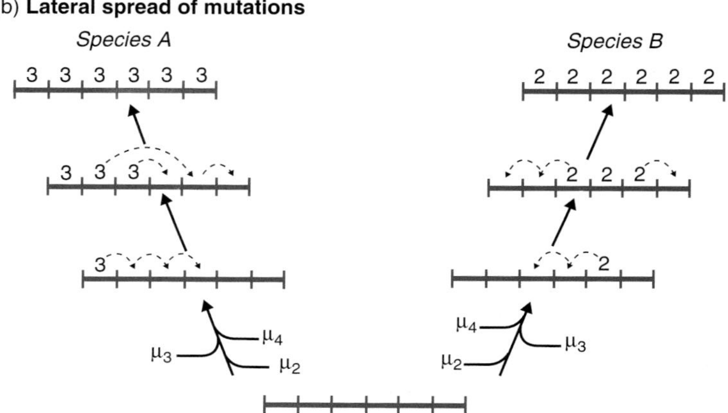

Figure 10.5 Concerted evolution. The different genes of a gene family tend to have similar sequences within a species, but are fixed for different sequences in different species. Thus, all genes in species *A* have sequence variant 3, whereas all the genes in species *B* have sequence 2. (a) This scenario could occur if many independent mutations arose at each locus, and selection favored the same sequence variant at every locus. (b) Alternatively, mutations might be able to spread horizontally.

explanation becomes increasingly incredible. When we reach the ribosomal RNA genes, with their innumerable tandem repeats, it is beyond belief that the same mutation could have occurred independently at each locus and been fixed by selection. In this case, it is generally accepted that some genetic mechanism must act to homogenize the different members of the gene family. One sequence somehow is copied horizontally from one gene to the others. Mutations are not occurring independently at the different loci.

Two main mechanisms have been suggested as causing concerted evolution: unequal crossing over and gene conversion. We encountered unequal crossing over when we discussed gene duplication (Figure 10.2). The same process generates chromosomes with different numbers and combinations of genes; these chromosomes can provide the mutational raw material for selection or drift to homogenize the gene family. Selection could directly favor more homogeneous gene clusters, or it could simply eliminate clusters with too many or too few copies of a type of gene (Figure 10.6).

In *gene conversion*, one of the alleles at a locus is converted into the other allele. Thus, when the heterozygote f_1f_2 segregated, instead of the Mendelian proportions

Figure 10.6 Unequal crossing over (Figure 10.2) can generate the variants needed for concerted evolution. The shuffling of genes among chromosomes can generate variants with more copies of one variant of a gene. Letters indicate genes, in a gene family, on a chromosome; arrows indicate the site of recombinational breaks. (a) Two successive unequal exchanges result in two chromosomes in which different variants have spread through two of the genes. (b) After one unequal crossing over, the chromosome with two *C* genes increases in frequency until an individual with two copies of the chromosome is formed several generations later. Another unequal crossing over could then form an offspring with three *C* genes. (a) and (b) represent just two courses among many possibilities.

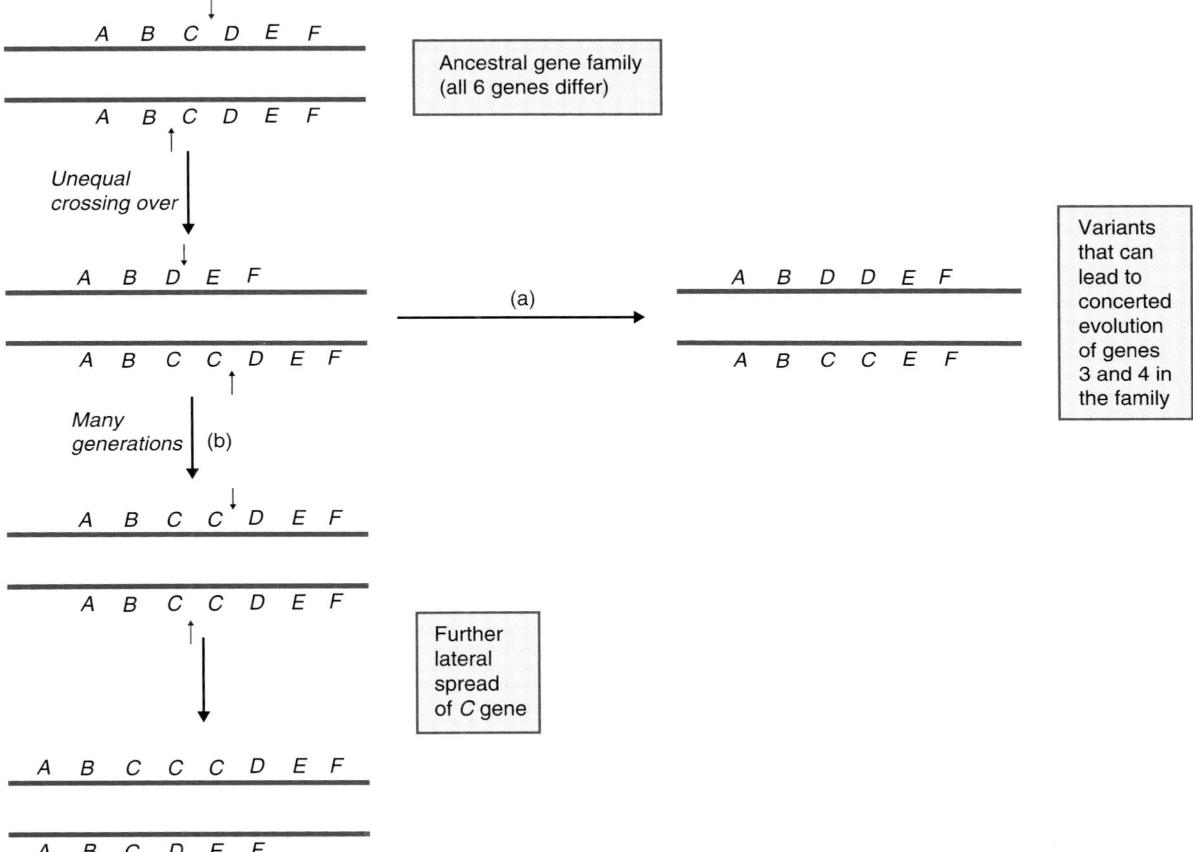

of one f_1 for one f_2, two f_1 or two f_2 (and none of the other allele) would emerge. The same process could take place horizontally between the members of a gene family, with the variant at one locus being copied into the other locus. The mutational raw material for concerted evolution would again be produced.

Gene conversion could be either biased or unbiased. Biased gene conversion means that it favors one variant of the gene rather than another. For example, two f_1 genes might be produced more often than two f_2. Biased gene conversion is a kind of directed mutation, and it increases the chance that a gene cluster will be homogenized. Unbiased gene conversion means that the production of two f_1 or two f_2 genes is equally likely.

Unbiased gene conversion and unequal crossing over do not by themselves produce concerted evolution, any more than undirected mutation alone produces other types of evolution. They account only for the origin of variants—that is, they cause some individuals of a population to have more homogeneous gene families than other individuals. Selection and drift are still needed to explain how one of the gene family variants becomes fixed in the population. In the case of drift, a lateral "march to homozygosity" would occur between loci much like the process for a single locus (section 6.6, p. 141).

The relative importance of unequal crossing over and gene conversion is unknown, but Hillis *et al.* (1991) found that the genes for ribosomal RNA showed concerted evolution even in an asexually reproducing lizard, which does not have crossing over. Gene conversion was therefore presumably responsible for the concerted evolution. Moreover, one sequence variant was favored, suggesting that gene conversion may be biased in this case.

So far we have considered concerted evolution in a single gene cluster. The process is not confined to gene clusters on one chromosome, however. The ribosomal RNA genes in primates (as we saw earlier) appear in clusters on five chromosomes, but the genes on different chromosomes show concerted evolution like genes on the same chromosome. This evolution is probably made possible when the regions of the chromosomes encoding the ribosomal RNA genes physically come into association. Unequal crossing over, or gene conversion, could then take place, just like that hypothesized to allow concerted evolution among genes on the same chromosome.

Concerted evolution is usually confined to only some of the genes in a gene family. For example, neither the α- and β-, nor the β- and γ-globins evolve in concert, even though they belong to the same gene family. The reason is that gene conversion takes place only between genes with some degree of sequence similarity. The chance of gene conversion is probably roughly proportional to the sequence similarity between two genes. A newly duplicated gene, therefore, might undergo one of two evolutionary courses. If gene conversion happens before the two genes have diverged dramatically, they may evolve in concert. If, however, they diverge too far for gene conversion to be possible, they will "escape" from one another, and their future evolution will happen independently. At a further stage, when they have diverged far enough, we should cease to classify these genes as members of the same gene family. Which of these two fates a pair of genes undergoes will be determined by the relative rates of gene conversion and of the mutations that break down sequence similarity. If the former is high relative to the latter, concerted evolution is likely; if the opposite is true, the genes will escape and diverge from one another.

The chance of gene conversion is probably not simply proportional to sequence similarity. The insertion of a mobile genetic element into a gene can prevent future conversion of that gene, or of part of it to one side of the insertion. Schimenti and Duncan have suggested that the insertion of an *Alu* element (see below) has prevented gene conversion, and concerted evolution, between certain genes in the β-globin families of cows and goats. The chance of concerted evolution is, therefore, controlled by both the sequence similarity of two genes and the insertion of a mobile genetic element in them.

In summary, the genes in gene clusters evolve in parallel. This concerted evolution is probably not caused by selection and drift operating on independent mutations at each locus; rather, it requires some mechanism for the same mutation to arise at the different loci. Gene conversion or unequal crossing over allow mutations to move between loci, and selection or drift could then fix the same mutation at all loci. In the case of biased gene conversion, the mutational mechanism itself will help to fix the favored variant. The result would be the observed pattern of concerted evolution.

10.5 *Not all DNA codes for genes*

So far we have been concerned with genes that code for proteins, or at least for RNA. However, much of the DNA in the organisms of many species probably does not code for anything. Consider humans as an example. The human genome contains about 3×10^9 nucleotide pairs, and geneticists disagree about what proportion of it is *coding DNA* (i.e., DNA that is transcribed to produce proteins or regulate the production of proteins). Estimates are fallible, but the figure may be about 10–25%; the highest estimates go up to 50%. Even if 50% of the genome does code for genes, a large proportion of the genome still does something other than coding for, or controlling the production of, proteins. DNA that does not code for genes is called *non-coding* DNA.

DNA re-association experiments allow an estimate of the proportion of non-coding DNA. In these experiments, the DNA is first heated and melted into single strands, and the single-stranded DNA is then allowed, in a cooler solution, to join together (re-associate). The rate at which the strands re-associate is controlled by their sequence similarity. Early work identified three classes of DNA. A part of the DNA re-associated quickly; it is mainly highly repetitive DNA, made up of large numbers of repeats of simple sequences. At the other extreme, "single-copy" DNA joined together relatively slowly; this part of the DNA probably codes for most of the genes. In between is a class of middle-repetitive DNA, which is repeated but not as much as the highly repetitive DNA. On the assumption that the single-copy DNA serves as the coding part of the genome, we can use the proportion of single-copy DNA (i.e., slowly re-associating DNA) as a first estimate of the coding DNA in the genome.

Table 10.1 lists the genome sizes and proportion of single-copy DNA for a number of species. These numbers suggest, in two ways, that a large quantity of non-coding DNA exists.

First, the figures for single-copy DNA are often much less than 100%. Only one-fifth of the DNA of the toad *Bufo bufo* is single-copy DNA, for example. If only single-copy DNA acts as the coding part, then four-fifths of *B. bufo*'s DNA is non-coding. This observation is generally true in eukaryotes, but not in prokaryotes. Bacterial DNA appears to be much more economically organized.

Second, large differences appear between species, and these differences can hardly all be due to variations in the numbers of coding genes. Fifteen times as many genes may be needed to build a human as to build a protozoan. In the words of Orgel and Crick, however, "it seems implausible that the number of radically different genes needed in a salamander is 20 times that in a man." The

TABLE 10.1

Amount of DNA and percentage of single copy DNA in various animal species. Amount of DNA is expressed as haploid DNA content, 1C, in pg (10^{-12}g). For conversion: 1 picogram (pg) = 0.98×10^9 base pairs (bp); 1 bp = 1.02×10^{-9} pg. Reprinted, by permission of the publisher, from John & Miklos (1988).

Invertebrates			Vertebrates		
Species	1C (pg)	% sc DNA	Species	1C (pg)	% sc DNA
Protozoa			Protochordata		
Tetrahymena pyriformis	0.2	90	*Ciona intestinalis*	0.2	70
Coelenterata			Pisces		
Aurelia aurita	0.7	70	*Scyliorhinus stellatus*	6.1	39
Nemertini (Rhyncocoela)			*Leuascus cephalus*	5.5	44
Cerebratulus	1.4	60	*Raja montagui*	3.4	47
			Rutilus rutilus	4.8	54
Mollusca			Amphibia		
Aplysia californica	1.8	55	*Necturus masculosus*	83.0	12
Crassostrea virginica	0.7	60	*Bufo bufo*	7.0	20
Spisula solidissima	1.2	75	*Triturus cristatus*	21.0	47
Loligo loligo	2.8	75	*Xenopus laevis*	3.1	75
Arthropoda			Reptilia		
Prosimulium multidentatum	0.18	56	*Natrix natrix*	2.5	47
Drosophila melanogaster	0.18	60	*Terrapene carolina*	4.1	54
Limulus polyphemus	2.8	70	*Caiman crocodylus*	2.6	66
Musca domestica	0.9	90	*Python reticulatus*	1.7	71
Chironomus tentans	0.21	95			
Echinodermata			Aves		
Strongylocentrotus purpuratus	0.9	75	*Gallus domesticus*	1.2	80
			Mammalia		
			Homo sapiens	3.5	64
			Mus musculus	3.5	70

obvious deduction is that the differences derive mainly from non-coding DNA. Thus, a genome consists of more than simply genes.

The experiments provide a second clue about the nature of the non-coding DNA—namely, that it is often repetitive. Geneticists distinguish a number of kinds of repetitive DNA. A first distinction is between *tandem* and *scattered repeats*—that is, whether the repeat sequences appear next to one another (in tandem) or are scattered through the genome. Repetitive DNA can also be classified by the length and number of the repeat units.

No generally agreed classification exists for repetitive DNA, although the suggested classifications are all similar. We will use the following classification developed by Charlesworth *et al.* (1994). There may be other kinds of repetitive DNA that are not readily classifiable in one of these categories—or there may not be: the question is a topic of current research. (The word "locus" features in the descriptions below. It is used here in a slightly different way from traditional genetics: a locus for a tandem repeat is the site in the chromosome where a whole series of repeats is present. Thus, it refers to the whole array, not just one unit within it.)

1. Tandem repeats
   1.1 Microsatellites. Repeats of short (2–5 base pairs) nucleotide sequences. The number of repeats varies between loci, but the average is on the order of approximately 100 repeats. Many microsatellite loci are scattered through the genome; an average human, for example, contains about 30,000 microsatellite loci. Microsatellites have also been detected in other vertebrates, insects, and plants.
   1.2 Minisatellites. Repeats of longer (approximately 15 base pairs) nucleotide sequences. The number of repeats varies among minisatellite loci, and many loci are scattered through the genome. The average length of any one locus is usually about 500–2000 nucleotides. Minisatellites have been studied in humans and other vertebrates, fungi, and plants.
   1.3 Satellite DNA. The size of the repeated unit varies in different cases, with some being as small (5–15 base pairs) as micro- and minisatellites, while others are larger (about 100 base pairs). They are often found in large blocks of 1000 or more repeats of the unit sequence in regions of the chromosome near the centromere or the telomere. They do not seem to be as variable in the number of repeats per site as are micro- and minisatellites.
2. Scattered repeats. Longer sequences (on the order of approximately 100 base pairs) that are distributed throughout the genome, usually in single copies bounded by other sequences rather than in tandem repeats. Three examples from humans are the sequences called *Alu* (which we will discuss further later in this chapter), *Kpn*, and poly (*C-A*); the three together make up about 20% of the human genome.

A unit sequence of any of the three kinds of tandem repetitive DNA can be found at more than one place in the genome. At all those sites they are found in the form of tandem repeats. Scattered repeats are also found at many sites, but usually with only a single copy of the sequence appearing at each site.

Two main evolutionary hypotheses have been suggested to explain the presence of this repetitive, non-coding DNA. One hypothesis says that such DNA is functional, even though it does not actually encode genes. It may be needed for some regulatory or structural reason, for example, or perhaps it keeps the genes apart or correctly configured in the DNA molecule's three-dimensioned shape. Alternatively, the repetitive DNA may be selfish DNA—either neutral "junk" DNA, or parasitic DNA.

We will concentrate on the "selfish DNA" hypothesis here, for two reasons. First, it is widely accepted as the explanation for much of the repetitive non-coding DNA. Second, it is a new evolutionary concept. If the repetitive DNA is functional, it will evolve just like genic DNA and requires no special discussion. If the DNA is non-functional for the organism that contains it, something different is occurring. It is also worth noting that comparisons of non-coding repetitive DNA made between different species have often shown the presence of concerted evolution, as in the gene families of coding DNA discussed earlier. The mechanisms presumably operate in much the same way in the two cases, and we will not discuss concerted evolution here.

10.6 *Repetitive DNA other than in gene clusters may be selfish DNA*

In 1980, Doolittle and Sapienza, and Orgel and Crick, made more explicit an idea that many other geneticists had also pondered—the idea that some, or all, repetitive DNA may have no use for the organism. It may instead be *selfish DNA*. Selfish DNA is non-transcribed and non-coding, and contributes nothing to the well-being of the organism. In most cases, such DNA is selectively neutral except for the energetic burden of replicating it. If it was excised, the organism would suffer no disadvantage. After a stretch of selfish DNA originates, it is passively replicated and passed on from parent to offspring. Changes in its frequency in the population would occur by drift. If a non-coding sequence was not neutral—perhaps because it interfered with the construction or metabolism of the organism—it would be selected against. Likewise, if it accumulated to such an extent that the cell cycle was slowed by the need to replicate the entire DNA sequence, selection would probably act to reduce it. Provided the quantity of a particular sequence is not excessive, is not transcribed, and accumulates in parts of the genome where is does not interfere with genic regulation and transcription, there is no reason why selfish DNA should not evolve.

Selfish DNA might be either "active" or "passive." The sequence itself might not influence the chance that it spreads in the DNA and is retained. It is then a passive kind of selfish DNA and could accumulate as "junk DNA" in the genome. Alternatively, a particular sequence might, for some reason related to its sequence, have a better than average chance of spreading through the DNA. These sequences would be a more active, parasitic kind of selfish DNA, and they would proliferate until checked by natural selection because they interfered with vital DNA functions or became so abundant that their replication imposed a significant cost on the organism. For the possible examples of selfish DNA we shall discuss here, we do not know for certain to which class they belong. This distinction should be kept in mind, however.

How might selfish DNA originate? In theory, any kind of mutation could give rise to selfish DNA. A point mutation could inactivate a gene, converting it into a pseudogene (section 7.11.3, p. 181). If the inactivated gene was redundant and the new pseudogene was not in a position to interfere with any important function, the mutation could be fixed by drift and the pseudogene retained as passive selfish DNA. For repetitive DNA, other kinds of mutations would be needed.

We will discuss tandem and scattered repetitive DNA separately, because they are believed to originate by different mechanisms. In this section, we will examine three ways of creating tandem repetitive DNA briefly. (In the next section we will look at the example of minisatellites in more detail before moving on to scattered repeat sequences.)

The three processes that have been suggested to create tandem repeats are called slippage, unequal crossing over, and rolling circle replication.

Slippage occurs when one DNA strand is being created by replication from another strand. The two strands may slip relative to one another by a few nucleotides, such that either a few nucleotides are missed or copied twice (Figure 10.7). Such an event is most probable in a region of short repeats, and the result

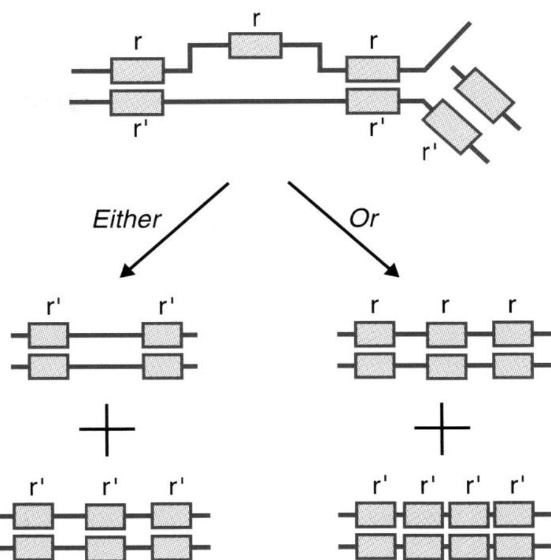

Figure 10.7 Slippage occurs when a sequence of DNA is copied twice, generating a new strand with an extra copy of that sequence, or missed out, generating a new strand with one copy less. As a result, the row of repeats may shrink or grow when the *R* and complementary *R'* sequences are misaligned. The molecular details are unknown, but may involve the formation of a loop. Short repeats may be particularly vulnerable to slippage.

is a new strand that has either more or less repeats than the original strand. Slippage occurs only over short distances, and is the most likely mechanism by which new length variants arise for microsatellite sequences. It may produce new length variants in minisatellites as well, although unequal crossing over is also important in this case.

Unequal crossing over (see Figure 10.2) can produce a strand with a greater or lesser number of repeats, and successive rounds of unequal crossing over can build up or break down long sequences of repeats. Unequal crossing over has not yet been confirmed to operate in minisatellites, but it is strongly suspected to play a role based on indirect evidence (one piece of which we shall see in the next section). While slippage could operate in the short unit satellites, unequal crossing over could operate in all of them.

The third mechanism, *rolling circle replication*, has also been suggested to affect all types of satellites. This method of replication has been well studied in bacterial plasmids and some viruses; the idea that it operates in eukaryotes is more speculative. It would require the excision of a sequence from the chromosome. The sequence would then have to form a circle, after which it could multiply by rolling circle replication. The replicated sequences would then have to be reinserted into the chromosome.

All of these processes are mechanisms for the origin of new variants. When they operate, one individual in the population will have a new length variant at a site of non-coding repetitive DNA. Drift or selection will then be needed for the variant to spread through the population. In the models of evolution we considered in chapters 5–9, evolution occurred when rare variants (produced, for example, by mutation) had their frequencies altered by selection or drift. Slippage, unequal crossing over, and rolling circle replication are all analogous to the "mutation" phase of those earlier models, and are thus only one part of an evolutionary explanation.

Under both slippage and unequal crossing over, new length variants of a unit sequence become more likely to arise if that sequence is present in more lateral copies in the genome. If only one copy of a sequence exists it is unlikely to misalign. If several repeats of the sequence are present, misalignment is more probable, as (for instance) copy two on one strand aligns with copy three on the other, or copy eight on one strand aligns with copy six on the other. The more copies of a sequence that exist, the more likely this misalignment becomes. If, therefore, a genome contained two sorts of neutral, functionless sequence—one in single copy and the other in multiple repeats—then (other things being equal) the repeated sequence would be more likely to grow laterally through the genome during evolution. If the highly repetitive rows of tandem repeats are actually selfish DNA, their lengths will fluctuate through evolutionary time. Unequal crossing over, slippage, and drift will constantly be causing them to shrink and grow. At any one time, some very long regions of repeats will be found, and these regions will be recognized as highly repetitive DNA. After a few million years, a currently long repeat could potentially have shrunk and some other, previously shorter series lengthened. The frequency distribution of lengths will likely remain probabilistically constant.

10.7 *Minisatellites are sequences of short repeats, found scattered through the genome*

While studying the human myoglobin gene in the early 1980s, Jeffreys discovered a short sequence of repeated DNA within an intron. He used the short sequence as a "probe" to see whether the same sequence was present anywhere else in the genome. The probe hybridized at several regions. Jeffreys then extracted the DNA from all these regions, and found that each DNA sample consisted of a unit sequence in a row of tandem repeats. The number of repeats at each site was highly variable between individuals; one site, for example, might contain 25 repeats in one individual, 5 in another individual, and 47 in another (Figure 10.8). Although the unit sequence itself is variable, a common core sequence has been characterized (Figure 10.9) and is 16 (or perhaps 8) bases long. The particular sequence shown in Figure 10.9 is not the only sequence that can form this sort of scattered short set of repeats, but it is the most studied. Jeffreys called

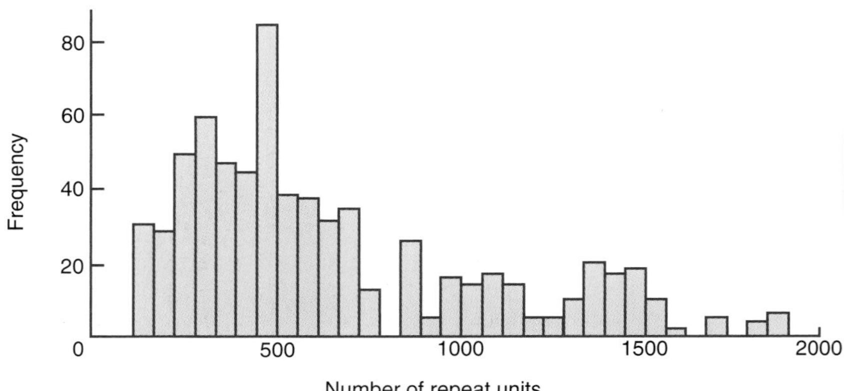

Figure 10.8 Frequency distribution of numbers of repeats at one minisatellite locus in a total of 344 individuals (688 gametes). Reprinted, with permission from *Nature*, from Jeffreys *et al.* (1988). Copyright 1988 Macmillan Magazines Limited.

Figure 10.9 The unit sequence of the human minisatellite. Different minisatellites do not all have identical copies of the sequence, but there is a consensus sequence of 16 nucleotides. Also shown for comparison is the sequence called χ in the bacterium *E. coli*. Genetic recombination occurs at higher than average frequency at the location of the χ sequence in *E. coli*, and recombination is also known to have elevated frequency at minisatellites. These events may or may not be a coincidence. Reprinted, with permission from *Nature*, from Jeffreys *et al.* (1985). Copyright 1985 Macmillan Magazines Limited.

Clone	Repeat length	Number of alleles	Number of repeats	DNA sequence	
				GGAGGTGGGCAGGAAG	Myoglobin probe
1	62	6	40–20	AAGGGTGGGCAGGAAC	Isolated clones
2	32	1	6	GGAGGTGGGCAGGAAX	
3	64	5	18–10	TGGGGAGGGCAGAAAG	
4	17	1	14	GGAGGYGGGCAGGAGG	
5	37	8	25–12	GGAGGAGGGCTGGAGG	
6	41	1	5	GGA-GTGGGCAGGCAG	
7	33	1	3	GGTGGTGGGCAGGAAG	
8	16	2	41, 29	AGAGGTGGGCAGGTGG	
				GGAGGTGGGCAGGAXG	Common core
				GCTGGTGG	E. coli χ

these sequences minisatellites; the name (like that of microsatellites and satellites) derives from the way they can be detected when DNA is centrifuged.

Why do minisatellites vary so much in the number of repeats? We saw earlier that slippage and unequal crossing over are thought to generate new length variants, and we can now look at two predictions of this idea. Both concern the heterozygosities of different minisatellite loci. The alleles at a minisatellite locus are the different length variants, which differ in their number of repeats of the unit sequence. We can measure the frequencies of the different length variants at each locus, and calculate the heterozygosity for each by the standard formula (section 6.6, p. 144).

The first relation to consider exists between the heterozygosity of a locus and the similarity of the unit sequence. So far we have described repetitive DNA as if the unit sequence were identical in all repeats. In practice the sequences can differ slightly between repeats, and they differ more at some loci than others. When Stephan and Cho plotted the heterozygosity against the sequence similarity for 10 minisatellite loci in humans, they found a strong negative relation (Figure 10.10a). The loci that are most variable in numbers of length variants possess the most similar unit sequences. The relation is expected if new variants arise by unequal crossing over, for two reasons. First, unequal crossing over is most likely if the unit sequences are more similar. Second, unequal crossing over causes the same variant to be spread through the DNA at a site. If unequal crossing over has happened more frequently at one locus than at another, the unit sequences should look more similar at those loci where the process is more common. If unequal crossing over has not occurred much at a locus, the sequences of the repeats will have diverged more.

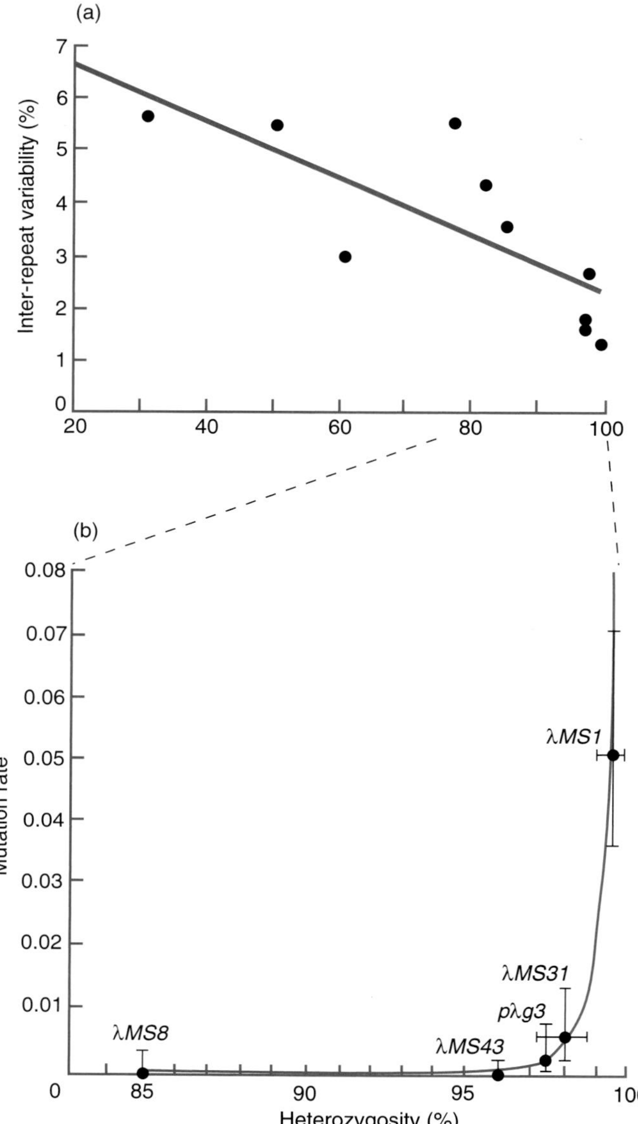

Figure 10.10 Two relations of the variation (heterozygosity) of different minisatellite loci in humans. (a) The sequences of the units are more similar at loci that are more heterozygous. (b) Mutation rates (i.e., rate of production of new length variants) are higher at more variable loci. Both relationships make sense if length variants at minisatellites mainly originate by unequal crossing over. The five loci in (b) are also the five loci at the right-hand end of (a). Reprinted, with permission from *Nature*, from Stephan and Cho (1994) and Jeffreys *et al.* (1988). Copyright 1988 Macmillan Magazines Limited.

Figure 10.10b shows another relation with heterozygosity for different human minisatellite loci. The more heterozygous loci have higher mutation rates. The mutations in this case form new length variants. If a locus is fixed for one length variant, unequal exchange is unlikely to operate. The more length variants present at a locus, the more likely it is that two different ones will cross over unequally and produce a new variant. That is, the more variety introduced into the system, the more variety it is likely to generate (Figure 10.10b). If the mutations arise by a more conventional genetic mechanism, there would be no reason to predict this trend; if they arise by unequal crossing over, we would expect this result. The mutation rates were estimated by tracing minisatellites through human pedigrees, to see when new variants arose. Notice that the rate is high, up to 5% per gamete.

The high mutation rates of minisatellites allow them to be used in *genetic fingerprinting*. New variants arise at a high enough rate for every individual (except monozygous twins) to have a unique frequency distribution (or "profile") of minisatellites. Genetic fingerprints are more forensically useful than real fingerprints. Not only is an individual's genetic fingerprint unique and identifiable from small quantities of body substances, but it is also heritable. The profiles of a father and his children, for instance, are more similar than two random members of the same population. Genetic fingerprinting can, therefore, be used in paternity testing as well as in straightforward identification. Minisatellites similar to those in humans have been found in many other species, and the genetic fingerprinting probe has become an important method of tracing paternity in behavioral ecology. Microsatellites are also used in genetic fingerprinting. (Another kind of genetic fingerprinting uses restriction enzymes to cleave an individual's DNA; this technique is quite different from the use of minisatellites and microsatellites.)

10.8 *Scattered repeats may originate by transposition*

Scattered repetitive DNA is thought to originate mainly by *transposition*. Certain sequences of DNA can, under appropriate circumstances, copy themselves elsewhere in the genome. These mobile genetic sequences are called transposable elements, or (more informally) jumping genes. The molecular biology is complex, but we can distinguish two types of transposition (Figure 10.11); these processes differ at a molecular level according to whether an RNA intermediate is present. The difference is evolutionarily interesting because transposition by an RNA intermediate may be more likely to increase the number of copies of the sequence in the DNA.

We can make a second distinction between *replicative* and *conservative* transposition. In Figure 10.11a, two copies of the sequence exist after the transposition, whereas only one was found before transposition. In Figure 10.11b, however, only one copy exists both before and after transposition. (It is possible for transposition by DNA intermediate to be replicative. For instance, the excised sequence may jump from one of the two strands produced after the DNA has replicated itself in the cell cycle, to a position upstream of the replication fork, where the DNA has yet to be replicated.)

Replicative transposition could theoretically produce a scattered repeat sequence. After its origin in one individual, it could then increase in frequency (by selection or drift) to result in the pattern of scattered repeats we now observe. Did real scattered sequences evolve in this way? The best evidence that a scattered repeat sequence had originated by transposition would be found by observing the sequence transposing. A more general method is to study the characteristics of the DNA sequences that are known to transpose, and of the scattered repeats, to see whether the latter resemble the former. Transposable elements have some

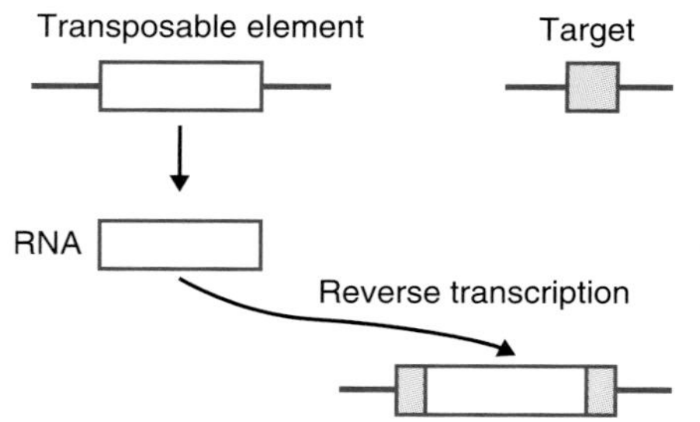

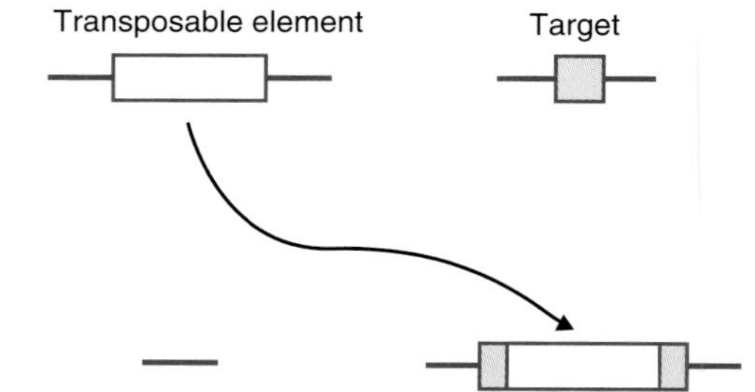

Figure 10.11 Two main kinds of transposable elements. (a) Retroelements, which transpose by the reverse transcription of an RNA intermediate. (b) Transposable elements, which transpose by excising from one site and inserting elsewhere, without an RNA intermediate.

characteristic features, and the similarity between these and at least some well-studied scattered repeat sequences is enough to make a compelling case for their origin by transposition. Let us look first at some known (or strongly suspected) transposable elements, and then at an example of a scattered repeat.

The simplest transposable elements lack the gene for the enzyme reverse transcriptase. They possess only a gene for a "transposase" enzyme together with one or more other genes, and are transposed by a DNA intermediate (Figure 10.11b). Such elements usually have inverted repeat sequences at either end; the same sequence is found at both ends, with one sequence forming a mirror image of the other. The most famous examples are the first transposable elements to have been described—the ones in maize discovered by McClintock. Other transposable elements have the gene for reverse transcriptase as well, and they are classified as retroelements. Some retroelements can live only in the genome; the element called *copia* in the fruitfly *Drosophila* is an example of this phenomenon. Other retroelements can live independently or within a host genome; these elements include the retroviruses, such as HIV, the agent of AIDS.

Which sequences in our DNA may have been produced from these kinds of transposable element? Transposition has been most studied in bacteria, and our knowledge for multicellular organisms is less certain, but some likely examples have been identified. The *Alu* sequence is a hot candidate in mammals. *Alu* is the name for a characteristic sequence recognized by the restriction endonuclease *Alu 1*. The sequence is about 281 bases long and is found throughout the human genome; each genome contains about half a million copies. About 5% of human DNA consists of the *Alu* sequence. Few, if any, of the sequences are transcribed, and the *Alu* sequence may have no function.

Where does the *Alu* sequence originate? One favored hypothesis is that it ultimately comes from the gene for a functional ribosomal RNA molecule called 7SL RNA. The gene for the 7SL RNA could have first been picked up by a retroelement and copied back into the DNA by reverse transcription; the many *Alu* sequences are believed to have been formed as second-order derivatives of the 7SL RNA gene, by recurrent rounds of retroelemental reverse transcription from a number of reverse-transcribed copies of the gene itself. The *Alu* sequence itself is not strictly a retroelement because it lacks the gene for reverse transcription, but it looks like a sequence that is derived from reverse transcription. Perhaps it is picked up by a retroelement, and the gene for reverse transcription is lost when it is copied back into the DNA. Although the sequence has not been proved to be functionless, a strong suspicion exists that it is an example of selfish DNA. The large numbers of copies of *Alu* suggest that something about its sequence makes *Alu* particularly likely to be copied; it would then be an example of active selfish DNA.

Transposable elements can have interesting evolutionary consequences, even if they evolved originally by neutral drift. The insertion of an *Alu* sequence, as we saw in section 10.4, can prevent gene conversion between a pair of genes. Concerted evolution between the two genes then becomes impossible. As a consequence, duplicated genes in which an *Alu* sequence is inserted would be more likely to show independent evolution than genes without such an insertion. A second suggested consequence involves chromosomal evolution. Chromosomes may be particularly likely to rearrange around *Alu* sequences, and the numbers and positions of the sequence may, therefore, influence the rate and form of chromosomal evolution. Finally, at the level of the DNA sequence, when a transposon inserts into a new gene, it causes a mutation. The recipient gene has a new sequence after receiving the transposon, which may (or may not) cause a mutation at the phenotypic level. McClintock first showed that mutations can be caused by transposition in maize. Transposition probably accounts for a substantial proportion of observed mutations. It may differ from other kinds of mutation in that it may occur more at some sites in the genome than at others. Mutations caused by nucleotide changes may be more randomly distributed.

10.9 *Selfish DNA may explain the C-factor paradox*

Let us return to the observation (section 10.5) that more DNA is present in the cells of some species, particularly eukaryotes, than is needed to code for their genes. The degree of excess is controversial, but no one doubts that at least some excess DNA exists in at least some species. (Likewise, no one doubts that at least some of the species differences in DNA content (Table 10.1), such as between humans and bacteria, are due to differences in phenotypic complexity.) The apparent excess of DNA is sometimes called the C-factor paradox. What purpose does the excess DNA serve?

Selfish DNA is an attractive, if hypothetical, solution to this question. The excess DNA would then have no use or function, but would be selectively neutral, and passively replicated from generation to generation. The assumptions of this hypothesis are plausible—unequal crossing over, gene conversion, transposition, and neutral drift all do happen. Moreover, if the hypothesis is correct, it would explain why this apparently functionless DNA actually *is* functionless. It could also explain why the degree of excess DNA differs so much among species, because the amount of repetitive DNA will fluctuate through time as it evolutionarily grows by mutation and drift (or selection) and shrinks when selection acts against excessive accumulations of repeats. Under this scenario, different species would represent the still-frames in a moving picture. While this idea is simply a hypothesis, it is, however, a theoretically attractive hypothesis. At the present time, the best available explanation for the apparent excess of DNA in our cells is that it results from the passive accumulation of selfish DNA.

10.10 *Conclusion*

Molecular biology is discovering many previously unsuspected genetic curiosities within the genome. The DNA of an organism is definitely not a string of structural cistrons and regulatory genes. Evolutionary biologists want to explain why the other kinds of genetic elements exist, and why the genomes have their current architecture. They also aim to explore the evolutionary consequences of such non-Mendelian processes as transposition and unequal crossing over. The phenomena are sufficiently new and imperfectly described, and the kinds of theories they require are sufficiently novel, that favored hypotheses and the details of the fit between fact and theory are changing almost weekly. It is an exciting area of the science.

SUMMARY

1. Genes are usually arranged in clusters (called gene clusters) of related genes on the chromosome. The cluster may consist of tandem repeats, like the ribosomal RNA genes, or a linked group of related genes, like the globin genes.
2. Much of the non-coding DNA consists of repeated sequences. Several different kinds of repetitive DNA have been recognized.
3. Gene families originate by gene duplication, which itself takes place by unequal crossing over or polyploidy.
4. The different genes in a gene family often show concerted evolution. That is, the genes at separate loci within a species are much more similar than the homologous copies of a gene in different species.
5. Concerted evolution among large numbers of genes is practically impossible to explain if mutations arise independently at each locus; it requires some mechanism for concerted mutations at all loci.
6. Concerted mutation can occur by unequal crossing over or gene conversion. Concerted evolution happens when the more homogeneous variants produced by these processes are fixed by selection or drift. Gene conversion

may be biased in favor of some sequences rather than others; the favored sequences would then proliferate by a form of lateral mutation pressure.

7. Gene conversion can occur between genes of similar sequence. When two genes have diverged more than a certain amount, concerted evolution will become unlikely.
8. Much of the genome does not consist of coding genes. It consists of various classes of repetitive DNA.
9. Repetitive DNA is classified as being either tandem or scattered repeat sequences. Tandem repeats, in turn, are distinguished by the size and number of the repeats, as being microsatellites, minisatellites, or satellite DNA.
10. The large quantities of repetitive, non-coding DNA may be selfish DNA. Such DNA may be non-transcribed, have no function for the organism, and be replicated from generation to generation like a passive parasite. It would change in frequency in the population mainly by random drift.
11. Tandem repeats probably originate by unequal crossing over and slippage (especially for short repeats). In contrast, scattered repeats probably originate by transposition.
12. Minisatellites are sequences consisting of a variable number of repeats of a characteristic short sequence; they are probably examples of selfish DNA. They have mutation rates as high as 10^{-2} per generation, as minisatellites with new numbers of repeats arise (probably by unequal crossing over) in high frequency.
13. Selfish DNA may explain the C-factor paradox—that is, the paradox that many eukaryotic organisms contain more DNA than appears to be necessary.

FURTHER READING

Maynard Smith (1989, ch. 11) and Li and Grauer (1991, ch. 6–8) introduce the subject. The volume edited by Selander *et al.* (1991), and the special issue of the journal *Current Opinion in Genetics and Development*, vol. 4 (1994), pp. 797–938 (including a huge bibliography), contain relevant papers. For background genetics, see a molecular genetics text such as Alberts *et al.* (1994), Lewin (1994), or John and Miklos's (1988) more specialized book.

Gene families. Bodmer (1983) is a good place to begin. On gene duplication, see Clark (1994). See Arnheim (1983), Jeffreys *et al.* (1983), and Hardison (1991) for concerted evolution in ribosomal RNA and globin genes; see Ohta (1988) and Walsh (1987) for the theory.

Selfish DNA and repetitive DNA. Charlesworth *et al.* (1994) is a recent review, containing many references. Doolittle and Sapienza (1980) and Orgel and Crick (1980) are the classic references. See also Queller, Strassmann, and Hughes (1993) on microsatellites; Britten (1994) on the *Alu* sequence; Neufeld and Colman (1990) and Avise (1994) on genetic fingerprinting; and Keller (1983) and Federoff and Botstein (1993) on McClintock.

The evolution of simple into complex genetic systems has inspired a series of questions, and this chapter examined only a few of these issues. Maynard Smith and Szathmáry (1995) look at many other questions. There is the question of the evolution of chromosomes (see also Shaw (1994) on centromeres and Marx (1994) on telomeres—both of which contain explicit or implicit evidence for selective constraints on non-coding repetitive DNA, including a selective advantage for concerted evolution), separate nuclei (Lake and Rivera 1994), mitosis, diploidy from haploidy (Kirkpatrick 1994), and meiosis (Haig 1993).

STUDY AND REVIEW QUESTIONS

1. The diagram below shows a family of four genes, in two species. The numbers indicate differerent variants of the gene.

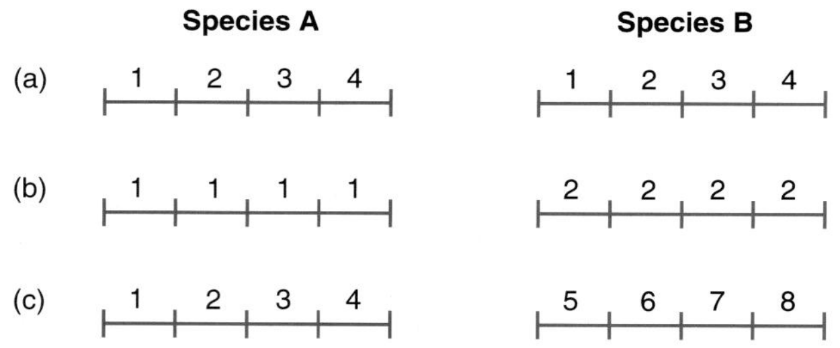

   Which of these genes show concerted evolution?

2. Suppose that a gene duplicates just before a species splits into two species, such that two copies of the gene appear side by side in both descendant species. Now suppose you measure the difference (i) between the two genes within a species and (ii) between either one of the genes from the two species. If evolutionary change is independent in the two genes, what can you predict about the relative values of (i) and (ii)?

3. (a) Review how unequal crossing over can cause (i) gene duplication and (ii) the lateral spread of a variant through a set of tandem repeats. (b) If unequal crossing over has operated as in (a), what other factors are needed to explain concerted evolution and the evolution of selfish DNA?

4. Describe the main classes of repetitive non-coding DNA. What genetic processes give rise to new genetic variants in each class?

5. (a) What relation do you predict between the mutation rates and heterozygosities of different minisatellite loci?

   (b) Look at the relation between mutation rate and heterozygosity in Figure 10.10b. Could any of the standard mechanisms of mutation (see section 2.4, p. 26) account for this relation?

PART 3

Adaptation and Natural Selection

How can we find out what (if any) advantage an organism gains from possessing some characteristic—whether in its anatomy, physiology, or behavior? That is, how can we study adaptation? For many characters, such as muscles or digestive systems, it is well understood how they function as adaptations. For others, the issue is less well understood, or not understood at all.

Chapter 11 begins by explaining the problem, and discusses three main methods for dealing with it. The main part of the chapter covers three case studies that illustrate these methods; all of the case studies are drawn from the subject of sexual reproduction. The most perplexing problem—why sex exists—is still mainly at the hypothesis development stage. The theory of sexual selection is well developed, and the crucial empirical work has begun. In the theory of sex ratio, a good match exists between theoretical prediction and empirical tests.

In chapter 12, we move on to ask in more theoretical terms which entity benefits from the evolution of adaptations. Evolution by natural selection happens because adaptations benefit something, but what is it exactly—genes, whole genomes, individual organisms, groups of organisms, or species? In other words, what is the unit of selection? Adaptations, the chapter suggests, usually benefit organisms, but a deeper criterion can be used to understand the exceptions as well as the rule—more fundamentally, adaptations evolve for the benefit of genes. Only genes last long enough for natural selection to be able to adjust their frequencies over evolutionary time. Organismal adaptations usually result because gene reproduction is more closely tied to the reproduction of organisms than to any other entity, and gene reproduction is maximized if adaptations occur at the organismal level.

Thus far we have accepted that adaptations evolve by natural selection. Chapter 13 focuses on certain conceptual problems in the study of adaptation and begins by asking whether any explanations for adaptation other than natural selection are possible. Although the chapter argues that no other theories fit adaptation, it notes that some characteristics have probably evolved by processes other than natural selection (though they are not adaptations). Not all evolution proceeds by natural selection, but all adaptive evolution does. The chapter considers various constraints that can influence the characteristics that evolve in a species, and describes how to test whether a constraint or natural selection is at work in a real case. The chapter also discusses the level of perfection achieved by adaptations of living species. Natural selection acts to improve adaptation, but various reasons exist to explain why a state of perfection is not reached.

CHAPTER 11

The Analysis of Adaptation

Many of the characters of organisms may superficially appear to make no difference to its life—to be non-adaptive. Closer study often reveals that they do matter, however. We begin here by looking at an example of a subtle adaptation that confirms this point. We then briefly examine the main methods of studying adaptation: the relation between the predicted and the observed form of a character, experiment, and inter-species comparisons. The chapter mainly concentrates on three related research questions—the questions of how sex, sex differences, and sex ratios are adaptive. More progress has been made with the sex ratio issue than with the explanation of the existence of sex, but together the three questions illustrate the methods used at various stages of research.

11.1 The way organisms are adapted may not be obvious

Cepaea nemoralis is a land snail, and the background color of its shell may be either yellow or some darker shade of brown or pink. A number (usually between zero and five) of dark bands are visible on top of the background color. The snails are, therefore, highly variable in their external appearance; an individual snail may have any combination of background color and banding pattern. Why do *Cepaea* vary in this way? One possibility is that the external appearance does not matter; as Mayr remarked in 1942, "there is no reason to believe that the presence or absence of a band on a snail shell would be a noticeable selective advantage or disadvantage."

We have since discovered that good reason does exist for this variation. So many selective factors have been shown to influence the pattern and coloration of *Cepaea* shells that, 35 years after Mayr's remark, the polymorphism was called "a problem with too many solutions." A series of ecological geneticists, particularly Cain and Sheppard, have studied this question over many years. The first suggestive observation was that the proportions of the different shell types vary from place to place. Initially, the geographic distribution of the shell types had seemed to be random, but it was soon found that the banded forms tend to be found in diversified habitats, such as mixed hedgerows, and the unbanded snails against more uniform backgrounds, such as those of dense woodlands. Perhaps the banding pattern is an adaptation for camouflage. Cain and Sheppard duly measured the rate at which the different snail types were eaten by birds. Birds, such as the thrush (*Turdus philomelos*), use stones as anvils to break open the snails, and it is possible to count the proportions of different snail types taken by the birds from the shell debris found around an anvil; these proportions can then be compared with those in the local habitat. At one site near Oxford (United Kingdom), for example, where the habitat is relatively uniform, Cain and Sheppard found that in the area as a whole 264 of a sample of 560 (47.1%) snails were banded, but 56.3% (486 out of 863) of the snails taken by thrushes were banded. The difference is statistically significant. In this habitat, the thrushes were taking the banded snails disproportionately. Similar results for other habitats also support the idea that differences in the color pattern and banding number serve as camouflage to protect the snails from bird predators.

The appearance of *Cepaea* is also influenced by the need for thermoregulation. The darkness of a shell affects how quickly the snail warms up in sunshine. It can be a matter of life and death for a snail if it has too many bands and lives on a sunny hill-slope. Given this factor, and the issue of camouflage, it can no longer be doubted that the color and banding patterns of *Cepaea* are adaptive, and adaptive in relation to more than one property of their local environments.

In 1942, as Mayr's remark shows, shell pattern in *Cepaea* was not recognized as an adaptation. It was certainly not obvious what purpose the adaptation might serve. In this chapter and the next two chapters, we shall be concerned with how we can find out why natural selection favors the particular characters that organisms possess. The first method, illustrated by *Cepaea*, is simply to look more closely at the character and see what consequences it has in the animal's life. Other methods may be used as well, and we shall be discussing them in relation to three questions about sex:

1. Why do organisms reproduce sexually?
2. Why do the sexes differ?
3. What sex ratio is favored by natural selection?

These questions are good examples because they are the subjects of active investigation, and because they reveal how evolutionary biologists actually carry out research on adaptation. Before we discuss these questions, let us look briefly at the abstract form of the methods they will illustrate.

11.2 *Three main methods are used to study adaptation*

The study of adaptation proceeds in three conceptual stages. The first stage is to identify, or postulate, what kinds of genetic variant the character can have. Sometimes, as in *Cepaea*, this identification is made empirically; other characters do not vary genetically and it may be necessary to postulate appropriate theoretical mutant forms. For example, if we are studying why sex exists, we postulate a mutant form that reproduces asexually.

The second stage is to develop a hypothesis, or a model, of the function of the organ or character. Cain and Sheppard's first hypothesis about snail banding, for instance, was that it functioned as camouflage. Hypotheses vary in quality, but the poorer sort can be improved as work proceeds. A good hypothesis will predict the features of an organ exactly, and the predictions will be testable. In morphology, these predictions are often derived from an engineering model. For example, hydrodynamics is used to understand fish shape, while construction engineering is used for shell thickness in a mollusc. In the latter case, the costs of building a thicker shell must be weighed against the benefits of reduced breakage, whether by wave action or predators. This sort of research touches on all levels, from the simple and qualitative through sophisticated algebraic modeling.

Stage three is to test the hypothesis's predictions, for which three main methods are available. One method is simply to see whether the actual form of an organ (or whatever character is under investigation) matches the hypothetical prediction. If the observation does not match the prediction, the hypothesis is wrong.

A second method, performing experiments, is useful only if the organ, or behavior pattern, can be altered experimentally. Almost any hypothesis about adaptation will predict that some specified form of an organ will enable its bearer to survive better than some other forms, but the alternatives are not always feasible. We cannot, for example, make an experimental pig with wings to see whether flight would be advantageous. When they are possible, experiments offer a powerful means of testing ideas about adaptation. Animal coloration, for instance, has been studied in this way. Color patterns in some butterfly species are believed to act as camouflage by "breaking up" the butterfly's outline. Silberglied *et al.* working at the Smithsonian Tropical Research Institution at Panama, experimentally painted out the wing-stripes of the butterfly *Anartia fatima*. The butterflies with obscured wing-stripes showed similar levels of wing damage (which is produced by unsuccessful bird attacks) and survived equally well as control butterflies (Table 11.1); the wing-stripes are, therefore, not adaptations to increase survival. They may have some other signaling or reproductive function, although that hypothesis would need to be tested by further experiments.

The third method of studying adaptation is the *comparative method*. It can be used if the hypothesis predicts that some kinds of species should have forms of an adaptation that are different from those observed in other kinds of species. Darwin's classic study of the relation between sexual dimorphism and mating system is an example we shall discuss later in this chapter. Some hypotheses

TABLE 11.1

The wing stripe of some butterflies was painted over, and controls were painted with transparent paint that did not affect their appearance. The numbers with intact wings at different times after the treatment was measured. The frequency distributions are not significantly different. Reprinted with permission from Silbergleid *et al.* (1980). Copyright 1980 American Association for the Advancement of Science.

Age at Capture (week)	Last Week in Which Wings Were Intact			
	Experimental		Control	
	N	%	*N*	%
0	81	83.5	88	90.7
1	14	14.4	6	6.2
2	2	2.1	2	2.1
3	0	0	1	1.0
4	0	0	0	0
5	0	0	0	0

predict that different kinds of species will have different adaptations, while others do not. Thus, Darwin's theory of sexual dimorphism makes such predictions. On the other hand, an optical engineer's model of how the eye should be designed might specify just a single best design, with the implication that all animals with eyes should have that design. In that case, the comparative method would not be applicable.

In summary, the three main methods of studying adaptation are to compare the predicted form of an organ with what is observed in nature (and perhaps also to measure the fitness of different forms of organism), to alter the organ experimentally, and to compare the form of an organ in different kinds of species.

11.3 Example 1: the function of sex

11.3.1 Sexual and asexual reproduction should be distinguished

In sexual reproduction, a new organism is formed by the fusion of two gametes. In contrast, in asexual reproduction females produce offspring without any male contribution; the female's gametes develop directly into female offspring. In sexual reproduction, the sex cells—eggs and sperm or pollen—are produced by a reduction division and the gametic fusion restores the original chromosome complement. In asexual reproduction, there is either no reduction division (apomixis) or the product cells of one individual's reduction division fuse together again (automixis). In addition, there are other, evolutionarily less interesting, methods of restoring diploidy, such as the doubling of the haploid chromosome set to produce a purely homozygous asexual offspring. The question to be discussed here is why sexual reproduction is so common. Why has it not been replaced by asexual reproduction more often?

The simplest way in which sex could be lost would be for a mitotic division to be substituted for meiosis, with the elimination of gametic fusion. The mitotically produced egg could then develop like a normal zygote, and no complicated mutation would be needed. All of the necessary mechanisms of mitosis already exist, and have much in common with meiosis. Substitute one process for the other is not a remarkably complex evolutionary step. If the loss of sex would be relatively easy, why is sex maintained? How is sexual reproduction adaptive? (The answer to these questions is also likely to answer the question of why genetic recombination exists; see section 8.11, p. 212.)

11.3.2 Sex has a 50% cost

Although some species reproduce asexually, the majority, at least of multicellular organisms, reproduce sexually. Thus, sex probably has a selective advantage in most species. The advantage, moreover, must be large, because it must overcome an automatic twofold *cost of sex*, relative to asexual reproduction.

What is this cost of sex? It is easy to understand by comparing the reproduction, over time, of a group of sexual females with a group of asexual females. We could start with a group of 100 asexual females and another group of 100 sexual females (if the sex ratio were 1:1, the sexual population would then have 200 members). There are 300 individuals in all, and one-third of them are asexual females, one-third sexual females, and one-third males. The members of the two groups, we suppose, are identical in all other respects: sexual and asexual individuals are equally good at finding food, avoiding enemies, and staying alive; they produce the same number of offspring, and those offspring have an equal chance of survival. We are considering only whether natural selection favors sexual or asexual reproduction.

Suppose, for simplicity, that each female produces two offspring. After one generation, the asexual group will have grown to 200 individuals, as each mother produces two daughters. The 100 sexual females will also produce 200 offspring, but only 100 of these offspring will be daughters; the size of the sexual group will remain constant. Now we have 400 individuals in all, and the proportion of asexual females has increased from one-third to one-half. After another generation, there will be 400 asexual females, 100 sexual females, and 100 males; the proportion of asexual females will have grown to two-thirds. It will not be long before asexual reproduction has completely taken over. The clone of offspring from an asexual female multiplies at twice the rate of the progeny descended from a sexual female, and a sexual female has only 50% of the fitness of an asexual female. Various provisos can be made about this argument, but it applies at least approximately in a wide variety of conditions—probably in the conditions of most sexually reproducing species.

Fifty percent is a large cost. The problem of explaining sex is to find a compensating advantage of sexual reproduction that is large enough to make up for its cost. This selective advantage must be extraordinarily large, as most evolutionary events are thought to involve selective advantages of a few percentage points at most, and more often 1% or less. To understand why, consider that a female who has survived to adulthood and is ready to reproduce must be fairly well adapted to her environment. If she were to reproduce asexually, she would just make a copy of herself and produce a daughter as well adapted to the conditions of the next generation as she would be herself. If she reproduces sexually instead, she discards half of her genes and produces an offspring by mixing the remaining half with other genes drawn from a stranger. If sex is to outweigh its twofold cost, the sexual female must by this procedure expect to produce a daughter who will be twice as fit as a simple copy of herself. The problem, therefore, is not trivial. Indeed, G. C. Williams has described it as "the outstanding puzzle in evolutionary biology." Let us look at some of the possible solutions.

11.3.3 *Sex can accelerate the rate of evolution*

A population of sexually reproducing organisms can, under some conditions, evolve faster than a similar number of asexual organisms. Sexual reproduction can greatly increase the rate at which beneficial mutations, at separate loci, can be combined in a single individual (Figure 11.1). Suppose, for example, that a sexual and an asexual population are both fixed for genes A and B at two loci. In the environment where the two populations live, mutations A' and B' are advantageous. A' and B' mutations would be likely to arise initially in different individuals. The asexual population will eventually consist of $A'B$ and AB' individuals, because the A' mutant cannot spread into the AB' clone, or vice versa.

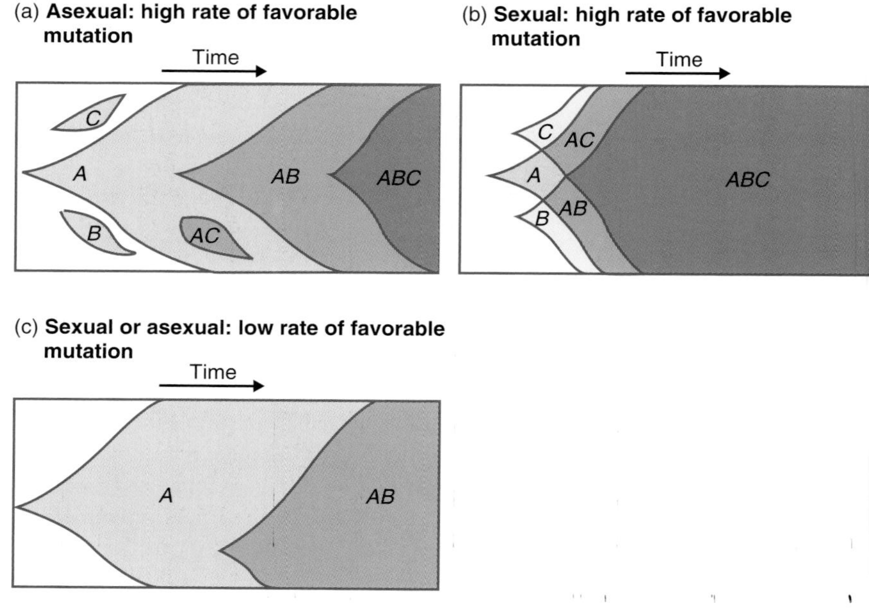

Figure 11.1 Evolution in (a) asexual and (b) sexual population. The mutations *A*, *B*, and *C* are all advantageous. In the asexual population, an *AB* individual can arise only if the *B* mutation arises in an individual that already has an *A* mutation (or vice versa). In the sexual population, the *AB* individual can be formed by breeding of a *B* mutation-bearing individual with an *A* mutation-bearing individual; the second mutation of *B* is unnecessary. (c) If favorable mutations are rare, each will have been fixed before the next mutation arises, and sexual populations will not evolve more rapidly. The relative rates of evolution in asexual and sexual populations depend on the rate at which favorable mutations arise.

$A'B'$ individuals cannot appear until an A gene mutates to A' within the AB' clone (or B to B' in the $A'B$ clone).

In the sexual population, evolution proceeds much more rapidly. After A' and B' have arisen in different individuals, they can soon combine in a single individual by sex without waiting for the mutations to occur twice. Natural selection can, therefore, take the population from the state AB to $A'B'$ more quickly than under asexual reproduction. This argument was first put forward by Fisher, who concluded that sexual populations have a more rapid rate of evolution than would an otherwise equivalent group of asexual organisms.

Fisher's conclusion depends on the rate of mutation, however. If favorable mutations are rare, each one will have been fixed in the population before the next one arises (Figure 11.1c). Sexual and asexual populations then evolve at the same rate. New favorable mutations will always arise in individuals that already carry the previous favorable mutation: they must, because the previous favorable mutation already appears in every member of the population. In the example given above, the B' mutation will arise in an $A'B$ individual in both sexual and asexual populations. If favorable mutations arise more frequently, Fisher's argument works: the sexual population evolves more rapidly. Each new favorable mutation will usually arise in an individual that does not already possess other favorable mutations; the greater speed with which the different favorable mutations combine together causes the sexual population to evolve more quickly. The higher the rate at which favorable mutations arise, the greater the evolutionary rate of a sexual relative to an asexual population.

11.3.4 *Is sex maintained by group selection?*

Perhaps the most common answer to the question of why sex exists is that it speeds up the rate of evolution. This "group selection" theory of sex accepts that sex is disadvantageous for the individual, because of its 50% cost, but claims that the faster evolution of the sexual population (or group) more than compensates for the cost. The sexual population would accumulate superior adaptations more rapidly and be able to outcompete an asexual population.

Group selection is a controversial subject, and we will discuss it at length in chapter 12. Here it is enough to note that most evolutionary biologists distrust theories that rely on advantages to whole populations. When individual advantage and group advantage conflict, individual selection is usually more powerful. We should not expect to find adaptations that prove disadvantageous for the individual even if they do benefit the group. Although sexual populations may evolve more rapidly than asexual ones, it is still true that sexual *individuals* reproduce more slowly than asexual individuals; asexuality, once it has arisen, will tend to take over sexual groups. Asexuality can arise in a group by either mutation or immigration, and neither of these processes is likely to be, on an evolutionary time scale, very rare. Asexuality should, therefore, arise at a fairly high rate. Group selection raises suspicion because it requires the rate at which asexual females arise in sexual groups to be very low.

The argument against group selection can, at best, show that adaptations for the benefit of the group are theoretically unlikely, not that they are theoretically impossible (see chapter 12). Consequently, we should also look at the facts. What do they suggest? We have seen that the effect of sex on the rate of evolution theoretically depends on the mutation rate, but what about populations in nature? Do natural populations occupy the "rare mutation" extreme in which sexual and asexual populations evolve at the same rate, or do favorable mutations arise frequently enough to prompt more rapid evolution of sexual populations? The question cannot be answered directly, because we do not know the rate of favorable mutation, but the taxonomic distribution of asexual species allows an indirect inference.

It has been known since an 1886 paper by Weismann that asexual reproduction has a peculiar taxonomic distribution. It is mainly confined to small twigs in the phylogenetic tree (Figure 11.2). That is, asexual reproduction is typically found in an odd species, or perhaps a whole genus, within a larger taxonomic group that mainly reproduces sexually; only very rarely is it found throughout a larger phylogenetic group. The only exceptions are a suborder of rotifers (the Bdelloidea) and of gastrotrichs (Chaetonoidea). All the former, and most of the latter, are believed to be asexual. The spindly taxonomic distribution of asexual reproduction suggests that asexual lineages have a higher extinction rate than sexual lineages—that asexual lineages usually do not last long enough to diversify into a genus or higher taxonomic level. If the difference in extinction rates is due to the slower evolution of asexual forms, then natural rates of favorable mutation must be high enough for Fisher's argument to apply. However, this conclusion is uncertain because the higher extinction rate of asexual lineages might also be cased by deleterious mutations, as discussed below (section 11.3.5).

Saying that sexual populations have a lower extinction rate than asexual populations is not the same as saying that sex exists because of its lower extinction

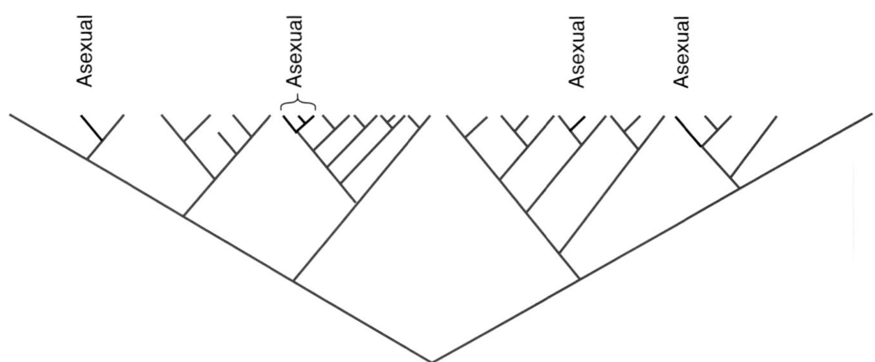

Figure 11.2 The taxonomic distribution of asexual reproduction is spindly; it is found in odd, isolated taxa.

rate (which is a much stronger statement). It could be that sex exists in sexual species because sex is advantageous to the individuals of those species, and asexual reproduction exists in asexual species because it is advantageous to the individuals in those species. The different extinction rates would then be species-level consequences of different individual adaptations in the two types of species. By analogy, carnivores could have higher extinction rates than herbivores, but that would not mean that herbivory was disadvantageous to individual herbivores and maintained because of its advantage to the group. The taxonomic distribution of asexuality, therefore, although it is consistent with the group selectionist theory of sex, does not confirm it. The same pattern could have arisen if sex had an individual advantage.

Williams has put forward a general argument—the *balance argument*—that casts doubt on the importance of group selection in the maintenance of sex. Some species, such as many plants, aphids, sponges, rotifers, and water fleas (Cladocera), can reproduce both sexually and asexually depending on the conditions. These species are called heterogonic. Many heterogonic species time their sexual reproduction for periods of environmental uncertainty, and reproduce asexually when conditions are more stable. Here we are concerned only with the individual's ability to reproduce in either way. Williams' point is that when an heterogonic individual like an aphid reproduces sexually, it must be advantageous to that individual; otherwise the aphid could have reproduced asexually. Both sexual and asexual reproduction must have "balanced" advantages to maintain them in the species' life cycle, or the inferior method would be lost.

The group selectionist, you will recall, believes that sex is disadvantageous to the individual, and only advantageous to the group. In aphids and other heterogonic species in which individuals have a "choice," sex must provide an individual advantage. The argument can be extended. If sex is advantageous in aphids, it is probably also advantageous to the individual in non-heterogonic species. We have no good reason to think that sex is exceptional in aphids, or that special factors favor sex in heterogonic species. If we must find an individual advantage for sex in aphids, that same advantage will probably also exist in other species. If group selection can be ruled out for aphids, it can probably also be ruled out for other species.

Williams' argument is powerful, but not decisive. In most heterogonic species, the asexual and sexual propagules differ in other respects besides being asexual and sexual. For example, the cladoceran sexual offspring form special winter eggs that are adapted for winter survival. Any cladoceran that gave up sex would also lose this wintering stage. In practice, the loss of sex while retaining the winter egg would need two mutations, one for the loss of sex and the other for transferring the winter egg phenotype to asexual eggs. Thus, the balance argument is not perfectly clearcut.

In summary, group selection will tend to favor sexual reproduction over asexual reproduction because sexual populations will have a lower rate of extinction. The taxonomic distribution of asexuality suggests that asexual populations tend to go extinct relatively quickly in evolution. However, biologists doubt whether group selection explains why sex exists, for two main reasons. One is a general disbelief in group selection; the other is Williams' balance argument. Neither of these objections is completely convincing, and group selection cannot be finally ruled out. However, the objections are strong enough to have inspired biologists to look for a short-term, individual advantage to sex. We will consider two of the most influential modern ideas in the following discussions.

11.3.5 *Sexual reproduction can enable females to reduce the number of deleterious mutations in their offspring*

We saw earlier (section 9.12, p. 247) that slightly deleterious mutations arise in each generation. They are normally removed by selection, but if selection is experimentally prevented then deleterious mutations accumulate and the fitness of the average member of the population declines over time. Here we will consider how effectively selection removes deleterious mutations, depending on whether reproduction is sexual or asexual. We will do so in two stages. First, we consider the case in which mutations at different loci have independent deleterious effects on an organism's fitness; in this case sex makes no difference. Next, we turn to the case in which an individual with two mutations has a lower fitness than would be expected from the fitnesses of individuals with one of the mutations (section 7.7.2, p. 167). In this case sexual reproduction can be advantageous because it purges the deleterious mutations more efficiently. We will then have defined the theoretical conditions for natural selection to favor sex, and we can examine whether the conditions are realistic.

The effect of the theory, which was mainly developed by Kondrashov, can be seen in its simplest form in a haploid model with two loci. We symbolize the normal versions of the genes by 1 and slightly deleterious mutants by 0. An individual with unmutated genes at two loci is symbolized by 11. Individuals with one mutant and one unmutated gene are 01 or 10. Double mutants are 00.

Let us begin with mutations that have independent effects on fitness (Figure 11.3, line b). The following is an example of independent fitnesses:

genotype	11	10	01	00
fitness (independent)	1	0.9	0.9	0.81

Figure 11.3 Three different relations between the fitness of an organism and the number of deleterious mutations it carries. The *y*-axis is logarithmic. (a) Multiple mutations have an increasingly damaging effect on the organism. (b) Independent, or multiplicative, fitness effects. (c) Multiple mutations have a decreasingly damaging effect. The text considers only (a) and (b). Sex can be advantageous with (a) and makes no difference with (b), but it is easy to generalize the arguments in the text to see that sex is disadvantageous with (c). These graphs are mirror-images of Figure 7.6 (p. 167), in which the number of good (rather than deleterious) genotypes was plotted on the *x*-axis. The decreasing returns graph for the number of good genotypes in Figure 7.6 corresponds to the "increasing returns" effect of deleterious genotypes in line (a) here.

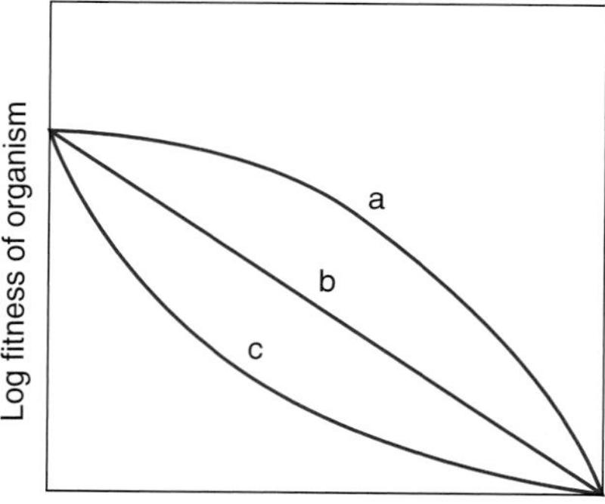

Each successive mutation decreases survival by the same multiple, regardless of the number of mutations already present in the individual. This scenario is most likely to be true if the loci have independent effects on the phenotype—for example, when locus 1 influences survival in the first year of life and locus 2 influences survival in the 10th year of life. The fitness of a double mutant would then equal the product of the fitnesses of the single mutants, as in the numerical example just given.

Natural selection will work against the deleterious mutations, but does sex make any difference? When mutations have independent effects, the genotype frequencies are on average the same in an asexual and a sexual population (the frequencies are those of linkage equilibrium between the loci—section 8.8, p. 207). Thus, it makes no difference to the average female whether she breeds sexually or asexually; the average number of slightly deleterious mutations in her offspring will remain the same. Asexual females who copy themselves have, at equilibrium, the same average number of mutations in their progeny as in themselves—the number of new mutations balances the number removed by selection. The frequency distribution of mutations is also the same in sexual individuals, male or female. When a female performs meiosis, she removes half of her deleterious mutations as she forms her haploid gamete. The mutations introduced by the sperm, however, have the same frequency distribution as those excised in meiosis. The resulting frequency distribution of mutations in the offspring is the same as if the females had copied themselves.

Contrast this situation with the case in which fitness effects of deleterious mutations show the "sloping down" pattern (Figure 11.3, line a). Now the fitness of a double mutant is lower than expected from the single mutants, and the fitnesses might take the following form:

genotype	11	10	01	00
fitness	1	0.9	0.9	0.7

In section 7.7.2 (p. 167), we saw that this may be a common occurrence because an organism can cope with one or two defects, but as the number of defects increases they eventually overwhelm the back-up mechanisms and send the organism's fitness into a serious decline. When the double mutants have this sort of lowered fitness, the strong selection against 00 haplotypes depresses their frequency to a level below that seen in the multiplicative case. Fewer 00 and 11 organisms and more 01 and 10 organisms than expected from the frequencies of the two individual 0 mutations will be produced. Thus, sex does make a difference in this case. In a sexual population, recombination between the 01 and 10 individuals generates new 00 and 11 individuals. In consequence, more 00 and 11, and less 01 and 10, individuals appear in a sexual population than in an asexual population. Sex acts to convert single mutant individuals into a combination of double and zero mutant individuals. Although the average number of mutations per individual before selection acts does not change, the distribution is modified.

The change in the distribution matters when we consider selection. Let us compare the progeny of females who have one mutation on average and who either reproduce sexually (and so produce some 00 and 11 progeny) or asexually (all of whose progeny are single mutants). At the zygote stage, before selection:

	Sexual Females		Asexual Females
genotypes	00	11	10/01
frequency in zygotes	50%	50%	100%
average number of mutations per offspring	1		1

At this stage sexual reproduction appears to have made things worse. In terms of the example introduced in section 7.6, is it better to produce one two-eyed and one blind offspring, or two one-eyed offspring? The answer is two one-eyed offspring; their average fitness is higher than the combination of one two-eyed and one zero-eyed offspring. However, this case is only at the zygote stage, and most of the zygotes will inevitably die. The sexual females make use of these deaths to improve the average quality of their offspring. In this simple example, the average number of deleterious mutations per offspring cannot be improved in the asexual females, but selection will improve it in the sexual females. We may now have the following situation:

	Sexual Females		Asexual Females
genotypes	00	11	10/01
frequency in adults	25%	75%	100%
average number of mutations per offspring	0.5		1

(The 75% and 25% are merely for illustration; the exact numbers would depend on the exact fitnesses.) The average number of deleterious mutations in the progeny of a sexual female after selection is less than for a comparable asexual female. The sexual females present the mutations to selection in their vulnerable double form, enabling selection to remove the mutations more powerfully than in the asexual population (the power derives not only from the lower fitness of the double mutants but also from the loss of two mutations when one double mutant dies). A sexual female who switched to asexual reproduction would experience the deterioration immediately, via her progeny, giving sex an immediate advantage.

The advantage to sex arises because recombination increases the frequency of double mutants, double mutants have disproportionately lower fitness, and therefore the creation of double mutant individuals effectively removes deleterious mutations from the progeny. The power with which the deleterious mutations are removed depends (when the fitness graph slopes down) on serving the mutations up to selection in multiple form; sex targets the deleterious mutations at a higher rate. Although this example focuses on the simple case of two genetic loci, the theory can operate in all genes in a genome, and it becomes even more powerful as more loci are considered.

How plausible is this theory? It has two requirements. First, the mutation rate must be high enough. A female gains the advantage whatever the deleterious mutation rate, but the advantage increases as the mutation rate increases. An interesting theoretical question is what deleterious mutation rate is needed to outweigh the twofold cost of sex. The answer depends on the details of the theoretical model, but a total mutation rate of about one per individual is probably enough. If the deleterious mutation rate is that high, a sizeable proportion of females in the population will potentially produce offspring with multiple mutations.

How realistic is this number? We do not really know, but a total mutation rate of one is higher than has traditionally been supposed. The traditional view enjoyed both factual and theoretical support. In theory, a mutation rate of one was thought to imply an intolerably high genetic load (section 7.3, p. 157). In fact, if we take a standard per locus mutation rate of 10^{-6}, and assume that practically all mutations are deleterious, then organisms with 10^4–10^5 loci will have total deleterious mutation rates that are 10–100 times less than 1. Moreover, the mutation rate implied by Mukai's experiment with fruitflies (section 9.12, p. 247), although much higher than the traditional estimates, implied a total mutation rate of about 0.5 per individual per generation, which is less than half that required for sex to be favored. However, neither of these arguments is conclusive. Intolerable mutational loads are incurred only if selection acts independently on each locus, and (as we saw in section 7.7.2, p. 167), a much higher mutation rate becomes plausible if selection eliminates inferior genotypes at several loci at once. This elimination occurs in the mutational theory of sex when the multiple mutations are ghettoized by recombination. (The argument can also be stated in the reverse. If the deleterious mutation rate is about one per individual, the fitness interactions of the mutations must show the sloping down form or the whole population would be dead.) Moreover, the measurements of mutational rates are uncertain, and the traditional view of them could easily be in error.

No strong factual evidence exists to rule out mutation rates as high as are required for sex to be favored. The estimate of 0.5 per individual from Mukai's work was a lower bound estimate, and his results are also compatible with a figure of more than one. Other ways of looking at the evidence for mutation rates make it reasonable to infer a total rate of deleterious mutation per individual of one or more.

The second theoretical requirement is that multiple deleterious mutations do not have independent effects on fitness. They interact, such that an individual with two mutations suffers a greater reduction in fitness than twice the average of the two single mutations. The plausibility of this requirement was discussed before (section 7.7.2, p. 167), when we saw that no conclusive evidence has been put forth but that a reasonable theoretical case can be made for this dependence. Therefore, one important conclusion from this discussion is that we urgently need further experimental work of the kind done by Mukai to measure both total mutation rates and the form of fitness interaction between different mutations. Meanwhile, we cannot conclude with certainty whether deleterious mutations alone are the reason why sex exists, but it remains one of the two most popular modern theories. Under this theory, the advantage of sex is that it enables females to eliminate more deleterious mutations from their progeny than could be accomplished by asexual reproduction.

11.3.6 *The coevolution of parasites and hosts may produce rapid environmental change*

The second theory we shall consider ignores the effect of deleterious mutation and concentrates on external environmental change. Sex is more likely to be advantageous if environments change rapidly. The problem is to work out *how* environments could possibly be changing rapidly enough. It is not difficult to believe that environments might change quickly enough to make sex advantageous every few hundred years, but how could they be changing at a rate that makes sexual reproduction advantageous every generation? Remember, the environment would have to change so rapidly that an average sexual female's daughters would be twice as fit as those of an average asexual female. We cannot take it for granted that ordinary environmental change will be enough: if we are to explain the existence of sex by environmental change, we have some work to do.

A recent, promising suggestion is that the coevolution between parasites and hosts may generate environmental change at a speed that renders sex advantageous in the short term. The "environment" for the parasite is the host's resistance mechanism; for the host, the environments consist of the parasite's method of penetrating its defenses. Several authors have suggested that *parasite-host coevolution* may be important in the maintenance of sex; Hamilton is the best known.

The theory can be made more explicit by examination of a simple model. In some parasite-host relationships, gene-for-gene matching systems operate such that one host genotype is adapted for resisting one parasite genotype, another host genotype for another parasite genotype, and so on. The best understood example involves wheat and parasitic rusts; similar selection may operate in the human HLA system (section 8.6, p. 203).

The simplest genetic model for host-parasite coevolution is haploid, with two alleles appearing in each of the host and parasite species. One parasite allele is adapted to penetrate hosts with one of the host alleles, the other parasite allele penetrates the other (Table 11.2). Selection of this sort generates cyclic changes in gene frequency (Figure 11.4); as a genotype increases in frequency, its fitness (after a time lag) decreases. If parasite genotype P_1 is more common, then host genotype H_1 will be favored and increase in frequency. The fitness of P_1 then declines as more hosts become resistant to it. As P_2 becomes more common, the fitness of H_1 decreases. When H_1 becomes rarer, the frequency of P_1 will, in turn, increase again. Cycles of gene frequency are driven by corresponding cycles of gene fitness.

TABLE 11.2

A simple model of gene-for-gene matching in a pair of host and parasite species. The numbers in the table are the fitnesses of the genotypes.

(a) Fitness of parasite genotype in two types of host			
		Host genotype	
		H_1	H_2
Parasite genotype	P_1	0.9	1
	P_2	1	0.9
(b) Fitness of host genotype against two types of parasite			
		Parasite genotype	
		P_1	P_2
Host genotype	H_1	1	0.9
	H_2	0.9	1

Figure 11.4 Frequency changes of host and parasite genotypes. (a) As H_2 becomes more common, selection acts to increase the frequency of P_2, which in turn selects against H_2, and H_1 increases in frequency. (b) Plotted against time, the frequency of each genotype oscillates cyclically.

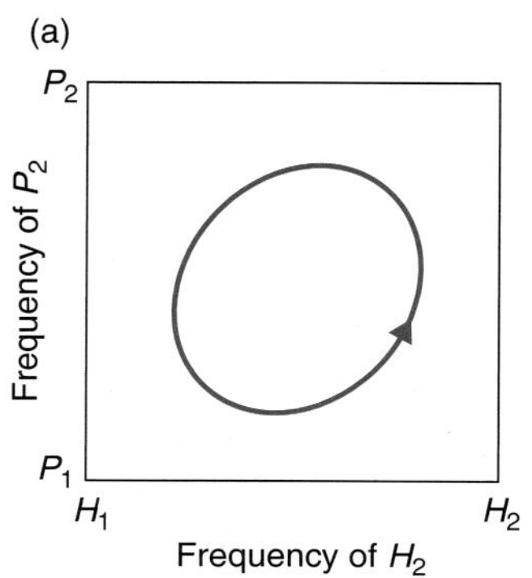

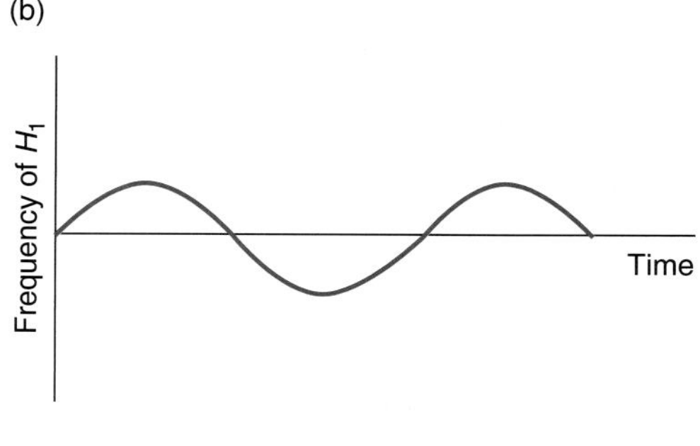

We need a more complex—and probably more realistic—model to produce an advantage for sex. Imagine now that resistance and counter-resistance are controlled by two loci. Again, a haploid model is simplest. With two loci and two alleles at each, there are four haplotypes: *AB*, *Ab*, *aB*, and *ab*. Complementary sets will appear in the host and the parasite; if A_HB_H, A_Hb_H, a_HB_H, and a_Hb_H are the host genotypes, then we could write the parasite genotypes as A_PB_P, A_Pb_P, a_PB_P, and a_Pb_P. A_HB_H and A_PB_P are analogous to H_1 and P_1 in the previous model. Imagine that A_H and B_H control two cell-surface molecules used by the parasite to penetrate the host. Hosts with allele A_H are efficiently penetrated by parasites with a_P, but not A_P; B_H hosts are penetrated by b_P parasites but not B_P. b_P parasites are, therefore, favored if the hosts are mainly B_H; likewise, b_H is a gene for resistance to b_P parasites. Haplotype frequencies at both loci will oscillate for the same reasons as did H_1 and P_1 in the simpler model. In the sexual parasites and the sexual hosts, alleles at the two loci can recombine; in asexual parasites and hosts, they cannot. A third locus determines whether reproduction is sexual or asexual.

How can sex be advantageous? As the frequencies of the four haplotypes oscillate through time, any one of them could potentially be lost at the trough of its frequency cycle. Suppose, for example, that the frequency of A_Hb_H is driven so low in one cycle that it is lost from both the asexual and sexual populations. In the asexual population it has been lost forever, whereas in the sexual population it will be recreated by recombination between the other three genotypes. As the frequency of the parasites that specialize in attacking A_Hb_H hosts increases again, sexual reproduction will offer an advantage because of its more frequent association with the resistant genotype. Thus, sex provides an advantage because it maintains in reserve an ability to recreate multi-locus genotypes that have been disadvantageous, but may be needed again. The cycles of host-parasite coevolution are exactly the kind of circumstances in which this ability is favored. If the environmental change is more erratic, or open-ended, such that once a genotype has been eliminated it is unlikely to be useful again, sex is not advantageous.

A number of empirical studies have found that, in taxa in which both sexual and asexual reproduction operate, sexual reproduction is more common where parasite pressure is higher. This work tentatively supports the theory, although alternative interpretations of the results may be made. The parasitic, like the mutational, theory requires certain preconditions, which need to be tested as well. It requires parasitism to be found in almost all species; it requires a certain kind of multi-locus gene-for-gene parasite-host relationship; and it requires that the frequencies of these genes oscillate through time. We know that parasite-host coevolution is almost universal in nature. Humans are parasitized by a long list of worms, flies, protozoa, bacteria, and viruses; all species, including bacteria, are parasitized by viruses. However, little is known about the genetics of host-parasite relations in most cases, and no evidence has been found of cycles in the frequencies of these genes. That lack of information is hardly surprising because the relevant genes have not been identified in most species, and no one has tried to measure their frequencies through time. Meanwhile, the parasitic theory of sex remains an attractive, and coherent hypothesis, but it is uncertain whether it identifies the main factor maintaining sex in nature.

11.3.7 *Conclusion: it is uncertain how sex is adaptive*

Both deleterious mutation and parasite-host coevolution are reasonable theories of sex, but it has not been conclusively shown that either really explains how sex is maintained in nature. They remain afloat in evolutionary biology today as hypotheses—as stimulating hypotheses, that are inspiring much research. They are not mutually exclusive ideas, and both factors could turn out to contribute to the selective advantage of sex. Other hypotheses have been posited as well—some of them highly ingenious.

Today, the question of why sex exists remains an "outstanding puzzle." Evolutionary biologists are not confident the question has been satisfactorily answered. Perhaps, as Maynard Smith once said, "some crucial aspect of the problem has been overlooked," and we need some radically new idea that has not yet been put forward or lies unappreciated. Alternatively, the gist of the answer may be hidden in the theories we have discussed and the problem involves showing how they apply in nature.

When the question is ultimately answered, some other deep beliefs in evolutionary biology may have to be surrendered. If sex is an adaptation maintained by group selection, ideas on the importance of group selection (chapter 12) will need to be modified. Deleterious mutation, or continual, rapid, cyclical coevolution between parasites and hosts may prove more common than is customarily supposed. In addition, both of these theories suggest that offspring are so different from their parents, whether because of external change or internal mutational deterioration, that a genetic copy of the parent would have half the fitness of a set of offspring produced by mixing the mother's genes with those of another individual. That form of evolution consists of dramatically rapid change, much more rapid than current evidence demonstrates, but it may yet turn out to be plausible. The subject has developed dramatically in the past 20 years and the hope now is that this crucial question will be answered in the next decade or so of research. Important related questions wait upon the answer.

11.4 *Example 2: sexual selection*

11.4.1 *Sexual characters are often apparently deleterious*

For the most part, the characters of organisms are adaptive: they increase the organisms' chances of surviving to reproduce. However, some characters have the opposite effect, and (as Darwin was well aware) natural selection does not explain why these characters exist. If a population contains some types with higher survival than other types, natural selection will fix the former and eliminate the latter.

Characters that reduce survival can be called "deleterious" or "costly." One large class of apparently costly characters are those found usually only in males and which Darwin called secondary sexual characters. The primary sexual characters are organs like genitalia that are needed for breeding. The secondary sexual characters are not actually needed for breeding, but they function during reproduction. The peacock's "tail" (or, more exactly, train) is an example of a secondary sexual character. In many other bird species, the males also possess tails or other extravagantly developed, brightly colored structures. A peacock could inseminate a female just as well without his remarkable tail, and in that sense it is a secondary, not a primary, sexual organ. The peacock's tail almost certainly reduces the male's survival, however (although this advantage has never actually been demonstrated). On the other hand, it reduces maneuverability, decreases powers of flight, and makes the bird more conspicuous; its growth must also impose an energetic cost. Why are these costly characters not eliminated by selection?

11.4.2 *Sexual selection acts by male competition and female choice*

Darwin's solution was his theory of sexual selection. He defined the process by saying that it "depends on the advantage which certain individuals have over other individuals of the same sex and species, in exclusive relation to reproduction." Thus, a structure produced by sexual selection in males exists not because of the struggle for existence, but because it gives the males that possess it an advantage over other males in the competition for mates. Darwin's idea is that the reduced survival of the peacocks with long, colorful tails is more than compensated by their increased advantage in reproduction.

Darwin discussed two kinds of sexual selection. One process involves the competition among males for access to females. *Male competition* can take the form of direct fighting, or it can be more subtle. Some male insects, for instance, while copulating have various means of removing any sperm the female possesses from matings with previous males; one example is the damselfly *Calopteryx maculata* in which Waage showed that the male uses brushes and hooks on the penis to scrape out the sperm. Many other delightful examples are known of male competition, but we shall not discuss them here because they do not pose deep theoretical questions. The situation is different for Darwin's other mechanism—*female choice.*

A structure like the peacock's tail cannot plausibly be explained by male competition. It would have no use in fighting—indeed, it would reduce the male's fighting power—and no one has ever thought up a more subtle competitive function for the tail. Darwin suggested that the tail exists instead because females preferentially mate with males that have longer, brighter, or more beautiful tails. If they do, the mating advantage of males with longer tails will compensate for a corresponding amount of reduced male survival.

Darwin's main argument for the importance of sexual selection was comparative. Sexual selection should operate more powerfully in polygynous than in monogamous species. In a polygynous species, in which several females mate with one male (and other males do not breed at all), a single male can potentially breed with more females than under monogamy; selection in favor of adaptations that enable males to gain access to females (whether by male competition or female choice) is proportionally stronger. Consequently, Darwin reasoned that secondary sexual characters would be more developed in polygynous, than monogamous, species. Polygynous species should have stronger *sexual dimorphism.*

Darwin's book *The Descent of Man, and Selection in Relation to Sex* (1871) contains a long review of sexual dimorphism in the animal kingdom. It remains the best (and classic) demonstration that sexual dimorphism is mainly found in polygynous species. In polyandrous birds, such as phalaropes, sexual selection is reversed: females compete for males, and the females are the larger and brightly colored sex. There are exceptions, such as monogamous ducks that are sexually dimorphic; Darwin had an additional theory for them. However, Darwin's principal evidence for sexual selection came from a comparison of large numbers of species. This comparison showed that the species with brightly colored, large, or dangerously armed males are more often polygynous and the species in which males and females are more similar were more often monogamous.

11.4.3 *Females may choose to pair with particular males*

For Darwin, female choice among males was an assumption; he was mainly concerned to show that, if it exists, it can explain extraordinary phenomena like the peacock's tail. He did not fully explain why the female preference should ever evolve. Selection can work on a female preference just like on any other character. If females with one type of preference produce more offspring than females with another, selection will favor the more productive preference. The difficult case involves an extreme example like the peacock, in which the form of female choice appears to be disadvantageous to the female. Peahens pick males that possess a costly character that will be passed on to their sons. Thus, the female preference seems to cause the females to produce inferior sons.

We can spell the problem out more fully, in terms of selection on a mutant non-choosy female. Suppose that peahens do prefer peacocks with dazzling tails, and a mutant female, who does not prefer these males, arises. She might mate at random, or prefer some other sort of male. How does selection affect this mutation? The mutant female will produce sons that do not possess the costly character, at least in so extreme a form; her sons will, therefore, survive better than average. Thus, the mutant should be favored, the female preference should be lost, and the extreme male forms should disappear.

Or should it? The mutant female will, indeed, produce sons that survive better than the population's average. As Fisher first realized, this greater than average survival is not enough to guarantee that the mutation will spread. When the mutant female's sons grow up, they will be rejected as mates based on their inferior tails. The mutant female is rare in a population where the majority of females prefer males with long tails, and this majority preference will work against the mutant's sons. Despite their superior survival, they will be condemned to celibacy. The randomly mating mutation, therefore, may not spread.

Fisher also discussed how the preference for a costly character could initially evolve. After the long male tail has evolved, it is costly. At an earlier evolutionary stage, before the female preference arose, things might have been different, however. Male tails would have been shorter at that time. Suppose that, before some mutant female arose who picked long-tailed males, most females picked their mates at random. Suppose also that a positive correlation existed between male tail length and survival at that point (Figure 11.5a). Selection would then favor a mutant female with a preference for males with longer tails; she would produce sons with longer than average tails and the associated higher survival. As the mutation spread, the males with longer tails would start to acquire a second advantage. The population would contain increasing numbers of females who prefer to mate with longer-tailed males, and the males so endowed will not only survive better but also enjoy an advantage in mating. The evolution of longer tails in males, and a mating preference for them in females thus come to reinforce one another, in what Fisher called a *runaway* process.

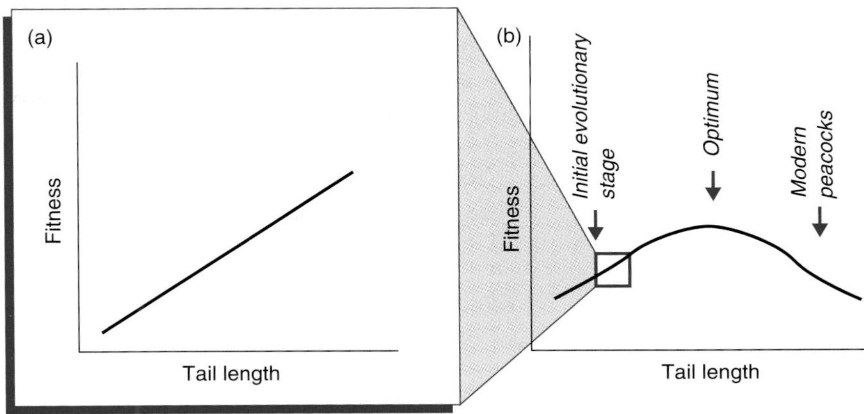

Figure 11.5 (a) Early stage in the evolution of a bizarre character such as the peacock's tail. Before females preferred to mate with long-tailed males, a positive correlation might have existed between tail length (then much shorter than in their descendants) and male fitness. (b) Full relation between the degree of exaggeration of character (tail length) and survivorship. An intermediate optimum exists as well. Modern species like the peacock lie toward the right of the graph.

Technically, these characters reinforce one another because the genes encoding them are in linkage disequilibrium (section 8.5, p. 200). The offspring of a female who mates preferentially with longer-tailed males will possess both their mother's genes for choice and their father's genes for long tails. As these two kinds of genes become non-randomly associated, the genes for female choice increase in frequency by hitchhiking with the advantageous genes for long tails in males.

If we consider a sufficiently wide range of tail lengths, the full relation between tail length and male survival presumably shows some increase and decrease on either side of an optimum (Figure 11.5b). Eventually, powered by female choice, the average tail length in the population will reach the optimum; evolution does not stop there, however. As the population evolves toward the optimal tail length, the longer-tailed males are still preferred. At this point, the female preference will have spread through the population and the majority of females will prefer longer-tailed mates. Now the mating preference alone drives the evolution of longer tails. The preference may have become strong enough to compensate lower male survival, and evolution will proceed into the interesting zone in which the male character, in a complete reversal of the original selective forces, evolves to become increasingly costly to its bearers.

Thus, three stages occur in the evolution of the long tail. In the initial stage, long tails have only a survival advantage. In the second stage, the survival advantage is supplemented by a mating advantage. As female choice grows more common and the tail length extends past the optimum for survival, the relative importances of the two advantages change places, until we reach a third stage at which further elongation is driven purely by female choice.

As the population evolves past the point of optimum tail length, the selective forces at work have become almost absurd. The males are evolving ever-longer tails even though the original selective advantage has ceased, and then reversed. The female mating preference will also continue to strengthen, even though it now favors a character that decreases survival. The runaway process will come to a stop only when the death rate of males, due to their feathery excess, is so high that their success in mating no longer compensates for the disadvantage. The tail length will then reach an equilibrium. That equilibrium, according to Fisher, is what we are now observing in birds like peacocks and birds of paradise.

The original problem was to explain the evolution of a set of apparently deleterious characters. Darwin's solution involved their maintenance by female choice. He did not, however, explain why females should come to choose males with deleterious characters, nor why the choice would not be lost by natural selection. In Fisher's theory, when the choice first evolved, the male character was much smaller and choice then favored males with higher survival. Genes for choice could, therefore, increase in frequency from being rare mutants to being the majority form in the population. Once nearly all females in a population choose mates in a certain way, mutant females that pick some other sort of male are selected against (because of the effect on the kind of sons they produce). The cost of the male character at the final equilibrium serves no function for the female. Rather, it is maintained by female choice but is the useless end-product of an initially useful process.

11.4.4 *Females may prefer to pair with handicapped males, because the male's survival indicates his high quality*

We now turn to a second theory—Zahavi's *handicap theory*—in which the costliness of the male character is positively useful to the female. ("Handicap" is Zahavi's term for what we have been calling a costly or deleterious character—a handicap is a character that reduces survival.) The argument takes the following form. Suppose that the males in the population vary in their quality. We shall concentrate on species, like peacocks, in which males contribute only sperm; in them, quality must mean genetic quality, because nothing else is transferred. We assume that some males have genes that confer higher fitness ("good genes") than do other males (who have "bad genes"). In practice, gene quality could exist in all degrees between good and bad, but the point can be explained in the simple dichotomous case.

If a female mates at random, her mates will have good and bad genes in the same proportions as the good and bad genes found in the entire population. Thus, if half of the males in the population have good genes and half have bad, then 50% of the female's mates will have good genes and 50% bad. Now suppose that some of the males in the population possess a handicap—a character that reduces their survival. If only males with good genes can survive possessing a handicap, a female who mates preferentially with handicapped males will mate with only males with good genes (Table 11.3). The choice will be favored by selection if the advantage through the superior genes outweighs the cost of the handicap. In that case, the net quality of the choosy female's offspring will be higher than those of the randomly mating female.

In this way, the handicap acts as an indicator of genetic quality. But why does the indicator have to be costly? The cost guarantees that the indicator will be reliable. A male's genetic quality is not written on him; instead, it must be inferred. If females inferred the genetic quality from an inexpensive signal, selec-

TABLE 11.3

The handicap principle. If only males with good genes can survive the possession of a handicap, females who mate with handicapped males will mate with only males who possess good genes. A female who mates with males lacking a handicap will mate with males possessing good and bad genes in their population proportions.

Males with:	Bad genes	Good genes
No Handicap	alive	alive
	in population proportions	
Handicap	dead	alive

tion would favor males that cheat. If females preferentially mated with males who merely said "I have good genes" (or, in a non-human species, sent out a signal analogous to this phrase) and rejected those that said "I have poor genes," then mutant males who said the former independently of their true genetic quality would be favored. Words (and their analogues) are cheap. On the other hand, if the criterion favored by females is costly, such as growing a long and ostentatious tail, then selection will less automatically favor cheats. In particular, if the cost of growing a handicap is less for a truly high-quality male than for a low-quality male, handicaps will be grown only by high-quality males and will serve as reliable signals for females. (This condition was met in the simple example in Table 11.3, in which the cost of the handicap for the males with bad genes was far higher than for males with good genes.)

As a result, the reason behind the costliness of the male character is completely different in Fisher and in Zahavi's theories. In Fisher's theory, the cost arose as the end-product of a runaway process. Initially, long tails were not costly, but as an open-ended female preference for males with longer tails was selected into the population, the tails evolved past their optimum and ended up reducing the survival of their bearers. In Zahavi's theory, the male character was costly from the start, and remains costly as the female preference spreads. The function of the chosen male character is to indicate genetic quality at other loci, and it must be costly to be reliable.

11.4.5 *Female choice in Fisher's and Zahavi's models must be open-ended, and this condition can be tested*

We have two crucial means by which Fisher and Zahavi's ideas can be tested. The first concerns the exact kind of female preference that they require. For the theories to succeed, the preference must be open-ended. We can distinguish between absolute preferences, which take the form "mate preferentially with males whose tails are 30 cm long," and open-ended preferences, such as "mate preferentially with the male who has the longest tail you can find."

In Fisher's theory, at the initial stage when the male character was positively correlated with survival, either an absolute or an open-ended preference could be favored. The average tail length might then have been 5 cm, and the longest tails in the population might have been 30 cm. If the mutant that was selected happened to encode an absolute preference for males with 30 cm tails, then evolution would proceed until the average tail length was 30 cm and then come to a stop. Only if the preference is open-ended can an equilibrium arise with a costly male character. At the equilibrium, the lower survival of males with longer than average tails must be offset by a higher frequency of mating. It is not enough for the females to prefer average males, or mate at random; they must actively prefer males with longer than average tails.

Likewise, in the handicap theory, females must prefer males with the most costly handicaps. If the greater cost paid by the higher-quality males is not offset by higher mating success, a less costly handicap will evolve. Therefore, both theories require that the female choice be open-ended in a species with a costly male ornament. If we find evidence for such preferences, it suggests the male character is maintained by female choice, but it does not tell us which theory is at work.

The prediction has been tested in more than one species. One example to illustrate the test procedure is Møller's study of barn swallows (*Hirundo rustica*). The two sexes are similar in the swallow, except for the outermost tail-feather, which is about 16% longer in the male than in the female. Møller tested whether females open-endedly prefer males with longer tails by experimentally shortening the tails of some males by cutting them off with a pair of scissors; he also elongated the tails of other males by sticking those severed tail feathers onto other intact males, using superglue that hardened in less than 1 second. He then measured how long it took the different males to find a mate. Møller found that males with elongated tails mated faster (Figure 11.6a), resulting in higher reproductive success (Figure 11.6b).

Møller also confirmed that the male character is costly. Swallows molt in the fall and grow a new tail for the following breeding season. A male's new tail is, on average, approximately 5 mm longer than in the previous year. Males whose tails were elongated grew a tail in the following year that was shorter than before the experimental treatment, however (Figure 11.6c). (Møller did not tamper with their tails in the year after the experiment.) Those males had enjoyed a good year during the experiment, but the extra effort of flying with an elongated tail exacted a physiological cost. In the next year, the cost was paid: the males took longer to find a mate and their reproductive success decreased. In summary, Møller has shown that the sexually dimorphic tail-feathers of swallows are maintained by female choice, that the choice is open-ended, and that the character chosen is costly.

11.4.6 *Fisher's theory requires heritable variation in the male character, and Zahavi's theory requires heritable variation in fitness*

In Fisher's runaway theory, females choose males with long tails at the final equilibrium point because a mutant female who mated at random would have lower fitness as her sons would have shorter tails and be rejected as mates. This statement is true only if male tail length is heritable. If all variation in tail length were environmental, and its heritability were zero (section 9.6, p. 235), the tail

Figure 11.6 Barn swallows with longer tails are preferred by females, but the character is costly. Møller experimentally shortened some males' tails and elongated others; as one control he cut the males' tails off and then immediately stuck them back on again (control 1) and as another he left the males untreated (control 2). Males with elongated tails (a) obtain mates more quickly, (b) have higher reproductive success, but (c) grow a shorter tail in the following year, while the other males grow a longer tail. (Møller also measured both the mating advantage and the cost of longer tails by other criteria, and those results support the results illustrated here.) Reprinted, with permission from *Nature*, from Møller (1994). Copyright 1994 Macmillan Magazines Limited.

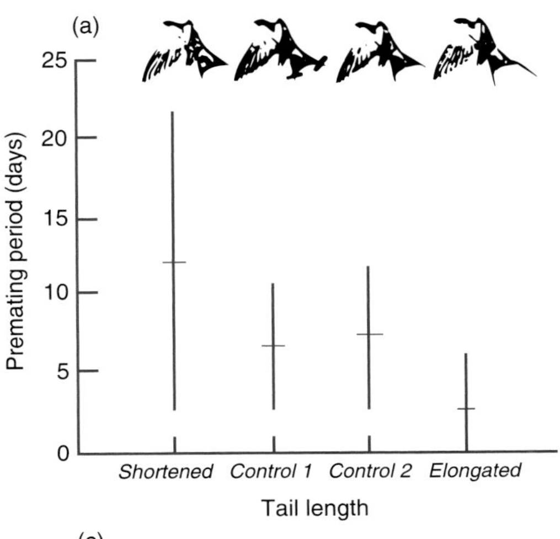

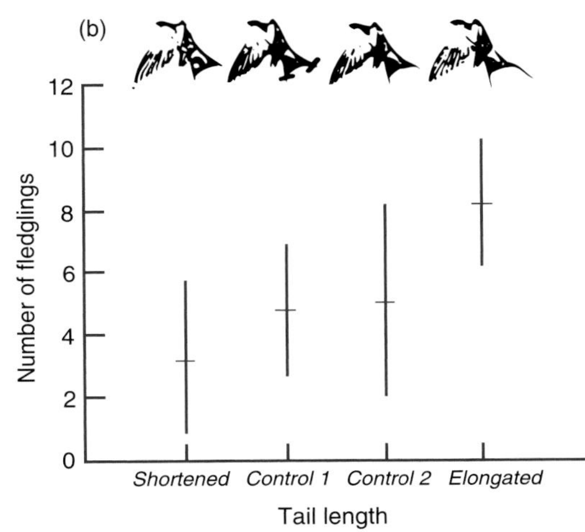

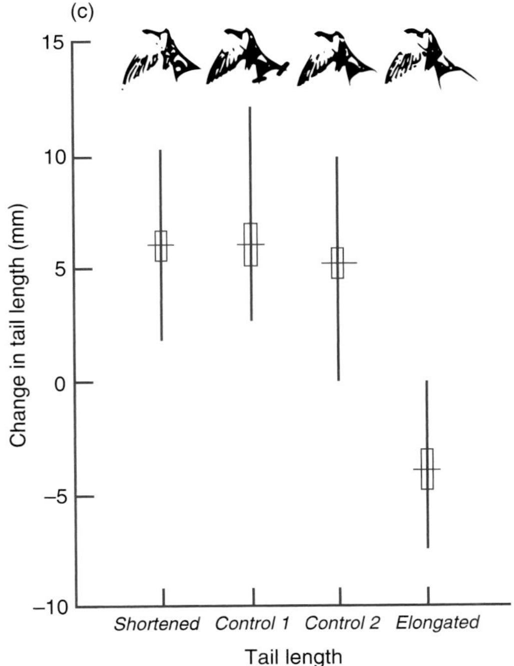

length of the mutant female's sons would be no shorter on average than those of the choosy females. If mate choice imposed any cost on a female, the randomly mating mutant would spread. Tail length, therefore, must be heritable or selection will favor the female who mates at random. This condition is testable, but has never been tested in a species with a costly and extravagant character like the peacock's tail.

In Zahavi's theory, the advantage of female choice does not depend on the inheritance of the male character, and choice could be maintained even if the heritability of tail length went to zero. This theory has an analogous condition. In species in which males transfer only sperm, female choice relates to male genetic quality. Male genetic quality must, therefore, vary, with some males having good genes, and others bad genes. This condition, which is called *heritability of fitness,* dictates that individuals of higher than average fitness (i.e., that produce more offspring than average) produce offspring who also have higher than average fitness. If the high-quality males do not produce high-quality offspring, then there is no point in picking them as mates.

The conditions in the two theories face a common difficulty. While selection operates on any character, it reduces its heritability (see the Illinois maize experiment, Table 9.2 and Figure 9.7). In a population in which some individuals possess good genes and others bad genes, selection acts to fix the good genes—and once fixation occurs, no variation in genetic quality will be left. Under those circumstances, Zahavi's theory would not work.

In fact, the amount of variation in genetic quality for a species in nature is unknown, and no firm conclusion can be drawn. However, several arguments have been made about this point. One is the possibility just noted, that Zahavi's theory may not work in species in which males transfer only sperm because of inadequate variation in genetic quality.

Alternatively, adequate variation may be present and the process may operate for either of two reasons. First, although mutation is relatively ineffective at producing variation at a single locus (section 5.11, p. 115), it is much more important when we consider all genetic loci in an organism (section 9.11, p. 247), and mutation alone might create new bad genes as fast as selection eliminates them. A similar argument can be made in Fisher's theory if the male character, such as tail length, is influenced by a large number of genetic loci; again, mutation might create new genes for short tails as fast as selection removes them. Second, environmental change may create new variation. If the environment changes, a "good" gene may become a "bad" gene after a few generations, and selection may never finally fix any of the genes. As with the function of sex, one particularly important type of environmental change may be that related to the coevolution of parasites and their hosts.

11.4.7 *Females may choose to pair with healthy, unparasitized males*

Earlier, we discussed the conditions for coevolution between parasites and hosts that cause unending cycles in the frequencies of genes for parasite resistance in the host population (Figure 11.4). At any one time, some individuals will be genetically resistant to the common parasites; they have good genes. Other individuals will be more easily parasitized, and have bad genes. Selection could then favor females who picked healthy, parasite-free males as mates, because they would pass on the males' genes for resistance to their offspring. Female choice of healthy males is a special case of female choice for good genes. This idea has been particularly supported by Hamilton.

If we acknowledge that females might choose males that are resistant to parasites, how does it relate to the evolution of brightly colored males? Hamilton and Zuk have suggested that these characters enable females to pick resistant mates. Bright plumage is a sign of health, because a parasitized male cannot produce as attractive a plumage as can an unparasitized male. A similar argument can be made for other sexual displays and courtship behavior patterns. As Hamilton and Zuk say,

> How could animals choose resistant mates? The methods used should have much in common with those of a physician checking eligibility for life insurance. Following this metaphor, the choosing animal should unclothe the subject, weigh, listen, observe vital capacity, and take blood, urine, and fecal samples. General good health and freedom from parasites are often strikingly indicated in plumage and fur, particularly when these are bright rather than dull or cryptic. The incidence of bare patches of skin, which may expose the color of blood in otherwise furry or feathered animals, and the number of courtship displays involving examination of male urine, are of interest in this regard. Vigor is often conveyed by success in fights and by frequently exhausting athletic performances of many displaying animals.

The parasitic theory of sexual selection is a recent hypothesis and few tests of it have been performed so far, but it is proving a remarkably rich source of ideas. We will look at two tests, one comparative and the other experimental.

Hamilton and Zuk reasoned that, perhaps for external ecological reasons, some bird species will suffer more from parasites than others. This parasitism will then spur selection on females to choose resistant mates, and the selection will be stronger in species suffering higher parasite loads than in species suffering lower loads. If male secondary sexual characters enable females to choose resistant males, we should expect to find the relatively brightly colored males in the more strongly parasitized species. Hamilton and Zuk collected evidence from the published literature of parasite loads and relative coloration of males and females in 7649 individuals from 109 bird species. They found a relation between the two (Figure 11.7) that supports Hamilton's theory; bird species that suffer more from parasites do have brighter male coloration.

The experimental test was made by Møller, again using the barn swallow. We saw earlier that female swallows prefer males with longer tails. In Hamilton's theory, tail length should indicate health, and Møller measured the health of male swallows by counting the numbers of a blood-sucking mite (*Ornithonyssus bursa*) found on the birds' heads. He found that males with longer tails do, indeed, carry fewer parasitic mites. Can the male's health be passed on to his offspring? Møller placed 50 mites in the nests of swallows with all tail lengths, and counted the number of mites on the offspring in each nest at a later date. He found that offspring with longer-tailed fathers carried fewer mites (Figure 11.8a). He was also able to confirm that the effect was genetic, by two excellent controls.

A simple correlation between a father's tail length and offspring health could have a non-inherited cause. It could arise, for instance, if longer-tailed fathers

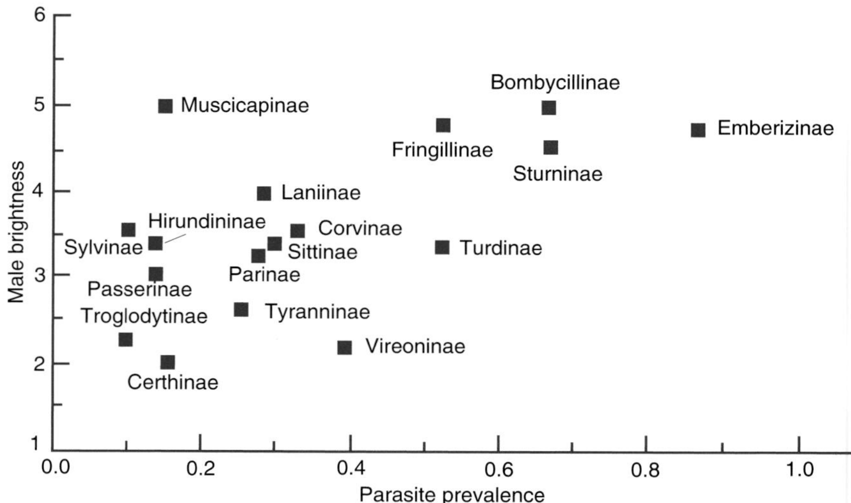

Figure 11.7 Bird species subject to higher degrees of parasitism tend to have more brightly colored males. The graph shows the relation for 17 subfamilies of North American passerines (based on data for 114 species). The scale of male brightness is a subjective ranking from 1 to 6. Reprinted, by permission of the publisher, from Read (1988).

Figure 11.8 Male swallows with longer tails produce healthier offspring. Møller put 50 mites in many nests during the egg-laying period, and then counted the number of mites on the offspring at seven days of age. He moved hatchling swallows between nests, to study the relations between (a) the genetic father's tail length and offspring parasite load, when the offspring remain in his nest. (b) The relation between the genetic father's tail length and offspring parasite load, when the offspring are reared in another nest. (c) The relation between the foster male's tail length and offspring in his nest, when the offspring were experimentally moved there. Reprinted, by permission of the publisher, from Møller (1994).

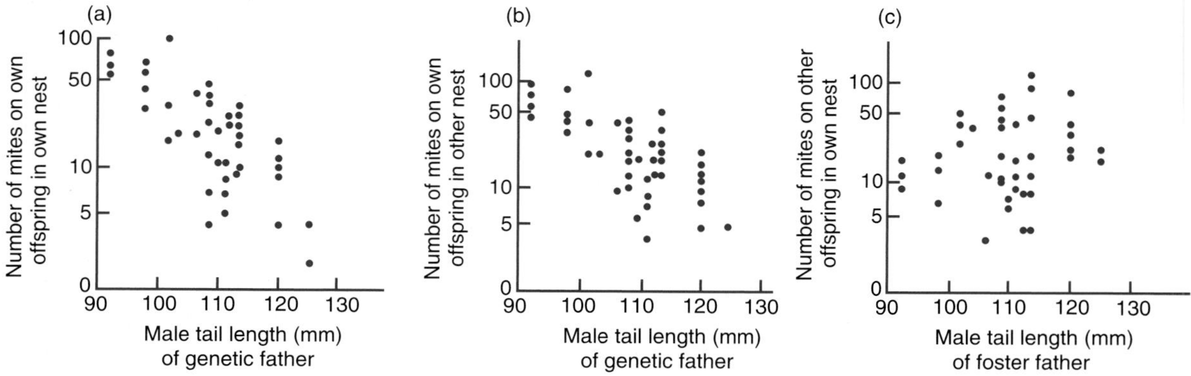

were better at spotting parasites and removing them from their offspring or if they were better at foraging and better-fed offspring resist parasites more effectively. To investigate this issue, Møller cross-fostered baby swallows, moving them between nests immediately after hatching (the adult swallows treat the foster-offspring like their own). No relation was observed between the parasite load of young swallows moved to another nest and the tail length of the foster father at their nest (Figure 11.8c); health in young swallows, as measured in this experiment, was not caused by paternal skill. Moreover, the correlation between the genetic father's tail length and the offspring's heath remained even when the offspring were reared in another nest (Figure 11.8b). Møller's results provide strong evidence that parasite resistance in swallows is inherited. In summary, Møller has shown that female swallows prefer males with longer tails, that tail length is costly for males, that natural tail length is an indicator of health, and that a male's health (or parasite resistance) is genetic and passed on to his offspring.

Møller's result (Figure 11.8) supports Zahavi's theory rather than Fisher's theory in the barn swallow. In Fisher's theory, the only advantage of female choice at equilibrium is the mating advantage of the sons. Sons with longer tails have lower survival, while female choice produces no effect in daughters (because they lack the long tail). In Zahavi's theory, the good genes indicated by the male handicap will be passed on to both daughters and sons. As Møller's investigation showed, the fortunate young swallows descended from long-tailed males were, in fact, more resistant to parasites. If Fisher's theory was the only process operating, we would not expect to see a correlation between the health of a female's offspring and the character of the male that she chooses.

11.4.8 *Conclusion: the theory of sex differences is well worked out but incompletely tested*

The theory of sexual selection is at a more advanced stage than the theory of why sex exists. The models, such as those of Fisher, Zahavi, and Hamilton, may be correct, and some work has been done to test them. The tests, however, remain at an early stage. Several pieces of evidence point to open-ended female choice in species with extravagant, costly male characters. This evidence suggests that Darwin's explanation of those characters as relating to female choice is correct. Less work has been conducted, however, on the other crucial theoretical variable—the inheritance of genetic quality. Møller's fine experiment suggests that such inheritance can occur in one species, but skeptics will wait for repeats and further work.

If further work supports Zahavi's and Hamilton's ideas, another set of questions will be opened up. For when we turn to points of detail, we rarely know how sexual selection has operated. The innumerable subtle details of sex differences in color, morphology, and behavior patterns are probably due, in an abstract sense, to sexual selection. But what do today's male characters indicate, and why? For the swallow's tail, we now know it indicates (perhaps among other things) the number of ectoparasitic mites, but why has tail length evolved to reveal this condition? And what about those more baffling male characters, such as the peacock's tail, or the ornaments of male birds of paradise? What do they reveal to the females of those species?

11.5 *Example 3: the sex ratio*

11.5.1 *Natural selection usually favors a 50:50 sex ratio*

The sex ratio is one of the most successfully understood adaptations, thanks to the work of Fisher. In most species, the sex ratio at the zygote stage is about 50:50. Fisher explained the 50:50 sex ratio as an equilibrium point—if a population ever deviates from this ratio, natural selection will drive it back.

At first glance, the 50:50 sex ratio might seem inefficient. Most species do not have parental care and are not monogamous; a male can fertilize several females. It would be more efficient for the species to produce more females than males, as the extra males are not needed to fertilize the females of the species and do not increase its reproductive rate. (This idea represents another "group selection" argument; see section 12.2.5, p. 325.) However, imagine what would happen to a population with a persistently female-biased sex ratio—one with four females for every one male, for instance. Each male in the population will fertilize, on average, four females. This condition could not be stable for long in evolution, because an average male is producing four times as many offspring as an average female. Being a male provides an advantage, as does being a female who produces extra sons because sons have a higher reproductive success than daughters.

If a mutant female arose who produced only sons, the total reproductive success of her offspring would be 20/8 times that of an average female (the mutant produces five males, each with relative reproductive success 4, for every one male and four females produced by the average female). The mutant would spread. As it did, the population sex ratio would become increasingly less female-biased. The same argument works in reverse for a population with a male-biased sex ratio. The reproductive success of the average female is then higher than that of a male, and natural selection will favor mutant females that produce more daughters than sons. Only when the sex ratio is equal are the relative reproductive successes of the two sexes equal. At that point, no advantage is associated with producing more of one sex than the other. Thus, over evolutionary time the population will gravitate toward, and then stay at, the 50:50 sex ratio. Any population that deviates from the 50:50 sex ratio will be shifted back to it by natural selection.

The fundamental reason why the 50:50 sex ratio is stable is that every organism has one father and one mother. All females jointly contribute the same number of genes to the next generation as all males together. When members of one sex are in short supply, their average success must increase.

A number of points should be noted about Fisher's argument. The argument usually applies to the sex ratio of zygotes, and this ratio is not affected by differential mortality. (It is not affected at all by differential mortality in species without parental care; in species with parental care, any sex difference in mortality during the period of care will affect the ratio the parents should produce.) If males die at a higher rate, the adult populations will have more females than males, and will deviate from the 50:50 ratio. Among the individuals that survive to reproduce, the average male will have a higher reproductive success than the average female.

Selection will not operate to produce extra sons, however. As far as a mother is concerned, the extra reproductive success of her surviving sons exactly balances the zero reproductive success of those that die before reproducing. When she produces a son, she cannot "know" in advance whether he will be a survivor who will have higher than average success, or die and not reproduce at all. At birth, a male can only be expected to have the success of an average male, and the average male has the same reproductive success as the average female. The offspring of any one parent will suffer any sex differences in mortality in the same proportion as the population as a whole, so producing more of the sex that will (in the adult stage) be in the minority offers no advantage. Any parent who did so would simply increase the average mortality rate among her progeny.

Thus, Fisher's argument explains the normal case of the 50:50 sex ratio. Now let us look at a number of cases in which, suitably modified, it can explain deviations from that ratio.

11.5.2 *Local mate competition causes deviation from a 50:50 sex ratio*

Fisher's theory assumes "population-wide" competition for mates. A typical male can expect to produce an average number of offspring equal to the total number of offspring produced by the population divided by the number of males in the population; the equivalent is also true for females. This expectation is correct if an individual competes for mates with the whole of the rest of the population. In some cases, however, that condition does not apply.

Hamilton (1967) first pointed out that the sex ratio will become biased in species in which an individual competes against only a limited part of the rest of its population. The condition is called *local mate competition*. The pyemotid mites are an extreme case such competition. In this group, brothers inseminate their sisters while they are still inside their mother. As a result, a mother needs to produce only one or two males in a brood to inseminate all of her daughters; the sex ratio is female-biased. In *Pyemotes ventricosus*, for instance, the sex ratio of a brood contains on average 4 males and 86 females. The males do not compete against the whole population for mates. Instead, they compete against only their brothers, and a mother can reduce the fruitless competition by producing fewer sons.

Local mate competition is particularly common in parasitic Hymenoptera. Parasitic wasps insert their eggs in the early stages of other insects. In "gregarious" species, the wasp lays more than one egg in a host, which often leads to female-biased sex ratios. Natural selection most strongly favors a female-biased sex ratio when only one female lays her eggs in a host. If a second female also later lays her eggs in the same host, the competition among males becomes less "local." (In the theoretical extreme case in which all females in a wasp population lay all of their eggs in the same host, Fisher's condition of population-wide competition for mates would be restored.)

Hamilton accordingly predicted that the second parasite of a host should produce more sons than the first parasite. The prediction has been tested, even with quantitative precision. Parasitic wasps can tell (using chemical senses in the ovipositor) whether an individual host has already been parasitized. It is not difficult to calculate the exact optimal sex ratio for each female wasp, although we will not work through the calculation here. Suffice it to say that the second female's best sex ratio depends on the relative number of eggs laid by herself and by the first female. If the second female contributes only a few eggs, most of them should be male. As she lays further eggs, an increasing proportion should be female.

Werren has tested this theory in *Nasonia vitripennis*, a parasite of maggots of common blow-flies. These wasps mate almost immediately after emerging from the maggot; the males wait on the outside for their sisters to emerge and then copulate with them. A female *N. vitripennis* lays about 30 eggs in a host,

The graph for sex ratio should cross the 50% males threshold at that social rank at which a hind produces, on average, fitter sons than daughters. Clutton-Brock's data are consistent with this prediction, but the amount of the scatter prevents us from saying that the theory has been quantitatively confirmed.

The relation between dominance and sex ratio, and local mate competition, are but two examples in which deviations from a 50:50 sex ratio have been successfully predicted. We will not discuss other examples here, but these two should be adequate to illustrate how the theory of adaptive sex ratio can be remarkably successful in explaining both the normal 50:50 sex ratio and deviations from it. The theory has suggested various kinds of tests; it is quantitative in its predictions; and the key variable—sex ratio—is easy to measure. The sex ratio will, therefore, probably remain in the vanguard of evolutionary research for some time to come.

11.6 *Different adaptations are understood in different levels of detail*

We have looked at the function of sex, sexual selection, and sex ratio, as three related examples of research on adaptation. In each case, the research has advanced to a different stage.

The problem of sex is still more or less unsolved. The main work is taking place at the imaginative level; the aim is to conceive of a reason why natural selection could favor sex and to build a model that shows that the hypothetical advantage to sex is large enough to outweigh its twofold cost.

In the case of sexual selection, the main ideas—of Fisher and Zahavi—provide a satisfactory abstract solution and are at least not known to be false. The full repertory of techniques—model building, experiment, and the comparative method—are being used to tackle this problem. At a detailed natural history level, many questions lack answers. We do not know whether the abstract ideas correctly explain the full natural variety of sexual behavior and dimorphism, and the ideas are not easy to test.

The theory of sex ratio is even further advanced. The relation between facts and theories is good. We have a general abstract theory as for sexual selection, and this theory also makes quantitative predictions and suggests a number of types of tests in special cases. Follow-up work has been instigated for several of the tests, and the fit of results to predictions suggests that the theory stands a good chance of being correct.

SUMMARY

1. For many characters, it is not obvious how (or whether) they are adaptive.
2. Adaptation can be studied by comparing the observed form of an organ with a theoretical prediction, by experimentally altering the organ, and by comparing the form of the organ in many species.
3. Sexual reproduction has a 50% fitness disadvantage relative to asexual reproduction.
4. Sexually reproducing populations will evolve faster than a set of asexual clones, provided that the rate of favorable mutation is high enough.

5. The taxonomic distribution of asexual reproduction suggests that asexual forms have a higher extinction rate than sexual forms. It is generally doubted that sex is maintained by group selection, however.
6. Two modern theories of why sex exists propose that it is favored by (1) the large numbers of deleterious mutations, which are more efficiently removed by sexual reproduction than by asexual reproduction, and (2) the coevolutionary "arms race" of parasites and hosts. The problem of why sex exists has not been finally solved.
7. Males in many species have bizarre and deleterious secondary sexual characters; the peacock's train is an example.
8. Darwin explained the evolution of strange secondary sex characters by sexual selection. In this theory, the characters reduce their bearers' survival, but increase their success in reproduction. Sexual selection in most species works by male competition and by female choice.
9. The greater sexual dimorphism of polygynous species than monogamous species suggests the importance of sexual selection.
10. The preference of females for males with deleterious characters is theoretically puzzling. It may be explained by Fisher's theory, in which deleterious characters were formerly advantageous and are maintained by majority preference, or by a Zahavi's handicap theory, in which the costly character indicates superior genetic quality.
11. In the barn swallow, females choose males with longer tail feathers, and tail length indicates health. Males with longer tails produce healthier, more parasite-resistant offspring.
12. The sex ratio is usually 50:50 because the reproductive success of all males in a population must equal the reproductive success of all females. If the population sex ratio deviates from 50:50, natural selection favors individuals that produce more offspring of the rarer sex.
13. The theory of sex ratio has correctly predicted when the ratio should differ from 50:50. It has been tested quantitatively and experimentally in the case of local mate competition.
14. The function of sex, sexual selection, and the sex ratio are three of the most important areas of research on adaptation. They have reached different stages of theoretical advance.

FURTHER READING

Curio (1973), Harvey and Pagel (1991), Maynard Smith (1978b), Parker and Maynard Smith (1990), and Rudwick (1964), are reviews, general and particular, of the methods of studying adaptation. On *Cepaea*, see Ford (1975), Sheppard (1975), Jones *et al.* (1977). Bulmer (1994, ch. 10–12) introduces the theory for sex, sexual selection, and sex ratio.

On sex, Williams (1975), Maynard Smith (1978a), and Bell (1982) are classic works; Michod (1995) is a recent popular book that argues for the importance of deleterious mutation but in a rather different way from this chapter. Williams and Maynard Smith contain an important exchange on the taxonomic argument (section 11.3.4) for high extinction rates in asexual species; Weismann's 1886 paper was translated in vol. 1 of Weismann (1891–1892). On the rate of evolution, see also Kirkpatrick and Jenkins (1989). A special issue of *Journal of Heredity*, vol. 84 (1993), pp. 321–414, contains a number of recent papers, including a classification of 14 theories by Kondrashov (1993).

On the mutation theory, the main references are Kondrashov (1988) and Charlesworth (1990); see also Redfield (1994). The theory was worked out by Kimura and Maruyama (1966) but they did not apply it to the existential question about sex. Chapter 9 gives references to background material on deleterious mutation. As described in the text, it makes no difference whether reproduction is sexual or asexual when mutations have independent effects on fitness. However, this concept holds true only in a deterministic, or infinite population size, model. In a stochastic finite population model, it does make a difference: the effect is known as "Muller's ratchet" and it is described in Maynard Smith (1978a) and a number of papers in the issue of *Journal of Heredity* cited above; Kondrashov's (1988) theory described in the text is sometimes confused, by imperfectly educated persons, with Muller's ratchet.

The other main theory is parasitic. Hamilton (1991, 1993) and Hamilton *et al.* (1990) are recent statements. Several relevant papers appear in the special issue of *Philosophical Transactions of the Royal Society of London* B 346 (1994), pp. 167–385. Howard and Lively (1994) modeled mutation and parasitism jointly and Lively has studied the parasitic theory empirically, using fish (Lively *et al.* 1990) and snails (e.g., Lively 1987). Kelley (1994) discusses other evidence, from plants. Chapter 22 contains further references on parasite-host coevolution generally, but on gene-for-gene matching see Thompson and Burdon (1992) and Briggs and Johal (1994) on the facts, and Parker (1994) on whether they fit the models of sex.

For both sexual selection and the sex ratio, Trivers (1985), Dawkins (1989), and Krebs and Davies (1993) are good introductions. For sexual selection, Andersson (1994) is a comprehensive review; Cronin (1991) is another clear introduction and is good on history and the broader context. The influence of parasites on mate choice is the topic of a conference organized by G. Hausfater and R. Thornhill published in *American Zoologist*, 30 (1990), pp. 225–352, including a paper by Hamilton. Møller (1994) describes his work on swallows, and Smith and Montgomerie (1991) confirm his experiment on female choice in North American barn swallows. Grafen (1991) discusses the evolution of costly signals of quality. On sex ratio, Fisher (1930) is the classic source; Bull and Charnov (1988) is a more recent review; and Gomendio *et al.* (1990), Hurst (1993), and Sundström (1994) discuss some other deviations.

STUDY AND REVIEW QUESTIONS

1. What is the cost of sex in a species in which the sex ratio at birth is (a) 1 male : 2 females, and (b) 2 males : 1 female?

2. (a) What condition is required, relative to the rate of deleterious mutation, for natural selection to favor sexual reproduction over asexual reproduction? Does reason and evidence suggest that the condition is met naturally? (b) Draw the relation between the number of deleterious mutations in an organism and its fitness that is required for sex and recombination to be advantageous (include a specification of the *y*-axis). What form is arguably true in reality?

3. What would you investigate to determine whether sex is favored by host-parasitic coevolution?
4. Why does a male character have to be costly (a handicap) to signal genetic quality?
5. In Fisher's runaway theory, what maintains the female preference for extreme males? Why does the preference not evolutionarily disappear?
6. If one male can fertilize several females in a species, why do parents not produce a sex ratio of many daughters per son?
7. In terms of the modes of selection (or fitness regimes) discussed in chapter 5, what kind of selection operates in Fisher's model of (a) female choice, and (b) the sex ratio?

The Units of Selection

ADAPTATIONS clearly benefit something in the living world, but to understand exactly why they persist we need to know in theory for whose benefit adaptations evolve. The chapter begins by explaining the problem. First, we consider a series of adaptations that benefit increasingly higher levels of organization of life. We begin with adaptations that benefit only a small cluster of genes, and move through the cell, organism, and family levels, up to possible adaptations that benefit whole groups. The examples illustrate the conditions for adaptations to evolve to benefit the different hierarchical levels, and reveal why adaptations at most levels other than the organism (and family group) are rare, although not non-existent. We finish this section of the chapter by establishing a general criterion that an entity must satisfy to evolve adaptations: it must show heritability. Next, we focus on the more fundamental question of the entity on which natural selection operates, and describe an argument to suggest that the entity is the gene, although "gene" is defined in a special sense.

12.1 For the benefit of which level in the biological hierarchy of levels of organization does natural selection produce adaptations?

It is a familiar idea that life can be divided into a series of levels of organization, from nucleotide to gene, through cell, organ, and organism, to social group, species, and higher levels. On which, if any, of these levels does natural selection act, and for which levels does natural selection produce adaptations? In a fairly superficial analysis, the answer does not matter. If an adaptation benefits an individual organism, it will often also benefit its species at a higher level and, at a lower level, all of the parts that make up the individual. Conflicts can exist between these levels, however. In some cases, what benefits an organism may not also benefit its species, and in these cases the evolutionary biologist needs to know which level natural selection most directly benefits. Thus, the question matters when we are studying particular adaptations. If we are seeking to understand why an adaptation evolved, we need to know for the benefit of which entities adaptations in general evolve. The question also has a more general, almost philosophical, interest: the theory of evolution should include a precise, and accurate, account of why adaptations evolve.

The issue can be made clearer in an example. Let us consider the adaptations that can be seen when lions go on a hunt. Lions often hunt alone, but they can improve their chance of success by hunting in a group. Bertram described a group of hunting lions in the following manner:

> When prey have been detected, a wildebeest herd perhaps, the lions start to stalk towards them. As they get close, they take different routes, some going on straight ahead, and some to the sides, so the prey herd is approached by lions stalking them from different directions . . . Eventually one lion gets close enough to make a rush at a wildebeest, or else a lion is detected by the prey.

The trap is then sprung. Panicked wildebeest run in all directions, some of them into the reach of other lions. The cooperative behavior of the hunting party seen here is an adaptation for catching food; but it is not the lion's only feeding adaptation. The lions' muscular jaws and limbs, their teeth and five senses, all contribute to the success of the hunt. Lions are well adapted for feeding. Although some hunts are unsuccessful, and individual lions may starve to death, the lions in the Serengeti Plains of Kenya spend about 20 hours per day in rest or sleep, and only one hour per day, on average, in hunting. Visitors tend to think lions are lazy.

When a lion hunt is successful, benefits affect all but the highest biological levels of organization. The individual lions obviously benefit, as does the pride. Each time a hunt is successful, a small incremental increase is made in the species *Felis leo*'s chance of survival, or avoidance of extinction. The survival probability will also be increased, if by a smaller amount, for the genus *Felis* and the cat family Felidae. The hunt's effect at higher levels will depend on exactly what prey the lions caught. Almost all Serengeti lions' food consists of other mammals, so when we reach the class Mammalia, the effect of the hunt has probably become neutral. That is, the lion's gain is the wildebeest's or zebra's loss, and the chance of survival of the class Mammalia is more or less unaffected. The beneficial effect of the hunt spreads downward as well as up from the individual lion. As the survival of the lion is increased, a corresponding effect is seen for the survival of its constituents. The lion's organs, cells, proteins, and genes all become more likely to survive every time a lion hunt proves successful. (If we trace the effect down through the nucleotides and their constituent atoms, it again disappears. A lion's atoms survive just as well whether it is alive and well fed or dead due to starvation.)

The levels of organization—from gene through individual lion to Felidae—are, to a large extent, bound together in their evolutionary fate, and what benefits one level will usually also benefit the other levels. However, this case is not always true. Male lions can join a pride only by forcibly evicting the incumbent males. In the fight, lions may be killed or wounded, and in any case lions have a low rate of survival after they have been evicted from a pride. These fights have losers as well as winners, and the benefit of winning is confined to the individual (or male coalition) level and below; the lion species as a whole does not benefit. The survival of the species may be affected little by the death of male lions, because the mating system is polygynous and has plenty of males to spare; but the effect is not necessarily positive.

Different adaptations, therefore, have different consequences for units in nature. At one extreme, some adaptations—an improved DNA replication mechanism perhaps—could benefit all life. More typically, adaptations will benefit only a smaller subsample of living things. Because the levels of living organization are bound together, if natural selection produces an adaptation to benefit one level, many other levels will benefit as a consequence. The question in this chapter is whether natural selection really acts to produce adaptations to benefit one level, with benefits at other levels being incidental consequences, or whether it acts to benefit all levels. If it benefits only one level, which is it? In evolutionary biology, this question is expressed as "What is *the unit of selection?*"

We will seek to answer this question by examining a series of adaptations that appear to benefit different "levels" in the hierarchy of biological organization. Some adaptations appear to be in the interests of individual genes, at the organism's expense; others benefit organisms at the group's expense; still others may benefit higher levels. After considering the example, we can discuss generally which of the types of adaptation we should expect to see most often in nature.

12.2 *Natural selection has produced adaptations that benefit various levels of organization*

12.2.1 *Segregation distortion benefits one gene at the expense of its allele*

With normal Mendelian segregation at a genetic locus, on average half of an organism's offspring inherit one of the alleles and the other half inherit the other allele. Mendelian segregation is, therefore, "fair" in its treatment of genes—genes emerge from Mendelian segregation in the same proportions as they entered. In some curious cases, however, Mendel's laws are broken: one of the alleles, instead of being inherited by 50% of a heterozygote's offspring, is consistently overrepresented. The *segregation distorter* gene of *Drosophila melanogaster* is an example of this phenomenon, which is also called *meiotic drive.*

The segregation distorter gene was first found in *Drosophila* stocks from Wisconsin and Baja California. The gene is symbolized by *sd*, and we can call the other, more normal alleles at the locus " + "; a heterozygote for the segregation distorter is then *sd* / +. The majority (90% or more) of offspring from male heterozygotes have the *sd* gene because the sperm containing the + gene fail to develop. Female heterozygotes have normal Mendelian segregation.[1]

A segregation distortion gene can confer a great selective advantage. The allele that is passed on to more than half of a heterozygote's offspring will automatically increase in frequency and should spread through the population. Once the allele becomes fixed, the effect would disappear, because segregation is normal in homozygotes. In the case of the segregation distorter gene complex in *Drosophila*, however, other things are not equal. The abnormal sperm are

1. The genetics are, in reality, more complex. Two tightly linked loci mainly control the process—one is a recognition locus, the other creates the distortion. A chromosome that produces the distortion in favor of itself will have a particular recognition sequence at the recognition locus and a distorter allele at the other. The system could work as follows. The distorter allele could produce a protein that, when it binds to the "distorter" recognition sequence, has no effect; when it binds to the other (+) recognition sequence, the protein prevents the chromosome from entering sperm. A third locus can influence the strength of the effect. For simplicity, however, we shall treat the case as if it was controlled by one locus, and write the heterozygote for segregation distorter as *sd* / +.

infertile, and the total fertility of male heterozygotes is accordingly lowered. The fertility of an *sd* / + male is about half that of a normal male. The effect of the lowered fertility on selection at the *sd* locus is complex, and depends on whether the reduction in fertility is more or less than 50% and the effect that the reduced fertility has on the number of offspring produced. At least some segregation distorter alleles produce enough copies of themselves in heterozygotes to have an automatic selective advantage, however. They produce more copies of themselves than would a normal Mendelian heterozygote; their increase in frequency up to fixation then becomes inevitable.

Segregation distorter sets up an interesting selection pressure in the rest of the genome. On average, all other genes at other loci suffer a disadvantage because of segregation distortion. In any case, one gene has only a 50% chance of getting into a particular gamete, because gametes are haploid whereas the individual is diploid. In addition, segregation distortion produces a further reduction of about 50% in the chance that genes at other loci are passed on. In a *sd* / + heterozygote, if a gene at any other locus is included in a sperm, there is a 50% chance it is an *sd* sperm, which will be fertile, and a 50% chance it is a + sperm, which will be infertile (we are ignoring linkage between the *sd* locus and the other locus under consideration). Genes at other loci are all net losers. If they appear in the same sperm as the favored *sd* allele, things are normal; if they are in the shrivelled, disfavored sperm, they will die. Selection at other loci will favor genes that suppress the distorters and restore the status quo.

The *sd* genetic system in the fruitfly is only one of several known examples of this phenomenon. Any system for distorting meiotic segregation is theoretically likely to require at least two linked genes (one for recognition, one for distortion), and those systems that have been investigated genetically do have this structure. The genes must be linked to spread. If recombination separated the effecter gene from the recognition sequence, it would be liable to work against itself. This argument has two evolutionary consequences. First, recombination might itself have evolved to break up cooperating sets of distorter genes; genes elsewhere in the genome could suppress the distorters simply by increasing the general recombination rate. Second, segregation distortion systems are particularly likely to be found in regions of the genome in which the recombination rate is low, or zero. One such place is the sex chromosomes, and sex chromosomes are therefore vulnerable to invasion by distorter systems. (We meet one uncertain implication in Box 15.2, p. 407, and another in section 16.10, p. 456.)

We can imagine the following series of evolutionary events. At each locus, selection favors mutant alleles that can enter more than 50% of the offspring; whenever such an allele arises, it has an automatic advantage. Once this allele has arisen, the next stage will depend on the fertility of its bearer. If it has no effect on, or increases, fertility, its frequency should increase to one. If it reduces fertility, alleles at other loci that suppress the effect will be favored. In this case, the distorter gene will either be suppressed or increase to a fixation, depending on how quickly a suppressor allele arises. In either event, segregation distortion will be evolutionarily short-lived. The allele responsible will either be fixed (and segregation distortion will then disappear) or suppressed. At any one time, only a few cases should exist (which probably explains why only a few examples have been found). Meiosis does appear to operate fairly, since alleles usually are equally represented in the offspring. It is important to remember, however, that this fairness exists despite selection at each locus for alleles that subvert the Mendelian ratios.

12.2.2 *Selection may sometimes favor some cell lines at the expense of the rest of the body*

In organisms like humans, a new individual develops from an initial single-cell stage, and that single cell derives from a special cell line, the germ line, in its parents. This kind of life cycle is called *Weismannist*, after the German biologist August Weismann, who first expounded on the distinction between germ and somatic cell lines. In a "Weismannist" organism, most cell lines (the soma) inevitably die when the organism dies. Reproduction is, therefore, concentrated in a separate germ line of cells.

The separation of the germ line limits the possibilities for selection at the suborganismic level, between cell lines. One cell may mutate and become able to outreproduce other cell lines and (like a cancer) proliferate through the body. This "adaptation" will not be passed on to the next generation, however, unless it has arisen in the germ line. Any somatic cell line comes to an end with the organism's death. For this reason, cell selection is not important in species like humans.

However, Buss has pointed out that Weismannist development is relatively exceptional among multicellular organisms (Table 12.1). We tend to think of it as the norm because vertebrates, as well as the more familiar invertebrates like arthropods, develop in a Weismannist manner. However, more than half of the taxa listed in Table 12.1 have the capacity for somatic embryogenesis—that is, a new generation may be formed from cells other than those in specialized reproductive organs. The most striking examples are from plants. Steward, for instance, in a famous experiment in the 1950s, grew new carrots from single phloem cells taken from the root of an adult plant.

In a species in which new offspring can develop from more than one cell lineage, selection between cell lines becomes possible. At conception, the organism will be a single cell, and for the first few rounds of cell division the organism will probably remain genetically uniform. No selection can take place between cell lines if they are all genetically identical. Eventually a mutation may arise in one of the cells. If the mutation increases the cell's rate of reproduction, the cell line will cancerously proliferate at the expense of other cell lines in the organism. In a Weismannist species, that cell line will die when the organism dies. If any cell line stands a chance of producing offspring, however, the mutant cell line would increase its chance of being passed on to an offspring. The process would be detrimental to the organism in such circumstances—but it could be more useful. Whitham and Slobodchikoff have argued that in plants this kind of somatic selection enables the individual to adapt to local conditions more rapidly than would be possible with strictly Weismannist inheritance.

At present, the process represents more of a theoretical possibility than a confirmed empirical fact, but it may well have importance in non-Weismannist species. It may also have been important in the non-Weismannist ancestors of such modern Weismannist forms as arthropods and humans. Buss has developed the idea that cell selection can explain certain features of embryology in Weismannist species.

TABLE 12.1

The modes of development in different groups of living things. In the cellular differentiation column, + means that it is present in all the species that have been studied in the group and +/− means it is present in some species and absent in others. In the developmental mode column, *s* means new organisms can develop from the "somatic" cells of their parent; *e* means epigenetic development; *p* means preformationistic development; and *u* means unknown. Reprinted, by permission of the publisher, from Buss (1987).

Taxon	Cellular Differentiation	Developmental Mode	Taxon	Cellular Differentiation	Developmental Mode
Protoctista			Animalia (continued)		
Phaeophyta	+/−	s	Mesozoa	+	p
Rhodophyta	+/−	s	Platyhelminthes	+	s, e, p
Chlorophyta	+/−	p	Nemertina	+	e
Ciliophora	+/−	s	Gnathostomulida	+	u
Labyrinthulamycota	+/−	s	Gastrotricha	+	p
Acrasiomycota	+/−	s	Rotifera	+	p
Myxomycota	+/−	s	Kinorhyncha	+	u
Oomycota	+	s	Acanthocephala	+	p
Fungi			Entoprocta	+	s
Zygomycota	+	s	Nematoda	+	p
Ascomycota	+	s	Nematomorpha	+	u
Basidiomycota	+	s	Bryozoa	+	s
Deuteromycota	+	s	Phoronida	+	s
Plantae			Brachiopoda	+	u
Bryophyta	+	s	Mollusca	+	e, p
Lycopodophyta	+	s	Priapulida	+	u
Sphenophyta	+	s	Sipuncula	+	u
Pteridophyta	+	s	Echiura	+	u
Cycadophyta	+	s	Annelida	+	s, e, p
Coniferophyta	+	s	Tardigrada	+	p
Angiospermophyta	+	s	Onychophora	+	p
Animalia			Arthropoda	+	e, p
Placozoa	+	s	Pogonophora	+	u
Porifera	+	s	Echinodermata	+	e
Cnidaria	+	s	Chaetognatha	+	p
Ctenophora	+	p	Hemichordata	+	s, e
			Chordata	+	e, p

12.2.3 *Natural selection has produced many adaptations to benefit organisms*

We do not need to consider an example of organismal adaptation here. Most of the adaptations described elsewhere in the book—from the beaks of the woodpecker (section 1.2, p. 5) and the Galápagos finches (section 9.1, p. 222), to the color patterns of *Biston betularia* (section 5.7, p. 103), the mimetic *Papilio* (section 8.1, p. 195), and *Cepaea* (section 11.1, p. 281)—benefit the individual organism. It can hardly be doubted, therefore, that organismal adaptations exist, and that natural selection can favor them.

12.2.4 *Natural selection working on groups of close genetic relatives is called kin selection*

In species in which individuals sometimes meet one another, such as in social groups, individuals may be able to influence each other's reproduction. Biologists call a behavior pattern *altruistic* if it increases the number of offspring produced by the recipient and decreases that of the altruist. (Notice that the term in biology, unlike in human action, implies nothing about the altruist's intentions: it is a motive-free account of reproductive consequences.) Can natural selection ever favor altruistic actions that decrease the reproduction of the actor? If we take a strictly organismic view of natural selection, it would seem to be impossible. Yet, as a growing list of natural observations records, animals behave in an apparently altruistic manner. The altruism of the sterile "workers" in such insects as ants and bees is one undoubted example. In such cases, the altruism is extreme, as the workers do not reproduce in some species.

Altruistic behavior often takes place between genetic relatives, where it is most likely explained by the theory of *kin selection*. Let us suppose for simplicity that we have two types of organism, altruistic and selfish. A hypothetical example might be that, when someone is drowning, an altruist would jump in and try and save him or her, whereas the selfish individual would not. The altruistic act decreases the altruist's chance of survival by some amount which we call c (for cost), because the altruist runs some risk of drowning. The action increases the chance of survival of the recipient by an amount b (for benefit). If the altruists dispensed their aid indiscriminately to other individuals, benefits will be received by other altruists and by selfish individuals in the same proportion as they exist in the population. Natural selection will then favor the selfish types, because they receive the benefits but do not pay the costs.

For altruism to evolve, it must be directed preferentially to other altruists. Suppose that acts of altruism were initially given only to other altruists. In such a case, what would be the condition for natural selection to favor altruism? The answer is that the altruism must take place only in circumstances in which the benefit to the recipient exceeds the cost to the altruist. This relation will hold true if the altruist is a better swimmer than the recipient, but it does not logically have to be true (if, for instance, the altruist were a poor swimmer and the recipients were capable of looking after themselves, the net result of the altruist's heroic plunge into the water might merely be that the altruist would drown). If the recipient's benefit exceeds the altruist's cost, then a net increase occurs in the average fitness of the altruistic types as a whole. This condition has only theoretical interest. In practice, it is usually (maybe always) impossible for altruism to be directed only to other altruists, because they cannot be recognized with certainty. It may be possible, however, for altruism to be directed at a class of individuals that contains a disproportionate number of altruists relative to their frequency in the population. For example, altruism may be directed toward genetic relatives. In this case, if a gene for altruism appears in an individual, it is also likely to be in its relatives. We will define r (for relatedness) as the probability that a new rare gene that arises in one individual also appears in another individual. This probability lies between 0 and 1, depending on the other individual concerned; the appropriate r can be deduced from Mendel's rules. If the new mutation arises in a parent, there is a ½ chance it will be in its offspring; likewise, there is a ½ chance that a gene in an individual is also present in its brother or sister.

Under what condition will natural selection favor altruism? The altruist still pays a cost of c for performing the act, while the recipient receives a benefit b. However, the chance that the altruistic gene is in the recipient is r. When rb exceeds c, the average fitness of the altruists will realize a net increase. The number of copies of the gene for altruism will increase because the loss of copies from the excess death of the individuals who actually perform acts of altruism is more than compensated for by the excess survival of the individuals who receive the benefit (and contain the gene for altruism). The condition for natural selection to favor altruism among relatives is that it should be performed if

$$rb > c$$

The theory of kin selection states that an individual is selected to behave altruistically provided that $rb > c$. The condition itself is called Hamilton's rule, after W. D. Hamilton, who primarily invented the theory of kin selection.

Hamilton's rule is testable. For an act of altruism we can measure the benefit and the cost, and r can be deduced if the pedigree relationship between altruists and recipients is known. The details of how b and c are estimated depends on the example. Here we shall look at one example involving "helpers at the nest" in the Florida scrub jay (*Aphelocoma coerulescens*, Color Plate 4).

The scrub jay is distributed widely across the western United States, and also has an isolated population that breeds in the shrinking areas of oak scrub in central Florida. It has been continuously studied in those regions since 1969 by Woolfenden and his colleagues. A breeding pair of Florida scrub jays may be helped by as many as six other birds. Woolfenden knows the pedigrees of the birds and has, therefore, been able to show that the majority of these helpers are either full or half siblings of the young they are helping. Thus, r is known for this example, but how can we estimate b and c?

"Benefit" and "cost" properly refer to the change in the lifetime reproductive successes of altruist and recipient, relative to the result if the act had not been performed. The true b and c cannot, therefore, be measured, because they refer to a situation that does not exist (namely, if the act had not been performed). They can, however, be estimated. Mumme experimentally removed the helpers from 14 nests in 1987 and 1988, and measured the reproductive success both in these experimental nests and in 21 untreated control nests in the same area. An average of almost 1.8 helpers appeared at the experimental nest, which we can round up to two helpers per nest in the calculations given below; a similar number of helpers were observed at the control nests.

The removal of the helpers significantly reduced the survival of the offspring (Table 12.2). Mumme found that the contribution of helpers mainly involves defending the nest against nest predators such as snakes and other birds. A nest with a helper is more likely to have a sentinel bird present at the nest at any time than is a nest without helpers, and nests with helpers can "mob" predators more effectively. ("Mobbing" is a kind of group defense of birds against predators, in which the birds dive at and harrass the predator. It is most commonly seen in birds such as crows, which often mob domestic cats.) The young in nests with helpers are also fed more and (probably in consequence) survive better after fledging.

Mumme's result enables us to estimate the benefit (b) of helping as the difference between the survival rate of young in nests with and without helpers. In Table 12.2, the survival rate was increased somewhere between two- and five-fold if helpers are present. Let us use the total figure for survival to day 60 to

TABLE 12.2

The survival of the young at nests of the Florida scrub jay, in nests at which the helpers had either been experimentally removed or left undisturbed. The offspring in experimental groups have lower survival during and immediately after the period of parental care, but not at the egg stage. These results are for 1987; the post-fledgling difference was similar in 1988, but the prefledging difference was not. Modified from Mumme (1992).

	Experimental Groups (helpers removed)	Control Groups (helpers present)
Initial sample size	45	63
% Survival from egg to hatching	67	68
% Survival from hatching to fledging	30	63
% Survival to day 60 after fledging	33	81
% Survival from egg to day 60	7	35

calculate whether natural selection favors helping. The survival of an average young scrub jay to day 60 is increased from 7% to 35% if helpers are present: the difference is $35 - 7 = 28\%$. We divide this value by 2 to find the benefit of helping per helper: $28/2 = 14\% = b$.

The cost of helping is more difficult to calculate. It consists of the reproductive success a helper would have had if it had not helped. We can make an upper- and lower-bound estimate. The lower-bound estimate is zero, in the case in which the helper is bred independently. This estimate may be close to the true value of c in the saturated habitat of Florida scrub jays. One of the main advantages of staying at the parental nest is thought to be the chance of inheriting the territory or budding off another territory at the edge of it; most new territories are formed in this way. A young jay may have to stay at home to be able to breed independently. An alternative, higher-bound estimate of the cost is the reproductive success of pairs without helpers. The justification of this estimate is that if the helper had bred by itself it would lack helpers (which are mainly derived from earlier clutches) and thus achieve the success of an unhelped pair. The cost of helping is then 7%, or the chance of survival to day 60 for an egg in a nest without helpers.

To apply Hamilton's rule in this case, we must notice that the helper's choice is between producing sibs and producing its own offspring. It should help its sibs if

$$r_{\text{sib}}b > r_{\text{off}}c$$

where r_{sib} is the relatedness to a sibling and r_{off} is the relatedness to its own offspring, both of which are ½ if the siblings are full siblings. (The small difference from the $rb > c$ version given above arises because we had previously imagined changes to the survival of altruists and recipients; the relatedness of an altruist to itself is necessarily 1 and the cost is implictly multiplied by this. Here the altruism affects the numbers of two kinds of offspring—its own and its parents'—and we must weigh each kind of offspring by the chance it shares a gene with the helper.)

For the two methods of estimating cost, the inequalities have the following approximate values:

lower-bound estimate: $\frac{1}{2} \times 14 > \frac{1}{2} \times 0$

higher-bound estimate: $\frac{1}{2} \times 14 > \frac{1}{2} \times 7$

Either way, natural selection favors helping behavior in young Florida scrub jays. The estimates of both b and c are fallible, however, and the test is uncertain. Nonetheless, it does illustrate how we can attempt a quantitative test of the theory of kin selection.

12.2.5 *The issue of whether group selection ever produces adaptations for the benefit of groups has been controversial, though most biologists now think it is only a weak force in evolution*

A group adaptation is a property of a group of organisms that benefits the survival and reproduction of the group as a whole. Adaptations produced by kin selection—such as helping in family groups of birds—will satisfy that definition, but we are concerned here with group adaptations that did not evolve by kin selection. If any adaptations exist, they will have come into existence by selection between groups. Groups possessing the group adaptation would have gone extinct at a lower rate, and sent out more emigrants, than groups lacking it. The group adaptation would have been favored by the differential reproduction of whole groups.

Many characters not only are beneficial at the group level, but also benefit all individuals within the group. For example, after an improvement in the hunting skill of a lion has spread by individual selection, all individuals in the group, and the group as a whole, will be better adapted. That concept is just a restatement of the earlier point that adaptations that have evolved for the benefit of one level of organization can incidentally benefit higher levels. The controversial group adaptations are those that benefit the group but not the individual. A hypothetical example is Wynne-Edwards' theory, put forward in *Animal Dispersion in Relation to Social Behaviour* (1962), that animals restrain their reproduction to avoid overconsumption of the local food supply. If all individuals in a group reproduce at the maximum rate, their offspring might overeat the food supply, and the group would then go extinct. They could avoid this fate by collectively restraining their reproduction, to maintain the balance of nature. Natural selection on individuals does not favor reproductive restraint. An individual that increases its reproduction will automatically be favored relative to individuals that produce fewer offspring. Within a group, if some individuals produce more offspring than others, the former will proliferate. In that case, can individual selection within the group be overcome by selection between groups?

This question is highly important, both conceptually and historically. It is important historically because vague group selectionist thinking—particularly in the form of statements like "adaptation X exists for the good of the species"—was once common. It is now more usual (although by no means universal) for biologists to believe that group selection is a weak and unimportant process. Both theoretical and empirical reasons underlie this belief.

Empirically, no definite examples exist of adaptations that need to be interpreted in terms of group advantage. Williams (1966) argued that the characters that Wynne-Edwards had suggested evolved to regulate population size can be explained as adaptations that benefit individuals. Individuals generally reproduce at the maximum rate they can; the only obvious exceptions concern genetically related individuals, and can be explained by kin selection. Moreover, living things have characteristics that contradict the theory of group selection. The 50:50 sex ratio, which we discussed in section 11.5 (p. 307), is a case in point. In polygynous species, it is inefficient for the population to produce 50% males, most of whom are not needed. The widespread existence of the 50:50 sex ratio suggests that group selection has been ineffective on this trait.

Group selection is also implausible in theory. Consider a population containing two genotypes, one for an altruistic, group adaptive trait like reproductive restraint (call them *A* individuals) and another selfish (*S*) type that reproduces as fast as it can. The population is made up of a number of groups, which can contain any proportion of *A* and *S* individuals. Groups with mainly altruistic members we call altruistic groups, and those with mainly selfish members we name selfish groups. The altruistic groups will go extinct at a lower rate, because group selection favors altruism. Individual selection favors the selfish individuals within all groups, because *S* individuals increase in frequency within each group. An altruistic group may temporarily contain no selfish members, but as soon as it is "infected" with one, the selfish trait will proliferate and become fixed in the group. What result should we expect to find? It depends on the balance between the two processes. In theory, we can imagine a rate of group extinction so high that altruists will predominate. Imagine, for the sake of argument, what would happen if all groups with more than 10% selfish types instantly went extinct. All groups would clearly have at least 90% altruists. Although that scenario is only an imaginary experiment, a more interesting question is what we should expect to happen naturally.

Most biologists suppose that group selection is a weak force in opposition to individual selection because of the slow life-cycles of groups as compared with individuals. Individuals die and reproduce once per generation, and many individuals can move between groups within a generation. As a result, groups go extinct at a much slower rate. The amount of time it will take for selfish individuals to infect and proliferate in a group represents only a small part of the group's lifespan. At any one time, therefore, individual adaptations will predominate.

Many models have been constructed of group and individual selection, but Maynard Smith has suggested that they can all be reduced to a common form (Figure 12.1). The groups are now supposed to occupy "patches" in nature. As before, some patches are occupied by altruistic (*A*) and by selfish (*S*) groups; empty patches (*E*) also appear in the landscape. A selfish group in the model drives itself extinct by overeating its patch's resources. The result of the model depends on whether a selfish group can infect an *E* or *A* patch before becoming extinct itself. Maynard Smith defines the number m as the number of successful migrants produced by one *S* group, on average, between its origin and extinction. (Successful means that the migrant establishes itself in another group and breeds.) If $m = 1$ the system will be stable; if $m < 1$ the selfish groups will decrease in

Figure 12.1 Maynard Smith's formulation of group selection models. E = empty patch, S = patch of selfish individuals, A = patch of altruists. A patch may change state in the direction of the arrows. Reprinted, by permission of the publisher, from Maynard Smith (1976). Copyright 1976 University of Chicago Press.

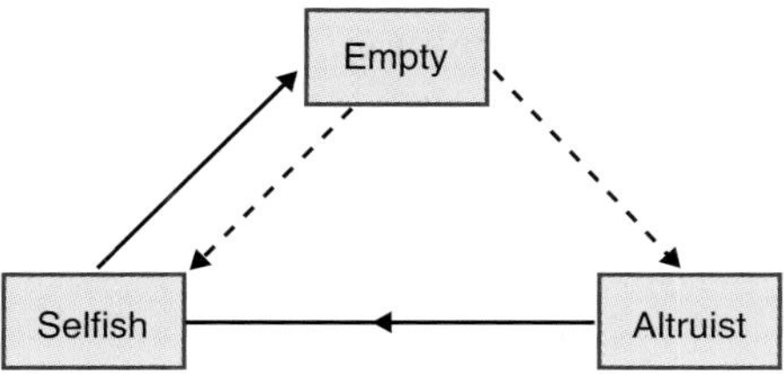

number; if $m > 1$ the selfish groups will increase. In other words, a selfish group needs to produce only more than one successful emigrant during its existence for the selfish trait to take over. This number is so small that we can expect selfish individual adaptations to prevail in nature. Group selection, we conclude, is a weak force. It works only if migration rates are implausibly low and group extinction rates implausibly high. It is also not needed to explain the facts.

Although the case against group selection has been stated in stark terms, this discussion is meant only to make the arguments clear; the matter has not been settled finally, and group selection probably operates on some occasions. The process is theoretically possible. Moreover, group selection can have evolutionary consequences even if it never overrides individual selection. In chapter 21 we shall discuss how individual selection can establish different adaptations in different groups (or species, in the case of sex) and those different adaptations could result in different rates of group extinction or expansion. No conflict exists between individual and group selection because in all groups (or species), individuals act in their own selfish interest. However, the kinds of group (or species) that go extinct less often will increase in frequency, and the increase is due to selection between groups. This idea is widely accepted; the "group selection controversy" in evolutionary biology was concerned solely with whether group selection could ever cause individuals to sacrifice their own reproductive interests to those of its group.

In nature, group selection is rarely likely to override individual selection and thereby establish individually disadvantageous behavior. In the lab, however, conditions can be made extreme enough for this inversion to occur. Let us finish by looking at such a case: Wade's experiment on flour beetles, *Tribolium castaneum*. The life cycle of the flour beetle takes place, from egg to adult, in stored flour. These beetles are pests, but they have also become a population biologist's standard experimental animal, particularly in Chicago. Wade set up an experiment to illustrate Wynne-Edwards' hypothesis of reproductive restraint. The experiment had the following design (Figure 12.2a). Each of three experimental treatments contained 48 different colonies of *Tribolium*. Each colony was allowed to breed for 37 days. Then 48 new colonies, of 16 beetles each, were set up from the progeny of the old ones. Wade artificially selected for groups that had showed a low (or high) fecundity (the third treatment was a control). He selected for low fecundity by forming a new generation of colonies from *Tribolium*

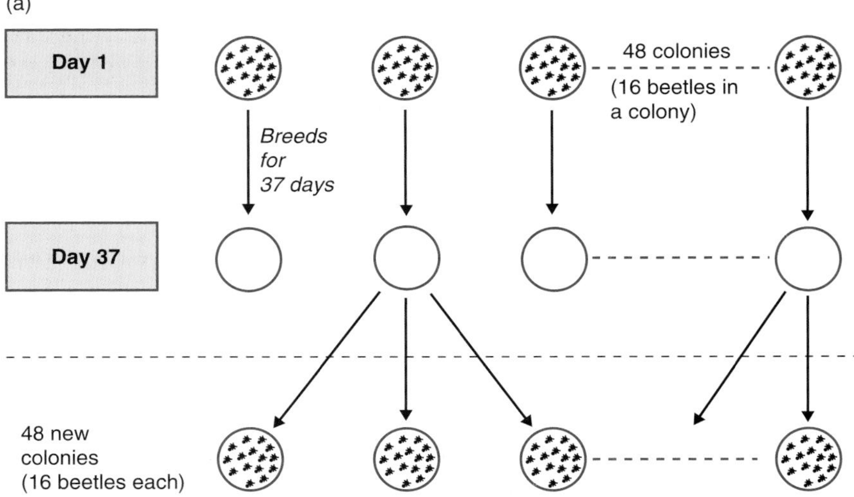

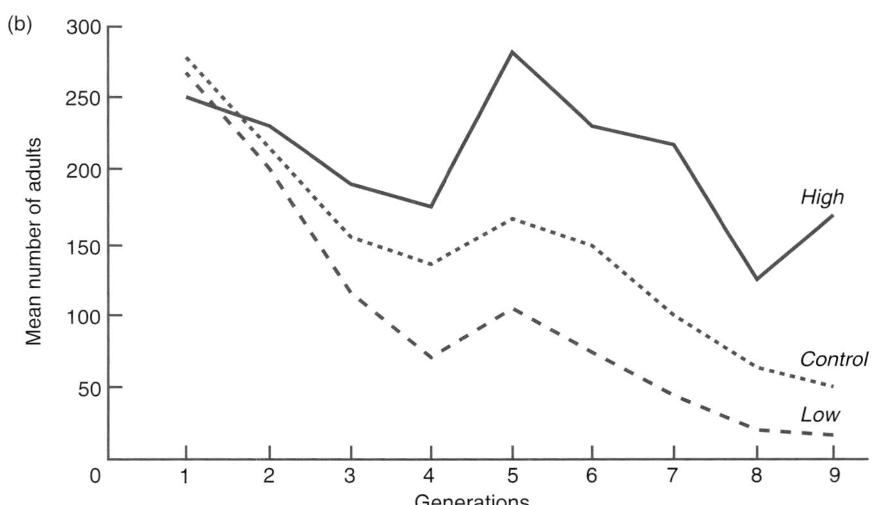

Figure 12.2 Wade's experiment with *Tribolium* beetles. (a) Experimental design. In the experiment, 48 colonies were bred for 37 days, and then a new round of colonies was formed from the colonies that had grown to a low (or high) population density. (b) Results. Population densities in lines selected for high or low population densities and in unselected controls. Results are means for 48 colonies. Reprinted, by permission of the publisher, from Wade (1976).

colonies that had a low population density at the end of the 37 days; he selected for high fecundity by forming each new generation from colonies that had high population densities. Wade repeated a number of rounds of the process.

Not surprisingly, the reproductive rate in Wade's "low" lines decreased, while the rate increased in the "high" lines (Figure 12.2b). The decrease in the low lines is due to group selection. Presumably, within the 37 days of any one cycle, the beetle types with high fecundity were increasing within each colony relative to the less fecund beetles. However, between cycles, Wade's group selection for low fecundity more than outweighed the individual selection, and average fecundity declined. The group selection was strong enough to work.

In a way, the group selective structure of Wade's experiment is superfluous. It would be perfectly possible to breed selectively from beetles with lower fecundity; the artificial selection would then reduce fecundity without the rigmarole of keeping the beetles in groups for 37 days. Wade's purpose was to illustrate group, not artificial, selection, however. His experiment, therefore, has alternating rounds of individual and group selection, and the experimental group selection is strong enough to produce the effect that Wynne-Edwards thought to be common in nature.

The fact that group selection can be carried out in an experiment does not have implications for its role in nature. Biologists doubt group selection for theoretical reasons, and because of the kinds of adaptations seen in nature. Wade's experiment had no impact on these doubts. The experiment is instructive, however, as it shows what group selection means.

12.2.6 *The level in the hierarchy of levels of organization that will evolve adaptations is controlled by which level shows heritability*

Adaptations can exist for the benefit of genes, cells, organisms, kin, or groups of unrelated individuals. Genic adaptations like segregation distortion are rare. Cell-line adaptations are very rare, at best, in Weismannist species, but may be found in non-Weismannist species such as plants. On the other hand, organismic adaptations are common. The number of examples of kin-selected adaptations is increasing, while group adaptations are probably rare.

Why do adaptations mainly appear at the organismic level, with a few additional cases for groups of kin? Earlier in this chapter, we discussed the answer to this question, but a more general answer can now be given. Maynard Smith has pointed out that the units in nature that show adaptations are the units that show heritability (section 9.6, p. 235). Mutations that influence the phenotype of a unit (whether a cell, organism, or group) must be passed on to the offspring of that unit in the next generation. In such a case, natural selection can act to increase the mutation's frequency.

Organisms show heritability in this sense. A finch with an improved beak shape, caused by a genetic change, will usually produce offspring with the improved beak shape. Natural selection can work on individual finches.

On the other hand, groups do not usually show this sort of heritability. A genetic variant that increases a group's chance of success tends not to be inherited by future groups. Immigration contaminates the group's genetic composition, such that heritability from generation to generation is low. Thus, altruistic groups do not exclusively generate descendant altruistic groups, and selfish groups generate selfish groups. Migration from selfish groups causes altruistic groups to become selfish. A group in one generation will be genetically correlated with the group of its offspring in the next generation only when practically no migration occurs. At that point, group selection works.

The same point can be made about kin selection and selection among cell lines. Kin selection operates because an "offspring" kin group genetically resembles the "parental" kin group. Cell selection tends not to operate in Weismannist species because somatic cells, although they are inherited down a cell line during one organism's brief life, are not passed on from an organism to its offspring. When such inheritance occurs (in non-Weismannist species), cell selection and the evolution of cell-line adaptations become theoretically more plausible. For genic adaptations such as segregation distortion, the same basic argument applies.

In summary, we should expect to find adaptations existing for the benefit of those units in nature that show heritability. Adaptations will, therefore, usually benefit organisms. The cases of adaptations that benefit higher or lower levels of organization can be understood in the same general terms, because they evolve only in circumstances when groups, or parts, of organisms show heritability from one generation to the next.

We can now give our first answer to our question about the unit of selection. The general answer is that the unit of selection is "that entity that shows heritability"; more specifically, it is usually the organism, with some interesting exceptions. This first answer specifies the units in nature that should possess adaptations.

12.3 *Another sense of "unit of selection" is the entity whose frequency is adjusted directly by natural selection*

Natural selection over the generations adjusts the frequencies of entities at all levels. We have implicitly seen this adjustment in the example of the lion hunt. If the lions of one pride become more efficient at hunting, perhaps because of some new behavioral trick, natural selection will favor them. If the trick is inherited, that type of lion will increase in frequency relative to other types of lion. All things associated with the trick will increase in frequency as well. The type of lion, its type of neurons, of proteins, and their encoding genes would all increase in frequency relative to their alternatives. When the hunting success of lions as a whole increases, the frequency of lions in the ecosystem will probably increase too. Over geological time, lions might come to replace other competing predators on the plains. The question in this section focuses on whether natural selection directly adjusts the frequency of any of these units—nucleotides, genes, neurons, individual lions, lion pride, or lion species?

The answer was most clearly given by Williams, in *Adaptation and Natural Selection* (1966), and Dawkins, in *The Selfish Gene* (1976). It is at least implicit in all theoretical population genetics (and in the previous section of this chapter). For natural selection to adjust the frequency of something over the generations, the entity must have a sufficient degree of permanence. You cannot adjust the frequency of an entity between times t_1 and t_2 if between the two times the entity has ceased to exist. For a character to increase in frequency under natural selection, therefore, it must be inherited.

We can work through this argument using the example of an improvement in lion hunting skill. (We will express the improvement in terms of selection on a mutation; the same arguments apply when gene frequencies are being adjusted at a polymorphic locus.) When the improvement first appeared, it was a single genetic mutation. At a physiological level, the mutation would produce its effect by making some minor change in the lion's developmental program. After the mutation has appeared, a "pool" of two types exists—the new mutation, and the rest of the population (i.e., all alleles of the mutation, and the behavior patterns they produce). Genetic variation will, of course, develop at loci other than the one where the mutation arose, but that variation can be ignored because it will be randomly distributed among the mutant and non-mutant types. The lions with the mutation will survive better and produce more offspring. Natural selection is starting to work. Now we can ask whose frequency natural selection is adjusting. Is it lions? Lion genomes? Or the mutation?

Williams' and Dawkins' answer is the gene—more specifically, the mutation that produces improved hunting. Natural selection cannot work on whole lions because lions die; they are not permanent. Nor can it work on the genome. The mutant lion's offspring inherit only genetic fragments, not a copy of a whole genome, from their parents. Meiotic recombination breaks the genome. In Williams' expression, "meiosis and recombination destroy genotypes [i.e., genomes] as surely as death." What matters, in the process of natural selection, is that some of the lion's offspring inherit the mutation. These offspring, in turn, produce more offspring, and the gene increases in frequency. The gene can increase in frequency because it is not fragmented by meiosis (like the genome) or returned to dust by death (like the phenotype). The gene, in the form of copies of itself, is potentially immortal, and is at least permanent enough to allow its frequency to be altered in successive generations.

Objections may be raised that recombination breaks genes as well as genomes. Recombination strikes at almost random intervals in the DNA and, therefore, could strike within the mutation with which we are concerned. A little reflection, however, shows that issue to be irrelevant. The information of the gene, not its physical continuity, is what matters. Consider the length of chromosome containing the gene and its mutant form; there will usually be a number of polymorphic loci around the mutant locus (Figure 12.3a). Now consider what happens when recombination strikes in either a neighboring gene or the gene itself. Nearby recombination breaks the information in the chromosome—which simply means

Figure 12.3 (a) Three genes along a chromosome. Loci *A* and *B* are heterozygous. The * indicates where the nucleotide differs from the other allele. (b) Intragenic recombination has no effect on the *A* genes that are produced. (c) Likewise, recombination in the neighboring *B* gene has no effect on the *A* genes produced. The same pair of *A* and *a* gene sequences are produced by recombination as were present before.

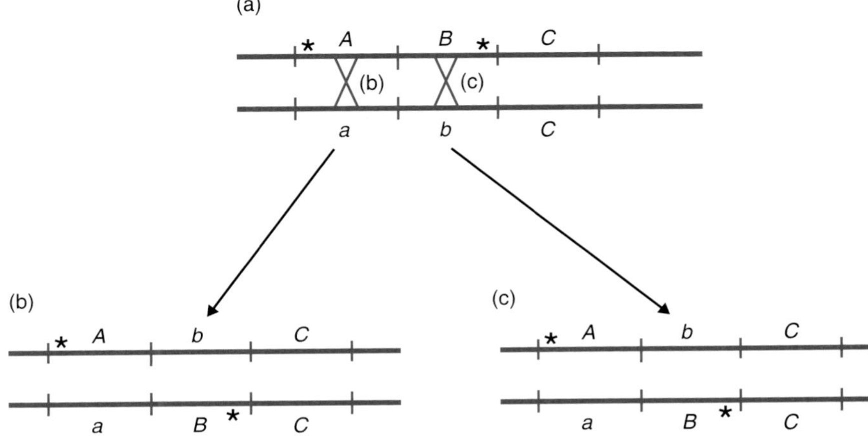

that recombination destroys the genome (Figure 12.3b). Recombination within the gene does not usually alter the outcome (Figure 12.3c). If the locus was homozygous before the mutation, all of the genes except for the mutant base pair will be identical in the original and mutant forms. Thus, intragenic recombination produces exactly the same result within the gene as no recombination, as it merely alters the combinations of genes.

Intragenic recombination can destroy the heritable information in a gene in one special circumstance. If the locus was heterozygous before the mutation, and recombination occurs between the mutant site and the site that differs between the two strands, the products of recombination will differ from the initial strands (Figure 12.4). When such an event occurs, the length of DNA whose information is inherited is shorter than a gene. For this reason, if we take a long enough view, the only finally permanent units in the genome are nucleotide bases, because recombination does not alter them. However, this long view holds little interest for us, as we are concerned with the time scale of natural selection. It takes a thousand or so generations for a mutation's frequency to be significantly altered (section 5.6, p. 100) and, over this time, genes, but not genomes or phenotypes, will be practically unaltered. Genes will then act as units of selection—they will be permanent enough to have their frequency altered by natural selection.

Williams defined the gene to make it almost true by definition that the gene is the unit of selection. He defined the gene as "that which segregates and recombines with appreciable frequency." According to this definition, the gene need not be the same as a cistron (i.e., the length of DNA encoding one protein, or polypeptide). Rather, it is the length of chromosome that has sufficient permanence for natural selection to adjust its frequency. Longer lengths are broken by recombination and shorter lengths have no more permanence than the gene (for the reason shown in Figure 12.3). The gene on Williams definition is what Dawkins calls the *replicator*. In practice, the replicator (or Williams' gene) does not consistently correspond to any particular length of DNA.

When selection takes place at one locus, a cistron at a neighboring locus will to some extent (depending on the amount of recombination) have its frequency adjusted as a consequence. In a population genetic sense, this hitchhiking (section 8.9, p. 211) builds up linkage disequilibrium between genes. The same will be true of loci further down the DNA from the selected locus. Although the hitch-

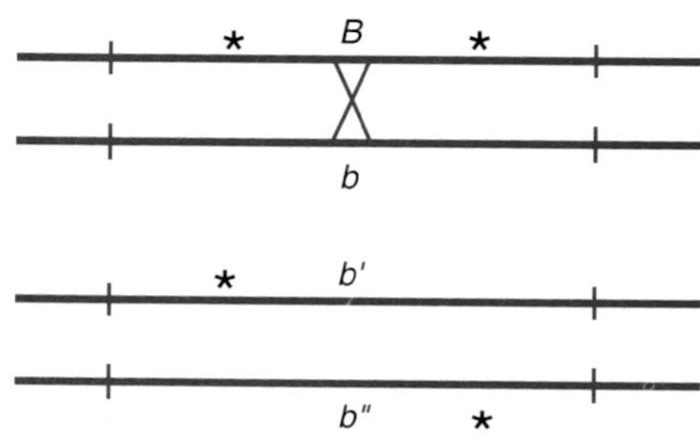

Figure 12.4 When intragenic recombination occurs between two alleles with different heterozygous nucleotides, it breaks the gene structure. The * indicates the nucleotide sequence differs from the allele. The gene sequences produced by recombination differ from the initial sequences.

hiking effect is gradually reduced with distance by recombination, no clear cut-off has been distinguished. This situation poses no problem for Williams' definition of the gene. The neighboring allele that is hitchhiking with the selected mutation is, in Williams' definition, part of the gene that is having its frequency altered.

Williams' "gene" has a statistical reality, because shorter lengths of DNA are more permanent and longer lengths less permanent. The random hits of recombination will generate a frequency distribution of genome lengths lasting for different periods of evolutionary time. The average length that survives long enough to undergo the effects of natural selection has been defined by Williams and Dawkins as the gene. Population geneticists have scolded Williams and Dawkins from time to time for assuming a one-locus zero linkage disequilibrium view of evolution, but the dispute is a matter of definition, not substance. These critics identify "gene" with "cistron." It would be interesting to know whether the gene in Williams' sense is also a physical cistron, but this is a secondary question and it is unrelated to the fundamental logic of Williams' and Dawkins' argument.

We must discuss one other matter before considering the significance of the genic unit of selection. Critics, such as Gould, have objected that gene frequencies change between generations only in a passive, "bookkeeping" sense. The frequency changes provide a record of evolution, but are not its fundamental cause. True natural selection, the critics would say, happens at the level of organismic survival and reproduction. For instance, the actual selection in the lion example happens when a lion catches, or fails to catch, its prey. The differential hunting success drives the gene frequency changes, and it is a mistake to identify the gene frequency changes as causal. Williams and Dawkins, however, do not deny that the ecological processes causing differential organismic survival produce gene frequency changes within a generation. What they deny is that this ecological interaction of organisms means that natural selection directly adjusts the frequencies of organisms over the evolutionary time scale of many generations.

An easy philosophical method has been developed for deciding whether natural selection works on genes, or larger phenotypic units. We can consider a phenotypic change such as a new hunting skill, and ask whether natural selection can work on it if it is produced genically and if it is produced non-genically. In the lion's case, the skill is produced genically—the advantageous new hunting behavior was caused by a genetic mutation. Now suppose that the same advantageous phenotypic change was caused by a non-heritable phenotypic change, such as individual learning or some developmental accident in the lion's nervous system. The thought-based experiment provides a test case between the organismic, phenotypic, and genic accounts of evolution. In the genic case, we know that natural selection favors the improved hunting type and the gene for it increases in frequency. But what happens in the phenotypic case? The individual lion with improved hunting ability will survive and produce more offspring than an average lion, but no evolution, or natural selection in any interesting sense, will occur. The trait will not be passed on to the next generation. Natural selection cannot work directly on organisms.

The change in gene frequency over time, therefore, is not just a passive "bookkeeping" record of evolution. Genes are crucial if natural selection is to take place. The need for inheritance, and the fact that acquired characters are not inherited, gives the gene a priority over the organism as a unit of selection. Whenever a gene is being selected, it produces a phenotypic change and the frequency of different organismal types will change concurrently with the gene frequency. The change in organism frequency is a consequence of the change in gene frequency, however. That is, natural selection actually works on the gene frequency. For this reason, Williams and Dawkins maintain that the gene represents the unit of selection.

The argument is more a matter of logic than a testable claim that could be refuted by facts or experiments. By no means do all biologists accept the argument. The main controversy has been concerned with the unit of selection in the first sense (see sections 12.1–12.2), but Williams and Dawkins' argument itself has also been criticized for the two reasons we have discussed. We believe that these criticisms have been misdirected. Confusion about the definition of "gene" has spurred arguments that hitchhiking and linkage disequilibrium mean that selection adjusts the frequency of larger units than the gene. In addition, the claim that natural selection "really" works on organisms rather than genes overlooks the importance of heredity.

Why does the argument matter? Its importance is to tell us for the benefit of which entities adaptations exist. Evolutionary biologists work on particular characters (like banding patterns in snails, and sex), trying to explain why the characters exist. The ultimate, abstract answer is that any adaptation exists because it increases the reproduction of the genes encoding it, relative to that of the alleles for alternative characters. The genes that exist in nature are the genes that in the past have outreproduced alternative alleles. Natural selection will always favor a character that increases the replication of the genes encoding it.

It is important to recognize the ultimate beneficiaries of adaptations. When we are attempting to explain the existence of particular characters, we need to know whether a proposed explanation is correct. The argument that genes are units of selection provides the fundamental logic behind relevant tests of these explanations. We imagine different genetic forms of the character, and the correct explanation must specify how the genes for the observed form of the character will outreproduce other genetic types. In practice, several possible hypotheses may be put forward, and they can be tested by the methods described in chapter 11. Before those methods are applied, however, we must ensure that the hypotheses make theoretical sense. We can rule out a hypothesis about adaptation before the practical testing stage if it contradicts the theory of gene selection.

12.4 *The two senses of "unit of selection" are compatible; one specifies the entity that generally shows phenotypic adaptations, the other specifies the entity whose frequency is generally adjusted by natural selection*

We have now defined the unit of selection in two different senses. These two specifications have sometimes been confused, but many evolutionary biologists now appreciate the distinction. The two senses have been given names; Hull, for instance, distinguishes between interactors and replicators, while Dawkins notes the distinction between vehicles and replicators. It is most important, however, to realize that there are two distinct issues and to understand the arguments used in the two cases.

Adaptations evolve because the genes encoding them outreproduce the alternative genes. In this sense, adaptations can evolve only if they benefit replicators. Genes do not, however, exist nakedly in the world. The kinds of adaptations that evolutionary biologists seek to understand, such as social behavior, beak shape, or flower coloration, are not simple properties of genes. Rather, they are phenotypic properties of higher-level entities (whole organisms, or societies). Consequently, we must also ask which higher-level entities should benefit from the natural selection of replicating genes. This train of thought leads to the discussion of section 12.2, and we have seen that adaptations should benefit entities that show heritability—which are usually organisms. This conclusion follows automatically from the fact that selection works on replicating genes, because the heritability results from the propagation of genes. In other words, those entities will show adaptations that propagate genes efficiently.

SUMMARY

1. Adaptations evolve by means of natural selection. When natural selection acts, it alters the frequencies of entities at many levels in the hierarchy of biological levels of organization. It also produces adaptations that benefit entities at many levels.
2. The discussion of units of selection aims to find out the level on which natural selection acts directly, and which levels it affects only incidentally.
3. Evolutionary biologists are interested in the unit of selection so as to understand why adaptations evolve and so that, when they study adaptations, they can concentrate on theoretically sensible hypotheses.
4. We can find out which level of organization shows adaptations by considering a series of adaptations at genic, cellular, organismic, and group levels and asking which evolves most often.
5. Segregation distortion is an adaptation of a gene against its allelic alternatives. Examples of this kind are rare.
6. In Weismannist organisms, which have separate germ and somatic cell lines, selection between cell lines is a weak force. Many species do not have separate germ lines, however; in these species, we expect cell lines to evolve adaptations enabling them to proliferate at the expense of other cell lines. No clear examples are known, but Buss has suggested that the embryology of modern Weismannist species can be explained by a history of cell selection.
7. Adaptations are common at the level of organisms. When genetic relatives interact, adaptations may evolve for the benefit of kin groups (kin selection).
8. Group selection, in which selection produces adaptations for the benefit of groups of unrelated individuals, is thought to be a weak force.
9. Adaptations are possessed by the levels in the hierarchy of life that show heritability, in the sense that genetic changes are inherited by the progeny at that level. Group selection is weak because of the low genetic correlation (heritability) between succeeding generations of groups.
10. Natural selection adjusts the frequencies of only entities that are sufficiently permanent over evolutionary time. Thus, it fundamentally adjusts the

frequency of small genetic units. This small genetic unit is called the replicator. The gene can be so defined to be the unit of selection, but it is then not necessarily always a cistron in length.

11. Adaptations evolve because they increase the replication of genes. The replication of genes, in the real world, is enhanced by adaptations that benefit entities that show heritability.
12. The question of whether natural selection adjusts the frequencies of genes or of organisms is distinct from the question of the relative power of individual, kin, and group selection.

FURTHER READING

The chapter follows the work of Dawkins (1982, 1989), Maynard Smith (1987b), and Williams (1966, 1985, 1992), who also refer to most of the prior literature. On segregation distortion, see Crow (1979) and a special issue of the *American Naturalist*, vol. 137 (1991), pp. 281–456; Haig and Grafen (1991) and Haig (1993) discuss the relation with meiosis; Dawkins (1982, 1989) discusses the argument that the fairness of meiosis is necessary for the evolution of complex life forms, a theme not explored here. On cell selection, see Buss (1987), and Maynard Smith *et al.* (1985, pp. 281–282) for a possible example in the *bobbed* mutation of *Drosophila*; Cosmides and Tooby (1981), Eberhard (1980, 1990), and Nordström and Austin (1989) discuss various subcellular selective phenomena that would bridge the gap between this chapter's sections on genic and cellular selection. One example is the cytoplasmic factor that is currently spreading through *Drosophila simulans* in California (Turelli and Hoffmann 1995). On kin selection, the fundamental works are by Hamilton (1964, 1972). For more introductory accounts see Dawkins (1989), Grafen (1984), Krebs and Davies (1993), Trivers (1985). See Woolfenden and Fitzpatrick (1984) on the scrub jays. Wilson and Sober (1994) review most of the vast literature on group selection. Critics of gene, or replicator, selection are discussed by Dawkins (1982) and Sober (1984); Gould (1980a, ch. 8; 1983b) is an example.

STUDY AND REVIEW QUESTIONS

1. Give examples of adaptations that benefit (a) both the individual organism and the species to which the organism belongs, (b) the individual organism, but at a cost to its species, (c) a local group of organisms, at a cost to its individual members, and (d) a small genetic system, at a cost to the organism containing it.

2. What are the main theoretical factors in models of group versus individual selection that determine whether individual or group adaptations tend to evolve?

3. In the measurements made by Woolfenden and Fitzpatrick (1984), the average number of young birds produced by a nest of scrub jays with helpers is 2.2 and the average number by a nest without helpers is 1.24. The average number of helpers present, for the nests with helpers, is 1.7. What values for b and c

can be estimated from these data? If the helpers are brothers or sisters of the individuals they are helping, does kin selection favor helping?

4. In both kin selection and pure group selection, adaptations often evolve that benefit the local group. What key difference is there between the kinds of groups, and the plausibility of the two processes?
5. What is the unit of selection, in the sense of a replicator, in (a) a species that reproduces asexually, and (b) a species in which there is no recombinational crossing over at meiosis?

CHAPTER 13

Adaptive Explanation

THIS chapter considers a series of points about adaptation. Many of them have inspired controversy in the professional literature and the chapter aims to explain the main positions in those controversies, and to provide the background to understanding them. We look first at the argument to show that natural selection is the only known explanation of adaptation. Next, we consider whether natural selection can explain all adaptations, including such complex organs as the eye. We then look at two definitions of adaptation, in terms of historical and modern factors. Most of the rest of the chapter gives various reasons why adaptations may not be perfect. In many cases, adaptations may be out of date because they have been constrained by genetics, developmental mechanisms, historical origins, or trade-offs between multiple functions.

13.1 Natural selection is the only known explanation for adaptation

The fact that living things are adapted for life on earth is sufficiently obvious that philosophers did not have to wait for Darwin to point it out. Adaptation was already a familiar fact, and much use was made of it by the school of thought called *natural theology*. Natural theologians explained nature's properties theologically (i.e., by the direct action of God). They were highly influential from the eighteenth century until the publication of Darwin's work—John Ray and William Paley were two important thinkers of this type.

Darwin himself was much influenced by the examples of adaptation, such as the vertebrate eye, discussed by Paley. Paley explained adaptation in nature by the creative action of God: when God miraculously created the world and its living creatures, he or she miraculously created their adaptations too. Natural theology was influential as a way of understanding adaptations in nature, but its main influence—beyond biology—was as an argument to prove that God exists. The "argument from design," for example, was invented by thinkers in ancient Greece, and the natural theologians updated it for the eighteenth and nineteenth centuries. Part of the reason why Darwin's theory was so controversial was that it wrecked one of the most popular (at that time) proofs for the existence of God. The key difference between natural theology and Darwinism is that the former explains adaptation by supernatural action, and the latter by natural selection.

Natural theology and natural selection are not the only two explanations that have ever been put forward for adaptation. The inheritance of acquired characters ("Lamarckism") suggests that the hereditary process produces adaptations automatically. Other theories suggest that the hereditary mechanism itself produces designed, or directed, mutations and adaptation results as the consequence. In Darwinism, variation is not directed toward improved adaptation; in these theories, however, mutation is undirected and selection provides the adaptive direction in evolution (section 4.8, p. 85).

One of the most fundamental claims in the Darwinian theory of evolution is that natural selection provides the only satisfactory explanation for adaptation. The Darwinian, therefore, must show that the alternatives to natural selection either do not work or are scientifically unacceptable.

Let us consider the natural theologians' supernatural explanation first. We can accept that an omnipotent, supernatural agent could create well-adapted living things—in that sense the explanation works. However, it has two defects. First, supernatural explanations for natural phenomena are scientifically useless (section 3.14, p. 65). Second, the supernatural Creator is not explanatory. The problem requires us to explain the existence of adaptation in the world, but a supernatural Creator already possesses this property. Omnipotent beings are themselves well-designed, adaptively complex, entities. Thus, the thing we are trying to explain has been built into the explanation. Positing a God merely invites the question of how such a highly adaptive and well-designed thing could, in its turn, have come into existence. Theological sophistry about the perfect simplicity of God and the inexplicability of the First Cause can be ignored here: the problem is to *explain* adaptive complexity. The first alternative to natural selection, therefore, is a viciously circular argument, and unscientific.

The "Lamarckian" theory—the inheritance of acquired characters—is not unscientific.[1] Since Weismann, in the late nineteenth century, it has generally been accepted that acquired characters are, as a matter of fact, not inherited. However, facts are always liable to revision and some factual exceptions to Weismann's general doctrine may exist. Even if no factual exceptions appear, it is still important to know whether the Lamarckian theory can account for adaptation in principle.

Can it explain adaptation? Consider the adaptations of zebras to escape from lions. Ancestral zebras would have run as fast as possible to escape from lions. In doing so, they would have exercised and strengthened the muscles used in running. Stronger legs are adaptive as well as being an individually acquired character. If the acquired character was inherited, the adaptation would be perpetuated. Superficially, this statement looks like an explanation, whose only defect is that as a matter of fact acquired characters are not inherited.

Now let us imagine (for the sake of argument) that acquired characters are inherited, and look more closely at the explanation. The adaptation arises because zebra, within their lifetimes, become stronger runners. However, muscles do not by some automatic physical process become stronger when they are exercised. They might just as well become weaker, because they are used up. Muscle strengthening in an individual zebra requires explanation; it cannot be taken

1. Quotation marks are placed around "Lamarckian" because, as we saw in chapter 1, the inheritance of acquired characters was not especially important in Lamarck's own theory, nor did he invent the idea. However, the inheritance of acquired characters has generally come to be called Lamarckism; we can conveniently follow normal usage, outside purely historical discussion.

for granted. Muscles, when exercised, grow stronger because of a pre-existing mechanism that is adaptive for the organism. But where did that adaptive mechanism originate? The Lamarckian has no answer for this crucial failing of the theory. For a complete explanation, it would be necessary to resort to another theory, such as God or natural selection. In the former case, it would run into the difficulties we discussed earlier, and in the latter case natural selection, not Lamarckism, provides the fundamental explanation of adaptation. Lamarckism could work only as a subsidiary mechanism, because it could bring adaptations into existence only in so far as natural selection had already programmed the organism with a set of adaptive responses. Pure Lamarckism does not, by itself, explain adaptation.

All theories of directed or designed mutation have the same problem. To be a viable alternative to natural selection, a theory of directed mutation must offer a mechanism for adaptive change that does not fundamentally rely on natural selection to provide the adaptive information. Most alternatives to natural selection do not explain adaptation at all. For example, earlier this century some paleontologists, such as Osborn, were impressed by long-term evolutionary trends in the fossil record. The titanotheres are a classic example. Titanotheres are an extinct group of Eocene and Oligocene perissodactyls (the mammalian order that includes horses). In a number of lineages, the earlier forms lacked horns, whereas later ones had evolved them (Figure 13.1). Osborn, and others, believed that the trend was *orthogenetic*—that it arose not because of natural selection among random mutations, but because titanotheres were mutating in the direction of the trend.

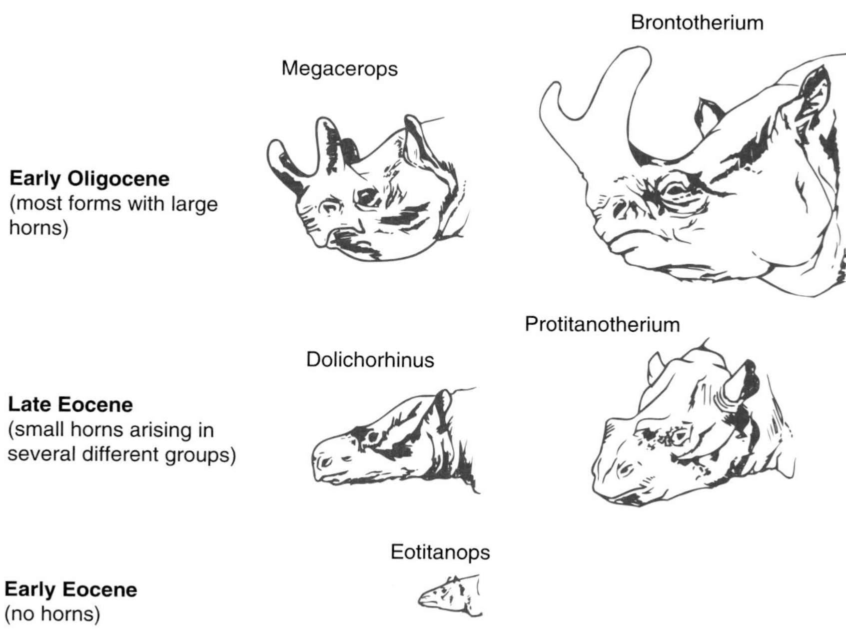

Figure 13.1 Two lineages of titanotheres, showing parallel body size increase and the evolution of horns. Only two of many lineages are illustrated. Reprinted, by permission of the publisher, from Simpson (1949).

This idea could explain a simple, adaptively indifferent trend. If horn size made no difference to a titanothere, then a trend toward larger horns might be generated by directed mutation. In fact, the horns almost certainly were adaptive. For trends toward undoubted adaptive complexity, such as the evolution of mammals from mammal-like reptiles (section 21.1, p. 582), we must ask how the "orthogenetic" mutations could continue occurring in the direction of adaptive improvement. If it is replied that variation just happens to be that way, then adaptation is being explained by chance—and chance alone cannot explain adaptation, almost by definition. Thus, theories of directed variation generally boil down to supernatural intervention or natural selection. As a result, they are either unscientific or infeasible alternatives to Darwinism. Directed mutation and the inheritance of acquired characters not only do not operate in fact, but they are also unsatisfactory in theory. We can, therefore, conclude that natural selection is the only available explanation for adaptation.

13.2 *Pluralism is appropriate in the study of evolution, not of adaptation*

We have now established that natural selection is the only explanation for adaptation. It should be emphasized that this statement applies only to *adaptation*—not to evolution as a whole. Biologists, such as Gould and Lewontin, have pointed out that Darwin did not himself rely exclusively on natural selection, but incorporated other processes into his theory as well. They urge that we should accept a "pluralism" of evolutionary processes, rather than relying exclusively on natural selection. For evolution as a whole, this approach is a sensible idea. In chapter 7, for instance, we saw that many evolutionary changes in molecules may take place by neutral drift. The molecular sequences among which drift occurs are then not different adaptations. Rather, they are different variants of one adaptation, and natural selection has nothing to say about why one organism has one sequence variant, and another organism has another. We need drift as well as selection to form a full theory of evolution.

The fact that processes other than natural selection can cause evolutionary change does not alter the argument presented in section 13.1. It simply shows that not all evolution need be adaptive. In that case, we should be pluralists about evolution, but concentrated on natural selection when we are studying adaptation. Critics might agree, but retort that some evolutionary biologists exaggerate the amount of evolution that is adaptive rather than non-adaptive, and mistake cases of the latter for the former. We shall have more to say about this point later. Here we focus on natural selection's role as the only known explanation of adaptation, rather than on how common adaptations are.

13.3 *Natural selection can, in principle, explain all known adaptations*

The argument we have made so far has been negative. That is, we have ruled out the alternatives to natural selection, but we have not made the positive case for natural selection itself. We saw earlier (chapter 4) that natural selection can explain adaptation. Now we will ask a stronger question: Can it explain *all* known adaptations?

This question has been important historically, and it often rises even in today's discussions of evolution. The case against selection would take the following form. Natural selection undoubtedly explains some adaptations, such as camouflage. The adaptation in this case, as well as in other famous examples of natural selection, are all simple, however. In the peppered moth, adaptation is simply a matter of adjusting external color to the background. The problem arises in complex characters, which are adapted to the environment in many interdependent respects. Darwin's explanation for complex adaptations is that they evolved in many small steps, each analogous to the simple evolution in the peppered moth; that is what Darwin meant when he called evolution "gradual." Evolution must be gradual because it would take a miracle for a complex organ, requiring mutations in many parts, to evolve in one sudden step. If each mutation arose separately, in different organisms at different times, the whole process becomes more probable (Box 7.3, p. 182).

Darwin's "gradualist" requirement is a deep property of evolutionary theory. The Darwinian should be able to show for any organ that it could, at least in principle, have evolved in many small steps, with each step being advantageous. If exceptions are found, it would cast serious doubt on the entire theory. In Darwin's words, "if it could be demonstrated that any complex organ existed which could not possibly have been formed by numerous successive slight modifications, my theory would absolutely break down."

Darwin argued that all known organs indeed could have evolved in small steps. He took examples of complex adaptations and showed how these examples could have evolved through intermediate stages. In cases such as the eye (Figure 13.2), these intermediates can be illustrated by analogies with living species; in other cases, they can only be imagined. Darwin had to show only that the intermediates could possibly have existed. His critics had the more difficult task: they had to show that the intermediates could not have existed. It is very difficult to prove negative statements. Nevertheless, many critics suggested that natural selection cannot account for various adaptations. These types of adaptation can be considered under two headings.

1. Coadaptations

Coadaptation here refers to complex adaptations, the evolution of which would have required mutually adjusted changes in more than one part. (Coadaptation is a popular word. It has already been used in a different sense in chapter 8, and will be used in a third sense in chapter 20!) In a historic dispute in the 1890s, Herbert Spencer and August Weismann discussed the giraffe's neck as an example. Spencer supposed that the nerves, veins, and bones and muscles in the neck were each under separate genetic control. Any change in neck length would then require independent, simultaneous changes of the correct magnitude in all of the other parts. A change in the length of the neckbones would malfunction without an equal change in vein length, and evolution by natural selection on one part at a time would be impossible. The example is unconvincing now because of the obvious retort that the lengths of all parts could be under common genetic control.

Figure 13.2 Stages in the evolution of the eye, illustrated by species of molluscs. (a) A simple spot of pigmented cells. (b) Folded region of pigmented cells, which increases the number of sensitive cells per unit area. (c) Pin-hole camera eye, as is found in *Nautilus*. (d) Eye cavity filled with cellular fluid rather than water. (e) The eye is protected by adding a transparent cover of skin, and part of the cellular fluid has differentiated into a lens. (f) Full, complex eye, as found in octopus and squid. Reprinted, by permission of the publisher, from Strickberger (1990).

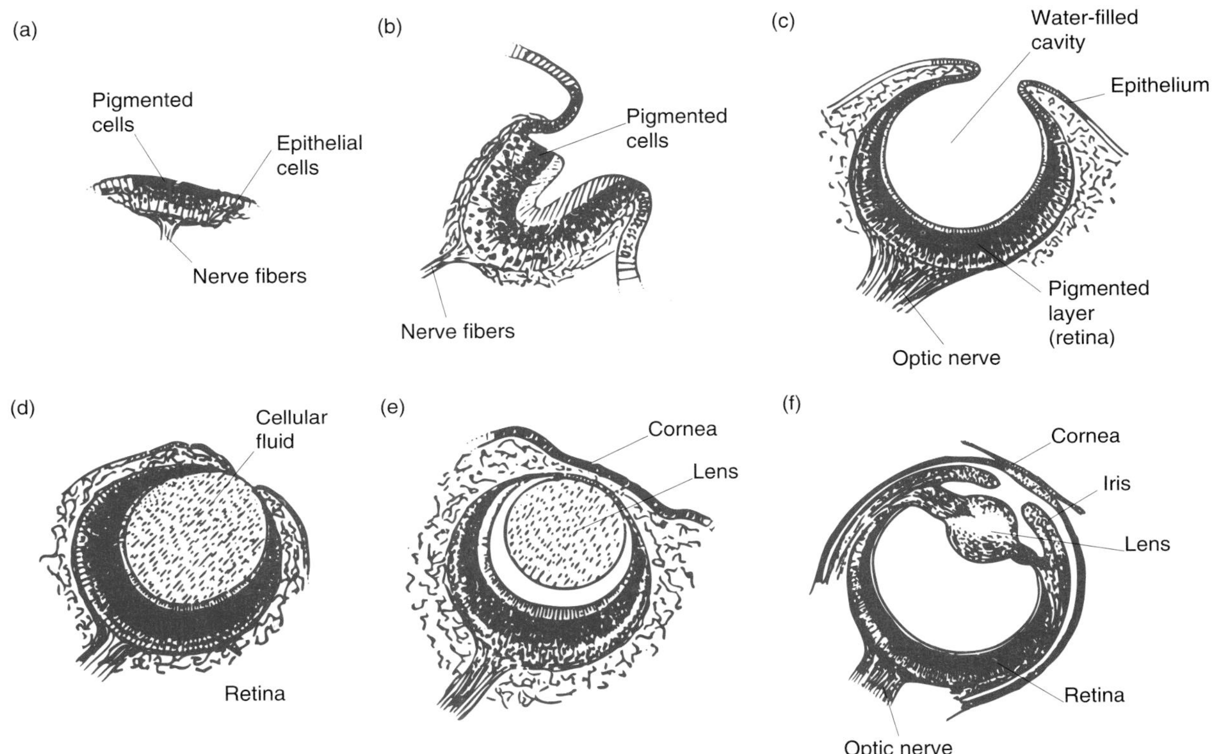

The other standard example of a complex coadaptation was the eye. When one eye part, such as the distance from the retina to the cornea, changes during evolution, changes in other parts, such as lens shape, would (it is said) be needed at the same time. According to this argument, the improbability of simultaneous correct mutations in both parts at the same time would mean that a complex, finely adjusted engineering device like the eye could not have evolved by natural selection. The Darwinian reply (illustrated in Figure 13.2) was that the different parts could evolve independently in small steps. It is not necessary for all the parts of an eye to change at the same time in evolution.

A recent computer model study by Nilsson and Pelger illustrates the power of Darwin's argument. Although the eye of a vertebrate or an octopus looks so complex that it can be difficult to believe it could have evolved by natural selection, light-sensitive organs (not all so complex) have actually evolved 40–60 times in various invertebrate groups—which suggests either that the Darwinian explanation faces a problem that is 40- to 60-fold more difficult than that presented by the vertebrate eye alone, or that evolution may not be so difficult after all. Nilsson and Pelger's simulation modeled the eye to find out how difficult its evolution really is. Their simulation begins with a crude light-sensitive organ—a layer of light-sensitive cells sandwiched between a darkened layer of cells below and a transparent protective layer above (Figure 13.3a). The simulation, therefore, does not cover the complete evolution of an eye. It initially takes light-sensitive cells as given (which is an important but not absurd assumption, because many pigments are influenced by light) and at the other end it ignores the evolution of advanced perceptual skills (which are more a problem in brain, than eye, evolution). The model concentrates on the evolution of eye shape and the lens, which is the problem that Darwin's critics have often pointed out, because they think it requires the simultaneous adjustment of many intricately related parts.

From the initial simple stage, Nilsson and Pelger allowed the shape of the model eye to change at random, in steps of no more than 1% change at a time. This small change fits in with the idea that adaptive evolution proceeds in small, gradual stages. The model eye then evolved in the computer, with each new generation formed from the optically superior eyes in the previous generation; changes that made the optics worse were rejected, as selection would reject them in nature.

The particular optical criterion used was visual acuity—the ability to resolve objects in space. The visual acuity of each eye in the simulation was calculated by the methods of optical physics. The eye is particularly well suited to this kind of study because optical qualities can readily be quantified. That is, it is possible to show objectively that one model eye would have better acuity than another.

Figure 13.3 Eight stages in the evolution of the eye in a computer model. The initial stage has a transparent cell layer, a light-sensitive cell layer, and a dark-pigmented bottom cell layer. It first improves its optical properties by buckling in (up to stage d–e); (e) approximately corresponds to the pin-hole camera eye (see Figure 13.2c). It then improves by (f) the evolution of a lens. The lens shape then changes, and the iris flattens, to improve the focussing properties. *The lens* has the best optical properties when the focal length of the lens (*f*) = the distance from lens to retina (*P*). This feature gradually improves in the final three phases (f–h). The normalized diameter of the eye (*d*) indicates the change in shape. Reprinted, by permission of the publisher, from Nilsson and Pelger (1994).

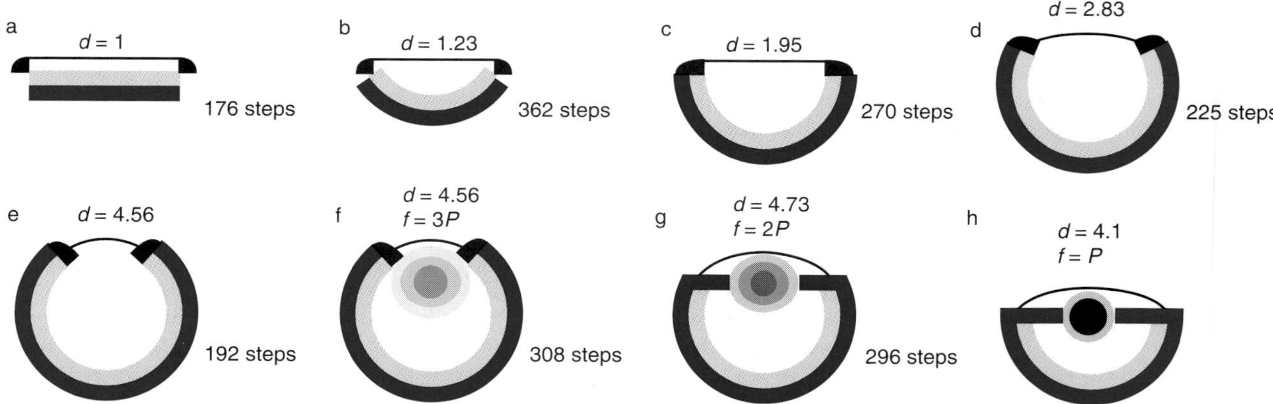

tion could then produce a Hb^+Hb^s chromosome, and that chromosome should be able to achieve anything that an Hb^+/Hb^s heterozygote can. The chromosome would also breed true, once it had been fixed. We might expect that the existing Hb^+/Hb^s system will evolve to a pure Hb^+Hb^+/Hb^sHb^s system. Some "dosage compensation" might be needed after the gene had duplicated, but that requirement should present no difficulty because regulatory devices are common in the genome. The apparent ease of this evolutionary escape from heterozygous advantage and segregational load is one possible explanation for the (apparent) rarity of heterozygous advantage. In any event, the existence of some cases of heterozygous advantage suggests that natural populations can be imperfectly adapted because a superior mutation has not arisen.

13.9 *Developmental constraints may cause adaptive imperfection*

A discussion (Maynard Smith *et al.* 1985) of *developmental constraints* by nine authors resulted in the following definition: "a developmental constraint is a bias on the production of variant phenotypes or a limitation on phenotypic variability caused by the structure, character, composition, or dynamics of the developmental system." The idea is that different groups of living things evolved distinct developmental mechanisms, and that the way an organism develops will influence the kinds of mutation it is likely to generate. A plant, for example, may be likely to mutate to a new form with more branches than would a vertebrate, because it is easier to produce that kind of change in the development of a plant (indeed, it is not even clear what a new "branch" would mean in the vertebrate—perhaps it might be extra legs, or having two heads). The rates of different kinds of mutation—or of "production of variant phenotypes" in the quoted definition—therefore differs between plants and vertebrates.

Developmental constraints can arise for a number of reasons. *Pleiotropy* is an example. A gene may influence the phenotype of more than one part of the body. A trivial instance would be that genes influencing the length of the left leg probably also influence the length of the right leg. The growth of legs probably takes place through a growth mechanism controlling both legs. This mechanism does not have to be inevitable for a constraint to exist. Perhaps some rare mutants do affect the length of only the right leg. A developmental constraint exists whenever there is a tendency for mutants (in this example) to affect both legs, and the tendency is due to the action of some developmental mechanism.

Pleiotropy exists because a one-to-one relation is not present between the parts of an organism that a gene influences and the parts of an organism that we recognize as characters. The genes divide up the body in a different way from the human observer. Genes influence developmental processes, and a change in development will often change more than one part of the phenotype. Much the same reasoning lies behind a second sort of developmental constraint. New mutations often disrupt the development of the organism. A new mutant, with an advantageous effect, may also disrupt other parts of the phenotype. The disruptions will probably be disadvantageous, but if the mutant has a net positive effect on fitness, natural selection will favor it. In some cases, the disruption can be measured by the degree of asymmetry in the form of the organism. In a species with bilateral symmetry, for example, any deviation from bilateral symmetry in an organism is a measure of how well regulated its development was. Mutations can, therefore, cause *developmental asymmetry*.

The Australian sheep blowfly *Lucilia cuprina* provides an example. It is a pest, and farmers spray it with insecticides. The flies, as we should expect (section 5.8, p. 109), soon respond by evolving resistance. The evolutionary pattern has repeated with a series of insecticides, followed by the development of resistance genotypes in the flies, for which McKenzie has studied a number of cases. When the resistance mutation first appears, it produces developmental asymmetry as a by-product. Presumably, the disruption of development is deleterious, although not so deleterious that the mutation is selected against: the advantage in insecticide resistance more than compensates for a little developmental disruption. As a result, the mutation increases in frequency. Selection will then start to act at other loci, favoring genes there that reduce the new mutation's deleterious side effects while maintaining its advantageous main effect. That is, selection will make the new mutation fit in with the blowfly's developmental mechanism. The genes at the other loci that restore symmetrical development, while preserving the insecticide resistance, are called *modifier genes*, and the type of selection is called *canalizing selection*. Over time, in the sheep blowfly, the resistance mutation was modified such that it no longer disrupted development (Figure 13.6).

Figure 13.6 Developmental asymmetry in genotypes of the Australian sheep blowfly (*Lucilia cuprina*) that are, or are not, resistant to the insecticide malathion. (a) Developmental asymmetry in genotypes when the resistant gene *RMal* first appeared, soon after malathion was first used. The original, non-resistant genotype is +. *RMal* disrupts development, producing greater average asymmetry, and is selectively disadvantageous in the absence of malathion. (b) Developmental asymmetry of *RMal* flies after modifiers (*M*) have evolved to reduce the developmental disruption; it is now reduced near to the level of the original +/+ flies, and in the absence of malathion *RMal* has little selective disadvantage or is neutral relative to +. The sample size is 50 flies for each genotype. Reprinted, by permission of the publisher, from McKenzie and O'Farrell (1993).

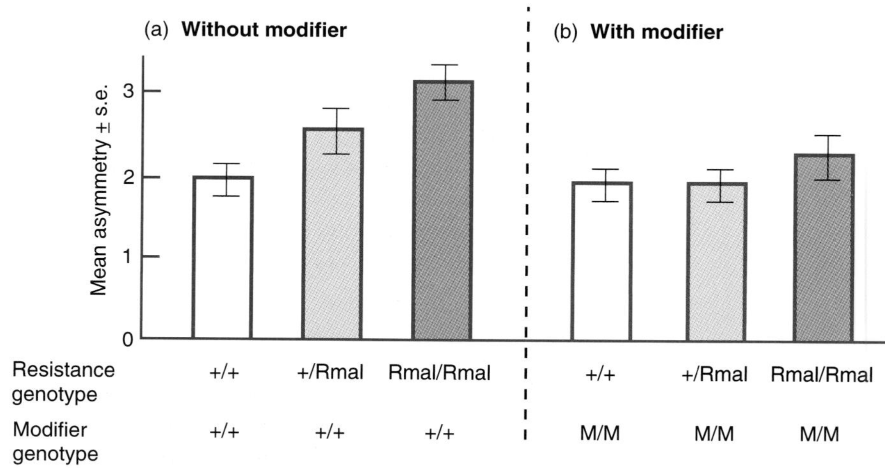

McKenzie was able to show that the modification was caused by genes at loci other than the mutation-carrying locus. (This distinction is important because, just as selection occurs at other loci to reduce the deleterious side effects of the mutation, so selection at that locus will favor other mutations that can produce insecticide resistance without harmful side effects.) It is probably common, given the extent of genetic interaction in development, for new mutations to disrupt the existing developmental pattern. Canalizing selection, which restores developmental regulation with the new mutation, is therefore likely to be an important evolutionary process.

Another sort of developmental constraint can be seen in the "quantum" growth mechanism of arthropods. Arthropods grow by molting their exoskeleton and then growing a new, larger one. They do not grow while the exoskeleton is hard. The arthropod growth curve shows a series of jumps, often with a fairly constant size ratio of 1.2–1.3 before and after the molt. Various models have been developed to show how body size can be adaptive. Body size, for example, influences thermoregulation, competitive power, and what food can be taken. None of these factors can plausibly explain the jumps in the arthropod growth curve, however. If, for example, the body size of an arthropod was adapted to the size of food items it fed on, it would hardly be likely that the distribution of sizes of food items in its environment created a selection pressure for quantum growth. The explanation for the quantum jumps is a developmental constraint. Growth by molting is dangerous, and growing with a smooth curve would require frequent risky molts. In this case, it is better to molt more rarely and grow in jumps.

Developmental constraints have been suggested as an alternative explanation to natural selection for two main natural phenomena. One is the persistence of fossil species for long periods of time without showing any change in form (section 20.6.2, p. 567). The other is the variety of forms to be found in the world. We can imagine plotting a *morphospace* for a particular set of phenotypes and then filling in the areas that are, and are not, represented in nature.

Raup's analysis of shell shapes is an elegant example of this technique. Raup found that shell shapes could be described in terms of three main variables: translation rate, expansion rate, and distance of generating curve from coiling axis (Figure 13.7). Any shell can be represented as a point in a three-dimensional space, and Raup plotted the regions in this space that are occupied by living shells (Figure 13.8).

Large parts of the shell morphospace in Figure 13.8 are not occupied. Two general hypotheses have been suggested to explain why these forms do not exist: natural selection and constraint. If natural selection is responsible, the empty parts of the morphospace are regions of maladaptation. When these shell types arise as mutations, they are selected against and eliminated. Alternatively, the empty parts could be regions of constraint: the mutations to produce these shells have never occurred. If the constraint was developmental, some reason would exist to make it developmentally impossible (or at least unlikely) for these kinds of shells to grow. The nonexistent shells would be embryological analogies for animals that disobey the law of gravity; that is, they are shells that break the (unknown) laws of embryology. The absence of these shells would no more be due to natural selection than is the absence of animals that break the law of gravity.

Figure 13.7 The shape of a shell can be described by three numbers. The translations rate (T) describes the rate at which the coil moves down the coiling axis. $T = 0$ for a flat planispiral shell, and is an increasingly positive number for increasingly elongated shells. The expansion rate (W) describes the rate at which the shell size increases; it can be measured by the ratio of the diameter of the shell at equivalent points in successive revolutions. $W = 2$ in the figure. The distance from the coiling axis (D) describes the tightness of the coil; it is the distance between the shell and the coiling axis, and in the figure it is half the diameter of the shell. See Figure 13.8 for many theoretically possible shell shapes with different values of T, W, and D. Reprinted, by permission of the publisher, from Raup (1966).

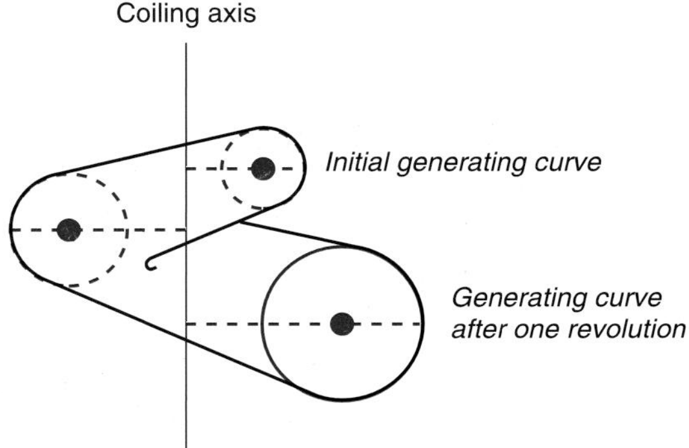

As well as being hypotheses to explain the absence of any form from nature, natural selection and constraint can both hypothetically explain the forms that are present. Faced with any form of organism, we can ask whether it exists because it is the only form that organism possibly could have (constraint), or whether selection has operated in the past among many genetic variants to favor the form we now observe. If the form of an organism is the only possible option, an analysis that treated it as an adaptation would be misdirected. In some cases we can be more certain that variation is strongly constrained than in others. If the constraint is the law of gravity, adaptation is a fanciful hypothesis. On the other hand, if the constraint is a conjectural piece of embryology, adaptation is worth investigating.

How can we test between selection and constraint? Maynard Smith and his eight co-authors listed four general possibilities: adaptive prediction, direct measure of selection, heritability of characters, and cross-species evidence.

The first test is the use of adaptive prediction. If a theory of shell adaptation predicted accurately and successfully the relation between shell form and environment—which forms should be present, and which absent, in various conditions—then, in the absence of an equally exact embryological theory, it would favor adaptation rather than developmental constraint. Conversely, a successful, exact embryological theory would be preferred to an empty adaptive theory.

The second test is a direct measure of selection. In the case of the shell morphospace, this test would involve somehow making the naturally nonexistent shells experimentally, and testing how selection then worked on them (section 11.2, p. 282). We would resort to observation to see whether negative selection works against these forms.

Third, we can measure the character's heritability. If a constraint prevents mutation in a character, it should not be genetically variable. Genetic variability

Figure 13.8 The three-dimensional cube describes a set of possible shell shapes. Around the outside of the figure, 14 possible shell shapes are illustrated; they were drawn by a computer. Only four regions in the cube are actually occupied by natural species: they are the regions *marked A, B, C,* and *D*. All other regions in the cube represent theoretically possible, but naturally unrealized shell shapes. The space is called a morphospace. Reprinted, by permission of the publisher, from Raup (1966).

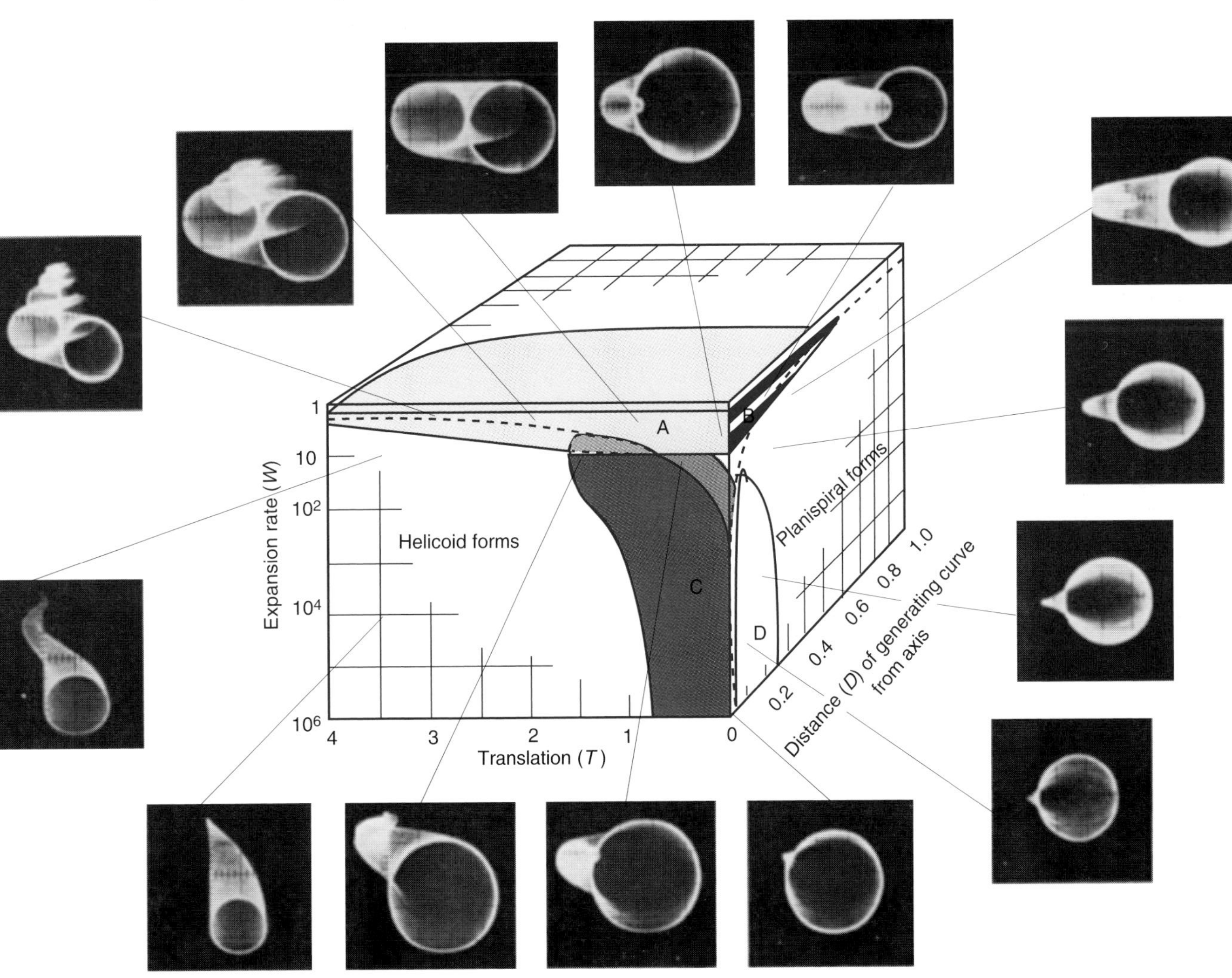

can be measured, and the constraint hypothesis refuted for any character that shows significant heritability. As it happens, this kind of evidence suggests that the gaps in the shell morphospace are not caused by developmental constraint. The heritability of a number of shell properties has been measured, and significant genetic variation found. Shell shape, therefore, is unconstrained to some extent.

Finally, cross-species evidence may be useful, particularly with pleiotropic developmental constraints. When more than one character is measured, and the values for the two characters in different organisms are plotted against one another, a relation is nearly always found. (This idea is true whether the different organisms are all members of the same species, or from different species.) The graphs have been plotted most often for body size and another character, and the relations are then called *allometric* (Darwin referred to it as the "correlation of growth"). Allometric relations are found almost every time that two aspects of size are plotted against one another graphically. A graph of brain size against body size for various species of vertebrates, for example, shows a positive relation. Such graphs function are two-dimensional morphospaces, and are analogous to Raup's more sophisticated analysis for shells.

The observed distribution of points might, once again, be due either to adaptation or to constraint. It might be adaptive for an animal with a large body to have a large brain. Alternatively, the size of an animal's brain might make no difference, and changes in brain size would simply be the correlated consequences of changes in body size (or vice versa). In that case, mutations altering one of the characters would be constrained to alter the other as well. Huxley was an influential early student of allometry, and he liked to explain allometric relations by the hypothesis of constraint: "whenever we find [allometric relationships], we are justified in concluding that the *relative size* of the horn, mandible, or other organ is automatically determined as a secondary result of a single common growth-mechanism, and *therefore is not of adaptive significance.* This provides us with a large new list of non-adaptive specific and generic characters" (Huxley 1932).

Some of these four classes of evidence are more persuasive than others. The allometric relations, in particular, do not provide strong evidence of developmental constraint. We can use the third kind of evidence (genetic variability) to see whether allometric relations are embryologically inevitable, or whether they can be altered by selection. Investigations have shown allometric relations to be as malleable as any other character.

Figure 13.9 illustrates an artificial selection experiment by Wilkinson on the rather strange Malaysian fly *Cyrtodiopsis dalmanni.* These flies have their eyes at the ends of long eye-stalks (Figure 13.9a, and Color Plate 5). The eye-stalks are particularly elongated in males and the character probably evolved by sexual selection. The important point here is that body and eye-stalk lengths are found to be correlated when they are measured in a number of individuals (Figure 13.9c). The ratio of eye span to body length in the natural population was 1.24 (that is not a misprint—the eye-stalks really are longer than the entire length of the body!). Wilkinson selected for increases or decreases in eye span relative to body length in two experimental lines and was able to alter the allometric relation in both directions (Figure 13.9). The allometric relation, therefore, is not a fixed law of embryonic development. Results such as those found by Wilkinson suggest

Figure 13.9 Artificial selection to alter the allometric shape of the stalk-eyed Malaysian fly *Cyrtodiopsis dalmanni*. (a) A silhouette of a fly, with arrows to indicate how eye span and body length were measured. (b) Results of one set of experiments, on males. Circles are experimental lines in which males with high ratios of eye span to body length were selected to breed; squares are experimental lines in which males with low ratios of eye span to body length were selected to breed; triangles are unselected control lines. Two replicates were done in each condition and they are distinguished by whether or not the symbol is filled in. (c) Another illustration of the allometric change in which there are four sets of points. The top two points are for males; the bottom two are for females. The filled-in symbols are individuals of the high line after 10 generations of selection for increased relative eye span; the open symbols are individuals of the low line after 10 generations of selection for decreased relative eye span; the dashed lines indicates the allometry in the unselected control line. The male points correspond to replicate 1 (open circle) in (b). Note the response to selection, showing allometric relations are changeable. The important change is in the *slope* of the lines in (c), which is more easily visible as a change in the ratio, in (b). From Wilkinson (1993). Reprinted with the permission of Cambridge University Press, © 1983.

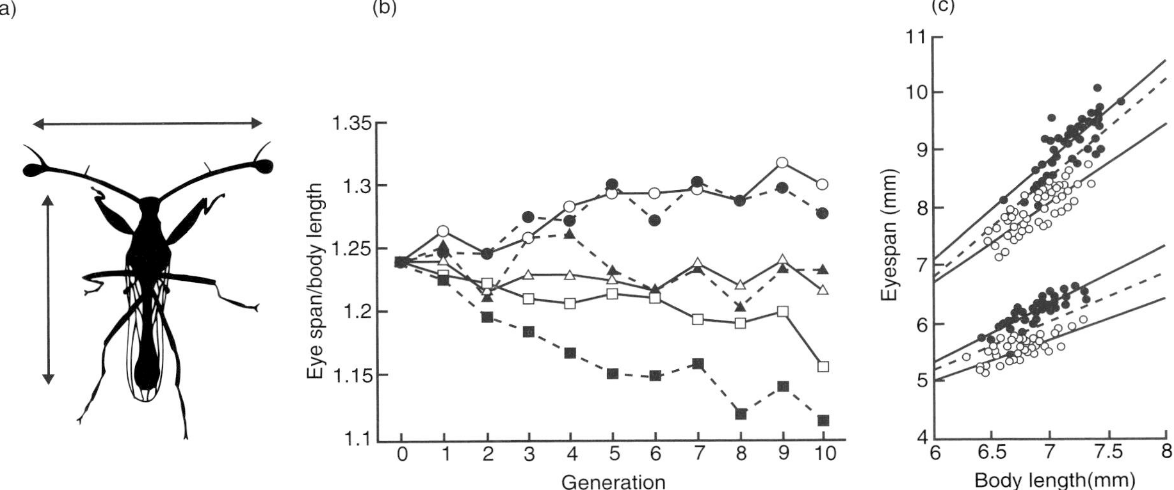

that allometric relations will have been tuned by natural selection in the past to establish a favorable shape in each species.

In conclusion, not much is known about how embryology constrains mutation, but the general idea is plausible. The way an organism develops will influence the mutations that can arise in some of its characters. The interesting problems begin when we try to move from this general claim to an exact demonstration in a real case; the attempts to date, as in the example of allometry, have not been finally convincing. In particular cases, we can test between the alternatives of selection and constraint.

13.10 *Historical constraints may cause adaptive imperfection*

As we said before, evolution by natural selection proceeds in small, local steps; each change must be advantageous in the short term. Unlike a human designer, natural selection cannot favor disadvantageous changes based on the knowledge that they will ultimately work out for the best. As Wright emphasized in his shifting balance model (section 8.13, p. 217), natural selection may climb to a local optimum, where the population may be trapped because no small change proves advantageous, although a large change could be. As we saw, selection itself (when considered in a fully multidimensioned context) or neutral drift, may lead the population away from local peaks. It may also have the opposite effect. Some natural populations now may be imperfectly adapted because the accidents of history pointed their ancestors in what would later become the wrong direction (Figure 13.10).

The recurrent laryngeal nerve provides an amazing example. Anatomically, the laryngeal nerve is the fourth vagus nerve, one of the cranial nerves. These nerves first evolved in fishlike ancestors. As Figure 13.11a shows, successive branches of the vagus nerve pass, in fish, behind the successive arterial arches that run through the gills. Each nerve takes a direct route from the brain to the gills. During evolution, the gill arches have been transformed; the sixth gill arch has evolved in mammals into the ductus arteriosus, which is anatomically near to the heart. The recurrent laryngeal nerve still follows the route behind the (now highly modified) "gill arch." In a modern mammal, therefore, the nerve passes from the brain, down the neck, round the dorsal aorta, and back up to the larynx (Figure 13.11b).

In humans, the detour looks absurd, but is only a distance of a foot or two. In modern giraffes, however, the nerve makes the same detour, but it passes all the way down and back up the full length of the giraffe's neck. The detour is almost certainly unnecessary and probably imposes a cost on the giraffe (because it has to grow more nerve than necessary and signals sent down the nerve will take more time and energy). Ancestrally, the direct route for the nerve was to pass posterior to the aorta. As the neck lengthened in the giraffe's evolutionary lineage, the nerve was led on a detour of increasing absurdity. If a mutant arose in which the nerve went directly from brain to larynx, it would probably be

Figure 13.10 Historical change in adaptive topography leaves a species stranded on a local peak. (a) Initially, a single optimum state exists for a character, and the population (x) evolves to that peak. (b) As the environment changes through time, the adaptive topography changes. The species has now reached the optimum. (c) The topography has changed, and a new global peak has arisen. The species is stuck at the local peak, because evolution to the global peak would traverse a valley. That is, natural selection does not favor evolution toward the global peak.

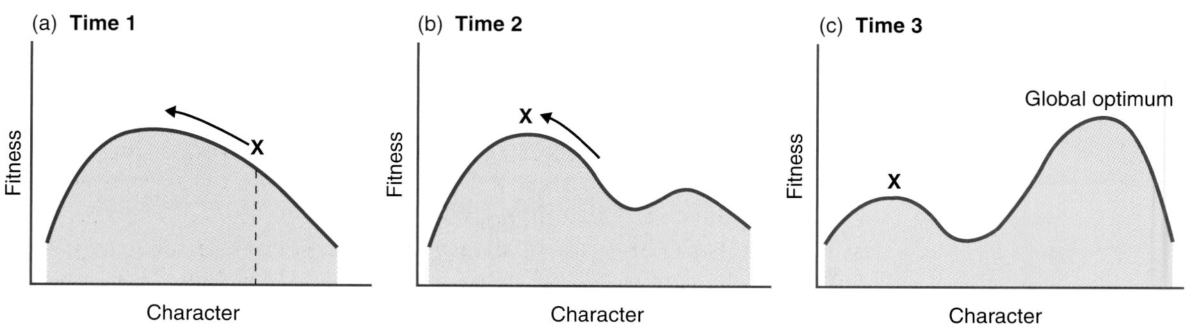

Figure 13.11 Evolution of the recurrent laryngeal nerve. (a) In fish, the vagus nerve sends direct branches between successive gill arches. (b) In mammals, the gill arches have evolved into a very different circulatory system. The descendant nerve of the fish's fourth vagus no passes from the brain, down to the heart (in the thorax) and back up to the larynx. Reprinted, by permission of the publisher, from Strickberger (1990), modified from De Beer.

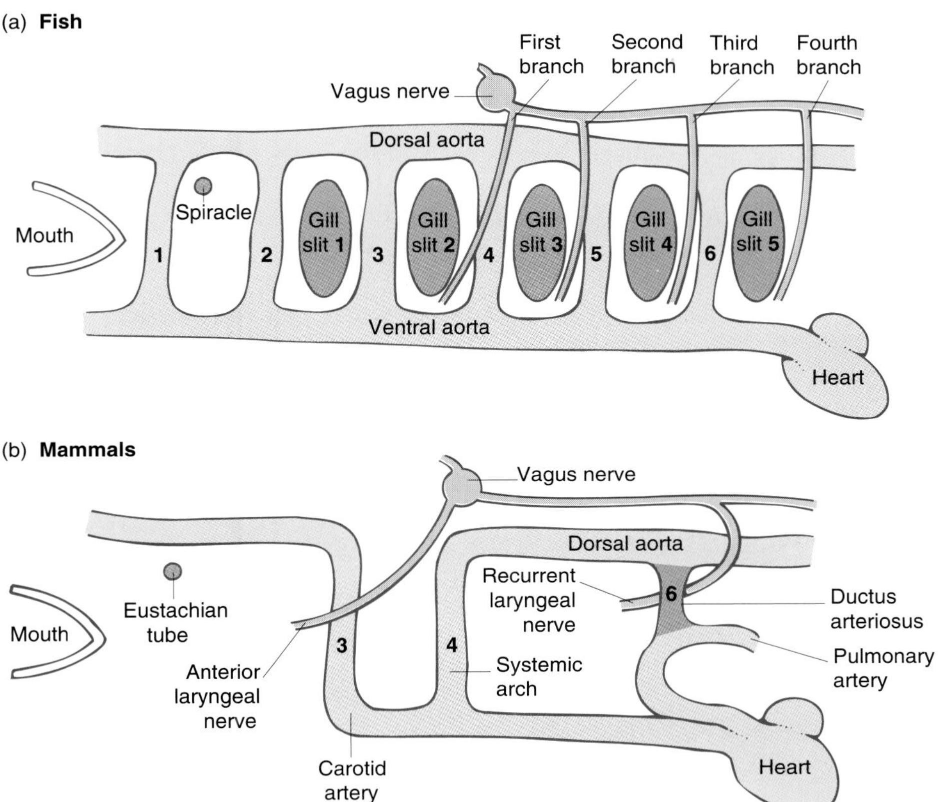

favored (although the mutation may be unlikely if it would require a major embryologic reorganization); the imperfection persists because such a mutation has not arisen (or it arose and was lost by chance). The imperfection arose because natural selection operates in the short term, with each step taking place as a modification of what is already present. This process can easily lead to imperfections due to historical constraint—although most will not be as dramatic as the giraffe's recurrent laryngeal nerve.

A similar historical contingency may produce not actual imperfection, but differences between populations or species that are not adaptively significant. In an adaptive topography with several adaptive peaks, more than peak of similar height may occur. The giraffe's laryngeal nerve looks like a case in which a local peak is clearly lower than the global peak, and it is, therefore, recognizably an imperfect adaptation. If several peaks of similar height existed, one would not be recognizably inferior to the others. Imagine now that the ancestors of a number of different populations started out near different future peaks. If they then experienced the same external force of selection, each one would still evolve to its nearest peak. The different populations would evolve different adaptations, because of their different starting conditions, rather than as a result of adaptation to different environments (Figure 13.12).

Kangaroos and a placental herbivore such as a gazelle have different methods of locomotion, and Maynard Smith suggested they are a possible example of such adaptation. The two forms are ecologically analogous. Kangaroo hopping is no better or worse a way of moving than running on four legs. The lineage leading to kangaroos improved one method of moving, while that leading to gazelles concentrated on the other. The difference is probably mainly a historical accident. If the argument is correct, the distant ancestors of kangaroos faced different selective conditions from those of gazelles; the adaptations fixed in those ancestors then influenced subsequent evolution such that the mutations influencing locomotion that are favored in the two groups are completely different, even though kangaroos and gazelles occupy similar ecological niches.

This example illustrates a different idea from the giraffes' recurrent laryngeal nerve. Neither kangaroo nor gazelle is claimed to be imperfectly adapted; the difference between the two may simply be a historical accident. In the giraffe lineage, a similar kind of historical accident has generated actual imperfection in its laryngeal nerve. Whether historical accident leads to imperfection, or a neutral difference between lineages, depends on whether a global peak remains a global peak during evolution or evolves into a local peak. In either case, past

Figure 13.12 Different starting conditions lead to two species occupying different, but equivalent, adaptive peaks. (a) The adaptive topographies for two species differ, with each species evolving to its own peak. (b) The adaptive topographies now change, until (c) they become identical for the two species. Each species remains on its own peak. At stage (b) the species difference was adaptive; it was better for species 1 to be on its peak, and species 2 at its own peak. By (c) the species difference is non-adaptive; either species would be equally well adapted on either peak.

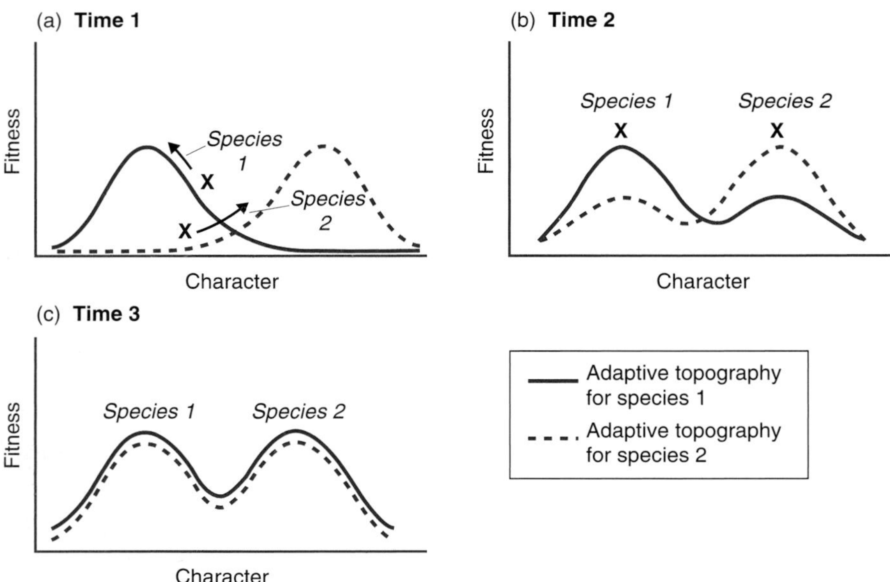

evolutionary events can lead to the establishment of forms that cannot be explained by a naive application of the theory of natural selection. Adaptation must be understood on a historical basis.

13.11 *An organism's design may be a trade-off between different adaptive needs*

Many organs are adapted to perform more than one function, and their adaptations for each function represent a compromise. If an organ is studied in isolation, as if it were an adaptation for only one of its functions, it may appear poorly designed.

Consider how the mouth is used for feeding and breathing in different groups of tetrapods (amphibians, reptiles, birds, mammals). In mammals, the nose and mouth are separated by a secondary palate, and the animal can chew and breathe at the same time. The earliest tetrapods, some modern reptiles, and all modern amphibians lack a secondary palate and have only a limited ability to eat and breathe simultaneously. A boa constrictor, for example, must stop breathing while it goes through the complex motions of swallowing its prey—a process that can take hours. The mouth of any such species that cannot breathe while it is feeding may, if it is judged only as an adaptation for feeding, appear inefficient compared with the mammalian system. The snake's mouth is simply a compromised adaptation for feeding. Of the reptilian groups, only crocodiles have a full secondary palate like mammals (it is presumably useful in crocodiles as it enables them to breathe air through the nose while the mouth is underwater), and reptilian feeding systems can be understood as compromised in varying degrees by the need to breathe.

Trade-offs do not exist only in organ systems. In behavior, an animal has to allocate its time between different activities, and the time allocated to foraging might, for example, be compromised by the need to spend time on other demands. Trade-offs exist over the whole lifetime as well. An individual's life history of survival and reproduction from birth to death is a trade-off between reproduction early in life and reproduction later on. At any one time, an animal may appear to be producing less offspring than it could. That characteristic does not mean the animal is poorly adapted, because it may be conserving its energies for extra reproduction later.

In summary, the adaptations of organisms represent a set of trade-offs between multiple functions, multiple activities, and the possibilities of the present and the future. If a character is viewed in isolation it will often seem poorly adapted. The correct standard for assessing an adaptation is, therefore, its contribution to the organism's fitness in all functions in which it is employed, through the whole of the organism's life.

13.12 *Conclusion: constraints on adaptation*

We can now draw two conclusions about the study of adaptation. First, we must realize the distinction between an adaptively insignificant difference between species and an imperfect adaptation within one species. The distinction corresponds to the two main ways of studying adaptation itself: evolutionary biologists are concerned with understanding both why different species have different adaptations and how adaptations function within each species. Let us consider how the different kinds of imperfection could upset the methods of analyzing adaptation that we discussed in chapter 11.

The comparative method could be misled by cases of adaptively insignificant differences between species. If the different forms of the adaptation are selectively neutral, or are equivalent locally adaptive peaks that different species evolved by historical accident, then attempts to correlate the differences with ecological circumstances should be unsuccessful. The fact that several equivalently good forms of an adaptation may exist does not, however, disturb the study of the character. The possibility of multiple adaptive forms should emerge from the analysis. If an optimal form of an enzyme exists, then it is no less an optimal form if 100 different amino acid sequences can realize it in practice. Instead, the problem for studies of particular adaptations within a species comes from the other source of imperfection. If the perfect form of the character has not arisen for reasons of history, embryology, or the genetic system, or because the environment has changed recently, then the character itself will be imperfectly adapted. If we try to predict the form of the character by an analysis of optimal adaptation, the prediction will be wrong.

What should the investigator do when a prediction proves wrong? An analysis made purely in terms of adaptation runs the risk of producing spurious results. Any particular character could have evolved as an adaptation for any of a large number of reasons. Body size, for example, may be adaptive for thermoregulation, storing food, subduing prey, fighting other members of the same species, or other factors. If we assume that body size is an adaptation, we should begin by picking one factor, then build a model and see whether it predicted body size correctly. If the model fails, we could move on to another factor. If, for instance, we started with a thermoregulatory model and it failed, we could move on to a diet model, and so on. This method, if carried far enough, will almost inevitably find a factor that "predicts" body size correctly. Eventually, by chance, a relation will be found, if enough other factors are studied. One will be found even if body size is a neutral character.

The solution to the problem—which is the chapter's second conclusion—can be stated in a conceptually valid, but not always practically useful, form. The methods of studying adaptation work well *if we are studying an adaptation.* If the character under study is an adaptation, then it must exist because of natural selection and it is correct to persist in looking for the particular reason why natural selection favors it. If body size is an adaptation, a correct adaptive model will exist. If the character (or different forms of it) is not favored by natural selection, however, the method breaks down. Methods of studying adaptation should, therefore, be confined to characters that are adaptive. In practice, they probably are. Adaptation can be a self-evident property of nature, and it would be absurd to claim that no properties of living things are adaptive. While research concentrates on obvious adaptations, it should be philosophically noncontroversial.

That conclusion leaves plenty of room for controversy, however. Biologists do not agree on how widespread, and how perfect, adaptations are in nature. Some biologists believe that natural selection has fine-tuned the details, and established the main forms, of organic diversity. Others think that the main forms may be historical accidents and the fine details due to neutral drift. Not surprisingly, evolutionary biologists who study adaptation tend to be among the former camp, and those who criticize it among the latter. This difference of opinion does not relate to the fundamental coherence of the methods, but focuses on the range of their application. History has—so far—been more on the side of the adaptationists. One character after another that had been written off as non-adaptive has turned out, following proper analysis, to be controlled by natural selection. Allometry is one case in point; snail shell banding is another (section 11.1, p. 281). The evidence of history, although it may encourage the study of adaptation, certainly does not prove that all characters of organisms should be assumed to be adaptations.

13.13 *How can we recognize adaptations?*

The study of adaptation could be made foolproof if we had a criterion to distinguish in advance between adaptive and non-adaptive characters. Criteria for recognizing adaptations do exist. Adaptations can be recognized as characters that appear too well fitted to their environments for the fit to have arisen by chance; they are characters that help their bearers to survive and reproduce; they are "purposive" and often complex; they are the sorts of characters that before Darwin would have suggested the existence of God. No one will doubt that some adaptations, like the vertebrate eye, fit this definition. Doubtful borderline cases can create problems, however. Small differences in a character in different species or individuals can be particularly troublesome to classify. For example, little doubt exists that the vertebrate brain is an adaptation, but it may make no adaptive difference whether an individual has a brain of 250 or 300 cm^3. The criteria given above will not tell us whether we should assume that such a difference is adaptive.

In theory, adaptation is a clear and objective concept. If a character is an adaptation, then natural selection will work against mutant alternatives. This criterion is theoretically objective, because a mutant either will or will not spread, but its use is mainly theoretical. Only rarely can we study the fate of mutants. The criterion can sometimes be applied in practice by measuring the fitness of variants of a character. If variation in a character is significantly related to an organism's survival and well-being, natural selection must automatically be operating on it; the favored forms of the character are adaptations.

When the method can be used, it is unambiguous. It is not always possible to use this method, however. Measuromg reproductive success in a character that does vary requires a great deal of effort. Moreover, some characters do not vary in an easily measurable way. The vertebrate eye is undoubtedly an adaptation, but no one has ever correlated variation in its optical properties with survival and reproductive success. A third problem is that a character could still be adaptive even if its relation with reproductive success is statistically undetectable. Natural selection can theoretically work on a character over millions of years and produce major changes through selection coefficients of 0.001 or less; it would be practically impossible to detect this amount of selection in a modern population with the normal resources of an evolutionary biologist. Forces that are important in evolution can, in some cases, be impossible to study directly because they are so small. A direct measurement of reproductive success is most likely to demonstrate that a character is adaptive if the selection coefficient is large, but such characters will tend to be "obvious" in any case. The method will be less useful for characters whose adaptive status is controversial.

If no detailed study of the relation between a character and fitness has been performed, or if one is not possible, then we must examine the character itself. The eye is a clearly adaptive: the criteria (beneficial, purposive, and so on) reveal this fact, without any need to measure fitness. The criteria are not foolproof, however, because of the borderline cases noted earlier. The ambiguity in these borderline cases is not just a practical problem. Rather, it is inherent in the concept of adaptation itself, because natural selection can favor simple as well as complex characters—and simple characters can also arise by chance.

We might make an analogy with the uncertainty in the definition of "design" in human fabrications. If we were to travel around the world and guess which objects were brought about by human design, we should see many obvious cases, such as architecture and engineered objects, and many nonobvious cases, such as heaps of earth. Earth could have been heaped up for a special purpose, such as for a burial mound, or it could have just accumulated there by natural accident. We should not expect the distinction between designed and nondesigned entities to be clear in either the case of natural adaptation or of human fabrications. Heaps of earth can be brought about by natural landslides or by human agency. Thus, we cannot always tell which cause operated just by looking at the result. An objective distinction can be made between the two causes, but it is historical: either the heaps of earth were constructed by human agency or they were not. The history is unobservable, however, and making the distinction purely using modern observable evidence will lead to difficult borderline cases.

Likewise, body coloration may be a simple adaptation, brought about by natural selection. Alternatively, it may be non-adaptive and brought about by chance, as may be the case for the red color of the sediment-dwelling worm *Tubifex* (visual factors are not important in the sediment at the bottom of the water column). Again, either natural selection favors the body coloration or it does not. If we try to decide whether which solution is correct just from looking at the character, the answer may not be clear. We have a clear theoretical concept of what an adaptation is; but that concept implies that adaptation cannot have a universal, foolproof, practical definition.

SUMMARY

1. Three theories have been put forward to explain the existence of adaptation: supernatural creation, Lamarckism, and natural selection. Only natural selection works as a scientific theory.
2. Natural selection is not the only process that causes evolution, but is the only process that causes adaptation.
3. Natural selection, at least in principle, can explain all known adaptations. Examples of coadaptation and useless incipient stages have been suggested, but they can be reconciled with the theory of natural selection. The vertebrate eye could have evolved rapidly by small advantageous steps.

4. Adaptation can be defined either historically or by current function. We can say either that an adaptation is a character that evolved by natural selection to perform its modern function, or one that evolved by natural selection regardless of whether its modern function is the same as the one it first evolved to perform.
5. Not all of the effects of an organ will have evolved as adaptations by natural selection. Some will be inevitable consequences of the laws of physics.
6. Adaptations cannot be simultaneously optimal for all levels of organization in life. What is optimal for the organism may not be optimal for its population.
7. Adaptations may be imperfect because of time lags. A species may be adapted to its past environment because it takes time for natural selection to operate.
8. Adaptations are imperfect because the mutations that would enable perfect adaptation have not arisen. The imperfections of living things are due to genetic, developmental, and historical constraints, and to trade-offs between competing demands.
9. For particular characters, adaptation and constraint can be alternative explanations. Likewise, differences in the form of a character between species may be due to adaptation to different conditions or to constraint. Forms that are not found in nature may be absent because they are selected against or because a constraint renders them impossible.
10. Adaptation and constraint can be tested between by several methods: the use of predictions from a hypothesis of adaptation or constraint; direct measures of selection; seeing whether the character is variable, and whether the variation is heritable and can be altered by artificial selection; and examination of comparative trends.
11. The methods of analyzing adaptation are valid when applied to adaptive characters and interspecific trends; they might be misleading for non-adaptive characters and trends.
12. Biologists disagree about how exact, and how widespread, adaptation is in nature.
13. Criteria to distinguish adaptive from non-adaptive characters have been developed. Measurement of selection provides an objective criterion, but is not always practical; other criteria, such as non-randomness and purposiveness are often useful, but may become subjective in borderline cases.

FURTHER READING

Williams (1966) is a classic work on adaptation. Gould and Lewontin (1979) is an influential paper that criticizes the way adaptation has often been studied; Cain (1964) vigorously argues the opposite. Reeve and Sherman (1992) is a stimulating paper about adaptation. Dawkins (1982, 1986, 1996) argues that only natural selection can explain adaptation; the 1986 and 1996 books are written for a wide audience. The natural theologian's argument from design was philosophically undermined by Hume (1779) in his *Dialogues Concerning Natural Religion*, which are in print in various paperback editions and (unlike some of Hume's other philosophical writings) readily intelligible. However, Hume's abstract argument did not convince people, and Darwin's mechanistic theory of natural selection historically toppled that long tradition of thought. See Simpson (1944, 1953) on orthogenesis. Wake and Roth (eds, 1989) discuss complex adaptations. Dawkins (1996) includes a popular account of Nilsson and Pelger's (1994) paper. For further material on eyes, see their references and Nilsson (1989). On preadaptation, see also the popular essay by Gould (1977b, ch. 12). Gould and Vrba (1982) is the reference for exaptation, on which see also Reeve and Sherman (1992); Pinker and Bloom (1990) interestingly discuss whether human language is exaptive or adaptive.

Dawkins (1982, ch. 3) and Krebs and Davies (1993, ch. 15) introduce constraints on perfection. See also Alexander (1985) and Williams (1992). On fruits, see Janzen and Martin (1982) in particular, and Janzen (1983) in general. Diamond (1990b), Cooper *et al.* (1993), and Givnish (1994) discuss some related possible ghost adaptations in New Zealand and Hawaiian plants, to the extinct moa and extinct flightless ducks and geese respectively. Macgregor (1991) reviews the remarkable genetic constraint in the crested newt and refers to earlier work. On developmental constraint, Maynard Smith *et al.* (1985) is the major review. McKenzie and Batterham (1994) discuss the insecticide resistance example (and see also the further reading in chapter 5). Macnair (1991) discusses why major gene changes are often involved in evolution of resistance to anthropogenic toxins; chapter 15 discusses the related example of resistance to heavy metal pollutants. There is a special issue of *Genetica*, vol. 89, pp. 1–317, 1993, on developmental stability; see also Leary and Allendorf (1989) and Hoffmann and Parsons (1991). Harvey and Pagel (1991) contains an account of, and references to, recent work on allometry, and Price and Langen (1992) discuss evolution of correlated characters. Chapter 9 has further references for canalizing selection, and chapter 21 for studies like Raup's on snails.

STUDY AND REVIEW QUESTIONS

1. What difficulties do theories of Lamarckian inheritance and directed mutation encounter when they are used as general theories of evolution, independent of (or in the absence of) natural selection?
2. Outline the main stages by which the verebrate or the octopus eye might have evolved, with successive stages showing improvements in the optical properties of the eye.
3. If natural selection is maintaining a character, against mutant alternatives, in a population, but the character has changed its function in the evolutionary historical past, is it an adaptation?
4. Consider a morphospace, such as the one for shell morphologies, or a brain-body size allometric graph. (a) There are regions in the space that contain no natural representatives. What are the two main theories to explain the absence of these forms? (b) How can we test whether a particular interspecies pattern in morphospace is caused by one theory or the other?
5. If organisms in a polygynous species produce a 50:50 sex ratio in their offspring, is it a perfect adaptation from the viewpoint of (a) the individual organism, and (b) the group of organisms? What general moral about the perfection of adaptation does the example illustrate?

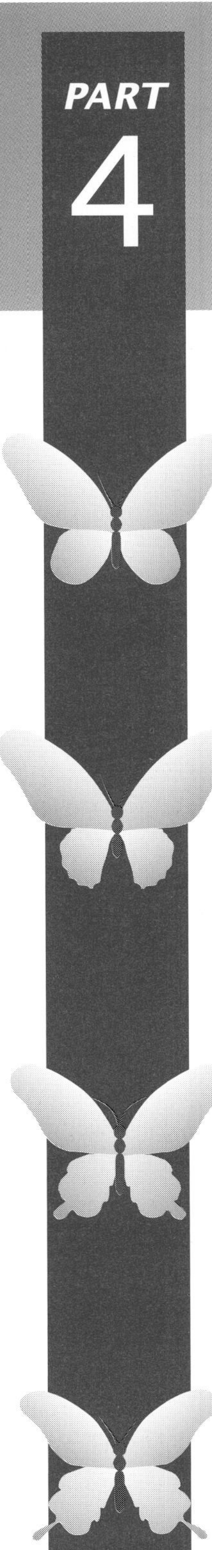

PART 4 Evolution and Diversity

Darwin closed *The Origin of Species* with the following words:

> There is grandeur in this view of life, with its several powers, having been originally breathed into a few forms or into one; and that, whilst this planet has gone cycling on according to the fixed law of gravity, from so simple a beginning endless forms most beautiful and most wonderful have been, and are being, evolved.

Part 4 of this book is about how the theory of evolution can be used to understand the diversity of life or, in Darwin's words, the "endless forms most beautiful." We begin with the principles of classification. Classification might seem a dry subject, but the abstract principles behind it raise deep theoretical issues. The chapter considers three main schools of classification and argues that classification is most objective when it represents only the branching relations (or phylogeny) of species and ignores how similar the species look to one another.

We then turn to the question of what a species is in chapter 15. In evolutionary biology, species can be understood as gene pools—sets of interbreeding organisms. These gene pools are important units because, in the theory of population genetics, natural selection adjusts the frequency of genes in such pools. The millions of species now inhabiting this planet have, as Darwin said, evolved from a common ancestor, and the multiplication in the number of species has been generated as single species have split into two. Speciation (chapter 16) requires special circumstances, and we shall consider both the geographical and genetic conditions in which it takes place.

Chapter 17 describes how the phylogenetic relations of species, and higher taxonomic groups, can be reconstructed. The history of species cannot be simply observed, and phylogenetic relations must be reconstructed from clues in the molecules, chromosomes, and morphology of modern species (and in the morphology alone of fossils). Phylogenetic reconstruction is a crucial part of classification if a classification aims to represent phylogenetic relations, and chapters 14 and 17, therefore, depend on one another.

Finally, the theory of speciation, classification, and phylogenetic reconstruction are all needed in evolutionary biogeography (chapter 18). This discipline involves using evolutionary theory to understand the geographical distribution of species.

CHAPTER **14**

Evolution and Classification

BIOLOGICAL classification is concerned with distinguishing and describing living and fossil species, and with arranging those species into a hierarchical, multilevel classification. The theory of evolution has a strong influence on classificatory procedures. This chapter will focus on classification above the species level; the next chapter (chapter 15) looks at classification of species. We begin here by looking at the two principles—phenetic and phylogenetic—that have been used to classify species hierarchically into groups (such as genera, families, and higher-level categories), and see how the three main schools of biological classification put them to use. We then examine the conditions in which the two principles give the same, or differing, classifications of a set of species. The main question of the chapter is which (if any) of the two principles and three schools is best justified by the evidence. The answer comes from an argument that phylogenetic classification at its best is objective, whereas phenetic classification always suffers from subjectivity. We then investigate some consequences of the strict use of phylogenetic relations to classify species into groups. We finish by considering why real evolution has resulted in a tree-like diverging pattern of relations among species.

14.1 *Biologists classify species into a hierarchy of groups*

Biologists have so far described approximately one million species of living plants and animals, and perhaps a further quarter of a million extinct fossil species. Estimates vary for the number of species that exist but have not yet been described—there may be between 10 and 35 million of them. Describing a species is a formalized activity, in which the taxonomist compares specimens from the new species and other, similar species, and then explains how the new species can be distinguished; the description also must be published.

Describing species is the most important task carried out by taxonomists, but it has no particular connexion with evolutionary biology. The evolutionary interest of classification begins at the next stage. Biologists do not think of their million or so described species simply as a long list, beginning with the aardvark, working through buttercup, honeybee, and starfish, to end with zebra. Since Linnaeus, species have been arranged in a hierarchy like the one we saw in Figure 3.5 (p. 46) for the wolf. Species are grouped in genera; the gray wolf species *Canis lupus* and the golden jackal *Canis aureus*, for example, are grouped in the genus *Canis*. Genera are, in turn, grouped into families. Thus, the genus containing dogs and wolves combines with several other genera, such as the fox genus *Vulpes*, to make up the family Canidae. Several families combine to make up an order (Carnivora, in this example), orders to make a class (Mammalia), classes to make up a phylum (Chordata), and phyla to create a kingdom (Animalia).

Each species, therefore, is a member of a genus, a family, an order, and so on. The problem of biological classification above the species level is how to group the species into these higher categories. The problem has both a practical and a theoretical side. Any number of practical problems can arise in deciding into which group to put a species (genus 1 or genus 2?) and what level particular groups should have (genus or family?). Even before these questions are considered, however, we must answer the question of what procedures should be used, and into what sort of hierarchy we should be trying to classify the species. If we take a million species and seek to arrange them into a classification, a large number of ways exist in which the arrangement could be made. A classification does not even have to be hierarchical. Chemists, for example, classify elements based on the periodic table, which is not hierarchical.

Why biological classification should be hierarchical is an interesting question in itself, and we shall consider it later in this chapter. Initially, however, we shall assume that the classification is hierarchical and ask what the exact form of the hierarchy is. The chapter will tackle is therefore about the theoretical question of what procedures classification ought to apply, rather than what happens when we try to apply the procedures, with greater or lesser facility, at the museum workbench.

14.2 *There are phenetic and phylogenetic principles of classification*

In biology, two main methods are used to classify species into groups: the *phenetic* and the *phylogenetic* methods. (Some biologists would prefer to substitute "phenotypic" for "phenetic" throughout this chapter.) The phenetic method groups species according to their observable phenetic attributes. If two species look more like each other than either resembles any other species, they will usually be grouped together in a phenetic classification. The full classification consists of a hierarchy of levels, such that the members of different groups at higher and higher levels have decreasingly similar appearances. A wolf and a dog (same genus) look phenetically more alike than do a wolf and a dolphin (same class).

In formal classification, phenetic similarity must be measured. Almost any observable attributes of organisms can be used for this purpose. Fossil vertebrates can be classified phenetically by the shape of their bones; modern species of fruitflies by the pattern of their wing venation; and birds by the shapes of their beaks or the color pattern of their feathers. Species can be grouped according to the number, shape, or banding pattern of chromosomes, by the immunological similarity of their proteins, or by any other measurable phenotypic property.

Nothing needs to be known about evolution to classify species phenetically. The species are grouped by their similarity with respect to observable attributes alone, and the same principle can be applied to any sets of objects, non-living or living, whether or not they were produced by an evolutionary process. Thus, this method could be applied to languages, furniture, clouds, songs, and styles of art and literature, as well as biological species.

The phylogenetic principle, however, is evolutionary. Only entities that have evolutionary relations can be classified phylogenetically. The clouds in the sky, for instance, cannot be classified phylogenetically (unless any of them were formed by the division of ancestral clouds). The phylogenetic principle classifies species according to how recently they share a common ancestor. Two species that share a more recent common ancestor will be put in a group at a lower level than two species sharing a more distant common ancestor. As the common ancestor of two species becomes ever more distant, the species are grouped even further apart in the classification. In the end, all species are contained in the inclusive phylogenetic category—the set of all living things—which contains all descendants of the most distant common ancestor of life.

In most real cases in biology, the phylogenetic and phenetic principles give the same classificatory groupings. If we consider how to classify a butterfly, a beetle, and a rhinoceros, the butterfly and beetle are more closely related both phenetically and phylogenetically (Figure 14.1a). The beetle and butterfly both

Figure 14.1 The phenetic and phylogenic principles of classification may (a) agree, or (b–c) disagree.

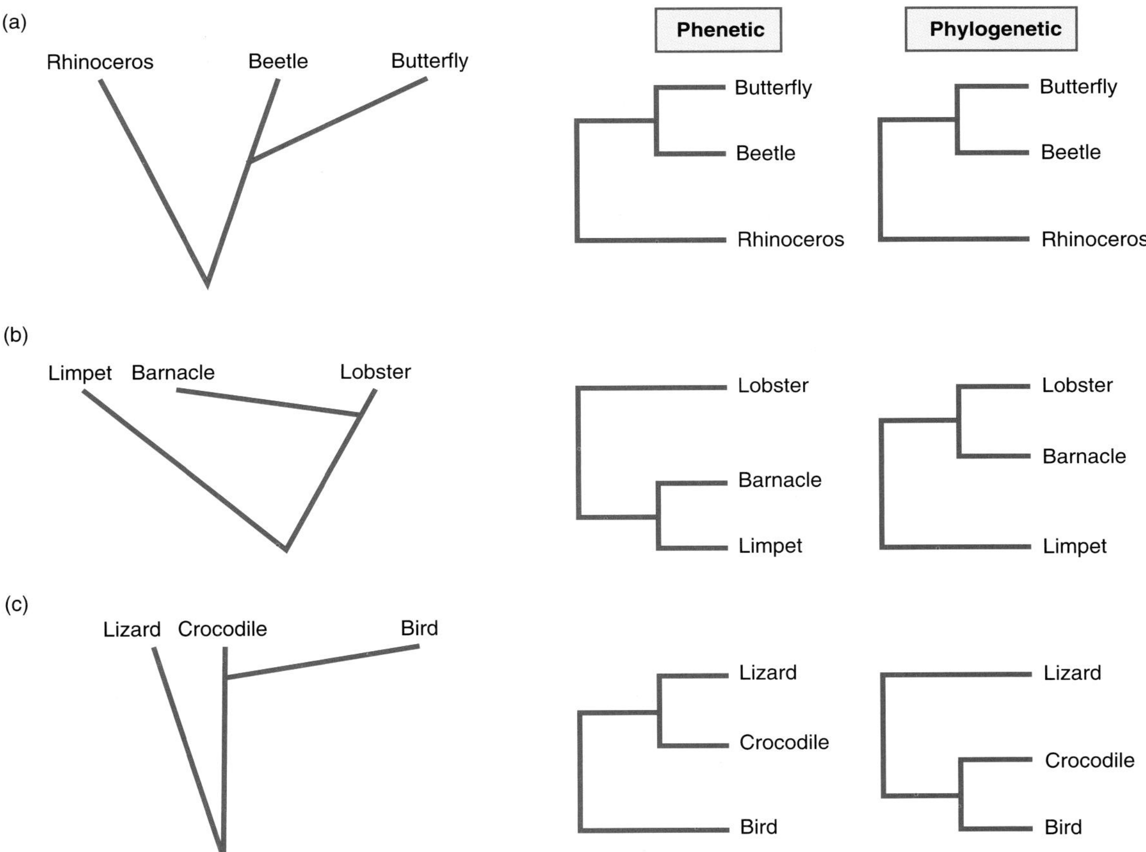

look phenetically more alike, and share a more recent common ancestor with one another than either does with the rhinoceros.

In other cases, the principles can disagree, for two main reasons. One reason is evolutionary convergence. Adult barnacles, for example, superficially resemble limpets. If we were to classify an adult barnacle, limpet, and lobster phenetically, we might well put the barnacle and limpet together even though the lobster and barnacle share a more recent common ancestor and are grouped together phylogenetically (Figure 14.1b). The other reason is illustrated by groups like reptiles. The phenetic and phylogenetic classifications of the reptilian groups differ because some descendants (such as birds) of the common ancestor of the group have evolved rapidly and left behind quite distantly related groups that resemble one another phenetically (Figure 14.1c). We shall discuss these two problematic cases later in this chapter. For now, the illustrations in Figure 14.1 are adequate to introduce the three main schools of classification.

14.3 *There are phenetic, cladistic, and evolutionary schools of classification*

The phenetic and phylogenetic principles are the two fundamental types of biological classification, but three main schools of thought exist about how classification should be carried out. This chapter will work toward an understanding of the three schools, and Table 14.1 (p. 381) summarizes their main technical properties.

The most influential modern school of phenetic classification is *numerical taxonomy*, which has been particularly defended by Sokal and Sneath. The terms phenetics, numerical phenetics, and numerical taxonomy are used almost interchangeably in modern biology.

Phylogenetic classification has been defended by the German entomologist Hennig and his followers. Hennig called it *phylogenetic systematics*, but *cladism* (from the Greek κλαδοζ for a branch) is now the more common term.

The third school uses a synthesis, or mixture, of phenetic and phylogenetic methods and is often called *evolutionary taxonomy*. In the reptilian example (Figure 14.1c), evolutionary taxonomy prefers the phenetic classification; in the barnacle example (Figure 14.1b), the phylogenetic. This school's best known advocates include Mayr, Simpson, and Dobzhansky.

14.4 *A method is needed to judge the merit of a school of classification*

How should we decide which school of classification, if any, is the best? To do so, we need a criterion against which to judge them, and we shall use the *objectivity* criterion for this purpose. An objective classification is one that represents a real, unambiguous property of nature; it is to be contrasted with subjective classification, in which the classification represents some property arbitrarily chosen by the taxonomist. You might arbitrarily choose, for instance, to classify species into one group if you discovered them on a Monday or Tuesday and another group if you discovered them between Wednesday and Friday. The classification would then be subjective because you would have no method of justifying the choice—except personal whim or convenience. If challenged about why you did not instead have one group for days beginning with the letter "T" and another group for all other days, you would have no principled argument with which to defend those classifications. The underlying classificatory principle—time of discovery—is ambiguous because it could be applied in a number of equally valid ways that would give differing classifications. It is also unreal because there is no inherent property shared by the organisms discovered on Monday and Tuesday and not shared by those discovered on Tuesday and Thursday. The objectivity test, therefore, asks whether a classificatory system has some compelling justification, external to the method it uses and the practitioners who practice it, for its method of partitioning objects.

A second distinction is between *natural* and *artificial classification*. To understand the difference, we must further distinguish between the characters of the organisms that are used to construct a classification and the characters that are not. In abstract terms, a natural classification then becomes one in which the members of a group resemble one another not only in the characters that define the group (as they must, by definition) but also for many other non-defining characters as well. An artificial classification is one in which the members of a group resemble each other only in the defining characters; they show no similarities for non-defining characters (Figure 14.2 gives an imaginary example). The advantage of natural classification lies in its ability to predict the distribution of other characters from the classificatory groupings alone.

Objective and natural classifications are preferable to subjective and artificial ones. If classification is objective, then rational people, working independently, should be able to agree that it offers a logical way to classify group members. The results should then be relatively stable and repeatable. Different, rational, independently working people would never agree on a stable classification if they all defined groups in some way related to the order in which the species were discovered. Likewise, if different taxonomists working on the same group are all satisfied with artificial classifications, they will probably all differ; conversely, if taxonomists all aim for natural classification, the results are more likely to be the same.

Now let us see how well the three classificatory schools meet the criteria of objectivity and naturalness. To carry out this examination, we will look in more detail at how the three schools actually operate.

14.5 *Phenetic classification uses distance measures and cluster statistics*

The modern forms of phenetic classification are numerical and multivariate, and they were developed in reaction to the uncertainties and imprecision of evolutionary classification. Evolutionary classification, whether the pure cladistic kind or the mixed evolutionary taxonomy of Mayr and others, requires a knowledge of phylogeny. We will discuss in detail in chapter 17 how phylogenies can

Figure 14.2 Natural and artificial classifications differ according to whether the members of a group tend to share characters that have not actually been used to define the group. (a) Natural classification. Suppose a group of species 1 and 2 have been defined by characters a_1 and b_1, and a group of species 3 and 4 by characters a_2 and b_2. The members of the groups share similar states in the other characters. (b) Artificial classification. The groups made up of species 1 and 2 and of species 3 and 4 are again defined by characters a_1 and b_1, and a_2 and b_2, respectively. In this case, the members of a group do not share the same states for other characters.

(a) **Natural classification**

Species	Characters and character states
1	a_1 b_1 c_1 d_1 e_1 f_1
2	a_1 b_1 c_1 d_1 e_1 f_2
3	a_2 b_2 c_2 d_2 e_2 f_3
4	a_2 b_2 c_2 d_2 e_2 f_4

(b) **Artificial classification**

Species	Characters and character states
1	a_1 b_1 c_1 d_1 e_1 f_1
2	a_1 b_1 c_2 d_2 e_1 f_2
3	a_2 b_2 c_1 d_1 e_2 f_3
4	a_2 b_2 c_2 d_2 e_1 f_4

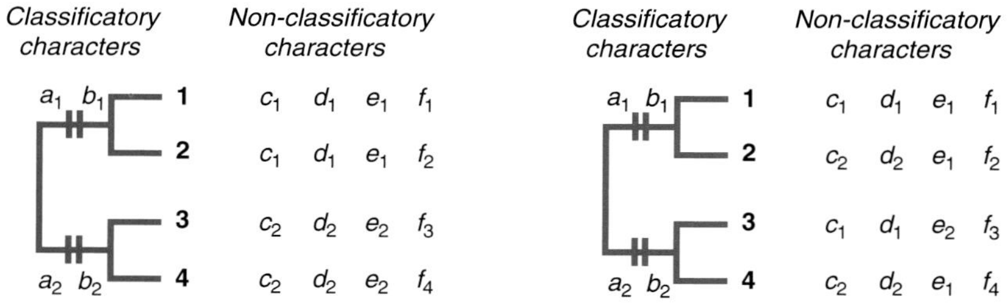

be inferred. Here, we need only know that, although the phylogenetic relations between species can often be inferred, the inferences can sometimes be highly uncertain. Phylogenetic knowledge is subject to change, as additional evidence emerges, and a classification of a group based on its phylogeny is liable to be unstable—not because the phylogeny itself is unstable but because our knowledge of it is constantly changing. For many groups of living things, little is known about their phylogeny, and an "evolutionary" classification of such a group will inevitably be poorly supported by evidence. Numerical phenetics aimed to avoid the evolutionary uncertainty by classifying only by phenetic relations, and by using quantitative techniques to measure those relations. The classification would follow automatically, and therefore (it was thought) objectively, from the phenetic measurements. Let us consider the methods in some more detail, and see how well these aims can be achieved.

The simplest kind of phenetic classification is defined by only one or two characters. We might classify the vertebrates, for example, by the number of their legs, to form groups with 0, 2, or 4 legs. Unfortunately, this procedure is likely to be subjective in the same way as classifying species by their order of discovery. Different individual characters show different distributions among species and, therefore, tend to produce different classifications. Consider the birds, and some reptilian groups, such as crocodiles, lizards, and turtles (Figure 14.3). Crocodiles are more similar to lizards and turtles than to birds if we look at their external surfaces, number of legs, and cold-blooded physiology; but crocodiles and birds have anatomically more similar skulls than either have

Figure 14.3 Character conflict in the phylogeny of birds and reptiles. The gait and the anatomy of the skull link crocodiles and birds; leg number, physiology, and external surface group link the reptilian groups. The anapsid skull has no openings, apart from the eye socket. The key feature of the diapsid skull is a single upper temporal opening, although most diapsids have an additional lower opening. Archosaurs and lepidosaurs differ in their skulls (to be exact, lepidosaurs lack a lower temporal arch).

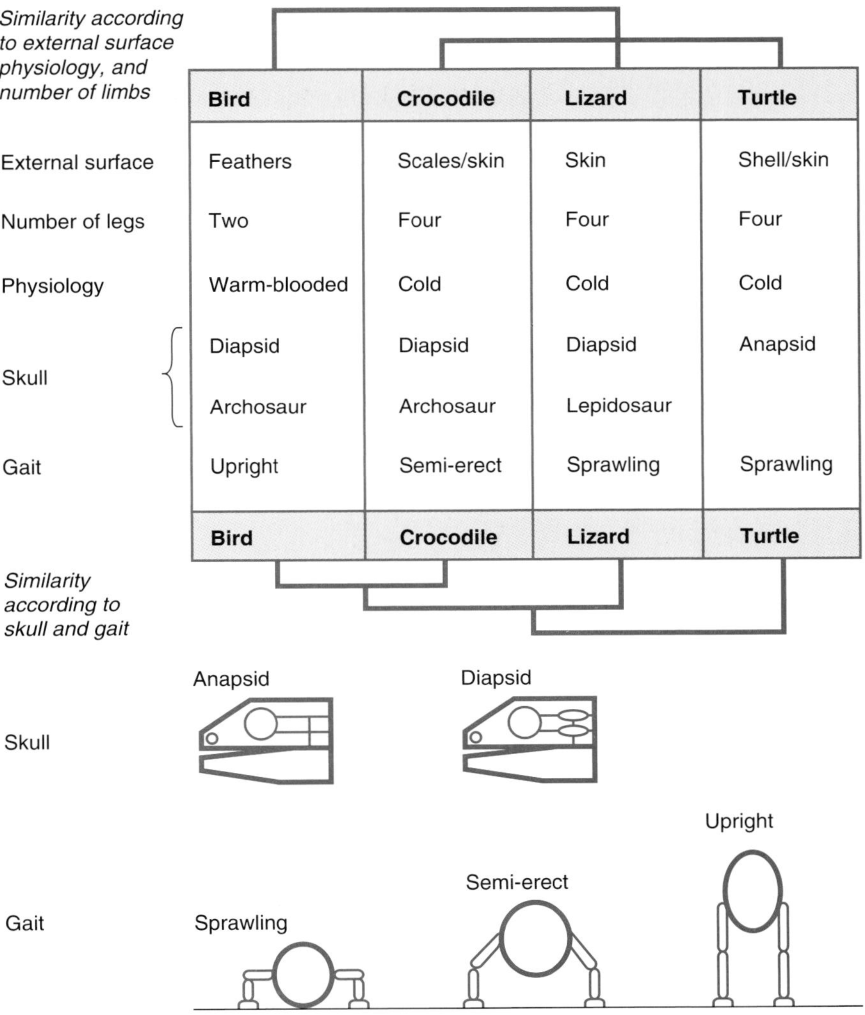

with lizards and turtles. The characters conflict. This discrepancy is a universal problem, not one confined to the vertebrates. A taxonomist working with one sample of characters will produce one classification, while one working with another sample will produce a second classification. As long as the principle of classifying with a small number of phenetic characters is followed, there is no way to decide which of the many classifications is the best.

The next step is to define the classification not by a few characters but by many. This technique became possible in the 1950s and 1960s as the statistical and computational apparatus became available for aggregating large numbers of phenetic measures into one grand measure of phenetic similarity. The aim of the numerical phenetic school was to measure so many characters that the idiosyncracies of particular samples should then disappear. The resulting classification groups the units according to their whole phenotype.

How do we aggregate a large number of measures into a single, combined measure of phenetic similarity? Several methods exist, and we can illustrate one of them in a graph. We shall start with the simple case of two characters (the extension to further dimensions is easy). Suppose that we wish to classify a group of fly species, and we have measured two characters, such as the length of a certain wing vein, and the length of the tibia of the hind-leg. The average for each species can be represented as a point (Figure 14.4). For any pair of species,

Figure 14.4 (a) The phenetic similarity between species can be expressed graphically. Suppose five species have been measured for two characters, length of a wing vein and length of tibia. The *x*-axis is the measurement of each species for length of a wing vein, and the *y*-axis is the measurement for length of tibia. The distance between two species on the graph is the phenetic distance between them. (Notice that the distance on the graph is different from the measure of mean character distance in Table 14.1.) (b) The phenetic classification by the nearest "nearest neighbor" technique puts species 3 with the group (cluster A) that has the nearest individual neighboring species (species 2). (c) Classification by the nearest "average neighbor" technique puts species 3 with the group (cluster B) that has the nearest average for all of its species. Species 4–5 have a nearer average distance. (The average is simply the average of the distances of species 1–2 and of species 4–5 from species 3.)

(a) **Phenetic measurements for five species**

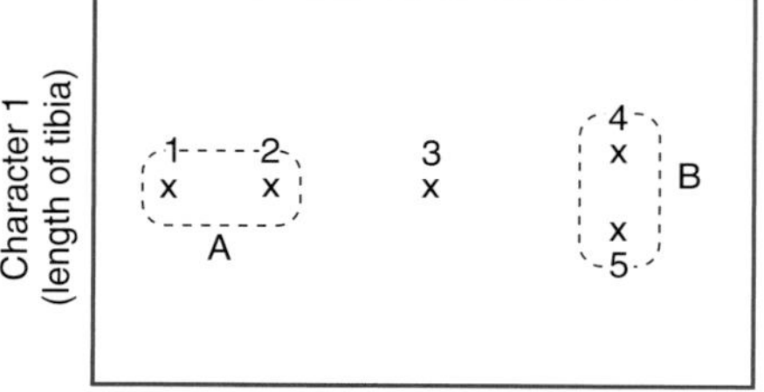

(b) **Nearest neighbor**

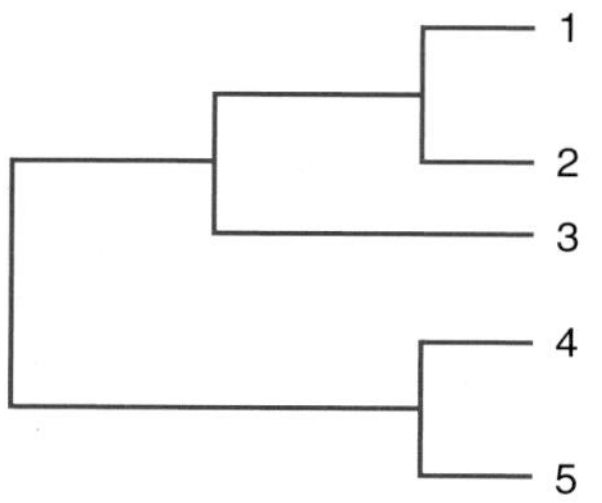

(c) **Average neighbor**

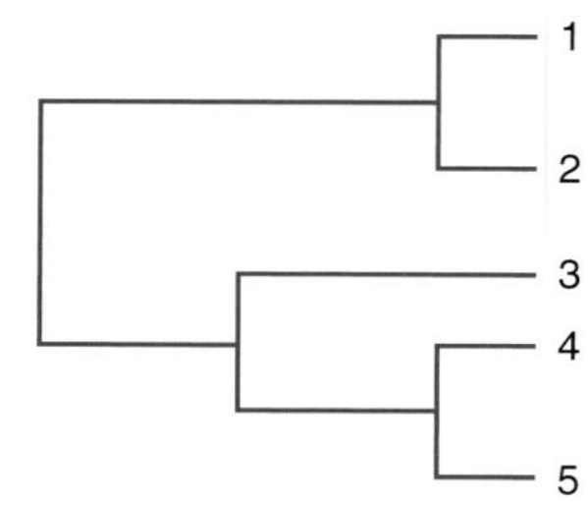

the average difference in their wing vein lengths is the distance between their points on the x-axis, and the difference between the lengths of their tibiae is the distance on the y-axis. If we used either character by itself, the classifications would differ; species 1 and 3, for instance, have identical tibial, but different wing vein, lengths. The aggregate difference for the two characters can be measured simply by the *distance* between the two species in the two-dimensional space. The species are then classified by putting each with the species, or group of species, to which it has the shortest distance.

If we measured a third character, such as pulse interval between the sounds in the courtship song, it could be drawn as a third dimension into the paper. Each species would then be represented by a point in the three-dimensional space, and the aggregate distance between the species could be measured, as before, by the distance between the species' points. We could likewise measure dozens of characters and measure the distance between the species by the appropriate line through hyperspace. Numerical taxonomists recommend measuring as many characters as possible—even hundreds—and making classifications according to the aggregate similarity for all of those characters. The more characters that are measured, the more likely it is that peculiar individual characters will be averaged out, and the better founded and more natural the classification will be.

Are phenetic classifications artificial or natural? The school of numerical taxonomy has pursued natural classification so far that it almost explodes the distinction. In the extreme case of a classification based on all possible characters, the classification would be both completely natural and completely artificial. The distinction under this system refers to the difference between defining and non-defining characters, and as more and more characters are used to define the groups, the number of non-defining characters decreases. However, for characters that were not used in the original classification, we have no reason to suppose they will fit in the same groups (see section 17.4, p. 468). A numerical phenetic classification is, therefore, natural in the sense that it is true of a very large number of characters; it must be, because it is defined by all characters for which it is true. On the other hand, the classification may well be artificial for other characters that have not yet been studied.

Is a numerical phenetic classification objective or subjective? Objective classifications, as you will remember, must represent some unambiguous property of nature. The phenetic classification itself represents the measure of aggregate morphological similarity for large numbers of characters. The question, then, is whether some property of nature—some hierarchy of "real" phenetic similarity—that the measurements of aggregate morphological similarity may reasonably be said to be representing.

We can begin by looking more closely at the statistical methods used in numerical taxonomy. Figure 14.4 included five species with two characters. To form Figure 14.4a, we grouped each species with its phenetically nearest neighbor. Two clusters—of species 1-2 and 4-5—immediately formed. To which of these clusters should we join species 3? The nearest species is 2. If we join species 3 to the cluster with the nearest neighbor (Figure 14.4a), we put it with cluster A (the nearest neighbor to species 3 in cluster A is species 2, whereas in cluster B species 4 and 5 are equally close). However, if we had calculated the average distance of each cluster as a whole, the answer is the opposite (Figure 14.4b); it follows from the geometry of Figure 14.4 that cluster B, rather than cluster A, has the nearer average neighbor to species 3.

The nearest neighbor and average neighbor methods are both examples of cluster statistics (other methods are used as well). Using these methods, we have, within the phenetic philosophy, managed to produce two different classifications. If the numerical phenetic claim to repeatability and stability is to be upheld, it must have some way of deciding which of the two is the correct phenetic classification—that is, the system must include some higher criterion to distinguish between possible classifications. The problem is that no such criterion exists. The higher criterion would presumably be a hierarchy of aggregate morphological similarity, but such a hierarchy does not exist in nature independently of the statistics that measure it. And—as Figure 14.4 shows—different statistics produce different hierarchies.

Thus, the phenetic philosophy incorporates an essential degree of subjectivity. If its classifications are to be consistent, it must pick on one statistic, such as the average neighbor statistic, and stick to it. Classification would then be repeatable, but at a price. The consistency does not follow from the phenetic system itself; it is imposed by the taxonomist—subjectively. In practice, numerical taxonomists have never been able to agree on which statistic to use, which partly explains why the school has lost much of its influence since its origin in the early 1960s.

Moreover, the choice of cluster statistic is not the only subjective choice in phenetic classification. The measurement of distance poses an analogous problem. The measure used in Figure 14.4 is *Euclidean distance,* or the straight line between two points; in two dimensions, this distance is measured by Pythagoras' theorem. Other distance measures also exist, such as *mean character distance* (MCD); MCD is the average distance between the groups for all characters measured. Thus, in two dimensions, if species 1 and 2 differ by x units in character A and y units in character B, then MCD $= (x + y)/2$ and Euclidean distance $= \sqrt{x^2 + y^2}$. The different measures of distance can give different hierarchies, once again forcing pheneticist to make a subjective choice.

Phenetic classification, therefore, even in its modern numerical form, is not objective. It can produce classifications, but the classifications lack a deep philosophical justification. Let us see how the introduction of evolution into classification can help with the problem. We will first consider purely phylogenetic, cladistic classification before we move on to examine the synthetic evolutionary school.

14.6 Phylogenetic classification uses inferred phylogenetic relations

14.6.1 Hennig's cladism classifies species by their phylogenetic branching relations

Phylogenetic classifications group species solely according to their most recent common ancestor. When a species splits during evolution it will usually form two descendant species, called *sister species*, and a cladistic classification groups these sister species together. The branching hierarchy of ancestral relations is a unique hierarchy, extending back to the beginning, and including all of life. The phylogenetic hierarchy is easily converted into a classification (Figure 14.5).

Figure 14.5 A simple relation exists between the phylogenetic (cladistic) classification of a group of species, and their phylogenetic tree. (a) The evolutionary history of seven species. (b) Their cladistic classification. (c) The formal Linnaean classification for species 5. This particular classification is only an example; depending on the detail in a particular case, different Linnaean levels might be used.

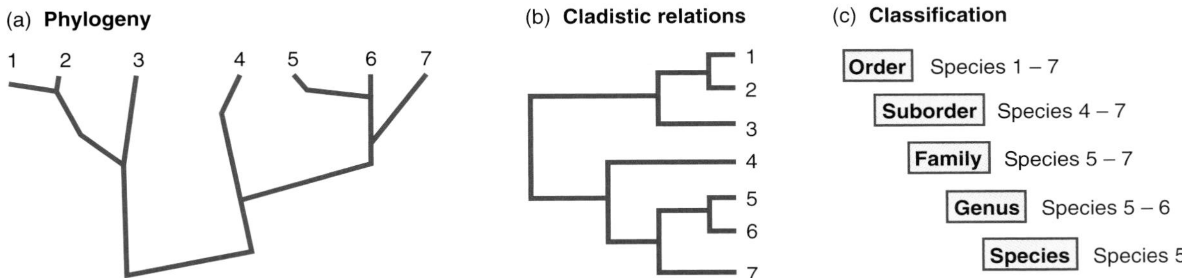

The advantage of the phylogenetic system should be apparent. The phylogenetic hierarchy exists independently of the methods we use to discover it, and it is unique and unambiguous in form. When different techniques for inferring phylogenetic relations disagree, we can always appeal to the external reference point. When we cannot work out the phylogeny of some group or other, we at least know that a solution exists and can aim for that goal. With the phenetic system, on the other hand, no such external solution is available, because there is no single, natural phenetic hierarchy analogous to the phylogenetic hierarchy.

When a pair of species, like 5 and 6 in Figure 14.5b, are classified together cladistically, it means that they share a more recent common ancestor than with any other species. Cladistic relations are fundamentally ancestral relations. In practice, the inference of ancestral relations (i.e., the phylogeny in Figure 14.5a) can prove difficult. Under this system, we need to know only that the inferences can be made, and the inferred phylogeny then becomes the starting point for phylogenetic classification.

We must, however, refer forward to make one important point (chapter 17 discusses it in detail). The main evidence for phylogenetic relations comes from a particular kind of characters called shared derived characters. The characters shared between species can be divided into three types (Figure 14.6 and Table 14.1). A first division is into *analogies* and *homologies*. A homology is a character

TABLE 14.1

Phenetic, evolutionary, and cladistic classification can be distinguished by the characters they use to defir groups, and the kinds of groups they recognize.

	Groups Recognized			Characters Used		
					Homologies	
Classification	Monophyletic	Paraphyletic	Polyphyletic	Analogies	Ancestral	Derived
Phenetic	Yes	Yes	Yes	Yes	Yes	Yes
Phylogenetic	Yes	No	No	No	No	Yes
Evolutionary	Yes	Yes	No	No	Yes	Yes

Figure 14.6 Different kinds of characters and taxonomic groups. Homologies are characters shared between species that were present in the common ancestor. They can be derived or ancestral. (a) Shared derived homologies are found in all descendants of the common ancestor, and are distributed in monophyletic groups. (b) Shared ancestral homologies are found in some, but not all, descendants of the common ancestor, and are distributed in paraphyletic groups. (c) Analogies (convergent characters) are characters shared between species that were not present in the common ancestor; analogies fall into polyphyletic groups. See Table 14.1 for the way the different characters are used by the different schools of classification. The crucial difference between paraphyletic and polyphyletic groups, in terms of the shaded gray zone, appears at the bottom, and not at the top, of the tree; the paraphyletic group contains the ancestor, whereas the polyphyletic group does not. The pattern at the top, in which the polyphyletic group seems to miss out a species between the included species, whereas the paraphyletic group seems to contain a set of contiguous species, is an accident of the way the pictures are drawn. By revolving appropriate nodes, it would be possible to make the species in the polyphyletic group contiguous or to intrude a gap in the paraphyletic group.

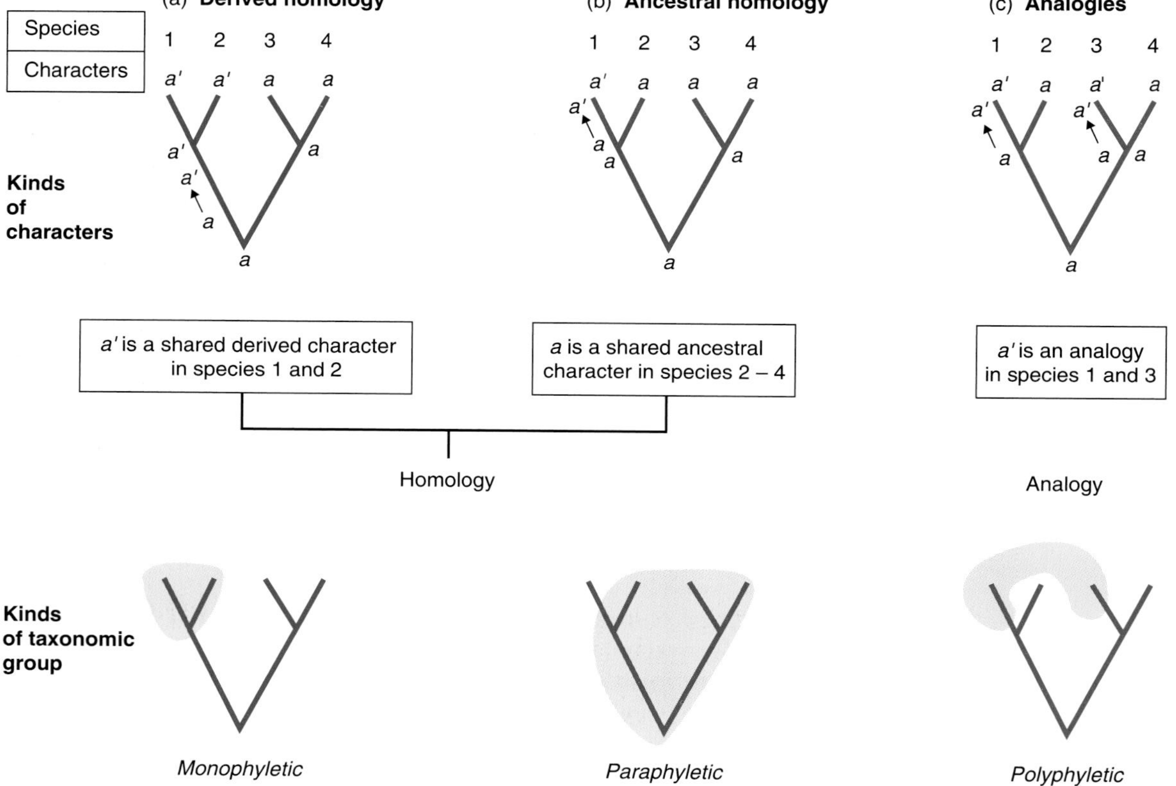

shared between species that was also present in their common ancestor; an analogy is a convergent character—one that is shared between species but that was not present in their common ancestor. (Note that this meaning differs from the pre-Darwinian meaning of chapter 3.) Homologies, in turn, are divided into shared derived homologies and shared ancestral homologies. A derived homology is unique to a particular group of species (and their ancestor), while a shared ancestral homology is found in the ancestor of a group of species but only in some of its descendants. Of these three kinds of shared characters, only the shared derived homologies indicate phylogenetic groups. Analogies indicate convergent groups and homologies indicate not only phylogenetic groups but a special kind of non-phylogenetic group as well.

The importance of this distinction between character types is to reveal the relation between cladistic and phenetic classification. Numerical phenetic classification groups species using as many characters as possible, and averaging them regardless of their evolutionary meaning. Cladists select, from all characters shared between species, one special class of characters—the shared derived characters—and use only those characters to group the species; all other kinds of characters are ignored in cladistic classification.

Are cladistic classifications natural? In a natural classification, the members of a group resemble one another in terms of both non-defining and defining characters. To answer the question, we should distinguish shared derived characters from the other kinds of characters (i.e., shared ancestral and convergent characters). The cladistic classification should be natural for all shared derived characters. The shared derived characters not formally used to define the classification should also fall into the same groups, because all evolutionary changes occur in the same phylogenetic tree. If two sets of shared derived characters seem to fall into contradictory groups, then at least one of them must have been mistakenly analyzed; correctly analyzed shared derived characters cannot contradict one another (section 17.7, p. 473). Shared ancestral and convergent characters need not show the same pattern, however, and the cladistic classification will, therefore, not be natural with respect to them. Cladistic classifications are natural for cladistic characters, but may be artificial for non-cladistic characters.

The cladistic philosophy is beguilingly simple, and it may be worth pausing to consider its implications. The complete shift from phenetic to phylogenetic classification produces results that, at first sight, can seem strange. Cladistic classifications are peculiar in two main ways: they exclude paraphyletic groups and they reclassify species even when scarcely any change of gene pool or phenetic appearance is observed. To understand the first point, we need to know some technical terms.

14.6.2 *Cladists distinguish monophyletic, paraphyletic, and polyphyletic groups*

The groups of cladistic classifications are *monophyletic* in the sense that they contain all of the descendants of a common ancestor—that is, each group has a common ancestor unique to itself (Figure 14.6a). Cladism rejects *paraphyletic* and *polyphyletic* groups. A paraphyletic group contains some, but not all, of the descendants from a common ancestor (Figure 14.6b). The included members are the forms that have changed little from the ancestral state; the excluded species are those that undergo more radical modifications. A paraphyletic group, therefore, contains the rump of conservative descendants from an ancestral species. Polyphyletic groups are formed when two lineages convergently evolve similar character states (Figure 14.6c). The key difference between paraphyletic and polyphyletic groups is that paraphyletic groups contain their common ancestor, whereas polyphyletic groups do not.

As Figure 14.6 shows, the different kinds of groups are defined by the different kinds of characters. Shared derived homologies fall into monophyletic groups;

shared ancestral homologies fall into both monophyletic and paraphyletic groups; analogies fall into polyphyletic groups. Paraphyletic groups are defined when some descendants of a common ancestor retain their ancestral characters, while others evolve new derived character states. The difference between paraphyletic and polyphyletic groups arises because ancestral characters would have been present in a group's common ancestor, but convergent characters would not.

Cladistic classifications include only monophyletic groups because these groups satisfy the requirement for having the unambiguous hierarchical arrangement of the phylogenetic tree. Only monophyletic groups are formed in the cladistic conversion of a phylogenetic tree into a classification (Figure 14.5). Such groups are defined unambiguously by their branching relations. They contain all of the branches below a given ancestor, but nothing need be said (or even known) about the phenetic evolution of species within each branch. To define paraphyletic and polyphyletic groups, however, we need to recognize phenetic similarity. We must decide which species to include and which to exclude, and such decisions are made by including only phenetically similar species in paraphyletic or polyphyletic groups. Because of the subjectivity of measures of phenetic similarity, the tree of life cannot be unambiguously divided into paraphyletic, or polyphyletic, groups. (Figure 14.10, p. 390, illustrates this point.)

Taxonomists had known about evolutionary convergence and polyphyletic groups for a long time before cladism, but paraphyletic groups proved a more insidious problem. Hennig was the first to recognize them clearly, but his work did not become widely known until it was translated into English in 1966.

We can see how paraphyletic groups tend to crop up by examining our earlier example, reptiles (see Figure 14.1c). The phylogenetic relations of the main tetrapod groups probably consist of those shown in Figure 14.7 (we return to this example in chapter 17). In conventional classifications, mammals, birds, and reptiles are given equal taxonomic rank as classes. To form these classes, two groups—mammals and birds—have independently undergone relatively rapid phenetic evolution and have come to have very different appearances from those of reptiles. The different reptilian lineages have changed more slowly and now look more like one another than like birds or mammals. Crocodiles and lizards, for instance, are cold-blooded, have scales and four legs, and walk with a reptilian gait; birds are hot-blooded, have feathers, two legs, and two wings, and fly. Yet crocodiles share a more recent common ancestor with birds than with lizards.

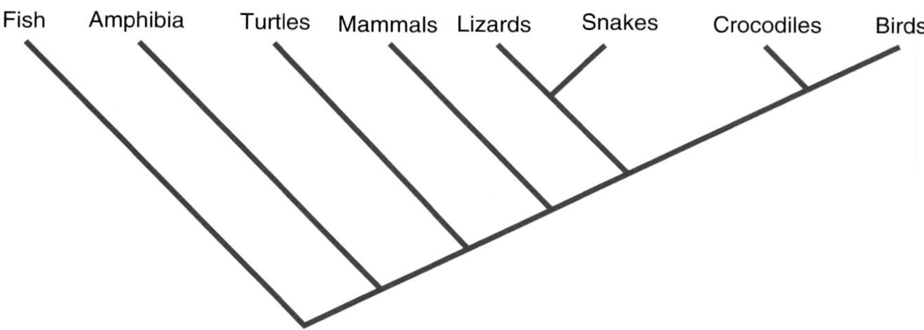

Figure 14.7 Phylogeny of main vertebrate groups. Reptiles are a paraphyletic group, made up of turtles, lizards, snakes, and crocodiles in this picture.

The characters of crocodiles and lizards (scales and so on) are ancestral for the group as a whole; the paraphyletic group was formed because ancestral characters were used for definition. Paraphyletic groups become a possibility whenever one or more subgroups have evolved relatively quickly and left their former relatives behind. If the cladistic philosophy is accepted, paraphyletic groups must be ruled out. They are defined phenetically (by shared ancestral characters) and their recognition is inevitably subjective.

The cladistic classification of the tetrapods is quite unfamiliar—even counter-intuitive. The Reptilia were recognized in almost every formal classification before cladism, but cladism rules them out. Many other examples of paraphyletic groups in non-cladistic classifications abound, including fish. The tetrapods evolved from one particular group of fish, the lobe-finned fish (section 13.4, p. 346). If we consider the relations of any tetrapod (such as a cow), any lobe-finned fish (such as a lungfish), and any ray-finned fish (such as salmon), the cow and the lungfish share a more recent common ancestor than do the lungfish and the salmon. The category "fish" (containing the lungfish and salmon, but excluding the cow) does not exist in cladism.

Some cladists have been rather fanatical in their insistence on ruling out Pisces and Reptilia. In a way, how we handle these cases matters little. Their paraphyletic status is well known, and can do little damage. In less well-known cases, it becomes more important to avoid paraphyletic groups. If a classification contains an unspecified mixture of monophyletic and paraphyletic groups, the evolutionary information in the classification becomes muddled. The beauty of a purely phylogenetic classification is that no doubt exists about the branching relations of the classificatory groups (Figure 14.5). If taxonomists define some relations phenetically and others phylogenetically, however, it is no longer possible to state the meaning of any relation with certainty. The branching relations are obscured and become lost.

Cladism has other controversial (and counter-intuitive) consequences besides its classificatory ban on paraphyletic groups. For instance, a group can change its name if it splits and one of the new lineages has hardly changed phenetically from its ancestor, or even not at all; this topic is discussed in the next chapter (section 15.3, p. 418). Another consequence arises when a new group originates by hybridization between two species, as occurs in hybrid speciation (sections 3.6 and 16.7, pp. 50 and 446); the shape of the phylogeny then does not imply a hierarchical classification of the Linnaean form. (Suggestions have been made about how to proceed in this case, but we shall not look into them here.) Yet another problem is how to reconcile the number of levels in a Linnaean classification with the number in a fully resolved phylogenetic tree. We will turn to that problem now.

14.6.3 *A strictly cladistic classification could theoretically have an impractically large number of levels*

If we could count all of the branch points between the earliest common ancestor of all modern species and the present, the number would be very high. It would be far higher, by orders of magnitude, than could be fitted conveniently into the Linnaean system. The Linnaean hierarchy has seven main levels (kingdom, phylum, class, order, family, genus, and species), and these levels can be multiplied by adding in super-, sub-, and infra-levels (e.g., superfamily, suborder, and so on; some are conventionally common, others are not) and extra levels, such as the tribe (between genus and family). Nevertheless, such a system clearly could not accommodate the thousands of levels that the complete cladistic hierarchy would require.

Figure 17.17 (p. 494) shows the phylogeny of the Hawaiian picture-wing fruitflies. A cladistic classification of one part of it—such as the *Drosophila adiostola* species group—would need the invention of several new levels (five in this case) in the Linnaean hierarchy between genus and species (Figure 14.8).

In practice, this problem is rarely acute, because we know very little about the phylogenetic relations of living things. The situation in the Hawaiian fruitflies is exceptional, while the crabs (Brachyura) are closer to the rule. Schram's *Crustacea* distinguishes about 50 families of crabs, but so little is known about their phylogeny that a classification can do little more than list the families alphabetically within about four tentative groupings. In the present state of our ignorance, only four taxonomic levels are needed: infraorder (Brachyura), three sections, two subsections, and 50 families. Only when we find out more about the relations of the families will extra levels be needed.

Figure 14.8 Possible classification of the *adiostola* group of species of fruitflies. (Taken from the larger phylogeny of Hawaiian *Drosophila* in chapter 17, Figure 17.8.) These 14 species are only a small part of the Hawaiian *Drosophila*, which are themselves only a (large) part of the worldwide *Drosophila* fauna. What we know about their phylogeny would require at least five new levels between the genus and species level.

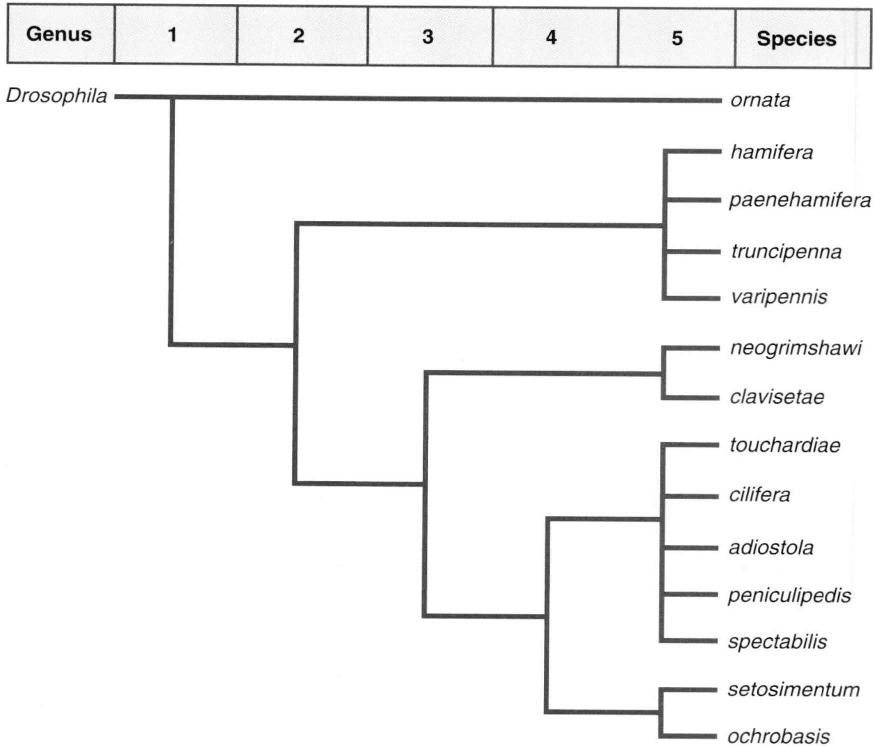

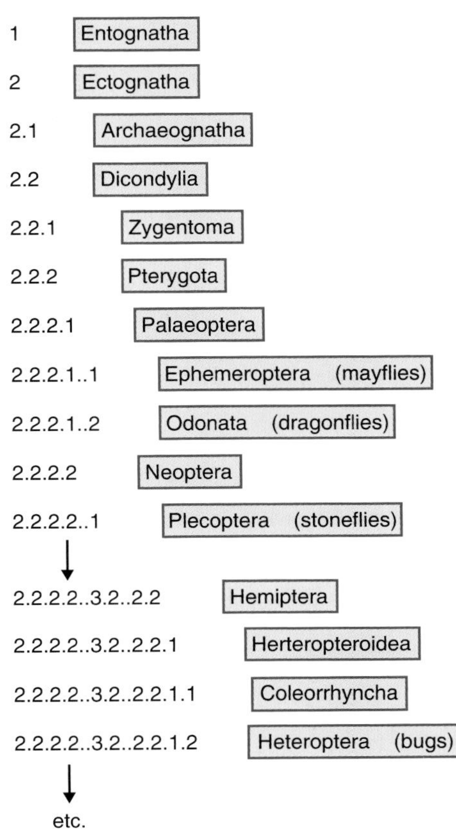

Figure 14.9 Part of Hennig's (1981) classification of the insects. The state of phylogenetic theory for the insects is too advanced for the limited number of Linnaean terms. Extra Linnaean terms could be invented, but Hennig instead classified each group by a series of numbers. The last number in the series identifies sister groups (e.g., 2.2.2.2..1 and 2.2.2.2..2, the mayflies and dragonflies, share a common ancestor). The order of splitting can then be traced back through the series of numbers. Only an illustrative part of Hennig's full scheme is shown here.

When we do know—or have some idea about—the fine branching of phylogeny, it is perfectly possible to devise a classificatory scheme to represent it. Hennig, for instance, in his final work on the phylogeny of the insects, produced at least hypotheses down to a fine branching level in several groups. He represented the phylogeny by a numerical scheme (Figure 14.9). All of the phylogenetic information in Figure 14.9 cannot be represented in a simple Linnaean system, although it can be represented partially, by collapsing several numerical levels into each rank in the Linnaean system. Linnaeus devised his system for approximately 10,000–20,000 named species. Roughly one million species have now been named, or 50–100 times the figure known in the late 18th century. The expansion of our knowledge requires an expanded classificatory scheme to accommodate it.

14.7 *Evolutionary classification is a synthesis of the phenetic and phylogenetic principles*

Evolutionary classifications incorporate both phenetic and phylogenetic elements. The school describes itself as synthetic, drawing on the advantages, and avoiding the shortcomings, of the two purer schools. However, for the same reason it is liable to be criticized for doing the opposite—for retaining the philosophical shortcomings of phenetic classification and adding to them the practical uncertainties of phylogenetic inference.

But first, what is evolutionary taxonomy? The school recognizes both paraphyletic and monophyletic groups. In terms of the kinds of characters used to infer phylogeny (Figure 14.6, and sections 17.5–17.7, p. 469), it forms groups by homologies rather than analogies, but it does not distinguish between ancestral and derived homologies. In Figure 14.1, evolutionary classification picks the phenetic classification in the reptilian case (Figure 14.1c) but the phylogenetic where convergence appears (Figure 14.1b). (In Figure 14.1a, evolutionary taxonomy makes no difference, and can be said to pick either one or the other or both.)

Is a classification defined by homologies natural or artificial? Homologies can be either ancestral or derived, and, as we saw, different shared derived homologies cannot (if correctly identified) contradict one another. Different ancestral homologies can, however, end up distributed in contradictory groupings. As a result, an evolutionary classification will be natural in so far as shared derived homologies were used to construct it and to test its naturalness. If it contains shared ancestral homologies, or if defining shared derived homologies are compared with non-defining shared ancestral homologies, the classification will be less natural. We have no reason to suppose that non-defining analogies will fall into the same groups as an evolutionary classification. An evolutionary classification is natural, therefore, in that it follows cladistic rules.

How can the evolutionary taxonomist's synthesis of phenetic and phylogenetic principles be justified? The evolutionary school predates both modern phenetics and cladism, and the original discussions of the school did not address the phenetic and cladistic arguments directly. Moreover, no complete modern evolutionary taxonomic defense against numerical and cladistic taxonomy exists. As such, the school differs from the two competing schools, which were conceived partly in opposition to evolutionary taxonomy and made their objections to it clear. Nevertheless, a case can be made to justify this school's veracity.

Evolutionary taxonomists disagree with phenetic classification for much the same reason as we discussed earlier, although they express the argument differently. They criticize phenetic systems for being *idealistic*—that is, for supposing that a phenetic classification represents some "ideal" phenetic relationship between species. The ideal relationship would be some "idea" or "plan" in nature. An example of this system might be the pre-Darwinian theory that classifications represent the thoughts of God. The idea according to which the idealist tries to classify then becomes an idea existing in the mind of God but manifesting itself in (and inferrable from) His creations. A modern scientist finds it difficult to make much sense of these old arguments, but "divine" taxonomy at least provides a concrete example of what idealism means. Other versions of idealism supposed that it was possible to deduce the existence of fundamental forms or plans of nature from a purely scientific analysis of species' morphology.

Notice that idealism could, in principle, solve the problem of subjectivity in phenetic classification. The plan of nature would provide an objective, external reference point for the phenetic classification. The only snag is that no plan of nature, in the idealist sense, exists. That absence prompted modern numerical phenetics to drop the idealist philosophy of earlier phenetic classifications.

Section 14.5 argued that, in the absence of any natural hierarchy of phenetic similarity, phenetic classification lapses into subjectivity. The idealist error represents the other side of the same coin. Idealists believe that a phenetic hierarchy exists, but offer no valid reason for their belief. When evolutionary taxonomists criticized phenetic classification for committing the error of idealism, they meant that no real phenetic hierarchy exists in nature, and a system that assumes such a hierarchy will be fundamentally subjective. Phenetic classifications try to group species according to a relationship—the ideal morphological system—that evolution does not produce. Evolution does not produce one particular privileged phenetic hierarchy that is more real than all other phenetic hierarchies.

Phenetic idealism can be avoided if taxonomists represent evolution, not phenetic similarity. But what does "evolution" mean? Evolutionary taxonomists exclude polyphyletic, but not paraphyletic, groups from classification. How do they interpret evolution to allow both paraphyletic and monophyletic groups? To see why, we must look at what evolutionary taxonomists say about cladism. Evolutionary taxonomists criticize cladism for its unnecessary puritanism. Cladism, as we have seen, leads to what might appear to be bizarre conclusions, such as the destruction of the Reptilia and the renaming of apparently unchanged species, because of its distinction between paraphyletic and monophyletic groups. If both kinds of group are allowed in classification, the cladistic novelties do not arise. Evolutionary classification duly allows paraphyletic groups, and thus seeks to avoid the apparent perversity of cladism.

That is how evolutionary classification sees itself. But cladists and pheneticists see the situation rather differently than evolutionary taxonomists. According to a cladist, the argument that evolutionary taxonomists accept against the pheneticist's polyphyletic groups works just as well against paraphyletic groups. Accepting paraphyletic groups requires accepting polyphyletic ones as well, or the system becomes inconsistent. Paraphyletic groups, like polyphyletic groups, are defined phenetically. If some of the descendants of a common ancestor will be excluded from a group, a decision must be made as to how many such descendants will be left out; according to cladists, this decision is phenetic and arbitrary (Figure 14.10). Paraphyletic groups are not formed by phylogenetic relations—as Figures 14.5 and 14.6 illustrated earlier. The standard phenetic problem then becomes apparent. According to some measures of phenetic similarity, one paraphyletic group will seem appropriate; according to another measure, a different group will be chosen. The choice between the groups is subjective—or idealist. When paraphyletic groups are admitted, the argument against phenetic classification is lost. If phenetic criteria can be used in the case of paraphyletic groups, why not for polyphyletic ones? Paraphyletic groups presuppose idealism just as much as polyphyletic groups. (Box 14.1 discusses a method by which we might objectively define non-monophyletic groups.)

Evolutionary classification presents another problem. Classifications that mix more than one type of information are less descriptive than classifications that represent only a single property. Consider an analogy with a library card index system. Suppose that a diligent librarian had read all of the books in the library and devised a coded system for abstracting the facts they contain. Imagine also that this coding system resembled closely the shelf mark codes. An evolutionary classification is like a card index that gives, for each book, a coded reference

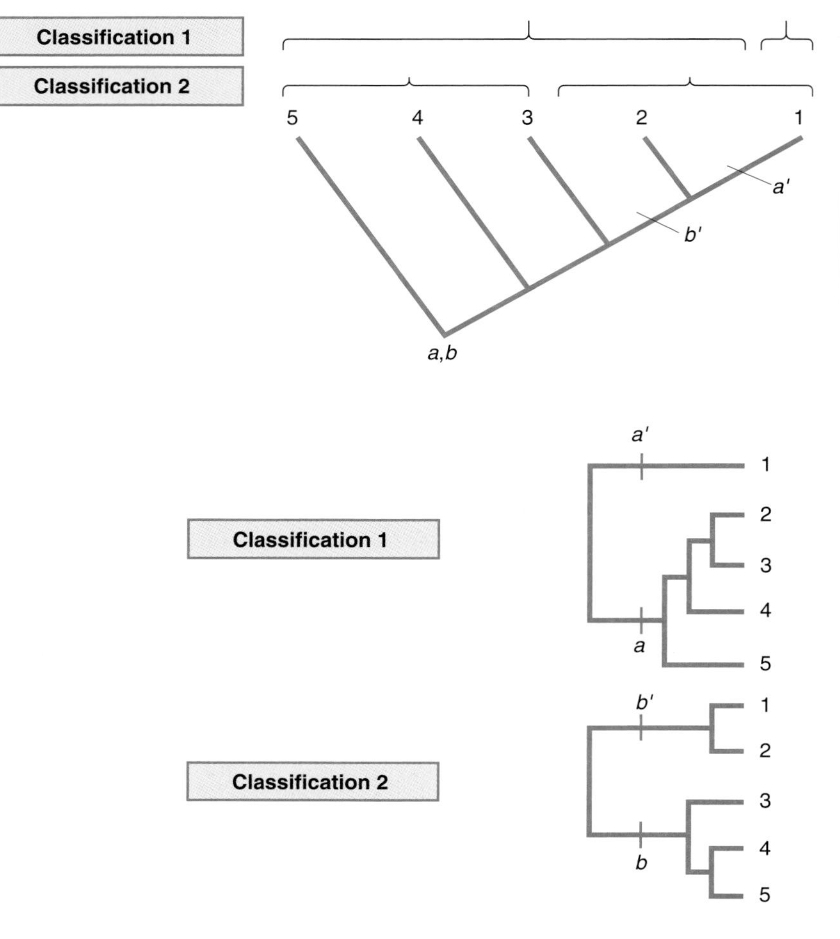

Figure 14.10 Paraphyletic groups contain some, but not all, descendants of a common ancestor. When they are defined, a decision must be made about which descendants to exclude. The decision is phenetic. In the figure, *a*, *a*′, *b*, and *b*′ are character states; *a*′ and *b*′ are derived from *a* and *b*, respectively. The question is whether to exclude species 1 (and define the paraphyletic group of 2–5 by the ancestral character *a*) or species 1–2 (and define the group 3–5 by ancestral character *b*). Notice there is no guarantee that shared ancestral characters can define such paraphyletic groups as 2–5 and 3–5, because *a* or *b* could, in principle, have undergone any number of changes within the branches leading to species 2–5 and 3–5.

number, but does not tell you whether the numbers represent the shelf mark or the abstracting code. The card index reference code itself tells you neither where to find the book on the shelf, nor what the book's contents are. Given the reference code, further work is mandatory before it may be interpreted. You would probably disagree if the librarian explained that, by adding the subject codes, the card indexes had been made more informative. While it is true that more information has been included, the cards are not necessarily more informative.

Likewise, an evolutionary classification is constructed with both phenetic and phylogenetic information. If you are given only the classification, you do not know which groups are phenetic and which phylogenetic. Classification by a pure system does not have this drawback.

Evolutionary taxonomy mixes phenetic and phylogenetic methods, but in a consistent and principled manner. It defines groups by homologies and excludes analogies. As a result, the classification system does not recognize polyphyletic groups, but it allows phenetic groups such as reptiles and fish. Such classifications in biology have a strong practical purpose, and a case can be made not to disband

BOX 14.1

Adaptationist Classification

When evolutionary taxonomists define paraphyletic groups, the decision about which groups (like birds and mammals) to split off from paraphyletic rumps (like the reptiles) is most easily understood as a phenetic decision. In such a case, the general criticism of phenetic classification applies. If this decision is not made on purely phenetic grounds, evolutionary taxonomy may escape that criticism. One rather speculative criterion has been suggested that comes from the idea of adaptation.

Birds do not differ from reptiles only phenetically. Birds have evolved a set of adaptations for flying. Thus, they have achieved what is called an *adaptive breakthrough*, in the sense that birds have "broken through" during their evolution to a different way of life from reptiles. Adaptation is a more promising concept than phenetic difference for an objective system of classification. Adaptations fit ecological niches, and ecological niches exist — outside the classificatory system — in nature. If some method could be found to measure adaptive differences, it might also prove useful in classification. Groups could then be defined by shared adaptations, rather than shared ancestry or phenetic similarity. Fish might then be defined as "the group of vertebrates with adaptations for swimming." (Some other criterion would have to be added to exclude swimming reptiles and mammals, but if some definition along these lines is possible, it could be used in adaptationist classification.)

At present this idea remains simply an interesting concept. To make it more concrete, we would need some method of measuring (and distinguishing) adaptive similarity. No formal method has been proposed, and it might turn out to be impossible. We should note, however, that the abstract proposal exists, and that it is distinct from phenetic and phylogenetic classification. It can be called *adaptationist classification*.

The use of shared adaptations to define groups has been most often proposed by evolutionary taxonomists, such as Mayr. He suggested it could be used to define paraphyletic higher taxonomic groups, such as birds. "In the history of the vertebrates [Mayr has written] we know many such cases of the formulation of new grades, such as the sharks, the bony fishes, the amphibians, reptiles, birds, and mammals." (A *grade*, in Mayr's words, is a classificatory group "characterized [= defined] by a well-integrated adaptive complex.") By itself, this argument is not an effective defense—it works just as well for the polyphyletic groups of phenetic classification as for the paraphyletic groups of the evolutionary taxonomist.

Convergence is usually caused by adaptation. For example, sharks and dolphins have converged on the same hydrodynamic adaptations for the same environment. Classification by shared adaptations, if it can be made practical, is as likely to end up justifying a future version of phenetic, as of evolutionary, classification.

these long-established groups as long as a convincing reason can be found for keeping them. However, it is questionable whether the evolutionary taxonomist's reason is convincing enough.

14.8 *The principle of divergence explains why phylogeny is hierarchical*

All three schools of classification—phenetic, cladistic, and evolutionary—aim at constructing a hierarchical classification. In the case of cladism, that goal is unsurprising. The phylogenetic tree is a hierarchy, so phylogenetic classification will be hierarchical as well. It is less obvious whether a phenetic classification must be hierarchical. An infinite number of phenetic patterns are found in nature. Some are nested hierarchies, but others are overlapping hierarchies or non-hierarchical networks. If we aim at a phenetic classification, we have no strong reason to classify hierarchically. Biological classifications are hierarchical because evolution has produced a tree-like, diverging, hierarchical pattern of similarities among living things.

Why should this pattern emerge? The question has an important place in the history of Darwin's thinking. Darwin thought up natural selection in the late 1830s, as a natural explanation for adaptation and evolution. As environments change, and competing species change, species will evolve new adaptations. By itself, this theory does not account for the tree-like, divergent course of evolution. Darwin was well aware that evolution had steered such a course. After all, the hierarchical structure, which places groups within groups in classification, had been established as fact in the early nineteenth century—by (among others) Geoffroy St. Hilaire's morphological, and Milne-Edwards's embryological, work. So striking a fact had to be fitted into the theory. As Darwin recalled in his autobiography, in the early (1844) version of his theory:

> I overlooked one problem of great importance, and it is astonishing to me how I could have overlooked it and its solution. This problem is the tendency in organic beings descended from the same stock to diverge in character as they become modified. That they have diverged greatly is obvious from the manner in which species of all kinds can be classed under genera, genera under families, families under suborders and so forth. . . . The solution occurred to me long after I had come to Down. The solution, I believe, is that the modified offspring of all dominant and increasing forms tend to become adapted to many and diversified places in the economy of nature.

This idea was Darwin's *principle of divergence* (Figure 14.11). Why should it be that species apparently push one another apart in evolution? Darwin suggested that it resulted from the relative strengths of competition imposed by closely related individuals on the one hand, and more distantly related individuals on the other. An individual of a species will compete strongly against other members of its own species, fairly strongly against members of other species in its own genus, and then more weakly against members of more distantly related groups. Little or no competition exists at the taxonomic extremes—between an average plant and an average animal, for example. The competition within a species will be strongest because the individual will encounter more members of its own species, and they will be more similar to it, exploiting more similar resources. One way to avoid competition is to become different from the competitors, and similar competing types will therefore tend to be pushed apart in evolution. Competition between similar individuals will promote the evolution of new adaptations in each species that reduce the intensity of competition, and divergence will result. In the next chapter we will consider, in the phenomenon of character displacement (section 15.2.5, p. 410), some evidence that competition can cause divergence between closely related species. The process is not inevitable, however, it depends on the contingencies of competition in particular cases. Provided that more similar individuals typically compete more closely, however, divergence will take place.

Figure 14.11 The divergent pattern of evolution. From Darwin's *Origin*.

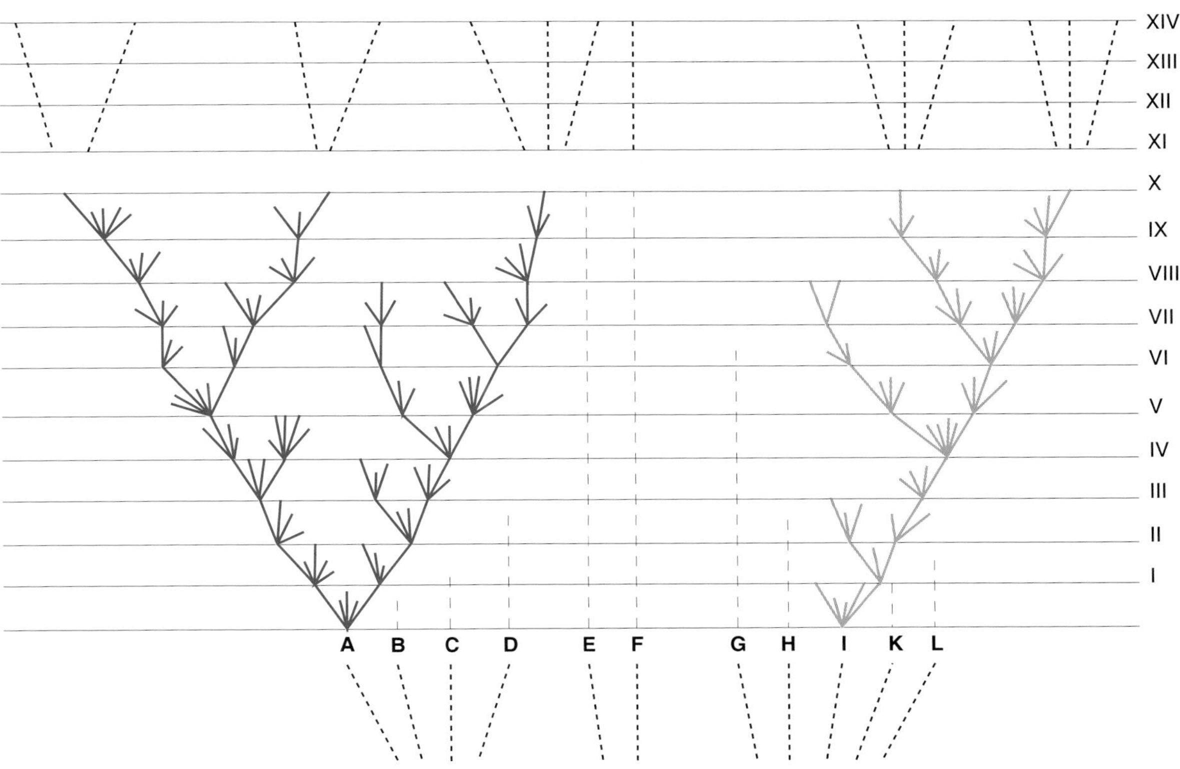

A second Darwinian principle—gradualism—helps to complete the explanation. Evolution proceeds in relatively small stages. If evolution often proceeded by large jumps, we might not see a smooth pattern of divergence. Species compete most often against close relatives, and progressively less against increasingly more distant relatives, as a result of the need to evolve in small stages. A species cannot rapidly escape the sphere occupied by its present competitors. A macromutational jump could lift a species out of the orbit of competition with its close relatives, and it could then evolve independently of them. In such a case, it might be as likely to converge as to diverge (Figure 14.12).

The contingent, exploratory nature of evolution is also influential. Even if two species were not driving one another apart by competition, they could still be expected to diverge. They will encounter different environmental challenges, and generate different favorable variants (because favorable variants are rare, the same one would be unlikely to arise coincidentally in both). As the different variants become fixed, the species will move apart.

Finally, the divergence of species can take place by a simple random walk of neutral evolution. Darwin may not have appreciated this possibility, for he wrote in the *Origin* that

> mere chance, as we may call it, might cause one variety to differ in some character from its parents, and the offspring of this variety again to differ from

Figure 14.12 (a) Gradual evolution is more likely to be tree-like and divergent than is (b) evolution with macromutational jumps. The macromutation in species 1 takes it out of competition with the species ancestral to 2 and 3. The species is then as likely to evolve toward its former competitor as it is to evolve away from it.

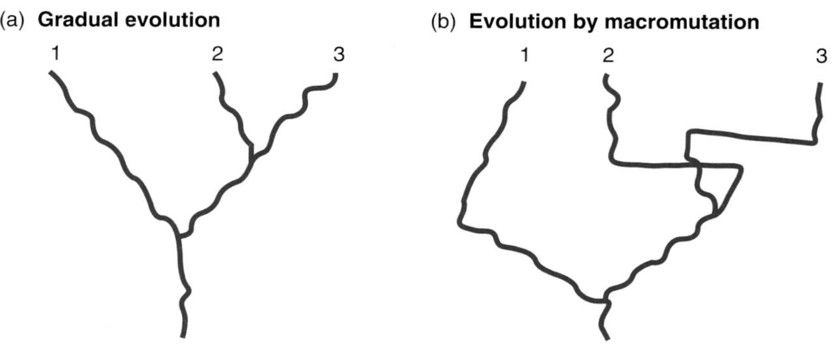

> its parent in the very same character and in a greater degree; but this alone would never account for so habitual and large an amount of difference as that between varieties of the same species and species of the same genus.

Two samples taken from a set of forms and allowed to evolve independently at random will soon diverge, in the way that evolution occurs in the neutral theory of molecular evolution (see also Figure 21.4, p. 589). Again, the reason is that the same mutations are not likely to occur in both lineages; random drift will tend to fix different amino acids in different lineages. Random divergence could have played only a part in evolutionary divergence, rather than being the primary force, because it cannot account for adaptation and much evolution is clearly adaptive. The branching pattern with which Darwin was concerned existed at the level of gross morphology. At that level, the microscopic scale and rarity of evolutionarily significant mutations, combined with the stronger competition among more similar forms, remains the most plausible principle of divergence.

14.9 *Conclusion*

If the arguments presented in this chapter are correct, cladism is theoretically the best justified system of classification. It has a deep philosophical justification that phenetic and partly phenetic systems lack. Cladism is objective, and objective classifications are preferable to subjective ones. Moreover, it is the only school that, in theory, will produce perfectly natural classifications. Under this theory, all sets of correctly identified shared derived characters must fall into the same phylogenetic hierarchy. Different sets of both shared ancestral homologies and of analogies can be distributed in contradictory groups of species, and classifications that use these characters may, therefore, be artificial.

Despite these theoretical advantages, cladism can run into practical problems. The uncertainties of phylogenetic inference make cladistic classifications liable to frequent revision. True cladists are not very worried about that possibility, and remark that all healthy theories are modified as new facts become known.

Many biologists, however, would draw a different conclusion. These biologists are less anxious about the philosophical coherence of a classificatory school, and more concerned either that its practical procedures should match its theoretical ambitions (in which case they may prefer phenetic classification, with its clearly articulated techniques) or that not too much violence should be done to existing classification (in which case they will prefer evolutionary classification). The most important point is to understand the arguments that have been used for and against each school. The exact conclusion we draw here is less important, as no orthodoxy appears forthcoming among evolutionary biologists about the best kind of classification.

SUMMARY

1. There are two main principles, and three main schools, of biological classification: phenetic and phylogenetic principles, and phenetic, cladistic, and evolutionary schools. The schools differ in how (if at all) they represent evolution in classification.
2. Phenetic classification ignores evolutionary relations and classifies species by their similarity in appearance. Cladism ignores phenetic relations and classifies species by their recency of common ancestry. Evolutionary taxonomy includes both phenetic and phylogenetic relationships.
3. Phylogenetic inference is uncertain, and phenetic classification has the advantage that it is not subject to revision when new phylogenetic ideas are put forward.
4. Phenetic classification is ambiguous because more than one way of measuring phenetic similarity exists and the different measures can disagree.
5. Cladism is unambiguous because only one phylogenetic tree of all living things exists.
6. Evolutionary taxonomy avoids some of the extraordinary properties of cladism. It suffers from the ambiguity of phenetic taxonomy, however, and its argument for excluding one kind of phenetic relation (convergence) works equally well against the kind of phenetic relation (differential divergence) that it includes.
7. Living things show a diverging, tree-like pattern of relationships probably because competition is stronger between more similar forms, evolution proceeds in small stages, and variation is undirected.

FURTHER READING

I have previously discussed many of the points in this chapter, at an introductory level and at greater length, in Ridley (1986). Sneath and Sokal (1973) is the standard work on numerical taxonomy; Sokal (1966) is a clear introduction. See Johnson (1970) for the main criticism. Hennig (1966) is the classic work on cladism—but he is not an easy read! Wiley *et al.* (1991) is an introduction to cladism. De Queiroz and Gauthier (1994) discuss the relation between phylogeny and formal nomenclature, and give references to the primary sources. Mayr

(1976, 1981, and Mayr and Ashlock 1991), Dobzhansky (1970), and Simpson (1961b) are key works by key evolutionary taxonomists. See the authors in Hall (1994) on homology, and on idealism. Osopovat (1981) discusses Darwin's principle of divergence historically.

The subject has been the topic of some interesting philosophical work. Hull (1988) gives an excellent history of the modern phenetic and cladistic movements, to illustrate his evolutionary philosophy of science. Beatty (1994) uses the controversy among schools of systematics to discuss a broader philosophical question: why does evolution have so many controversies about relative frequencies, or relative significance, in contrast with the Newtonian paradigm of physics, and some other areas of biology, in which questions are asked that have a single answer? Sober (1994) contains some other philosophical chapters.

STUDY AND REVIEW QUESTIONS

1. Match the kinds of taxonomic groups to the schools that allow them:
   Groups: polyphyletic, monophyletic, paraphyletic
   Schools: evolutionary, cladistic, phenetic
2. Give (a) a phenetic, (b) an evolutionary, and (c) a cladistic classification of the cow, the lungfish, and the salmon.
3. Given below are measurements of two characters in three species.

	Species 1	Species 2	Species 3
character *a*	2	5	2
character *b*	2	2	6

Calculate the differences for each character and then calculate (*a*) the Euclidean distance, and (*b*) the mean character distance, for the three species pairs. You could write the distances in the following matrix, (*a*) above the diagonal, and (*b*) below it.

	Species 1	Species 2	Species 3
species 1			
species 2			
species 3			
	MCD		

4. Here are the pairwise distances (either MCD or Euclidean) among five species:

Species	2	3	4	5
1	1	2.31	4.28	5.27
2		2.31	4.28	5.27
3			2	3
4				1

The average distances from species 3 to species 1 and 2 is 2.25, and from species 3 to species 4 and 5 is 2.5. (a) What is the phenetic classification of the five species? (b) What does the result suggest about the objectivity of numerical phenetic taxonomic classification?

5. Devise a classification with Hennig's numerical system like Figure 14.9 for some subsection of the Hawaiian fruitflies, such as species 3–16 (in Figure 14.8), of another group of species from the phylogeny in Figure 17.1 (p. 463).
6. Supporters of phenetic classification have sometimes replied to criticism that the system can be ambiguous by saying that if enough more characters are measured, the ambiguity will be resolved. (In Figure 14.4, if further characters were measured in the species, the average neighbor and nearest neighbor statistics would come to agree.) Is this a good reply?
7. (a) Why are biological species classified hierarchically whereas chemical elements are classified non-hierarchically in the periodic table? (b) Why does evolution usually have a divergent branching pattern in the form of a tree, rather than some other pattern (in the form, perhaps, of a row of telegraph poles, or of erratic zigzags)?

CHAPTER **15**

The Idea of a Species

EVOLUTIONARY theorists have suggested a number of reasons why biological species exist, and controversy has arisen about which of the reasons is most important. The chapter focuses on species concepts, and the controversy among them. We begin by seeing how species are recognized in practice, and then move on to the theoretical ideas. Our review covers, in order, the phenetic, reproductive (biological and recognition), and ecological concepts, all of which aim to define species at a point in time. We concentrate on two properties of each species concept: (1) whether it theoretically identifies natural units, and (2) whether it explains the existence of the discrete phenetic clusters we recognize as species. While looking at the biological species concept, which defines species by interbreeding, we also consider isolating mechanisms, which prevent interbreeding between species. We examine some test cases, from asexual organisms and from genetic and phenetic patterns in space. We then turn to cladistic and evolutionary species concepts that can supplement the non-temporal concepts and define species through time. We finish by considering the philosophical question of whether species are real categories in nature, or nominal ones.

15.1 In practice, species are recognized and defined by phenetic characters

Biologists almost universally agree that the species is a fundamental natural unit. When biologists report their research, they identify their subject matter at the species level and communicate it by a Linnaean binomial such as *Haliaeetus leucocephalus* (bald eagle) or *Drosophila melanogaster* (fruitfly). However, biologists have not been able to agree on exactly what a species is, or how species should be defined in the abstract. The controversy—which occurs on a theoretical, rather than a practical, level—no doubt exists about how particular species are defined in practice. Taxonomists practically define species by means of easily identifiable morphological characters. If one group of organisms consistently differs from other organisms, it will be defined as a separate species. To give the species a formal definition, the taxonomist searches for reliable characters—characters that are possessed by all (or as many as possible) members of the species to be defined, and not by members of other species.

Almost any phenotypic character may be useful in the practical recognition of species. Figure 15.1, for example, shows the adult bald eagle (*Haliaeetus*

Figure 15.1 Adult bald eagle (*Haliaeetus leucocephalus*) (a) and the golden eagle (*Aquila chrysaetos*) (b), seen from underneath. The species can be distinguished by their pattern of white coloration.

leucocephalus) and the golden eagle (*Aquila chrysaetos*). A bird guide, such as Peterson's *Birds*, gives a number of characters by which the two species can be distinguished. In the adult, the bald eagle can be recognized by its white head and tail, and its massive yellow bill. More of a problem occurs when trying to distinguish the immature bald eagle from the adult golden eagle; the immature bald eagle usually has enough white color in its wing linings for it to be recognizable. Therefore, we can define the bald eagle in North America as an eagle with a white head and tail in the adult. (Strictly speaking, the characters used to recognize species are often "diagnostic" rather than "defining" characters. Box 15.1 explains the distinction.)

Practical problems of species recognition occasionally arise. Based on the theory that species have evolved from a common ancestor, we should expect some awkward intermediate cases in which it is unclear whether two populations have diverged enough to count as separate species. Variation poses most of the practical problems. Phenotypic characters often vary between the individuals of a population at any one place, and between populations living in different places. The taxonomist's aim, in looking for characters to recognize species, is to find a character that does not vary much between individuals within a species, and is found only in that species and not in members of closely related species. For most species it has been possible to find a reasonably reliable character (or group of characters), although it may require an expert to recognize them.

For some closely related species, however, no reliable characters have been identified. If the species vary geographically, a good character for species recognition in one place may become useless in another place. In ring species, in which the geographic extremes of a species overlap because the geographic distribution is ring-shaped, two species appear where the ranges overlap, but they are connected by a continuously varying series of intermediates (section 3.5, p. 47). It

BOX 15.1

Description and Diagnosis in Formal Taxonomy

The example of the two eagle species is intended merely to demonstrate that species are defined in practice by particular characters. We should also notice a terminological formality, because the "definition" of a species can have two meanings. We must distinguish a formal *description* of a species from a *diagnosis*. The formal description of a species must satisfy certain rules and the characters that, in a formal sense, "define" the species are those characters mentioned in the first formal description. In theory, the characters mentioned in the formal definition could be very difficult to use on a practical basis, but they still define the species in a legalistic sense.

The characters in a species' formal definition could, for example, include arcane details of its genitalia, which can be observed through a microscope on a dead specimen in a museum. Taxonomists do not purposely pick obscure characters to include in their definitions, but if such characters were the only distinct characters noticed by the species' first taxonomist then they will provide its formal definition. This method can prove inconvenient, at least, and subsequent taxonomists will then try to find other characters that are easier to work with. These useful characters, if they are not in the formal description, provide a "diagnosis." A diagnosis does not have the legalistic power of a description to determine which names are attached to which specimens, but it is more useful in the day-to-day taxonomic task of recognizing the species to which specimens belong.

As research progresses, better characters (i.e., more characteristic of the species, and more easily recognized) may be found than those in the first formal description. The formal definition then loses its practical interest, and the characters given in a work like Peterson's *Birds* are more likely to be diagnostic than formally defining.

When an evolutionary biologist discusses the definition of species, the formal distinction between description and diagnosis is immaterial. All that matters is that phenotypic characters are used to recognize species, as in the eagles. The distinction has importance, however, because it allows us to avoid unnecessary muddles, and for other reasons—taxonomic formalities are important in the politics of conservation, for instance. Nevertheless, strictly speaking, we can ignore the distinction in the species controversy of evolutionary theory.

would be practically impossible to produce a non-arbitrary phenotypic character to divide the ring into two species. It would also be theoretically meaningless—there really is a continuum, not a number of clearcut, separate species.

These practical difficulties have brought us to the theoretical question of the species concept. Species are, in practice, mainly recognized by phenotypic characters, but the interest of the subject for evolutionary biologists lies elsewhere. In this chapter we will discuss whether some deeper theoretical concept underlies the individual characters used to recognize individual species. Is the bald eagle just the set of eagles that have white heads and tails, or is there more to it? Should we first ask what it would mean for the species of bald eagles to have both a superficial, practical and a deeper, theoretical definition?

A thought experiment should make the relation clear. Imagine a bald eagle that lacked its diagnostic characters, but was otherwise unchanged; imagine that a parental pair of bald eagles with white heads and tails produced a nest of eagles of some different color pattern. Would they have given birth to a new species? If the color of the head and tail defined being a member of *Haliaeetus leucocephalus*, then the answer would clearly be yes. However, if a more fundamental definition of the species was developed and the coloration was picked only as a practically useful marker, then the answer would be no. Indeed, the birth of the brood without the white coloration would mean that that character had become a less useful character, and taxonomists should search for some other characters to recognize the species.

15.2 *Some species concepts define species at a point in time; others define species through evolutionary time*

A first distinction among species concepts is between temporal and non-temporal concepts. Most work has been done on non-temporal concepts, which aim to identify the members of a species at a particular time. The world contains millions of organisms, and we wish to know which ones belong to the same species. This kind of non-temporal definition is practically useful in the study of modern species. For example, a field naturalist wishes to know which eagles are members of *Haliaeetus leucocephalus* now, and is not much interested in eagles that lived a million years in the past or that will exist in the future. Conversely, temporal concepts take account of the time dimension of evolution, and aim to define which individuals are members of a species not only at one time, but throughout all time.

Three main theoretical kinds of non-temporal species concepts have been suggested. One type of concept suggests that species exist because of interbreeding; in the following sections, we will discuss two versions of this concept, the biological and recognition species concepts (we can use the term "reproductive species concept" to refer to them jointly). A second concept suggests that species exist because selection maintains variation in the forms we recognize as species; we will discuss this idea as the ecological species concept.

We will only mention the third possibility here. It suggests that species exist because of genetic constraints on variability (section 13.8, p. 352). According to this theory, living things exist in discrete clusters because the mutations that would produce intermediates cannot arise. No one has seriously developed this sort of structuralist species concept, and it is only mentioned here for completeness.

In addition to these theoretical definitions, it has also been suggested that the species concept should not be tied to any particular theory. This suggestion brings us to the phenetic species concept.

15.2.1 *The phenetic species concept*

The *phenetic species concept* applies phenetic classification (section 14.5, p. 375) to the species category. Informally, the phenetic species concept defines a species as a set of organisms that resemble one another and are distinct from other sets. More formally, it would specify some exact degree of phenetic similarity, and similarity would be measured by a phenetic distance statistic. An exact definition would, therefore, sound more like "a species is a set of organisms not more than x phenetic distance units apart" or "a species is the set of organisms separated by a phenetic distance of at least y units from the nearest distinct set," and the value of x or y could be chosen to produce species much like those defined by earlier methods.

In practice, the phenetic concept uses measures of as many characters as possible in as many organisms as possible, and recognizes phenetic clusters by multivariate statistics. The species then consist of the smaller clusters that match the level of similarity typical of what would have been called a species before numerical techniques became available. Species can undoubtedly be defined in this way, as the techniques of phenetic measurement exist and are known to work.

The phenetic concept can be regarded as a simple extension of the way species are recognized in practice. Species are recognized by morphological characters; the pheneticist then says that species are groups of individuals with certain morphological characters. The phenetic species is properly defined by a large number of characters. The diagnostic characters like "white head and tail" that are used for rapid identification then serve as proxies for the underlying multi-character clusters. After the clusters have been found by multivariate numerical techniques, it will always be possible to find particular characters that discriminate conveniently between the clusters. These characters can, in turn, be used in practical species identification.

The phenetic concept can also be viewed as an updated, numerical form of the earlier *morphological* or *typological species concept* (this interpretation is particularly emphasized by the concept's critics). Particularly before the modern neo-Darwinian synthesis, species had traditionally been defined by reference to a morphological type. Under this system, a species would have been imagined to possess a standard "type" or "ideal" form, and the members of the species could then be defined as those individuals sufficiently similar to the type.

The phenetic species concept suffers from the general difficulty of all phenetic classification. We have worked through the argument before (chapter 14) and need only repeat it here in outline. The phenetic concept lacks a sound philosophical basis, and can be forced to make subjective and arbitrary decisions. For any set of organisms, one phenetic measure may divide them up into one pattern of species, while another measure results in a different pattern. If the phenetic concept was sound, it would then have a criterion for deciding which pattern was correct. No such criterion exists, however, because it would require a real phenetic pattern of species to be present in nature before the phenetic measurements were made. If we pursue the argument, any purportedly real pattern will probably turn out to be some pattern of morphological types. Neo-Darwinians dismissed typological classification because there is no reason to suppose that a real pattern of morphological types exists in nature (sections 14.5 and 14.7, pp. 375 and 387). Evolution does not make any variants in a population more typical or more real than others. Thus, the theory of evolution provides no deep support for the phenetic species concept. Identifying species phenetically is done only for practical convenience; the underlying theoretical concept of species embraces something other than a phenetic cluster.

Although it may be impossible (or at least very difficult) to provide an objective phenetic criterion of species, we do not deny that nature contains groups of organisms that we recognize as looking phenetically similar, and that we describe as species. Our perceptual systems operate a sort of cluster statistic, such that we recognize some members of a set of objects as being more similar, and some less similar. Moreover, different observers agree on the pattern of similarity in most (if not all) cases. By this criterion of simple observation, we see that organisms come in discrete, recognizable phenetic clusters in nature. To our eyes, "human," "common chimpanzee," and "gorilla" represent discrete species, not a continuum from one species to the next. The phenetic concept is generally rejected because it provides no criterion on which to fall back in ambiguous cases, but the existence of phenetic clusters in nature remains an important observation. As we move on to the theoretical concepts, we can ask two questions about these clusters. First, do they make internal theoretical sense? That is, do they identify a fundamental unit in evolutionary theory? Second, do they explain the existence of phenetic clusters in nature? A successful species concept should satisfy both conditions.

15.2.2 *The biological species concept*

The *biological species concept* defines species in terms of interbreeding. Mayr, for instance, defined a species as follows: "species are groups of interbreeding natural populations that are reproductively isolated from other such groups." The concept actually predates Darwin—It was the species concept used by John Ray in the seventeenth century, for instance—but it was strongly advocated by several influential founders of the modern synthesis, such as Dobzhansky, Mayr, and Huxley, and it remains the most widely accepted species concept today, at least among zoologists. The biological species concept is important because it places the taxonomy of natural species within the conceptual scheme of population genetics. A community of interbreeding organisms is, in population genetic terms, a gene pool. In theory, the gene pool is an abstract conception of a set of reproducing genetic units, within which gene frequencies can change. In the biological species concept, gene pools become more or less identifiable as species. The identity is imperfect, because species and populations are often subdivided, but that detail does not nullify the concept as a whole.

The biological species concept explains why the members of a species resemble one another, and differ from other species. When two organisms breed within a species, their genes pass into their combined offspring; as the same process is repeated every generation, the genes of different organisms are constantly shuffled around the species gene pool. Different family lineages (of parent, offspring, grandchildren, and so on) soon become blurred by the transfer of genes between them. The shared gene pool gives the species its identity. By contrast, genes are not (by definition) transferred to other species, and different species take on a different appearance.

The movement of genes through a species by migration and interbreeding is called *gene flow*. According to the biological species concept, the sexual shuffling around of genes gives a species its phenetic coherence—causing it to form a phenetic cluster. Species differ from other species because the genes of each species are confined to their own species' gene pool. Moreover, the constant shuffling around of genes sets up a selection pressure favoring genes that interact well with genes at other loci to produce an adapted organism; a gene that does not fit in with the workings of other genes will be selected against. When we look at an organism today, we see the cumulative effects of selection, and we expect to see genes that interact well together within a species. The same is not true of genes at different loci in two separate species. These genes have not been mixed together and sifted by selection, and we have no reason to expect them to interact well. When combined in a single body, they may produce genetic chaos. Sexual interbreeding within a species, therefore, produces what Mayr (1963) calls "cohesiveness" in the species' gene pool.

Based on the biological species concept, how should the taxonomist's method of defining species be interpreted? Taxonomists actually identify species by morphology, not reproduction. According to the biological species concept, the taxonomist's aim should be to define species reproductively; the justification for defining species morphologically is that the morphological characters shared between individuals are indicators of interbreeding. Taxonomists should study interbreeding in nature whenever possible and define the arrays of interbreeding forms as species. With dead specimens in museums, morphological criteria should be sought that mimic the interbreeding principle. Taxonomists should, therefore, seek morphological criteria that define a species as a set of forms having the kind of variation expected in an interbreeding community. The morphological characters of species then serve as indicators of interbreeding, as estimated by the taxonomist. Thus, eagles with white heads and tails are one interbreeding unit; eagles with the color pattern of the golden eagle are another.

Taxonomists can estimate the degree of morphological variation corresponding to a reproductive species. Interbreeding gives a species a degree of morphological uniformity, and someone experienced with a group can estimate the typical degree of uniformity for a reproductive species in that group. Members of a species are by no means all uniform, however. Instead, biological species are *polytypic*—they have many (or no) morphological types. Different biological species may have different degrees of morphological variation, making the biological species concept practically as well as conceptually distinct from the phenetic species concept.

The discrete phenetic clusters exist, according to the biological species concept, because of interbreeding. It is also possible for species to differ reproductively but not morphologically, which case they are called *sibling species*. The classic example of a pair of sibling species is *Drosophila pseudoobscura* and *D. persimilis*. The two are phenetically almost indistinguishable. If a numerical phenetic cluster analysis was performed on them, they would appear to be a single phenetic species. Indeed, they *are* a single phenetic species. On a reproductive basis, however, they are separate: if flies from a *persimilis* line are put with flies from a *pseudoobscura* line, they do not interbreed.

Supporters of the biological species concept such as Dobzhansky (1970) used sibling species to criticize the principle of phenetic classification, but the criticism should not be well taken. Sibling species merely show that phenetic and biological species are not the same. To criticize the phenetic species concept for failing to distinguish *D. pseudoobscura* from *D. persimilis* assumes that the biological species concept has priority and the phenetic concept is to be judged by how well it recognizes biological species. Pheneticists would disagree, turning the argument around to criticize the biological species concept for ruining a perfectly good phenetic species by splitting it in two.

The main point is that phenetic and reproductive units can differ in nature. For the biological species concept, phenetic characters matter only in so far as they indicate interbreeding. In fact, species do often form phenetic clusters, and the biological species concept has, in gene flow, an explanation for this situation.

15.2.3 *Interbreeding between species is prevented by isolating mechanisms*

Why is it that closely related species, living in the same area, do not breed together? Interbreeding is prevented by *isolating mechanisms*. An isolating mechanism is any property of the two species that stops them from interbreeding. Several main types of isolating mechanisms are distinguished; Table 15.1, taken from Dobzhansky, gives one classification. Species may not interbreed because they live in different habitats, have different courtship, their gametes do not fuse, or hybrid offspring fail to reproduce. Examples are known of all of the mechanisms listed in Table 15.1, with the possible exception of mechanical isolation. No definite case is known of a pair of species that do not interbreed because copulation is mechanically impossible. (In many cases, of course, copulation between species is impossible, but some prior factor always seems to isolate the species before any mechanical difficulties come into play.) Let us examine three examples.

The first example illustrates isolation by time of breeding. Two closely related toads, *Bufo fowleri* and *B. americanus*, have overlapping ranges in central and eastern United States. *B. americanus* breeds earlier than *B. fowleri*, and little interbreeding takes place between them through most of the range. However, in Michigan and Indiana, they do interbreed and a third type (a hybrid) is found.

TABLE 15.1

Dobzhansky's classification of reproductive isolation mechanisms. From Dobzhansky (1970). Copyright © 1970 by Columbia University Press. Reprinted with permission of the publisher.

1. *Premating* or *prezygotic* mechanisms prevent the formation of hybrid zygotes
   (a) *Ecological* or *habitat isolation.* The populations concerned occur in different habitats in the same general region
   (b) *Seasonal* or *temporal isolation.* Mating or flowering times occur at different seasons
   (c) *Sexual* or *ethological isolation.* Mutual attraction between the sexes of different species is weak or absent
   (d) *Mechanical isolation.* Physical non-correspondence of the genitalia or the flower parts prevents copulation or the transfer of pollen
   (e) *Isolation by different pollinators.* In flowering plants, related species may be specialized to attract different insects as pollinators
   (f) *Gametic isolation.* In organisms with external fertilization, female and male gametes may not be attracted to each other. In organisms with internal fertilization, the gametes or gametophytes of one species may be inviable in the sexual ducts or in the styles of other species
2. *Postmating* or *zygotic* isolating mechanisms reduce the viability or fertility of hybrid zygotes
   (g) *Hybrid inviability.* Hybrid zygotes have reduced viability or are inviable
   (h) *Hybrid sterility.* The F_1 hybrids of one sex or of both sexes fail to produce functional gametes
   (i) *Hybrid breakdown.* The F_2 or backcross hybrids have reduced viability or fertility

The hybrid is fertile and as vigorous as either parent species; at a site in New Jersey, only this hybrid is found. One interpretation, based on the work of Blair and others, is that recent human-induced changes have created new habitats that both species live in, whereas previously there was some separation by habitat. The toads have now been attracted to a common habitat, where they interbreed. The normal isolation would then have been partly temporal and partly by habitat.

The time of breeding represents an example of a *prezygotic isolation mechanism*. Such a mechanism prevents the two species from breeding together initially. In other cases, two species may interbreed but will still be isolated from one another if the viability or fertility of the hybrid offspring is so low that the interbreeding is practically fruitless. The species are then maintained by a *postzygotic isolation mechanism*.

One common means of incomplete postzygotic isolation is the phenomenon called *Haldane's rule*. Haldane (1922) observed that, in crosses in which only one sex of offspring is inviable or infertile while the other sex develops as normal, the heterogametic sex usually has reduced viability or fertility (the heterogametic sex is the one with two different sex chromosomes—i.e., in mammals, the males, whose genotype is XY, whereas the female genotype is XX). Since the tendency was noticed by Haldane, it has been supported by a large amount of evidence (Table 15.2). For example, 20 mammalian crosses between closely related species result in offspring in which only one sex is infertile or inviable, and in 19 of them that sex is the male. The genetic mechanism of Haldane's rule is the topic of active modern research, because it is likely to reflect the key genetic changes that lead to speciation (Box 15.2).

In particular cases, isolation is not likely to be solely due to a single factor from Dobzhansky's list. It may be caused by a mix of several prezygotic and postzygotic factors. Moore estimated the relative importance of four different types of isolating mechanisms among six congeneric North American frog (*Rana*) species. His results are given in Table 15.3. *Rana sylvatica*, for instance, is 100% isolated from *R. clamitans* by hybrid inviability and 30% by different ecology. The totals add up to more than 100% because species are isolated by more than one mechanism—some isolatory overkill occurs. For some pairs of species, the isolation of 1 from 2 differs from the isolation of 2 from 1. The geographic range of one species may be contained within another, for example. The more local

TABLE 15.2

Support for Haldane's rule. "Asymmetry" in the column "Hybridizations with asymmetry" means that one sex is affected more than the other with respect to the trait such as fertility. Many species of butterflies, moths, and mosquitos are also known to follow the same rule. Reprinted, by permission of the publisher, from Coyne and Orr (1989b).

Group	Trait	Hybridizations with Asymmetry	Number Obeying Haldane's Rule
Mammals	Fertility	20	19
Birds	Fertility	43	40
	Viability	18	18
Drosophila	Fertility and viability	145	141

BOX 15.2

What Is the Genetic Mechanism of Haldane's Rule?

Haldane's rule states that when only one sex of hybrids between closely related species shows reduced viability or fertility, it is the herogametic sex (that is, males in mammals and in fruitflies). The rule has extensive factual support, and only about 2% of observations are exceptions (in which the homogametic hybrids have lower fitness). What is its explanation? A first hypothesis might suggest that it is something to do with sex, with male hybrids being more vulnerable for some reason. However, the hypothesis is wrong, because in groups, such as birds and butterflies, in which females are the heterogametic sex, it is the female hybrids that show reduced fitness (in cases in which one sex is affected more than the other). The explanation for the rule must lie in an individual's sex chromosomes, not its sex. Several hypotheses have been suggested, but none of them accounts for all the observations.

One kind of hypothesis explains the rule by the interaction of sex chromosomes and autosomes in hybrids. In mammals and fruit flies, a male hybrid has a set of chromosomes like A_1, A_1, . . . X_1/A_2, A_2, . . . Y_2, and a female hybrid has one like A_1, A_1, . . . X_1/A_2, A_2, . . . X_2. Female hybrids have a full haploid set of chromosomes from both parental species, but the male lacks the X chromosome of one species while possessing autosomes from that species. (There is a similar mismatch for the Y chromosome, but the Y chromosome is not thought to contribute much to the observations underlying Haldane's rule. The Y chromosome has few active genes and genetic evidence also shows that it has little influence on hybrid fitness.) Genes on the X chromosome probably interact with genes on the autosomes. These interactions should work normally in an individual of either parental species and in female hybrids, whereas in a male hybrid a gene on an A_2 (or an A_1) may not interact well with a gene on the X_1 (or X_2). However, an experiment by Coyne placed in doubt any hypothesis invoking interactions between autosomes and sex chromosomes. He used some genetic tricks to breed experimental hybrid female fruitflies that had a similar mismatch between autosomes and sex chromosomes as is found in a normal hybrid male. These females did not show the reduction in fitness seen in hybrid males. It therefore seemed that Haldane's rule was not due to the interaction between sex chromosomes and autosomes.

A second kind of hypothesis considers the sex chromosomes alone. Genetic evidence suggests that the reduction in hybrid fitness is almost wholly caused by genes on the X chromosome (see section 16.9, p. 452, for a related observation). Why should this be? One possibility is that the X chromosome may evolve more rapidly than other chromosomes, because an advantageous recessive mutation will be expressed more on an X chromosome than would a similar mutation on an autosome. One difficulty facing this hypothesis is that the large effect of the X chromosome seems to be confined to hybrid fitness. If we look at species differences in other characters, such as in courtship behavior or in morphology, we do not find that the X chromosome has a disproportionate influence. However, if the X chromosome evolves rapidly, we should expect to see a large X effect on all characters: or at least, we have no reason to suppose its influence will be confined to hybrid viability and fertility. Another possibility is that the reduced recombination in part of the X chromosome makes the spread of "driver" genetic elements (section 12.2.1, p. 318) particularly likely. Finally, we can consider a hypothesis that does not rely on a relatively rapid rate of evolution in the X chromosome. The hypothesis suggests that the genes on the X chromosome that cause reduced hybrid fitness tend to be recessive. A deleterious recessive gene on an X chromosome is expressed in a male but not in a female. The most recent discussion of Haldane's rule, by Turelli and Orr, concludes in favor of this hypothesis. However, awkward observations remain. Turelli and Orr's version of the hypothesis cannot comfortably explain results like Coyne's experiment discussed above. The rule is therefore still something of a mystery. Perhaps new ideas are needed, or perhaps some of the observations need to be reinterpreted in terms of the existing ideas. Haldane's rule has been studied so much because it is far more than a genetic curiosity. Evidence from fruitflies suggests that the majority of incipient species evolve through a "Haldane's rule" phase in which the fitness of heterogametic hybrids is reduced before they evolve lowered fitness in both sexes of hybrid. The rule therefore probably tells us something about the genetic events that often occur when a new species is beginning to evolve.

Further reading: Coyne and Orr (1989b), Coyne (1992, 1994b), Frank (1991), Turelli and Orr (1995).

TABLE 15.3

Moore's estimates of the relative power of geographic isolation (G), ecological isolation (E), seasonal isolation (S), and hybrid inviability (D) among six species of North American frogs. From Moore (1949).

	Males					
Females	*Rana sylvatica*	*Rana pipiens*	*Rana palustris*	*Rana clamitans*	*Rana catesbeiana*	*Rana septentrionalis*
Rana sylvatica		G 29	G 61	G 59	G 68	G 80
		E 70	E 40	E 30	E 70	E 60
		S 60	S 100	S 100	S 100	S 100
		D 100	D 100	D 100	D 100	D ?
Rana pipiens	G 59		G 74	G 67	G 62	G 88
	E 70		E 70	E 70	E 85	E 80
	S 60		S 40	S 100	S 100	S 100
	D 100		D 0	D 100	D 100	D 100
Rana palustris	G 13	G 0		G 3	G 23	G 72
	E 40	E 70		E 60	E 70	E 50
	S 95	S 40		S 95	S 100	S 100
	D 100	D 0		D 100	D 100	D ?
Rana clamitans	G 28	G 0	G 24		G 18	G 22
	E 30	E 70	E 60		E 30	E 20
	S 100	S 100	S 95		S 50	S 0
	D 100	D 100	D 100		D 100	D 100
Rana catesbeiana	G 51	G 0	G 47	G 29		G 93
	E 70	E 85	E 70	E 30		E 30
	S 100	S 100	S 100	S 50		S 50
	D 100	D 100	D 100	D 100		D ?
Rana septentrionalis	G 0	G 0	G 37	G 36	G 79	
	E 60	E 85	E 50	E 20	E 30	
	S 100	S 100	S 100	S 0	S 50	
	D ?	D 100	D 100	D 100	D 95	

Complete isolation = 100; absence of isolation = 0.

species will then be 0% geographically isolated from the widely distributed species, but the widely distributed species will be isolated somewhere in the 0–100% range from the local species. From the table, it appears that 74% of the range of *R. palustris* is outside that of *R. pipiens*, but none of *pipiens* is outside *palustris*.

15.2.4 *The recognition species concept*

Within a single habitat in the United States, as many as 30 or 40 different species of crickets may be breeding. The male crickets broadcast their songs and are approached by females. Interbreeding is confined within a species because each species has its own distinctive song and females approach only males that sing their species song. In experiments, female crickets in the wild will approach loudspeakers playing cricket songs, and the crickets turn out (on identification) to belong to the species whose song is being played. A still better experiment is to put female crickets on a kind of Y-maze (Figure 15.2); the song of the female's own species can be played from one side and the song of another species from the other side. The females almost invariably turn toward the song of their own

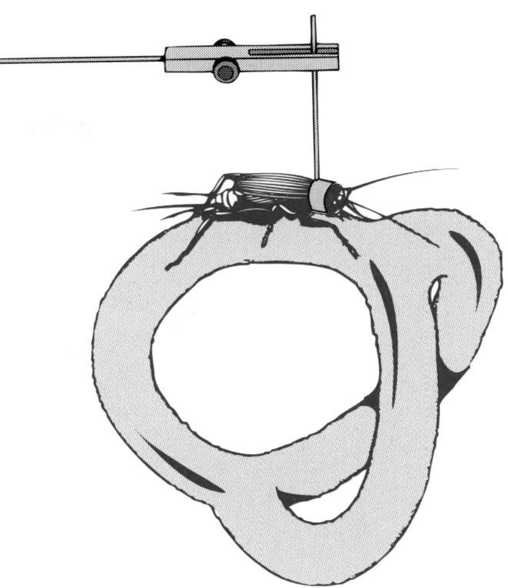

Figure 15.2 A "Y-maze" for testing the acoustic preference of a female cricket. As the cricket walks, the Y-maze moves beneath her feet. Loudspeakers can play the songs of different species from either side. Reprinted, by permission of the publisher, from Bentley and Hoy (1974).

species. (The direction from which the two species song are played can be reversed, to control for any tendency of crickets to turn to the left or right.)

The female cricket recognizes the song of males of her own species and will breed only with a male who sings that song. The song, and the female recognition of it, constitutes a mate recognition system. Thus, the species has a specific mate recognition system (SMRS). Paterson has particularly emphasized that species can be defined as a set of organisms with a common method of recognizing mates (a shared SMRS). This idea is known as the *recognition species concept.* In practice, the recognition concept should define very similar, or identical, species to the biological concept and it is possible to think of these ideas as being two versions of a general "reproductive" species concept. An isolation mechanism to keep species apart and a recognition mechanism to ensure breeding takes place within a species are, to a large extent, two sides of the same coin.

Paterson has suggested that defining species by shared mate recognition systems rather than interbreeding and reproductive isolation can offer a number of advantages. One advantage is practical. Mate recognition systems can often be observed in a few, even dead, specimens, whereas interbreeding is more difficult to observe and is usually inferred indirectly. A second possible advantage is more theoretical. The recognition concept may represent more accurately what happens when a new species originates. The crucial event for the origin of a new species, according to Paterson, is the evolution of a new mate recognition system. He doubts whether the evolution of isolating mechanisms is so important. Dobzhansky argued that isolating mechanisms will evolve under natural selection to prevent interbreeding among incipient species, so that the isolating mechanisms then become a part of the causal process leading to new species. Paterson doubts whether isolating mechanisms are ever directly favored during speciation, and suspects they generally evolve as incidental by-products of evolutionary changes taking place for other reasons during the divergence that leads to speciation, or later. Thus, Paterson's preference for his recognition concept stems in part from his view of the relative importance of mate recognition systems and isolating mechanisms in speciation. He sees the former as causal and the latter as more of a consequence. The recognition concept then better represents the causal process of speciation.

In chapter 16, we will return to the question of whether isolating mechanisms evolve during speciation. For our purposes here, we can consider isolating mechanisms and specific mate recognition systems to be two ways of defining species as reproductive communities; the biological and recognition concepts are, therefore, two parts of a more general "reproductive" species concept. It is worth noting that the biological species concept does not necessarily have to be defined in terms of isolating mechanisms, although both Mayr (in some publications but not in others) and Dobzhansky took this approach. For the fundamental biological species concept, the crucial factor is interbreeding, and whether interbreeding is brought about by a shared mate recognition system or prevented by an isolating mechanism is a secondary matter.

15.2.5 *The ecological species concept*

The forms and behavior of organisms are, at least to some extent (chapter 13), adapted to the resources they exploit and the habitats they occupy. According to the *ecological species concept*, populations form the discrete phenetic clusters that we recognize as species because the ecological and evolutionary processes controlling the division of resources tend to produce those clusters. Half a century of ecological research, particularly regarding closely related species living in the same area, has abundantly demonstrated that the differences between species in form and behavior are often related to differences in the ecological resources that the species exploit. The set of resources and habitats exploited by the members of a species form that species' ecological *niche*, and the ecological species concept defines a species as the set of organisms exploiting a single niche.

Why should ecological processes produce discrete species? Parasites provide good examples. Imagine that parasites exploit two host species. The host species will differ in certain respects—perhaps in where they live, or the times of day they are active, or their morphology—and the parasites will evolve appropriate adaptations to penetrate one or the other host. The parasites will tend to become two discrete species, because their environmental resources (hosts in this case) exist in discrete forms. A strong version of the ecological species concept would claim that all (or nearly all) species form discrete units, or phenetic clusters, because of discontinuities in their environments.

The ecological species concept supposes that ecological niches in nature occupy discrete "zones," with gaps appearing between zones. The idea that nature is divided into *adaptive zones* was originally invented for larger categories than species. Simpson suggested that large taxa such as the Mammalia occupy an adaptive zone—that is, an abstract space, set by the resources and competitors in nature, within which the mammalian sort of body plan is adaptive. The mammalian adaptive zone is more inclusive than a species niche, but we can imagine a hierarchy of adaptive zones corresponding to the successive taxonomic levels. (Simpson developed the idea from Wright's more formally defined adaptive topography (section 8.12, p. 215; see also Raup's analysis of snail shells, Figure 13.8, p. 357).) According to this idea, some sets of adaptations work well in nature, but others do not; when this idea is applied to explain the forms and interbreeding patterns of species it gives the ecological species concept. If, for example, two ecological species, such as two parasites, interbreed, they may produce offspring that fall in a gap between two niches. Thus, the offspring may be adapted to nonexistent hosts. Selection will tend to favor interbreeding with another organism possessing adaptations to at least approximately the same niche.

Resources, however, may not all come in discrete units, in the manner of the hosts of parasites. Consider a number of species of birds, whose beaks are adapted to eat seeds of certain sizes, as we saw in the example of Darwin's finches (section 9.1, p. 222). Seed size may often form a continuous distribution (although not always, as we shall see in a later example). Even if seed sizes are not an example of a continuous resource, some habitat variables surely are.

Many species differ from close relatives in the regions of a habitat that they occupy. For example, in a classic study of five species of warblers in Maine in the 1950s, MacArthur showed that each species mainly exploited a particular subregion of the trees in which they all lived. Some species foraged higher, others lower; some foraged near the ends of branches, others nearer the center of the tree. These variables, like height in tree, are continuous. Why should discrete species form on such a continuous resource? Why does a continuum of forms of warbler, from one extreme to the other, emerge, instead of the five discrete warbler species each of which interbreeds only with its own kind? The answer most probably lies in the processes we have already studied in relation to the biological species concept: sex and interbreeding. It is evolutionarily difficult for a sexually reproducing species to produce an array of forms that match the proportions of available environmental resources (asexual warblers might manage this task, but here we are taking sex as given). Rather, the system evolves to produce clusters of forms, within which interbreeding takes place, that are adapted to exploit a certain range of the resources. Selection will again work against interbreeding between individuals from different clusters because of the danger of producing maladapted offspring.[1]

The explanation for discrete species based on continuous resources makes for a weaker ecological species concept. It explains species by sex, the fundamental factor in the biological species concept, in addition to more purely ecological processes. Both ecological and reproductive processes likely matter when trying to understand species in nature, and the two concepts are closely related.

Ecological processes undoubtedly influence the array of forms within a species, and how discretely they are clustered. Let us consider the example of an array of species exploiting resources that form a single resource axis, such as seeds of varying sizes (Figure 15.3). An individual suffers intraspecific and interspecific competition for food. Selection within a species might favor individuals at the extremes, as they suffer less competition. Limits exist as to how far the process

1. The force becomes stronger as time passes, because different genes will be substituted in the various species. If individuals of different species interbreed after the species have been separated for millions of years, the DNA of the two will likely have diverged so far that the hybrids become nonfunctional, independently of adaptation to ecological resources. For example, a mule is a hybrid between a male ass (*Equus africanus*) and a female horse (*Equus caballus*) and is somatically vigorous, but sterile. The reason is presumably not the poor ecological adaption of mules, but the independent genetic divergence between asses and horses.

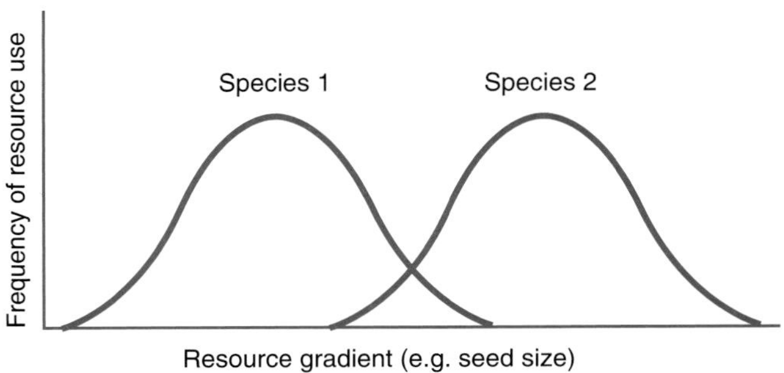

Figure 15.3 A species possesses a frequency distribution of phenotypes that exploit a certain range of resources. Two related species of birds in an area might exploit the larger and smaller seeds as shown here.

can go, however, because extreme individuals also suffer more competition from other species. The next species up the array will be better adapted (for instance, by possessing larger beaks) and a subpopulation from the smaller-beaked species would be unlikely to last long in competition with this larger-beaked species for larger seeds. The larger-beaked species will drive its competitor extinct. The event represents an example of *competitive exclusion*: if two species compete for the same resource, one of them will usually be driven extinct. Members of one species will exploit the resource more efficiently, and be able to survive on a lower level of the resource. When resources decline to a level below which a population of the less efficient species cannot survive, that species goes (locally) extinct.

The clearest evidence that interspecific competition influences the frequency distribution of phenotypes in a species comes from the phenomenon of *character displacement*. Imagine two closely related species with partly overlapping geographic ranges. These species live together in some parts of their ranges, but separately elsewhere. The species are technically said to be sympatric where they live together, and allopatric where only one of them is present. Character displacement means that the two species differ more sympatrically than allopatrically; it can be explained by the balance of intraspecific and interspecific competition. If the characters with respect to which the species differ more in sympatry are related to ecological competition, then character displacement will occur because of the advantage of avoiding competition with a better adapted species where it is present. In a place where only one species exists, selection from interspecific competition is relaxed and the species evolves to exploit a larger niche.

Let us look at one of the classic examples of character displacement: Darwin's finches on the Galápagos. Two of the finches, the larger *Geospiza fortis* (Color Plate 3b) and the smaller *G. fuliginosa*, both occupy the island of Santa Cruz. Only *G. fortis* lives on Daphne, and only *G. fuliginosa* lives on the small islands of Los Hermanos. In the 1940s, Lack had observed that the beak sizes of the two species had a similar distribution where they lived alone but they differed sympatrically. Figure 15.4 shows more recent measurements, with the beak size distributions appearing as the histograms at the bottom of the graphs. Is the character displacement for beak size caused by competition? It could simply result from differences in the sizes of seeds on the three islands, if the seeds on

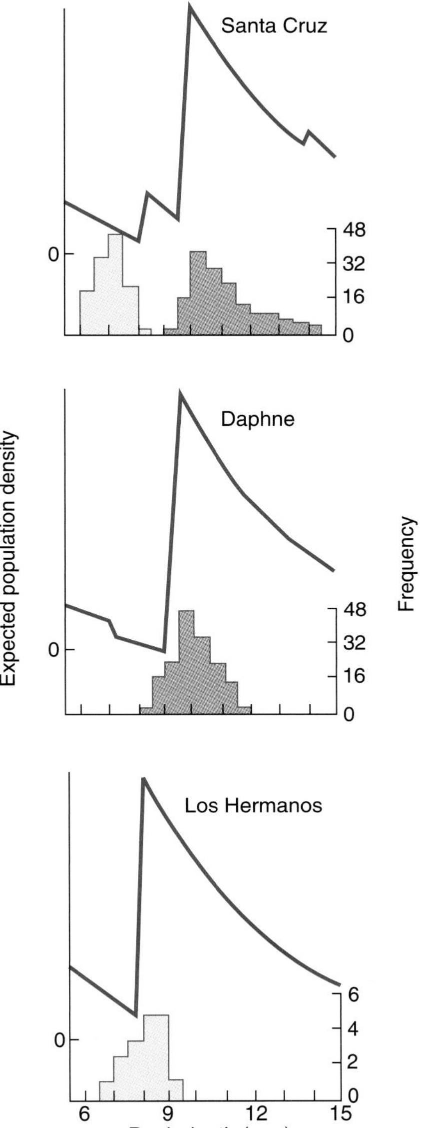

Figure 15.4 Character displacement in Darwin's finches. *G. fortis* (dark blue histograms) and *G. fuliginosa* (light blue histograms) are more similar allopatrically than sympatrically. Compare the histograms of beak sizes for Daphne and Los Hermanos (below) with those for Santa Cruz (above). The difference between the islands is not due to differences in seeds. The upper line shows the population densities of birds that would be predicted if they were determined by seed availability alone. Instead, the lines are similar in all the islands and do not explain the differences in the observed histograms for beak sizes. Reprinted with permission from Schluter *et al.* (1985). Copyright 1985 American Association for the Advancement of Science.

Santa Cruz have a bimodal distribution while those on the other two islands are unimodal and have a similar average size.

Schluter *et al.* took the analysis a stage further in the 1980s. They measured the seed sizes on the islands, and the density of the birds. These two factors should determine the degree of competition. The upper line in Figure 15.4 shows the population density of birds that would be predicted from the distribution of seed sizes. This measurement was similar in all three populations, although it shifted to the left on Los Hermanos. Differences in the available seeds, therefore, do not explain the difference between allopatric and sympatric beak size distributions.

What has happened is that the population of *G. fuliginosa* on Los Hermanos has evolved larger beaks and begun to exploit the larger seeds from which it is excluded on Santa Cruz, where it eats, and is adapted to eat, smaller seeds. The measurements strongly suggest that the presence or absence of interspecific competition determines the distribution of beak sizes. Ecological competition, therefore, influences the phenotypic frequency distribution of phenotypes in a species. (Character displacement is also important in understanding divergence (section 14.8, p. 391), and the radiation of higher taxa (chapter 23).)

Although ecological processes can explain the existence of phenetic clusters, and we will see in later examples that ecology explains certain phenetic patterns better than interbreeding, the soundness of the ecological species concept remains questionable. This concept includes a stronger claim than simply saying that ecological processes shape phenetic patterns. Instead, it says that the members of a species can be defined by sharing a common ecological niche. As G. C. Williams (1992) has recently argued, the different life stages of an organism commonly have utterly different ecologies. A marine larva in the plankton, for example, does not exploit the same niche as a sessile adult form attached to the rocks or to the sea bottom. Perhaps the concept could be salvaged by defining each species in terms of a unique set of niches, but the work has not been done and the complexity of the concept looks as if it might multiply in the manner of the Ptolemaic epicycles. For ecology to provide a full species concept, we must do more than show that ecological processes influence the phenetic and genetic patterns of species.

A controversy has arisen about the processes that explain phenetic patterns in nature. Some supporters of the biological species concept have argued that interbreeding is the main explanation, whereas some ecologically minded critics have suggested that selection, due to ecological patterns, provides the basis for such process. Many biologists, however, would reject the party lines and look at the empirical evidence.

15.2.6 *Selection and gene flow can both explain the integrity of species*

The reproductive and ecological aspects of species are probably correlated in nature. If the members of a species are interbreeding with too wide a range of mates, and producing maladaptive offspring, selection should adjust their mating preferences. Because of the correlation, it is difficult to test between the ideas, to find out whether gene flow or selection provides a better explanation of species integrity. Gene flow (migration) can rapidly unify the gene frequencies of separate populations if selection is weak (section 5.15, p. 125). On the other hand, if selection is strong, it can (in theory) keep populations distinct. The importance of the two processes in nature is an empirical question. Here we will examine a few test cases in which the processes conflict. The reproductive and ecological concepts make different predictions when interbreeding is present but selection favors divergence, and when interbreeding is absent but selection favors uniformity. Ehrlich and Raven discussed examples of each in a classic paper in 1969, in which they strongly criticized the idea that species are held together by gene flow.

1. Selection can produce divergence despite gene flow

Bradshaw carried out a major ecological genetic study of the plants, particularly the grass *Agrostis tenuis*, on and around spoil-tips in the United Kingdom. Spoil-tips are deposited from metal mining and contain high concentrations of such poisonous heavy metals as copper, zinc, or lead. Few plants can grow on them. Some plants have colonized those sites, however, including grass *A. tenuis*. *A. tenuis* has colonized the spoil-tips by means of genetic variants that can grow where the concentration of heavy metals is high; around a spoil-tip, therefore, one class of genotypes grows on the tip itself, and another class inhabits the surrounding area. Natural selection works strongly against the seeds of the surrounding forms when they land on the spoil-tip, and these seeds are poisoned. Selection also acts against the resistant forms at locations away from the spoil-tips. The reason behind this situation is less clear, but some cost, or adaptive trade-off, probably results from possessing a detoxification mechanism. Where the mechanism is not needed, the grass thrives better without it.

Populations of *Agrostis tenuis* show divergence, in that markedly different frequencies of genes for metal resistance appear among populations found on and off the spoil-tips. The pattern is clearly favored by natural selection, but how is gene flow affected? The biological species concept predicts that gene flow will be reduced; otherwise the divergence could not have taken place. In fact, gene flow is large. Pollen blows in clouds over the edges of the spoil-tips and interbreeding between the genotypes is extensive. In this case, selection has been strong enough to overcome gene flow.

The conditions on a spoil-tip are exceptional, so the example illustrates a possibility rather than determines the normal pattern in nature. Most other examples of populations that have diverged in the face of gene flow also involve exceptional conditions, as we would expect. If gene flow and selection come into conflict, immediate selection works to alter reproductive behavior. Signs indicate that this reproductive selection is already happening in *Agrostis tenuis*. The flowering times of the resistant and normal types already differ significantly, which will reduce the amount of interbreeding. If the selection for and against heavy metal tolerance in close proximity continues, we should expect reproductive isolation to evolve and the grass to speciate. The conflict between gene flow and selection will be short-lived, because either the gene flow pattern or the selection regime will change.

2. Selection can produce uniformity in the absence of gene flow

Ehrlich and Raven also discussed a number of cases in which a species is subdivided into several geographically separate populations. The gene flow between the populations was arguably trivial or zero, yet the populations were remarkably uniform. Unfortunately, all of the evidence was unquantitative. The recurring pattern in one example after another was suggestive at least—even persuasive for some—but each example could be criticized.

The checkerspot butterfly *Euphydryas editha*, which Ehrlich has extensively studied, illustrates the point. In Ehrich and Raven's (1969) words:

> Colonies of the butterfly occur scattered throughout California, many of them separated by distances of several kilometers and some by gaps of nearly 200 kilometers. It has been demonstrated that there is almost no gene flow in this species over as little as 100 meters. For this reason, there seems no possibility that gene flow "holds together" its widely scattered

populations.

This argument may well be true. It should be noted, however, that a skeptic would want more exact evidence about morphological similarity, gene flow, and selection in relation to morphology before reaching a conclusion about gene flow in *E. editha*.

Another example comes from the work of Ochman *et al.* on the snail *Cepaea nemoralis* in the Pyrenees. The snail rarely lives above 1500 meters in the mountains, and never above 2000 meters, because of the cold. In the Pyrenees, it lives in neighboring river valleys separated by mountains. Where those mountains are higher than 1500 meters, gene flow between valleys will be absent—and little gene flow probably occurs even between the valleys with lower mountains. If gene flow is required to maintain the integrity of the species (i.e., the similarity of gene frequencies), populations in different valleys should have diverged.

Ochman *et al.* observed the shell morphologies for *Cepaea* (which vary in banding pattern; see section 11.1, p. 281) and measured gene frequencies by gel electrophoresis in 197 populations of the snail. Shell morphology showed little geographic differentiation; the frequency distribution of color forms and banding patterns was approximately the same in all populations. For protein polymorphisms, the situation was more complex. The frequencies of different protein forms had diverged between different areas, but in a pattern that transcended the mountainous barriers to gene flow. Ochman *et al.* recognized three main areas, within which protein form frequencies were relatively constant; but each of the areas contained several valleys separated by mountains. Gene flow cannot explain the uniformity within each of the three areas, and particularly not within the high central area (Figure 15.5). *Cepaea* in the Pyrenees, therefore, is another case in which selection maintains uniformity between populations when no gene flow occurs.

Further evidence comes from asexual species. There is no reason why only sexual, and not asexual, forms should inhabit niches, and selection should, therefore, maintain asexual species in integrated clusters much like sexual species. If gene flow is more important in holding a species together, asexual forms should have blurred edges, unlike the discrete species of related sexual forms. After all, nothing should stop asexual species from blurring into a continuum. Unfortunately, the evidence published so far is indecisive. Many authors—especially critics of the biological species concept, Simpson being an example—have asserted that asexual species form integrated phenetic clusters just like sexual

Figure 15.5 (a) Map of the Pyrenees, showing sites where the snail *Cepaea nemoralis* was sampled, and river valleys. Rivers are separated by high ground and mountains, and the shaded area running from left to right indicates regions where altitude exceeds 1500 meters. (The stippled area in the middle indicates the region around which gene frequencies are differentiated.) (b) Shell morphology (in this case, background color) shows little geographic variation, whereas (c) protein polymorphism falls into three main areas. The map relates to the location of the four alleles of one enzyme, indophenol oxidase (Ipo-1). Three or so regions can be seen from left to right, with characteristic gene frequencies: to the left, there is more of allele 130, in the center more of allele 100, to the right more of allele 80. These regions transcend the high ground shown in (a) and similarity therefore within an area is unlikely to be maintained by migration (gene flow). Reprinted, by permission of the publisher, from Ochman *et al.* (1983).

(a)

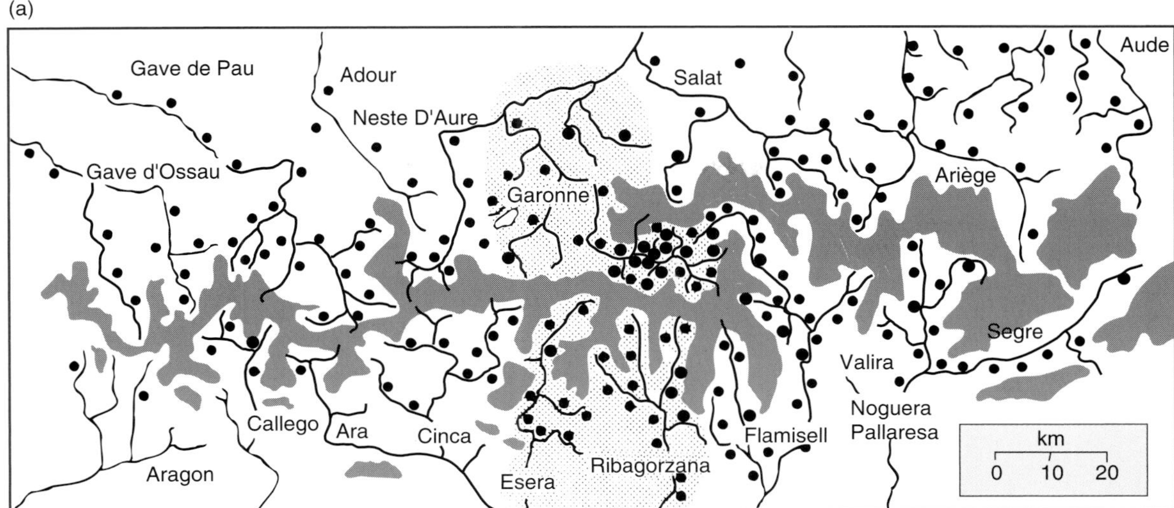
Gave de Pau
Adour
Salat
Aude
Neste D'Aure
Gave d'Ossau
Garonne
Ariège
Segre
Valira
Noguera
Pallaresa
Callego
Ara
Cinca
Flamisell
Ribagorzana
Aragon
Esera
km
0
10
20

(b)

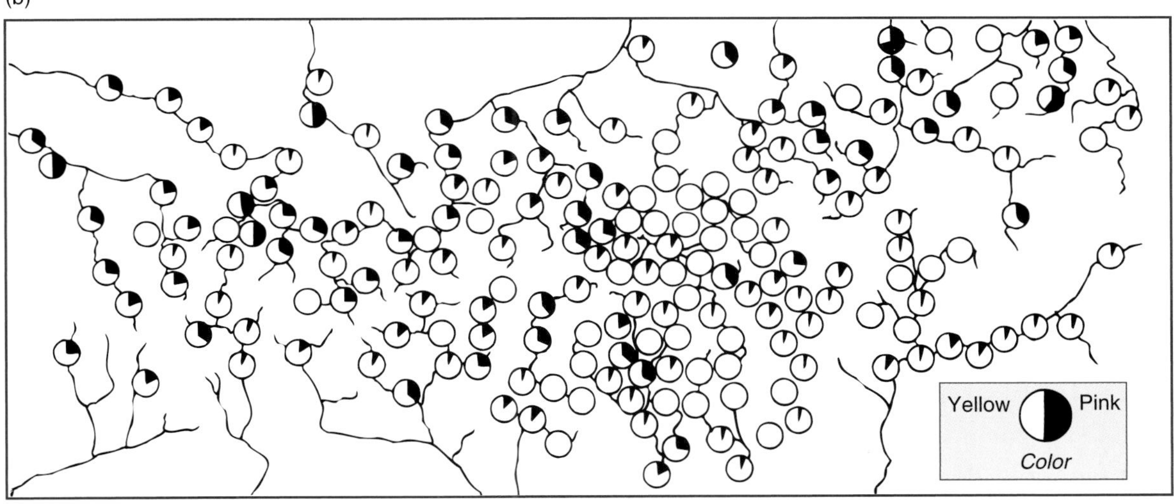
Yellow
Pink
Color

(c)

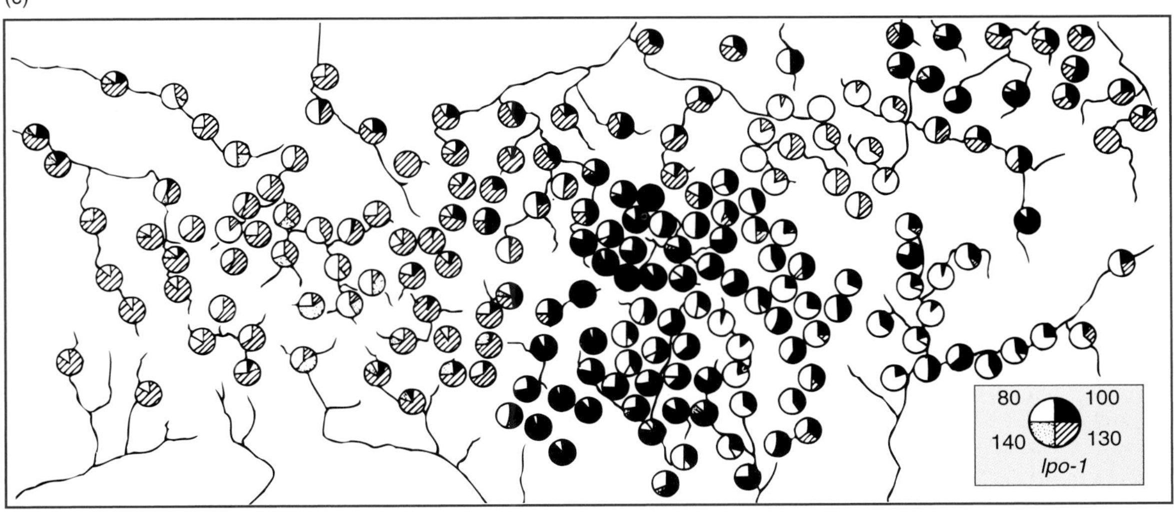
80
100
140
130
Ipo-1

species. However, it can be argued for many of these cases that they have only recently become asexual, and they simply retain the discreteness of their sexual ancestors. Likewise, although many bacteria seem to form discrete species and usually reproduce asexually, they may have enough genetic exchange to maintain species of a similar kind to those of eukaryotes.

Moreover, counter-examples may exist. Maynard Smith (1986) has pointed to the example of hawkweed (*Hieracium*). It reproduces asexually and is highly variable, such that taxonomists have recognized many hundreds of "species"; no two taxonomists agree on how many forms exist. In contrast, an analysis of rotifers, by Holman, seems to show the opposite. He compared bdelloid rotifers, which are asexual, with monogont rotifers, which can reproduce sexually, and argued that species of the asexual bdelloids are more consistently recognized by different taxonomists than are the species of the sexual monogonts. Asexual species, therefore, present a potentially interesting test case, but the evidence assembled to date does not point to a definite conclusion.

In most instances, selection and gene flow probably do not represent opposed forces in nature. Gene flow, which tends to break down differences between two forms, and natural selection, which works on reproductive behavior, should normally unify the ecological and reproductive species concepts. The few examples we have looked at here may not be representative of nature, but they do show that natural selection can be powerful enough to overcome gene flow, and to produce integrity without the help of gene flow. (Similar issues have been raised in relation to sympatric and allopatric, and parapatric and allopatric speciation, which we discuss in chapter 16.)

15.2.7 *A pluralistic species concept*

Two possible conclusions can be drawn about the controversy between ecological and reproductive species concepts.

First, we could accept that one or the other of the processes is fundamentally more important. If it were shown, for example, that species of a similar degree of integrity are formed in the absence of gene flow and in its presence, or, on the other hand, that populations diverge in the absence of gene flow even when selection favors uniformity, then the ecological or reproductive concept respectively would be supported.

Alternatively, we might conclude that both factors work in different degrees in different taxa to produce the kind of phenetic clusters we call species. Some species may be more ecological, others more reproductive. We should then need more than one species concept. Cain, for example, in 1954 suggested that we need at least three. More recently, Mishler and others have favored a pluralistic species concept, which would explicitly recognize that no single concept accounts for all species.

In any event, the main processes at work probably have been identified, even if their relative importance remains undecided. Interbreeding and selection within adaptive zones are probably the reasons why species form phenetic clusters in nature.

15.3 *The cladistic species concept defines species throughout their*

evolutionary history

The species concepts we have discussed so far aim to define how, and explain why, individuals should be grouped into species at one instant. Species last for many generations, however, and another species concept may be needed to include all members of a species throughout its existence. As shown in Figure 15.6, the species concepts we have discussed so far have been "horizontal." They seek to define which individual organisms belong in which line at any one time—that is, which organisms are in species 2 and which are in species 3 in Figures 15.6a and b. The question now is "vertical": what length of line should be included in one species?

The horizontal concepts are unambiguous in Figure 15.6. For all five (phenetic, biological, recognition, ecological, pluralistic) concepts, the line 1–2 represents only one species because the individuals throughout the lineage are a phenetic, ecological, and reproductive unit. (Let us ignore the pedantic point that individuals from different times cannot interbreed; we are interested only in whether they could interbreed if they were contemporary. Formally, this assumption implies that the reproductive concept is defined in terms of isolation or mate recognition systems rather than interbreeding.) In Figure 15.6b, the lineage 1–2 would, under all five concepts, probably contain two species, because at the time of the split the population evolved to a different phenetic (and therefore reproductive and ecological) form. Thus, the pattern in Figure 15.6a differs from that in Figure 15.6b for the horizontal concepts. The former figure includes two species, while the latter figures indicates three species. The fact that a species has branched off between 1 and 2 in Figure 15.6a is irrelevant to the species definition for all of the horizontal concepts.

Now let us turn to a gradually changing lineage (Figure 15.6c). How do the horizontal concepts divide this lineage? All five will divide the lineage into more

Figure 15.6 Three patterns of evolution. Time goes up the page. The *x*-axis has been drawn as representing phenetic distance, but it represents all five horizontal concepts if only certain sets of phenetic forms can interbreed and those sets fit neatly into the available ecological niches. (a) The ancestral species does not change phenetically (or reproductively or ecologically) after a daughter species evolves. Species 1 and 2 are phenetically identical but cladistically different. (b) The ancestral species changes after the evolution of the new species; species 1 and 2 are both phenetically and cladistically different. (c) A species that changes gradually through time. Is the lineage one species, or more? If it is more than one, how should it be divided?

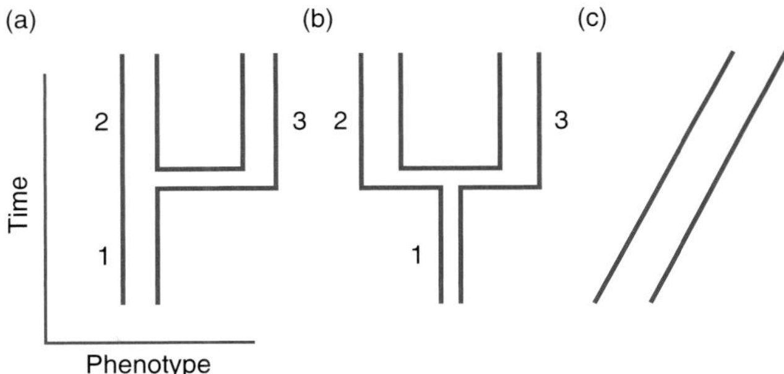

than one species, if the lineage is long enough. The longer the lineage is, the more species they will recognize. Difficulties arise, however, when we ask exactly how the division is carried out. The divisions into species will inevitably be arbitrary because the lineage shows continuous change.

One way to avoid the arbitrariness is to supplement the horizontal concepts with a vertical concept. Hennig's cladistic principle of classification (chapter 14) provides one such concept. The lineages between branch points exist objectively; we run into the problem of arbitrary division only when we try to divide up a single branch. We can define as a *cladistic species* that set of organisms in a lineage between two branch points.[2] Therefore, in Figures 15.6a and b, the two patterns are cladistically identical. Both designate three species, and species 1 gives birth to two new species at the branch point in both instances. Whereas the horizontal concepts give two species in Figure 15.6a and three species in Figure 15.6b, the cladistic concept leads to three species in both cases. This scheme results because the cladistic concept recognizes species by branch points, independently of how much change occurs between them. For the same reason, Figure 15.6c has only one cladistic species, regardless of the extent of phenetic, reproductive, or ecological change.

The division of the constant lineage in Figure 15.6a and the non-division of the changing lineage in Figure 15.6c are both controversial features of the cladistic species concept. No orthodoxy has emerged among evolutionary biologists about whether the cladistic treatment is correct in these two cases, although many people hold strong views favoring one opinion or the other.

Simpson's *evolutionary species concept* is another temporal concept. It defines a species as "a lineage (an ancestral-descendant sequence of populations) evolving separately from others and with its own unitary evolutionary role and tendencies." The definition is vague, but of all concepts presented in this chapter it is probably closest to the ecological concept, with an added time dimension.

In summary, we need both a definition of species at a time instant, and a definition for all evolutionary time. At any one time, species may be defined by interbreeding (biological and recognition concepts) or by a shared ecological niche. If these concepts are applied through time, they lead to arbitrary divisions of lineages. Consequently, they are best supplemented by a temporal concept so that they can also be applied though time.

15.4 *Taxonomic concepts may be nominalist or realist*

15.4.1 *The species category*

When we classify the natural world into units such as species, genera, and families, are we imposing categories of our own devising on a seamless natural continuum, or are the categories real divisions in nature? This problem is an old one. It applies to all taxonomic categories, but has particularly been discussed in the case of the species category. The idea that species are artificial divisions of a

2. To be more exact, we need to take account of extinction and the possibility that a species is still alive. A fuller definition of a cladistic species is the set of organisms between two branch points, or between one branch point and an extinction event, or that is descended from a branch point.

natural continuum is called *nominalism*; the alternative, that nature is itself divided into discrete species, is called *realism*.

The question is often posed in phenetic terms, but the answer depends on the species concept that we are using. If we use a phenetic species concept, the correct questions to ask become "If we measure many individual organisms phenetically, do they fall into clusters or are they evenly scattered through the morphospace? Do the species exist in the frequency distribution for individual variation, or must they be imposed on a continuum?" Clusters are, indeed, usually found, especially if we sample from only a limited geographic area. (Discrete clusters would not be found, for example, in a sample from the full geographic range of the ring species of salamanders *Ensatina klauberi* and *E. eschscholtzii* on the Pacific coast.)

The most striking evidence that species exist as phenetic clusters comes from "folk taxonomy." People working independently of western taxonomists usually have names for the species living in their area, and we assess whether they have identified the same species as have western taxonomists working with the same raw material. Some people, it seems, do classify species much like western taxonomists. The Kalám of New Guinea, for instance, recognize 174 vertebrate species, all but four of which correspond to species recognized by western taxonomists. People elsewhere do not agree quite so closely with museum classifications. A reasonable generalization is that the classifications for large vertebrates are usually the same, but those for smaller, and economically unimportant, species more often are not.

The fact that independently observing humans can see much the same species in nature does not show that species are real rather than nominal categories. The most it shows is that all human brains are wired with a similar perceptual cluster statistic. That cluster statistic, however, has no objective priority over any other cluster statistic, and it would be easy (for the reasons discussed in chapter 14) to invent another statistic that classified the birds of New Guinea differently from the Kalám or the taxonomists of a Natural History Museum. In this sense, the question of whether phenetic species exist nominally or in reality is meaningless. For the same set of observations, one cluster statistic might sense a continuum, while another cluster statistic indicated discrete clusters. In practice, phenetic taxonomy certainly reveals clusters, and cross-cultural anthropology supports these clusters in some cases. Those clusters have no firm theoretical justification, however, and suffer from the general problem of phenetic classification.

The other species concepts have sounder theoretical foundations, and the question of nominal and real classification can be answered more conclusively for them. For the reproductive species concepts (both the biological and recognition versions), the question deals with whether species exist as reproductive communities. A gradual blurring and disappearance of interbreeding could occur, for instance, as we move to ever more distant forms. In practice, although mating is often non-random within a species, interbreeding within a species is typically extensive and hybrids between species are rare. For example, in a sample of 725 copulating pairs of two similar cicada species (*Magicicada*) breeding at a site near Chicago, Lloyd and Dybas found only seven heterospecific pairs. Even those seven pairs probably did not produce offspring. The reproductive criterion is clearcut; real species do exist in the reproductive sense.

The same is probably true for the ecological species concept. In this case, however, we usually have no external measure of the adaptive zones in nature and we cannot say for sure whether they are discrete and separated by gaps, or blur into one another. Actually, the strongest reason to think that niches are discrete is that species form phenetic clusters—but this argument could only persuade someone who already accepts the ecological species concept. Skeptics might interpret the evidence differently.

The controversy between realistic and nominalistic concepts applies more directly to instantaneous than temporal classification. A strong argument can be made that species really exist in the cladistic sense, however. The reality of ecological and reproductive lineages at any one time will guarantee that the phylogenetic hierarchy contains real branches, and that cladistic species are identified. Indeed, to deny that real cladistic species exist would be to deny the existence of a phylogenetic tree of life. We can, therefore, conclude that species, in the reproductive, (probably) ecological, and cladistic senses are real, and not simply nominal units in evolutionary theory.

15.4.2 *Categories above the species level*

Evolutionary biologists who support the biological species concept characteristically differ from those who support the ecological species concept in their attitude toward the reality of taxa above the species level. The biological (and recognition) species concepts can apply to only one taxonomic level. If species are defined by interbreeding, then genera, families, and orders must exist for some other reason. Mayr has been a strong supporter of the biological species concept and (in 1942, for example) duly reasoned that species are real, but that higher levels are defined more phenetically and have less reality—that is, higher levels are relatively nominalistic. Dobzhansky and Huxley held a similar position.

Simpson, however, favored a more ecological theory of species. The ecological concept can apply in much the same way at all taxonomic levels. If the lion occupies a narrow adaptive zone, the genus *Felis* is associated with a broader zone and the class Mammalia with a still broader zone. Adaptive zones could have a hierarchical pattern corresponding to (and causing) the taxonomic hierarchy. All taxonomic levels could then be real in the same way. The relative reality of the species, and of higher taxonomic levels, is therefore part of the larger controversy between the ecological and reproductive species concepts.

15.5 *Conclusion*

In evolutionary biology, the interesting questions about species are theoretical. The practical question of which actual individuals should be classified into which species can on occasion be awkward, but biologists do not tie themselves in knots about it. The majority—perhaps more than 99.9%—of specimens can be fitted into conventionally recognized species and do not raise even practical problems. Other specimens can be identified after a bit of work—or even left unclassified until more is learned about them.

The more interesting question is why variation appears in nature arranged in the clusters we recognize as species. We have now seen several possible answers. Different horizontal (non-temporal) species concepts follow from different ideas about the importance of interbreeding (or gene flow) and natural selection. Although it is sometimes possible to test between those concepts, the results to date have not confirmed any one concept (or any plurality of concepts) decisively. General agreement has been reached that phenetic distinction alone is not an adequate concept, and that the key explanatory processes are interbreeding and the pattern of ecological resources.

SUMMARY

1. In practice, species are redefined by easily recognizable phenotypic characters that reliably indicate the species to which an individual belongs.
2. The phenetic species concept defines a species as a set of organisms that are sufficiently phenetically similar to one another.
3. The biological species concept defines a species as a set of interbreeding forms. Interbreeding between species is prevented by isolating mechanisms.
4. The recognition species concept defines a species as a set of organisms with a shared specific mate recognition system. Different members of the species recognize one another (by an evolved method of communication) as potential mates.
5. The ecological species concept defines a species as a set of organisms adapted to a particular ecological niche. Discontinuities in resources, such as in the hosts of parasitic species, and interspecific competition can shape the phenetic distributions of species into discrete clusters.
6. The biological species concept explains the integrity of species by interbreeding (which produces gene flow), the ecological concept by selection. The two processes are usually correlated, but it is possible to test between them in special cases. Selection can be strong enough to overcome gene flow, and selection can maintain a species' integrity in the absence of gene flow.
7. The cladistic species concept defines a species as the members of an evolutionary lineage between two branch points. If the biological species concept is applied temporally, it can lead to arbitrary divisions of a lineage.
8. The question of whether species exist as real, or only nominal, phenetic clusters has no deep theoretic meaning. Species, however, do exist as reproductive units, in the sense of the biological species concept.
9. According to the biological species concept, species can be real, but not the higher taxonomic levels. According to the ecological species concept, all taxonomic levels from species to kingdom can have a similar degree of realism.

FURTHER READING

Mayr (1963) is the classic account of the species in evolutionary biology; Mayr (1976, 1986, 1988) update his views on various details; see also Mayr and Ashlock (1991). Coyne (1994a) discusses species concepts, particularly in relation to Mayr's ideas. Dobzhansky (1970), Huxley (1942), Cain (1954), and Simpson (1961b) also contain classic material. See Raven (1976) and Levin (1979) on species in plants. Hull (1965) and Ghiselin (1984) explain the distinction between the theoretical definition and practical recognition of species; Sober (1994) is another philosophical source. Ereshefsky's (1992) anthology contains many of the important papers on species concepts.

On the phenetic species concept, see Sneath and Sokal (1973) and many of the references in chapter 14. Roth (1992) discusses numerical phenetic identification of fossil species, in elephants. On the biological concept, see the Mayr references. For isolating mechanisms see the books by Mayr and Dobzhansky. On plants, see Levin (1978). On the recognition concept, see Paterson (1993) and the authors in Lambert and Spencer (1994); Coyne *et al.* (1989) and Mayr (1988) criticize it.

On the ecological concept, see Van Valen (1976). On character displacement in Darwin's finches, chapter 10 of Weiner (1994) is a popular, and Schluter *et al.* (1985) the primary, source; see also Grant (1986). Grant (1975) discusses the other classic avian example; Abrams (1986b) is a major theoretical study; Schluter (1994) experimentally investigates it in fish (and see the exchange in *Science*, 268 (1995), 1065–1067); Taper and Case (1992) is a recent review; Brown and Wilson (1958) is the *fons et origo*. The controversy between selection and gene flow was stimulated particularly by Ehrlich and Raven (1969); see also Jackson and Pounds (1979) and Slatkin (1985). On heavy metal tolerance in plants, see Antonovics *et al.* (1971), Bradshaw (1971), Ford (1975), and Antonovics *et al.* (1988) for more recent related work. (See also the references to the related topic of insecticide resistance in chapter 13, including the paper by Macnair (1991).) See Ridley (1989) on the cladistic concept. Dykhuizen and Green (1991) discusses microbial species (on which see also Maynard Smith *et al.* 1993), Van Valen (1971) adaptive zones, and Berlin (1992) and Gould (1980a) folk taxonomy.

STUDY AND REVIEW QUESTIONS

1. Review the main arguments for and against the phenetic, biological, recognition, ecological, and cladistic species concepts.
2. In a pair of "sibling species," how many species are there under the (a) phenetic, (b) biological, and (c) ecological species concepts?
3. Describe the kinds of prezygotic and postzygotic isolation mechanisms, and review Haldane's rule.
4. Devise a genetic hypothesis to explain Haldane's rule.
5. Do asexual organisms form species like sexual organisms, and what consequences does the answer have for our concept of species?
6. In Figure 15.6, how many cladistic species are present in (a), (b), and (c)?
7. Under the (i) phenetic, (ii) biological, (iii) recognition, and (iv) ecological species concepts, are (a) species and (b) higher taxonomic categories real or nominal entities in nature?

CHAPTER 16

Speciation

WE begin by defining three possible kinds of geographic relationships between speciating populations: allopatric, parapatric, and sympatric. The first task of the chapter is to consider the theory of, and evidence for, each of these types of speciation. The theory of allopatric speciation has several versions: dumb-bell and peripheral isolate, with or without a founder effect and genetic revolution. Parapatric speciation occurs via a stage called a hybrid zone, and we examine studies of hybrid zones in nature. We then turn to some controversial examples of sympatric speciation in animals, and some uncontroversial examples from hybrid speciation in plants. In the discussion of the three geographic modes of speciation, we meet the important concept of reinforcement—that is, natural selection to increase reproductive isolation between incompletely formed species. The second main purpose of the chapter is to understand the arguments and evidence for and against the process of reinforcement. The discussion of reinforcement at the end of the chapter leads into some inferences about the genetics of speciation, and about how long speciation takes.

16.1 How can one species split into two reproductively isolated groups of organisms?

The crucial event, for the origin of a new species, is reproductive isolation. The members of a species usually differ genetically, ecologically, and in their behavior and morphology (i.e., phenetically) from other species, as well as in the organisms with which they will interbreed. None of these criteria may ultimately provide a universal species definition, applicable to all animals, plants, and microorganisms (chapter 15). Many species, if not all, differ by being reproductively isolated, however. To understand in these cases how a new species comes into existence, we need to understand how a barrier to interbreeding can evolve between the new species and its ancestors.

Evolutionary biologists have studied the circumstances and process of speciation in a number of ways. They have considered whether new species evolve in geographically isolated populations—whether species become reproductively isolated while geographically separated, or whether the process only starts there and is completed if the incipient species later re-encounters its ancestors. They have also investigated whether the populations that give rise to new species are peculiar in any way—for instance, by being small. The genetics of speciation has been studied, to see what kinds of genetic changes give rise to reproductive isolation. In addition, work has been done on how long it takes for one species to split into two. All these topics are considered in this chapter.

16.2 A newly evolving species could theoretically have an allopatric, parapatric, or sympatric geographical relation with its ancestor

In the abstract, the following series of events must happen in the origin of a new species. We start with a single species, made up of a set of interbreeding organisms. A genetic variant (or several variants) must spread through part of the species, and the bearers of this genetic variant must mate only (or preferentially) with other bearers of the same genetic variant. If the reproductive isolation starts as a mating preference, it must eventually narrow to become exclusive breeding with its own type. Once this event occurs, the species will have split into two. That is, two separate interbreeding populations will have evolved from one initial population. At some stage along the way, further phenetic, ecological, and behavioral differences associated with the reproductive isolation may evolve.

The major question inspired by speciation is why, and under what circumstances, the genetic variant causing reproductive isolation should evolve. The first matter to consider is whether the new species evolves in geographic isolation from its ancestor (*allopatric speciation*), in a geographically contiguous population (*parapatric speciation*), or within the geographic range of its ancestor (*sympatric speciation*) (Figure 16.1).

16.3 Geographic variation is widespread and exists in all species

Johnston and Selander measured 16 morphological variables in 1752 house sparrows (*Passer domesticus*) sampled from 33 sites in North America. The 16

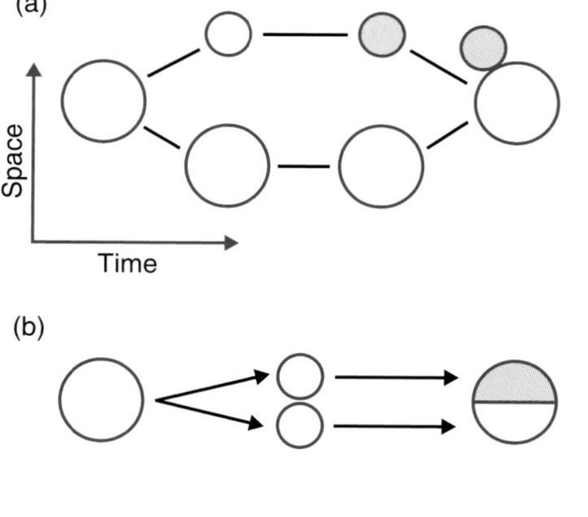

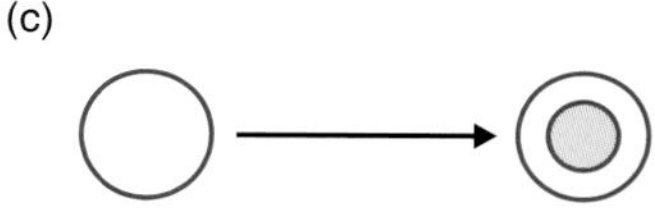

Figure 16.1 Three main theoretical types of speciation are distinguished, according to the geographic relations of the ancestral species and the newly evolving species. (a) In allopatric speciation the new species forms geographically apart from its ancestor. (b) In parapatric speciation, the new species forms in a contiguous population. (c) In sympatric speciation a new species emerges from within the geographic range of its ancestor.

characters could be reduced to a single abstract character of "body size" (to be statistically exact, this character was the first principal component). Two things should be noticed about the map plotting the average body size of these house sparrows (Figure 16.2).

First, and most important for our purposes, is simply that the characters vary in space. House sparrows from one part of the continent differ from those in other parts. Almost every species that has been studied in different places has been found to vary in some respect. Not all characters vary, of course (humans have two eyes everywhere), but populations always differ in some characters. Moreover, if we look at enough species, and enough populations of those species, we can usually find a case in which any particular character will vary geographically, especially if we concentrate on metrical variation. Different populations differ in size, color, physiology, chromosomes, and behavior and courtship patterns. Geographic variation has even been found at the genetic level for gel electrophoretically assessed gene frequencies.

Mayr, most powerfully in his book *Animal Species and Evolution* (1963, ch. 11), has collected more evidence about geographic variation than any other

Figure 16.2 Size of male house sparrows in North America. Size is measured as a "principal component" score, derived from 16 skeletal measurements. The score of 8 is for the largest birds, the score of 1 is the smallest. The study described in section 3.2 (p. 42) is an earlier relative of this research. Reproduced, with permission, from Gould and Johnston (1972), corrected from Johnston and Selander (1972). © 1972 by Annual Reviews Inc.

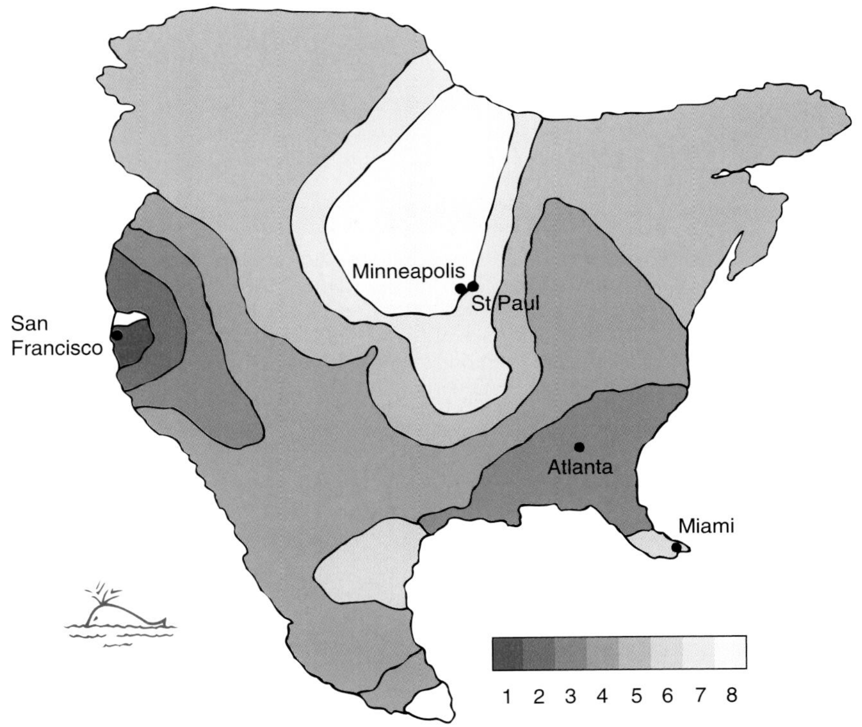

researcher. He concludes that "every population of a species differs from all others," and "the degree of difference between different populations of a species ranges from almost complete identity to distinctness almost of species level." Moreover, "all characters employed to distinguish species from each other are known also to be subject to geographic variation." That is, species differ by body size and proportions, and these measurements also vary intraspecifically; species differ in their chromosomes, and these vary intraspecifically; species differ in their courtship, and genital morphology, and these also can vary within a species from place to place.

The second point to notice in Figure 16.2 is that the form of the geographic variation is explicable. House sparrows are generally larger in the north, in Canada, than in central America. The generalization is imperfect (compare, for instance, the sparrows of San Francisco and Miami), but where it applies, it illustrates *Bergman's rule*. Animals tend to be larger in colder regions, presumably for reasons of thermoregulation. Geographic variation in this species is therefore adaptive. The form of the sparrows differs between regions because natural selection favors slightly differing shapes in different regions.

If we drew a line on Figure 16.2 from Atlanta to Minneapolis/St. Paul, or from Minneapolis/St. Paul to San Francisco, and looked at the size of sparrows along it, we should have an example of a cline. A *cline* is a gradient of continuous variation, in a phenotypic or genetic character, within a species. Clines can arise for a number of reasons. In the house sparrows, a cline probably exists because natural selection favors a slightly different body size along the gradient, and sparrows are continuously adapted to their local environments (Figure 16.3a). A cline can also arise if two forms are adapted to two different environments separated in space, and migration (gene flow) takes place between them (Figure 16.3b,c). We shall return to the theory of clines later (section 16.5.1), but the important point at this stage in the argument is that geographic variation exists and often in the form of a continuous cline.

Figure 16.3 A cline can arise (a) in a continuous environmental gradient. The house sparrow example (Figure 16.2) probably has polygenic inheritance. The *y*-axis would more appropriately express the proportion of genes for larger body size than the average for the United States. (b) When natural selection favors different genotypes in different discrete environments, gene flow (migration) occurs between the species. (c) This example is like (b) except that the environment changes gradually rather than suddenly.

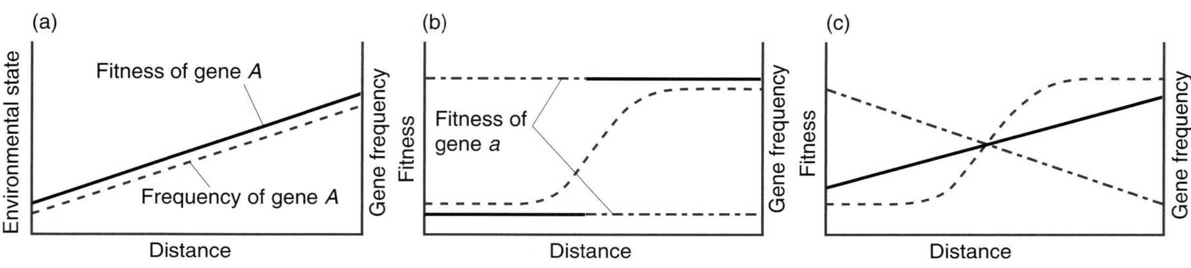

16.4 *Allopatric speciation*

16.4.1 *Allopatric speciation may occur when a barrier is intruded within the continuous geographic variation*

The extremes of geographic variation within a species can be almost as different from one another morphologically as are two closely related species, as the passage quoted above from Mayr remarked. Normally, however, we do not know whether the extremes can interbreed, because they never meet. An interesting test case is provided by ring species (section 3.5, p. 47). In ring species, the populations of a species are arranged geographically in a ring, and the end points do meet. When they meet, the two forms behave as perfectly good, reproductively isolated, biological species. Neither the Pacific coast salamanders discussed in chapter 3, nor any of the other classic examples of ring species, are thought to have evolved simply by one population gradually spreading out through space until it met itself again. The history of the subpopulations is difficult to reconstruct and probably complicated, but ring species show that geographic variation within a species can be extensive enough to account for the origin of new species.

One species could split into two if a physical barrier divided its geographic range. The barrier could be something like a new mountain range, or river, cutting through the formerly continuous population. If the barrier is large enough, gene flow between the two groups would cease and the two separate populations would evolve independently. Over time, different alleles would be fixed in the subpopulations, whether because of mutation and drift, or because selection favored different characters in the two. Whether two full species would evolve depends on whether the populations persist long enough for them to diverge to that degree.

Suppose that the two geographically isolated populations have diverged to approximately the difference in genetics and morphology that we see between two species. The two might at this point meet again, either because the barrier between them disappeared or because of migration. If the groups had diverged enough, they might not recognize one another as members of the same species, and they would not interbreed. Thus, these subpopulations would have speciated, and be isolated by a prezygotic isolation mechanism. Two species would exist where formerly only one was found. (The same process could occur with any kind of isolation mechanism (section 15.2.3, p. 405); the species might, for instance, mate but the offspring not develop. Isolation is then postzygotic.) Alternatively, the two populations might meet again before they had diverged as far as separate species. They might recognize each other as conspecifics, interbreed, and their offspring develop normally; the two populations would then soon merge back into a single continuous population. On the other hand, one of the two new groups might be better adapted than the other, and drive the second population extinct.

16.4.2 *Laboratory experiments illustrate how separately evolving populations of a species tend incidentally to evolve reproductive isolation*

When two geographically separate populations are evolving independently, different genes will be fixed in each, whether by drift or adaptation to different environments. The theory of allopatric speciation suggests that two such populations will also, at least sometimes, evolve some degree of reproductive isolation in consequence. If the populations simply diverged without becoming reproductively isolated to any degree, when they met again they would tend to fuse and no speciation would occur.

This idea has been tested experimentally, by seeing whether two populations tend to acquire any degree of reproductive isolation after they are allowed to evolve separately. Dodd, for example, performed the experiment with fruitflies (*Drosophila pseudoobscura*). The flies had originally been caught in Utah. They were then taken to a lab at Yale University and divided into eight populations. Four of these populations were placed on a starch-based food medium; the other four were exposed to a maltose-based medium. The populations were reared on these different resources for a number of generations. After a while the flies showed certain electrophoretically detectable enzyme differences—differences that were almost certainly adapted to the different resources. Thus, the populations had diverged under the influence of selection to live on different resources in the laboratory.

Dodd exploited these populations to see whether any reproductive isolation had evolved as an incidental consequence of the divergence. She placed recently emerged males and females from the starch and maltose populations in a cage after marking all members of one population. She then measured which flies mated, and found that the "starch" flies preferred a "starch" mate, and the "maltose" flies preferred a "maltose" mate (Table 16.1). Some reproductive isolation had evolved—in this case, prezygotic isolation. It presumably evolves because the changes that have occurred in the population influence reproductive behavior in some way.

Rice and Hostert, in a recent review, listed 14 experiments that measured whether prezygotic isolation emerged between populations that had been experimentally isolated. They found that in 11 of these experiments, it did; in the other experiments no significant change was observed. In summary, good experimental evidence suggests that reproductive isolation tends to evolve in populations that are geographically separated for many generations.

16.4.3 *When the diverged populations meet again, reproductive isolation may be reinforced by natural selection*

When the two geographically isolated populations come back into contact, the reproductive isolation between them might be complete or incomplete. If it is complete, speciation has occurred. If it is incomplete, the two forms might exert a selection pressure on one another, and another theoretical possibility arises. The two forms might initially prefer to mate with their own type, but occasionally mate with the other type. Some hybrids would be produced. If the hybrids had lower fitness than either parental form, selection would act to increase the reproductive isolation because each form would do better not to mate with the other and form the disadvantageous hybrids. Speciation might then be speeded up by selection in sympatry. The process is called *secondary reinforcement.* It is "secondary" if the reproductive isolation has partly evolved allopatrically, and is then reinforced when the two populations come into secondary contact.

TABLE 16.1

Prezygotic isolation emerges between populations that have adapted to different conditions. Four population of fruitflies were kept on starch medium; four others on maltose medium. After a number of generations, the tendency of the flies to mate with other like themselves was measured.

In the experimental series, 12 females from a maltose population and 12 more from a starch population were put in a cage with 12 males from a starch and 12 more from a maltose population; the numbers of the four kinds of mating couples that formed were counted. One such experiment was done for each of the four starch populations with each of the four maltose populations, making 16 experiments in all. In (a) one of the 16 experiments is shown as an example. In (b) the average isolation in all 16 experiments is given. The isolation index is calculated as (number of matings to same type − number of matings to different type) / total number of matings. It varies from 1 (for complete reproductive isolation) to 0 (for random mating, or zero isolation) through to the theoretical extreme of −1 if matings were exclusively between opposite typed flies. (The same isolation index is used again in Figure 16.14.)

In the control series, 12 females and 12 males from one of the four starch populations were put with another 12 males and 12 females from another of the four starch populations; the same controls were done with the maltose populations as well. Again, one example is given: a pair may be formed between a male and a female from the same starch population, or a male from one of the four starch populations may pair with a female from a different starch population. The average is then given for all 16 controls.

Notice the higher value of the isolation index for the experimental rather than the control cages. Prezygotic isolation had evolved between the populations that were isolated, but on the same media. Reproduced, by permission of the publisher, from Dodd (1989).

(a) Examples of results:

	Experimental Cage			Control Cage	
	Female			Female	
	Starch	Maltose		Same	Different
Male			Male		
Starch	22	9	Same	18	15
Maltose	8	20	Different	12	15
Isolation index = $\left(\frac{42-17}{59}\right) = 0.42$			Isolation index = $\left(\frac{33-27}{60}\right) = 0.1$		

(b) Average isolation indexes for all 16 crosses:	Average isolation index
Starch × maltose population crosses	0.33
Control crosses	0.014

The process by which selection increases reproductive isolation, independently of the history of the populations, is simply called *reinforcement*. Reinforcement may occur when two forms coexist, and the hybrids between them have lower fitness than crosses within each form. (It is also possible for selection to act to reduce the partial reproductive isolation, if the hybrids are as fit—or fitter—than the two original forms.)

Reinforcement is a controversial theory. Many influential biologists, from Wallace to Fisher and Dobzhansky and their followers, have argued that it often operates during speciation. Reinforcement is certainly possible in theory. If an individual has a choice between two sorts of potential mates, and would produce more offspring by mating with one sort, selection undoubtedly will favor a mating preference. Selection will therefore act to reinforce the prezygotic reproductive isolation.

Moreover, the process can be simulated by artificial selection, through the simple matter of selecting for assortative mating. Figure 16.4 illustrates the results of an experiment by Kessler. *Drosophila pseudoobscura* and *D. persimilis* are sibling species (section 15.2.2, p. 404), and their degree of prezygotic isolation depends in the laboratory on temperature. At low temperature, females of each species will mate with males of the other (heterospecific) species as well as males of their own (homospecific) species. Kessler watched the behavior of the females closely. He selected for increased isolation by breeding from females that he saw reject a heterospecific male and then mate with a homospecific male; he selected for decreased isolation by the reciprocal procedure. Kessler obtained significant

Figure 16.4 Artificial selection in female *Drosophila pseudoobscura* for increased (low isolation, solid circles) and decreased (high isolation, open circles) tendency to mate with male *D. persimilis*. The *y*-axis is an index of frequency of mating with a member of the other species (heterospecific mating). When the index is positive, females are more likely to mate with heterospecific males than are control females; when it is negative, the females are less likely to mate with such males. Reproduced, by permission of the publisher, from Kessler (1966).

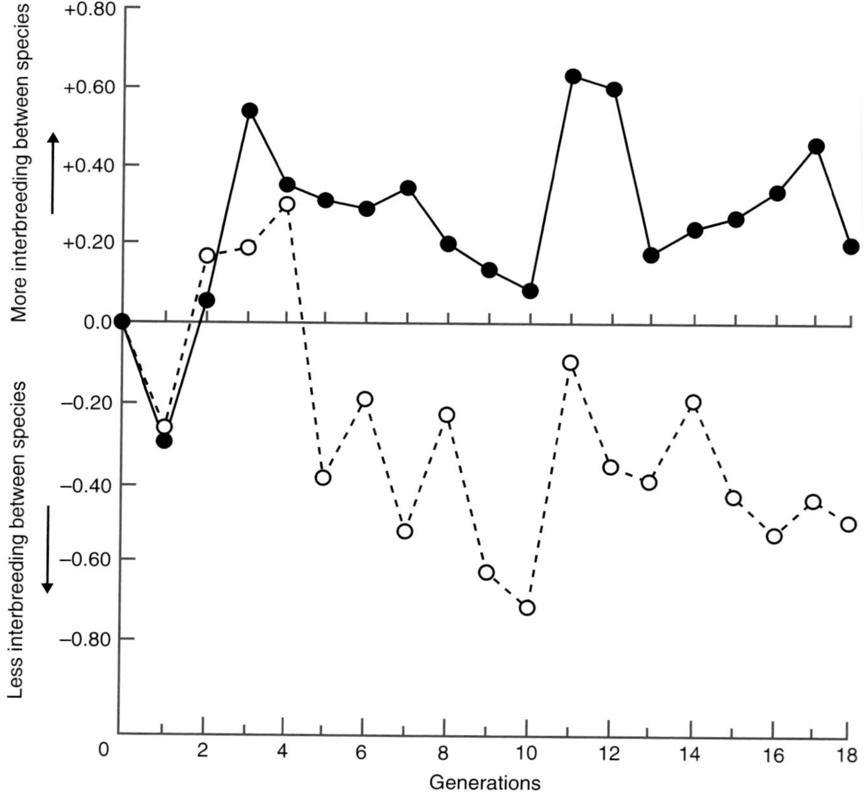

changes in prezygotic isolation in 18 generations. Rice and Hostert's recent review paper identified eight other experiments like Kessler's, all of which showed much the same result. Thus, artificial selection can alter the mating preferences of females for different kinds of males.

The theoretical conditions for speciation to take place by reinforcement are difficult. Our ability to devise a process in theory, and even to enact that process in an artificial selection experiment, does not mean that the process of reinforcement is generally important in nature. In real cases, a time window exists after secondary contact during which reinforcement must happen.

Consider, for example, the simplest case in which selection favors reinforcement. Imagine that we have two alleles *A* and *a* at a locus, and *AA* and *aa* have higher fitness, while *Aa* is selected against. Natural selection then favors *AA* types that mate preferentially with other *AA* types, and *aa* types that mate only with other *aa*. The result could be speciation, with one species having genotype *AA* and the other species *aa*. Before speciation is complete, however, the selection against heterozygotes will also act to remove the rarer allele (if *a*, for example, is rare it will tend to find itself more in *Aa* than in *aa* individuals and will accordingly be selected against). Positively frequency-dependent selection occurs (section 5.13, p. 121), such that natural selection favors the more common genotype. In the evolutionary race between speciation and the loss of the allele, the latter process will often occur more rapidly because it needs no new genetic variation, whereas speciation cannot happen without genetic variation for mating preferences. A more complex model might consider genetic differences like *AA* and *aa* at many loci, but here again selection will tend to remove the rarer genotypes until the population becomes uniform. Moreover, recombination between the genes controlling the mating preference and the genes under selection will frustrate the evolution of reinforcement.

In the case of two populations that have diverged partly in allopatry and then re-encounter each other, gene flows becomes an additional factor. Selection might reinforce any isolation between the populations. Until isolation is complete, however, gene flow will act to equalize their gene frequencies. Once the two populations become genetically identical, selection will no longer decrease breeding between them. These theoretical considerations make Kessler's experiment an imperfect test of reinforcement. No hybrid offspring were allowed to contribute to the next generation in the experiment, and the two species were, therefore, kept experimentally distinct for the 18 generations while reproductive isolation was being reinforced. In a natural case, the two species would be blurring together during those 18 generations, by gene flow, and one might be driving the other extinct. Such an action would reduce the power of selection for reinforcement.

Selection for reinforcement is likely to be strongest immediately after the two populations meet. If the necessary genetic variation in mating preferences is present, and selection is strong enough, the two species may completely split. On theoretical grounds, a gradual backslide into a single species is also possible. We shall see later in this chapter (sections 16.5.2, 16.8–16.9) that, in addition to the difficult theoretical conditions, the evidence does not always favor the theory of reinforcement. A theory of speciation, therefore, can avoid a theoretical and empirical minefield if, while not excluding the possibility of reinforcement, it nevertheless does not depend on it.

The allopatric theory has this virtue. In the allopatric theory, reproductive isolation might evolve only in allopatry (and when it does not, the two populations simply collapse back into one when they meet); alternatively, partial reproductive isolation might sometimes evolve allopatrically and then be reinforced on secondary contact. Either possibility fits in with the general theory of allopatric speciation. The two alternatives—parapatric and sympatric speciation—depend strongly on the theory of reinforcement; indeed, they require it, and if reinforcement does not operate, neither do they.

16.4.4 *Allopatric speciation may take place in peripherally isolated populations*

The model of allopatric speciation discussed above proposed that some barrier physically divided the species' geographic range. Bush calls this process speciation by subdivision, and Mayr terms it the "dumb-bell" model; its characteristic is that the ancestral population is divided into two fairly equal halves (Figure 16.5a). This concept can be contrasted with the *peripheral isolate* model. Here the idea is that a small population, at the extreme edge of the species' range, becomes separated from the main group (Figure 16.5b). The same sequence of divergence and, it may be, subsequent meeting of the two populations could then take place as in speciation by subdivision.

What are the relative frequencies of allopatric speciation via peripheral isolates and via subdivision? It has been argued, particularly by Mayr, that the process via peripheral isolates has occurred much more commonly in evolution. One reason is that it may be physically more probable that a small population would be isolated at the edge of a species range than that a barrier would divide the whole of a species range. A second reason comes from the pattern of geographic variation in modern species. Isolated populations at the edge of a species range

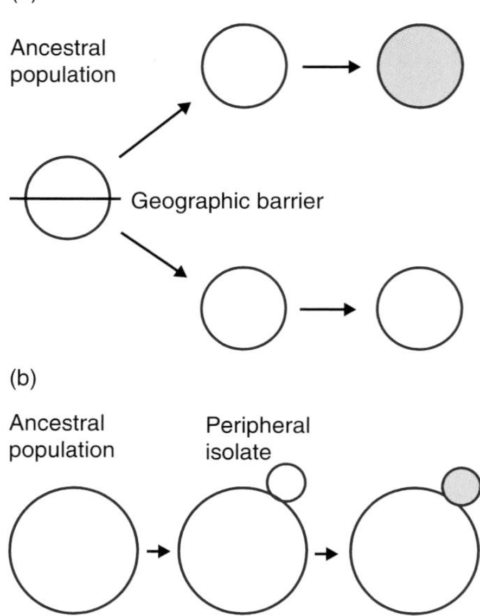

Figure 16.5 Two models of allopatric speciation. Stippling indicates evolution of new species. (a) The "dumb-bell" model in which the ancestral species is divided into two roughly equal halves, each of which forms a new species. (b) The peripheral isolate model, in which the new species forms from a population isolated at the edge of the ancestral species range.

frequently show distinct forms, whereas the individuals in the main part of the species range show less variation. Distinct peripheral isolates have often been observed on islands. Mayr (1954) described one such example:

> Let us look, for instance, at the range of the Papuan kingfishers of the *Tanysiptera hydrocharis-galatea* group (Figure 16.6). It is typical of hundreds of similar cases. On the mainland of New Guinea three subspecies occur which are very similar to each other. But whenever we find a representative of this group on an island, it is so different that five of the six Papuan island forms were described as separate species and four are still so regarded.

The kingfishers on the peripheral islands have diverged more than would be predicted from the degree of variation found on New Guinea. We should bear in mind that the phenomenon of phenotypically distinct peripherally isolated populations has been illustrated only by examples, and has not been supported by systematic study. Nevertheless, we can provisionally accept its reality. The phenomenon has two possible explanations. First, local adaptation may emerge if the conditions on the island are such that natural selection favors a distinct phenotype there. Second, isolation (the explanation favored by Mayr) results in reduction of gene flow from the rest of the species on the island, allowing the population there to diverge. Whatever its explanation, if peripherally isolated populations are particularly likely to diverge from the form of their ancestors, they may indeed be a common stage on the way to speciation.

A number of more controversial conjectures have been made about how the process of speciation operates via peripherally isolated populations. The speciating population is likely to be small and at the edge of the species' range. It may live in relatively extreme physical conditions. Mayr, and his followers, have accordingly

Figure 16.6 Kingfishers of the *Tanysiptera hydrocharis-galatea* group in the New Guinea region. The three forms on mainland New Guinea (1, 2, 3) are almost indistinguishable. The five island forms (4–8) are strikingly distinct and most were originally described as separate species. The Aru Islands (H1) and southern New Guinea once formed an island, on which the form *hydrocharis* evolved. When southern New Guinea (H2) joined the main island (dotted line), the southeast New Guinea subspecies (3) invaded the area, but did not interbreed with *hydrocharis*. *Tanysiptera hydrocharis* had, therefore, evolved into a full species while isolated. From Mayr (1942). Copyright © 1942 by Columbia University Press. Reprinted with permission of the publisher.

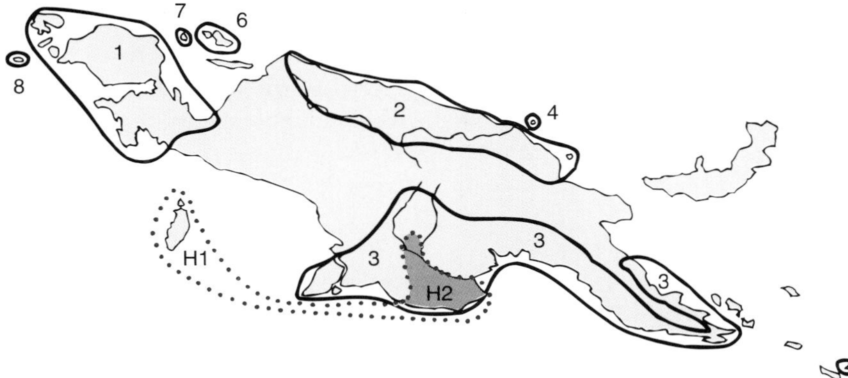

suggested that the speciating population may evolve rapidly (because the population is small) and by drift as well as selection. Because the peripheral isolate has been cut off accidentally, it may contain a non-representative sample of the ancestral species' genes. In the *founder effect* (section 6.3, p. 136), the peripheral population is formed by a small number of founder individuals, who lack some of their ancestor's genes. This idea remains controversial, because it implies that speciation might be non-adaptive. One long-standing criticism of the theory of evolution by natural selection is that many species differences, which concern apparently minor morphological details, may be non-adaptive. In such a case, a theory of non-adaptive speciation would be needed. Founder effect speciation, with random genetic loss and drift, would be one such theory. Hypotheses of non-adaptation are fallible (section 11.1, p. 281), however, and no convincing evidence exists that any differences between species are non-adaptive.

A final suggestion made by Mayr is that speciation might be accompanied by a *genetic revolution*. This extensive reorganization of the gene pool would take place in the extraordinary genetic and environmental conditions of the peripherally isolated population. In terms of Wright's shifting balance theory of evolution (section 8.13, p. 217), the genetic revolution might occur when a population evolved suddenly from one peak in the adaptive topography to another; this process is referred to as speciation by a "peak shift."

All of these ideas have been influential (see, for example, the theory of punctuated equilibrium in section 20.4, p. 560), but they remain largely hypothetical. Theoretical work has clarified the conditions required for them to operate, but the crucial evidence needed is factual. Unfortunately, the ideas refer to events that mainly occurred in the past, at the time of origin of new species. It is, however, possible to study modern species that are either in the process of splitting or have recently split. The conclusions from work of this kind are controversial, but have provided little or no support for the idea that speciation occurs via the founder effect, or by genetically revolutionary peak shifts. The peripheral isolate model, on the other hand, does not require these exotic genetic processes. It could be (and this idea is a conventional claim in evolutionary biology) that speciation does often occur via peripherally isolated populations, but the process is then driven by a fairly normal mix of natural selection and drift.

16.4.5 *Allopatric speciation: conclusion*

Allopatric speciation represents an important evolutionary process. Ring species provide clear evidence for its feasibility, but the evidence that the process is common comes from the widespread existence, and extent, of geographic variation in all species, and the theoretical possibility that reproductive isolation could evolve as part of the genetic divergence in an isolated allopatric population. Geographic variation also provides the best evidence for the peripheral isolate version of the theory of allopatric speciation; observations of isolated populations that differ from the rest of the species suggest that peripheral isolates may be a common route to speciation. The main controversies about the role of geography in speciation concern not whether allopatric speciation happens, but about the details of how it operates. In addition, the question of whether it is the exclusive mode of speciation, or whether speciation can also take place parapatrically, or even sympatrically, remains unresolved.

16.5 *Parapatric speciation*

16.5.1 *Parapatric speciation begins with the evolution of a hybrid zone*

In parapatric speciation, the new species evolve from contiguous populations, rather than completely separate ones (as in allopatric speciation). The crucial evidence concerns a kind of geographic variation called a hybrid zone. A *hybrid zone* is an area of contact between two noticeably different forms, or races, of a species, at which hybridization takes place. The forms on either side of the zone may be different enough to merit classification as separate species.

The carrion crow (*Corvus corone*) and hooded crow (*Corvus cornix*) in Europe are a classic example of the development of a hybrid zone (Figure 16.7). The hooded crow is distributed more to the east, the carrion crow to the west; the two species meet along a line in central Europe. At that line—the hybrid zone—they interbreed and produce hybrids. The hybrid zone for the crows was first recognized phenotypically, because the hooded crow is gray with a black head and tail, whereas the carrion crow is black all over. The two species are now known to differ in many other respects as well. The fact that the crows interbreed in the hybrid zone mean that speciation between them is incomplete.

Figure 16.8 illustrates another example of a hybrid zone, from Harrison's work on the crickets (*Gryllus*) of Connecticut. In the crickets, the hybrid zone

Figure 16.7 Hybrid zone between the carrion crow (*Corvus corone*) and hooded crow (*C. cornix*) in Europe. Reprinted, by permission of the publisher, from Mayr (1963), Cambridge, Mass.: Harvard University Press. Copyright © 1963 by the President and Fellows of Harvard College.

Figure 16.8 Mosaic hybrid zone in crickets *Gryllus firmus* and *G. pennsylvanicus* in Connecticut. (a) Collecting localities, (b) average ovipositor length, (c) frequencies of the *Est*$^{-10}$ allele, (d) frequencies of two mitochondrial DNA genotypes called *A* and *B*. The samples are almost pure *G. firmus* (long ovipositor, high frequency of *Est*$^{-10}$ allele, *B* mtDNA) at the coast and *G. pennsylvanicus* (short ovipositor, low frequency of *Est*$^{-10}$ allele, *A* mtDNA) inland. A mosaic pattern, rather than a simple line, of hybrid zones appears between the species. Reprinted, by permission of the publisher, from Harrison and Rand (1989).

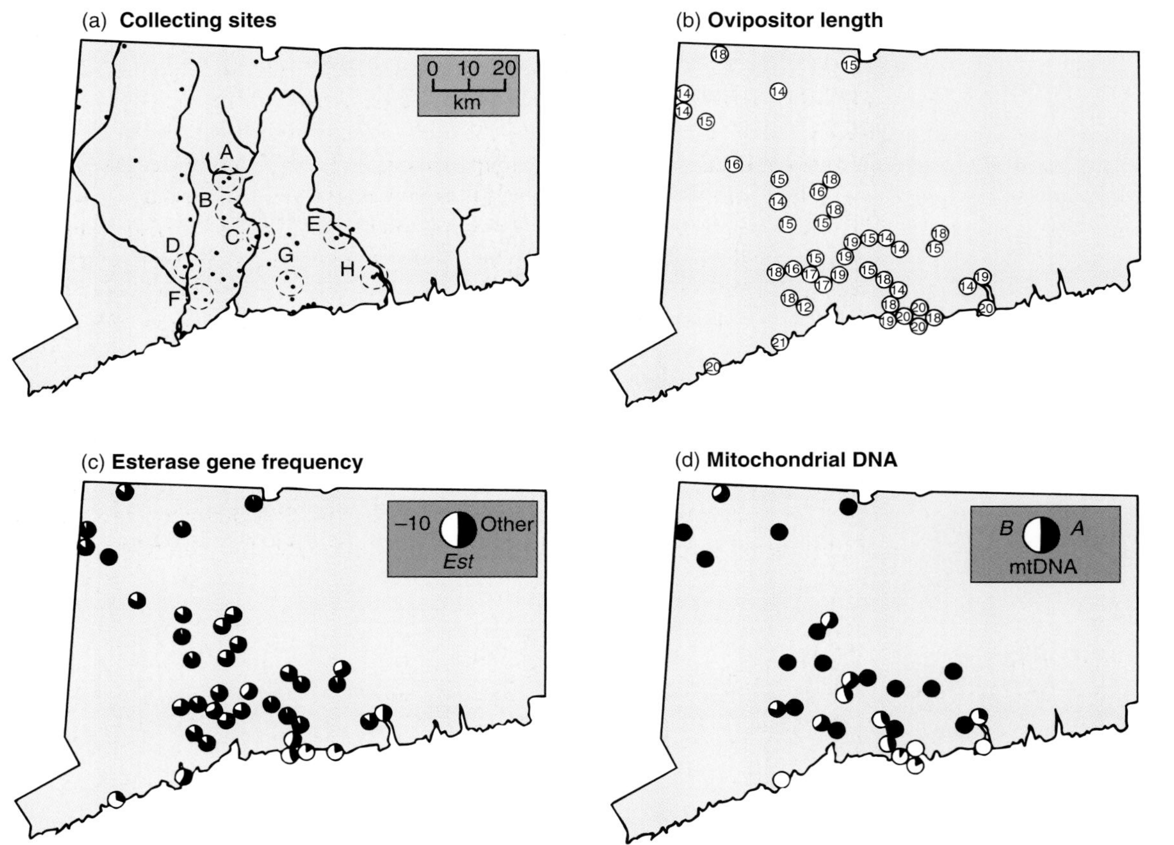

is not a simple cline as it is in the crows; rather, a mosaic of zones appears where hybrids meet.

Hybrid zones have been described in all main groups of living things that have been studied by population geneticists, and for gel electrophoretic, chromosomal, and phenotypic characters. The actual zones vary considerably in form in different cases. Some, like the hybrid zones occupied by the composite weed *Gaillardia pulchella* in Texas, are narrow and the transition takes place in a few meters. Others, like the hybrid zone between forms of the bishop pine *Pinus muricata* in California, are a few kilometers wide.

Hybrid zones are examples of *stepped clines*. We saw earlier that continuous clines are a common kind of geographic variation, and that they will evolve when different genotypes are favored in different places and individuals migrate between the areas (Figure 16.3b,c). A simple model would be as follows. Suppose there are three genotypes at a locus, *AA*, *Aa*, and *aa*, and *AA* is increasingly favored in one direction on the geographic gradient and *aa* in the other (Figure 16.3c). The heterozygote is, on average, selected against. The cline will be "steeper" (or narrower, depending on your perspective) if the selection against heterozygotes is stronger, and if less migration occurs. To be exact, if h is the selection against *Aa* heterozygotes, and if s is the standard deviation of the distance to which genes disperse per generation (i.e., the average distance between an average reproducing offspring and its parent), then the "width" of the cline $= s/h$. (Width can be defined as 1/maximum slope of the graph of gene frequency against space. If the change over in gene frequencies is sudden, the slope is steep, and the cline is narrow.)

This formula can be tested against natural observations. It almost always holds true (as discussed below) that hybrid heterozygotes are selected against in hybrid zones, although the exact disadvantage is not known in most cases. We can test for the predicted positive relation between degree of dispersal and width of cline, on the assumption that different intensities of selection will contribute only noise rather than spoiling the trend. Figure 16.9 illustrates the relation for 26 hybrid zones and shows that species with higher dispersal rates actually form wider hybrid zones. The kind of cline that will evolve depends on the rate of dispersal and the pattern of the environment.

We can now see how these ideas relate to the theory of speciation. Suppose that a population initially existed in an area to which it was well adapted, and that it then started to expand into a contiguous area in which the environment favored a different form. If the transition between the two environments was

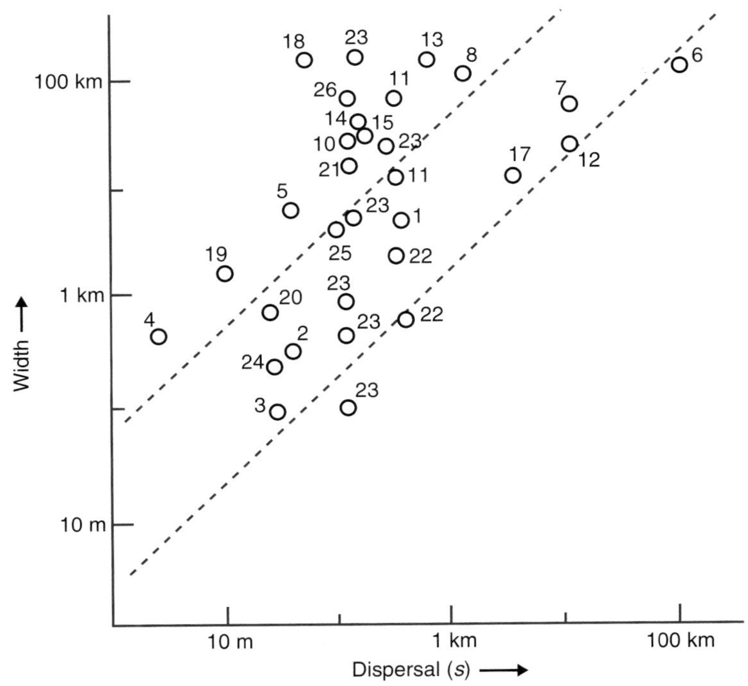

Figure 16.9 Species with higher rates of dispersal form wider hybrid zones. The graph shows the relation between dispersal (*s*) and width (*w*) for hybrid zones in 26 genera. Different points with the same number mean more than one species has been studied in a genus. For the exact meaning of *s* and *w*, and the theoretically predicted relation, see section 16.5. The two dashed lines are for selection against the heterozygote of $h = 10^{-5}$ and 10^{-1}. Reproduced, with permission, from Barton and Hewitt (1985). © 1985 by Annual Reviews Inc.

sudden, a stepped cline will evolve at the border. As selection worked on the population in the new area, different genes would accumulate in it and the two populations would diverge to become adapted to their respective environments. If they diverged almost to become different species, the border would be recognized as a hybrid zone. The two populations would have separated while they were geographically contiguous, along an environmental gradient. This procedure is the first stage of parapatric speciation.

The two populations on either side of the hybrid zone, according to this interpretation, have diverged without any period of geographic separation. The contact at the zone is called primary. The allopatric theory of speciation would have a different interpretation of the observed hybrid zones. It suggests that the two forms first diverged allopatrically; their ranges then changed, and they spread up to each other's borders. Prezygotic reproductive isolation, however, had not evolved, and the two populations could interbreed and form hybrids. Hybrid zones then comprise cases of secondary contact—they exist because populations have diverged allopatrically and subsequently come back into contact.

Neither mechanism is theoretically controversial. A stepped cline can evolve along an environmental gradient and populations can diverge in allopatry and then change their ranges. Empirical controversy can arise about particular hybrid zones, however, and it has turned out to be difficult, or even impossible, to tell whether the races of most real hybrid zones diverged allopatrically or while in contact.

One piece of evidence that has been argued to favor a "secondary" explanation of hybrid zones is the coincidence of clines in many characters and genes at the same place. At a typical hybrid zone, the frequencies of many allozymes, morphological markers, and mitochondrial DNA genotypes all change through the same region (e.g., as shown in Figure 16.8). If selection were acting on each independently (the argument says), it would be unlikely to favor a change in all of these characters at exactly the same site. On the other hand, if the populations diverged in all of these characters allopatrically, they would all change in space at the same hybrid zone. This argument is unconvincing, because an abrupt change in the environment at the location of a hybrid zone will place selection on many features of the organisms at the same point in space.

Although the evidence is indecisive, many hybrid zones probably are secondary. The ice ages may explain the crows, and other examples. Species ranges generally changed at that time (section 18.5, p. 513), and it is possible that populations would have been isolated in refuges, and then expanded later when the climate became warmer.

16.5.2 *Hybrid zones may evolve into species barriers by reinforcement*

Will reinforcement operate in a hybrid zone? If the hybrids are not selected against, reinforcement will not appear, and the hybrid zone could remain stable indefinitely. If the hybrids are disadvantageous, however, natural selection will act to reinforce the reproductive isolation between the two forms. If reinforcement is effective, two full species can evolve. The full process of parapatric speciation, therefore, calls for a cline to evolve first, without any allopatric phase, as far as a hybrid zone; the hybrids are selected against in this zone. The two different forms are then selected to mate only with others of their own type.

The difficulty with reinforcement, we saw earlier, is that it must occur in a limited time. If we have two populations, containing genotypes that when crossed produce inferior hybrids, reproductive isolation is favored only while the populations remain distinct. If they are not yet isolated, natural selection will either eliminate one of the genes or gene flow will merge the two populations. In the theory of parapatric speciation, this problem is solved by the stable cline. A cline like Figure 16.3b will be maintained indefinitely by natural selection, and reinforcement has a much longer time to operate. The stable cline is an essential feature of parapatric speciation; without it, not enough time would be available for reinforcement to occur.

Thus, the theoretical possibility of reinforcement exists. What evidence supports this theory in modern hybrid zones? We must first establish that the hybrids in hybrid zones are actually disadvantageous, and the evidence suggests that they usually are—though with some exceptions. Barton and Hewitt had in 1989 reviewed 170 reported hybrid zones and concluded that most of them were "tension zones." A *tension zone* is a hybrid zone in which the hybrids are selected against, and in which the two races on either side of it are adapted to different environments. The hybrid zone exists because of the dispersal of individuals out of the area to which they are best adapted.

In such a case, natural selection should have favored mating preferences. Within the zone, the mating preference should be stronger (because of reinforcement) than outside it where only one form is found. Curiously, however, little evidence has been found of reinforcement in such locations. A general mating preference exists in most cases; in the crickets studied by Harrison, for example, each species prefers a mate of its own species. Evidence that the mating preferences are any stronger in the hybrid zone is in short supply. In a 1985 review of 32 hybrid zones that supported partial isolation, Barton and Hewitt found evidence of reinforcement in only three cases. It is not known why reinforcement of prezygotic isolation is not more powerful. Perhaps hybrid zones are not tension zones at all, and the apparent evidence that hybrids are selected against is misleading in some way. In some cases, as Moore (1977) has argued, hybrids may be favored.

The fact that interbreeding persists probably explains another property of hybrid zones: their stability over time. Three chromosomal races of the grasshopper *Warramaba* meet in hybrid zones on Kangaroo Island, South Australia, and the hybrid zones are thought to have been stable for 8000–10,000 years.

In summary, the theoretical argument for the two stages of parapatric speciation—divergence along an environmental gradient, followed by reinforcement—is persuasive. Its application to natural evidence is less complete, however. It can usually be argued that real hybrid zones are secondary, not primary, and further work is needed to sort out how reinforcement operates (or does not operate) within them.

The same evidence of geographic variation used to support the theory of allopatric speciation can apply to parapatric speciation as well. Both processes require geographic variation; in both, the source material is converted into speciation. In the case of particular modern species, however, it is difficult to decide whether any given pair of species evolved by parapatric or allopatric speciation. The key difference between the two processes involves the geographic distribution during a short phase in the past, and that evidence has now been lost through the stream of history. Modern geographic distributions can be suggestive, but not decisively so. The parapatric theory predicts that recently formed pairs of species will have contiguous geographic distribution, whereas the allopatric theory does not particularly predict this pattern (although it does not rule it out). The relative importance of the two processes is unlikely to be settled by biogeographic evidence. A solution is more likely to come from theoretical work, and further research on hybrid zones.

16.6 Sympatric speciation

16.6.1 Sympatric speciation is theoretically possible

In sympatric speciation, a species splits into two without any separation of the ancestral species' geographic range. It has been a source of recurrent controversy whether sympatric speciation ever happens. Mayr has particularly cast doubt on this theory, which has, in turn, stimulated others to work out conditions under which it may be possible. Although many models have been proposed, we will consider only one general model, by Seger, which clarifies why some models apparently permit sympatric speciation whereas others do not.

We have already encountered the main problem with this idea. Speciation is most likely to be favored when the population contains two genotypes, and cross-mating between the genotypes produce deleterious hybrids. In that case, selection will favor individuals that mate only with other individuals of the same genotype. (A population in which individuals preferentially mate with others of their own genotype shows *assortative mating*; this situation contrasts with random mating.) Natural selection will eliminate all but one of the genotypes and gene flow will act to unify a population genetically and to break down any incipient differences between subpopulations. The conditions for speciation become easier if selection is forcibly maintaining the genetic diversity in the population (as it does in the stable stepped cline of a hybrid zone). Under these circumstances, natural selection and gene flow will not preempt the speciation process. If the mutations needed to cause reproductive isolation can arise, speciation will be possible. In short, we need a stable polymorphism, and a stable polymorphism in which assortative mating would be favored by selection.

Seger suggests we imagine a bird species in which (like the Galápagos finches, section 9.1, p. 222) beak size determines the type of food the bird can eat. If the food consists of seeds, then the environment will include some distribution of seed sizes. Suppose also that several genotypes influence beak size in the bird population. The fitnesses of the genotypes will be negatively frequency-dependent (section 5.13, p. 121) because, as more birds arise with a certain size of beak, they will compete with one another for food and lower one another's fitness. A stable polymorphism of beak genotypes will be found. The question is how the seed size distribution will influence the evolution of the bird population.

The simplest case occurs when the seed sizes have a normal distribution (Figure 16.10a). The bird population can then match the resource distribution by random mating among the genotypes controlling beak size. The population will evolve to a state at which all the beak sizes have equal fitnesses; in that

Figure 16.10 Two theoretical resource distributions. (a) Seed sizes have a normal distribution. Random mating among bird genotypes will result in a distribution of beak sizes that matches the resource distribution. (b) Seed sizes have a flat distribution. Random mating among bird genotypes will produce too few birds with extreme beak sizes, and those birds will have a higher fitness because of reduced competition for food.

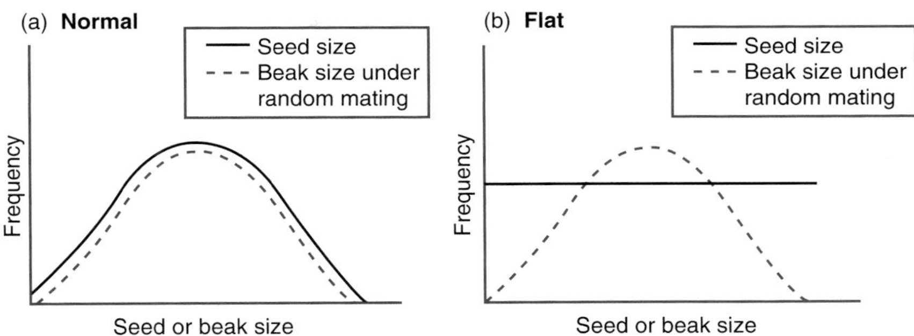

case, no beak size will have an advantage over any other beak size, and the population will be stable.

The interesting cases occur when the resource distributions are not normal. Seger first considers the case of a flat distribution, in which there are equal quantities of all seed sizes (Figure 16.10b). With random mating of birds, the beak size distribution will be normal as before, but now the different genotypes will not have equal fitness. Birds with extreme beak sizes will experience less competition for food, and have higher fitness. In this circumstance, selection will favor assortative mating for beak size. By mating assortatively, birds at the edge of the distribution would produce more offspring like themselves, and therefore with higher fitness, than if they mated at random. As selection fixes assortative mating, the population will split into two new species—one with long beaks, the other with short.

When random mating can produce as good, or a better, fit of beak sizes to the seed resource distribution than can assortative mating, then selection will not produce speciation. Sympatric speciation becomes particularly unlikely when the resources to which the genotypes are adapted are normally distributed. If assortative mating produces a better fit, then a condition for sympatric speciation exists. In that case, selection favors assortative mating and can be strong enough to produce full speciation.

We have discussed Seger's model in terms of seeds and beaks, but it could apply to any resource and to any genotypically controlled character related to the resource. When the phenotypic distribution can be brought more into line with the resource distribution by assortative mating, sympatric speciation becomes a possibility.

The conditions producing a multiple niche polymorphism (section 5.14, p. 123) and the conditions of Seger's model are closely related. A multiple niche polymorphism will be stable within a species provided that either no genes enable assortative mating, or random mating generates genotypes in the same frequencies as their niches exist in the environment. If the niches are in frequencies that cannot be matched by random mating, then selection will tend to produce sympatric speciation. The general conditions for Seger's model to apply are that the genotypes are adapted to different resources and the limited resources generate frequency-dependent selection. If the resources do not match the frequencies of the genotypes generated by random mating, sympatric speciation becomes a possibility. How commonly these conditions are found in nature is not the issue here; rather, we have sought only to clarify the conditions needed for sympatric speciation to take place.

16.6.2 *Two species of green lacewings may have split sympatrically*

A study by Tauber and Tauber provides a possible example of sympatric speciation, in which the genetics are also understood. The Taubers worked on the green lacewings *Chrysoperla carnea* and *C. downesi*. *C. carnea* and *C. downesi*, which live in North America, are similar in appearance, but show subtle differences in their coloration. *C. carnea* is light green in spring and early summer and changes to brown in the fall; *C. downesi* is a darker green year-round. Their colors are adapted, as camouflage, to their habitats. *C. carnea* lives in fields and meadows in the summer and moves to deciduous trees in the fall, whereas *C. downesi* mainly lives on conifers. The geographic ranges of the two species are sympatric: the range of *C. downesi* is completely contained within that of *C. carnea*. The Taubers know of no evidence that the two species' ranges were ever separated in the past.

The species are separated by breeding season as well as habitat. *C. carnea* breeds in winter and again in the summer, while *C. downesi* breeds only in the spring. The breeding time is controlled by a photoperiodic response. Thus, by bringing the lacewings into the laboratory and keeping them on artificial light-dark cycles, the Taubers were able to make the two species breed at the same time. The species could hybridize successfully.

From genetic experiments, the Taubers found that the species differed in three main genetic loci. One locus controlled color: the *C. carnea* color phenotype was produced by one homozygote (G_1G_1) and the *C. downesi* phenotype by another homozygote (G_2G_2); heterozygotes (G_1G_2) were intermediate in color. The fact that no lacewings with the hybrid coloration were found in nature suggests that the two are "good" species that do not interbreed naturally. The other two loci controlled the photoperiodic response, and dominance appeared at both. A lacewing with two double recessive homozygous conditions breeds (like *C. downesi*) in the spring; a lacewing with a dominant allele at at least one of the loci breeds in the winter and summer (like *C. carnea*).

The Taubers suggest that speciation in these lacewings took place as follows. The first stage was the establishment of a polymorphism within a species. One morph was adapted to live on conifers and the other in the deciduous habitat. Thus, the lacewings developed multiple niche polymorphism. The stable polymorphism is necessary, or selection and gene flow would reduce the species back to a single genotype. Once the polymorphism existed, selection would favor assortative mating, because crosses between the two morphs would not be well camouflaged to either habitat. The divergence in breeding times could have been the mechanism by which assortative mating was brought about. The result would be the two species we now see.

Henry subsequently showed that reproductive isolation between the species is effected by their courtship songs, as well as the separation of breeding seasons and habitats that the Taubers identified. If the Taubers' theory is correct, the differences in courtship songs would have evolved after the establishment of the habitat polymorphism. Courtship song, along with breeding season, would have caused assortative mating. In theory, the divergence of song could have occurred before, after, or at the same time as the divergence in breeding season.

The Taubers' interpretation is theoretically plausible and fits in with all the evidence. However, it is impossible to confirm it with certainty. We cannot prove that the two species never had an allopatric phase in the past and an advocate of the allopatric theory of speciation need not feel compelled to accept the sympatric interpretation. However, the question of what the evidence suggests is at least as interesting as the amount of evidence necessary to force a fanatic to accept this concept. There is no evidence for an allopatric phase, whereas the species' sympatry is now observable. All the conditions for sympatric speciation seem to be met. The Taubers' study gives about as complete a set of evidence for sympatric speciation as is possible.

16.6.3 *Phytophagous insects may split sympatrically by host shifts*

Rhagoletis pomonella is a tephritid fly and a pest of apples. It lays its eggs in apples, and the maggot then ruins the fruit. It did not always follow this path, however. In North America, *R. pomonella*'s native larval resource is the hawthorn. Only in 1864 were these species first found on apples. Since that time, the fly has expanded through the orchards of North America, and has also started to exploit cherries, pears, and roses. These moves to new food-plants are called *host shifts*. In the host shift of *Rhagoletis pomonella*, speciation may be happening before our eyes.

The *R. pomonella* found on the different hosts are currently different genetic races. Females prefer to lay their eggs in the kind of fruit in which they grew up. Thus, females isolated as they emerge from apples will later choose to lay eggs in apples, given a choice in the laboratory. Likewise, adult males tend to wait on the host species in which they grew up; mating takes place on the fruit before the females oviposit. Thus there is assortative mating. Male flies from apples mate with females from apples, males from hawthorn with females from hawthorn.

The races are presumably about 100 generations old (given that they first moved on to apples a century ago). Is this period long enough for genetic differences between the races to have evolved? Gel electrophoresis shows that the two races have evolved extensive differences in their enzymes. They also differ genetically in their development time; maggots in apples develop in about 40 days, whereas hawthorn maggots develop in 55–60 days. This difference also acts to increase the reproductive isolation between the races, because the adults of the two races are not active at the same time.

Apples and hawthorns differ and selection will, therefore, probably favor different characters in each race; this selection may be the reason behind their divergence. If so, selection may also favor prezygotic isolation and speciation. If flies from the different races are put together in the lab, however, they mate together indiscriminately. Either reinforcement has not operated when expected, or, alternatively, the differences in behavior and development time in the field may be enough to reduce interbreeding to the level natural selection favors. Selection would then not act to reinforce the degree of prezygotic isolation. To determine which interpretation is correct, we would need to know more about the forces maintaining the genetic differences between the races.

In the case of host shifts, unlike the Taubers' lacewings, we can be practically certain that the initial host shift, and formation of a new race, has happened in sympatry. The shift took place in historic time. It does not provide a full example of sympatric speciation, however, because the races have not fully speciated. Indeed, we do not know whether they will, or whether the current situation, with incomplete speciation, is stable.

How general is the process of sympatric speciation by host shifts? A definite answer cannot be given. In truth, it has not even been confirmed that sympatric speciation ever takes place by host shifts; but there are interesting hints that the process might be important (section 22.2, p. 612). Several phytophagous insect taxa have undergone extensive phylogenetic radiations on plant host taxa. For example, about 750 species of fig wasps are known, each of which breeds on its own species of fig. In Great Britain, the dipteran family Agromyzidae alone includes 300 species of leaf miners, and 70% of those species each feed on only one plant species. It is easy to imagine how these groups could have radiated from a single common ancestor, as successive new species arose by host shifts like the one taking place in the apple maggot fly in the United States. If phytophagous insect species consisted of an odd species scattered through the phylogeny of the insects, and feeding on unrelated kinds of food plants, the process would probably have not been operating. The existence of whole large taxa of host-plant-specific phytophages, on the other hand, suggests that speciation by host shifts could have contributed to their diversification.

16.7 *Some plant species have originated by hybridization and polyploidy*

We first encountered the origin of plant species by hybridization in section 3.6 (p. 50), where we saw how a new species of primrose and a natural species of *Galeopsis* were artificially produced by hybridization. Interspecies hybrids are typically sterile, usually because the chromosome pairs, which consist of one chromosome from one species and another chromosome from the second species, do not segregate regularly at meiosis. For a new species to evolve, this sterility must be overcome, and the common mechanism is polyploidy. If the chromosome numbers are doubled, each chromosome pair at meiosis contains two chromosomes from one species, and regular segregation is restored. Polyploidization is encouraged by applying the chemical colchicine in the commercial production of new species, but it can also occur naturally at a low rate. A new hybrid species may then evolve. The polyploidy hybrids are interfertile among themselves, but reproductively isolated (by the mismatch in chromosome numbers) from the parental species; they are therefore well-defined new species.

The simplest cases to identify are those, like *Primula kewensis*, in which the new species is a simple 50:50 hybrid, produced from two parental species, with 50% of its genes coming from one parental species, and 50% from the other. Within the past century, the natural evolution of four new species of this sort has been recorded, two in Britain and two in North America.

The two North American examples belong to the genus *Tragopogon*. *Tragopogon* is an Old World genus, but three species have been introduced to North America: *T. dubius*, *T. pratensis*, and *T. porrifolius* (whose common name is salsify, and whose roots can be eaten as a vegetable). All three species are found together in regions of east Washington and Idaho, and they first became established there in the first 2–3 decades of this century. By 1950, Ownbey discovered that two new species had appeared in this region, *T. mirus* and *T. miscellus*. Both species continue to thrive, and samples 40 years later by Novak *et al.* (1991) showed that *T. miscellus* is now a common weed of roadsides and vacant lots in and around Spokane, Washington, and to the east.

Ownbey showed that *T. mirus* and *T. miscellus* (each with 12 pairs of chromosomes) are tetraploid hybrids of pairs of the three introduced species (which are diploid and have six pairs of chromosomes). The forms of the chromosomes in the species, as well as other characters such as flower color, revealed that *T. mirus* is derived from *T. dubius* and *T. porrifolius*, and *T. miscellus* from *T. dubius* and *T. pratensis*.

Ownbey found many interspecies hybrids in nature, but they were all diploid and sterile. Presumably hybrids crop up from time to time in nature, and at some stage a tetraploid mutant occurred among them, and gave rise to the new species. The tetraploid hybrids are fertile, and reproductively isolated from the parental species. Subsequent work has used more discriminating genetic markers; Roose and Gottlieb used gel electrophoresis, and Soltis and Soltis used restriction enzymes on nuclear and cytoplasmic DNA. Both studies showed that the two new species are both derived from more than one origin. That is, the tetraploid mutants must have arisen more than once. Roose and Gottlieb's gel electrophoretic study also hinted at the possible basis of the success of the hybrids in their new environment. It revealed that the hybrids had novel forms of certain enzymes, due to combinations of components from the enzymes from each parental species.

The diploid hybrids of *Tragopogon* are sterile, but in other cases the initial hybrids are partly interfertile with one or both parental species. The hybrids then backcross to the parents. This gene flow from parental species into hybrid population is called *introgression*. Many outcomes are possible from introgression, depending on the degree of interfertility with the parents. It is believed that some degree of interbreeding occurs between hybrids and parents, and the hybrid population builds up a complex mixture of the genes of the two parental species. At some point, perhaps after a polyploid mutation, the hybrid population becomes reproductively isolated from the parental species. At that point, it has evolved into a new species (Figure 16.11).

Many cases of hybrid speciation in plants probably involve a number of generations of introgression, rather than an instantaneous speciation event. The difference between introgression and simple hybridization is that in introgression the new species will have a complex mix of parental genes, according to the history of backcrosses between hybrid and parental species during the origin of

Figure 16.11 Hybrid speciation by introgression. The initial hybrid individuals are interfertile with one or both parental species and backcross with them, producing a hybrid population with various mixtures of genes from the two parental species. At some stage, the hybrid population may evolve far enough to be reproductively isolated fom the parental species; it is then recognized as a new species. Reprinted from Rieseberg and Wendel (1993) by permission of Oxford University Press.

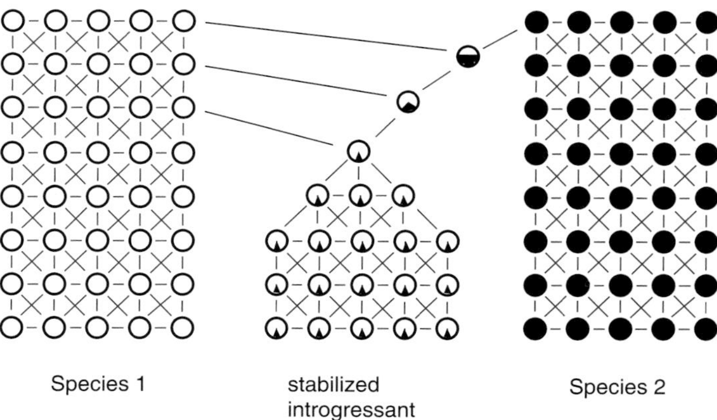

the new species. In contrast, a simple hybrid receives 50% of its genes from one parental species and 50% from the other. Rieseberg and Wendel recently reviewed introgressive speciation in plants; they listed 165 cases in which it had been suggested, and they judged that the evidence for introgression was good in 65 of them.

One of the best examples of introgression comes from the wetlands of south Louisiana (Color Plate 6). A number of species of attractive irises grow in the swamps and rivers in that region, and Arnold and his colleagues have been using genetic markers to reconstruct their origin. Color Plate 6a illustrates the three parental species in this example. Two of them are *Iris fulva*, with tawny colored flowers, and *I. hexagona*, whose flowers are violet with yellow crests. Both species are widespread in streams in the southeast, and in southern Louisiana they live in the water channels called bayous that are derived from the Mississippi River. The third parental species is *I. brevicaulis*, which is colored like *I. hexagona* but has a different growth habit. *I. hexagona* grow up to 4 feet, whereas *I. brevicaulis* tend to lie flatter on the ground and curve upward.

Unlike *I. fulva* and *I. hexagona*, *I. brevicaulis* lives in drier habitats such as hardwood forests. Where a bayou happens to flow into an *I. brevicaulis* habitat, the three species come into near contact and hybridize (Figure 16.12). Natural hybrids are not uncommon, though they are formed at low rates. Hybrids may, to some extent, cross back to the parental species, producing a complicated mix of genotypes in the populations where the species meet.

In the 1960s, a new species of iris was detected in the region where hybrids are found, and it was named *Iris nelsonii* (Color Plate 6b). The new species has a morphology, including flower color, and chromosomal complement that shows similarities to *I. fulva* and *I. hexagona; I. brevicaulis* also contributed genes to its

Figure 16.12 Where bayous flow into swamps in southern Louisiana, the bayou-dwelling *Iris fulva* and *I. hexagona* come into contact with the swamp-dwelling *Iris brevicaula*. In the intermediate regions, the hybrid species *I. nelsonii* has evolved by introgression. Reprinted from Arnold and Bennett (1993), after Viosca. Copyright © 1993 by Oxford University Press, Inc. Reprinted by permission.

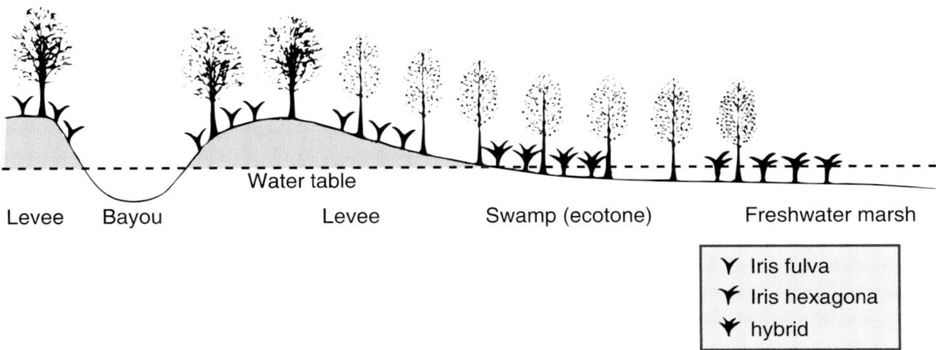

origin, however. Genetic markers suggest that *I. nelsonii* mainly resulted through repeated back-crossing into *I. fulva*, rather than from one simple hybridization event in the manner of the Kew primrose (*Primula kewensis*). *I. nelsonii* is not polyploid.

We have concentrated on the problem of how a new reproductively isolated hybrid genotype can evolve. Two other problems are also likely to arise in the evolutionary transition from a rare new hybrid genotype to a full hybrid species. One is finding a mate. When a fertile polyploid hybrid first arises, it is one hybrid (or perhaps one of a small number) within two large populations of the parental species. It may simply be infertile with both parental species because of the chromosomal difference. The situation may be worse if parental species pollen fertilizes the hybrids' eggs and they then fail to develop or reproduce. The hybrid's interfertility with other hybrids like itself can only be expressed if other hybrids exist. Natural selection on the hybrid, therefore, has a kind of positive frequency dependence (section 5.13, p. 121). When the hybrid is rare its fitness is lower because of the difficulty of finding a mate. As a result, it may have to reach some threshold of abundance before natural selection favors it. (Strictly speaking, this is number, rather than frequency, dependence, but frequency-dependent selection may occur in at least an informal sense.)

This problem probably explains why hybrid speciation has been much more common in some groups of plants than others. It is much easier for a new hybrid to cross the difficult transition stage, in which it is rare, if it has alternative reproductive options besides sexual cross-fertilization. Stebbins has shown that hybrid speciation is more common in groups in which asexual reproduction or self-fertilization are possible. *Iris nelsoni*, for example, can reproduce asexually by rhizome runners, in addition to sexual cross-fertilization via pollen that is carried by bumblebees.

The other problem is ecological competition. We saw earlier (section 15.2.5, p. 410) that two species with identical ecological needs will usually not be able to coexist. If one of the species can survive on a lower level of resources, it will drive the other locally extinct. When the hybrid arises, it will occupy in the same place as the parental species, and it is likely to have ecological needs that overlap with those of its competitor. The hybrid species that survive and become established are, therefore, thought to be those that are adapted to different ecological niches from the parental species. Hybrids that were ecologically too similar to the parental species would have been lost, and we do not observe them today. The survival of *Iris nelsonii* may have been helped, for example, by the way it lives in habitats in between the other three species (Figure 16.12).

Hybrid speciation is an example of sympatric speciation, but it is much less controversial than sympatric speciation in animals. In an earlier section (16.7) we encountered the problem that reinforcement may not be able to operate before the incipient species go extinct or are drowned by gene flow. The same problem could arise in a prolonged case of introgression, but the mechanism of polyploidy can quickly isolate a hybrid from its parental species. There will therefore be problems, as we have seen, that will arise in finding mates and in ecological competition. The problem of gene flow from the parental species has been cut off, however. Hybrid speciation in plants, therefore, demonstrates that sympatric speciation can occur, but its existence does not provide evidence that it can occur easily in animals, in which polyploidy is much rarer (for reasons we shall not discuss here).

16.8 *Reinforcement is suggested by greater sympatric than allopatric prezygotic isolation between a pair of species*

Drosophila mojavensis and *D. arizonae* are two closely related species of fruitfly that coexist in Sonora, Mexico, while each species lives on its own elsewhere in the southwest. *D. mojavensis*, for example, lives in Baja California where *D. arizonae* is absent, and *D. arizonae* lives in other regions of Mexico without *D. mojavensis* (Figure 16.13). When a male of one species comes into contact with a female of the other species, they are less likely to mate than are a pair from the same species. That is, the two species show a degree of prezygotic reproductive isolation. Wasserman and Koepfer measured the degree of mating discrimination in populations taken both from where the two species coexist and from where only one of the species lives. They found that discrimination against potential mates from the other species was stronger in the flies from regions where both species are found (Figure 16.13c); this finding provides another example of character displacement (section 15.2.5, p. 410). Character displacement occurs when two species differ more in sympatry than in allopatry. The term can refer to any character, and *Drosophila mojavensis* and *D. arizonae* show reproductive character displacement—to be exact, character displacement for prezygotic reproductive isolation. These two species represent only one example among several in which this result has been found.

Wasserman and Koepfer's observations suggest that reproductive isolation has been reinforced in sympatry. When the two species do not encounter one another (i.e., allopatrically), no selection will have acted on them to produce discrimination against mates from the other species. In sympatry, where interbreeding may produce hybrids of reduced fitness, selection will have favored

Figure 16.13 (a) A study of character displacement requires two species with partly overlapping ranges. (b) Distributions of *Drosophila mojavensis* and *D. arizonae* in the southwest. They coexist in part of Sonora, Mexico, and are found alone in other areas, including large regions of Mexico for *D. arizonae* and Baja California for *D. mojavensis*. The dots on the map for *D. arizonae* are the collecting sites for the experiment in (c), within a fairly continuous distribution. (c) Experimental demonstration that reproductive isolation is higher between the two species in sympatry than in allopatry. The experiments give a female of one species a choice of mating with males of the other two species, or vice versa. In the experiment, the number of matings with members of the same (H_s) and with the other species (H_o) and total number of matings (N) were measured, and the isolation index $I = (H_s - H_o)/N$, as explained in Table 16.1. In the table, the top left number means that *mojavensis* females, taken from a place where there are no *arizonae* (Baja California)—that is, allopatric *D. mojavensis*—when put with males of both species, show an isolation index of only 0.3. The same format is followed for the other seven conditions. Reprinted by permission of the publisher. Map (b) from Koepfer (1987), results (c) from Wasserman and Koepfer (1977).

(a) Simplified system to test for character displacement

(b) Distribution of *Drosophila arizonae* and *D. mojavensis* in the southwest

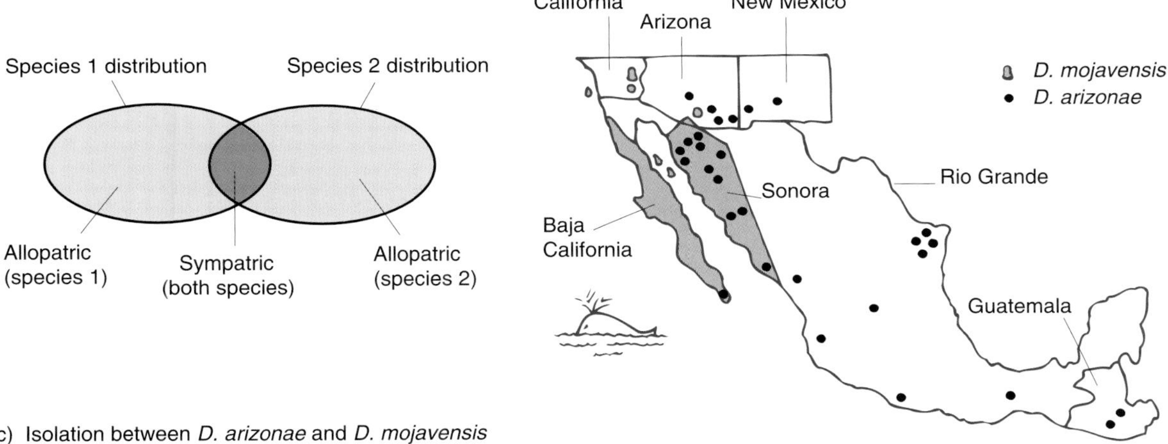

(c) Isolation between *D. arizonae* and *D. mojavensis*

D. mojavensis female with *arizonae* male		*D. arizonae* female with *mojavensis* male	
D. mojavensis female		*D. arizonae* female	
Allopatry	$I = 0.3$	Allopatry	$I = 0.9$
Sympatry	$I = 0.94$	Sympatry	$I = 0.8$
D. arizonae male		*D. mojavensis* male	
Allopatry	$I = 0.54$	Allopatry	$I = 0.78$
Sympatry	$I = 0.6$	Sympatry	$I = 0.92$

mechanisms to prevent cross-breeding. Wasserman and Koepfer's observations give particularly persuasive evidence of reinforcement because the comparison takes place between populations within species. We have only limited evidence of this kind, but far more evidence of a less persuasive kind has been gathered. When reproductive isolation is measured by mating experiments between pairs of closely related sympatric species and between pairs of closely related allopatric species, this measurement is often found to be higher in the sympatric species pair than in the allopatric species pair (we shall see some examples of such cases in the next section). This finding could again be due to reinforcement, but, as Templeton pointed out, the same result could occur without reinforcement.

Fusion is one alternative (and not the only one). Suppose that some pairs of species, while the two species are living allopatrically, evolve higher degrees of prezygotic isolation than do other pairs. The geographic ranges within the various pairs of species then change, and the members of each pair come into contact. In those cases in which reproductive isolation is low, the pair would be more likely to fuse back into a single species than would more isolated pairs. The same process will not operate in allopatry, where fusion is impossible. Sympatric species pairs would then show higher reproductive isolation than allopatric pairs, because the allopatric pairs include all kinds of species, whereas the pairs of sympatric species include only those pairs that have higher isolation—those with lower isolation have fused and no longer exist as "pairs" to use in a comparison.

Evidence in which whole species are compared is, therefore, unconvincing when considering the possibility of reinforcement. Comparisons, like Wasserman and Koepfer's, between populations within a species serve as stronger tests of reinforcement. Templeton's argument can explain why sympatric species pairs have higher reproductive isolation than allopatric species pairs, but it cannot easily explain why sympatric populations within a species should have higher reproductive isolation than allopatric populations of the same species.

16.9 *A study of speciation in* Drosophila, *by Coyne and Orr, provides evidence about reinforcement and other interesting results*

A subtle comparison of reproductive isolation between allopatric and sympatric species pairs in *Drosophila*, by Coyne and Orr, also suggests that reproductive isolation has been reinforced. Their result emerges from a comparison of prezygotic and postzygotic isolation between many pairs of species; the species pairs differed in how closely related they were. (Box 16.1 explains how the key variables were quantified.) Some of the pairs of species lived sympatrically, in at least part of their geographic range, while others were completely allopatric. Coyne and Orr found 42 independent pairs of species of *Drosophila* for which all four variables (prezygotic and postzygotic isolation, genetic distance, and geographic distribution) had been measured.

Figure 16.14 shows the relation between genetic distance and reproductive isolation. Reproductive isolation increases with time (that is, with genetic distance). This result makes good evolutionary sense. As time passes after the original split between the two species, genes are substituted in each lineage, genetic distance increases, and as the two species diverge they will come not to recognize, or meet, one another (prezygotic isolation). If they do mate, the genetic differences between them will tend to result in inviable or sterile hybrids (postzygotic isolation).

BOX 16.1

Quantitative Measures of Genetic Distance, Prezygotic Isolation, and Postzygotic Isolation

Genetic distance is a standard concept, of which the most common measure is Nei's index D. We calculate D by taking two populations and measure the frequency in both of the alleles at several loci. Consider an allele A_i at a locus in population 1 and 2, and define its frequency as p_i in population 1 and q_i in population 2. (By the usual convention, i has as many values as there are alleles at a locus; if there are two alleles, $i = 1$ for one of them and $i = 2$ for the other.)

First, calculate $j_1 = \Sigma p_i^2$, $j_2 = \Sigma q_i^2$, and $j_{12} = \Sigma p_i q_i$. In this case, j_1 is the probability that two randomly chosen genes in population 1 are identical, j_2 is this probability for population 2, and j_{12} is the probability that two genes, one drawn randomly from population 1 and the other from population 2, are identical. (We may recognize j_1 more readily as the frequency of all homozygotes in population 1, and j_2 as the same frequency for population 2.) The same set of three numbers can be calculated for every locus; the next step is to take averages for all loci. Define the averages as follows: $J_1 = j_1$, $J_2 = j_2$, $J_{12} = j_{12}$. Finally, Nei's formula for genetic distance is:

$$D = -\log_e \frac{J_{12}}{\sqrt{J_1}\sqrt{J_2}}$$

D decreases as more alleles are shared between the two populations, and make the product J_{12} higher. In the extreme case of two samples from one population, $J_1 = J_2 = J_{12}$, and $D = -\log_e 1 = 0$. As shared alleles are lost, D increases. D has no mathematical upper limit (in the extreme case of two samples that share no alleles, $J_{12} = 0$, $D = -\log_e 0 = \infty$). In practice, it is found to have values of about 0.1–2 for different species, and of more than 1 for different genera. The formula can also be expressed $D = -\log_e I$, where $I = J_{12} / \sqrt{(J_1 J_2)}$. I is called the genetic identity between the two populations.

Because many proteins evolve at least approximately in the manner of a molecular clock (chapter 7), genetic distance increases approximately in proportion to the time since the two species diverged.

Postzygotic isolation. If two species can be crossed in the laboratory, some or all of their offspring may be inviable or sterile. We saw in Box 15.2 (p. 407) that the heterogametic sex is often inviable or sterile, a generalization known as Haldane's rule. There are two sorts of cross between two species: either a male of species 1 and a female of species 2, or a male of species 2 and a female of species 1. According to its sex and the type of cross, an offspring can be one of four types: male and from male species 1 × female species 2; female and from male species 1 × female species 2; and so on. For a species pair we count the number of the four types that are sterile or inviable and divide the result by four. The index can, therefore, have values 0, 1/4, 1/2, 3/4, 1 in order of increasing isolation. For a pair of species that fit Haldane's rule, the index = 1/2.

Prezygotic isolation. Prezygotic isolation is generally measured in a mate choice experiment in which an individual of one species has a choice of mates from its own, and another, species. If it mates with its own species, the mating is called homospecific; if it mates with the other species, the mating is termed heterospecific. As an index of prezygotic isolation, Coyne and Orr used the following calculation:

$$\text{Prezygotic isolation} = 1 - \frac{\text{frequency of heterospecific matings}}{\text{frequency of homospecific matings}}$$

In theory, the index can vary from 1 (complete isolation) to 0 (no isolation) through to $-\infty$ for complete disassortative mating. In practice, the index varies between 0 and 1 for most pairs of species. (In Table 16.1 and Figure 16.14 a slightly different isolation index was used.)

Figure 16.14 Strength of isolation in relation to genetic distance for pairs of species of *Drosophila*. (a) Prezygotic isolation. (b) Postzygotic isolation. Note that prezygotic isolation is higher for small distances than is postzygotic isolation. Reprinted, by permission of the publisher, from Coyne and Orr (1989b).

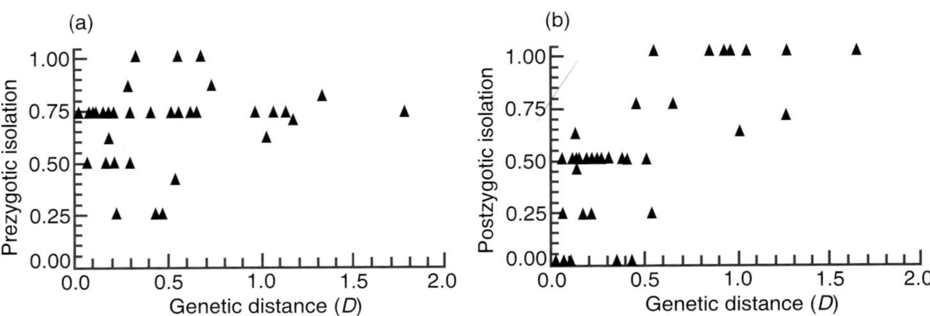

A second interesting result is less immediately visible in Figure 16.14. If we concentrate on closely related pairs of species—those with genetic distance (D) ≤ 0.5—prezygotic isolation is higher than postzygotic isolation. Both prezygotic and postzygotic isolation eventually reach 1 for high enough genetic distances, but prezygotic isolation does so at a lower value of D. Once the difference has been pointed out, it can be seen in Figure 16.14 (compare the large number of points with isolation about 0.75 in the prezygotic isolation figure with the large number at more like 0.5 in the postzygotic isolation figure), and it can be confirmed statistically. The result suggests that prezygotic isolation tends to evolve during speciation before postzygotic isolation. We shall look at a possible explanation later; here we need only notice the result.

Finally, we can compare the isolation of sympatric and allopatric pairs of species. Figure 16.15 shows the degree of prezygotic isolation between pairs of sympatric species and between pairs of allopatric species. Notice that the sympatric pairs are much more isolated from each other. The result is an interspecific

Figure 16.15 Strength of prezygotic isolation in relation to genetic distance for pairs of species of *Drosophila* (a) for allopatric pairs of species, and (b) for sympatric pairs of species. Note that prezygotic isolation is higher for sympatric pairs of species. Reprinted, by permission of the publisher, from Coyne and Orr (1989b).

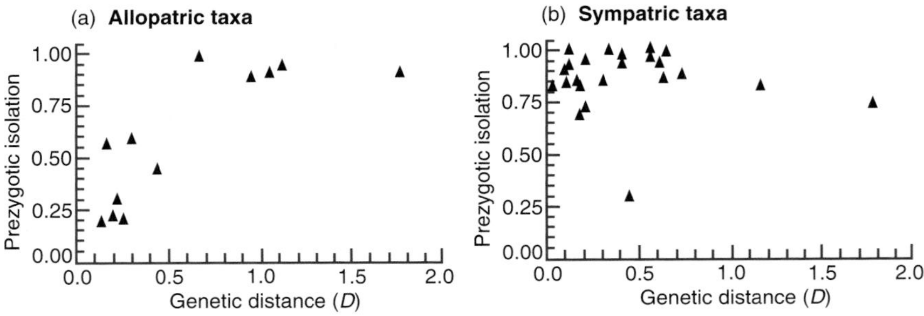

analogue of the character displacement of *Drosophila mojavensis* and *D. arizonae*. In that case, different populations of the same species were compared; here we are comparing different species. The result again suggests that sympatric species have evolved mechanisms to prevent interbreeding, although Templeton's objection could again apply.

The result suggesting that reinforcement has taken place during evolution is found in Figure 16.16. Prezygotic isolation is higher in sympatric than in allopatric species, but no such difference appears for postzygotic isolation. Templeton's argument that species with low isolation will tend to fuse in sympatry applies equally to postzygotic and prezygotic isolation; it does not predict that prezygotic isolation will be increased in sympatry while postzygotic will not.

By contrast, the theory of reinforcement can explain the result. Suppose that hybrids are initially selected against, though by some small amount such that their disadvantage is not detected in Coyne and Orr's index of postzygotic isolation. Selection would then favor preferential mating with conspecifics, and in sympatry prezygotic isolation will evolve. Postzygotic isolation, however, evolves only as an incidental consequence of the divergence between the two species; natural selection will not act to reinforce postzygotic isolation in fruitflies. As the species diverge, the genes within each will become less well coadapted to one another, and the hybrids become decreasingly viable. After prezygotic isolation has been reinforced, the two species' gene pools will diverge more rapidly, and the evolution of postzygotic isolation therefore follows that of prezygotic isolation.

This argument provides the explanation of Coyne and Orr's result by reinforcement. Its crucial prediction is that hybrids are disadvantageous even between closely related (low *D*) pairs of species. This disadvantage drives the entire system. If closely related species in which prezygotic isolation has evolved actually have hybrids as fit as crosses within a species, the argument fails. The prediction is testable, but has not yet been tested. Meanwhile, the argument is plausible, but uncertain.

Another interesting conclusion can be drawn from Coyne and Orr's study. The *x*-axis of the graphs have all been drawn as genetic distances. As is remarked in Box 16.1, genetic distance is approximately proportional to time for low values of *D*. Nei has estimated, from a number of studies, that it takes about 5,000,000

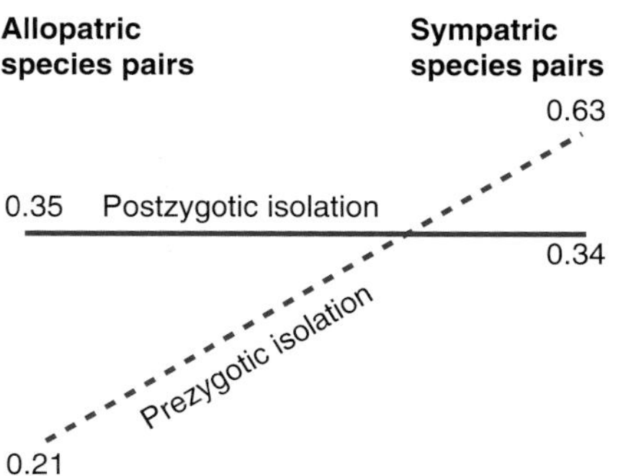

Figure 16.16 Average strength of prezygotic and postzygotic isolation between pairs of species of *Drosophila*, for allopatric and sympatric pairs. Note that prezygotic isolation is higher among sympatric than among allopatric pairs of species. Reprinted, by permission of the publisher, from Coyne and Orr (1989b).

years for D to increase from 0 to 1. We are now in position to calculate how long it takes for reproductive isolation to evolve between an average pair of *Drosophila* species. The average figure for isolation between sympatric pairs of *Drosophila* is about 0.9; the lower bound, from 95% confidence intervals, is 0.85. Isolation of 0.85 is, therefore, probably sufficient to keep two sympatric species apart. The relation between genetic distance and reproductive isolation suggests that isolation of 0.85 evolves at a genetic distance of $D = 0.53$ (for all data combined), or 0.66 (for allopatric species), or 0.31 (for the sympatric species). Reading these numbers back through Nei's estimate implies that it takes between 1,500,000 and 3,500,000 years for a typical new *Drosophila* species to evolve.

In a separate study, Coyne and Orr collected genetic results of crosses between closely related species of fruitflies. In these crosses, it is possible to make hybrid flies with various chromosomal combinations and find out where the genes responsible for hybrid sterility are located. Their striking result is that the genes are *always* concentrated on the X chromosomes (Table 16.2, which also includes results from four other taxa). This finding is probably related to Haldane's rule (Box 15.2, p. 407). If the genetic changes during the early stages of speciation occur mainly on the X chromosomes, we should expect to see difficulties in the hybrids with XY sex chromosomes (males in fruitflies) evolutionarily before we see them in those with XX sex chromosomes. The XY hybrids will have a mismatch between the X chromosome and one set of autosomes, and the interactions between these two might cause defects (see also Box 15.2).

In summary, Coyne and Orr's work suggests that selection acts in sympatry

TABLE 16.2

Experimental genetic interspecific crosses in which the sex chromosomes have a large effect on hybrid sterility. (Adapted, by permission of the publisher, from Coyne and Orr 1989b; see their work for the original references.)

Cross	Trait affected	Reference
Drosophilia species		
pseudoobscura/persimilis	Male and female fertility	Dobzhansky (1936); Orr (1987)
pseudoobscura/Bogota	Male fertility	Dobzhansky (1974); Orr (1989)
simulans/mauritiana	Male fertility	Coyne and Kreitman (1984)
simulans/sechellia	Male fertility	Coyne (1986)
mohavensis/arizonensis	Male fertility	Zouros *et al.* (1988)
micromelanica A/B	Male fertility	Sturtevant and Novitski (1941)
littoralis/virilis	Male fertility	Orr and Coyne (1989)
novamexicana/virilis	Male and female fertility	Orr and Coyne (1989)
texana/virilis	Male and female fertility	Orr and Coyne (1989)
lummei/virilis	Female sterility	Orr and Coyne (1989)
buzzattii/serido	Female sterility	Naviera and Fontadevila (1986)
mulleri/aldrichi-2	Female viability	Crow (1942)
texana/montana	Female viability	Patterson and Griffen (1944)
Glossina morsitans subspp. 1/2	Fertility or mating ability	Curtis (1982)
Anopheles arabiensis/gambiae	Fertility or mating ability	Curtis (1982)
Colias eurythreme/C. philodice	Female viability, fertility, mating vigor	Grula and Taylor (1980)
Cavia rufescens/C. porcellus	Male sterility	Detlefsen (1914)

to reinforce the prezygotic isolation of closely related species. It provides evidence of the rates (and relative rates) at which prezygotic and postzygotic isolation evolve, and demonstrates that isolation between species increases through time. It also shows that the earliest genetic changes in speciation are on the X chromosome.

16.10 *Chromosomal changes could potentially lead to speciation*

Species differ in their chromosomes as well as in their morphology, genetics, and behavior. The chromosomes of many taxa have been described and species are known to differ in their chromosomal numbers, lengths, and structure. Humans, for example, have 46 chromosomes, whereas chimps, gorillas, and the orangutan have 48 chromosomes, suggesting that a pair of chromosomes have fused in our ancestry. These sorts of observations would not inspire peculiar interest—given that species differ in all other respects—were it not for a theory suggesting that chromosomal changes may be exceptionally important in speciation.

Consider two forms of a chromosome, differing (for example) in an inversion. We can call the standard and inverted forms A and A'. We suppose, for simplicity, that A and A' have identical sets of genes and produce the same phenotype (Figure 16.17a). The two kinds of homozygotes (AA and $A'A'$) will have identical fitness. The heterozygote, however, is likely to be selected against, because recombinants between the two forms may have double sets of some genes, and lack others (Figure 16.17b), and organisms that lack genes will usually be selected against. Thus, the situation presents a precondition for the evolution of assortative mating. AA types should preferentially mate with other AA types and avoid crossing with $A'A'$, and vice versa. The usual difficulties with reinforcement

Figure 16.17 (a) A chromosomal inversion has a set of genes inverted. The letters represent genes along the chromosomes. (b) Recombination in a heterozygote for a chromosomal inversion can produce chromosomes that lack some genes and have others in double dose. These forms are probably selected against.

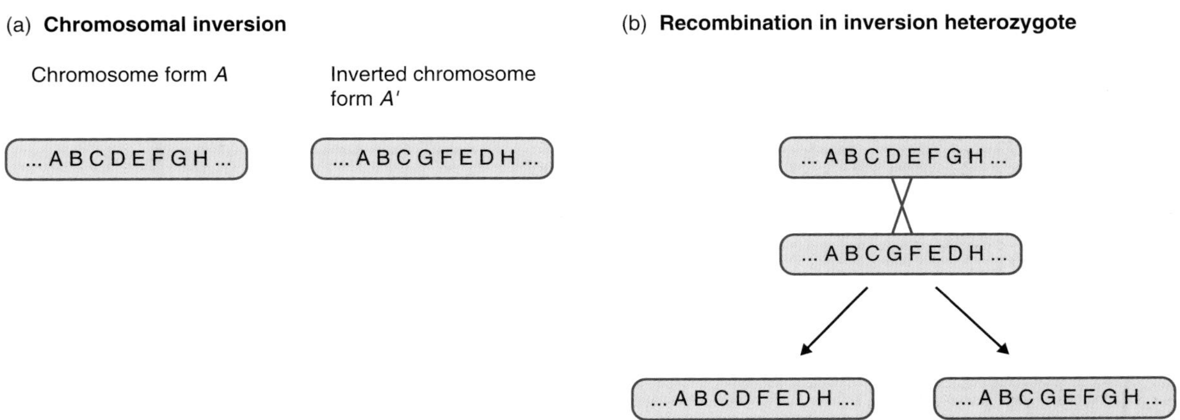

will apply, but a gene causing its bearer to mate assortatively with respect to chromosomal genotype will be favored. If assortative mating becomes established, speciation will result.

The argument works for a population that already possesses both chromosomes in reasonable frequencies. During evolution, one of the chromosomes would have existed first, and the second would have originated as a rare mutation. Suppose the population was initially fixed for A and A' arose as a mutation. A' would initially exist only in AA' heterozygotes. These heterozygotes are selected against, and the most likely evolutionary result would be the loss of A'.

Two ideas have been put forward to suggest how an initially rare A' chromosome could spread. One is that the chromosome could be a "driver" chromosome that is passed on to more than 50% of the offspring of heterozygotes and spreads automatically (section 12.3, p. 330). The other idea makes use of inbreeding to help the spread of the chromosome. The evolutionary problem is to produce $A'A'$ homozygotes. A rare AA' heterozygote's close relatives will tend to include other heterozygotes, and if they inbreed they will produce some homozygous $A'A'$ offspring. The spread of the chromosome may also be assisted by random genetic drift, which is more likely in an inbred population to produce one subpopulation with a new chromosomal form in high frequency. Either way, inbreeding may help the spread of the new chromosomal form.

Once the population possesses both homozygotes, two things can happen. Selection favors either the loss of one of the chromosomal forms or reproductive isolation between them. If one chromosomal form is lost, it could be the new form (in which case nothing would have changed) or the old (in which case a chromosomal evolutionary event would have occurred). If speciation proceeds, we should have two new species, one with the old chromosomal form and one with the new.

If the theory is correct, we can predict an association between the likelihood of the members of a taxonomic group having an inbred population structure, their rate of speciation, and the rate of chromosomal change. Some species live in small social groups, or otherwise have subdivided populations; the degree of inbreeding in these species will be higher than in a species in which individuals outbreed in a large population.

Bush *et al.* tested whether genera whose members live in small family groups have a higher rate of speciation and chromosomal evolution than panmictic genera. They collected evidence for 225 vertebrate genera. For each genus, they counted the number of species in it and its chromosomal diversity. The two are correlated, as would be predicted by almost any theory. They then argued that the taxa with higher speciation rates tended to have subdivided population structures. For example, mammals, especially primates and horses (though not whales), have high rates of speciation and chromosomal evolution; fish, amphibians, and reptiles have lower rates. Primates and horses are two taxa that usually live in social groups. Bush *et al.* did not systematically compare the variables in all taxa; but the result is still suggestive. Perhaps taxa with subdivided population structures do have higher speciation rates, and this elevated rate may be due to the way chromosomal evolution can proceed more rapidly in these kinds of species.

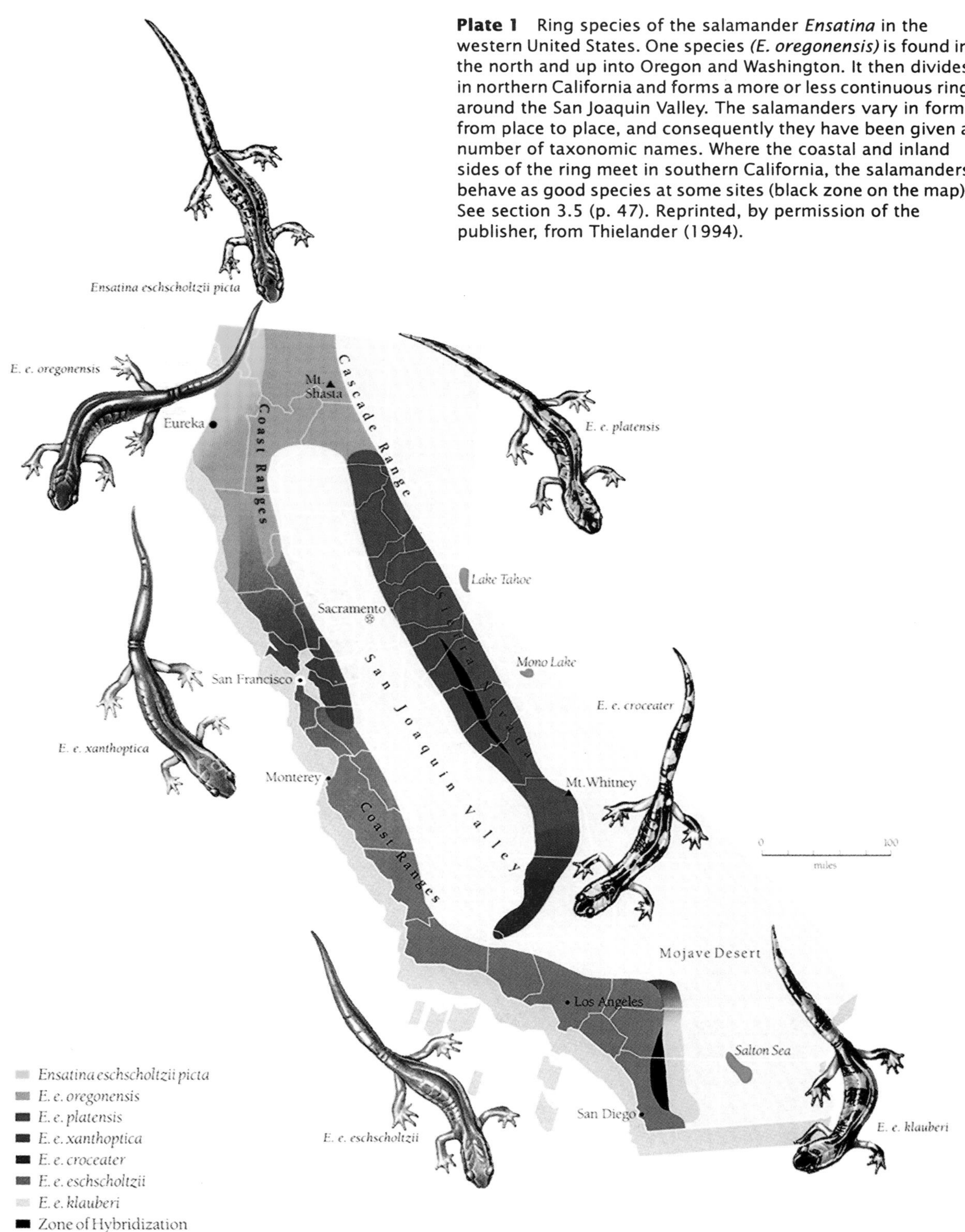

Plate 1 Ring species of the salamander *Ensatina* in the western United States. One species *(E. oregonensis)* is found in the north and up into Oregon and Washington. It then divides in northern California and forms a more or less continuous ring around the San Joaquin Valley. The salamanders vary in form from place to place, and consequently they have been given a number of taxonomic names. Where the coastal and inland sides of the ring meet in southern California, the salamanders behave as good species at some sites (black zone on the map). See section 3.5 (p. 47). Reprinted, by permission of the publisher, from Thielander (1994).

(a) (b) (c) (d) (e) (f)

(g) (h) (i) (j) (k) (l)

(m)

Plate 2 The lower row shows six of the many forms of *Papilio memnon*, beneath the model species that they may mimic. (a–f) Six suspected models: (a, b) two forms of female *Papilio coon;* (c) *P. aristolochiae;* (d) *Troides helena;* (e) *Troides amphrysus;* (f) *Papilio sycorax.* (g–l) Six forms of *Papilo memnon.* Three of the forms (g, h, i) mimic species (a, b, c) that have tails, three (j, k, l) mimic species (d, e, f) that lack tails. (m) Another form of *Papilio memnon,* the rare probable recombinant form anura, from Java. This form is like the normal mimetic form called achates (illustrated in g–i), but it lacks achates' tail. It may be a recombinant between achates and a tailless form such as j–l. See section 8.1 (p. 195). From Clarke *et al.* (1968) and Clarke and Sheppard (1969).

16.11 *Conclusion*

The question of the relative frequencies of different modes of speciation probably will not be settled by counts of individual cases. It is too difficult to establish conclusively the mode of speciation for any one case because of the many unobservable, historical, but crucial elements in the models. Some progress can be made in reconstructing individual cases, but not to the point of certainty. The question also must be studied theoretically, to establish the plausibility of speciation under different conditions. We have a good idea of the main ways in which speciation could take place, but plenty of room remains for more conclusive work on the question of how a new species can arise on earth—the question that John Herschel once described, in a phrase that inspired Darwin, as "the mystery of mysteries."

SUMMARY

1. The evolution of a new species occurs when one population of interbreeding organisms splits into two separately breeding populations.
2. It has been a matter of controversy whether new species evolve only from subpopulations that are geographically isolated (allopatric) from the ancestral population, or whether they can also evolve from subpopulations that are contiguous with (parapatric), or overlap (sympatric), the ancestral population.
3. At least in vertebrates, the majority of new species probably evolve allopatrically. All species show extensive geographic variation that could be converted into speciation, and there are no theoretical difficulties for allopatric speciation (whereas there are for parapatric and sympatric).
4. Experiments have demonstrated that reproductive isolation tends to arise incidentally between two populations that are kept separate from one another and allowed to evolve independently for a number of generations.
5. Allopatric speciation may occur by subdivision of the species range or by a peripheral isolate—a small population that becomes cut off at the edge of the species range.
6. Parapatric speciation could happen if a steep cline evolved into a hybrid zone and barriers to interbreeding then evolved.
7. Sympatric speciation is most likely if selection first establishes a stable polymorphism and then favors assortative mating within each polymorphic type.
8. Many new plant species have originated by hybridization of two existing species, followed by polyploidy of the hybrids.
9. Reinforcement is the enhancement of reproductive isolation by natural selection: forms are selected to mate with their own, and not with the other, type. Sympatric speciation requires reinforcement to happen. Parapatric speciation usually requires reinforcement. Allopatric speciation can take place with or without reinforcement. Reinforcement is theoretically difficult, which is one reason to suspect that allopatric speciation may be more common than sympatric and parapatric speciation.
10. Reinforcement may explain why pairs of closely related species show greater reproductive isolation where they coexist sympatrically, than where they are

allopatric populations. The same result could arise without reinforcement, however.

11. The amount of prezygotic isolation is greater between recently evolved sympatric pairs of species of fruitflies than it is between equivalent pairs of allopatric species. The amount of postzygotic isolation is similar in sympatric and allopatric pairs of species. Reinforcement is probably the explanation.
12. New species of fruitfly take on average about 1,500,000–3,500,000 years to evolve.
13. Chromosomal evolution might produce more rapid speciation in species with a subdivided population structure, in which inbreeding is more frequent. Some evidence suggests that chromosomal rates of evolution are correlated with social structure in vertebrate species.

FURTHER READING

The classic treatises on speciation by Mayr (1942, 1954, 1963) and Dobzhansky (1970) remain good, if dated, introductions. See Mayr (1982a, and Mayr and Ashlock 1991) for his more recent ideas; Coyne (1994a) discusses speciation, particularly in relation to Mayr's ideas. There have been a series of multiauthor books on the topics that are also good, if erratic, introductions: Atchley and Woodruff (1981), Barigozzi (1982), and Otte and Endler (1989). Review papers include Bush (1975), Templeton (1981), Barton (1988), Harrison (1991), and Coyne (1992). Rice and Hostert (1993) is an excellent review of experimental work.

Mayr's works are the references for allopatric speciation, including ring species. Murray and Clarke (1980) describe two other examples of ring species, in snails; Tudge (1992) is a popular account of these snails, and the ecological catastrophe they have recently suffered. Rice and Hostert (1993) review other experimental work like Dodd's and Kessler's; Macnair and Christie (1983) is a wild example, from heavy metal tolerance in plants, of the process in Dodd's experiment. On the peripheral isolate and founder effect theories, see Barton (1989), Giddings *et al.* (1989), and Lande (1980, 1986).

There are several good reviews of hybrid zones. Harrison (1993) is now the main source, from which most earlier work can be traced. See also Hewitt (1988) and Harrison (1990). On the European crows, see Cook (1975). The general works listed above also discuss the topic. On parapatric speciation see Endler (1977), which includes an important discussion of the biogeographical evidence.

On sympatric speciation, Mayr (1942, 1963, 1982) is the classic critic. Tauber and Tauber (1989) review the work on *Chrysopa*, the evidence for other insects, and give references to other theoretical models beside Seger's discussed here. Bush (1994) is a recent introductory review on host shifts and phylogenetic evidence. Lynch (1989) uses biogeographic evidence to assess the relative frequencies of the different geographic modes of speciation; Chesser and Zink (1994) follow up his method.

On hybrid speciation in plants, Abbott (1992), Arnold (1994), Arnold and Hodges (1995), and Thompson and Lumaret (1992) are shorter introductory works, and Lewis (1980) and Rieseberg and Wendel (1993) longer research-level reviews. Stebbins (1950) and Grant (1981) are classical and cover plant speciation in general. In addition to the two case studies described here, another good example is the sunflower *Helianthus* in the southwest; Rieseberg and Wendel (1993) discuss it.

On reinforcement generally, look at Howard (1993a) and Butlin (1987, 1989), as well as the general reviews and hybrid zones references cited above. For Coyne and Orr's study, see Coyne and Orr (1989a,b). The rate of speciation in other cases may be very different form those inferred for fruitflies. See, for example, Meyer (1993) on African lake cichlid fish. On chromosomal mechanisms of speciation, see White (1978), Lande (1979, 1985), and M. King (1993). Howard (1993b) discusses this and other circumstances in which inbreeding can influence speciation.

STUDY AND REVIEW QUESTIONS

1. Why do species show geographic variation?
2. What evidence from modern species suggests that speciation occurs particularly often via peripheral isolates?
3. Explain why hybrid zones exist according to the theory of (a) allopatric and (b) parapatric speciation.
4. What factors determine how steep the spatial change in gene frequency is at a hybrid zone?
5. What evidence can be used to infer that a species had a hybrid origin, and to identify its ancestry?
6. What reasons suggest that reinforcement may be a weak evolutionary force in nature?
7. In a mating experiment, an individual female of species 1 is given a choice of a male of species 1 and a male of species 2. The experiment is repeated with 100 females. Numbers of females who mated with each kind of male are given below. Calculate an index of prezygotic reproductive isolation (I) for the three cases.

Male mated:	
Species 1	Species 2
(a) 100	0
(b) 75	25
(c) 50	50

8. Two species have partly overlapping ranges. Females of species 1 taken from an area where only species 1 lives are given an experimental choice between males of the two species and mate indiscriminately. Females of species 1 taken from the area where both species live are given the same choice and mate preferentially with males of species 1. What is the phenomenon called? What are the two main evolutionary explanations for it?
9. Why are genetic changes on the sex chromosomes so influential in the early stages of speciation?

The Reconstruction of Phylogeny

A knowledge of the phylogenetic relations among species is essential for many other inferences in biology, and consequently a major effort has been made to reconstruct the tree of life. The fundamental principle used in most phylogenetic inference is the principle of parsimony, and a widely used version of parsimony requires a distinction between "unrooted" and "rooted" trees. We begin by defining these terms. We then look at the relevant techniques, in two stages: first, the techniques mainly used on the morphology of organisms, and then those mainly used with molecular sequences. We consider why different techniques have been used with morphological and molecular evidence, and finish with two case studies in which different kinds of evidence have come into conflict.

17.1 Phylogenies are inferred from characters shared between species

A *phylogenetic tree*, or *phylogeny*, or *tree*, for a group of species is a branching diagram that shows, for each species (or group of species), with which other species (or group of species) it shares its most recent common ancestor. A phylogeny implicitly has a time axis, and time usually goes up the page. Figure 17.1 shows some examples of phylogenies. In the figure, species *A* and *B* share a more recent common ancestor with one another than either shares with any other species (or group of species). This chapter discusses how phylogenetic relations are inferred from imperfect evidence. They must be inferred because we exist now, whereas the splitting events and common ancestors existed in the past and cannot be directly observed.

How do we know, for example, that humans and chimpanzees share a more recent common ancestor than either shares with the amoeba? The quick answer is that humans and chimpanzees look more similar. For phylogenetic inference, the vague expression "look more similar" must be made more precise. We can divide up the phenotypic appearance of an organism into a series of characters and character states. We can then say that humans and chimpanzees have similar appearances because they share many more characters in common than either does with the amoeba. Humans and chimpanzees share vertebrate characters such as brains and backbones, mammalian characters such as lactation, and great ape characters such as their distinctive molar teeth and absence of a tail. Humans and amoebas do have characters in common—many of the characters of cellular

Figure 17.1 A phylogeny shows, for a group of species, the order in which they share common ancestors with one another. (a) Species *A* and *B* share a more recent common ancestor with one another than either does with *C*; the group of species *A, B,* and *C* share a more recent common ancestor with one another than any of them do with species *D*. (b) In a phylogeny, any of the nodes can be rotated without altering the relation shown. (a) and (b) are identical. (c) differs from (a) and (b) because the order of branching is altered. (d) is a different phylogeny. (e) Phylogenies may be drawn with either right-angled or diagonal lines; the information is identical. Thus, (d) and (e) are the same phylogeny. The only information in the phylogeny is the order of branching; the *x*-axis does not necessarily represent phenetic similarity. In particular, (d) does not imply that species *B* and *C* show convergent evolution. Sometimes a phylogenetic diagram also displays phenetic similarity (e.g., Figure 14.1, p. 373), it may be explicitly drawn on as well. The vertical axis expresses the direction of time; it goes up the page. However, the axis is not usually exactly proportional to time. That is, (a) does not imply that the time between the successive branching was constant. Some phylogenetic diagrams do display absolute time, and then it is again made explicit (e.g., Figure 17.19, p. 497).

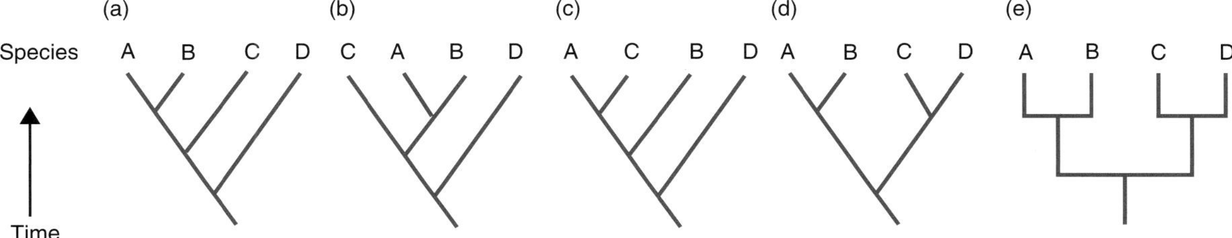

metabolism, for example—but these characters are equally shared with chimpanzees and do not suggest a separate grouping of human and amoeba.

A more fundamental principle underlies phylogenetic reconstruction than that of shared characters, however. A phylogeny is more plausible if it requires less, rather than more, evolutionary changes in character states. This is the principle of *parsimony*. Phylogenetic inference using parsimony proceeds in two stages, and to understand them we need to distinguish between a *rooted* and an *unrooted tree*. Full phylogenies like Figure 17.1 are "rooted" trees—that is, the position of the deepest ancestor, or "root" within the tree is specified. In Figure 17.1a, for example, the root is in the branch between species *D* and the group of species *A, B,* and *C*. An unrooted tree (Figure 17.2) shows the branching relations between the species but does not show the position of the deepest common ancestor. Thus, an unrooted tree consists of a phylogenetic tree with the time dimension removed. It can be rotated through any angle on the page without altering the information in it, because the only information it contains is the branching topology of the species. An unrooted tree is, therefore, less informative than a rooted tree.

A rooted tree is compatible with only one unrooted tree. In contrast, an unrooted tree is compatible with a number of possible rooted trees (of which at most one will be the true one), depending on where the common ancestor, or root, lies. The smallest number of species for which a nontrivial unrooted tree can be drawn is four. A group of three (or two, or one) species can have only one possible unrooted tree, and it is therefore trivial to write this tree down.

Figure 17.2 Unrooted and rooted trees. One unrooted tree for four species is compatible with five rooted trees. An unrooted tree is a timeless picture of branching relations and does not specify the location of the ancestor (or root) of the tree. The root could appear anywhere in the tree, and there are five topological possibilities, as drawn below. In general, any one unrooted tree of s species has $2s - 3$ internal branches and therefore $2s - 3$ possible rooted trees. (Here, as elsewhere in the chapter, we confine ourselves to strictly bifurcating trees.)

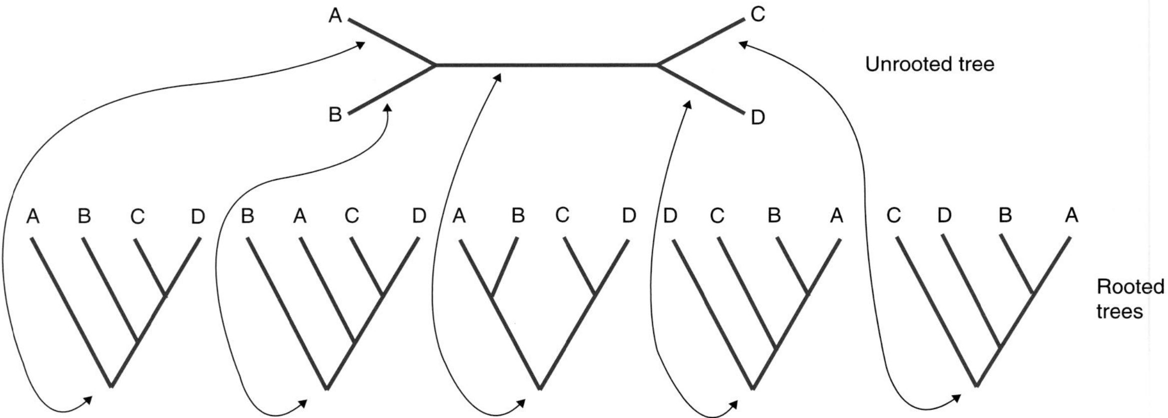

Four species have three possible unrooted trees, however, and it is nontrivial to assert that the true unrooted tree is any one of those alternatives.

The two stages in the method of parsimony are first to infer the unrooted tree for a set of species, and then to locate its root. How can we infer the unrooted tree? Let us consider the simplest case, in which we have four species. We can add one species, such as a magnolia tree, to the human, chimpanzee, and amoeba in the example above. The procedure is as follows (Figure 17.3). First, write out all possible unrooted trees for the species. In this case, and throughout the rest of the chapter, we shall confine ourselves to the case of bifurcating (or dichotomous) trees, so three trees are possible for the group of four species. (Phylogenetic research normally confines itself to trees with bifurcating branches, because most real trees probably have bifurcating, rather than multiple, branches.) Next, given a knowledge of the character states in each species, count the minimum number of evolutionary events (that is, changes in character states) implied by each tree. The best estimate of the true unrooted tree is the one requiring the least evolutionary change.

Which of the three trees in Figure 17.3 requires the smallest number of events? Suppose we know that 1000 characters are shared in all four species (all of these characters, such as the use of DNA, proteins, and many cellular structures, will be common to all eukaryotic life). We also know that each species has 10 characters unique to itself (leaves, wonderfully perfumed flowers, and so on in the magnolia, an adult brain of about 1500 cm^3 in humans, and so on). Likewise, we know that 100 characters are shared between humans and chimps, but are absent in amoebas and magnolias (backbone, vocal communication, and so on). No other kinds of character are known.

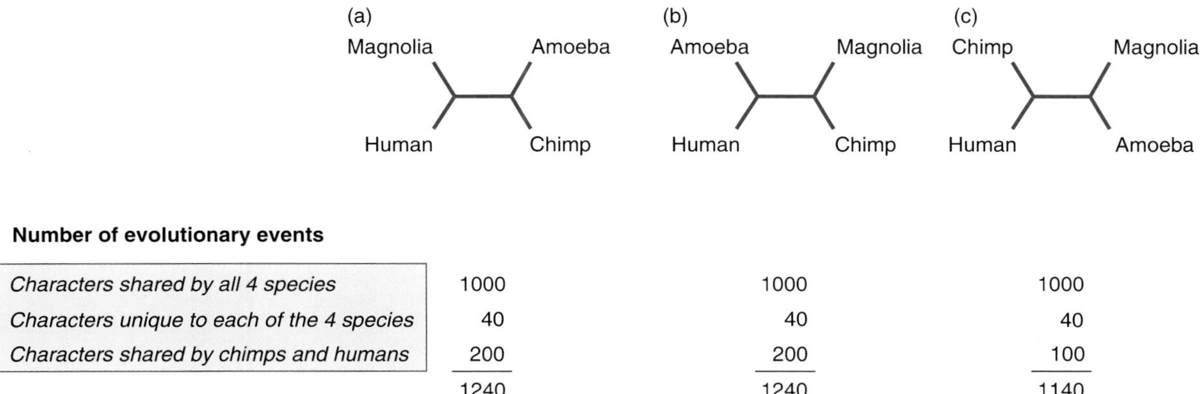

Figure 17.3 Possible phylogenetic relations of amoeba, chimpanzee, human, and magnolia. (a), (b), and (c) are three unrooted trees; they specify branching, but not ancestral, relations. The sums for the number of evolutionary events implied by each tree are made for 1000 characters shared by all four species, 10 characters unique to each species, and 100 characters shared by humans and chimps.

Number of evolutionary events

	(a)	(b)	(c)
Characters shared by all 4 species	1000	1000	1000
Characters unique to each of the 4 species	40	40	40
Characters shared by chimps and humans	200	200	100
	1240	1240	1140

We can now count the smallest number of evolutionary events implied by each unrooted tree. In Figure 17.3a, the 1000 common characters could have evolved once each in the common ancestor of all four species and been retained throughout the phylogeny. That possibility would require a total of 1000 evolutionary events (one evolutionary origination for each character). Each species' 10 unique characters can have evolved in the lineage leading to that species, making 40 evolutionary events in all. The 100 characters shared by humans and chimps are more puzzling. The smallest number of events producing the observed distribution of characters is 200. They could have evolved separately in the human lineage and in the chimp lineage (100 events in each, making 200). Alternatively, if the branch leading to the magnolia contains the common ancestor of the group of four species, all 100 could have evolved once between that common ancestor and the common ancestor of the amoeba, human, and chimp. They would then need to be lost in the lineage leading to the amoeba, making 100 gains and 100 losses, or 200 events in all. (The same argument can be made if the common ancestor of the four species appears in the branch leading to the amoeba; the characters then must be lost in the lineage leading to the magnolia.) We do not need to know the real pattern of character changes to count the number of events implied by the tree, because either way the total number—and therefore the minimum number—of events in Figure 17.3a is 1240.

The tree in Figure 17.3a is logically compatible with any number of events between 1240 and infinity. The minimum number was obtained by supposing the 10 characters unique to the magnolia evolved only in the lineage leading to the magnolia. Logically they could have evolved and then been lost any even number of times in any other lineage. Perhaps, in the human lineage, leaves were evolved and then lost, scented flowers were evolved and then lost, then leaves were evolved and lost again . . . and so on to infinity. The interesting number is the minimum number of events implied by the tree, not the number of events that logically could have occurred.

The same calculation in Figure 17.3b also implies 1240 events. In Figure 17.3c, however, the 100 characters shared by humans and chimps need to have evolved only once, in the common ancestor of humans and chimps, and would not have to be lost again. Thus, this tree only requires 1140 events. It is the most parsimonious unrooted tree, and (according to the principle of parsimony) represents the best estimate of the real unrooted tree.

The unrooted tree completes stage one of the inference. The second stage is to locate its root, to turn it into a phylogeny like Figure 17.1. We shall consider the evidence used to "root" the tree in section 17.8.

The example in Figure 17.3 illustrates another important distinction, between informative and uninformative characters (or evidence). An uninformative character, in phylogenetic inference, is one that is equally compatible with all possible unrooted trees for the species under analysis. In Figure 17.3, the 1000 characters shared by all four species are uninformative. They require the same number of events in (a), (b), and (c) and, therefore, do not help us calculate which tree is more likely to be correct. The 4×10 unique characters are also uninformative because they require the same 40 events in all three unrooted trees. All of the difference in the total number of events required by the unrooted trees is contributed by the characters shared between chimpanzees and humans. Thus, these characters are the only informative ones in Figure 17.3. They require fewer events in (c) than in (a) or (b) and allow the inference that (c) is most likely.

With this knowledge, we could have performed the same analysis in another way. We could first have examined all 1140 characters for which we have evidence and divided them into informative and uninformative characters. Then we could discard all the uninformative characters, and count the number of events implied by the unrooted trees for the informative characters alone. The result would be 200 events in (a) and (b) and 100 for (c). Using this method, Figure 17.3c again represents the most parsimonious tree. The result, and the underlying logic of the argument, is the same in the two methods. But the second method, in which all uninformative evidence is identified and discarded before the events are counted can often be quicker and clearer.

17.2 *The parsimony principle works if evolutionary change is improbable*

How can we justify the parsimony principle? Why is a phylogeny requiring less evolutionary events a more plausible inference than one requiring more? The parsimony principle is reasonable because evolutionary change is improbable. Suppose we know that a modern species and one of its ancestors both have the same character state (Figure 17.4). Parsimony suggests that all intermediate stages in the continuous lineage between the ancestor and modern species possessed that same character state. As we have seen, a large number of changes—indeed, an infinite number—could logically have occurred between ancestor and descendant. A change followed by a reversal of that change is an unlikely occurrence,

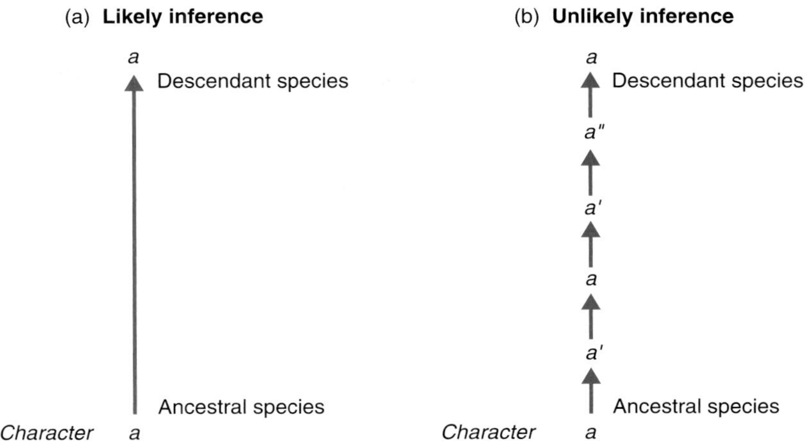

Figure 17.4 The same character is found in both a descendant species and one of its ancestors. It is more likely (a) that the character has remained constant and been passed on by inheritance than (b) that it has changed and reverted to its original state a number of times between ancestor and descendant.

however. Each change requires a gene (or set of genes) to arise by mutation and then to be substituted, either by drift (if the change is neutral) or by selection; both these processes are improbable. It is much more likely that the same character would have been continuously passed on, in much the same form, from ancestor to descendant by simple inheritance. We know this process is plausible because it happens every time a parent produces an offspring: the parental characters are passed on.

For the characters shared between humans and chimps, this argument is particularly powerful. Chimps and humans share whole complex organ systems like hearts, lungs, eyes, brains, and spinal cords. The initial evolution of each of these characters required improbable mutations, and natural selection operating over millions of generations. It is evolutionarily improbable to the point of near impossibility that the same changes would have evolved independently in the two lineages after their common ancestor. By contrast, there is nothing improbable about postulating that the characters could have been passed on via passive inheritance from the common ancestor of chimps and humans to the modern descendants.

For some characters other than the complex morphological characters shared between humans and chimps, this argument is less powerful. At the other extreme, if we find one nucleotide, at a particular site in the DNA, shared between two species, there is a 25% probability that it could be shared by chance, and the principle of parsimony does not strongly suggest that the nucleotide has not changed through all evolutionary intermediates between the two species.

While the argument's power varies from case to case, evolutionary change in all characters is improbable to some extent, as compared with simple inheritance. The principle of parsimony, therefore, has a sound evolutionary justification. In conclusion, it is more likely that a character will be shared by common descent than by independent, convergent evolution. For any set of species, a phylogeny requiring less evolutionary change is more plausible than one requiring more.

17.3 Phylogenetic inference uses two principles: parsimony and distance statistics

So far we have encountered two principles of phylogenetic inference, and we use these principles throughout this chapter. These principles are:

1. *Parsimony.* Species are arranged in a phylogeny such that the smallest number of evolutionary changes is required.
2. *Distance (or similarity).* Species are arranged in a phylogeny such that each species is grouped with the other species with which it shares the most characters.

In the case of the chimpanzee, amoeba, and human, the argument that humans and chimpanzees share a more recent common ancestor because they have more similar appearances implicitly used measurements of distance (section 14.5, p. 375; Box 16.1, p. 453). The counts of evolutionary events in the three unrooted trees for the human, chimp, amoeba, and magnolia used the principle of parsimony.

In easy cases like that of humans, chimpanzees, and amoebas, the distance and parsimony principles give the same result, and it might seem that it does not matter which is used. In other cases, however, results differ for these principles. The parsimony principle (as we shall see, in section 17.7) is then more reliable, because it has a better theoretical justification. Distance statistics remain an important method in phylogenetic inference, because under certain circumstances they are almost as reliable as parsimony—and they can often be collected more rapidly. To know when we can rely upon any given method, it is necessary to understand the underlying principle of phylogenetic inference.

17.4 In most real cases, not all characters suggest the same phylogeny

In the imaginary example of Figure 17.3, all of the characters implied the same phylogeny. Some characters were unique to each species, other characters were shared by all species, and still other characters were shared by only chimpanzees and humans. The 1140 characters "agreed" in the sense that we could divide them into two sets and still obtain the same answer. If, for example, we divided the 1000 + 40 + 100 characters into two sets of 500 + 20 + 50, the phylogeny inferred from each of those two sets would be the same. Now imagine that, in addition to the 100 characters shared between chimpanzees and humans (and absent from magnolias and amoebas), we have 50 characters shared by magnolias and humans (and absent from chimpanzees and amoebas). In this case, the characters do not agree. If we concentrate on the characters shared between humans and magnolias, the phylogeny in Figure 17.3a is correct; on the other hand, the characters shared between humans and chimps favor the phylogeny in Figure 17.3c.

Conflict between characters is typical in real phylogenetic problems. (In a case like the human, chimpanzee, amoeba, and magnolia, conflict might not present a problem, but the answer is obvious. It would be more accurate to say that the cases in which no character conflict occurs have all been worked out,

Figure 17.5 Two routes of phylogenetic inference. (a) Two-stage character analysis to reduce conflict, followed by parsimony (if needed). (b) Proceed straight to parsimony without character analysis; the polarity of at least one character change must be found to convert the unrooted tree into a rooted one. (a) is more often used with morphological evidence, (b) with molecular evidence.

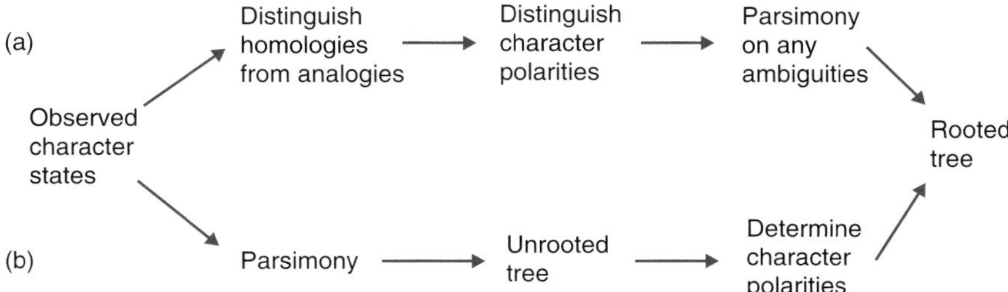

and the real problems left for phylogenetic research are the cases of conflicting evidence.) The reptiles and birds are an example of such conflict in characters (Figure 14.3, p. 378). Some characters, such as gait, suggest one phylogenetic grouping; other characters, such as skull anatomy, suggest another tree.

When characters conflict, how should we proceed? There are two main answers, and they are related rather than mutually exclusive. They almost form, in practice, two routes to infer a phylogeny (Figure 17.5).

One possibility is to resolve the conflict directly by parsimony. If humans and chimpanzees share 100 characters, and humans and magnolias share 50 characters, the phylogeny putting the humans and chimpanzees together will be more parsimonious; that is, it will require fewer evolutionary events. As we shall see in sections 17.10–17.13, this method is essentially the same as that used with molecular evidence.

The other possibility is to analyze the characters, to see whether some serve as more reliable phylogenetic indicators than others. This "classical" cladistic method has particularly been used with morphological evidence. In sections 17.5–17.9, we shall look at the theory and practice used to sift the reliable evidence from the unreliable evidence. The unreliable shared characters, once they have been identified, can be discarded. Any remaining conflict can then be resolved by parsimony.

We can, therefore, either go straight to the principle of parsimony, or sift the evidence first and then use the principle of parsimony. The fundamental evolutionary logic of the two methods is in many ways identical, and the different applications with morphology and molecules mainly derive from the level of biological understanding of the evidence (section 17.10).

17.5 *Homologies are more reliable for phylogenetic inference than are analogies*

The first stage in the top route of Figure 17.5, in which we analyze the characters to assess their phylogenetic reliability, is to distinguish homologies from analogies. We encountered this distinction when discussing classification (see Figure 14.7

and Table 14.2), but there we assumed that the character analysis had been performed, and focused instead on how to use the different characters in classification. In this case, we will consider how to analyze the characters. What is the difference between homologous characters, or *homologies*, and analogous characters, or *analogies*? A homology is a character shared between two or more species that was present in their common ancestor; an analogy is a character shared between two or more species that was not present in their common ancestor. Homologous similarity reveals a phylogenetic relationship, while analogous similarity does not.

A homologous character such as the heart, or lungs, of a human and a chimpanzee is easily recognized as the same character, and is presumably derived from a common ancestor that also possessed that character. In other cases, the similarity is less obvious. The five-digit limb of tetrapods is homologous even though its form varies (Figure 3.6, p. 54), and in extreme cases homologies are so subtle that it takes clever detective work to reveal them. The ear bones of mammals, for example, do not superficially resemble the gill arches of a fish. One of the classic discoveries of nineteenth century comparative anatomy was, however, a series of intermediates that was traced between them, and these characters were found to have a common embryonic origin—that is, they are homologous. Strictly, a homologous character does not have to be identical in all species possessing it, but there does have to be shared morphological information among them.

Convergence can also be impressively deceptive. Consider a classic example of convergence from the two major groups of mammals, the marsupials and placentals (Figure 17.6). The marsupial and placental saber-toothed carnivores both evolved long, gashing canine teeth. In addition, striking similarities in skull shape and body form are found in the marsupial and placental wolves. If we inferred the phylogeny of the marsupial wolf, the placental wolf, and the kangaroo from the pattern of phenetic similarity, we should obtain the wrong answer. The two wolves are phenetically more similar, even though the marsupial wolf is phylogenetically closer to the kangaroo than to the placental wolf. The phenetic similarity between the wolves is analogous and is not due to a close phylogenetic relationship.

Thus, analogous characters do not reliably indicate phylogenetic relationships because they can fall into any grouping of species; they are not constrained to fall into phylogenetic groups. Different analogies can fall into different and conflicting groupings, and the existence of analogies is, therefore, one reason for conflict among characters (see Figure 17.8, p. 474). (This point forms the first part of the full argument why phenetic similarity, as measured by distance statistics, is not a reliable guide to phylogeny. We shall complete the argument in section 17.7.)

17.6 *Homologies can be distinguished from analogies by several criteria*

Imagine that we have a list of 100 characters that show similarities within a group of species. The characters give conflicting indications about the true phylog-

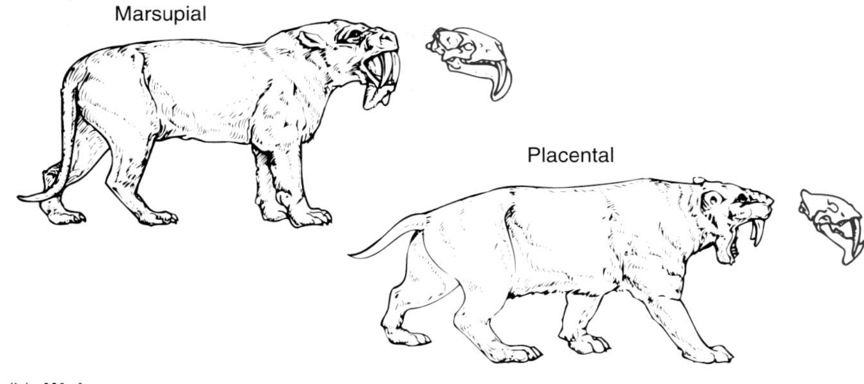

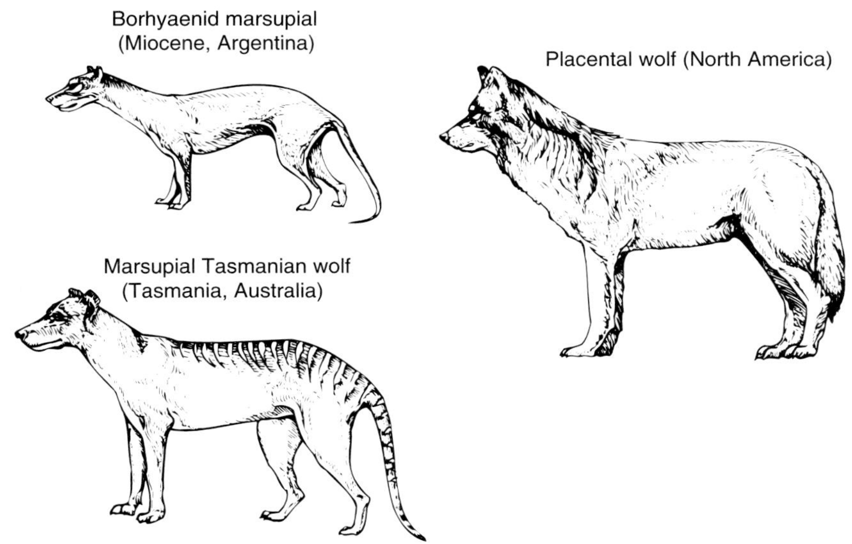

Figure 17.6 Convergence in marsupial and placental carnivores. (a) The reconstructed bodies, and skulls, of *Thylacosmilus*, a saber-toothed marsupial carnivore that lived in South America in the Pliocene and of *Smilodon*, a saber-toothed placental carnivore from the Pleistocene in North America. (b) *Prothylacynus patagonicus*, a borhyaenid marsupial from the early Miocene in Argentina; *Thylacinus cynocephalus*, the extinct marsupial Tasmanian wolf; and *Canis lupus*, the modern placental wolf. From Strickberger (1990).

eny, with perhaps 30 characters implying one phylogeny, 30 characters implying another tree, 20 still another, and the final 20 pointing to various idiosyncratic possibilities. The task is to shorten the list and reduce its ambiguity by eliminating the analogies. How can we detect them? If a shared character is a real homology (i.e., inherited from a common ancestor), it must be the same character in the two species, and we can test it by examining the character in detail to find out whether it is the same in the different groups.

First, if a character is homologous it must have the same fundamental structure. The wings of birds and bats, for example, are superficially similar, but they are constructed from different materials and supported by different limb digits (Figure 17.7). Thus, they probably evolved independently, and not from a winged common ancestor.

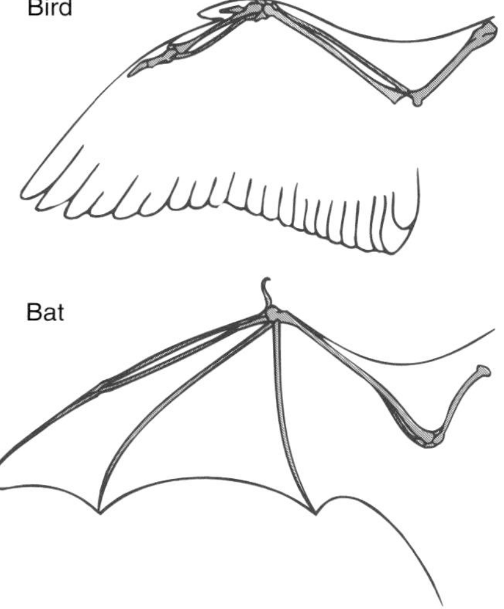

Figure 17.7 The wings of birds and bats are analogies. They are structurally different. The bird wing is supported by digit number 2, the bat wing by digits 2–5. The bird wing is also covered with feathers, the bat wing with skin.

Second, homologies must have the same relations to surrounding characters. Homologous bones, for example, should be connected in a similar way with their surrounding bones.

Third, the characters must have the same embryonic development in different groups. A character that looks similar in the adult forms, but develops by a different series of stages, is unlikely to be homologous. One example, which we saw earlier is the relation between a barnacle, a mollusc such as a limpet, and a crab (Figure 14.1, p. 373). At least superficially, the adult form of a barnacles resembles a limpet more closely than a crab. The relations of barnacles had been uncertain for centuries until John Vaughan Thompson discovered their larvae in 1830; these forms are very like the larvae of several groups of Crustacea, and unlike those of molluscs. Barnacles, therefore, share a more recent common ancestor with crabs than with limpets. The similarities between the adult barnacle and limpet, such as their hard external armor, attachment to rocks, and feeding through a hole in the shell, are all convergent.

Finally, some other criteria may sometimes be useful in distinguishing homologies from analogies. Convergence is caused by natural selection, when organisms in different evolutionary lineages face similar functional requirements (such as flying, in birds and bats). We have grounds for suspecting that a shared morphological form may be analogous, when the species that share it clearly need it for their way of life.

Thus, it is possible to analyze characters shared between species to see whether some are homologous or analogous. A homology can be recognized as a character that has fundamentally the same structure, relations with surrounding parts, and development in a set of species. Once the homologies are (often tentatively) identified, they can be allowed into the list of evidence used to infer the phylogeny. The analogies are removed. In this way, character conflict should be reduced.

17.7 *Derived homologies are more reliable indicators of phylogenetic relations than are ancestral homologies*

Character conflict may not, however, be reduced to zero. Of the 100 initial characters in our example, perhaps 45 could be ruled out as analogies. The 55 remaining (inferentially homologous) characters might consist of 25 characters that support one phylogeny, 15 that imply a second tree, 12 a third, and 3 that point to various other groupings. We now have a list of homologies that are shared in various combinations among the species. In most real cases, the list will contain conflict, though likely less conflict than in the list of raw, unanalyzed characters. Some of the remaining conflict may be due to mistakes—that is, some of the inferred homologies may really be analogies. However, conflict can still exist among genuine homologies because not all homologies reliably fall into phylogenetic groups.

The next stages in the analysis of the characters involve dividing the homologies into ancestral and derived homologies. An ancestral homology was present in the common ancestor of the two species, but evolved earlier and is also shared with other, more distantly related species. In contrast, a derived homology first evolved in the common ancestor of the two species and is not shared with other more distantly related species. If the two species were a human and a salmon, for example, the backbone is a derived homology; it first evolved in the common ancestor of the vertebrates—which is also the common ancestor of these two species. The presence of a distinct nucleus within all their cells is an ancestral homology; it was present in the common ancestor of the vertebrates, but evolved earlier, in the common ancestor of all eukaryotes. Thus, whether a homology is ancestral or derived depends on the set of species being compared. The backbone is a shared derived homology in a human and a salmon, but it is a shared ancestral homology in a human and a chimpanzee.

The complete analysis of a character has two stages: distinguishing analogies from homologies, followed by dividing the homologies. A character can, therefore, belong to any one of three types (Figure 17.8). The distinction matters because, of the three kinds of shared characters, only shared derived homologies constitute evidence that the two species share a more recent common ancestor with one another than with any other species under investigation. Analogies or shared ancestral homologies can be shared by species that do not form phylogenetic groups. We saw above how analogies are phylogenetically unreliable (see Figure 17.8a).

We can now complete the argument against inferring phylogenies from phenetic similarity (that is, similarity with respect to all three kinds of characters together). That phenetic similarity is misleading in the case of convergence is widely appreciated. Ancestral homologies cause the same problem, however, and in a more insidious fashion. We have already seen this problem in the reptiles: a crocodile looks more like a lizard than a bird, but is phylogenetically closer to a bird than to a lizard. The crocodile and lizard share ancestral homologies—characters that were present in the common ancestor of the bird, crocodile, and lizard—that have been lost in the rapidly evolving bird lineage.

The point of these examples is not that phenetic similarity never indicates

Figure 17.8 Shared characters divide into analogies, ancestral homologies, and derived homologies. (a) *a′* is an analogy: it is not in the common ancestor of the species that share it. (b) An ancestral homology: it is in the common ancestor of the species that share it, but has been lost in some descendants of that common ancestor. (c) A derived homology: it is in the common ancestor of the species that share it, and in all of its descendants. Notice that only derived homologies always indicate phylogenetic relationships. In addition, conflict may appear among analogies and among ancestral homologies, but not among derived homologies. Consider the four characters. In (a), *a′*, *b′*, *c′*, and *d′* are all analogies, and they fall in phylogenetically conflicting groups. In (b), *a*, *b*, *c*, and *d* are all ancestral homologies, and the group indicated by *d* (species 3 + 4) could be a subgroup of that indicated by a (species 2 + 3 + 4), but otherwise the characters contradict one another. In (c), *a′*, *b′*, *c′*, and *d′* are all derived homologies and do not conflict. The group indicated by *b′* is a subgroup of that indicated by *c′*, and similarly for *a′* and *d′*.

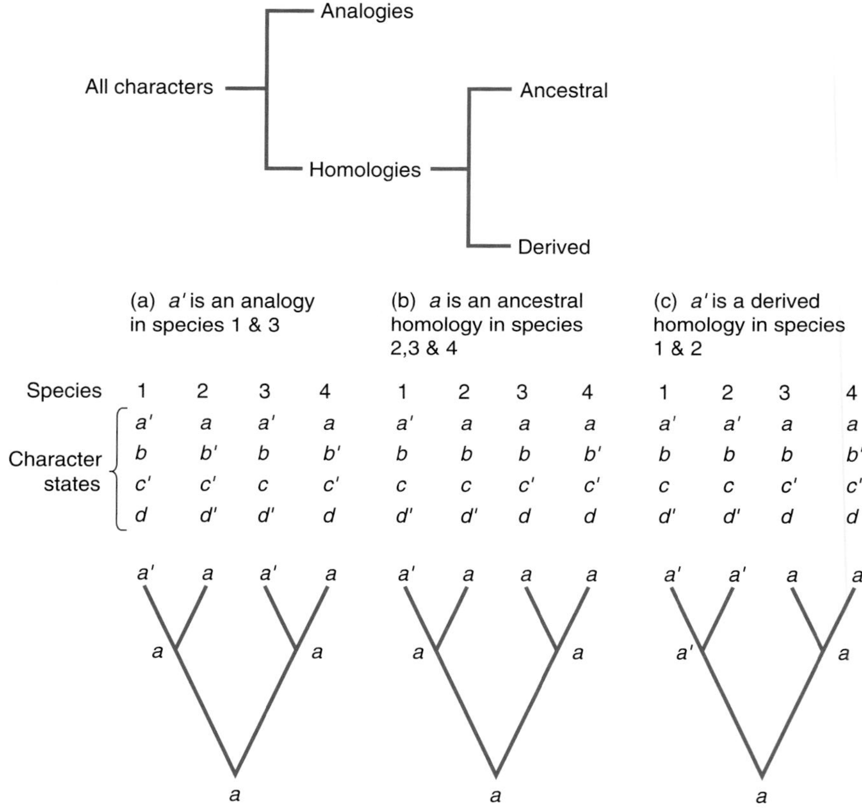

phylogenetic relationship, but that it is unreliable; in some cases, it gives the wrong answer. The unreliability of phenetic similarity gives us a reason to distrust distance statistics in phylogenetic inference, because distance statistics group species according to their phenetic similarity (exactly the same argument applies for genotypic similarity). If we had applied the informal reasoning used earlier in this chapter for the three species case of chimpanzee, human, and amoeba to the case of the lizard, crocodile, and bird, we would have blundered. Shared ancestral homologies can, like analogies, fall into conflicting groups of species (Figure 17.8b).

Only derived homologies eliminate the possibility of character conflict: they must all agree on the same phylogeny (Figure 17.8c). Because all characters have evolved in the same phylogenetic tree, all the shared homologies must fall into the same pattern of groups ("horizontal" transfer of characters between lineages is an exception). If the homologies and character polarities have been correctly identified in a number of characters, it is impossible for different shared derived characters to suggest incompatible phylogenies. A set of species can no more have multiple phylogenetic relations than a human family can have more than one family tree. If a human family has two family trees in its possession, at least one of them must be wrong. Likewise, if two characters suggest incompatible phylogenies, the analysis for at least one of them must be incorrect. The same is not true for analogies and ancestral homologies, however. Ten different analogies, or ten different ancestral homologies, can lead to ten different, and conflicting, groupings of species.

In summary, we have divided characters into three kinds and considered theoretically how each falls into groups of species in a phylogeny. Only shared derived homologies consistently reveal phylogenetic groups. But how can we distinguish in practice between ancestral and derived states of characters?

17.8 *The polarity of character states can be inferred by three main techniques*

The question of how to distinguish derived from ancestral homologies (also called the question of *character polarity*) has the following general form. A character has two states, which we can call a and a'. In this case, we need to know whether a evolved from a', or vice versa. In this section, we will discuss the three most important methods of discerning character polarity: outgroup comparison, the embryological criterion, and the fossil record.

17.8.1 *Outgroup comparison*

Amniotes are the group made up of reptiles, birds, and mammals. All of these animals possess an egg membrane, called the amnion, during their development. It is known that amniotes are a monophyletic group—that is, they all share a unique common ancestor. Here, we can assume that the amniotes are indeed known to be a good phylogenetic group but that we do not know the relations among the different amniotes. For instance, in a set of six amniotic species (such as a mouse, a kangaroo, a bird of paradise, a robin, a crocodile, and a tortoise), does the kangaroo share a more recent common ancestor with a mouse, a bird of paradise, or one of the other animals?

Suppose we have established homologies in various characters, including reproductive physiology. The kangaroo and mouse are viviparous, and the other four species are oviparous. Did the ancestor of the group of six species breed viviparously, in which case viviparity was ancestral and oviparity derived, or did it breed oviparously, in which case evolution occurred in the opposite direction? By the method of *outgroup comparison*, the answer is found by looking at a closely related species that is known to be phylogenetically outside the group of species under investigation. The character state in that outgroup is likely to have been ancestral in the group under consideration.

In this case, we might look at a salamander, or a frog, or even a fish. They are all near relatives of the amniotes, but not amniotes themselves. These "outgroup" species almost all breed oviparously. The inference by outgroup comparison, therefore, is that oviparity is ancestral in the amniotes. Viviparity, in the kangaroo and mouse, would then be a shared derived character and oviparity, in the other four animals, would be a shared ancestral character.

In the abstract, two species, species 1 and 3, might share homology a and two others, species 2 and 4, with homology a' (Figure 17.9). To find whether character a evolved into a', or a' into a, we would look at a closely related species and infer that the state there is ancestral in the group of four. If the outgroup had a, we would infer that species 2 and 4 share a more recent common ancestor with one another than with any of the other species; the relation of species 1 and 3 remains uncertain (as is discussed later).

Outgroup comparison works on the assumption that evolution is parsimonious. In Figure 17.9, if the character in the outgroup a is ancestral in the group of species 1–4, at least one evolutionary event must have occurred in the phylogeny: a transition from a to a' before the ancestor of species 2 and 4. If, having observed a in the outgroup, we had reasoned that a' was the ancestral state of species 1–4, at least two events must have taken place: a change from a' to a somewhere between the outgroup and species 1–4, and then a change from a' back to a in species 1 and 3. If the character state in the outgroup is ancestral, the fewest evolutionary events are required.

(a) **Observations**

Species	1	2	3	4	Outgroup
Character state	a	a'	a	a'	a

(b) **Phylogenetic inference**

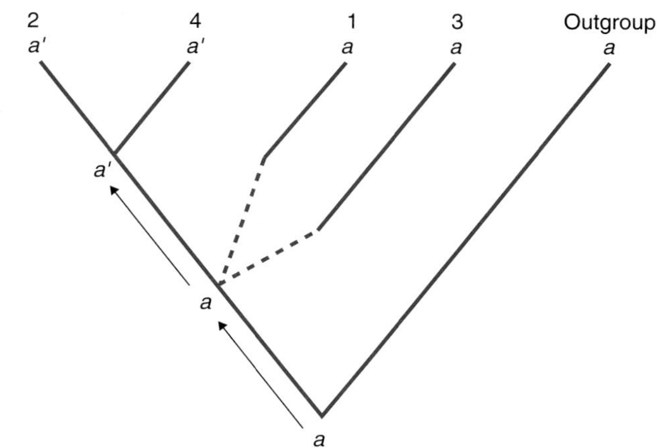

Figure 17.9 (a) Species 1–4 have the character states given. We wish to know whether a or a' was the state in their common ancestor. (b) We look at a closely related species, the outgroup. It has state a, and we infer that was the state in the ancestor of species 1–4. The dotted lines for species 1 and 3 indicate their branching relations remain uncertain.

Outgroup comparison, like all techniques of phylogenetic inference, is fallible. Sometimes, one possible outgroup will suggest that one character state is ancestral, but another possible outgroup will suggest that a different character state is ancestral. The result will then depend on which outgroup is used. The method is most reliable when the closely related species that could be used as outgroups all suggest the same inference, but the method may also lead us astray in particular cases. The inference should be treated with caution, and if possible tested against other evidence.

Returning to the case of the six amniotes, the inference about viviparity provides evidence that the mouse and kangaroo share a more recent common ancestor with one another than with any of the other four animals. It provides no evidence about the relations among the other four species. They share only an ancestral character. The evidence does not distinguish between the two types of phylogeny in Figure 17.10. The mammals could be a separate branch, or they could have branched off from any one of the other four taxa. The shared ancestral homology of oviparity does not provide evidence that the four share a more recent common ancestor with one another than with the mammals. We need more characters to sort out the four species with the ancestral character state.

Before we can use outgroup comparison, we must know something about the phylogeny. For example, we needed to know that fish and amphibians were

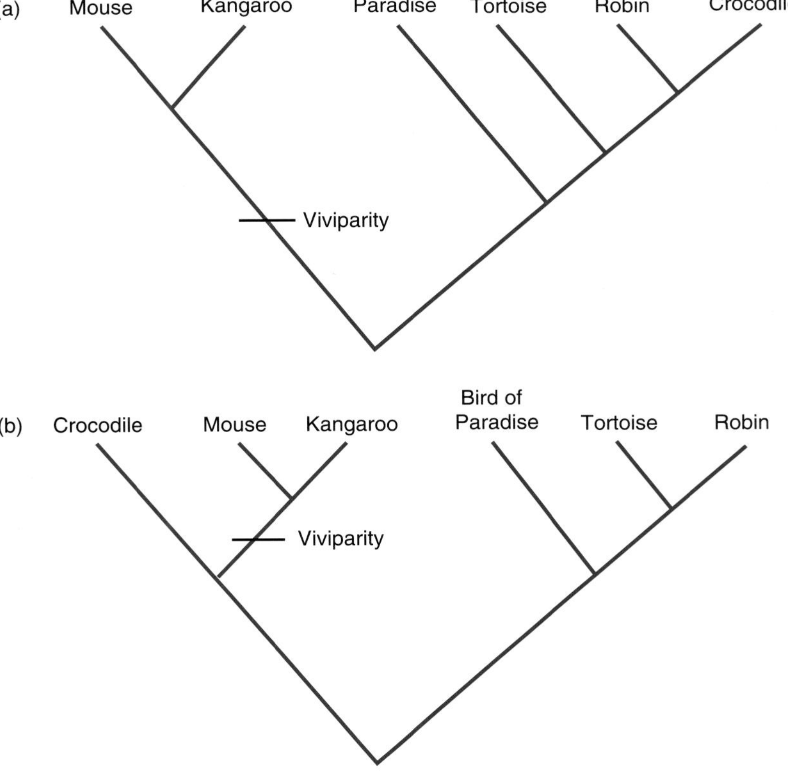

Figure 17.10 Two possible inferences for six amniote species. Once we know that mouse and kangaroo are more closely related, we still know nothing about the relations of the other four species. Nor do we know whether the mouse and kangaroo (a) share a separate branch, or (b) branched off from any one of the other taxa. Many more phylogenies would be compatible with the evidence.

outside the Amniota to use them as "outgroups." In practice, this is not a major problem. Outgroup comparison cannot be used when we are absolutely ignorant, but we can build on our knowlege about the phylogeny of a group (for example, that amphibians are not amniotes, but are closely related to them) to find out more (in this case, more about the phylogeny within the amniotes).

17.8.2 *The embryological criterion*

The pre-Darwinian embryologist Karl Ernst von Baer (1792–1876) described, from microscopic observations, the course of development of several animal groups. He summarized his observations in his embryological laws. For phylogenetic inference, von Baer's first law is most important. It states: "The general features of a large group of animals appear earlier in the embryo than the special features." Cartilage, for example, is found in all fish—in cartilaginous fish such as sharks, rays, and dogfish as well as in bony fish. Cartilage is a general character; bone is a special character, being found only in bony fish. Von Baer's law predicts that, in bony fish, cartilage will appear earlier in individual development, and will transform into bone. This prediction is correct (Figure 17.11).

Von Baer's law can be given an evolutionary interpretation and then put to phylogenetic use. The characters von Baer called "general" are, in evolutionary terms, ancestral, and his "special" characters are evolutionarily derived. Thus, the successive transformations from general to special forms of a character are evolutionary changes between ancestral and derived character states. By the embryological criterion, cartilage is inferred to be an ancestral state, and bone a derived state. The bone in bony fish, therefore, evolved from a cartilaginous ancestry. In general, if we have a group of species and a list of homologous characters, then (if they are the kind of characters that undergo development—i.e., not things like chromosomal bands) the relative ancestral and derived character states can be inferred from their order in development.

The embryological criterion works only when von Baer's law is correct (see section 21.5, p. 592). The law's scope has never been systematically studied. It

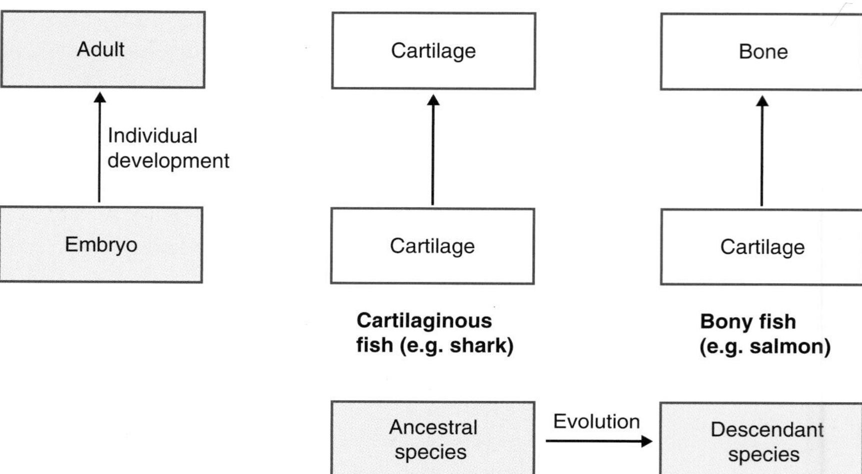

Figure 17.11 The character (cartilage) that appears earlier in the development of bony fish is a general character, found in the wider phylogenetic group of bony and cartilaginous fish. Bone is a more special character, being found only in bony fish. The character state "cartilage" is inferred to be ancestral, and the state "bone" derived.

is widely accepted to have some truth, and for this reason it can be used in phylogenetic inference. It is also known to have exceptions, however. Clearly, von Baer's law should be applied only where we can be reasonably confident that it is valid, and inferences made from it should, if possible, be tested against other classes of evidence.

17.8.3 *The fossil record*

In the evolution of the mammals from mammal-like reptiles, many characters changed over time (section 21.1, p. 582). Posture evolved from a "sprawling" to an "upright" gait; jaw articulation and circulatory physiology also changed. Some, though not all, of these characters leave a fossil record, and we can infer which character states were ancestral and which derived by studying the earlier fossils. This type of investigation is the *paleontological criterion* of character polarity.

The reasoning could hardly be more simple. The ancestral state of a character must have preceded the derived states in fact, and therefore the earlier state in the fossil record is likely to be ancestral. In the case of the mammal-like reptiles, the criterion is reliable, because the fossil record is relatively complete. If the record is less complete, a derived character could be preserved earlier than its ancestral state (Figure 17.12), and the paleontological inference will be the opposite of the truth.

For a whole fossil series like the mammal-like reptiles, we can be reasonably sure which states are ancestral. At the other extreme, of few fossils and a highly imperfect record, the evidence may be practically worthless. Most real cases lie between these extremes, and an intermediate level of confidence is appropriate.

17.9 *Any residual character conflict can be resolved by parsimony*

The division of shared morphological characters into analogies and homologies, and ancestral derived homologies, should reduce the conflict relative to an unanalyzed list of characters, for the theoretical reasons given in section 17.7. In an ideal case, the conflict should be reduced to zero, because real derived homologies cannot conflict. The actual level of conflict, however, is likely to be reduced

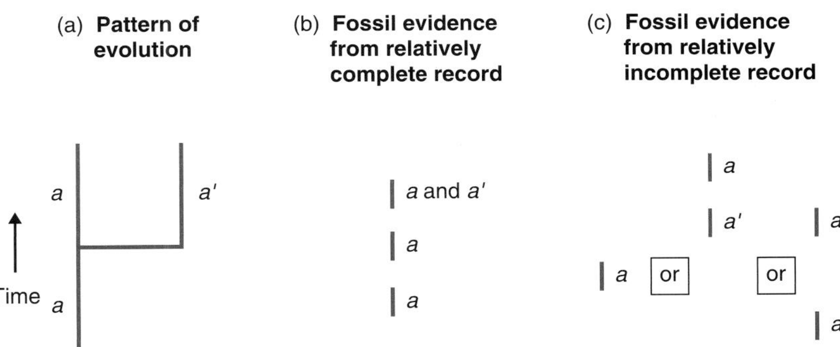

Figure 17.12 (a) The ancestral state of a character must have evolved before its derived state. (b) If the fossil record is relatively complete, the ancestral state will be preserved in earlier fossils. (c) If the fossil record is incomplete, the derived state may or may not be preserved earlier than the ancestral.

to something more than zero because the techniques can all yield mistakes. Convergence can be deceptively exact, and analogies can be mistaken for homologies. The criteria for determining character polarities may be inapplicable (if the character lacks a fossil record, or its embryonic development is unknown or nonexistent, or its phylogenetic relations with nearby outgroups are obscure), and even when they can be used they are still fallible. Moreover, the existence of more than one criterion may increase the uncertainty. If a character can be studied by multiple criteria, the results can be played against one another. If the criteria agree, it increases the plausibility of the conclusion; if they disagree, we face the problem of deciding which evidence to trust.

We have four options when faced with conflict in the evidence: we can scrutinize, and re-scrutinize, the contradictory results to test their reliability; we can suspend judgment; we can collect more evidence; or we can resolve the conflict by the parsimony principle (that is, pick the phylogeny that requires the fewest evolutionary events). The viability of the four will vary from problem to problem, but, after all character analysis is finished, any remaining conflict must be resolved by parsimony. Biologists vary in how much they prefer to persist with character analysis, before moving on (or resorting) to parsimony. The fact that parsimony may be used in the end does not mean the result will be the same as if it had been used initially.

Consider the example of the 100 initial characters, of which 30 pointed to phylogeny *a*, 30 to phylogeny *b*, 20 to phylogeny *c*, and 20 to other idiosyncratic arrangements. The exposure and elimination of analogies reduced the list to 25, 15, 12, and 3 (homologous) characters, respectively. The character polarities might reduce it, in turn, to 12, 3, 6, and 0 characters, respectively. The principle of parsimony is finally needed to resolve the conflict (in favor of phylogeny *a*, which is supported by 12/21 characters), but the inference becomes much more certain by virtue of the character analysis. The initial data were indifferent between phylogenies *a* and *b*, and only 30% of the data supported each of them. On the other hand, the sifted data unambiguously favored phylogeny *a*, which was supported by almost 60% of the characters.

A final point can be made about the recognition of analogies and homologies, and of character polarities. The methods we have examined so far worked before (*prior* to) we knew the phylogeny. It is also possible to recognize the character types after (*posterior* to) the phylogeny is known. Take the case in which 12, 3, and 6 characters, as analyzed by the "prior" methods, suggest phylogenies *a*, *b*, and *c*. They cannot all be derived homologies, because real derived homologies do not conflict. Thus, some of the characters must have been misidentified. We then apply the principle of parsimony to the conflict, and deduce that phylogeny *a* is correct. The conclusion implies that the 12 characters supporting *a* are derived homologies; but the other 9 characters are not. This is the "posterior"method. The distribution of the 9 characters on phylogeny *a* will reveal, by inspection, which are convergent and which are ancestral homologies. The "posterior" method cannot be used to infer the phylogeny, because it would create a viciously circular argument. A knowledge of whether a shared character is homologous or analogous, and ancestral or derived, can be useful for other purposes, however, and a complete list of the techniques of character analysis should include both the posterior and the prior techniques.

17.10 *Molecular sequences are becoming increasingly important in phylogenetic inference, and they have distinct properties*

The sequences of proteins and DNA are both used in phylogenetic inference. Proteins blazed the trail, with the first protein to have its amino acid sequence worked out—insulin—being sequenced by Sanger in 1954. Protein sequencing became an automated process through the 1960s, and the sequences of some proteins, such as cytochrome *c* and hemoglobin, became available in enough species for large-scale phylogenies to be inferred. DNA sequences followed approximately 20 years later. Sanger was also the first to sequence a decent sized portion of DNA, in this case the whole genome (containing 5375 bases) of the bacteriophage ϕX174, in 1977. Since that time, DNA sequencing has expanded, almost explosively, and most current molecular phylogenetic work is concerned with DNA sequences. Many of the methods and concepts of molecular phylogenetics were established for proteins, however, and here we shall consider the two kinds of molecules together.

The deep logic of phylogenetic inference is identical for molecular and morphological characters, but the two have distinct properties and the methods and concepts used for each can appear very different. The homology/analogy distinction, in particular, differs for the two. When confronted by apparently conflicting homologies for morphological characters (like the wings of birds and bats), we first reexamine the organs, and their embryology, in detail to see whether their similarity really is fundamental, or superficial and analogous. Homology represents a powerful concept for morphological organs such as wings. Wings are complex in structure and can take on an almost infinite variety of shapes; they have an embryonic development and morphological relations with the rest of the body. If the information in the structure and development of a wing in two species is the same, it is highly likely (for the reasons given in section 17.2) that the character evolved from a common ancestor who had similar wings.

The homology/analogy distinction is much less powerful for molecular evidence. Suppose a nucleotide is identical (as an *A*) in two species. Evolutionary changes take place among a very limited set of alternatives (the four bases *A*, *C*, *G*, and *T*), and it is fairly probable that the same informational state could independently evolve in the two species. The argument is relaxed for proteins, because 20 amino acid states are possible. Nevertheless, it still applies because 20 fixed states make up a small number compared with the variety of morphological forms. Thus, on the molecular level it is not so unlikely that similarity in the states of two species could have evolved independently. Moreover, the morphologist's methods are absurd for molecules. The amino acid found at site 12 of cytochrome *c* is methionine in humans, chimpanzees, and rattlesnakes, but glutamine in all other species—including many mammals and birds—that have been studied. We cannot dissect the rattlesnake's methionine, or trace its embryonic development, to see whether it is only "superficially" methionine and "more fundamentally" glutamine. It is a methionine molecule, and that is that.

Nor can we usually assess the reliability of different pieces of molecular evidence by considering how natural selection could have acted on them. When morphologists examine a similarity between the organs of two species, they also watch for functional convergences—such as the evolution of wings in species that fly. This kind of analysis is impossible if we do not understand the relation between the structure (the wing) and its function (flight). For molecules, we usually lack this understanding. If we knew, for example, that a change from glutamine to methionine at site 12 of cytochrome *c* made functional sense in certain kinds of animals, then the same kind of arguments applied in morphology could be used for the protein. Otherwise, we must treat molecules in the same way that a morphologist would treat an organ of unknown function.

As a result, research with molecular evidence makes little use of the distinction made in morphological research between homology and analogy. Homologous and analogous similarities can be distinguished for molecules, but by the "posterior" method, after the phylogeny has been inferred. With morphological evidence, we saw that it is sometimes possible to infer strongly that a particular shared character is homologous—and the character may then have great weight in phylogenetic inference. For molecular sequences, it cannot be firmly established in advance that an identical amino acid or nucleotide is homologous. We can observe similarity, but the similarity could just as well be homologous as analogous.

Molecular sequences have other distinctive properties. They provide a great deal of evidence. Cytochrome *c*, for example, has 104 amino acids that can be treated as 104 pieces of phylogenetic evidence. A typical morphological study might be based on perhaps 20 or so characters, and it is exceptional for many more than 50 characters to be used.

In addition, the recognition of independent units of evidence is apparently straightforward with molecular sequences. With morphological evidence, two apparently separate organs may really comprise a single evolutionary unit. At one extreme, non-independence is obvious; no one would think of treating the right leg and the left leg as two pieces of evidence. Less obvious correlations can also arise as a consequence of developmental processes, which makes the recognition of independence tricky. For nucleotides, the mutations down the DNA molecule are effectively independent; each site can evolve independently of each other site. (This is not true in fact, however, when there is no or little recombination between the sites—though how much of a problem this creates is the business of current research.)

Evolution at different amino acid and nucleotide sites is easily comparable. One change at one site is equivalent to one change at another site. How, then, can we say what amount of evolution in a knee bone is equivalent to any given change in a skull bone?

The four properties of DNA and protein sequence data—the impossibility of any deeper analysis of the character, the large amounts of evidence, the recognizability of independent units, and the comparability of evidence—have encouraged the development of statistical techniques to infer phylogenies. The same techniques are, in principle, equally applicable to morphological evidence, although in such cases it is always tempting to try to preempt statistical analysis and resolve the apparent conflicts by ever-deepening character analysis. Morphological data are also less readily divisible into neat character states for statistical analysis. Molecular phylogenetics therefore tends to follow the method shown at the bottom of Figure 17.5.

17.11 *Molecular sequences can be used to infer an unrooted tree for a group of species*

A number of statistical techniques can produce an unrooted tree from a set of molecular sequences. They broadly divide into "parsimony" and "distance" statistics. In this section, we will concentrate on parsimony.

The first, and most difficult, step is to align the sequences from the various species. With normal length sequences of more than 100 nucleotides, regions will usually have been deleted during evolution in some species and added to others, such that the sequences of the different species do not simply align, with nucleotide number 39 of species 1 corresponding to the nucleotide 39 of the other species. Alignment is a difficult art, and we shall not discuss it further here (more information on this process can be obtained through the Further Reading section on pp. 504–505).

Given a set of aligned sequences, the principle of parsimony with molecular sequences follows the same procedure as used in the simple example of the human, chimpanzee, magnolia, and amoeba presented at the beginning of this chapter. We write out all possible phylogenies and count the minimum number of events implied by each (Figure 17.3). Figure 17.13 illustrates a mechanical procedure for some simplified nucleotide data (although it looks at only three of the seven possible unrooted trees for five species). Four of the five sites in the figure are informative (only site 5 is uninformative, being equally compatible with all trees), a higher proportion than found in most real cases.

At Stage 4 in Figure 17.13, the stage at which we have the most parsimonious unrooted tree, the branches in which character changes occurred are known, but their polarities are not. In the figure, the branches with changes have lines drawn through them. In the top left branch of Figure 17.13a, for example, a change occurred between *A* and *T*, but we do not know whether $A \rightarrow T$ or $T \rightarrow A$. The polarities of the changes are needed at the final stage, at which we find the root of the unrooted tree. If the characters were not molecules, it might be possible to find the root using any of the criteria given in section 17.8. With molecular evidence, however, only outgroup comparison can be used. We therefore look for the sequence in a closely related species and see to which branch in the unrooted tree it corresponds most closely.

Two possibilities are shown in Figure 17.13. In one case, the outgroup sequence most parsimoniously branches off the bottom right-hand branch of the unrooted tree, and implies the rooted tree as drawn. In the other case, the root of the tree might be located in any of three branches. It is also possible that outgroup sequences might be unknown or so different from the five species that further evidence would be needed before we could locate the root of the tree.

17.12 *Different molecules evolve at different rates, and molecular evidence can be tuned to solve particular phylogenetic problems*

The art of using molecules to infer phylogenetic relations involves picking a molecule that evolves at a rate appropriate to the group of species in question.

Figure 17.13 Counting the minimum number of changes among a set of species. Here we have five species, and five variable bases have been sequenced in the DNA of each. The first step is to write out all possible unrooted trees; only three are given here. Take the tree (a) as an example. The procedure is as follows.

(1) Pick any species as a starting point (e.g., species 3). Trace back down its lineage to the nearest node (node ①). Deduce the possible sequence at the node that would minimize the number of changes below the node. Thus, if a base is the same in species 3 and 4, write that base at the node. For example, site 1 has *A* in both species and we write *A* at site 1 at the node. If a site varies, write both alternatives at the node. For example, site 3 has *C* and *T* and we write *C* or *T* at the node because the node could have had either and one change would be required in either case.

(2) Trace back to the next node (node ②). Use the sequence in species 5 and at node ① to deduce the sequence at node ② that would require the fewest changes. Again, if a site is invariant, write the base at the node; if it is variable, write both alternatives. Also, if a site is variable at node ① and one of the alternatives is in species 5, strike out the other alternative at node ①. For example, the *T* at site 3 in species 5 means that node ① must have had *T* (not *C*) at that site.

(3) Deduce node ③ from the sequences in species 1 and 2.

(4) Use the common bases between the possible sequences at nodes ② and ③ to strike out alternatives when one base is common to both. For example, site 3 of node ③ could be *A* or *T* from looking at species 1 and 2, but it is *T* at node 2. We therefore strike out the *A* at node ③.

We now have the most parsimonious sequences throughout the tree and can count the number of changes required. They are written as lines struck through the branches, and there are six. The same calculations for (b) and (c) give 8 and 9 changes. In fact, (a) requires the least changes for all possible trees connecting the five species, and is the correct unrooted tree.

The root can be found by examining the sequence in an outgroup. Two possibilities are shown. If the outgroup has the sequence *AATTA*, the phylogeny is uniquely determined because only one branch part of *a* contains that sequence. If it is *AATTT*, there are three regions with that sequence and the tree could have any of three forms. The outgroup logically could have located the root at any point in the unrooted tree.

Different proteins, and stretches of DNA, evolve at different rates (Table 7.1, p. 156, and Table 7.8, p. 184), and they can be used like clocks with hands that move at different rates. If you use a rapidly evolving molecule for an ancient group, the molecule will have "turned over" many times during the phylogeny. In such a case, once multiple changes at the same site become common, the phylogenetic information in the sequence similarity is lost. By analogy, a stop-watch with only a seconds hand would be of no use in comparing professors' lecture times. Likewise, slowly evolving molecules are useless for fine phylogenetic resolution because they will not have changed enough.

Ribosomal RNA genes are particularly valuable in phylogenetic reconstruction because they are found in almost all species; they are present in both mitochondrial and nuclear DNA. The mitochondrial genes evolve more rapidly than nuclear DNA (Figure 17.14a,b), and mitochondrial ribosomal RNA genes are useful for resolving phylogenetic problems in the 10–100 million year range. In contrast, the slowly evolving nuclear ribosomal RNA genes are useful in the hundreds of millions of years range.

Thus, when Milinkovitch *et al.* wished to resolve the phylogeny of dolphins and whales, which the fossil record suggests to have originated less than 35–40 million years ago, the mitochondrial rRNA genes were an appropriate choice.

Observed DNA sequences

	Base				
Site number	1	2	3	4	5
Species:					
1	G	C	A	T	T
2	G	C	T	T	T
3	A	A	C	G	T
4	A	A	T	G	T
5	A	A	T	T	A

	(a)	(b)	(c)
Three of the possible unrooted trees	1, 2, 3, 4, 5	1, 4, 3, 2, 5	1, 5, 3, 2, 4
Counting the changes **Step 1**	AA C/T GT, AACGT, AATGT, ①, ②, ③	③, ②, ①, GCTTT, GCTT, AATA, AATTA	③, ②, ①, GCTTT, GC T, AA T G T, AATGT
Step 2	AA C/T GT, ①, ②, ③, ATT TA/GT, AATTA	CG, AATTT, AACGT, ③, ②, ①, GCT T T, AA A	AA T/C GT, AACGT, ③, ②, ①, GCT T T, AA G
Step 3	GC A/T TT, GCATT, ③, GCTTT, AAT TA/GT	GCAG T, AATT, GCATT, ③, AATGT, CG, AATTT	GCA T A, AAT T, GCATT, ③, AATTA, AA T/C GT
Step 4	GCATT, AATGT, AACGT, AATGT, GCTTT, GCTTT, AATTT, AATTA	AAT G/T T, AAT G/T T, AACGT, GCATT, GCTTT, AATGT, AATTT, AATTA (changes drawn for G at site 4 in center branch)	GCATT, AATTT, AATGT, AACGT, ③, ②, GCTTT, AATTA, AATGT, AATGT
	Six changes	Eight changes	Nine changes

Outgroup	Rooted tree(s)
AATTA	1 2 3 4 5
AATTT	1 2 3 4 5 or 1 2 5 3 4 or 1 2 5 4 3

Figure 17.14 Matching the molecule to the phylogenetic problem. The ribosomal RNA genes in (a) the mitochondria evolve more rapidly than those in (b) the nucleus. The different points indicate species pairs, for which the date of their common ancestor can be estimated from fossils. The graphs tail off (at about 33% divergence) because of multiple substitutions at a site. (c) Phylogeny of dolphins and whales, using mitochondrial rRNA genes; the deepest root is about 35 million years ago. (d) Relations of major animal groups, as revealed by nuclear rRNA genes; the deepest root is probably over 600 million years ago. Reprinted, by permission of the publisher, from Mindell and Honeycutt (1990), Milinkovitch *et al.* (1993), and Lake (1990).

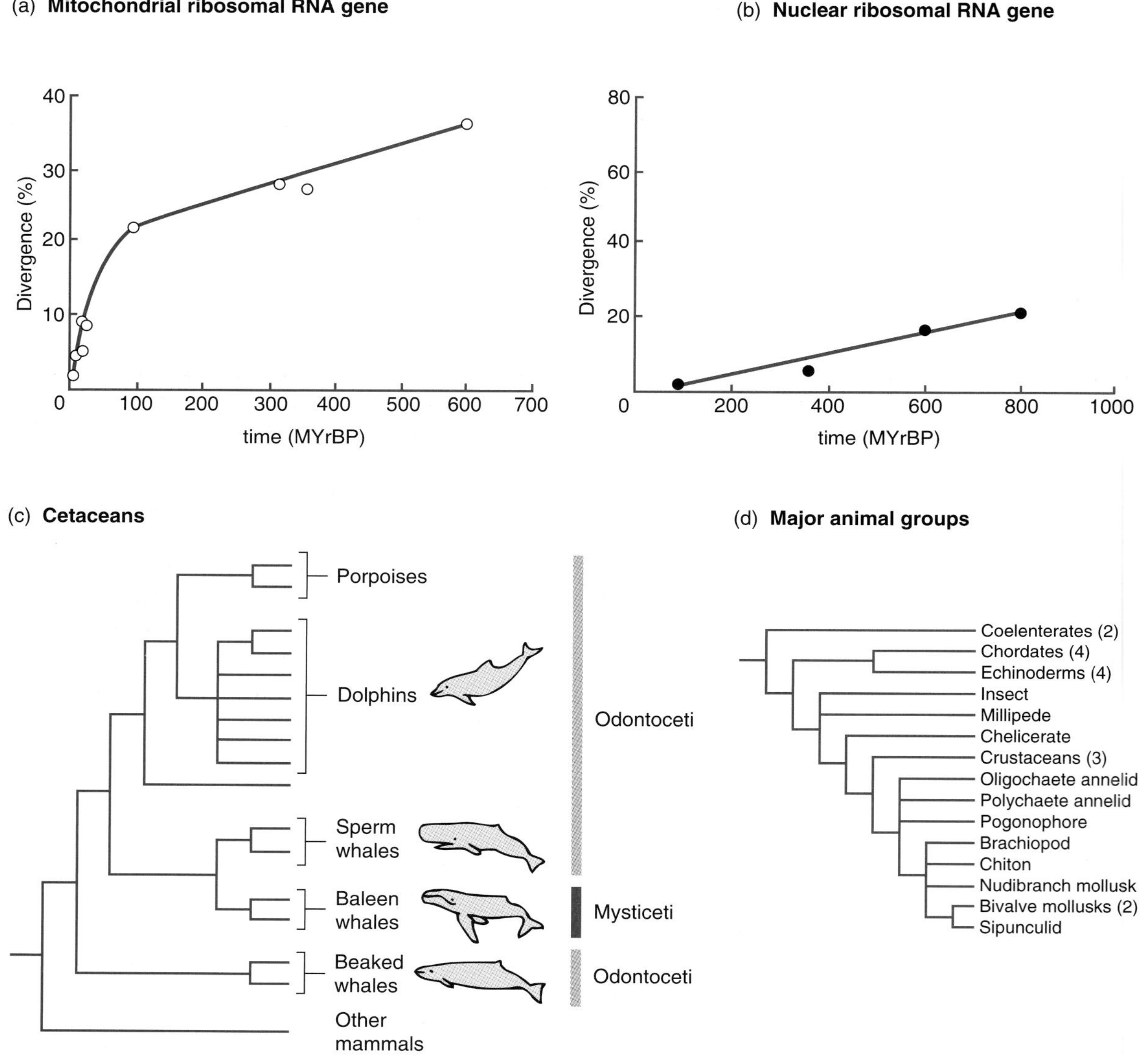

The result of their study is shown in Figure 17.14c. Much of it is uncontroversial, but one of the branches is surprising. Cetaceans are usually classified into two groups, Odontoceti (which contains the echolocating toothed whales and dolphins) and Mysticeti (which contains the filter-feeding baleen whales). Milinkovitch *et al.*'s mitochondrial DNA evidence suggests that the Odontoceti are paraphyletic (see Figure 14.10, p. 390) and that one group of odontocetes (the carnivorous sperm whales) shares a more recent common ancestor with baleen whales than with the other odontocete groups. The grouping will need further independent confirmation before it becomes generally accepted.

By contrast, Figure 17.14d shows the results of a study by Lake on the grand groups in the animal kingdom. These groups originated hundreds of millions of years ago, so the nuclear rRNA genes were the appropriate molecule for the problem. Careful inspection of the phylogeny reveals another zoological surprise—the arthropods may be paraphyletic. Lake's result, at least for the arthropods, has been challenged, but the main point here is that slowly evolving molecules are needed to infer phylogenetic relations of this degree of antiquity.

This point can be understood in terms of the distinction between informative and uninformative characters. As we saw, a site can be uninformative either because it is identical or because it is unique in all species. The former is true when the site changes too little because the phylogeny is too short relative to the evolutionary rate of the molecule; the latter is true when the site changes too much. For the time periods and evolutionary rates in between these extremes, the proportion of informative sites will be highest.

17.13 *Molecular phylogenetic research encounters difficulties when the number of possible trees is large and not enough informative evidence exists*

Research on molecular phylogenies runs into a number of snags, and we shall look at two of the main problems here. In section 17.11, we saw that it is necessary to search through all possible trees to find the most parsimonious one. The first potential difficulty is that the number of possible trees may be impossibly large. With four species, as we saw in our initial example of human-chimpanzee-magnolia-amoeba, three bifurcating unrooted trees are possible, and it was not difficult to count the number of events implied by them all. For five species, however, 15 trees are possible. The general formula for the number of possible unrooted bifurcating trees for s species is:

$$\text{number of possible unrooted bifurcating trees} = \prod_{i=3}^{s} (2i - 5)$$

The Π term means "product": we multiply all possible terms in the parentheses. For three species, $s = 3$ and we have only one term to take the product (from $i = 3$ up to s, which is also 3); the parenthetic term for $i = 3$ is $6 - 5 = 1$, and the number of possible trees is therefore 1. For $s = 4$, we must multiply that 1 by the parenthetic term for $i = 4$, which is 3. Thus, $3 \times 1 = 3$, the number of unrooted trees for 4 species. For $s = 5$, the product is $5 \times 3 \times 1 = 15$, and so on. The number of possible trees increases explosively as the number of species increases. For 50 species, there are about 3×10^{76} possible unrooted trees, and for the 30 million species that may be alive on earth today, the number is about $10^{300,000,000}$. No computer can search through that quantity of trees. Approximately 25 species is the practical upper limit for which we can search all possible trees.

Students of molecular phylogenies distinguish between "algorithms" and "optimality criteria." Parsimony is an example of an optimality criterion. It says that the best tree is the one requiring the least evolutionary change. An optimality criterion is a criterion against which all possible phylogenies can be compared, and the best estimate of the phylogeny is the one that is closest to the criterion.[1] Optimality criteria often encounter the problem of limited computer search capacity, because all trees must be compared with the criterion in this method. If the number of species is too large for all possible trees to be searched, the search must instead be carried out by means of an "algorithm." An algorithm is a rule about how to search from one tree to the next, and to assess which of the two trees is better; it eventually will find a tree that is better than any of the alternatives, but it searches through only a limited number of trees to reach that end.

The following analogy should clarify how a search algorithm works. Suppose you are in San Francisco and are giving someone instructions on how to find Los Angeles. An optimality criterion would be to say "find the city with the largest population in the United States." The unfortunate person who receives this direction must visit every city in the country, and measure their population sizes, to be sure he or she has found that destination. (We assume the searcher has no other source of information.) An algorithm might say "face south and, keeping the Pacific Ocean on your right-hand side, move forward until you arrive at a city with more than one million inhabitants." Now only a small proportion of the United States must be searched, and the conclusion will be satisfactory so long as no other cities exist that meet the criterion between the starting and finishing points.

With the principle of parsimony, the optimality criterion is known, and research is concerned with developing algorithms that search through the trees as efficiently as possible. The particular algorithms used in phylogenetic research have constantly improved in recent years, but remain vulnerable to becoming trapped on "local optima" when they search through the possible trees in a particular way. A local optimum is a tree that appears to be the best possible choice, by comparison with the other trees that the algorithm investigates, but is actually less parsimonious than other trees in a very different part of the space of possible trees. One practical response to the problem is to run the algorithm several times on a set of sequences, starting each run at a different starting point in the "tree space." If all runs converge on the same answer, it strongly suggests that the algorithm has located the most parsimonious tree. If the runs give conflicting results, however, it may suggest that the evidence is inadequate in some way.

A recent study of humans using mitochondrial DNA illustrates this problem. Figure 17.15 is a branching diagram for 135 human mitochondrial types. (A mitochondrial type is a particular mitochondrial sequence. Mitochondria were sequenced from 189 individual humans, and because the study had 189 humans

1. We could say, formally, that the optimality criterion of parsimony is zero evolutionary change, or the tree that comes closest to having zero change is the best. Notice that statement is not equivalent to saying we expect any tree to *have* zero change, because we know that evolution has happened. It is a formal logical criterion, not a theory of reality.

Figure 17.15 Phylogenetic relations within *Homo sapiens*, as revealed by mitochondrial DNA. Each of the 135 tips is a mitochondrial DNA type; the 135 types came from 189 individual human beings. The phylogeny suggests that humans originated in Africa and there have been successive colonizations from that source. The phylogeny is based on sequences of the control region within the mitochondrion, which evolves 4–5 times as fast as the average for the whole mitochondrion. The 135 types have the following ethnic sources: Western Pygmies (1, 2, 37–48), Eastern Pygmies (4–6, 30–32, 65–73), !Kung (7–22); African Americans (3, 27, 33, 35, 36, 59, 63, 100), Yorubans (24–26, 29, 51, 57, 60, 63, 77, 78, 103, 106, 107), Australian (49), Herero (34, 52–56, 105, 127), Asians (23, 28, 58, 74, 75, 84–88, 90–93, 95, 98, 112, 113, 121–124, 126, 128), Papua New Guineans (50, 79–82, 97, 108–110, 125, 129–135), Hadza (61, 62, 64, 83), Naron (76), and Europeans (89, 94, 96, 99, 101, 102, 104, 111, 114–120). The computational procedures for calculating the most parsimonious tree for 135 units are imperfect, and the tree shown is only one possibility among many. The tree was inferred using PAUP and rooted using the chimpanzee. Reprinted with permission from Vigilant *et al.* (1991). Copyright 1991 American Association for the Advancement of Science.

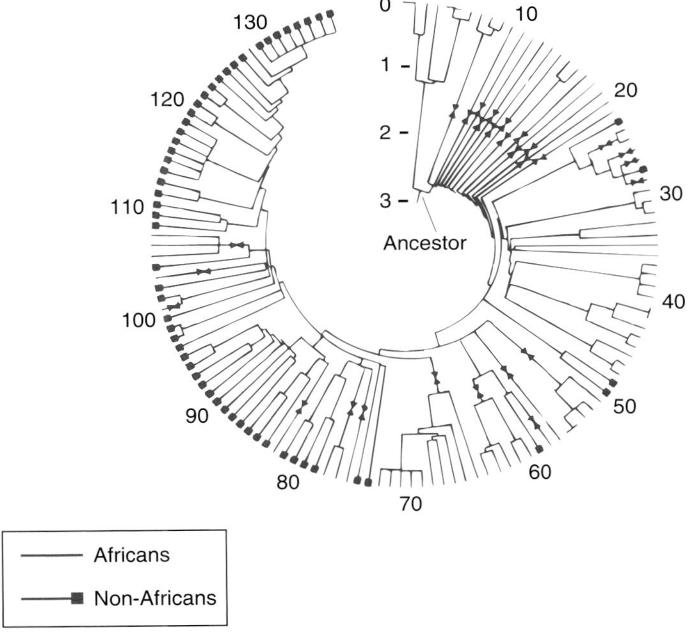

and 135 mitochondrial types, each tip of the phylogeny represents one, or a few, individual human beings.) The number of possible phylogenies with 135 tips is astronomic; they cannot all be searched. The result in Figure 17.15 is the output after one run of a parsimony algorithm, and it has a number of interesting properties. For example, it shows that the deepest branch is African; it has African mitochondrial types to one side and a mix of African and non-African mitochondrial types on the other side, implying that the root of the tree was an individual who lived in Africa.

Another interesting result is that the mitochondrial types do not fall into the groupings that might have been expected. Look, for instance, at the Yorubans. The Yoruban mitochondrial types are scattered through the phylogeny, even though they all live in Nigeria. Likewise, Papua New Guineans do not form a discrete group. It is possible that our naive expectations are incorrect—but the results more likely indicate that the tree is unreliable. Indeed, when the program was rerun, starting in different regions of tree space, many more trees were found that were more parsimonious than the one given in Figure 17.15. Some had African deep roots; others did not. In addition, the different trees showed all sorts of groupings of human populations.

The problem of finding the most parsimonious tree was compounded in this study by a further difficulty. It has 135 tips, but the sequence evidence contained only 201 variable sites and only 119 of those sites were informative. Thus, the evidence is insufficient to resolve the tree. As a rule of thumb, at least as many informative changes are needed as there are tips, and ideally many more informative changes should be included. The algorithm might still stop at a local optimum even if the evidence had contained sufficient changes (e.g., 1000 or so) to resolve the tree, but the danger is greater if the evidence is insufficient, because many trees are now equally compatible with the evidence. The deficiency of evidence explains the strange patterns in Figure 17.15.

The region of the mitochondrial DNA that was used to produce Figure 17.15 had not accumulated enough changes because it does not evolve rapidly enough relative to the time scale of the problem. The time scale is short, because human populations may have possibly diverged as recently as 100,000 years ago or less. More evidence is needed to resolve the tree, perhaps from a more intensive study of mitochondria, or by turning to other parts of the DNA that evolve even faster. Color Plate 8 is a human phylogeny that Bowcock *et al.* (1994) constructed using a set of DNA sequences called microsatellites (section 10.5, p. 265). Microsatellites are made up of a series of short, repeating sequences and they evolve rapidly (see section 10.7, p. 270). These high-speed sequences are well tuned to resolve human populations, and the arrangement in Color Plate 8 is a more probable representation of human relations than Figure 17.15. Color Plate 8, like Figure 17.15, suggests that humans originated in Africa.

In summary, we have considered two separate, but related problems. First, when the number of species (or other taxa) at the tips of the phylogeny is large, the number of possible phylogenies may be too large for all of them to be searched. The algorithms that are used to search among the trees are usually reliable, but not infallible. Second, when the evidence is insufficient, the algorithms are particularly likely (on any one run) to become stuck on a local optimum, and fail to find the most parsimonious tree or even the set of all equally parsimonious trees. (To this point we have treated parsimony as if it were a single technique. In fact, it has several versions and Box 17.1 considers how to choose among them.)

17.14 *Unrooted trees can be inferred from other kinds of evidence, such as chromosomal inversions in Hawaiian fruitflies or comparative anatomy in the mammal-like reptiles*

For some reason, an extraordinarily large number of species of fruitflies (*Drosophila*) are found in the Hawaiian archipelago. There are probably about 3000 drosophilid species in the world, and approximately 800 of them appear to be confined to this archipelago. The phylogeny of one subgroup of the Hawaiian fruitflies is better known than that of any other equivalently large group of living creatures. It was worked out, by Carson and his colleagues, from chromosomal banding patterns. Chromosome bands are clearly visible in fruitflies (section 4.5, p. 77).

The banding patterns differ between species, and it soon becomes obvious that regions of the chromosomes have been inverted during evolution. In fact, a segment of genes within a chromosome has been inverted as a whole. The important event, for phylogenetic inference, occurs when a second inversion takes place across the end of an earlier inversion (Figure 17.16). When this modification happens, we can infer with near certainty that the unrooted tree is 1↔2↔3, not 1↔3↔2. If species 1 had evolved directly into species 3, and then 3 into 2, the two inversions in Figure 17.16 would be needed for the evolution of species 3. To go to species 2, the exact same two breaks (one at each end) of the second inversion would have to happen again in reverse—which is much less probable than evolution in the order 1↔2↔3. As more species are added, with more overlapping inversions, the improbability of most alternative trees multiply to the point of practical impossibility.

To apply the technique, we must first identify the chromosomal banding patterns of the group of species. One species is then picked, more or less arbitrarily, as the "standard" against which the other species are compared. Starting with the species that have a chromosomal banding pattern most like the standard, we gradually work outward through the tree until all species have been included. Carson has concentrated on the "picture wing" group of Hawaiian drosophilids; this group contains about 110 species. A recent phylogeny—a marvellous piece of work—included 103 of these species, based on 214 inversions (Figure 17.17). We can get some idea of the certainty of the inference from the fact that none of the 214 inversions contradict the phylogeny; in other words, all of the characters agree.

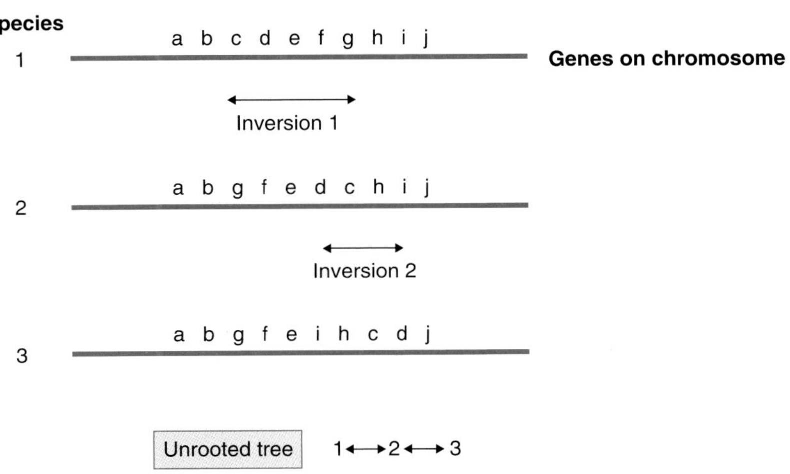

Figure 17.16 Overlapping inversions, in different species, can be used to infer their phylogenetic relations, in the form of an unrooted tree. With this pattern of inversions, the tree must be 1′ 2′ 3, and not 1′ 3′ 2 or 3′ 1′ 2.

BOX 17.1

Parsimony Is Not the Only Method of Inferring Unrooted Trees, and the Merits of Different Methods Are Under Investigation

We have looked at the principle of parsimony to illustrate how to infer phylogenetic trees, but it is not the only technique to accomplish this task. The other main class of techniques consists of distance techniques, in which the similarity between pairs of species is measured and each species is grouped with the other species to which it has the shortest distance (the techniques are formally much the same as the numerical phenetic techniques of classification, section 14.5, p. 375).

Parsimony itself has a number of versions. The kind of parsimony we discussed is called *unweighted parsimony*. We saw that the art of successful phylogenetic inference is to concentrate on informative evidence, which can be achieved by picking a molecule with the appropriate evolutionary rate for the species under study. As we look down the sequence of any one gene, however, some sites evolve faster than others. Consequently, it might also be possible to tune into the most informative evidence within a gene. For example, transitions have a higher evolutionary rate than transversions. Sites that have experienced transitional changes tend to become uninformative quickly, while the sites undergoing slower transversions continue to indicate the phylogenetic relations between species. One way of "weighting" the parsimony criterion is to weight transversions more heavily than transitions. When counting the total number of events implied by a tree, we might divide the number of transitions in each tree by some factor, such as 10. Although other ways exist to weight the events counted by the parsimony criterion, this method is an important one and illustrates the principle.

How well do the various techniques perform? We can try to find out by simulating evolution on a computer; the true (simulated) phylogeny is then known, together with all of the character states in the final (simulated) species. We can then provide the character states of the final species to the techniques and see how well they recover the phylogeny through which they evolved.

Color Plate 7 shows results of one such study, by Hillis *et al.* (1994). The axes express a range of evolutionary rates, from slow to fast, in two regions of the simulated phylogeny. The blue zones indicate where the technique accurately inferred the phylogeny, and the increasingly red zones indicate where the technique was increasingly likely to infer an erroneous phylogeny. Thus, the whole area would be blue for a perfect technique.

The results show a number of striking properties. First, parsimony clearly did better than the distance

Morphological homologies can also be used to produce an unrooted tree for a series of forms. For example, the fossil record suggests that mammals evolved from reptiles through a series of intermediates known as the mammal-like reptiles. The transition took place in gradual stages (section 21.1, p. 582), and the fossil series can be arranged in a row, leading from those that are more like reptiles to those that are more like mammals (Figure 17.18). The position of most of the species in Figure 17.18 has been inferred from their morphological similarity with the other species. *Diademodon* is placed between *Probelesodon* and *Cynognathus* rather than between (for example) *Probainognathus* and *Oligokyphus,* because it is more similar to *Probelesodon* and *Cynognathus* than to *Probainognathus* and *Oligokyphus.* The pattern of similarity tells us which species are most closely related, but it does not tell us the direction of evolution.

technique, and unweighted parsimony was the most accurate of all. (This result, however, can depend on the exact way that evolution is simulated in the computer; deciding the most realistic way to simulate evolution continues to generate controversy.) In some regions, weighting strongly improved the parsimony technique; these regions are, as we should expect, where the evolutionary rates are higher. Figure 17.21 is a real example in which weighting makes a difference.

Second, all the techniques do best along and around the 45° line. The 45° line is the zone where the evolutionary rate in all branches is equal; the phylogeny is then easiest to infer. As we move to the left and right of the line, the rate of change in some branches differs from others. This point is where the techniques begin to misbehave.

A third noteworthy feature is that all the techniques, and particularly the distance technique, do badly toward the top left of the graph. Indeed, in the red area above and to the left of the dashed line, the techniques are appallingly badly behaved. Only three unrooted trees are possible for this four-species problem, which means that a technique that picked a tree at random would find the right answer 33% of the time. The dashed line defines the area beyond which the techniques find the right answer less than one-third of the time. They do worse than chance, actively confusing the issue!

The reason for this behavior is that the top left of the graph is the region where two of the branches have evolved rapidly, leaving the other two species resembling one another even though they are phylogenetically unrelated. The groups of reptiles have evolved relatively slowly as compared with birds and mammals, and distance statistics, which measure phenetic similarity, tend to recognize reptiles as a phylogenetic group, with their own unique common ancestor, because of their high similarity to one another. The similarity is ancestral, however, and an unreliable guide to phylogeny. The figure shows that parsimony techniques are prone to the same error (although for a different reason).

In summary, simulation can test the merits of techniques of phylogenetic inference, such as different kinds of weighting. The research remains at an early stage, but it has already revealed how well the different techniques perform with a range of kinds of evidence, and the results make sense.

Further reading: Hillis *et al.* (1994).

Figures 17.17 and 17.18 are unrooted trees, constructed from homologies alone without any knowledge of polarities. An unrooted tree is valuable biological knowledge, but it becomes even more valuable if we can identify the position of the common ancestor. The criteria noted in section 17.8 can identify the root (although embryology is inapplicable to chromosomal inversions and fossil bones). The root in Figure 17.18 is easy to locate by the paleontological criterion. It is found at the location of the oldest fossils, at the bottom of Figure 17.18.

For the Hawaiian fruitflies, two independent lines of evidence have been used to locate the root. One is to look outside the archipelago for the nearest outgroup and see the banding pattern it includes. The fruitflies that are thought to be the nearest outgroup of the picture-wing group live in South America and

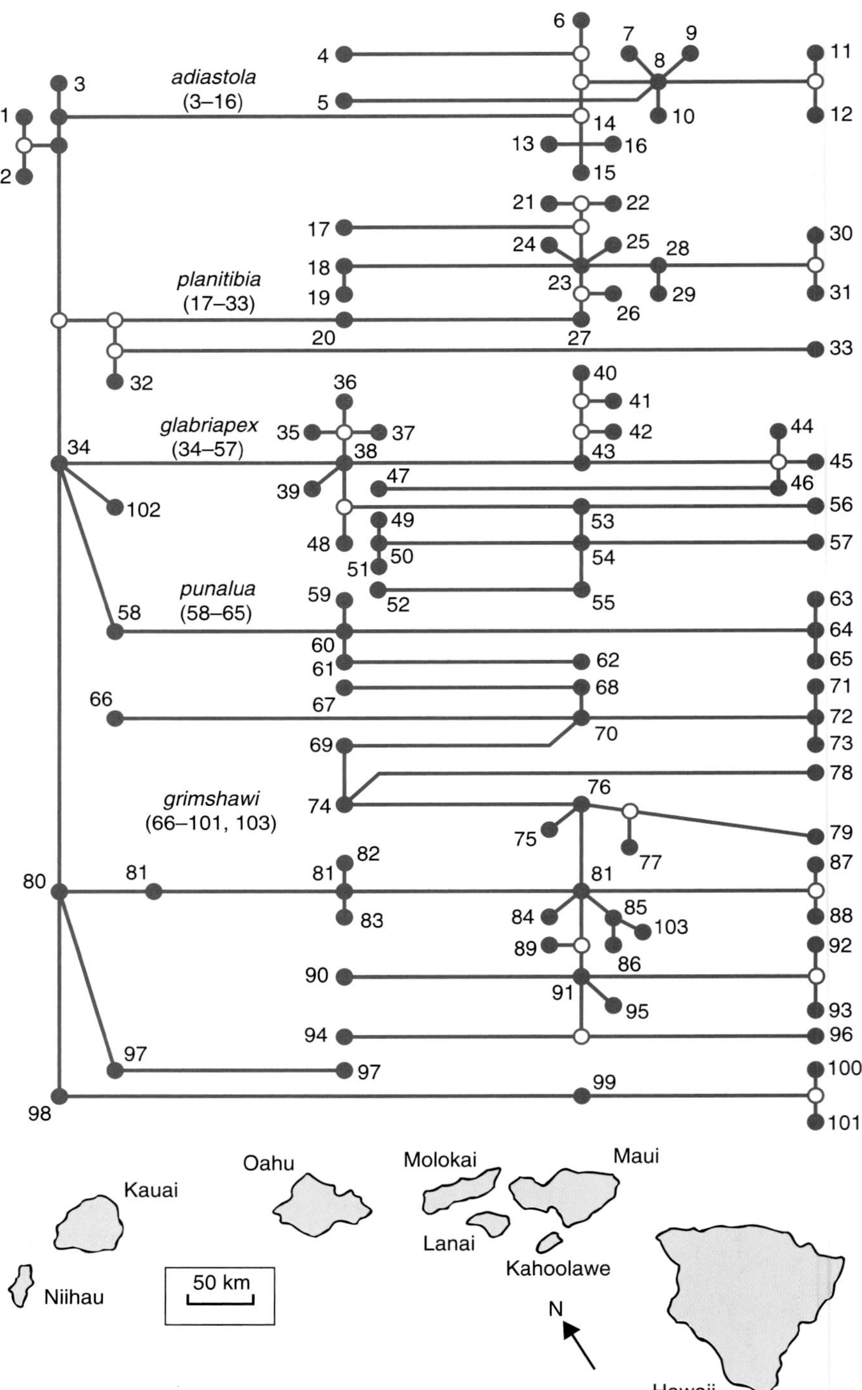

Figure 17.17 Phylogeny of 103 species of Hawaiian fruitflies (*Drosophila*) of the picture-wing group. The unrooted tree was inferred by patterns of chromosomal inversions. The tree can be rooted by geochronology of the islands, and comparison with closely related fruitflies in South America. Some details of the tree are inferred by biogeography rather than inversion patterns. Species shown as ancestors may actually be descendants of the (now lost) ancestor. Reprinted, by permission of the publisher, from Ridley (1986).

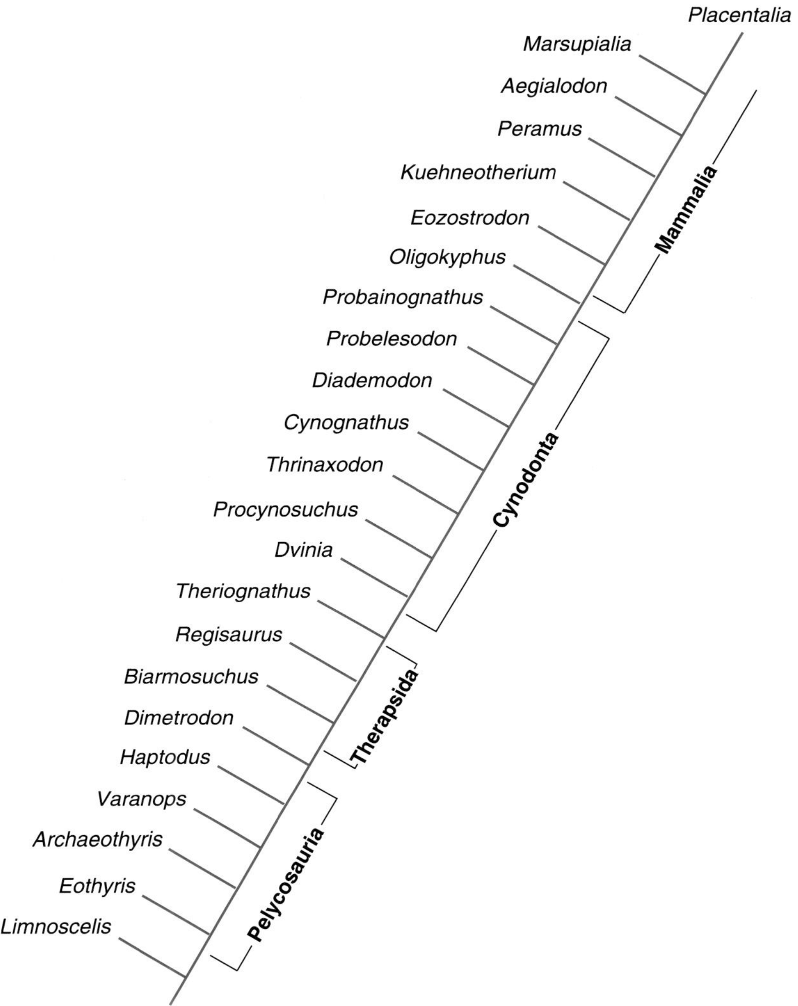

Figure 17.18 A morphoseries of fossil mammal-like reptiles. Each species is intermediate in form between the one below it and the one above it in a series. Figure 20.2 (p. 584) illustrates some of the forms. Reprinted, by permission of the publisher, from Kemp (1982a).

are most similar to *Drosophila primaeva* and *D. attigua* among the fruitflies in Figure 17.17. These species are, therefore, probably closest to the root of the tree. The inference is supported by the geological history of the archipelago. Kauai is the oldest island and Hawaii the youngest (Figure 17.17). The ancestor of the group would probably have colonized Kauai and, if it is still alive, it should still be on Kauai because almost every Hawaiian drosophilid is confined to a single island. *D. primaeva* and *D. attigua* live on Kauai. Indeed, much of the phylogenetic history of the picture-wing group consists of populations of species on the older islands moving to the younger islands, where they form new species by allopatric speciation. No examples are known of species on older islands that are derived from a species on a younger island. Thus, the youngest island, Hawaii, supports the most recently evolved species and the oldest islands in the west have the most ancient species (see Figure 18.11a, p. 524).

17.15 *Some molecular evidence can only be used to infer phylogenetic relations with distance statistics*

The rate at which molecules—both proteins and DNA—can be sequenced is increasing constantly, but we still have sequenced only small parts of the genome even in well-studied species. A phylogeny based on one or two genes may be biased, and methods that do not require exact sequences and use samples from a larger part of the genome can be useful. Several versions of these methods are employed. We will discuss one of the earlier methods, using immunological similarity, in the next section. A more recent method uses rates of DNA hybridization. The feature common to all methods is their inference of phylogenies by distance statistics. They deduce the phylogeny from the simple, unanalyzed similarities between pairs of species. No distinction is made between analogous and homologous similarity, or derived and ancestral similarity.

Normal double-stranded DNA can be "melted", to form single-stranded DNA, by heating it up. If single-stranded DNA is then put in (cooler) solution, it will join together again, or "hybridize" (the same method was discussed in section 10.5, p. 264, for another purpose). The strength of the bond between the two strands is proportional to the similarity of their sequences, and the similarity of the whole DNA of two species can be measured by allowing their DNA strands to hybridize and then measuring how much lower the melting temperature is for the hybrid DNA than the DNA of each species. The temperature reduction is proportional to the percentage difference of the DNA. If the DNA of the two species is very similar, it must be heated almost to the same temperature as the DNA of the two species before it melts apart. If the DNA is different for the species, however, the hybrid DNA will melt at a much lower temperature than the single-species DNA.

The relation between temperature and percentage similarity of the DNA is known. The lab measurements therefore result in a set of percentage similarities (or differences) for the pairs of species that can be used as distance statistics. The DNA of humans and chimpanzees, for example, differs on average at about 1 1/2 % of nucleotides. This finding is the evidence for the famous claim that the DNA of humans and chimps is 98.5% similar. The figure can be converted into a time to the common ancestor if a species pair can be found for which both the age of their common ancestor and the percentage difference in their DNA is known. The fossil record suggests that Old World monkeys and great apes shared a common ancestor about 30 million years ago, and this figure can be used to convert percentage differences into absolute times (Figure 17.19).

The problem with distance statistics is that they do not distinguish between the three kinds of shared characters. If shared ancestral similarities are more common between two species than are shared derived similarities, distance statistics confuse paraphyletic and monophyletic groups. If some members of a group

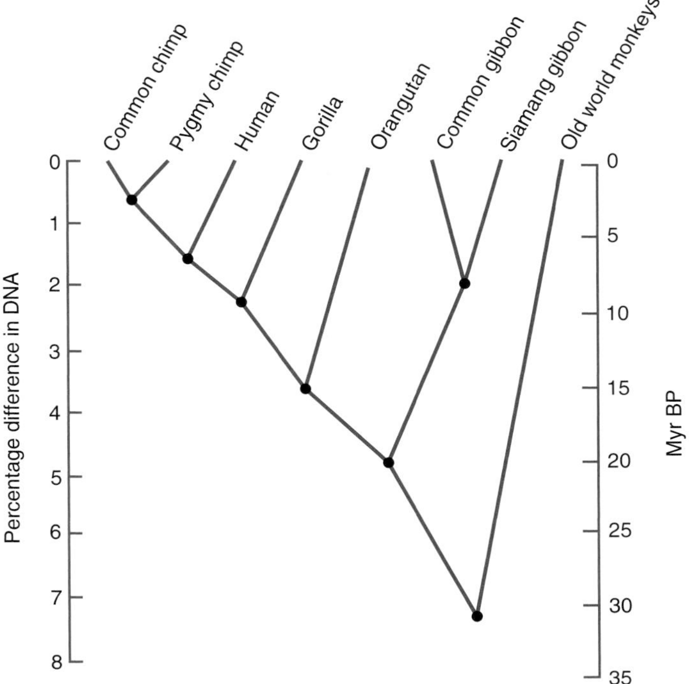

Figure 17.19 Phylogenetic relations of hominoids, as revealed by DNA hybridization. This result contains the evidence that the DNA of humans is 98.5% similar to that of chimpanzees. Reprinted, by permission of the publisher, from Sibley and Ahlquist (1987).

of species (like the reptiles) evolve slowly and other members of the group evolve rapidly, the slowly evolving species resemble one another even though they do not share a recent common ancestor. Thus, distance statistics group species by similarity and will recognize the disparate set of slowly evolving species as a phylogenetic group.

It can be argued that this general problem does not apply to DNA hybridization. DNA hybridization measures the percentage similarity of the whole (or a large part of) DNA of two species. The DNA sequences of species may well evolve at a relatively constant rate. Genes and proteins provide reasonably good molecular clocks (chapter 7), and the same is likely to be true for the DNA as a whole. In that case, the DNA hybridization method is justified. The problematic cases like reptiles arise when evolution proceeds more rapidly in one lineage than another. If that rapid evolution occurs only in some characters, such as observable morphology, while other characters, such as the changes in the whole DNA molecule measured by hybridization, evolve at a constant rate, then distance statistics can be considered reliable for the constantly evolving characters.

In summary, there are classes of molecular evidence, of which DNA–DNA hybridization is currently the most important, that can be used to infer phylogenies by distance statistics. Distance statistics suffer the general disadvantage that they do not concentrate on shared derived homologies. They are most reliable if molecular evolution is clock-like, which, in turn, is most likely if it is driven by neutral drift.

17.16 Comparing molecular evidence and paleontological evidence

17.16.1 *Molecular evidence successfully challenged paleontological evidence in the analysis of human phylogenetic relations*

It is always interesting when two independent lines of evidence, from very different fields, are applied to the same question. We shall finish by looking at two examples, in which older fossil evidence and more recent molecular evidence confict. In one example, the molecular evidence inspired a reinterpretation of the fossil evidence. In the other example, morphological evidence may be correcting a molecular inference.

The first example concerns the time of origin of the human evolutionary lineage. We can consider the fossil evidence first. "*Ramapithecus*" (which is now usually classified in the genus *Sivapithecus*) is a group of fossil apes that lived about 9–12 million years ago. Until the late 1960s, almost all paleoanthropologists thought that *Ramapithecus* was a hominid—that is, it was more closely related to *Homo* than to chimpanzees and gorillas (Figure 17.20a). (Hominoids (formally superfamily Hominoidea) are the group of all great apes, including humans; hominids (formally family Hominidae or subfamily Homininae) are the narrower group consisting of *Homo* and the australopithecines.) *Ramapithecus* and *Homo* apparently shared a number of derived characters. For example, *Homo* has a rounded, "parabolic" dental arcade, whereas chimpanzees have a more pointed dental arcade. *Ramapithecus*'s dental arcade was initially thought to be shaped more like that of *Homo*. *Ramapithecus*'s canine teeth were thought to be relatively diminished compared with its other teeth, as in *Homo* but unlike chimpanzees

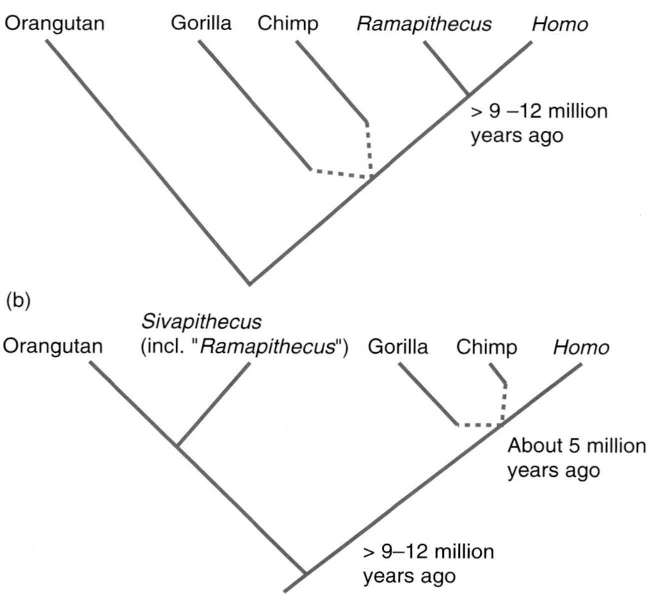

Figure 17.20 Relations of *Homo*, other great apes, and *Ramapithecus*, according to (a) original paleontological and morphological evidence, and (b) molecular (and revised paleontological and morphological) evidence. (Dotted lines imply uncertainty in the order of the human/chimpanzee/gorilla split.)

(in which the canines, especially in males, are large). Most importantly, *Homo* and *Ramapithecus* were thought to share, as a derived condition, a thickened layer of tooth enamel, unlike the thinner layer in other apes (and which was thought to be the condition in the ancestors of the Hominidae).

This morphological and paleontological argument for a relation between *Homo* and *Ramapithecus* has a classical form. A set of character states were shown to be shared uniquely by these two species, and the characters were derived within the larger group of Hominoidea. The corollary was that the human lineage must have split from the great apes at least 12 million years ago, because *Ramapithecus* is nearer to humans than to the great apes.

In the early 1960s, Goodman first demonstrated the molecular similarity of humans and other great apes. The molecular argument for a recent human-ape split was most influentially made, however, in a paper by Sarich and Wilson in 1967. Sarich and Wilson used an immunological distance measure. This method is similar in philosophy to DNA hybridization, but differs in the exact molecule used.

To measure immunological distance, Sarich and Wilson first made an antiserum against human albumin by injecting human albumin into rabbits (albumin is a common protein that circulates in the blood). They then measured how much that antiserum cross-reacted with the albumin of other species, such as chimpanzees, gorillas, and gibbons. The antiserum recognizes the albumins of closely related species, because they are similar to human albumin. It does not recognize them quite as efficiently as it does human albumin, however. The degree of cross-reactivity gives a measure of the immunological distance (ID) between a pair of species. ID increases among phylogenetically more distant relatives, and the relative rate test (Box 7.1, p. 172) suggests that ID increases at a constant rate through time. Thus, immunological distance serves as a sort of molecular clock. The clock can be calibrated using the fossil record for some of the studied species, and the ID can be used to estimate the divergence time for other pairs of species.

The results of this method suggest that *Homo* and the other great apes have too short an ID to fit with a pre-*Ramapithecus* divergence. Consequently, Sarich and Wilson suggested that humans and chimpanzees diverged only about 5 million years ago. Subsequent molecular work has supported their conclusion. The DNA hybridization results suggest a similar, if perhaps slightly older, figure (Figure 17.19), and other molecules suggest a figure of 3.75–4 million years. The corollary is that if *Homo* diverged from the chimpanzee and gorilla 5 million years ago, it cannot be more closely related to *Ramapithecus* than to the living great apes. Thus, the phylogeny must be more like that shown in Figure 17.20b.

Clearly, the molecular and fossil evidence disagreed. A controversy began, in which both the molecular and morphological evidence was challenged (often by experts in the other field). The controversy has now been settled (with a few dissenters) in favor of the original molecular evidence. The morphological characters previously believed to show a relation between *Homo* and *Ramapithecus* succumbed to reanalysis. The dental arcade of *Ramapithecus* had been wrongly reconstructed (originally by combining parts from different specimens), the canine teeth may appear reduced because the fossil *Ramapithecus* specimens were female, and Martin in 1985 finally removed the last important character—thickened enamel—by reinterpreting it as an ancestral character. Moreover, when *Ramapithecus* was compared with another fossil (*Sivapithecus*) that was generally accepted to be a close relative of the orangutan, and with the orangutan itself, it showed clear similarities to them. The specimens formerly classified as *Ramapithecus* are now usually included in the genus *Sivapithecus*, which in turn is thought to be a close relative of the ancestors of modern orangutans (Figure 17.20b).

In summary (simplifying things a little), molecular evidence helped to inspire a reanalysis of the fossil evidence for hominid origins—with the result that a figure of about 5 million years (or at least the range of 4–8 million years) is now widely accepted for the time of origin of the hominid lineage.

17.16.2 *Paleontological evidence may be correcting molecular evidence in the analysis of amniote relations*

Some would draw the conclusion from the previous example (and others like it) that molecular evidence is more reliable than morphological and paleontological evidence. The uncertainties in molecular phylogenetic inference that we encountered in this example would certainly cast doubt on that conclusion, however. In this section, we will consider another phylogenetic problem that is the subject of current research and in which the opposite process may be happening. We say "may be" because the controversy is current, and no conclusion has been reached as yet.

This problem concerns the relations of the amniotes (the group of mammals, birds, and reptiles). Mammals and birds are usually thought to have evolved separately, from different reptilian taxa (Figure 17.21a). The main evidence for this suggestion comes from fossils. A group of reptiles called the mammal-like reptiles (or synapsids) is known to form intermediates between the reptiles and the mammals (see Figure 17.18, and section 21.1, p. 582). Another set of fossils, including *Archaeopteryx*, connects the birds with the diapsid reptiles, a very different group. (Figure 14.3 illustrates a diapsid skull; the synapsids have a different form.)

In 1982, Gardiner suggested a different phylogeny, using the morphology of modern amniotes (Figure 17.21b). He considered 47 morphological characters

Figure 17.21 (a) Standard phylogeny of amniotes and their near relatives, based on skull anatomy and fossil evidence from mammal-like reptiles. (b) Gardiner's phylogeny. The numbers (1–47) indicate numbers of inferred derived homologies shared by each group; the characters are morphological and exist in living species. Note the unorthodox grouping together of birds and mammals. (c) Molecular evidence. α-globin supports, though not strongly, the grouping of mammals with birds; α-crystallin A does not. (α-crystallin A itself gives an unorthodox grouping of birds with lizards rather than alligators, however, and look at the newts in the α-globin tree! The evidence is not definitive.) 18S and 28S ribosomal RNA support the pattern in (b), though not strongly, in an unweighted analysis. (d) When improbable changes such as transversions are more heavily weighted than probable changes such as transitions, the ribosomal RNA genes support (a). Reprinted, by permission of the publishers, from Gardiner (1982); Bishop and Friday (1988); Hedges *et al.* (1990); and Lockhart *et al.* (1994).

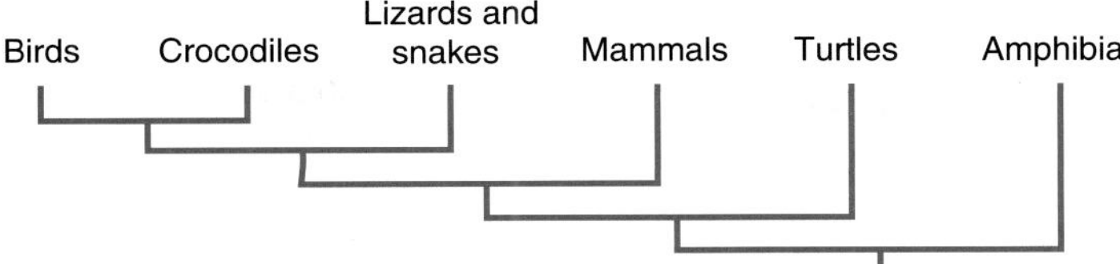
(a) Standard tetrapod phylogeny
Birds
Crocodiles
Lizards and snakes
Mammals
Turtles
Amphibia

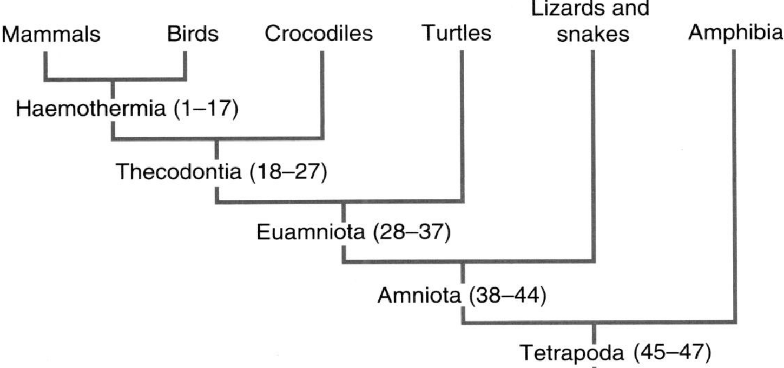
(b) Gardiner's tetrapod phylogeny, from modern morphological evidence
Mammals
Birds
Crocodiles
Turtles
Lizards and snakes
Amphibia
Haemothermia (1–17)
Thecodontia (18–27)
Euamniota (28–37)
Amniota (38–44)
Tetrapoda (45–47)

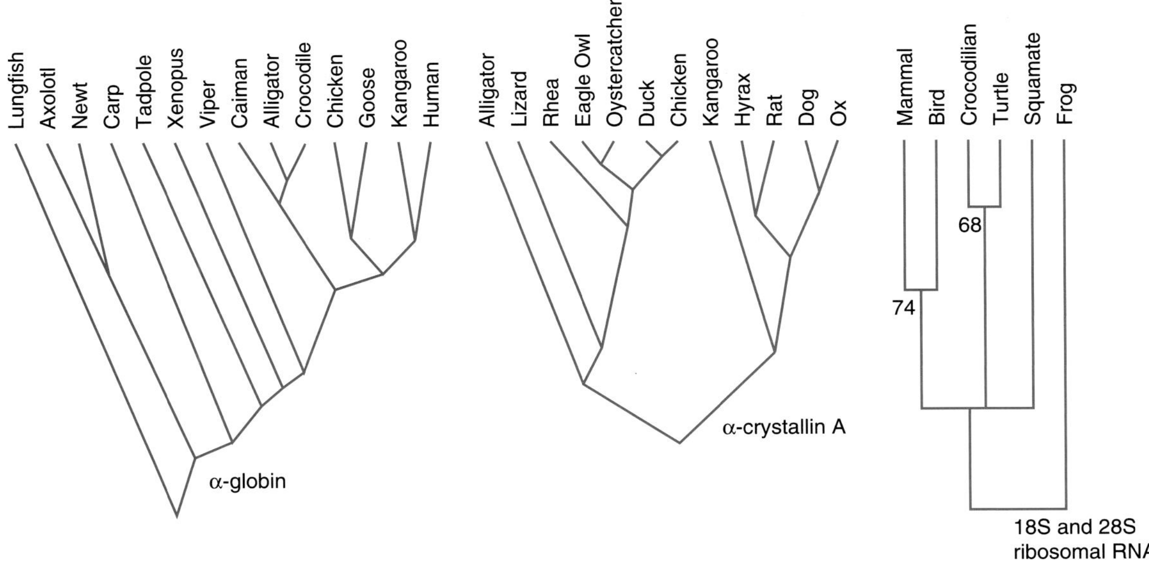
(c) Molecular phylogenies, using unweighted parsimony
Lungfish
Axolotl
Newt
Carp
Tadpole
Xenopus
Viper
Caiman
Alligator
Crocodile
Chicken
Goose
Kangaroo
Human
α-globin
Alligator
Lizard
Rhea
Eagle Owl
Oystercatcher
Duck
Chicken
Kangaroo
Hyrax
Rat
Dog
Ox
α-crystallin A
Mammal
Bird
Crocodilian
Turtle
Squamate
Frog
68
74
18S and 28S ribosomal RNA

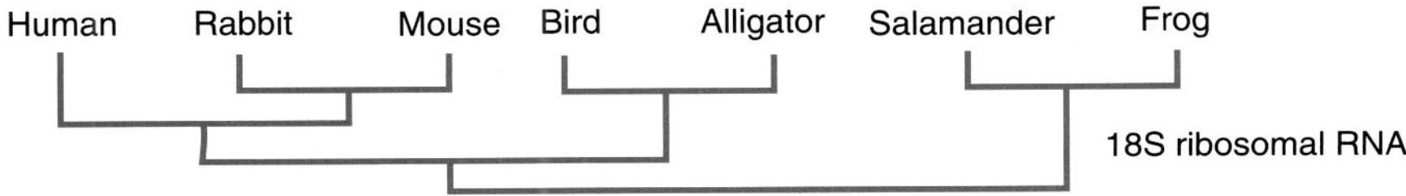
(d) Molecular phylogeny, using weighted parsimony
Human
Rabbit
Mouse
Bird
Alligator
Salamander
Frog
18S ribosomal RNA

and concluded that birds and mammals form a group (called the Haemothermia, because they are warm-blooded) that evolved from the reptiles. The Haemothermia would share a common ancestor subsequent to the reptiles, whereas in the pattern of Figure 17.21a their common ancestor would be found deep within the reptiles. If the fossil evidence and Gardiner's morphological evidence are accepted at face value (which not every one does), the two conflict. If the birds and mammals are closely related groups, the fossil evidence for their separate origins must be reinterpreted to show that it is misleading in some way. On the other hand, if the fossil evidence is accepted, we must show that the modern morphological evidence is misleading, perhaps because the independent evolution of a warm-blooded physiology by birds and mammals led to many other convergences. Gauthier *et al.* (1988) showed that the traditional grouping is supported when all evidence is included and analyzed by parsimony (though Gardiner, in a paper in 1993, disagrees).

What does the molecular evidence suggest? The answer is not simple, but the weight of evidence at least initially seemed to support Gardiner's controversial scheme. Figure 17.21c shows three molecules. α-globin supports Figure 12.21b; α-crystallin A does not. The ribosomal RNA genes support Figure 12.21b using one method of phylogenetic inference (but not when using another, arguably inferior, method). Perhaps this question represented yet another case in which dusty fossil evidence was to be brushed aside by go-go molecules. In this instance, the comparative anatomy of living species, according to Gardiner's analysis, supported the molecular evidence.

Although that assessment may yet turn out to be correct, current research does not appear to be heading that way. The fossil evidence in this case (unlike with *Ramapithecus*) is extensive and cannot easily be ignored or interpreted away. The force of the fossil evidence has driven people to look harder at the molecular evidence.

We saw earlier (Box 17.1) how parsimony can be used in various "weighted" and "unweighted" forms. The results of Figure 17.21c use a method corresponding to unweighted parsimony. The DNA of birds and mammals is known to be *G-C* rich, however. That is, it has a higher proportion of the nucleotides *G* and *C* as opposed to *A* and *T* when compared with reptiles and other taxa. The reason behind this variation may be the warm-blooded physiology of birds and mammals, and their similarity in *G-C* content would then be likely to be analogous rather than homologous. We can test whether the high *G-C* content of mammals and birds influences the result by analyzing the pattern suggested when we use a form of weighted parsimony in which *A-T* similarities are weighed more heavily than G-C similarities. Figure 17.21d demonstrates the results, and reverses the initial conclusion—now the molecular evidence supports the traditional grouping (Figure 17.21a).

These results will not be the final word in this case. It is not known, for example, whether the particular set of weights used in Figure 17.21d is correct, and ribosomal RNA is not the only molecule that can be investigated. We can draw the provisional conclusion that molecular evidence, appropriately weighted, may well support the traditional tetrapod phylogeny. It is worth noting that the weightings added to parsimony effectively reintroduce the morphologist's distinctions (as seen in sections 17.5–17.9) into molecular research. We noticed earlier (section 17.10) that such distinctions are often ignored for molecules, mainly because we do not know enough about molecules to distinguish reliable from unreliable evidence. As our knowledge improves, the distinctions can be reintroduced.

In summary, the phylogeny for amniotes suggested by fossil evidence was challenged by evidence from modern morphology and molecules in the 1980s. Instead of crumbling in the face of the newer analytical method, it helped to force a reanalysis of the molecular inference. The statistical technique used initially with the molecular evidence was probably misleading, and a new system of weights has been introduced. Although only time will prove which results are correct, it appears as if the evidence of dead bones may be correcting the molecular research on amniote relations.

17.17 *Conclusion*

Phylogenetic inference draws on all kinds of evidence, from molecular sequences, to chromosomal inversions, to morphology in modern and fossil forms. It employs (in some form or other) the parsimony criterion: the best estimate of the true tree is the arrangement requiring the least change. With morphological evolution, the most common (though not universal) course of research is to distinguish analogies from homologies, and try to infer the polarity of the homologous characters. The conflict between the characters shared among the species may then be reduced (ideally to zero) and a residue of reliable shared derived homologies used to infer the (rooted) tree. With molecules, individual characters are analyzed less and parsimony is used to infer the unrooted tree; the location of the root can then be inferred by other evidence. Molecular evidence can also be analyzed, and weighted parsimony used in an attempt to focus on the evidence from the more reliable parts of molecules. In other cases, molecules are thought to evolve in enough of a clock-like manner for distance methods to be used as a short-cut to the phylogeny.

It is easy to be deceived by discussions of phylogenetic inference into thinking that it represents an exceptionally uncertain, shaky kind of science. In a full discussion of an unsolved and controversial problem, there can seem to be an endless series of pieces of evidence pointing first one way and then another. One student (of Prof. C. F. A. Pantin of Cambridge University in England), when confronted with a classically recalcitrant problem in phylogenetic inference—the origin of the chordates—summed it up as "Paleontology is mute, comparative anatomy meaningless, and embryology lies." This impression could be misleading, however. Discussion (as well as research) naturally focuses on unsolved problems—and these issues tend to be the difficult ones. Many phylogenetic problems have been solved, and it is worth balancing our reasonable certainty in the human-chimpanzee-magnolia-amoeba case with a more slippery problem such as the amniotes. In addition, the phylogeny of the picture-winged fruitflies of Hawaii, deduced from 214 conflict-free, multiple overlapping, chromosomal inversions, the phylogenetic inference has a level of certainty that compares favorably with most of the facts of the natural sciences.

SUMMARY

1. Phylogenetic relations are inferred using the shared characters of species. The best estimate of the phylogeny of a set of species is the one requiring the smallest number of evolutionary character changes. This concept is called the principle of parsimony.
2. Phylogenies are often incompletely determined, in the form of an unrooted tree. An unrooted tree specifies the branching relations among species, but not the direction of evolution. The root can be identified if character polarities can be inferred.
3. When different characters imply the same phylogeny, phylogenetic inference is easy; when they do not, methods are needed to unravel the disagreement.
4. Theoretical arguments suggest that some kinds of shared characters indicate phylogenetic relations reliably, whereas others do not. Analogies (convergent characters) and ancestral homologies do not reliably indicate phylogenetic groups. Derived homologies indicate relations more reliably. Phylogenetic inference should rely on shared derived homologies.
5. Techniques exist to distinguish homologies from analogies, and ancestral homologies from derived homologies.
6. Character polarities can be inferred, with varying degrees of certainty, by outgroup comparison, the embryological criterion, and the fossil record.
7. For molecular characters, the kinds of character analysis used with morphology are often inapplicable. Phylogenies are usually inferred by statistical techniques that weigh all characters equally, though methods with "weighting" are sometimes employed.
8. The principle of parsimony can reconstruct an unrooted tree from a set of molecular sequences. The molecular similarity of species pairs can also be measured without the sequences (for instance, by DNA hybridization). The phylogeny can then be inferred provided that similarity is proportional to the recency of common ancestry—a condition that is violated when different lineages evolve at different rates.
9. The statistical methods used with molecular evidence, and the character analysis used in morphology, both reconstruct phylogenies from those characters that, according to the principle of parsimony, can be expected to reveal phylogenetic relations most reliably.
10. The unrooted tree of Hawaiian fruitflies is the most firmly established phylogeny of any large group of species. It has been reconstructed using chromosomal inversions.
11. When different classes of evidence, such as molecular and fossil evidence, indicate differing phylogenies, the evidence can all be rescrutinized to determine which evidence is most reliable. No single class of evidence has been shown to be universally more reliable than other classes.

FURTHER READING

The cladistic method, in which rooted trees are inferred using character polarities, often with morphological evidence, tends to be explained separately from the methods used more often with molecular evidence, which infer unrooted trees by parsimony or some other method.

1. *Cladistics.* For the cladistic method, see Wiley *et al.* (1991) and several chapters in Scotland *et al.* (eds) (1994); references can be traced from these sources. Hennig (1966) is the classic reference. Phylogenetic inference is increasingly a computer activity, e.g. Maddison and Maddison's (1992) program. MacClade is friendly to its users; the manual also introduces phylogenetic inference and the uses of trees in biology, and contains many references. Sober (1989) is a philosophical discussion of most of the main methods, and he has edited an anthology (Sober 1994) that reprints a number of relevant papers. Another philosophical question, not covered in this chapter, is whether phylogenetic reconstruction revolves a circular argument; see Hull (1967).

   On homology, see the authors in Hall (1994) for recent viewpoints. On convergence, Wake (1991) discusses the topic, not explicitly discussed in this chapter, of embryological influences; and Doolittle (1994) considers molecular convergence. For the use of fossils, see Donoghue *et al.* (1989), Norell and Novacek (1992), Benton (1994), Smith (1994), and many of the papers in the "molecules vs. morphology" section below, such as Gauthier *et al.* (1988).
2. *Molecular phylogenetics.* Li and Grauer (1991) and Stewart (1993) introduce techniques used with molecular evidence. The most authoritative general source is the book edited by Hillis, Moritz, and Mable (1996). The chapter by Swofford, Olsen, and Waddell (1996) is the best introduction to the theory. Penny *et al.* (1992) also discuss some of the statistical problems. For computer programs, MacClade (referred to above) can be used, and is the most friendly program, but is designed more for morphological than molecular characters. The most widely used programs in molecular phylogenetic research are PAUP (Swofford 1992) and PHYLIP (Felsenstein 1993). The problem about independence mentioned in section 17.10 can be partly followed up through the literature on hitchhiking (section 8.9) and the relation between polymorphism and species differences (section 7.13).

   Avise (1994) discusses the contributions of molecular evidence and has a large reference list. Werman, Springer, and Britten (1996) review the use of DNA hybridization and Sibley and Ahlquist (1990) apply it to birds. Diamond (1991, ch. 1) is a popular essay that discusses DNA hybridization. Wheeler *et al.* (1993) disagree with Lake's result in Figure 17.14d. A great deal of literature focuses on human mitochondrial DNA. Templeton (1993) summarizes most of the difficulties in a readable way, and gives references. See Lynch and Jarrell (1993) on the rate of mitochondrial DNA evolution (it does not generally appear to be faster than nuclear, except in primates).
3. *Molecular vs. morphological evidence.* Patterson *et al.* (1993) is a general review, and discusses both the hominid and amniotic cases. On hominid phylogeny, Klein (1989) is a general text, Diamond (1991, ch. 1) is a popular source, and Andrews (1995) introduces and refers to more recent paleoanthropological discoveries. Primary references include Kay and Simons (1983), Martin (1985), Pilbeam (1986), and Sarich and Wilson (1967). On amniote phylogeny, see Eernisse and Kluge (1993), Gardiner (1993), Gauthier *et al.* (1988), Kemp (1988), Lockhart *et al.* (1994, and see also Lake (1994) on the weights they used), and the more general books by Benton (1990) and Carroll (1988).

   On Hawaiian fruitflies, see Carson and Kaneshiro (1976), Carson (1983, 1990), Kaneshiro (1988), and M. Williamson (1981). For the phylogeny of other taxa not discussed in the chapter, see Fernholm *et al.* (eds) (1989) on everything, Doolittle and Brown (1994) on the deep relations of the major groups, Conway Morris (1993) on metazoan origins, Novacek (1992) on mammals, Hennig (1981) on insects, Nielsen (1994) on animals, Sytsma (1990) on plants, and appropriate further reading in chapter 19.

STUDY AND REVIEW QUESTIONS

1. Shown below is an unrooted tree for five species. Draw all of the rooted trees that are compatible with this tree.

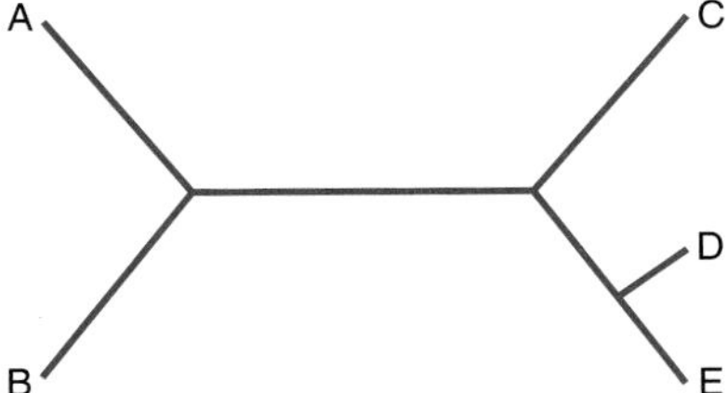

2. Shown below is a rooted tree. Draw all of the unrooted trees that are compatible with this tree.

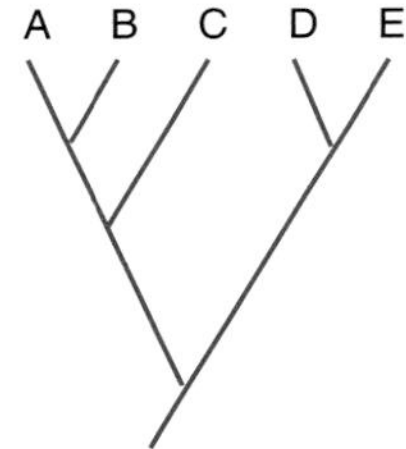

3. Imagine four species, *A*, *B*, *C*, and *D*. Each species has five unique character states; 25 characters are shared by *A*, *B*, and *C*; 15 characters are shared by *A* and *B*; and 100 characters are shared by all four species. What is the minimum number of evolutionary events implied by each of the possible unrooted trees for the four species?
4. Under what conditions (of evolutionary rates) are distance statistics reliable guides to phylogenetic relations?
5. What features would you consider, in some superficially similar structure such as the tail of a dolphin and a salmon, to decide whether it was homologous or analogous?

6. Given below are the character states of species 1–6.

(a)

Species	1	2	3	4	5	6
Character state	A	A'	A	A	A	A

(b)

Species	1	2	3	4	5	6
Character state	A'	A	A	A	A'	A

(a)

Species	1	2	3	4	5	6
Character state	A	A'	A	A'	A	A'

Which of *A* and *A′* is ancestral, and which derived, in the group of species 1 + 2? Assess the relative certainty of the inference in the three cases.

7. Compare the attributes of molecular and morphological evidence that make character analysis and the principle of parsimony more or less applicable to evidence of each kind.

8. Given three molecules:
Hemoglobin
Minisatellites or microsatellites
Histone

And three phylogenetic problems, for which to work out relations:
Populations within a species, such as different human populations
Members of different classes of vertebrate, such as toads, newts, and bats
Members of different kingdoms, such as fungi, bacteria, and roadrunners
Which molecule would you use for each phylogenetic problem?

9. What do you need to know to introduce "weighting" into parsimony?

10. Given below are the orders of genes (or some other markers) along the chromosomes of three species:
(1) adebcfg (2) abcdefg (3) abedcfg
What is the unrooted tree of the three species?

Evolutionary Biogeography

BIOGEOGRAPHY is the science that seeks to explain the distribution of species and higher taxa on the surface of the Earth. The chapter begins by describing the elemental facts of biogeography—the kinds of distribution. We move on to short-term processes, such as species ecology and movement in relation to climate, that explain distributions at the species level. We then examine grander biogeographic patterns and the longer-term processes, particularly continental drift, that produce them. We see how to study the relation between the phylogenetic history of a species and the geological history of the area it has occupied. We finish by looking at another evolutionary phenomenon resulting from continental drift: encounters between faunas when previously separated areas are tectonically brought together. The classic example of this phenomenon is the Great American Interchange.

18.1 Species have defined geographic distributions

The geographic distributions of species can be of a number of types. Consider Figure 18.1, which shows the distribution of three species of toucans in the genus *Ramphastos*, in South America. Two of the species, *R. vitellinus* and *R. culminatus*, have endemic distributions. That is, they are limited to a particular area. Endemic distributions can be more or less widespread and, in the extreme case of species that are found on all continents of the globe, are called *cosmopolitan*. The pigeon, for example, is found on all continents except Antarctica. By a strict definition, the pigeon might not be considered cosmopolitan, but the term is usually intended less strictly. Thus, the pigeon is called a cosmopolitan species. Other species, like *R. ariel* in Figure 18.1, are not confined to a single area, but are distributed in more than one region with a gap between them. These patterns are called *disjunct* distributions.

Maps like the ones for species in Figure 18.1 can be drawn for a taxonomic group at any Linnaean level. Just as species have geographic distributions, so too do genera, families, and orders. Biogeography aims to explain the distributions of both the higher taxa and species, and different explanatory processes are often appropriate at different levels. Short-term movements of individuals influence the distributions of populations and species, whereas slower-acting geological processes may control the biogeography of higher taxa. The distributions of the higher taxonomic levels are, obviously, more widespread than those of species, but some taxonomically isolated higher groups, with small numbers of species (usually living fossils; see section 18.5), have localized distributions. For example,

Figure 18.1 The natural distribution of three species of toucans in the genus *Ramphastos* in South America. *R. vitellinus* and *R. culminatus* have endemic distributions; *R. ariel*'s distribution is disjunct. An extensive hybrid zone exists between the species. Adapted, by permission of the publisher, from Haffer (1974).

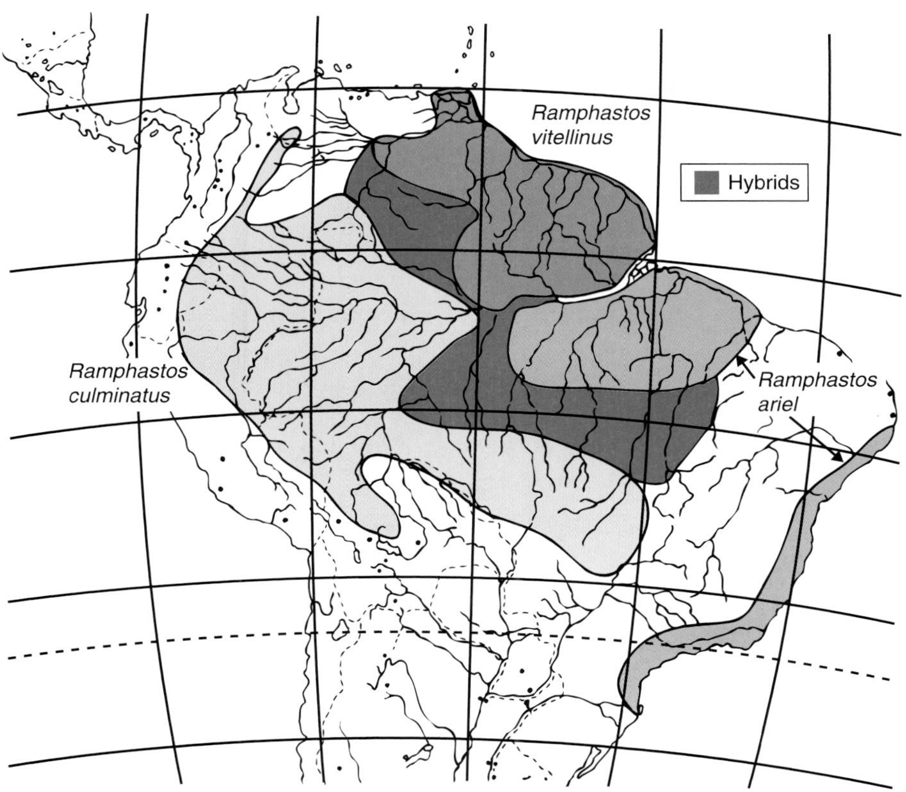

the tuatara *Sphenodon punctatus* is the only surviving species of a whole order of reptiles (or almost the only survivor—there may be more than one surviving species of *Sphenodon*). Of about 20 orders of reptiles, 16 are completely extinct and 4 have living survivors. Of those four, three contain the turtles and tortoises, lizards and snakes, and crocodiles, respectively. The fourth has only *Sphenodon*, which is now confined to some rocky islands off New Zealand.

When the biogeographers of the nineteenth century looked at the distributions of large numbers of species on the globe, they saw that different species often lived in the same broad areas. This discovery led to the suggestion that there are large-scale faunal regions on the Earth. The first map of these faunal regions was drawn for birds by the British ornithologist Philip Lutley Sclater (1829–1913). Alfred Russel Wallace soon generalized Sclater's regions to other groups of animals. The Earth was thus divided up into six main biogeographic regions (Figure 18.2). The regions are mainly defined by the distribution of birds and mammals, and might not have been recognized if other groups had been used. Botanists, for instance, tend to draw different lines on the map. They usually combine the Nearctic and Palearctic regions of Figure 18.2 in one larger region

Figure 18.2 The world has been divided into six main biogeographic areas, according to the similarity of their animals (particularly birds and mammals—see Table 18.1). This version was developed by Simpson (1983). The discontinuity between the Australian and Oriental regions is called Wallace's line.

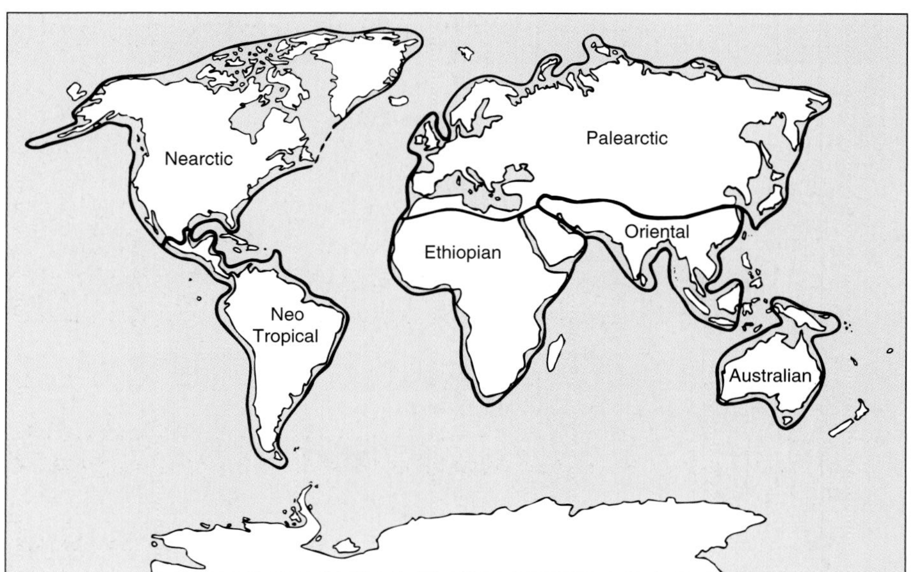

called the Boreal, or the Holarctic, and recognize a separate floral region, called the Cape, in southern Africa. Figure 18.2, therefore, does not illustrate a set of hard-and-fast facts; the regions are approximate. The regional terms—like Nearctic and Neotropical—are often used in biogeographic discussion.

The division into regions was made according to the degree of similarity between the lists of the species living in the various places. Biogeographic similarity can be quantified by various *indexes of similarity.* One of the simplest indexes is Simpson's index. If N_1 is the number of taxa in the area with the smaller number of taxa, N_2 is the number of taxa in the other area, and C is the number of taxa in common between the two regions, then Simpson's index of similarity between the two areas is

$$\frac{C}{N_1}$$

Table 18.1 gives the faunal similarities, for mammalian species, between several regions; these similarities are expressed as percentages (i.e., $(C/N_1) \times 100$). The indexes in the table show some of the justification for the division of the Earth into faunal regions in Figure 18.2. For example, the faunas of Australia and New Guinea are 93% similar, whereas those of New Guinea and the Philippines are only 64% similar. In fact, the Philippines have as high a similarity with Africa as they do with New Guinea. This Indonesian discontinuity, which can be seen in Figure 18.2, is known as *Wallace's line.* It was not properly understood until the discovery of continental drift.

TABLE 18.1
Indexes of similarity for the mammalian species of various regions. Data from Flessa *et al.* (1979).

	North America	West Indies	South America	Africa	Madagascar	Eurasia	South East Asian Islands	Philippines	New Guinea	Australia
North America										
West Indies	67									
South America	81	73								
Africa	31	27	25							
Madagascar	38	27	35	65						
Eurasia	48	27	36	80	69					
South East Asian islands	37	20	32	82	63	92				
Philippines	40	20	32	88	50	96	100			
New Guinea	36	21	36	64	50	64	79 —	64		
Australia	22	20	22	67	38	50	61	50	93	

18.2 *The ecological characteristics of a species limit its geographic distribution*

The distributional limits of a species are set by its ecological attributes. One way of understanding how ecological factors limit a species' distribution is in terms of a distinction, first made by Hutchinson and MacArthur in the 1950s, between the *fundamental niche* and *realized niche* of a species. A species will be able to tolerate a certain range of physical factors—temperature, humidity, and so on—and could, in theory, live anywhere these tolerance limits were satisfied. This area is its fundamental niche. Competing species will often occupy part of this range, however, and the competition may be too strong to permit both species to exist. Each species' realized niche will then be smaller than its physiology would make possible. That is, each species will occupy a smaller range than it could in the absence of competition. Much ecological research has been carried out to discover the factors—whether physical or biological—that act to limit particular species' distributions in any place.

Patterns in the distribution of species probably cannot be explained by ecological factors alone. It becomes difficult, after looking at the maps of several species' distributions, to accept that species are always excluded from all of the places where they are absent purely by ecological factors that have operated persistently in the past and present; it is also difficult to believe that the places where a species exists always differ in some crucial ecological respect from the places where the species is absent.

Another possibility is that historical factors have been at work. In some places, a species ecologically could be present, but it is absent because it has never arrived—that is, it never migrated and established itself. The same factor could work via an ecological competitor. Suppose that a species is equally well adapted to the physical conditions of two places, but is present in one area and absent from the other. We then find that the species is excluded by an ecological competitor from one place, but not the other. The distribution could then be determined by the historical accident of which of the two competitors established itself in a place first. The distribution of the species will then again be determined historically—by where the species was found at certain times in the past.

In what sense are ecological and historical factors alternatives? If we consider a particular distributional limit of a species, we can ask whether it lies at the limit of the species ecological tolerance, or whether the species could ecologically survive on the other side of the border but for some historical reason does not live there. It can, therefore, be meaningful to test between ecological and historical explanations. In most real cases, however, a complete account of a species distribution requires both ecological and historical knowledge. A species cannot live outside its ecological tolerance range. As a result, its biogeography cannot contradict its ecology. Within its ecological tolerances, historical factors may have determined where it is living and where it is absent. The two factors will then not be opposed, and the sensible method of analysis is to examine how ecology and history have combined to produce the species distribution.

18.3 *Geographic distributions are influenced by dispersal*

A species' range will be changed if the members of the species move in space—a process called *dispersal.* Individual animals and plants move, actively and passively, through space both to seek out unoccupied areas and in response to environmental change. (Plants move passively, at the seed stage.) When the climate cools, the ranges of species in the northern hemisphere moves southward, and tropical forests fragment into smaller forest patches. The range of a species might also change, when the climate changed, without the movement of individuals. For example, species in the colder regions might die off, and the range would shrink and move on average to the south. In practice, though, individuals would move southward as well, thereby extending the species range. If a species originated in one area and subsequently dispersed to fill out its existing distribution, then the place where it originated is called its *center of origin.*

Various dispersal routes might have been followed in the biogeographic history of a species. Simpson distinguished dispersal by means of *corridors, filter bridges,* and *sweepstakes.* Two places are joined by a corridor if they are part of the same land mass—Georgia and Texas, for example. Animals can move easily along a corridor, and any two places joined by a corridor will have a high degree of faunal similarity. A filter bridge is a more selective connection between two places, and only some kinds of animals will manage to pass over it. For instance, when the Bering Strait was above water, mammals moved from North America to Asia and vice versa, but no South American mammals moved to Asia and no Asian species moved to South America. The reason is presumably that the land bridges at Alaska and Panama were so far apart, so narrow, and so different in ecology that no species managed to disperse across both of them. Finally, sweepstakes routes are hazardous or accidental dispersal mechanisms by which animals move from place to place. The standard examples are island hopping and natural rafts. Many land vertebrates live in the Caribbean Islands, and (if their biogeography is correctly explained by dispersal) they might have moved from one island to another, perhaps being carried on a log or some other sort of raft.

Good evidence illustrates the power of dispersal. In 1883, for example, a volcanic eruption covered the small Indonesian island of Krakatau with ash and killed all of the plants and animals there. Biologists then recorded the recolonization of the island, particularly for birds and plants. The recolonization was astonishingly rapid. Fifty years later, the island was recovered with tropical forest, which supported 271 plant species and 31 bird species. Invertebrate animals, such as insects, had come as well, although their numbers were less closely monitored. The immigrants mainly came from the neighboring islands of Java (40 km away) and Sumatra (80 km away). The birds would have dispersed by active flight, and the plants would have been carried as seeds. In the right circumstances, dispersal can have a clear effect on the ranges of species.

18.4 *Geographic distributions are influenced by climate, such as in the Ice Age*

The current geological age is called the Quaternary; it began 2.5 million years ago (see section 19.2, p. 539, for geological time). During the Quaternary, the climate has mainly been cooler than in the preceding Tertiary, and continuous cycles of increasing and decreasing temperature have occurred. Many of the cooler times were glacial periods, and the warmer times interglacials. These climatic changes have happened recently enough for the fossil record in some cases to be revealingly complete. When the weather turns cool, animals tend to migrate southward and plant ranges contract. At any one site, the local ecology changes to one characteristic of the cooler climate. A change from a temperate to a tundra-type ecosystem, for example, has been well documented from pollen records in the northern temperate zone through recent Ice Ages.

The change can also be seen in the distribution of single species (Figure 18.3). The most recent Ice Age ended about 10,000 years ago, and Figure 18.3 shows how the geographic distributions of hemlock and beech trees moved north through the United States as the temperature warmed up and the ice cap retreated. The same factor has controlled the distribution of certain beetle species in Europe. Approximately 38,000 years ago, the climate was cooler than today. In British deposits dating from that time, Coope has found fossil beetles belonging to species that are now not found south of northern Scandinavia. Likewise, he has found beetles belonging to species that are now not found north of France in sediments beneath Trafalgar Square in London, dating from 120,000 years ago when the weather was warmer (Figure 18.3b). Thus, the beetles serve as indicators of past climates. In addition, note that the beetles do not show any evolutionary changes in their morphology when the climate changes. As the interglacials and glacials come and go, the beetles do not evolve—they simply move north and south.

North-South movements are one response to climatic change. Another is for ranges to expand and coalesce, or contract and fragment. Such events are crucial in the hypothesis of *glacial forest refuges*. The ranges of tropical forests may, for example, change through the glacial cycle (Figure 18.4). Tropical forests, like those of Latin America, are now continuous over large (though decreasing) areas.

Figure 18.3 (a) Changing American geographic distribution of beech (*Fagus*) and hemlock (*Tsuga*) as the polar ice cap retreated after the most recent Ice Age. (b) Modern distributions of two beetle species: *Diachila arctica*, which lives well north of Great Britain; and *Oodes gracilis*, which lives further south. *D. arctica* Fossils have been found from 38,000 years ago in England and *O. gracilis* Fossils from 120,000 years ago.

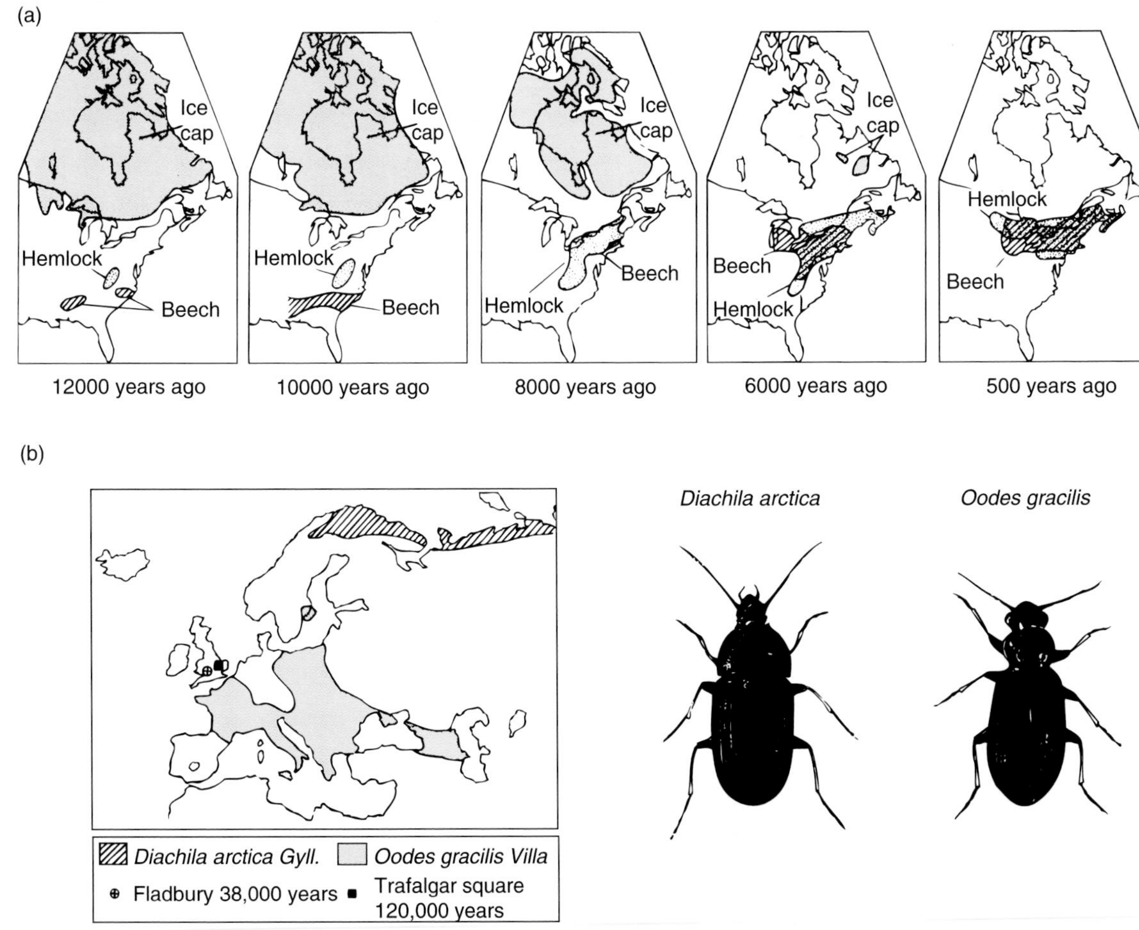

Figure 18.4 Possible cycle of forest from (a) glacial minima, to (b) present pattern, to (c) glacial maxima (ice age) with forest refuges. The positions of the refuges were inferred from several kinds of evidence and are uncertain. The cross-hatched region indicates the Andes (> 1000 m). The coastline has been adjusted in (a) and (c) to reflect changes in sea level relative to the present (b). From Lynch (1988). Reprinted with permission of Cambridge University Press.

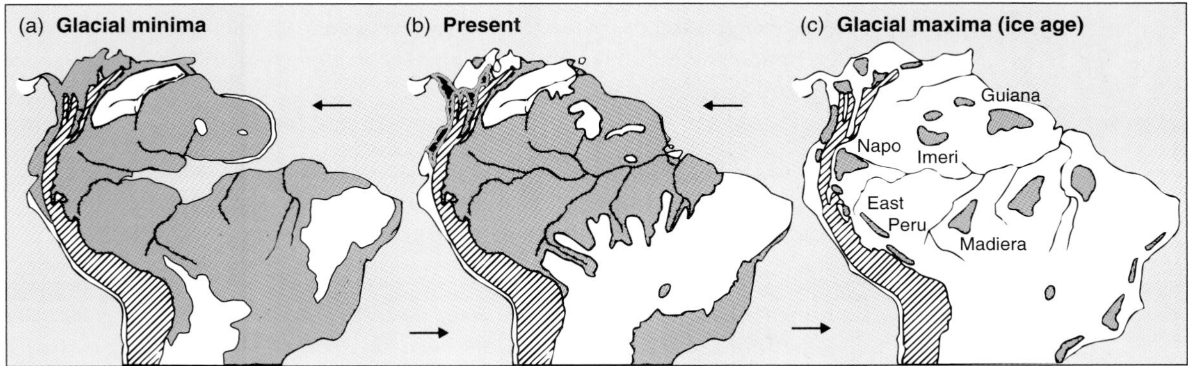

In the drier conditions of the Ice Age, they may have shrunk and fragmented into smaller areas.

This (possible) glacial cycle of range fragmentation has been used to explain a number of biogeographic facts. Haffer in 1969 suggested it explains the high diversity of living species in tropical forests. (He also used it to explain disjunct distributions such as the toucan *R. ariel* in Figure 18.1.) The fragmentation of ranges is exactly the kind of process that could cause allopatric speciation, and as each locality evolved its own species, different from that of other localities, the total diversity of species would increase.

This hypothesis is attractive, but controversial. One question is whether the forest actually did contract. The "refuges" drawn in Figure 18.4 were inferred from a number of kinds of indirect evidence; Haffer originally inferred their locations from areas of high modern bird species diversity and the pattern of modern rainfall. The most direct kind of evidence would be from pollen remains, however, and Colinvaux has argued that the pollen remains from one area that Haffer inferred to be dry and forest-free actually supported forest throughout the Ice Age. Perhaps the glacial forest was more continuous than the refuge hypothesis implies.

Another controversial question concerns the timing of speciation. Haffer's hypothesis predicts that speciation should have had a higher than average frequency in the Pleistocene, during the Ice Ages, and speciation in different groups should have happened almost simultaneously. Critics argue that the speciation events that generated modern species in the region occurred long before the last Ice Age. The detailed work on the phylogenetic and biogeographic history of Amazonian species needed to test the theory is being carried out, though it is not yet conclusive.

The hypothesis of glacial refugia has been applied broadly, to explain such phenomena as the Müllerian mimicry rings of *Heliconius* (section 8.3, p. 198) and the language groups of Native Americans. Nor is the principle confined to glacial events. The same principle operates, in inverted form, in species that now have local distributions but were more widely distributed in past climatic conditions. The Nevadan deserts contain the vestiges of former large lake systems, and the desert pupfish (*Cyprinodon*) sometimes live in the scattered remaining waterholes. The 20 or so isolated populations of these remarkable fish have diverged into a number of (perhaps four) species, and when the next pluvial period brings water to the desert, they may expand from their interpluvial refuges to encounter one another in a process that is analogous to the one suggested for Amazonian birds and butterflies.

18.5 *Geographic distributions are influenced by vicariance events, such as continental drift and speciation*

A second factor influencing geographic distributions is *continental drift*. The continents have moved over the surface of the globe through geological time. The positions of the main continents since the Permian have been reconstructed in some detail (Figure 18.5), and these maps immediately suggest the reason for many biogeographic observations. For example, when we looked at the faunal regions of the world (Figure 18.2), we saw the difference between the faunas of the northern and southern Indonesian Islands known as Wallace's line. As can be seen in Figure 18.5, the two regions have separate tectonic histories and have only recently come into close contact. The patterns of faunal similarity match, therefore, what we should expect given the fact of continental drift.

One of the main modern research programs that studies the relation between biogeography and continental drift is called *vicariance biogeography*. The drifting apart of tectonic plates is the sort of event that could cause speciation (section 16.4.1, p. 429). If the splitting of the land and of the species on it coincide, two species come to occupy complementary parts of a formerly continuous area that was inhabited by their common ancestor. This example represents a *vicariance event*. (Vicariance means a splitting in the range of a taxon.) In theory, continental drift is just one process that could split a species' range. A species could also be split by parapatric speciation, or some other process (like mountain building, or the formation of a river) that subdivided an area. In any such case, a pair or more of species would occupy the new subsections of the ancestral species range. According to the theory of vicariance biogeography, the distributions of taxonomic groups are determined by splits (or vicariance events) in the ranges of ancestral species.

We can contrast this idea with the theory that distributions are determined more by dispersal. Before continental drift was known, or at least accepted, the main process believed to alter biogeographic distributions was dispersal. Taxonomic groups were thought to originate in one confined area, called the center of origin, and then descendant populations dispersed away from it. Thus, the geographic history of a group could have taken the form of a series of splits within formerly larger ancestral ranges, a series of dispersal events, or some mixture of the two (Figure 18.6). Although such events hypothesized by the dispersal and vicariance theories took place in the past, they can still be tested by two methods.

Figure 18.5 Continental drift. (a) The movements of the continents during the past 200 million years. (b) The positions of the main tectonic plates today.

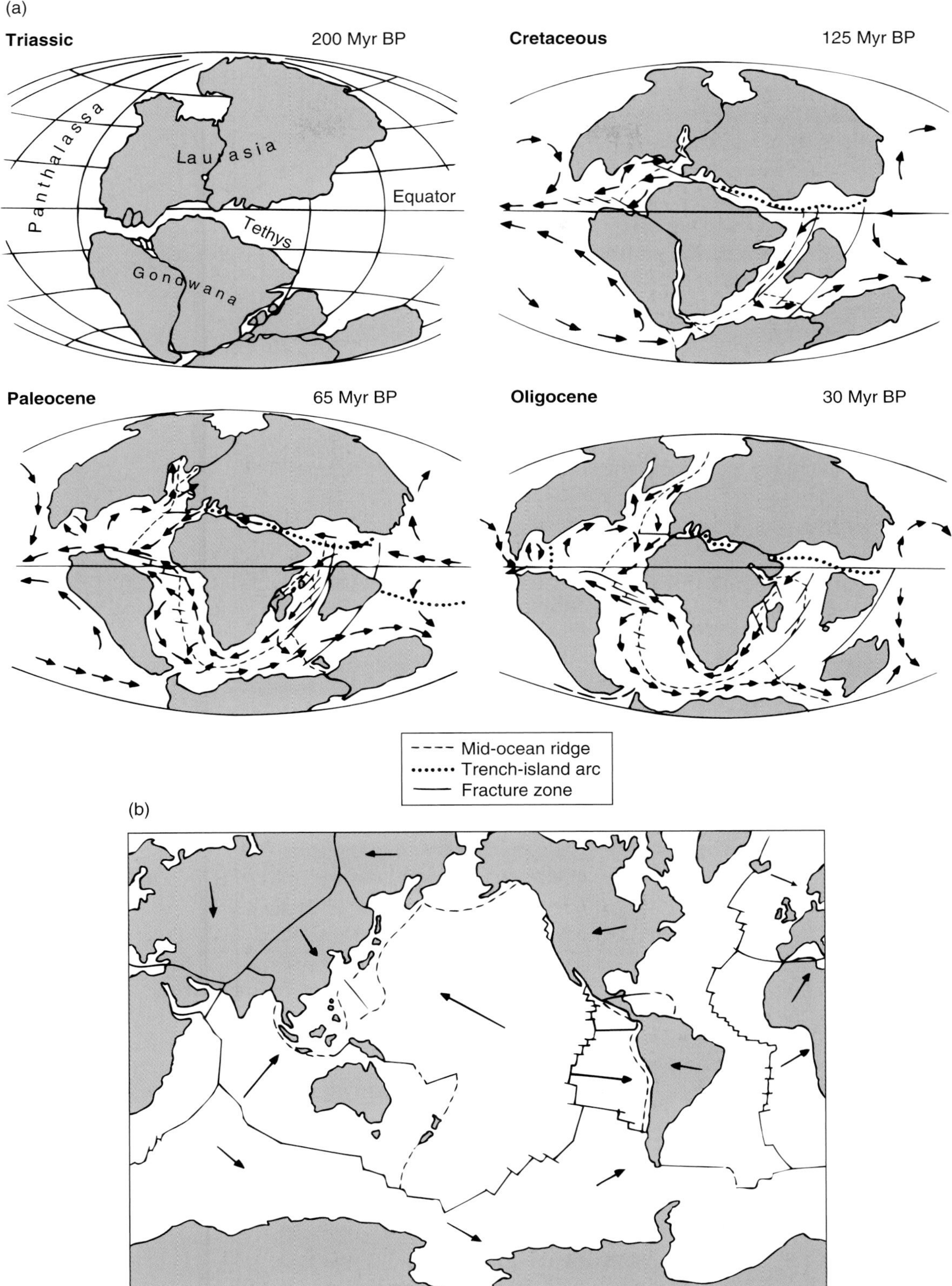

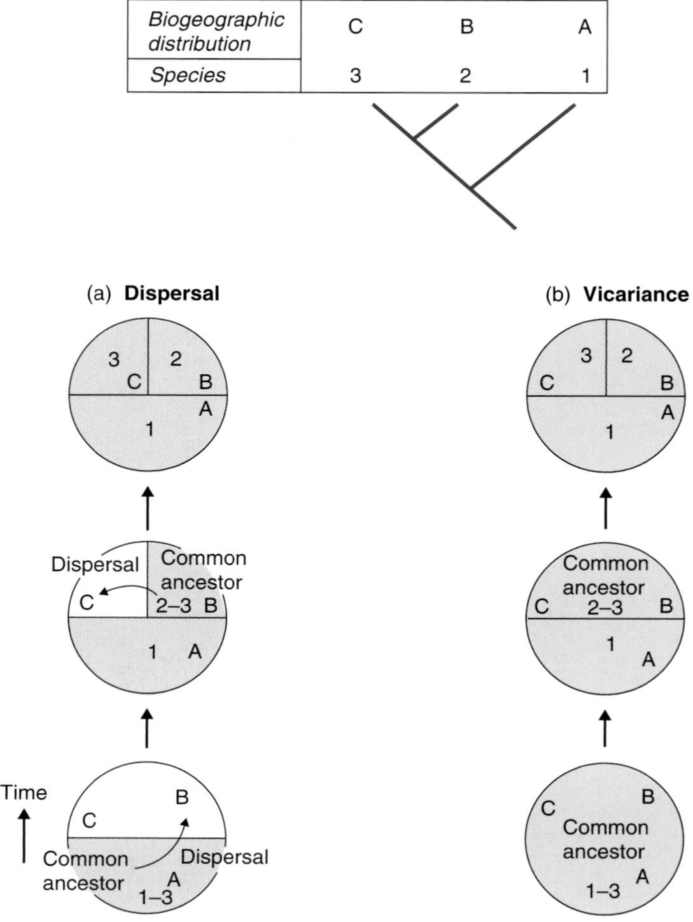

Figure 18.6 Dispersal and range splitting can be alternative hypotheses to explain the biogeography of a group. (a) An ancestral species with center of dispersal in area A dispersed first to area B and a descendant from there then dispersed to area C. (b) An ancestor occupying A + B + C had its range split first into A and B + C, and then the descendant in B + C had its range split.

The first method tests whether the pattern of splitting in one species matches the geological history of the region where it lives. The first major piece of vicariance biogeographic research, by Brundin in 1966, used this approach. Brundin studied the Antarctic chironomid midges, which are distributed around the southern hemisphere (Figure 18.7). He reconstructed their phylogeny by standard morphological techniques (see chapter 17) and then used the species' modern biogeographic distributions to draw a combined picture of their phylogeny and biogeography, called an *area cladogram* (Figure 18.8). If the successive splits in the phylogeny were driven by successive break-ups of the land, the phylogeny implies a definable sequence of tectonic events. Initially, the common ancestor of the modern forms would have occupied a large area made up of all of their modern distributional zones—which implies the existence, sometime in the past, of a southern supercontinent, Gondwanaland. Gondwanaland would then, Brundin's analysis predicts, have split in the following order. First, South Africa split from a combination of Australia, New Zealand, and South America; New Zealand split from South America and Australia; and finally Australia split

Figure 18.7 Biogeography of chironomid midges in the southern hemisphere. Reprinted, by permission of the publisher, from Brundin (1988).

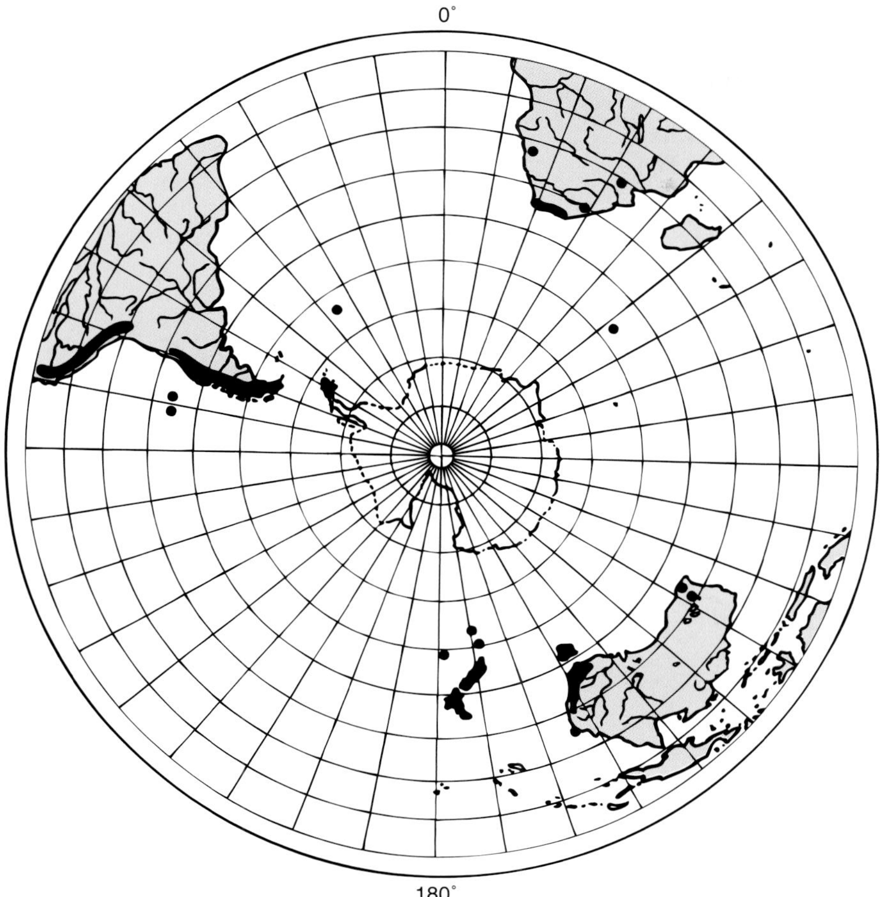

from South America. This prediction can be tested against the geological evidence, which turned out to fit Brundin's prediction (see Figure 18.5a, but more detailed maps are needed for a strong test).

Brundin's test concerns a single taxon. A second test is to compare the relation between phylogeny and biogeography in many taxa. As continents—or, in general, the habitats occupied by species—move in a particular pattern through time, all groups of living things found in an area will be affected in a similar manner. If members of each group tend to speciate when their ranges become fragmented, they should show similar relations between phylogeny and biogeography. Thus, their area cladograms should match.

This prediction can be tested, as shown in Figure 18.9. Figure 18.9a shows the phylogeny and biogeography of three species in a hypothetical Taxon 1. We can also look at another taxon inhabiting the same region and see whether it supports the same area cladogram. In technical language, vicariance biogeography predicts their area cladograms will be *congruent*. Congruence is a term that can

Figure 18.8 Area cladogram of chironomid midges. The diagram shows the phylogenetic relations between the midges from different areas. The midges of Australia, for example, are phylogenetically more closely related to those of South America, than those of South Africa. For Laurasia see Figure 18.5. Reprinted, by permission of the publisher, from Brundin (1988).

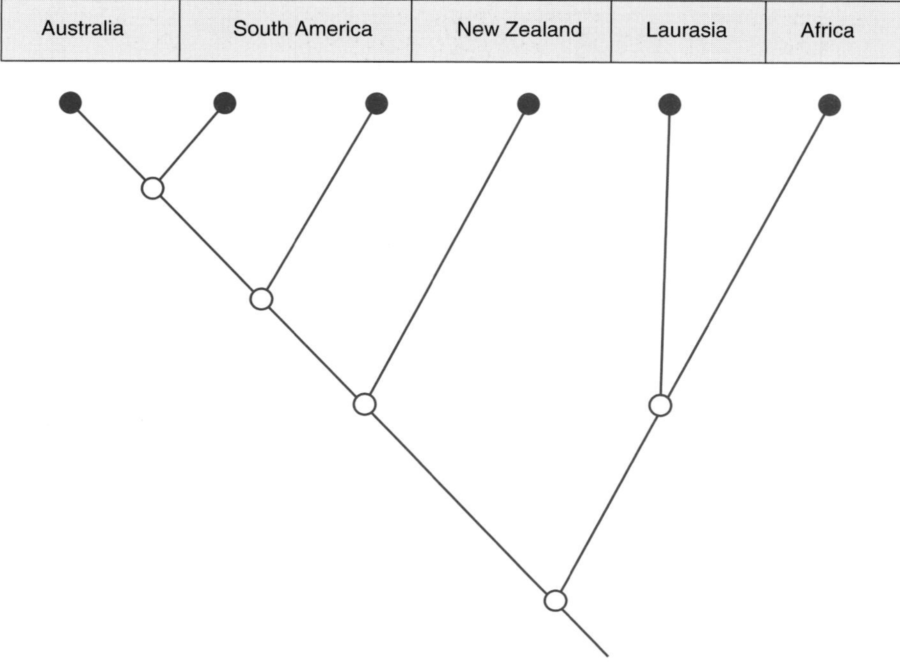

be applied to any sort of branching diagram (in phylogeny or biogeography). If two branching diagrams are congruent, the order of branching in the two do not contradict one another. The two diagrams do not have to be identical, because one place, or taxon, might be missing from one of the branching diagrams. The order of branching in the entities that are present in both must be the same, however. In the figure, the area cladograms for taxa 2 and 3 are congruent, and that of taxon 4 is incongruent, with taxon 1. If the land area A + B + C had first split into A + B and C, and then into A and B, the phylogeny of taxa 1–3 would fit with this series of geographic events. Thus, the phylogeny can be understood as a series of vicariant events. On the other hand, taxon 4 does not fit. If its common ancestor occupied the entire area, its first split suggests that the land first divided into A + C and B, not into A + B and C. The congruence of taxa 1–3 conform with the ideas of vicariance biogeography, but taxon 4 does not.

Before these methods were developed—indeed, before continental drift was widely accepted—the Venezuelan biogeographer Léon Croizat had established that different taxonomic groups often show correlated distributions. Croizat called these patterns "generalized tracks," with the distribution of any one species being its "track." He argued that, if different species independently dispersed

Figure 18.9 Testing vicariance biogeography by comparing area cladograms of four taxa. (a) Phylogeny and biogeography of four species. The species are symbolized by numbers (1, 2, 3, 4) and the places where the species live by letters (A, B, C). (b) Inferred vicariant history of distributions. (c) Taxa 2–3 have distributions that are congruent with species 1, but taxon 4 is incongruent. (d) Either dispersal events occurred during the history of taxon 4, or its phylogeny is wrong. Species 15 may have been wrongly classified, for example, because it has evolved rapidly (the group of species 12–14 in (c) is then an example of a paraphyletic group). The suggested history with migration is only one of a number of possibilities that are compatible with a range split in the order A+B+C → A+B / C → A / B / C.

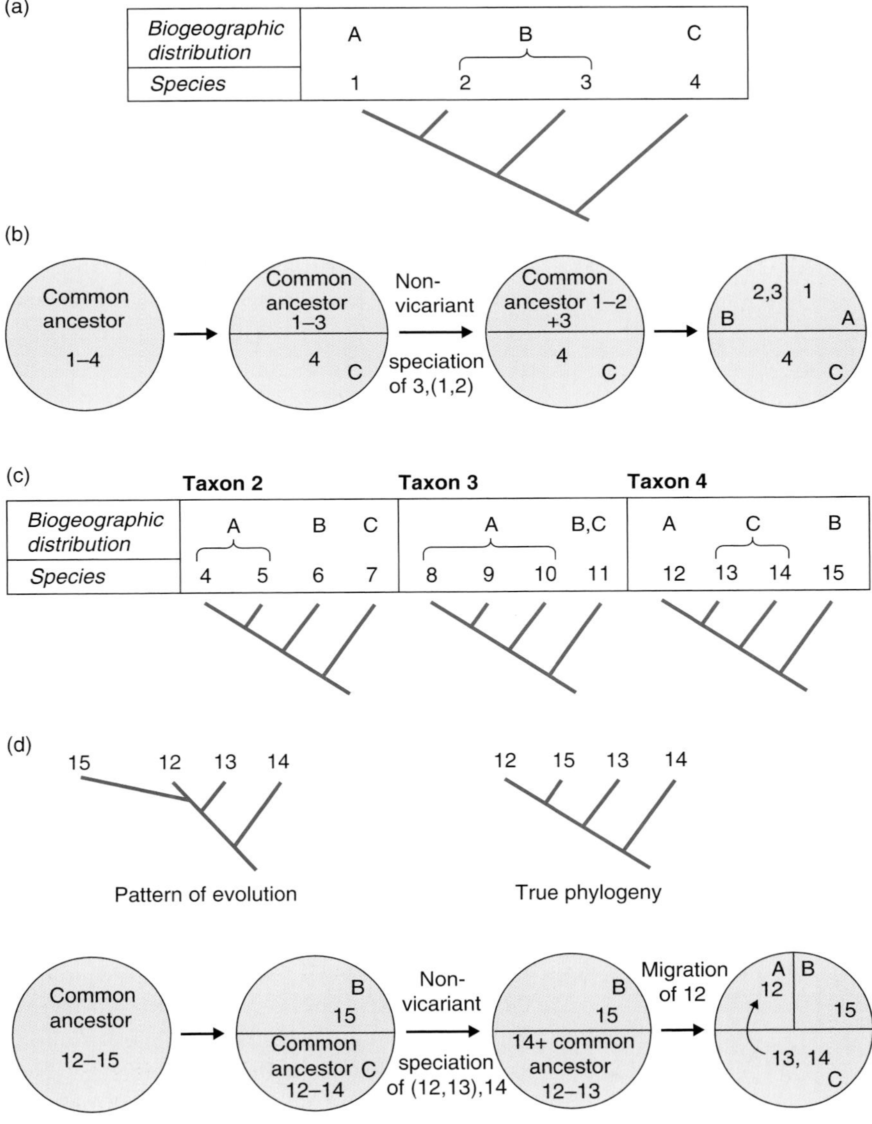

from centers of origin, they would not have correlated distributions. Correlated distributions are more likely to result from common vicariance events, such as continental drift, that split the ranges of several taxa in the same way. Modern vicariance biogeography has added to Croizat's ideas in two ways. We now know more about the details of continental drift. We also recognize the importance of using a realistic phylogeny when testing whether different taxa have congruent distributions.

The analyses in Figures 18.8 and 18.9 are possible only for taxa that are monophyletic in the cladistic sense (Figure 14.6, p. 382). If a set of phylogenetic groups have been classified into a mixture of mono-, para-, and poly-phyletic groups, then their area cladograms need not be congruent, even if they have experienced the same sequence of range subdivisions.

Consider Figure 18.9 once again. The area cladograms of taxon 4 and taxon 1 are incongruent. If taxon 4 has the phylogeny of Figure 18.9b—that is, if the groups are monophyletic—then the incongruency between taxa 1 and 4 implies some dispersal events must have occurred in the past (Figure 18.9d). On the other hand, if the classification of taxon 4 was paraphyletic or polyphyletic, the theory of vicariance biogeography no longer predicts that the area cladograms of the taxa will be congruent. We have no reason to expect different paraphyletic or polyphyletic groups to have congruent biogeographic patterns with one another, or with monophyletic ones. It is, therefore, essential for vicariance biogeographers to use cladistically defined taxa, which reflect the order of phylogenetic branching. If the classifications contain a mixture of phenetic and cladistic taxa, any general biogeographic study is liable to become meaningless.

Let us now turn to an example from part of a larger study by Patterson. His starting point was a probable area cladogram for the marsupials (Figure 18.10a). Recent marsupials live in Australia and New Guinea, and South and North America (where they are represented by the opossum *Didelphis*). Fossil marsupials can also be found in Europe, making five areas in the complete area cladogram for marsupials.

Marsupials have evolved on the same globe as all other species. If the modern distributions of vertebrates result from a history of range splitting, they should all share much the same geological history and their area cladograms should, therefore, all be relatively congruent. What do the area cladograms for the other vertebrates look like? Figure 18.10b reveals, for five other vertebrate groups, that the vicariance prediction is upheld—their area cladograms are congruent. The result could, in theory, result from the dispersal of all taxa in the same order and the same direction, but that would require an unlikely series of coincidences. More likely, the common pattern is simply due to a shared history of range splits by tectonic events.

Dispersal has probably had some influence on the history of the taxa in Patterson's study. The osteoglossine fish are found in South East Asia as well as Australia, New Guinea, and South America (Figure 18.10b). As it happens, none of the other four taxa are represented in South East Asia. Three explanations for the result are possible. First, all six taxa could have once lived in South East Asia and five of them could have since gone extinct there. Second, all six could have been originally absent from Asia and osteoglossine fish (in the form of *Scleropages*) could have arrived there by dispersal. The fossil record could, in

Figure 18.10 (a) Area cladograms of Recent and fossil marsupials. (b) Area cladograms of five other taxa with congruent biogeographic distributions. From Nelson and Rosen (1981). Copyright © 1981 by Columbia University Press. Reprinted with permission of the publisher.

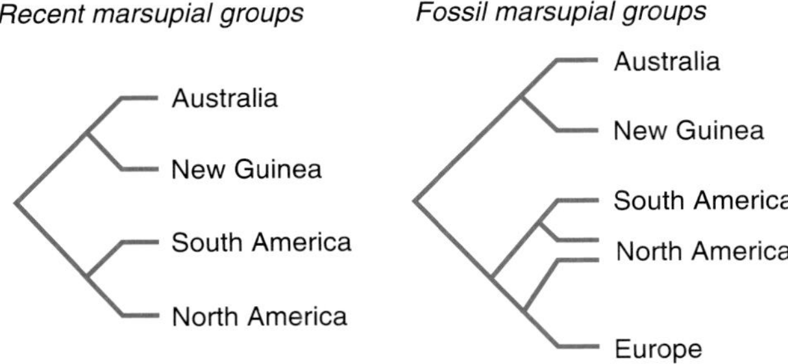

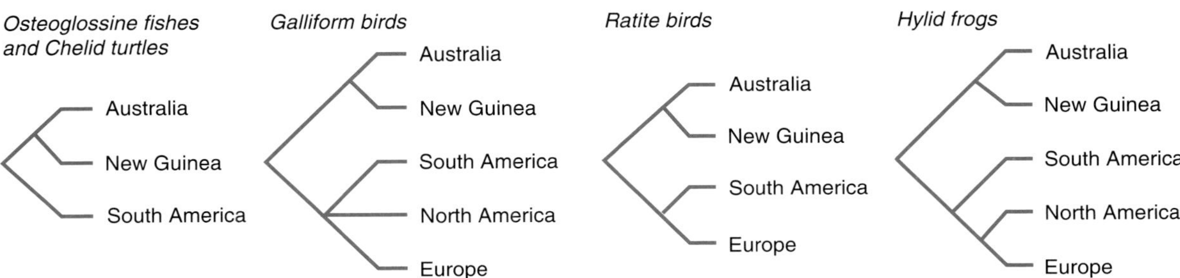

principle, be used to show that a taxon once lived in Asia but is now extinct. In the absence of any such evidence, Patterson reasons that it is more likely that one group (the osteoglossines) dispersed to Asia than that five groups went extinct there. Third, osteoglossines could have originally had a broader distribution than the other five vertebrate groups, and thus might have been ancestrally present in South East Asia. The vicariance of the osteoglossines would then have taken place within a larger range. Vicariance biogeography has been successful in finding a number of area cladograms that are mainly consistent between different taxa and also consistent with tectonic history.

Range splits are not the only process that explain area cladograms, however. In chapter 17, we discussed the phylogeny of the picture-wing fruitflies of Hawaii, and Figure 17.17 (p. 494) illustrated the biogeographic distribution of the species among the islands. The islands were geologically formed successively, perhaps as the tectonic plate moved from east to west over a volcanic "hot-spot" that threw up one island after another. The oldest islands are found in the west; the youngest island is Hawaii in the east. It is, therefore, likely that the picture-wing group originated as populations from the older islands colonized the younger islands, with these original populations emerging from the ocean floor and evolving into new species by the process of allopatric speciation. Most of the species in the phylogeny (Figure 17.17) appear to have evolved from ancestors on older islands. The pattern can be shown more clearly on a map (Figure

18.11a). Most of the colonizations took place from an older island to a younger one, but a minority went in the opposite direction. The same pattern can be found in other Hawaiian taxa, such as the tarweeds (Figure 18.11b).

The area cladogram of these groups was probably generated by dispersal; that explanation fits in with our knowledge of the geological history of the archipelago and with our understanding of speciation. It is highly unlikely that the ancestral fruitfly occupied all of the islands and gave rise to the group by successive rounds of range fragmentation. No known geological process could have produced the vicariance. The islands clearly originated as volcanic islands and not, for instance, by the sinking of a larger area, with waters dividing up a previously continuous dry land. Moreover, all of the islands have existed for only a short period; for much of the time, only some of them were above water. Finally, if the history of the groups were due to vicariance, the pattern of Figure

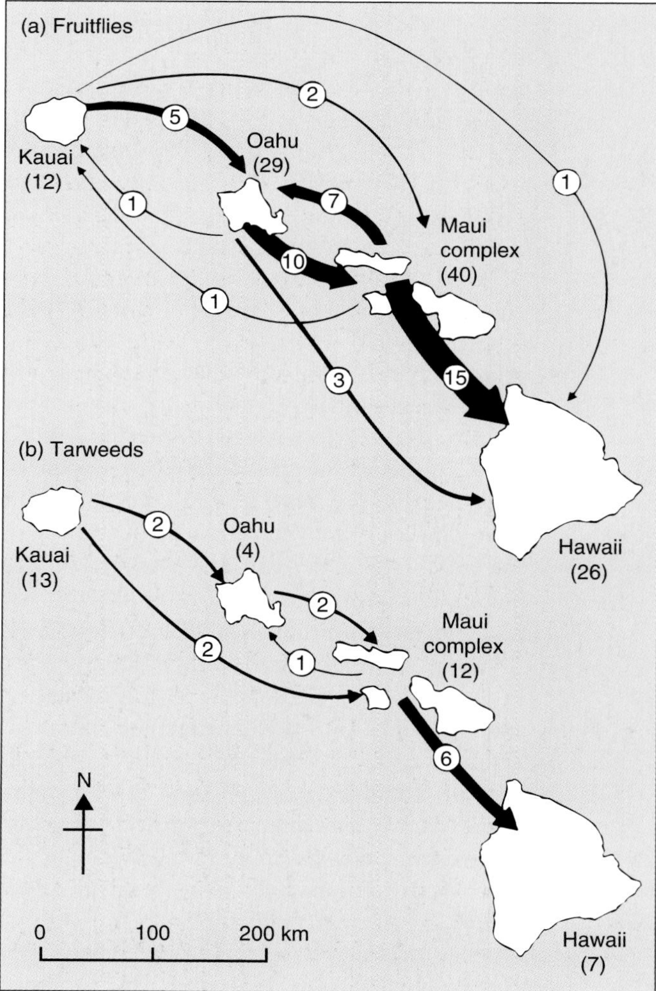

Figure 18.11 (a) The dispersal events suggested by the phylogeny of the Hawaiian picture-wing group of fruitflies (*Drosophila*). The phylogeny appears in Figure 17.17 (p. 494), but the numbers here are not exactly those implied there because this figure is more recent. Numbers in arrows are inferred number of dispersal events; parenthetic numbers by islands are numbers of endemic species living on that island. (b) Comparable figure for the tarweed plant. The geological history of the archipelago, in which the islands have been successively formed toward the east, has imposed the same biogeographic histories on the two groups. Reprinted, by permission of the publisher, from Carr *et al.* (1989).

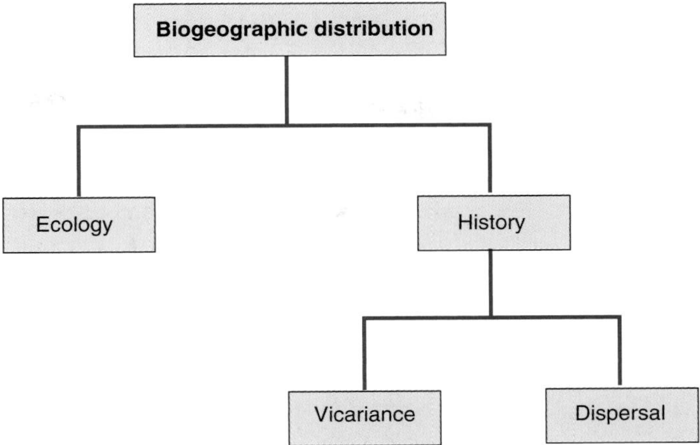

Figure 18.12 Relation between explanatory dichotomies.

18.11—ancestral species on older islands, and derived species on younger islands—could be explained only as a coincidence, which is improbable.

In summary, dispersal and vicariance form a pair of possible explanations one stage below the historical alternatives (Figure 18.12). Much the same point can be made about the distinction between dispersal and vicariance as was made for ecological and historical factors. It is possible that in any particular case, either dispersal or vicariance has been exclusively at work. The area cladogram of Brundin's midges was likely generated by vicariance, whereas that of the Hawaiian fruitflies and tarweeds was generated by dispersal among an emerging archipelago of volcanic islands. The two processes can also operate together. In that case, the challenge becomes working out the relative contributions of the two processes.

18.6 *The Great American Interchange*

The processes of continental drift and dispersal have both contributed to the events that take place when two previously separate faunas come into contact. Several of these events, which are called biotic interchanges, are known from the history of life. The most famous is the Great American Interchange. Its deep geological cause is probably connected to the tectonic processes that have been raising the Andean mountains for the past 15 million years. The rate of this mountain building has varied from time to time, but during a period between 4.5 and 2.5 million years ago, it intensified. At the same time—perhaps 3 million years ago—the modern Isthmus of Panama rose out of the sea and the South and North American continents reconnected. The connection had dramatic repercussions for the fauna, most noticeably the mammalian fauna, of the southern continent.

North and South America had been connected before, over 50 million years earlier. They may have had similar mammalian inhabitants, but the Cretaceous mammals of South America are too poorly known to be certain. Probably in the late Paleocene, the two halves of the American continent drifted apart. At that time, the modern orders of mammals—the groups such as horses, dogs, and cats that remain the dominant land vertebrates today—evolved in North America, Africa, and Europe. South America, on the other hand, shows no sign of possessing these forms, instead evolving its own distinctive mammalian fauna. The South American mammals of the Paleocene and Eocene are classified into three groups: marsupials, xenarthrans (armadillos, sloths, anteaters), and ungulates. Armadillos, tree sloths, and opossums still survive in South American forests, but they formerly lived along with many other curious, and now extinct, forms. Such forms included marsupial saber-tooth carnivores (see Figure 17.5), ground sloths (the group from which the giant ground sloth *Megatherium* of the Pleistocene evolved), and the most heavily armored mammals that ever lived—the glyptodonts (Figure 18.13), which were first described from Darwin's collections made during the *Beagle* voyage.

New arrivals came from outside South America, on rare occasions, beginning during the early Oligocene. They probably immigrated by waif dispersal, hopping from island to island before a continuous land bridge appeared between the continents. Rodents are a major group that first appeared in the Oligocene. Their origin remains so uncertain that experts still dispute whether the South American rodents are more closely related to African or North American species (although the latter is the more widely favored source). The South American rodents, like the other mammalian groups in that land, also evolved peculiar South American forms, including one called *Telicomys gigantissimus* (in the Pleistocene) that is the largest rodent ever to have lived—it was almost as large as a rhinoceros.

In the late Miocene, about 8–9 million years ago, further small additions to the fauna arrived. These new forms included the procyonids (racoons and allies), which came from North America, and the cricetid rodents. These animals almost certainly entered by waif dispersal. While it is possible that North and South America had tectonically wandered closer together at that time, the connection can have been neither close nor lasting, because another 6 million years passed before the South America mammal fauna encountered the full range of the outside world's mammalian types.

About 3 million years ago, the Bolivar Trough finally disappeared and the modern Panamanian land bridge formed. The vegetation on both sides of the

Figure 18.13 A reconstruction of *Doedicurus*, a Pleistocene glyptodont. The glyptodonts were a strange group of armored South American mammals, related to armadillos. From Simpson (1980). Copyright © 1983 by Scientific American Books. Used with permission of W.H. Freeman and Company.

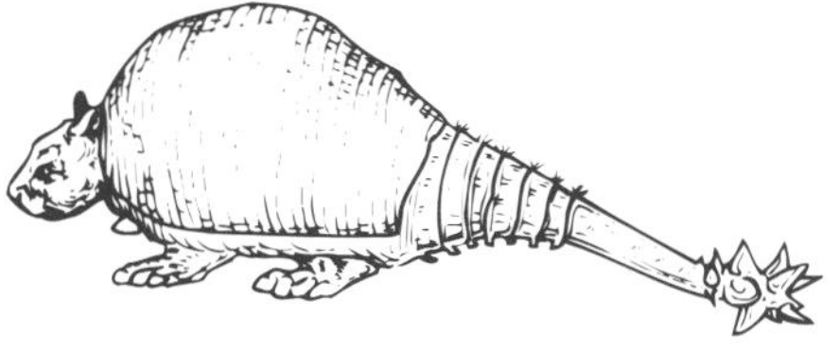

bridge was probably savannah, not the tropical rain forest of modern times. Mammals adapted to the similar vegetation of the two sides and could move freely both ways, allowing the mustelids (skunks), canids (dogs), felids (cats), equids (horses), ursids (bears), and camels to invade South America from the North, while the dasypodids (armadillos), didelphids (opossums), callithricids (marmosets), and edentate anteaters moved rather less dramatically in the opposite direction. In both cases, immigrating animals were accompanied by many other less well-known forms. This extraordinary clash and exchange of faunas is known as the Great American Interchange, and popular biology still portrays it as a competitive rout of the South American mammals by the superior northern forms. Although that idea contains some truth, the increasing quantity of fossil evidence is allowing a more detailed reconstruction of the events.

A study by Marshall, Webb, Sepkoski, and Raup has examined the time course of the Interchange in detail. These scientists counted the number of mammalian genera in South and in North America at successive times and divided the genera according to where they originally evolved. They then divided the immigrant genera into primary immigrants (genera that evolved in the South and immigrated to the North, or vice versa) and secondary immigrants (genera descended from primary immigrants). Their argument claims that the primary invasions were roughly equal in both directions, and that the takeover of the South by northern mammals had two other sources: weight of numbers and different rates of speciation after arrival.

Figure 18.14 shows the numbers of genera, expressed both as absolute numbers and as proportions, of mammals in North and South America. On both sides, after 2.5 to 3 million years ago, an increasing proportion of the mammal genera were immigrants (or descendants of immigrants) from outside the region. At present, about 50% of South American genera are descended from originally North American mammals; the proportion of southern mammals in the North is much lower, less than 20%.

The numbers become more revealing when we break them down even further (Table 18.2). We can begin by counting the total number of genera before the Interchange. The total number is higher in the North, perhaps because of the continent's greater area. (It is an important principle of island biogeography that a larger area supports a larger number of species.) Marshall and his colleagues then counted the numbers of North American mammals moving South, and South American mammals moving North, and expressed them both as proportions of the total pool. According to their calculations about 10% of available North American mammals invaded the South (for example, as Table 18.2 shows, about 100 endemic mammalian genera were found in North America and about 10 of them migrated south) and about 10% of South American mammals moved North (for example, about 60 South American mammals lived 3 million years ago, and the number of South American genera in the north increased by about 6 between 4.5 and 1 million years ago). Thus, although the proportions are the same, the greater absolute numbers moving South mainly result from the larger number of mammals initially present in the North.

The pattern of primary immigration is, therefore, similar in both directions. Roughly 10% of the genera from each side successfully invaded the other. When

Figure 18.14 Numbers (and percentages) of genera of land mammals in the last 9 million years in North and in South America. Immigrant and native genera are distinguished in both places. Note the wave of immigration after about 3 million years ago. Reprinted with permission from Marshall *et al.* (1982). Copyright © 1982 American Association for the Advancement of Science.

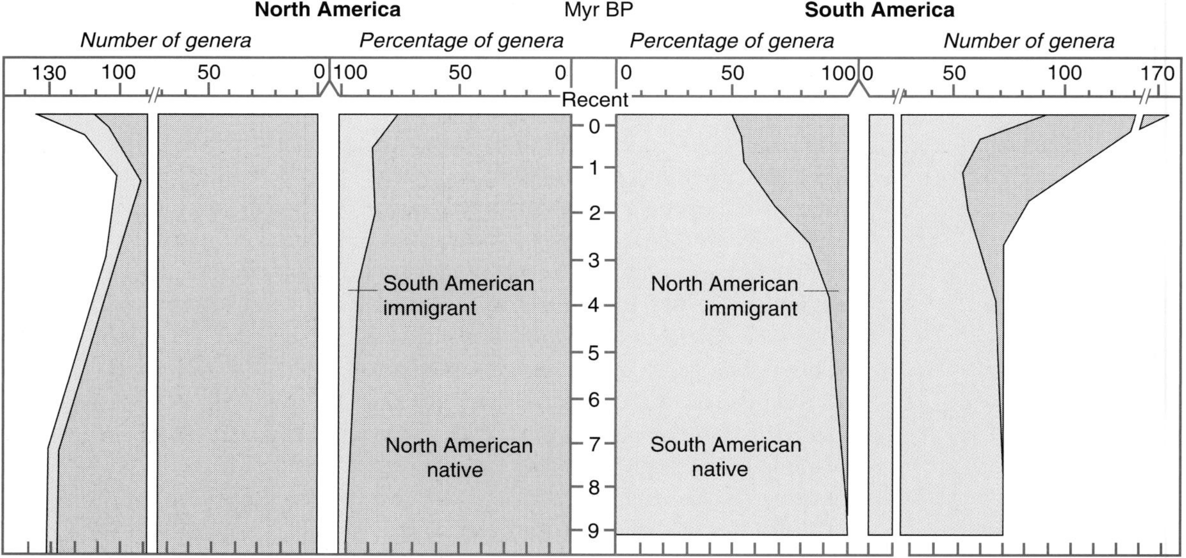

we look at the subsequent proliferation of the immigrants, the pattern diverges markedly (Table 18.2). By the Recent, a total of 12 (the nine species in Table 18.2 represent the number alive—three others had arrived and then gone extinct) immigrant southern mammal genera had produced only 3 new genera, while the 21 immigrant northern mammalian genera in the south produced 49 new genera. In the Recent, the trend has continued. The North American mammals showed their superiority, therefore, not in the original invasion, but in their relative success afterward.

Why did the North American mammals prove superior? The increase in number of originally North American genera can be seen across a wide range of mammal types, which suggests they possessed some general advantage. Several ideas have been suggested to explain this apparent advantage. One is that the North American mammals had lived a more competitive life, in a larger continent with more species, than the isolated southern mammals. The "arms race" of competition had moved further in the North. This idea can be illustrated by Jerison's study of brain size (section 22.7, p. 621). In North American mammals, brain sizes, relative to body size, increased with time in both predators and prey in the past 65 million years. Jerison's interpretation is that brain sizes increased as predators and prey grew increasingly intelligent, in an escalating improvement of offensive and defensive behavior; the pattern of brain evolution fits his interpretation (Figure 22.6, p. 621). In South American mammals, however, no such increase seems to have occurred (Table 18.3). Arguably, when the North American

TABLE 18.2

Pattern of faunal exchange between North and South America. The table gives the total numbers of genera of South or North American origin in each region (the numbers are plotted in Figure 18.14), and breaks down the immigrant genera according to whether they were "primary" (the genus itself immigrated) or "secondary" (the genus descended from a primary immigrant genus e.g., a secondary immigrant in North America evolved in North America but from a genus that itself evolved in South America). The total of the immigrant genera in the bottom rows equals the number of alien genera in the "number of genera" row. Note (1) the similar proportions of primary immigrant genera moving in each direction, and (2) the much greater numbers of secondary immigrants in South than in North America. Simplified, with permission from the publisher, from Marshall *et al.* (1982). Copyright © 1982 American Association for the Advancement of Science.

	South American Mammals						North American Mammals			
Time period	9–5	5–3	3–2	2–1	1–0.3	0.3–Recent	9.5–4.5	4.5–2	2–0.7	0.7–Recent
Duration (million)	4	2	1	1	0.7	0.3	5	2.5	1.3	0.7
Number of genera										
North American	1	4	10	29	49	61	128	99	90	102
South American	72	68	62	55	58	59	3	8	11	12
Total	73	72	72	84	107	120	131	107	101	114
Numbers of immigrant genera										
Primary	1	1	2	10	18	20	2	6	8	9
Secondary	0	3	8	19	31	41	1	2	3	3

mammals invaded the South, they had been prepared by 50 million years of more demanding competition. They possessed advanced armaments, probably not only in intelligence, that enabled them to overrun the southern mammals.

Alternatively, as Marshall and his coauthors suggest, the North American mammals may have enjoyed some advantage in the environmental change of the past 3 million years. The Andean upthrust sheltered the Americas from the Pacific, creating a rain shadow east of the mountains. In South America, dryer pampas or even semi-desert replaced the moist savannah and forest. Why such a change should benefit the North American mammals at the expense of the South American forms is unclear, but such a change would be likely to

TABLE 18.3

Relative brain sizes (expressed as encephalization quotient, which increases with increasing brain size; see section 22.7, p. 621) of North and of South American ungulates in the Cenozoic. Reprinted, by permission of the publisher, from Jerison (1973).

	Ungulate Brain Size (EQ)	
	South America	North America
Time (Million yr ago):		
65–22	0.44 $n = 9$	0.38 $n = 22$
22–2	0.47 $n = 11$	0.63 $n = 13$

benefit one of the two groups more than the other. The sheer magnitude of this change suggests that it was probably influential in the faunal replacements of the time.

The Great American Interchange is one of the most dramatic case studies in historical biogeography. The mammalian faunas of North and South America have been connected, by a narrow isthmus, for less than 3 million years. Yet 50% of the mammalian genera in the South are now of northern origin, and such wonderful animals as that rhinoceros-sized rodent, the giant ground sloth, and the saber-toothed borhyaenid were somehow involved in the general destruction of species during the interchange. It is clearly plausible that the events of the Interchange were at least partly due to competition; demonstrating that point is a taller task, however.

18.7 *Conclusion*

Evolutionary biogeographers have been particularly interested in the historical processes that have shaped the geographic distributions of species—though they by no means rule out the well-documented influence of modern ecology. They have mainly studied two sorts of historical processes: movement and range splitting. Species undoubtedly do move—by dispersal—and when a new corridor appears on the Earth allowing a new encounter of faunas, it can precipitate dramatic evolutionary events. The Great American Interchange is a famous example. It is not easy to disentangle the exact causes at work in this scenario, but the data allow a plausible inference that the faunal changes were substantially influenced by both weight of numbers and competition.

The recent biogeographic work on vicariance events illustrates well how a discipline can be invigorated by new, formal techniques. It is possible to write the modern distributions of large taxa on a phylogeny, and use the information to answer two questions. First, is the order of phylogenetic splitting consistent with the tectonic history of the areas occupied by the taxa? Second, do other large taxa show the same congruent biogeographic patterns? It is unlikely that vicariance events account for all biogeographic distributions, because we know that dispersal can be influential in cases like the recolonization of Krakatau and the movement of species during the Ice Age. The relative importance of vicariance and dispersal, and of ecology and history, are both topics that can, in principle, be settled by empirical research.

SUMMARY

1. Species, and higher taxa, have geographic distributions that biogeographers seek to describe and explain.
2. The similarity of the flora or fauna of two regions can be measured by indexes of similarity. The world can be divided up into six main faunal regions, based on the distributions of bird and mammal species. Different taxa imply different regional divisions.
3. The ecological properties of a species set limits on where it can live.

4. The distributions of species are influenced by historical accidents—that is, where species happened to be at certain times—as well as their ecological tolerances.
5. The ranges of species may be altered by dispersal (when a species moves in space) and by continental drift (when movement of the land subdivides the ranges of species). The splitting of a species range is called vicariance.
6. When climates cooled in the Ice Age, the ranges of species in the northern hemisphere moved to the south. The ranges of tropical forests may have been fragmented into localized refuges.
7. An area cladogram shows the geographic areas occupied by a group of phylogenetically related set of taxa.
8. Vicariance biogeography suggests that geographic distributions are determined mainly by splits in the ranges of ancestral species, not dispersal. It predicts that the area cladogram of a taxon should match the geological history of the area, and the area cladograms of different taxa in an area should have compatible (congruent) area cladograms. Both predictions are supported by evidence, although other evidence illustrates the importance of dispersal events.
9. In the encounter between the North and South American faunas when the Isthmus of Panama formed 3 million years ago, similar proportions of mammals initially moved in both directions. The immigrant North American mammals in the South proliferated at a greater rate, however.

FURTHER READING

Cox and Moore (1993) is an introduction, Brown and Gibson (1983) is a comprehensive textbook, and Myers and Giller (eds) (1988) is a symposium on modern research topics. The *American Zoologist*, vol. 22, pp. 347–471 (1982) contains another such symposium, including a general paper about ecological and historical factors by Endler.

Simpson (1983) explains how the great faunal regions of the Earth were discovered, as well as the importance of movement, and how to measure faunal similarity. The ecological influences on distribution can be read about in any ecology text, such as Begon *et al.* (1990). On niche concepts, see the entries by Griesemer and Colwell in Keller and Lloyd (1992). On Krakatau, see Bush and Whittaker (1991) and the narrative in Wilson (1992).

On Ice Age biogeography, see Pielou (1991), and the general references above, which include a chapter in Cox and Moore (1993). See Pease *et al.* (1989) for relevant theory; Graham and Grimm (1990) for the influence of climate; Huntley and Webb (1989) for North American trees, and Elias (1994) for Quaternary beetles. On refugia, Haffer (1969) is the original; Lynch (1988) is a more recent study; Colinvaux (1993) criticizes the Amazonian application; Turner (1976) applies the idea to the Heliconius discussed in section 8.3 (p. 198); and Rogers *et al.* (1991) apply it to Native American languages. On pupfish, see Brown (1971).

On vicariance biogeography, see Brundin (1988); Wiley (1981, 1988); Nelson and Rosen (eds) (1981), which contains the chapter by Patterson; *Systematic Zoology*, vol. 37, pp. 219–419 (1988); and Page (1994). For Croizat's biogeography, see Croizat *et al.* (1974). On Hawaii generally, see Carlquist (1970), the special issue of *Trends in Ecology and Evolution*, vol. 2 (1987), pp. 175–228, and Gillespie *et al.* (1994, popularized by Holmes and Harvey 1994) for a further phylogenetic study of Hawaiian biogeography, this time on spiders. Avise (1994) describes how molecular evidence can be used to reconstruct historical geography (or "phylogeography") at low taxonomic levels, such as within a species.

Vermeij (1991) is a general study of biotic interchanges; as is an issue of *Paleobiology*, vol. 17 (1991), pp. 201–324, which contains a paper by Webb on the Great American Interchange. Stehli and Webb (1985) is a book about the Great American Interchange; see also Vrba (1992). Simpson (1980) describes the South American mammals, and see the chapters in Goldblatt (1993) for South American biogeography generally. On the brain size difference, see Jerison (1973) and a popular essay by Gould (1977b, ch. 23). Part of another of Gould's popular essays (1983a, ch. 27) is about the Interchange.

STUDY AND REVIEW QUESTIONS

1. Review the geographic terms Boreal, Nearctic, Palearctic, Holarctic, and Neotropical.
2. Calculate the indexes of similarity between Areas 1 and 2:

Number of species in area 1	Number of species in area 2	Number of species common to areas 1 and 2	Index of similarity
10	15	5	
15	10	5	
10	10	5	
5	15	5	

3. Examine the phylogeny of Figure 17.17 (p. 494). How many dispersal events does it imply from younger to older islands, and from older to younger islands? (The ancestral species are numbers 1 and 2 at the top left. The oldest island is at the left, the youngest at the right. The species can be treated as four "columns" inhabiting the islands of Kauai, Oahu, Mauai, and Hawaii, respectively. The small deviations to the right and left on the page are biogeographically insignificant.) You could draw the dispersal events on the map at the bottom of the figure. What relevance does the answer have for vicariance and dispersalist theories of area cladograms?
4. Using the same phylogeny (Figure 17.17), draw an area cladogram in the form of Figure 18.16 for fruitfly species 1–15.
5. Given below are the geographic areas occupied by the species of two taxa:

Area	A	B	C	D
Species in taxon 1	1	2	3	4
Species in taxon 2	5	6	7	8

Three (rooted) phylogenies for the two taxa appear in the figure on the following page. Which pairs of area cladograms are congruent, and which are not?

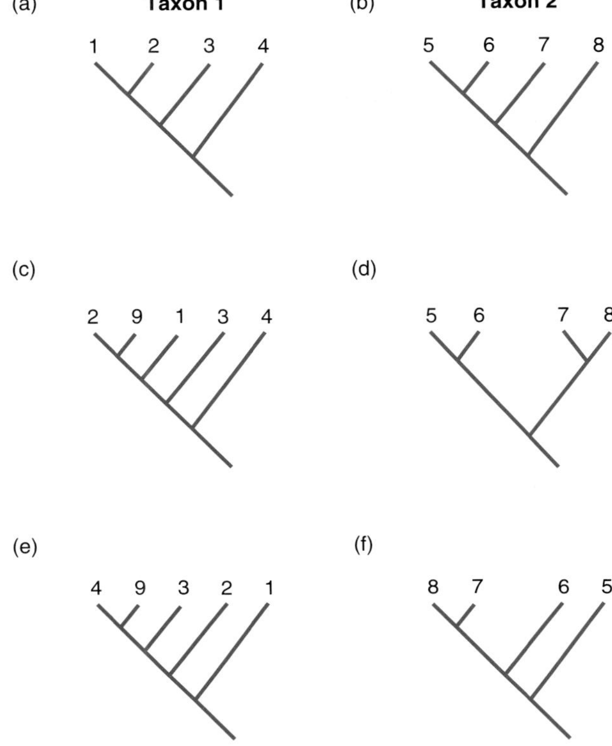

6. What are the main hypotheses to explain the proliferation of North American mammals in South America after the formation of the Isthmus of Panama?

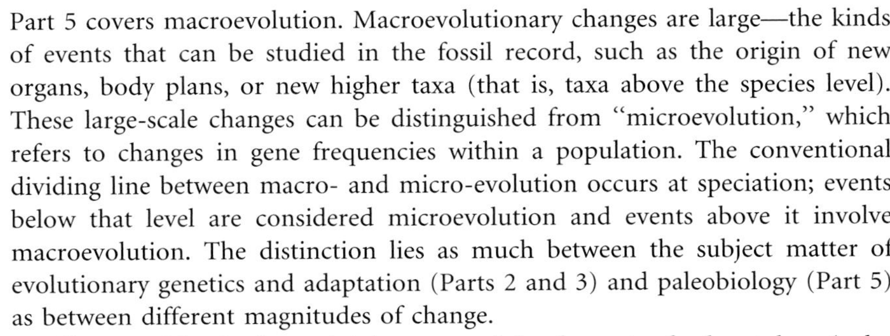

PART 5 Paleobiology and Macroevolution

Part 5 covers macroevolution. Macroevolutionary changes are large—the kinds of events that can be studied in the fossil record, such as the origin of new organs, body plans, or new higher taxa (that is, taxa above the species level). These large-scale changes can be distinguished from "microevolution," which refers to changes in gene frequencies within a population. The conventional dividing line between macro- and micro-evolution occurs at speciation; events below that level are considered microevolution and events above it involve macroevolution. The distinction lies as much between the subject matter of evolutionary genetics and adaptation (Parts 2 and 3) and paleobiology (Part 5) as between different magnitudes of change.

Chapter 19 begins by explaining some of the elements of paleontology (paleontology is the science that studies fossils) and briefly reviewing the history of life. We then move on to the study of evolutionary rates in chapter 20; we see how these rates are measured and consider one controversy—about the relative rates of evolution during, and between, speciation events—in detail. That controversy introduces us to a much larger question: Is macroevolution really microevolution extrapolated over a long time scale or does it take place by different, though not incompatible, mechanisms from microevolution? We stay with that issue in chapter 21, which discusses macroevolutionary change. The chapter presents evidence that higher groups, such as the mammals, evolved by natural selection. We then consider the different ways that development may change during evolution, because changes in developmental mechanisms probably represent the typical means by which large changes in morphology evolve. Chapter 22 focuses on coevolution, in which the evolution of one species is directed by evolutionary changes in the other species that make up its environment. Coevolution is studied both in living species, by the methods of sections 2 - 4 of this book, and in fossils; we look at examples of both kinds of study.

Chapter 23, the final part of this section, considers why extinctions happen. We look at how the longevity of ecological niches, and the characters possessed by species, influence the chance of extinction. We then turn to mass extinctions, which are more likely caused by physical catastrophes. Mass extinctions, by clearing the ground, are thought to have been important influences in the replacements of one higher taxon by another in the history of life.

CHAPTER 19

The Fossil Record

This chapter describes how fossils are formed, how their age is estimated, and how the completeness of the fossil record is calculated; it also includes a synoptic review of the history of life. The study of fossils—paleontology—is a broad science, and its scope extends far beyond the theory of evolution. This chapter introduces only those parts of paleontology that are most important in evolutionary biology, and a knowledge of which is assumed in chapters 20–23. The chapter serves a similar purpose for paleontology as chapter 2 did for genetics.

19.1 Fossils are remains of organisms from the past, and are preserved in sedimentary rocks

A fossil is any trace of past life. The most obvious fossils are body parts, such as shells, bones, and teeth; fossils also include remains of the activity of living things, such as burrows or footprints (called trace fossils), and of the organic chemicals they form (chemical fossils). For any organism to leave a fossil requires a series of events, each of which is unlikely. We can consider these events for the case of hard parts; analogous points apply for trace and chemical fossils.

When the organism dies, its soft parts will either be eaten by scavengers or be decayed by microbial action. For this reason, organisms that consist mainly of soft parts (such as worms and plants) are much less likely to leave fossils than are organisms possessing hard parts. The fossil record is biased in favor of species that possess skeletons. Only an organism's hard parts stand much chance of fossilization, although in most cases these parts will be destroyed as well. Hard parts may be crushed by rocks, stones, or wave action, or broken up by scavengers. If they survive, the next stage in fossilization usually is their burial in sediment at the bottom of a water column—only sedimentary rocks contain fossils. (Geologists distinguish three main rock types: igneous rocks, formed from volcanic action; sedimentary rocks, formed from sediments; and metamorphic rocks, formed deep in the Earth's crust by the metamorphosis of other rock types—when sedimentary rocks undergo metamorphosis, any fossils are lost.)

Animals that normally live within sediments are more likely to be buried in sediment before being destroyed, and these animals are, therefore, more likely to leave fossils than are species that live elsewhere. Likewise, species that live on the surface of the sediment (i.e., on the sea bottom) are more likely to be fossilized than are species that swim in the water column. Terrestrial species are least likely of all to be fossilized. The further a species lives from sediments, the less likely its fossilization becomes. For most of the delicate kinds of animals that live on the sea bottom, such as feather stars and worms, practically the only way they may come to leave fossils is by "catastrophic" burial, such as a slide of sediment from shallower water into the depths that carries with it and buries some soft-skeletoned animals. Feather stars, for instance, are known to decay into nothing within 48 hours of death on the sea bottom; they must be buried rapidly to have any chance of fossilization.

Once an organism's remains have been buried in the sediment, they may stay there for an indefinitely long period of time. As new sediment piles on top of older sediment, the lower sediments are compacted. In this process, water is squeezed out and the sedimentary particles forced closer together. The fossil hard parts may be destroyed or deformed in the process. As the sediments compact, they are gradually turned into sedimentary rock. They may subsequently be moved up, down, or around the globe by tectonic movement, and can be re-exposed in a terrestrial area. Any fossils they contain can then be picked up, or dug up, on land (a fossil is, etymologically, anything that is dug up). Sediments may also be lost by tectonic subduction and geological metamorphosis.

Any particular sedimentary rock will be made up of sediments that were deposited at a certain time, or through a certain range of times, in the geological past. The fossils contained within that rock will consist of organisms that lived at the time the sediments were deposited. It is possible to draw a geological map of an area, showing the ages of the rocks that are either exposed at the surface or are near the surface but concealed beneath the topsoil. Color Plate 9 presents a geological map of North America; maps of this kind, at varying levels of detail, have been produced for many areas of the globe. A geological map is a first guide to where it may be possible to find fossils of particular times. Dinosaurs, for example, lived in the Mesozoic, and we can read directly from the geological map of the United States that the pink and mauve regions—for instance in Texas and New Mexico, and north through Colorado, Wyoming, and Montana—are appropriate regions to hunt for fossil dinosaurs. Indeed, abundant dinosaur remains have been located at some sites there. The pattern of rock types found on a geological map can be understood in terms of the theory of plate tectonics.

Over geological time, the original hard parts of an organism will be transformed while lying in the sedimentary rock. Minerals from the surrounding rock slowly impregnate the bones, or shell, of the fossil, changing its chemical composition. Calcareous skeletons also change chemically. Carbonate comes in two forms: aragonite and calcite. Aragonite is less stable and becomes rare in older fossils. Calcite may be replaced in some fossils by silica or pyrite. In extreme cases, the calcite may be dissolved away completely, and the space filled in by other material. The fossil then acts as a mould, or cast, for the new material. The remains will subsequently reveal the shape of the organism's hard parts.

Fossilization is an improbable eventuality. It is more probable for some kinds of species than others, and for some parts of an organism than for other parts. After burial in sediment, the fossils slowly transform through time. The transformed remains, if they are preserved, can still tell us (after expert interpretation) much about the original living form.

19.2 *Geological time is conventionally divided into a series of eras, epochs, and periods*

19.2.1 *The successive geological ages were first recognized by characteristic fossil faunas*

Figure 19.1 shows the main time divisions of the geological history of the Earth during the past 600 million years. Earlier divisions are recognized as well, but

Figure 19.1 The geological time scale. Adapted with permission from SCOPE 47, Long-term Ecological Research: an international perspective, edited by Paul G. Risser, 1991, John Wiley & Sons, Chichester, UK.

Era	Period	Epoch	Myr BP (approx)
Cenozoic	Quaternary	Recent	0.01
		Pleistocene	1.8
	Tertiary — Neogene	Pliocene	5
		Miocene	24
	Tertiary — Paleogene	Oligocene	37
		Eocene	54
		Paleocene	65
Mesozoic	Cretaceous		144
	Jurassic		213
	Triassic		248
Paleozoic	Permian		286
	Carboniferous	Pennsylvanian	320
		Mississippian	360
	Devonian		408
	Silurian		438
	Ordovician		505
	Cambrian		590

the past 600 million years have the greatest paleontological importance because fossils are much less common before this time.

The time divisions in Figure 19.1 were recognized by nineteenth century geologists on the basis of characteristic fossil faunas. The times of transition between two eras represent transitional periods between different characteristic fossil faunas. The fossils of the Permian, for instance, characteristically differ from those of the Triassic; thus, a relatively sudden transition between the fossil faunas appears at the Permo-Triassic boundary. In the nineteenth century, it was not known whether the transition times corresponded to mass extinctions and sudden replacements, or to long gaps in the fossil record while a slower replacement was proceeding. Such transitions are now known to mark mass extinctions occurring over short periods of time. To demonstrate the feasibility of such events, techniques to establish absolute times were necessary. For the smaller divisions of geological time, such as epochs, disagreements continue to abound over the absolute dates, and more than one geological time scale exists.

19.2.2 *Absolute geological time is inferred by radioactive dating techniques*

Geologists date events in the past both by relative and absolute techniques. An absolute time is a date expressed in years (or millions of years); a relative time is a time relative to some other known event. The times given in Figure 19.1 are absolute; they were established from the radioactive decay of elements. For example, the isotope of rubidium, ^{87}Rb, decays into an isotope of strontium, ^{87}Sr. The decay is very slow, having a half-life of about 48.6 billion years. Thus, half of an initial sample of ^{87}Rb will have decayed into ^{87}Sr in 48.6 billion years (about 10× the age of the Earth).

Radioactive decay proceeds at an exponentially constant rate. Exponential decay means that a constant proportion of the initial material decays in each time unit. For example, suppose we start with 10 units and 1/10 of them decay per time interval. In the first time interval, one unit will decay, and we will have 9 units left. In the second time interval, a proportion equal to 1/10th of the remaining 9 units (i.e., 0.9 units) will decay; and we will be left with 8.1 units. In the third time interval, a further 1/10th of the 8.1 units will decay, leaving 7.29 units (8.1 − 0.81) at the beginning of the fourth time interval. . . and so on. In radioactive decay, the proportion of the isotope that decays each year is called the decay constant (λ), and for $^{87}Rb/^{87}Sr$ the decay constant is 1.42 × 10-11 per year. Therefore, for whatever the amount of ^{87}Rb is present at any time, a proportion equal to 1.42×10^{-11} of that amount will decay into ^{87}Sr in the next year.

To estimate the age of a rock by the radioisotope technique, we need to make two measurements and validate one assumption. The two measurements are the isotope composition of the rock now and when it was formed. The current proportions of ^{87}Rb and ^{87}Sr are obviously measureable. The composition of the rock was originally fixed when it crystallized as an igneous rock from liquid magma, and the ratios of ^{87}Rb and ^{87}Sr in modern magma can be measured. This ratio provides a good estimate of the isotope ratio when the rock first formed. The isotope ratio will slowly change from the original ratio as ^{87}Rb radioactively decays into ^{87}Sr. To estimate the age of the rock from the change in the isotope ratio, we must assume that all the change in the ratio is due to radioactive decay. For the case of ^{87}Rb/^{87}Sr, this assumption is probably valid. Neither isotope seeps into, or leaks out of, the rock, and the ratios are, therefore, solely determined by time and radioactive decay. For some other radioisotopes, this assumption is less solid. Uranium, for example, can be oxidized into a mobile form and move among rocks (although this problem can be dealt with by combining two uranium decay schemes, such that the time is inferred from the ratio of two lead isotopes and the concentration of uranium does not matter).

Table 19.1 lists the main radioisotopes used on geochronology. The decay of ^{40}K, for example, is a geochronologically useful decay scheme. When a volcano erupts, the heat volatilizes all of the ^{40}Ar from the volcanic lava and ash, but not the ^{40}K. When the volcanic dust cools, therefore, it contains (of ^{40}K and ^{40}Ar) only ^{40}K. The ^{40}K then decays into both ^{40}Ca and ^{40}Ar. In practice, so much ^{40}Ca remains in the rock from other sources that it is not convenient to use it for dating purposes, but all ^{40}Ar in the rock will have been produced by the decay of ^{40}K. The decay is so slow that it is not practical to use this dating method for rocks less than about 100,000 years old; other radioisotopes are typically used for such younger rocks. The decay of carbon-14 into nitrogen, for example, has a half-life of only 5,730 years.

The exact age of a rock is calculated as follows. Take ^{87}Rb/^{87}Sr as an example. Let N_o be the number of ^{87}Rb atoms in the sample of rock when it was formed, and N be the number today. Then $N = N_o e^{-\lambda t}$, where λ = the decay constant and t is the age of the rock. Take logs and $t = (1/\lambda)\ln N_o/N$. On a practical level, it is easier to measure the quantity of the isotope generated by decay. Therefore, let N_R = the number of ^{87}Sr atoms generated by radioactive decay up to any time. Because each ^{87}Sr atom has been generated by the decay of one ^{87}Rb atom, $N_R = N_o - N$. This expression can be substituted into the formula for time:

$$\frac{1}{\lambda} \ln \frac{N + N_R}{N}$$

TABLE 19.1

Radioactive decay systems used in geochronology.

Radioactive Isotope	Decay Constant ($\times 10^{-11}$/year)	Half-life (years)	Radiogenic Isotope
^{14}C	1.2×10^7	5.73×10^3	^{14}N
^{40}K	5.81 + 47.2	1.3×10^9	^{40}Ar + ^{40}Ca*
^{87}Rb	1.42	4.86×10^{10}	^{87}Sr
^{147}Sm	0.654	1.06×10^{11}	^{143}Nd
^{232}Th	4.95	1.39×10^{10}	^{208}Pb
^{235}U	98.485	7×10^8	^{207}Pb
^{238}U	15.5125	4.4×10^9	^{206}Pb

*^{40}K decays into both ^{40}Ar and ^{40}Ca, with the two decay constants given; the half-life is for the sum of the two.

For example, if 3% of the original rubidium-87 in a rock has decayed into strontium-87, then the age of the rock is calculated as

$$\frac{1}{1.42 \times 10^{-11}} \ln (100/97)$$

or about 2 billion years.

Note that the decay constant and the half-life of a radioisotope are simply related. When half of the original ^{87}Rb has decayed into ^{87}Sr, the number of ^{87}Sr atoms formed must equal the number of ^{87}Rb atoms that have decayed. That is, $N = N_R$. If we substitute that expression into the formula for time, and $t_{1/2} = \ln 2/\lambda = 0.693/\lambda$.

In summary, if we know the isotope ratio in the rock when it was formed and in a modern sample, and if we can reasonably assume that the change in the ratio between then and now was caused by only radioactive decay, then we can estimate the absolute age of the rock.

The radioisotope method can be used only for rocks that contain the specified isotopes. Some fossils contain carbon-14, because carbon is found in living material, and these fossils can be dated directly if they are not too old. The ratio of ^{14}C to ^{14}N can be substituted into the formula for radioactive decay to infer the age of the fossil. Some other radioisotopes are found in corals or shells, but many (such as ^{87}Rb or ^{40}K) are found solely in igneous rocks. To date a deposit of fossils, it then becomes necessary to infer that the fossils were laid down at about the same time as an associated igneous rock that can be absolutely dated. If the fossiliferous sediments are deposited on top of an igneous rock, it can be inferred that the fossils are no older than the date of the igneous rock (on the principle that younger rocks lie on top of older ones). If an igneous rock has been intruded into a sedimentary rock, it can be inferred that the sediments are older than the igneous rock (because igneous rocks only intrude into existing sedimentary rocks). In the best case, a fossiliferous sediment will lie on top of an older set of igneous rocks, and have another younger igneous rock intruded in it. The fossils can then be dated to a time between the age of the two igneous rocks.

19.2.3 *Relative geological time is inferred by correlating fossils between different rocks*

The inference of the age of fossils from that of surrounding igneous rocks is an example of relative time measurement. If we know the relative date of a rock, or fossil, it means we know its date relative to that of another rock, or fossil. Thus, we have a statement of the form "rock A was laid down before / at the same time as / after rock B." Several procedures may be followed to find relative times. At any one site, more recent sediments are deposited on top of older sediments. Fossils lower down a sedimentary column are, therefore, likely to be older (sometimes a large geologic convulsion, such as a volcanic explosion, may turn a sedimentary column upside down, but the aftermath of such an event is obvious). The date of any one fossil deposit relative to those at different sites can also usually be estimated by comparing the fossil composition of the site, for some common fossils such as ammonites or foraminifers, with a standard reference collection. For these reference fossils, the fossils deposited at one place and time will be much the same as those deposited at another place. Their presence shows that two sites had the same relative date. This kind of study is called correlation; the paleontologist is said to be "correlating" the two sites.

Magnetic time zones supply a similar principle. The magnetic field of the Earth has reversed its polarity at intervals through geological history. When the magnetic field has its current polarity (compasses point north), it is called normal; when it has the opposite polarity, it is called reversed. The history of the Earth is an alternating sequence of normal and reversed time zones. (The reason for the reversals is not known for certain—although hypotheses abound.) Polarity switches have been more common at some times than others; Figure 19.2 gives an idea of their frequency in recent times. All rocks found in the globe at any one time have the same polarity, and the polarity at the time rocks were formed can be detected. The polarities can then be used in fine-scale time resolution. If two rocks are known to have been formed at similar times, but we are not sure whether they are exact or only near contemporaries, magnetic polarities can provide the answer. If the rocks have different polarities, they cannot have been exact contemporaries.

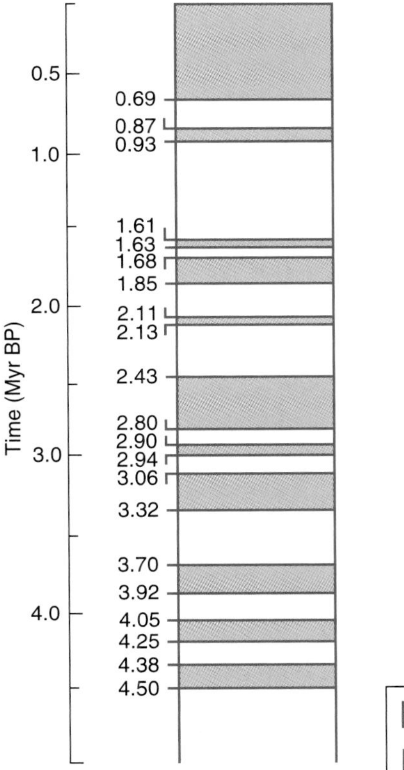

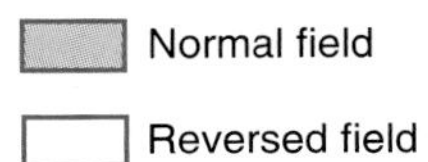

Figure 19.2 Polarity reversals of the Earth's magnetic field in the past 4.5 million years. The picture may not be complete. It is possible that further, shorter events have still to be discovered.

19.3 *The history of life*

19.3.1 *The origin of life*

The origin of life consisted of events on a molecular scale and cannot be studied in the fossil record, but the record provides valuable information about timing. The Earth itself is about 4.5 billion years old, and for the first few hundred million years it was far too hot to support life. The earliest fossil evidence for life is about 3.5 billion years old, and comes from some of the oldest rocks on the planet. Schopf (1993) has described filamentous microbes from rocks in Western Australia that are dated to 3465 million years ago. The rocks belong to the Archean Apex Chert, which is probably the oldest rock system that could retain fossil evidence of life. Although small amounts of older rocks are found on Earth, they are thought to have undergone too severe metamorphosis to possibly contain older fossils. The 3.5 billion-year-old fossils already are cellular and presumably descended from a long period of prior evolution. In Schopf's (1994) words, "it seems certain that life must have originated substantially (and probably hundreds of millions of years) earlier."

The striking fact about Schopf's evidence is that it suggests a very early time of origin for life. We can infer that life originated almost as soon as the physical conditions of the Earth permitted it. Various interpretations of this inference (if it is correct) are possible, including the idea that the origin of life, from non-living chemical raw materials, is an evolutionarily easy step.

Most of what we know—with suitable uncertainty—about the origin of life comes from modern research on the kinds of chemical reactions that may also have taken place on the Earth 3.5 billion years ago. Many of the molecular building blocks of life (such as amino acids, sugars, and nucleotides) can be synthesized from a solution of simpler molecules, of the sort that probably existed in the prebiotic seas, if an electric discharge or ultraviolet radiation is passed through the solution. Once the molecular building blocks exist, the next crucial step is the origin of a simple replicating molecule.

Although the earliest ancestral replicating molecule has not been positively identified, several lines of evidence point to RNA. RNA, which is single-stranded, is simpler than the double-stranded DNA. In addition, "prebiotic soup" experiments have more readily yielded the nucleotide *U* than *T*. In the 1980s, it was discovered that RNA can act not only as a replicator but also as an enzyme (in which capacity it is called a ribozyme), catalyzing certain kinds of reactions that contribute to replication. No fully autocatalytic replicating RNA system—one in which the RNA acts as a ribozyme to catalyze its own replication—has (at the time of writing) been discovered, but such a system would require no fundamentally new properties beyond those already identified. Research in this area is moving rapidly; the technology was developed in 1994 to select for new ribozymes *in vitro*, and the discovery of a fully autocatalytic system is probably only a matter of time.

19.3.2 *The origin of cells*

Nothing was inevitable about the evolution of cells; unadorned replicating molecular systems could have persisted, and the molecules being replicated as their component building blocks bonded to them and formed copies, or near copies, of the whole. Limits exist as to how complex a system can evolve in this molecular form. Indeed, experiments have been performed with RNA in which natural selection operates on the naked replicating molecules. In these experiments RNA of a fairly short optimal length soon evolves, and evolution then comes to a halt.

For the system to become more complex, it needs enzymes and metabolic systems that enable it to harvest resources more powerfully, or to exploit the resources better by converting them into the molecular units needed for replication. This step is difficult for at least two reasons. First, mutational error becomes increasingly damaging as the replicating molecule increases in size. Second, any advantageous innovation—for example, one that can produce useful molecules—will share its produce with all competing replicating molecules in the locality. A "selfish" replicating molecule that used resources manufactured by others but did not itself manufacture them would, therefore, have a selective advantage over other replicating molecules that both manufactured and used resources (section 12.2, p. 318).

This second difficulty was probably overcome by the evolution of cells. If the replicating molecules remain enclosed within cells, the products of their metabolism are confined to the cell that produced them and are not available for any selfish replicating molecules outside. Another advantage of cell membranes is that metabolic enzymes can be arranged spatially; a chain of metabolic reactions can then operate in an efficient sequence. Thus, the first cells were probably little more than replicating molecules either surrounded by, or arranged within, membranes. Modern prokaryotic cells are complex versions of this form of life.

Unicellular prokaryotic life exists in the fossil record approximately 3.5 billion years ago. The other kind of cell, the eukaryotic cell, does not appear in the record until 1.8 billion years ago. Eukaryotic cells are more complex, with a separate nucleus and organelles, and are usually larger than prokaryotic cells (see Figure 2.1, p. 22). The long time from the origin of life and of prokaryotic cells to the origin of eukaryotic cells—about 50% of biological history—suggests that it was an improbable, or evolutionarily difficult, transition. It may well have occurred in a number of stages, with the origin of the nucleus being a separate event from the origin of the organelles, and the origin of chloroplasts separate from the origin of mitochondria. Direct evidence of this process is lacking, however. Much research has been conducted on the origin of organelles, and it is widely (if not universally) accepted that they originated in a symbiotic union of ancestral prokaryotes. One cell may have entered another (either as engulfed food, or as a parasite), and instead of being digested or parasitizing its host, it may have evolved into a symbiotic mitochondrion. (It is, of course, most unlikely that it was a unique event; rather, it would often occur that one cell entered another, and all sorts of evolutionary possibilities could ensue.) If the mitochondrial ancestor was originally a parasite, the theory about evolutionary increases and decreases in virulence (section 22.5, p. 615) explains why it might have evolved into a symbiont.

An important event associated with the origin of eukaryotes is the evolution of photosynthesis, or of photosynthesis on a mass scale. Photosynthesis itself probably originated earlier—indeed, Schopf's 3.5 billion-year-old microbes may have been photosynthetic. Around the time that eukaryotic cells were evolving, however, the quantity of oxygen is known to have increased, suggesting that photosynthesis was becoming much more important. Oxygen concentrations can be inferred from the ionic nature of iron in rocks of various ages, and the concentration of oxygen was probably much lower in the early history of the Earth. By a time more than 2 billion years ago, the concentration began to increase, and by about 1.5 billion years ago it had reached the sort of concentration that exists today. The most likely reason for this transition is a greater abundance of photosynthesizing organisms that were pouring out oxygen as a by-product. Perhaps the chloroplast-containing cells of eukaryotes were more efficient photosynthesizers than the prior prokaryotes, which explains why the oxygen concentration increased at about the time when eukaryotes were evolving. Whatever the reason, when oxygen was first released in large amounts it was probably a poison to most existing forms of life, because they had evolved in environments with little oxygen; an ecological disaster may have ensued. Subsequent forms of life have mainly descended from species that evolved to tolerate, and then make use of, this chemical novelty.

19.3.3 *Multicellular life*

The earliest definite fossils of multicellular animals (Metazoa) come from the Ediacarian deposits in Australia. These deposits, and similar deposits elsewhere in the world, date to the period from 670 to 550 million years ago. The Ediacarian fossils are of soft-bodied aquatic animals such as jellyfish and worms (Figure 19.3). Most or all of them went extinct before the beginning of the Cambrian period about 550 million years ago. The Cambrian holds a great increase in the diversity and numbers of fossils—so much so that the main time periods of the fossil record (Figure 19.1) begin with Cambrian. Precambrian fossils were not discovered until the 1940s, and in the nineteenth century it was thought no fossils were developed before the Cambrian.

The great proliferation of fossil evidence since the Cambrian occurred because of the evolution of hard parts—shells in molluscs and skeletons in arthropods and echinoderms. Vertebrates are another major group of animals with skeletons. Vertebrates may have evolved as early as the Cambrian (a group called the conodonts, which are thought to be early vertebrates, exist in the Cambrian). On the other hand, no fossil fish have been observed before the Ordovician, and these animals do not become abundant until the Silurian.

19.3.4 *The colonization of land*

Terrestrial environments were probably first colonized by crusts, or mats, of microbes at the water's edge. Such microbes were most likely Cyanobacteria, although it is not known for sure because the fossil evidence is chemical—localized areas enriched in organic content. Terrestrial life persisted in this form until the evolution of the group that includes the vascular plants, probably in the middle Ordovician. Two key adaptations led to the origin of the vascular plants: a resistant spore, which probably originated in the taxonomic group containing the green algae, and the tracheid cells, with lignin-containing walls, that form tubes and are the crucial structural feature of vascular plants. Vascular plants originated in the Ordovician, and by the end of the Devonian the land was covered in plants in a way that architecturally (although not taxonomically) resembles the modern world.

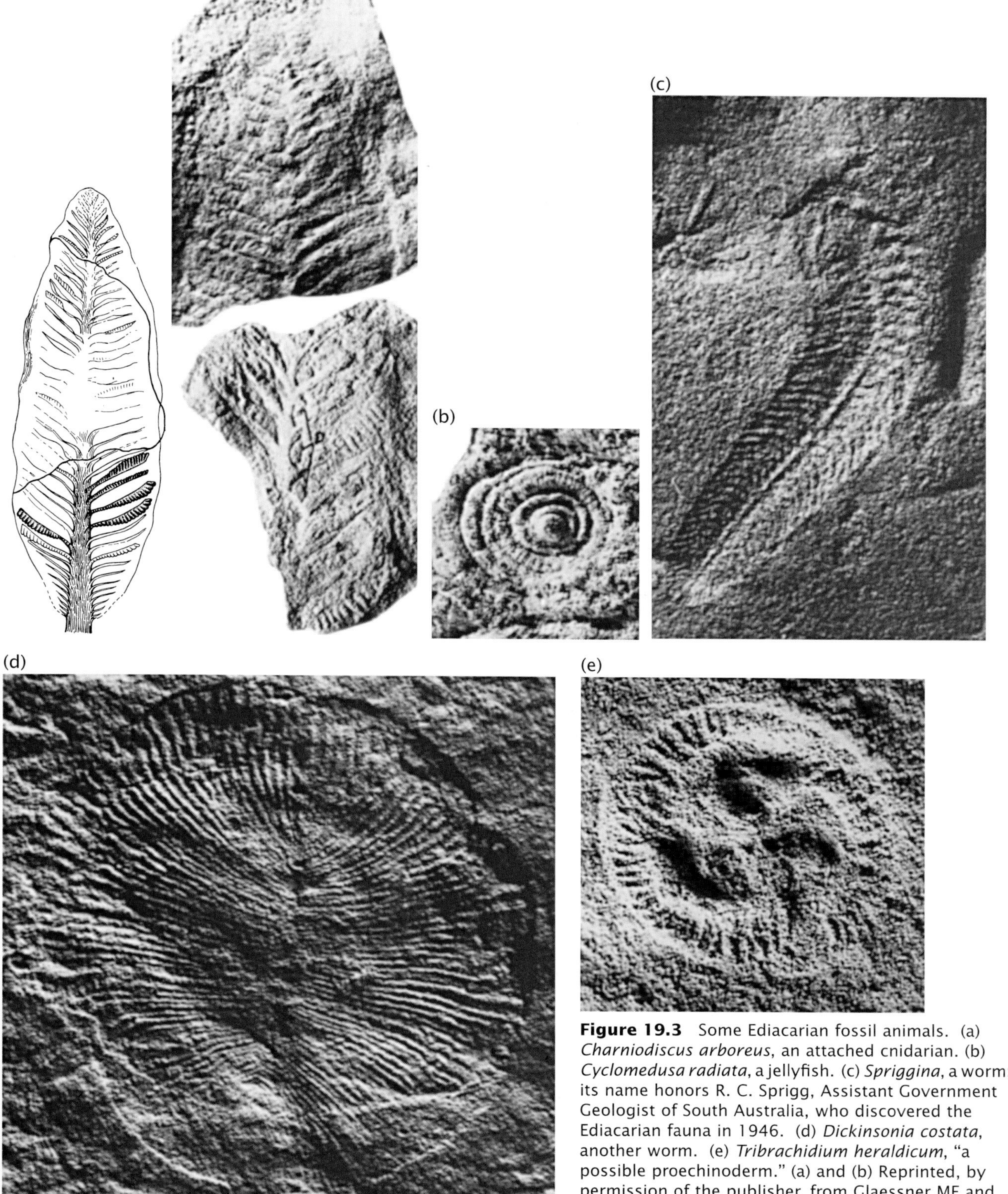

Figure 19.3 Some Ediacarian fossil animals. (a) *Charniodiscus arboreus*, an attached cnidarian. (b) *Cyclomedusa radiata*, a jellyfish. (c) *Spriggina*, a worm; its name honors R. C. Sprigg, Assistant Government Geologist of South Australia, who discovered the Ediacarian fauna in 1946. (d) *Dickinsonia costata*, another worm. (e) *Tribrachidium heraldicum*, "a possible proechinoderm." (a) and (b) Reprinted, by permission of the publisher, from Glaessner MF and Wade MJ, *Palaeontology* 9; 1996; and Wade MJ, *Palaeontology* 15; 1972. (c), (d), and (e) courtesy of MF Glaessner.

The Devonian also provides the earliest evidence, from body fossils, of terrestrial animals. Animals apparently took some time to evolve adaptations for herbivory life, because it was well into the Carboniferous before definite herbivores—in the forms of arthropods—appeared. The Carboniferous is also known for the presence of gigantic predatory insects, including dragonflies with wingspans of more than 60 cm.

Vertebrates show a similar pattern to arthropods. The first terrestrial tetrapod fossils, which date to the late Devonian, were predators (eating fish and insects); the first herbivorous tetrapods did not evolve until the end of the Carboniferous and the early Permian.

One of the important evolutionary steps enabling vertebrate life on land was the origin of the amniotic egg. Reptiles, birds, and mammals are amniotes, and members of these groups, unlike most amphibians, do not return to water for the early stages of the life cycle. The origin of egg types cannot be traced directly in the fossil record. The origin of the reptiles, however, is marked by changes in skeletal morphology as well as egg type, and good evidence is known for the former modifications. The reptiles probably evolved in the Carboniferous. An early definite reptile, for example, is the small lizard-like creature called *Hylonomus* that comes from deposits in Nova Scotia.

After the origin of the reptiles, the two main events in vertebrate evolution were the origin of flight in birds and the origin of the mammals. The late Jurassic fossil *Archeopteryx* is a good intermediate stage in the evolution of flight.

The origin of mammals is the fossil record's best-documented transition for the origin of a major taxon. Section 21.1 (p. 582) discusses this development as an example of a macroevolutionary change.

19.4 *The completeness of the fossil record can be expressed by various quantitative measures*

It is important for a number of questions in evolutionary biology to know how complete the fossil record is. When estimating rates of evolution (chapter 20), we need to know whether apparently sudden evolutionary changes really indicate bouts of rapid evolution or whether they just mean that a long series of intermediate stages left no record as fossils. In mass extinctions (chapter 23), we are interested in whether extinctions in the fossil record at different sites were exactly synchronous (i.e., they happened within a few tens of years of one another) or were more approximately synchronous (within a million or two years of one another).

Determining the completeness of the fossil record can be tackled in two stages. In a series of fossils from successive deposits at a site, geologists first look for any evidence of a break, or non-uniformity, in the sedimentary deposition. For instance, a change from one form of rock to another might suggest that the earlier rocks were laid down under different conditions from the later; a hiatus also might have appeared between them. It would be a reasonable hypothesis that changes in the form, or composition, of fossils between the different kinds of rock reflect changes in fossilizing conditions, and a possible gap in the record between the two rocks, rather than continuous evolutionary change. Within an apparently uniform sedimentary rock, the apparent events in the record are more likely to be real.

Even a fairly uniform rock will not contain deposits from every year when it was laid down. Gaps, of shorter or longer duration, are typically found. We can estimate how complete the record is if we have absolute dates for the top and bottom of the sequence. The rate of deposition of sediments in various environments, such as the ocean bottom or river estuaries, can be measured over a human time scale. We can then extrapolate the short-term rate over the time of the fossiliferous sedimentary sequence to estimate how complete it is:

$$\text{completeness} = \frac{\text{observed thickness}}{\text{short-term rate of deposition} \times \text{timespan of rock}}$$

Completeness is then a number between 0 and 1. If sediments were laid down continuously at the short-term rate, the rock sequence would have a completeness of 1. In so far as there were periods without deposition, completeness will be reduced below 1. The estimates must also be corrected for compaction.

For various studies that have been used to test between punctuated equilibrium and phyletic gradualism (chapter 20), Schindel has estimated the completeness of the record. His results appear in Table 19.2. Schindel used a compaction factor of 2 in his estimates (i.e., he assumed that compaction had reduced the thickness of the rock to one-half of its expected thickness if sediments had simply piled up at the short-term rate). The choice of a compaction factor is a matter of some uncertainty, however, and the estimates should, therefore, be treated as uncertain. The figures show how the completeness of fossil records can be low, even though these samples represent some of the most complete records available; several are less than 10%. Also, the completenesses of older fossil sequences tend to be lower. This background information will be useful when we use the actual fossil records to test theories about the pattern of evolutionary rates.

The completeness figures in Table 19.2 are not calculated exactly from the formula above. Rather, they are completenesses for a time interval of 100 years. Completenesses are usually measured relative to some unit time interval, and

TABLE 19.2

Completeness of some important fossil sequences. Simplified, with permission, from Schindel 1982. Copyright 1982 MacMillan Magazines Limited.

Subject	Authority	Sediment Thickness (m)	Temporal Scope (M yr)	% Stratigraphic Completeness
Eocene mammals	Gingerich	520	3.5	28
Permian forams	Ozawa	200	12	4
Neogene radiolarians	Kellogg	5	1.9	2
Neogene forams	Malmgren and Kennett	200	8.3	23
Jurassic ammonites	Raup and Crick	14	1+	3
Pennsylvanian snails	Schindel	1300	27	34
Plio-pleistocene molluscs (1)*	Williamson	265	0.4	45
(2)*		110	0.1	73

*(1) Entire Koobi Fora formation, (2) Lower Member, Koobi Fora formation

the size of the interval can be chosen to be appropriate for the problem under study. Specifying a time interval can be convenient for the following reason. The rate at which sediments are deposited fluctuates through time according to various geological factors. In a simplified example, sediments might potentially be deposited at a site for 100 years, followed by a 900-year hiatus; sediments might then be deposited for another 100 years, followed by another 900-year hiatus; and so on (Figure 19.4). Completeness as defined by the formula above will be 10%. Alternatively, if we define completeness relative to units of time, we could ask what percentage of 1000-year time intervals are represented in the record. The answer to this question is 100%: every 1000-year unit contains some representation in the record. If the time unit is 100 years, then only 10% of the time units are represented. That is, only 1 in 10 of all 100-year time intervals left deposits at the site. The completeness can be expressed relative to any time unit, and in this example it will be 100% for all units of 1000 years or longer and 10% for all units of 100 years of less.

Completenesses expressed relative to a certain time unit are useful figures when thinking about particular pieces of evidence. The best time unit will depend on the question being asked. If we are examining speciation over one million years, then we can obtain a very complete answer if the record contains samples from most 10,000-year intervals; if the speciation event is more rapid, we should

Figure 19.4 Completeness is expressed relative to a unit time interval. (a) and (b) Sediments are deposited for 100 of every 1000 years. (c) Completeness is 100% for 1000-year time units, but 10% for 100-year units.

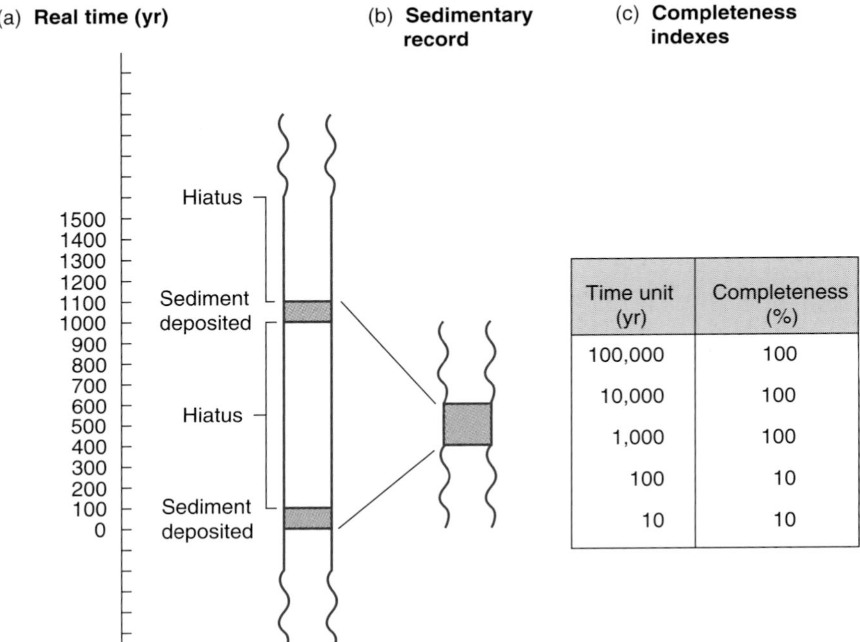

ask about the completeness for shorter time intervals. Provided that the record is complete in the sense of providing enough samples for each interval, it does not matter if sediments were deposited over only a small proportion of the total time. If sediments are deposited, as in the example, only 100 years of every 1000 years, the record can still be highly complete for a study of speciation. In such a case, the record provides the geological analog of a series of snapshots of the species as it evolves through time.

Although we have discussed completeness only in temporal terms so far, we have other ways to judge the completeness of the fossil record. We can measure, for instance, the geographical completeness of the record—that is, the percentage of the geographic range of a species that is represented by fossils. Completeness for single species is appropriate for evolutionary research, such as speciation study, that is concerned with single species. On the other hand, the fossil record is also used to study the grand pattern of change in the number of species through time. In this instance, it is interesting to know the proportion of species from the past that are represented in the fossil record, and how the proportion changes with time during the past 600 million years. Various estimates have been made of this proportion, and some of them are strikingly high. Valentine, for instance, estimated that some fossil record exists for 12% of all skeletonized species from the Cambrian to the present, and for 85% of skeletonized species from the Recent epoch. Although the fossil record is biased and incomplete, it can provide a large amount of evidence, and theories that are formulated in appropriate terms can be tested with convincing rigor.

SUMMARY

1. Fossils are formed when the remains of an organism are preserved in the sediment deposited at the bottom of the water column; the sediment may then form a sedimentary rock by compaction over time. If the sedimentary rock is later exposed at the surface of the earth, the fossils can be removed from it.
2. The history of the Earth is divided into a series of time stages. Most fossils are from organisms that lived in the past 600 million years. This time period is divided into three eras (Paleozoic, Mesozoic, and Cenozoic); the eras, in turn, are divided into successively into periods and epochs.
3. Rock ages can be measured absolutely using their radioisotopic composition, and relatively by correlating their fossil content with other rocks elsewhere. Magnetic time zones also provide useful chronological evidence.
4. The fossil record provides evidence of the history of life. Life originated more than 3.5 billion years ago. Prokaryotic cells already existed 3.5 billion years ago; eukaryotic cells originated maybe 1.8 billion years ago. At roughly the same time, photosynthesis expanded, and caused an increase in the oxygen concentration of the Earth's atmosphere. Multicellular life forms appear first, in soft-bodied animals, in the Ediacarian faunas of 670–550 million years ago; hard-bodied animals evolved about 550 million years ago and their appearance led to a much richer fossil record, starting with the Cambrian. Terrestrial ecosystems were colonized from the Ordovician onward, and the fossil record provides good evidence concerning the evolution of reptiles, and the origin of birds and of mammals from reptilian ancestors.

5. The completeness of the fossil record can be studied, at a gross level, by the geological continuity of the rock sequence, and, at a finer level, by estimating the percentage of the total time range represented by the rocks during which sediments were actually being deposited.

FURTHER READING

There are many books about the fossil record. For example, see texts and introductory books by Clarkson (1992), Cvancara (1990), Fortey (1991), Lane (1992), Simpson (1983), and Smith (1994); also the relevant chapter in Hoffman (1989). Donovan (ed, 1991, 1994) contains more advanced papers about fossilization and trace fossils. For British (and anglophile) readers, Fortey (1993) is a delightful popular introduction. On fossil evidence of early life see the three papers (by Schopf, Knoll, and Valentine) in *Proceedings of the National Academy of Sciences USA* vol. 91 (1994), pp. 6735–6757; Schopf and Klein (1992) contains many papers about early life and early environments. Glaessner (1984) is about the Ediacarian fauna; Forey and Janvier (1994) and Long (1995) about early vertebrates; Shear (1991) about the colonization of land in general; Graham (1993) about the first terrestrial plants; Ahlberg and Milner (1994) about early tetrapods (among which there have been important new findings). See Gould (1989) for the Burgess Shale; many of his popular essays, anthologized in Gould (1977b, 1980a, 1983a, 1985, 1991, 1993), are paleobiological. Benton (ed, 1993) provides authoritative summaries of the fossil records of individual taxa. Maynard Smith and Szathmáry (1995) discuss the main events in the history of life, but mainly non-paleontologically; see them for the origin of life. On completeness, McKinney (1991) and Benton (1994) are recent introductions that explain two approaches to the problem.

STUDY AND REVIEW QUESTIONS

1. Review (a) the events leading to fossilization, and (b) the divisions of geological time.

2. Fill in the age of a fossil that contains the following ratios of ^{14}C to ^{14}N. Assume all ^{14}N has been formed by decay of ^{14}C.

	$^{14}C : {}^{14}N$	Age
(a)	1 : 1	
(b)	2 : 1	
(c)	1 : 2	

3. A sedimentary rock is 10 m thick and was formed over a period of 100,000 years. Suppose the sediments were deposited at the rate of (a) 0.001 m per year for 1000 years followed by 9000 years of non-deposition, repeated 10 times, and (b) 0.005 m per year for 200 years followed by 9800 years of non-deposition, repeated 10 times. Calculate the completenesses for the following time units:

Time unit (yr)	Completeness (a)	Completeness (b)
100,000		
10,000		
1,000		
100		

Rates of Evolution

EVOLUTIONARY rates can be measured quantitatively for a character within a lineage, and this chapter begins by demonstrating how this measurement is conventionally made. We then look at a large compilation of more than 500 such measurements and ask whether fossil evolutionary rates conform with the theory of population genetics. Punctuated equilibrium is an influential modern idea about evolutionary rates in fossils, and we discuss how to test this theory, evidence for and against it, and its conceptual relation with the modern synthesis. We finish with two other measures of evolutionary rates: rates of change in arbitrarily coded character states—which we illustrate by a classic study of a living fossil (the lungfish)—and taxonomic rates, which are obtained from survivorship curves for fossil taxa.

20.1 How to measure the rate of evolution of horse teeth

The 6–8 modern species of the horse family (Equidae) are the modern descendants of a well-known evolutionary lineage in the fossil record. The record extends back through forms such as *Merychippus* and *Mesohippus* to *Hyracotherium*, which lived 55 million years ago and was once called *Eohippus*. Horses have characteristic teeth, adapted to grind up plant material, and fossilized teeth provide the main evidence used to trace the history of horses. Early members of the lineage were smaller on average than later forms, as Figure 20.1a illustrates. The Eocene ancestors of modern horses were about the size of a dog, and the smallest was the size of a cat. Their teeth were smaller as well, and had different shapes from modern horses. Because the ancestor-descendant relations of the equid species are known with reasonable, if imperfect, confidence, it is possible to estimate the rate of evolutionary change in the teeth by direct measurement.

Horse teeth are classic subject matter in the study of evolutionary rates, and the most comprehensive modern work on them is by MacFadden. He measured four properties of 408 tooth specimens, from 26 inferred ancestor-descendant pairs of species (Figure 20.1b–d). The measure of rate used by MacFadden, and many other paleontologists, was first suggested by Haldane (1949b). Suppose that a character has been measured at two times, t_1 and t_2; t_1 and t_2 are expressed as times before the present, in millions of years. Thus, t_1 might be 15.2 million years ago and t_2 might be 14.2 million years ago (t_2 is the more recent sample and has a shorter time to the present). The time interval between the two samples can be written as $\Delta t = t_1 - t_2$, which is 1 million years if $t_1 = 15.2$ and $t_2 = 14.2$. The average value of the character is defined as x_1 in the earlier sample

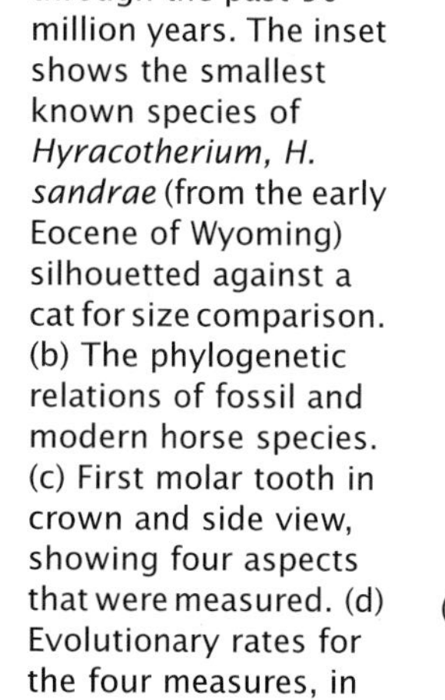

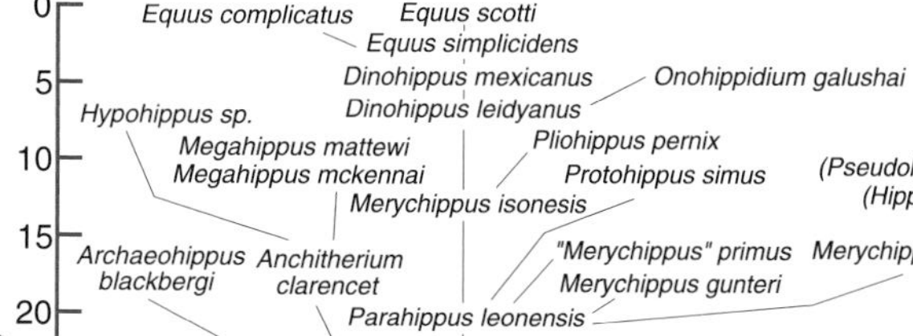

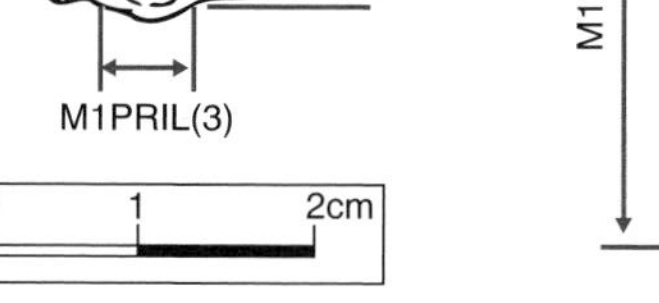

(d)

Species pair	Δt (Myr)	M1APL(1) (d)	M1TRNW(2) (d)	M1PRTL(3) (d)	M1MSTHT(4) (d)
Equus simplicidens-Equus complicatus	2.0	0.000	−0.014	0.115	0.054
Parahippus leonensis-Merychippus primus	3.0	−0.009	−0.026	−0.012	0.154
Parahippus leonensis-Protohippus simus	6.0	0.051	0.034	0.073	0.247
Miohippus quartus-Anchitherium clarencei	7.0	0.065	0.057	0.049	0.046
Mesohippus bairdi-Mesohippus barbouri	1.0	0.157	0.146	0.136	0.050
Equus simplicidens-Equus scotti	2.0	0.040	0.005	0.162	0.097
Dinohippus mexicanus-Equus simplicidens	2.0	0.064	0.081	0.088	0.142
Dinohippus leidyanus-Dinohippus mexicanus	2.0	−0.004	−0.026	0.074	−0.054
Dinohippus leidyanus-Onohippidium galushai	2.0	0.033	0.027	0.030	-0.074
Merychippus isonesis-Pliohippus pernix	2.5	0.072	0.101	0.185	0.180
Megahippus mckennai-Megahippus matthewi	2.0	0.036	0.022	0.064	0.219
Anchitherium clarencei-Megahippus mckennai	4.0	0.083	0.096	0.094	0.109
Anchitherium clarencei-Hypohippus large sp.	8.0	0.062	0.072	0.077	0.077
Parahippus leonensis-Merychippus insignis	3.0	0.053	0.011	0.038	0.228
Parahippus leonensis-Merychippus isonensis	3.0	0.030	0.014	0.028	0.339
Parahippus leonensis-Merychippus gunteri	2.0	−0.023	−0.107	−0.116	0.172
Parahippus tyleri-Parahippus leonensis	2.0	−0.062	−0.039	−0.179	0.088
Miohippus quartus-Archaeohippus blackbergi	7.0	−0.017	−0.018	−0.029	−0.021
Miohippus quartus-Parahippus tyleri	5.0	0.067	0.052	0.066	0.091
Mesohippus bairdii-Miohippus quartus	6.0	0.022	0.018	0.012	0.022
Epihippus gracilis-Mesohippus bairdii	14.0	0.023	0.037	0.032	0.036
Orohippus pumulis-Epihippus uitensis	4.5	0.047	0.052	0.042	−0.007
Orohippuspumulis-Epihippus gracilis	4.5	0.026	−0.010	0.006	0.020
Hyracotherium vaccassiense-Orohippus pumulis	2.5	0.011	0.029	0.012	0.068
Hyracotherium angustidens-Hyracotherium vaccassiense	3.0	0.010	0.026	−0.020	0.000
Hyracotherium angustidens-Hyracotherium tapirinum	3.0	0.101	0.106	0.091	0.096
Mean species pair evolutionary rate		0.045	0.047	0.0690	0.104

Figure 20.1 (a) Modern horses are descended from a group containing a number of lineages, and have increased in average body size through the past 50 million years. The inset shows the smallest known species of *Hyracotherium, H. sandrae* (from the early Eocene of Wyoming) silhouetted against a cat for size comparison. (b) The phylogenetic relations of fossil and modern horse species. (c) First molar tooth in crown and side view, showing four aspects that were measured. (d) Evolutionary rates for the four measures, in 26 inferred ancestor-descendant species pairs, expressed in darwins (d). It is not important to study the numbers in detail! They are meant only to illustrate the results that come out of a study of evolutionary rates. Reprinted, by permission of the publisher, from MacFadden (1992).

and x_2 in the later sample; we then take natural logarithms of x_1 and x_2 (the natural logarithm is the log to base e where $e \approx 2.718$, and it is symbolized by log or ln). The evolutionary rate (r) then is

$$\frac{\ln x_2 - \ln x_1}{\Delta t}$$

The rate is positive if the character is evolutionarily increasing and negative if it is decreasing. For many purposes, however, the absolute rate of change, independent of the sign, is the important factor. Haldane defined a "darwin" as a unit to measure evolutionary rates; one darwin is a change in the character by a factor of e ($e \approx 2.718$) in one million years. The formula above for r gives the rate in darwins provided that the time interval is in millions of years. If, for example, $x_1 = 1$, $x_2 = 2.718$, and $t = 10$ million years, then $r = 0.1$ darwins.

The reason for transforming the measurements logarithmically is to remove spurious scaling effects. If logarithms were not taken, the rate of evolution of a character would appear to speed up when it became larger even if its proportional rate of change remained constant. With logarithmically transformed measurements, it is also possible to compare rates of change in species of very different size, such as mice and elephants. Many people think intuitively about changes in terms of percentages rather than natural logarithms; thus, a change of 10% appears meaningful, a change of 0.1 natural logarithmic units less so. For short enough time intervals, an almost linear relation exists between time and percent change.

We can examine this relation by imagining a lineage that is evolving at 1 darwin. For times up to about 1000 years, the percent change will be roughly constant per year; after 1000 years the lineage will have changed by close to 0.1%, and by about 1/1000th of that amount (that is, 0.0001%) every year up to then. The nonlinear relation then becomes more influential. If the lineage continued to change by the same increment per year up to 1 million years, it would have increased by 100%, but at a rate of 1 darwin it will really have increased by 272%. (The best way to familiarize yourself with the meaning of darwins is to calculate a few; see the study questions at the end of this chapter.)

The 26 ancestor-descendant species pairs and 4 dental characters measured by MacFadden produced $4 \times 26 = 104$ estimates of evolutionary rates (Figure 20.1d). The different tooth characters show different patterns, with height (M1MSTHT in the figure), for instance, evolving rapidly between *Parahippus* and *Merychippus*, while the other characters were evolving at normal rates. The detailed pattern of the numbers is not important here; they could perhaps be interpreted in terms of the grinding functions of the teeth and the diets eaten by individual horse species. The approximate absolute values of the rates are worth bearing in mind. The values in Figure 20.1 are mainly about 0.05–0.1 darwins, or about a 15–30% change per million years. They are mainly positive, indicating that the lineage was on average increasing in size. Negative values also appear, as the horses in the lineage evolutionarily shrunk as well as expanded.

The values in Figure 20.1 are averages for a lineage connecting an ancestral-descendant species pair, and do not imply that evolution had a constant rate of 0.05 darwins every million years throughout that time. An average is not a constant, and the rates for short periods may have been very different from the

TABLE 20.1

Gingerich's summary of evolutionary rates. The summary is large but not complete; it is based on 521 different measurements. Gingerich divided the measurements into four classes. The importance of the column for time intervals will become apparent in section 20.3. Reprinted with permission from Gingerich (1983). Copyright 1983 American Association for the Advancement of Science.

		Evolutionary Rate (d)		Time Interval	
Domain	Sample Size	Range	Geometric Mean	Range	Geometric Mean
I Selection experiments	8	12,000–200,000	58700	1.5–10 yr	3.7 yr
II Colonization	104	0–79,700	370	70–300 yr	170 yr
III Post-Pleistocene Mammalia	46	0.11–32.0	3.7	1000–10,000 yr	8200 yr
IV Fossil Invertebrata and Vertebrata	363	0–26.2	0.08	8000 yr–350 Myr	3.8 Myr
Fossil Invertebrata alone	135	0–3.7	0.07	0.3–350 Myr	7.9 Myr
Fossil Vertebrata alone	228	0–26.2	0.08	8000 yr–98 Myr	1.6 Myr
I to IV combined	521	0–200,000	0.73	1.5 yr–350 Myr	0.2 Myr

d, darwins; yr, years; Myr, million years ago.

long-term average. As average figures, the values in Figure 20.1 are fairly typical for the fossil record, being neither exceptionally fast nor slow. We shall see (Table 20.1) that the average figure for a large set of evolutionary rates in vertebrates is about 0.08, and rapidly evolving vertebrates show rates more like 1–10 darwins over short periods. Simpson, who did more than anyone to stimulate the study of fossil evolutionary rates, noticed that rates vary between taxa, characters, and times, and he invented the terms bradytelic, horotelic, and tachytelic to describe slow, typical, and rapid evolution, respectively; horse evolution as such is horotelic.

20.2 *How do population genetic and fossil evolutionary rates compare?*

Rates of evolution in the fossil record have been measured for many characters, in many species, at many different geological times. A compilation by Gingerich in 1983 included 521 different estimates, of which 409 were for the fossil record. The estimates vary between 0 and 39 darwins in fossil lineages. The main problem of evolutionary rates is to understand why they differ between times and taxa in the patterns observed.

Before tackling that problem, we can ask a more general question. Are the rates of change seen in the fossil record consistent with the mechanisms of evolutionary change studied by population geneticists? Population genetics identifies two main mechanisms of evolution, natural selection and random drift, although it is questionable whether drift is important in morphological evolution (section 7.9, p. 170). For changes like those of tooth size in the history of horses, we cannot confirm directly that selection was the cause. We would have to show that larger-toothed horses gave rise to larger-toothed offspring than average (i.e., the character was inherited) and produced more offspring than average (i.e., selection favored it). That kind of study is usually impossible with fossils. We can, however, at least find out whether the results of research in the two areas are consistent. First, we ask whether any contradiction appears between the rates of evolution observed in population genetic work, such as artificial selection experiments, and those observed in fossils. If, for example, the fossil rates are significantly higher, it would suggest that selection alone cannot be the only cause of evolution; some other, more rapid factor would be needed. In fact, the rates of evolution in artificial selection experiments are far higher than those measured in fossils. Evolution under artificial selection has proceeded about five orders of magnitude faster than in the fossil record (Table 20.1). We can conclude that the known mechanisms of population genetics can comfortably accommodate the fossil observations.

Strictly speaking, this finding does not confirm that the fossil changes were driven by selection and (perhaps) drift. It does show, however, that the observations are consistent. For this reason, and because no other mechanisms of evolution are known, no one seriously doubts that the microevolutionary processes discussed in chapters 4–10—even if operating indirectly (tooth sizes might increase because of selection for larger body size, for instance)—are fundamentally responsible for all evolution in the history of life. As we shall see (section 20.6.6), it is in some ways debatable whether macroevolution can be reduced to microevolution, but those debates are not concerned with the most fundamental question of whether selection, mutation, and drift cause evolution.

20.3 *Why do evolutionary rates vary?*

Paleobiologists have studied a number of generalizations about evolutionary rates. For example, it has been suggested that species usually change more rapidly during, rather than between, speciation events; that structurally more complex forms evolve faster than simpler forms; and that some taxonomic groups evolve more rapidly than others—that mammals, for instance, evolve faster than molluscs (this old idea was one of Lyell's favorite generalizations). We shall examine the first of these issues in more detail later. Before addressing this suggestion, let us return to Gingerich's compilation of evolutionary rates and consider a general point about the study of evolutionary rates.

Gingerich observed, in his compilation of evolutionary rates, an inverse relation between the rate and the time interval over which it was measured. The observed cases of rapid evolution have tended to be for shorter intervals than the cases of slower evolution (Figure 20.2). The relation is unlikely to be due to any strong force in the evolutionary process itself. Nothing in evolutionary theory constrains rapid evolution, at these speeds, to take place in short intervals and slower evolution in longer intervals. At the molecular level, by way of comparison, the rates of evolution seem to be fairly constant over all time periods (e.g., Figure

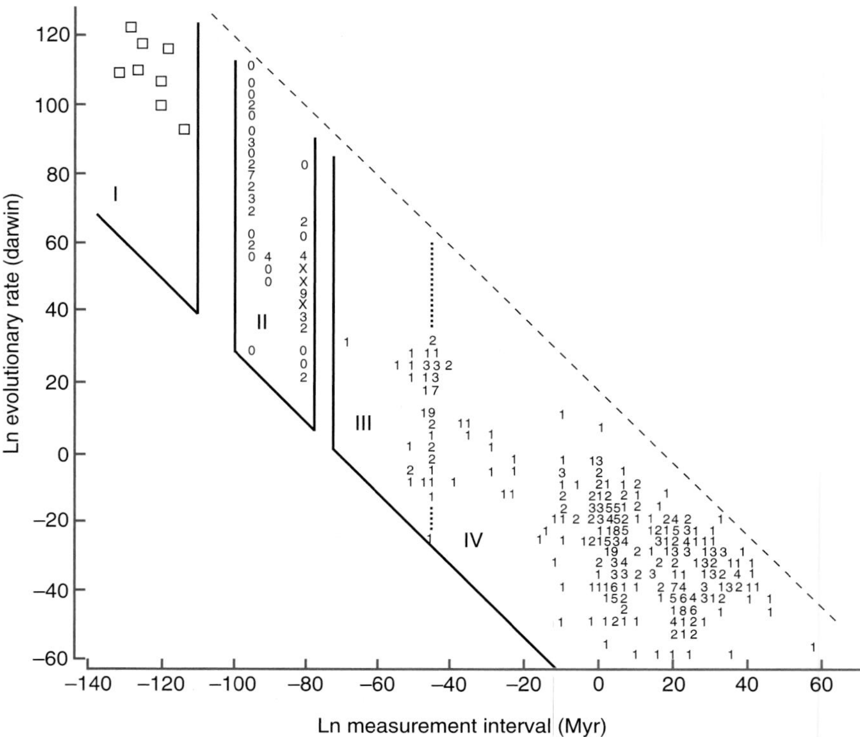

Figure 20.2 The relation between estimated logarithmically transformed evolutionary rate and the time interval used, for the 521 studies summarized in Table 20.2. The relation is negative. For the meaning of samples I, II, III, and IV see the caption to Table 20.1. The digits higher than 1 on the graph mean that number of cases fell on that spot (*x* for numbers higher than 9). Reprinted with permission from Gingerich (1983). Copyright 1983 American Association for the Advancement of Science.

7.7, p. 170). Gingerich's favored interpretation can be explained by reference to the evolutionary pattern we saw in Darwin's finches (section 9.1, p. 222).

The finches' beaks evolved to be larger in times of food shortage and smaller in times of abundance; the food supply fluctuated through time according to the weather, particularly the periodic El Niño disturbance. Imagine measuring the rate of evolution both within one of these cycles and over the cycle as a whole (Figure 20.3a). If the direction of evolution fluctuates, the rate of evolution measured over a short interval is inevitably higher than the rate measured over a longer time interval because the short-term changes cancel out. The pattern in Figure 20.3a is simplified, giving zero net change over a cycle. If any fluctuations in evolutionary direction occurred (Figure 20.3b), the rate measured over a shorter interval will be higher. Probably almost all evolutionary lineages show some reversals in the direction of change, and the pattern illustrated by Darwin's finches may be quite common. Hence, we have the general relation in Gingerich's compilation (Figure 20.2).

Other factors may contribute to this relation as well. For instance, the cases of rapid evolution over short time intervals are for artificial selection experiments (data set I) and natural ecological colonizations (data set II); these data sets could potentially be extraordinary events and have higher than average selection intensities. (Alternatively, it might be argued that the rates are high only because the measurement interval is short enough to catch evolution in its unidirectional

Figure 20.3 The inverse relation between measured evolutionary rate and time interval (Figure 20.2) will be found if the direction of evolution fluctuates through time. (a) Simplified cycle of evolutionary change; the rate of change measured for short time units is higher than for the cycle as a whole. No net change occurs over the cycle and the rate of evolution is zero. The numbers under "measurement intervals" in the table refer to the time intervals in the *x*-axis of the graph (the arbitrary beak size units can be thought of as logarithmic, to make the rates comparable with the formula for calculating rates in section 20.1). (b) With a more realistic pattern of evolution, the inverse relation between rate and measurement interval will still be found to some extent if any fluctuations occur in the direction of change.

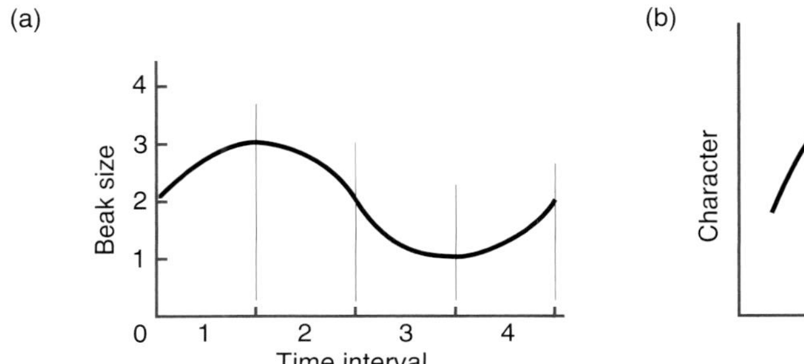

Measurement interval		Rate of evolution
1 unit	Within 1 or 2 or 3 or 4	1
2 units	1 + 2 2 + 3 3 + 4	0 1 } av. 1/3 0
3 units	1 + 2 + 3 2 + 3 + 4	1/3 1/3
4 units	1 + 2 + 3 + 4	0

phase, and not because the intensity of selection is peculiar. Opinions differ about how representative the selection intensities in data sets I and II are of those in the lineages making up data sets III and IV.)

This interpretation, if it is correct, matters for some kinds of generalizations about evolutionary rates, but not others. It does not invalidate the measurements themselves. In the 14 million years between *Epihippus gracilis* and *Mesohippus bairdii*, horse teeth undoubtedly evolved at a rate of 0.023–0.036 darwins (Figure 20.1). All questions about individual measurements, and comparisons between them, remain valid. It is when dealing with the more general patterns that Gingerich's result should make us suspicious. The generalization that mammals evolve faster than molluscs, for example, is reflected in Gingerich's data. He

have not, been made by Eldredge, Gould, and their followers. Here we shall discuss the independent ideas separately, concentratating on four such claims.

20.6.1 *Punctuated evolution may be due to allopatric speciation, or to other speciation mechanisms, such as the break-down of homeostatic constraints, or macromutation*

The original model of punctuations attributed the rapid change at speciation to allopatric events involving peripheral isolates. A number of more or less well worked-out suggestions have been made about how evolution might happen in a small peripherally isolated population. One possibility is that evolution might proceed as normal, under selection and drift, thereby generating the punctuated equilibrial pattern.

Another possibility is that the peripherally isolated population may evolve in an abnormal manner, because it may be living under abnormal conditions. What might happen in that instance? One hypothesis draws on the way stabilizing selection—in the form of canalizing selection in this case—favors the evolution of homeostatic mechanisms (section 13.9, p. 353). Under extreme conditions, outside the normal experience of the species, these mechanisms may break down and an unusual array of phenotypes may then be expressed. The generation of new phenotypes alone has no straightforward, permanent evolutionary effect; new genotypes would be needed to create such an effect. Perhaps genetic variation that was previously disguised by homeostatic mechanisms could be expressed in the new conditions. The course of selection could then be altered. Moreover, strange genotypic effects can arise under abnormal conditions. Recombination rates, for instance, can increase under stress (Figure 20.8).

The break-down of homeostatic constraints is not the only unorthodox hypothesis about punctuated equilibrium. Gould revived the idea, particularly associated with Goldschmidt, that major evolutionary breakthroughs could occur by macromutation. Macromutations may tend to arise in genes controlling regulatory, or developmental, processes rather than in structural genes, and new species and major groups might (according to the neo-Goldschmidtian view)

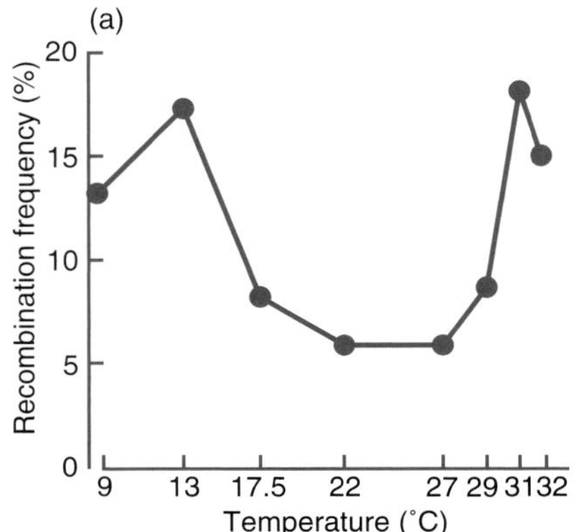

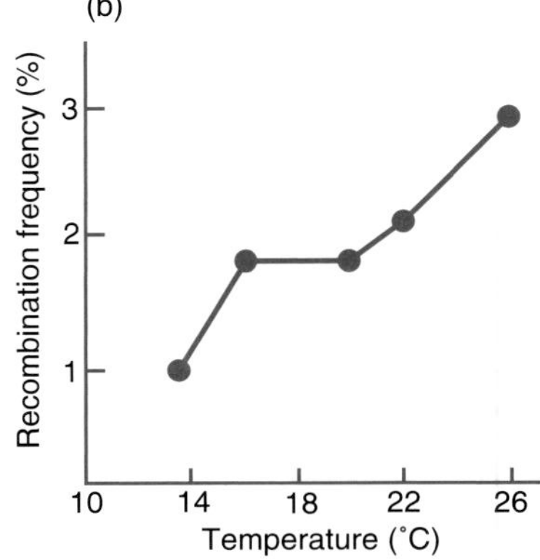

Figure 20.8 Organisms can be subjected to stress by keeping them at abnormal temperatures. This figure includes two examples of how recombination rate increases at stressful, unnatural temperatures: (a) recombination rate in the purple-black region of chromosome 2 in the fruitfly *Drosophila melanogaster*, (b) recombination rate between *dpy-5* and *unc-15* in the nematode worm *Caenorhabditis elegans*.

arise by regulatory macromutation, which is a different type of evolution from the natural selection of small mutations. Macroevolution would then proceed by non-adaptive mechanisms, and adaptive evolution would be confined to minor adjustments within populations. (We will return to this topic in chapter 21.)

These ideas are simply hypotheses at this point, but we should be aware of them. The main conjecture here is that natural selection and ordinary variation might produce only small-scale, microevolutionary change. Major changes might be achieved in the revolutionary crucible of small isolated populations. Only there might the new genotypes of the future be forged.

20.6.2 *Does rapid change always take place at speciation?*

Eldredge and Gould not only claimed that evolution proceeds in alternating moments of rapid change and stasis, but they also suggested that the rapid evolution would be accompanied by speciation. That is, when evolution is rapid, the lineage splits. For much, though not all, real fossil evidence, this process is virtually a taxonomic tautology. Few lineages will be preserved so completely that all intermediates survive, and whenever rapid change occurs a new species will usually be named whether or not the lineage has split. Even if the record is complete, the possibility that a sudden change is due to immigration must be ruled out. (Sheldon's trilobites are a possible exception. If fossils only from the beginning and end of one of the lineages had been found, they would probably have been put in different species.) What reason do we have to think that rapid evolution will always cause, or only be caused by, a split in the lineage?

Logically, evolutionary changes do not have to be concentrated in the times when a lineage splits. Eldredge and Gould, however, gave a more biological argument to suggest that such changes often will fit this pattern. Evolutionary change, they suggested, does not happen in whole populations; it happens in geographically isolated sub-populations. As the isolated subpopulation changes, it will split from the unchanged ancestral rump population because the two will now possess different phenotypes.

Is it true that large evolutionary changes cannot happen in whole, rather than isolated, populations? Examples of evolutionary transitions in large, central populations are known (chapter 5), and Turner has pointed to cases of rapid change, without splitting, in butterfly species with mimetic polymorphism. Perhaps it could be argued that these cases are exceptional. It is a safe general statement that we do not know the relative frequency of rapid evolution in small isolated sub-populations and in large, complete populations. As a result, it would be premature to draw any conclusion ruling out one or the other. Evolution in large populations is reasonable in theory and has been observed in fact.

Real speciation (in which the ancestral species splits) and phyletic speciation (in which a lineage changes enough to be given a new name but has not split) are usually indistinguishable in the fossil record. Eldredge and Gould's idea that rapid evolution is confined to the times of lineage splits is, nevertheless, an interesting hypothesis. It has some theoretical appeal, although as a limited rather than absolute generalization (because exceptions are known). Decisive evidence for or against it remains to be found.

20.6.3 *Is stasis due to stabilizing selection or constraints?*

In the fossil record, stasis—evolutionary lineages that persist for long periods without change—is common. Stasis is not a new discovery, but it has taken on a new importance since Eldredge and Gould's paper. Two main ideas—*stabilizing selection* and *constraint*—have been put forward to explain it. We originally encountered this pair of ideas in section 13.9 (p. 353).

Constraint means that species do not change because they lack the necessary genetic variation. It has been suggested that in the sorts of peripherally isolated populations where speciation may take place, latent, or new, genetic variation is expressed. Genetic constraint is then lifted and evolutionary change becomes possible.

We can thus distinguish two extreme theories to explain stasis and rapid speciation. On the one hand, species may remain constant because they are under stabilizing selection and then change rapidly when new selective conditions arise. On the other hand, the species may stay constant while they are constrained and change rapidly when those constraints break down under exceptional conditions. In the real world, any mixture of these processes could operate.

Stabilizing selection does not have to mean that the environment remains constant. In section 18.4 (p. 513), we saw how beetle populations migrate south and north as ice ages come and go. The beetles remain constant in form, but move about as the conditions change. Thus, the beetle population as a geographic whole could be experiencing stabilizing selection even though the environment at any one place is not constant.

The question of whether evolutionary constancy is caused by stabilizing selection or constraint is a particular form of the general question of what causes phenotypic patterns in nature, and the same general points can be made about the answer. The hypothesis that a character is not changing because it has no genetic variation can be tested. It predicts that the character's heritability is zero. The heritability of fossil characters cannot be measured, unlike that of similar characters in modern species. For example, a much-discussed possible case of punctuated equilibrium was found by Williamson in the Pleistocene snails of Lake Turkana, now in Kenya. Williamson suggested that these snails did not change for long periods because of some developmental constraint. Various population geneticists replied that heritabilities have been measured in snail characters like those studied by Williamson, and significantly positive values found. (The counter-argument, that Pleistocene snails had different genetics from their modern descendants, is desperate.) Stabilizing selection is well documented, whereas a lack of genetic variation has never been shown to explain why any character remains unchanged. Most characters show significant heritability; those that do not are puzzling, because the nature of the constraint is not understood.

The evidence of geographic variation provides another reason to doubt the importance of constraint. At any one locality, species show a limited range of variability, but they vary markedly from place to place. The existence of ring species shows that geographic variation within a species can be large enough to produce speciation (section 16.3, p. 426). Williamson, however, denied this argument. He suggested that geographic variation is limited, and inadequate to account for speciation, and he attributed the limitation to developmental constraint. Williamson did not support his case with appropriate evidence, however. Variation within a species must, as a necessary consequence of taxonomy, be narrower than the variation of a group of species. The case where populations of a species are arranged geographically in a ring provides the only clear test case for Williamson's claim. The fact that these rings lead to speciation in all known cases strongly suggests that ordinary geographical variation alone can explain speciation.

In conclusion, debate continues over why species can persist in the fossil record for long periods of time without changing. Stabilizing selection is a well-documented and plausible explanation. Constraint is an undocumented but still possible explanation that should not be ruled out before investigation begins.

20.7 *Darwin was not a phyletic gradualist*

We have now dealt with the main scientific questions in the punctuated equilibrium controversy. The full controversy also includes two historical issues. Gould has claimed that the theory of punctuated equilibrium contradicts both Darwin's own ideas and also those of neo-Darwinism. Darwin certainly did emphasize repeatedly that evolution is slow and gradual; Gould thought that Darwin was advocating the theory of phyletic gradualism instead of punctuated evolution. It is more likely, as Dawkins has argued, that Darwin meant something crucially different by gradual evolution. Darwin did not make his remarks about gradualism in the context of discussing evolutionary rates at and between speciation events. When he did discuss that subject, he said things that sound quite like punctuated equilibrium, such as (from *On the Origin of Species*):

> Many species once formed never undergo any further change. . . ; and the periods during which species have undergone modification, though long as measured by years, have probably been short in comparison with the periods during which they retained the same form.

Darwin's theory was not gradualist in the sense of demanding similar evolutionary rates at, and between, speciation events. Rather, his theory is strongly gradualist concerning the evolution of adaptations, particularly of complex adaptations. Adaptations cannot evolve in sudden jumps (section 13.3, p. 341). If they are to evolve at all, it must take place in a large number of small stages; this process will, therefore, be slow and gradual. Darwin's theory is gradualist about the evolution of complex adaptations, not about the pattern of evolutionary rates. Darwin's own statements consistently suggest that he was talking about adaptation, rather than speciation, when he described evolution as slow and gradual. Both the structure of Darwin's theory and the literary evidence from Darwin's writings count against Gould's claim. Darwin was not a phyletic gradualist—but he did believe that adaptations are built up in many small evolutionary stages.

20.8 *The theory of punctuated equilibrium does not render the modern synthesis "effectively dead"; the relation between microevolution and macroevolution is an open question*

Eldredge and Gould in 1972 originally accepted a neo-Darwinian theory of speciation and used it to update the paleobiological study of evolutionary rates. Around 1980, Gould moved on to a more radical position, which has been actively discussed ever since. He turned the theory of punctuated equilibrium back on the modern synthesis and (he claimed) toppled that synthesis off its throne. In 1980, he described the synthetic theory of evolution as "effectively dead."

According to Gould, the modern synthesis asserted that evolution at all levels takes place by the same process: natural selection among alleles of different fitnesses. Microevolution within populations, speciation, and the origin of higher groups, are all thought to be driven by natural selection among alleles within populations. In this sense, the modern synthesis is narrow and *extrapolationist* (or *reductionist*). It is narrow because it recognizes only one cause of evolution; it is extrapolationist because it explains all evolution by extrapolating from microevolution; it is reductionist, in a philosophical sense, because it "reduces" macroevolution to microevolution.

Let us consider an example. We saw earlier (section 9.1, p. 222) how natural selection operates on the beaks of Darwin's finches. The evidence in that case suggested natural selection within a species, and it demonstrated that individuals with larger beaks are favored when the seeds and fruit that they eat are large, whereas smaller beaks are favored when the food size is smaller. In 1976–1977 (and two subsequent periods), Grant measured the strength of selection on the finches' beaks, and its evolutionary results (Figure 9.9, p. 240). Today, the 14 Galápagos finch species known mainly differ in their beak and body proportions (Figure 20.9). We can, therefore, calculate whether the kind of selection Grant observed would account for the origin of all finches in the Galápagos in the time available.

How long would it take for the process studied by Grant in 1976–1977 to convert one species of finch into another? During the 1977 drought, the beak size of *Geospiza fortis* on the island of Daphne Major increased by 4%. *Geospiza magnirostris* is a close relative of *G. fortis*; the two species differ mainly in body and beak proportions and they coexist on many islands of the Galápagos. From the average difference in beak size between *G. fortis* and *G. magnirostris* on Daphne Major, Grant estimated that 23 bouts of evolution of the 1977 type would be enough to turn *G. fortis* into *G. magnirostris*. On other islands *G. fortis* is larger than on Daphne Major and, using one of the larger *G. fortis* populations as starting point, only 12–15 such events would be needed.

How much time is available? The Galápagos archipelago are volcanic islands. The oldest rocks are 4.2 million years old, and all the islands had appeared by 1–0.5 million years ago. (Recent geological work has suggested the islands may be older, although that difference would affect the argument that follows only if the finch ancestors arrived earlier than we will assume. In such a case, the argument would be strengthened.) It has been estimated that the common ancestor of Darwin's finches arrived from South America about 570,000 years ago.

Figure 20.9 Possible phylogeny of Darwin's finches, according to Lack. The dotted lines indicate uncertainty. Other phylogenies have been suggested as well. Was there time enough for the evolution of 14 species, by selection within a population, since the Galápagos were colonized by the ancestral finch 570,000 years ago? Reprinted, by permission of the publisher, from Lack (1947).

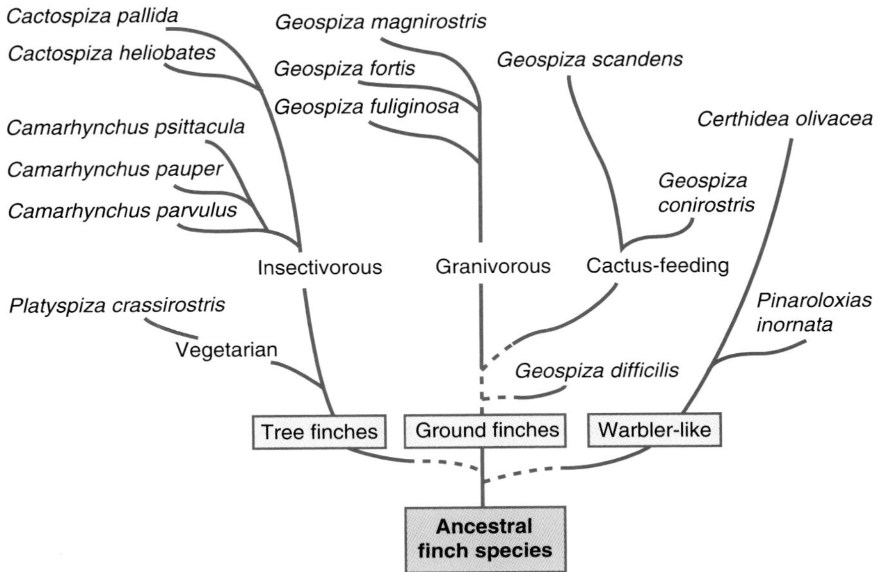

Thus, the radiation of the 14 finch species has occurred in about half a million years. Using either the low estimate of 12–15 events (where an "event" represents a bout of evolution such as occurred in 1977 on Daphne Major) or the higher estimate of 25 events, we can see that even if only one such event took place per century, the evolutionary divergence between the two species could still have been accomplished well within the time available. In fact, the real evolutionary transition probably did not happen that way. The 1982–1983 El Niño event reversed the evolution of 1977 and a steady transition from one population into another probably did not occur; instead, frequent reversals of accumulated small changes may have inspired the evolution. In any case, it is unlikely that *G. fortis* simply changed into *G. magnirostris* (or vice versa). Rather, they probably both diverged from another common ancestor.

This rough calculation is not intended to represent the exact history of the birds. Instead, it illustrates how we can extrapolate from natural selection operating within a species to explain the diversification of the finches from a single common ancestor about 570,000 years ago to the present 14 species. If the extrapolation is correct, the reason for the speciation in the finches was the same process as has been observed in the present—natural selection for changes in beak shape, which were probably due to changes in food types through time and between islands. Arguments of this kind are common in the theory of evolution. We shall meet another such issue in chapter 21, where natural selection over long periods will be used to explain the major evolutionary transition from the mammals to the reptiles. All of these arguments are extrapolationist.

The theory of punctuated equilibrium, in its more radical form, suggests that large-scale evolution is not simply microevolution extended over long periods. If speciation, for example, happens only under extraordinary, revolutionary circumstances, it would not involve a simple extrapolation of normal evolution within a population. Instead, speciation would be uncoupled from microevolution. The theory of punctuated equilibrium would then contradict the modern synthesis's "extrapolationism." Gould suggested that several evolutionary processes might operate, each with its own taxonomic level for its operations. He calls his idea *pluralist* (because it employs more than one causal process), and suggests that, once pluralism is accepted, the narrow modern synthesis would benefit from "hierarchical expansion" (as different processes would be studied at the micro- and macro-evolutionary levels).

We shall see in the next three chapters how paleobiologists have also challenged extrapolationism at higher levels than speciation. It has been suggested that the origin of major groups, and macroevolutionary trends, may be driven by non-adaptive processes (i.e., distinctive developmental reorganization) or by selection at higher levels than within a population. This line of research is both influential and substantial, but how much of a challenge does it pose to the modern synthesis?

Let us consider another interpretation of the modern synthesis. The modern synthesis, it can be argued, does not predict that evolution has only one, consistent rate. In his classic treatment of evolutionary rates in the modern synthesis, Simpson reasoned that rates may be slow, average, or fast (section 20.1). Thus, Simpson was a pluralist about evolutionary rates. In terms of the patterns in Figure 20.5, the modern synthesis arguably suggested that any combination of the patterns might be found, and not that evolution always proceeded in the form of the phyletic gradualist extreme. Seen against this background, the theory of punctuated equilibrium is novel mainly for being narrow. The modern synthesis did not predict that every speciation event involved punctuated equilibrium, but it did not deny that it could happen in an unspecified proportion of cases. The two alternative schools of thought are then not punctuated equilibrium and phyletic gradualism, but punctuated equilibrium and modern synthesis's pluralism. If punctuated equilibrium turns out to be empirically universally true, it will refute the pluralist as well as the gradualist view of rates; nevertheless, the way in which punctuated equilibrium differs from the modern synthesis is arguably not what Gould claims.

What about the extrapolationism of the modern synthesis? According to the alternative interpretation, the modern synthesis is extrapolationist only in explaining adaptations. For other kinds of (non-adaptive) evolution, we have no reason why several processes should not operate, and at various taxonomic levels. Gould's interpretation of the modern synthesis would then imply that it claimed all evolution to be adaptive. The modern synthesis never ruled out non-adaptive evolution, however. Individuals such as Cain did occasionally question its existence, which is different from ruling it out in theory. In any case, Cain's position was not considered mainstream. Concerning non-adaptive evolution, the modern synthesis was again pluralist. Nothing in the modern synthesis rules it out, and entire areas of research (like molecular evolution) are devoted to it.

TABLE 20.2
Two interpretations of the modern synthesis.

	Gould's 1980 Development of the Theory of Punctuated Equilibrium	Modern Synthesis: Gould's Interpretation	Modern Synthesis: Alternative Interpretation
Macroevolutionary changes in adaptations	pluralist, with uncoupling of micro- and macro-evolutionary levels	extrapolationist	extrapolationist
Macroevolutionary changes in all characters		extrapolationist	pluralist
Pattern of evolutionary rates	punctuated equilibrium	phyletic gradualism	pluralist

In summary, the two interpretations of the modern synthesis differ about the evolution of adaptive and non-adaptive characters and about the pattern of evolutionary rates (Table 20.2). According to one interpretation (that of Gould), the modern synthesis takes a narrow and reductionist view of all evolution and favors the phyletic gradualist theory of rates. According to the alternative view, the modern synthesis is reductionist only about adaptive characters and is pluralist both about non-adaptive evolution and about the pattern of rates.

20.9 *Evolutionary rates can be measured for non-continuous character changes, as illustrated by the evolution of "living fossil" lungfish*

The measurement of evolutionary rates in darwins is appropriate for metrical changes, such as a character evolving to be longer or shorter. For larger changes, such as from a leg to a wing, this method ceases to be useful (section 20.2). It is still possible, however, to measure rates of evolution for larger changes. The last two sections of the chapter describe two methods for taking such measurements. The first involves a famous early study of evolutionary rates: Westoll's work on lungfish.

Lungfish (Dipnoi) form one of the four main divisions of fishes (see Figure 13.4, p. 347) and are an ancient group dating back more than 300 million years; only six modern species are known. The modern forms are examples of living fossils—that is, species that have changed little from their fossil ancestors in the distant past. They should therefore show, at least recently, slow rates of evolution. Westoll investigated this question quantitatively. He distinguished 21 different characters of fossil Dipnoi; they are all skeletal characters. For each of the 21 characters, he distinguished a number of character states (like the character states discussed earlier for classification and phylogenetic inference). The 21 characters showed between three and eight different states. Character number 11, for example, was "degree of fusion of bones along supraorbital canal." Westoll distinguished five different states:

4. Irregular, more or less random fusions.
3. Tendency for fusions to be in twos, especially in some parts of canal.
2. Still stronger tendency to fusion, rarely in threes or fours in specific sections.
1. Three or four elements (K to M) generally fuse, but numerous irregularities.
0. Three or four elements (K–L_2 or K–M) always fuse.

(Letters like K and L_2 refer to particular, identifiable bones.) The highest state (4) is the most primitive condition of the character, and 3, 2, 1, and 0 are successively later, more derived states. Westoll made an analogous list of states for all 21 characters. These character states are not the sort of metrical changes for which evolutionary rates can be measured in darwins (section 20.1). Instead, the fusion of two bones into one is a discrete, not a continuous, evolutionary change.

For each fossil, Westoll calculated a total score, made up of its scores for all 21 characters. The most advanced possible lungfish, with 21 characters in the most advanced state, would have a total score of zero; the score for the most primitive possible lungfish, which had the highest scores for all 21, would have been 100. The rate of change in the score measures the rate of evolution of the group. The numbers assigned to the characters states are arbitrary, but they can still be used to portray evolutionary rates.[1] Westoll's result is shown in Figure 20.10. Dipnoi, it reveals, have not always been "living fossils." Around 300 million years ago, they were evolving rapidly, but their evolution has slowed considerably since about 250–200 million years ago, and their description as living fossils is accurate for the modern forms.

The obvious biological question is why evolutionary change almost stopped in lungfish 200 million years ago. Lungfish are not the only examples of living fossils; other examples include the brachiopod *Lingula* and the horseshoe crab *Limulus* (Figure 20.11). The supreme examples of living fossils are the Cyanobacteria (sometimes called "blue-green algae"); 3 billion-year-old fossils look much like forms living today. Many conjectures have been made about why these groups have changed so little, but no general theory has been proposed. The question is an instance of the general question of why evolutionary stasis occurs. The stability of living fossils may be due to stabilizing selection or absence of genetic variation. Some living fossil species live in relatively isolated habitats, with no apparent competitors, and if their habitats have been stable, they will have faced no pressure to change. No evidence exists that living fossils have peculiar genetic systems that might prevent evolutionary change—for instance by constraining the degree of new genetic variation. We can measure the amount of protein polymorphism in modern living fossil species by gel electrophoresis, and in *Limulus polyphemus* this measurement is not noticeably low.

The point of this example is methodological, not biological. It shows how evolutionary rates can be studied quantitatively in characters whose evolutionary changes are not simply metrical. The characters can be divided into discrete states; the states assigned arbitrary scores; and the changes in those scores measured through time.

[1]The lungfish in Westoll's study are not a simple sequence of ancestors and descendants, as were the horses in MacFadden's work. Westoll's rates apply to evolution within the Dipnoi as a whole and are not rates of change down a single evolutionary lineage; the whole groups of Dipnoi would have contained many lineages.

Figure 20.10 (a) Evolution in lungfish, shown as the total score for each fossil; the text explains the scoring. Score = 100 for the most primitive form, 0 for the most advanced. The rate of evolution is the slope of the graph. When the graph is flat, evolution is not happening. (b) Rate of evolution. The graph is derived from (a) and shows the rate of change of the score through time. Lungfish have been living fossils since about 200 million years ago. Modified from Westoll (1949).

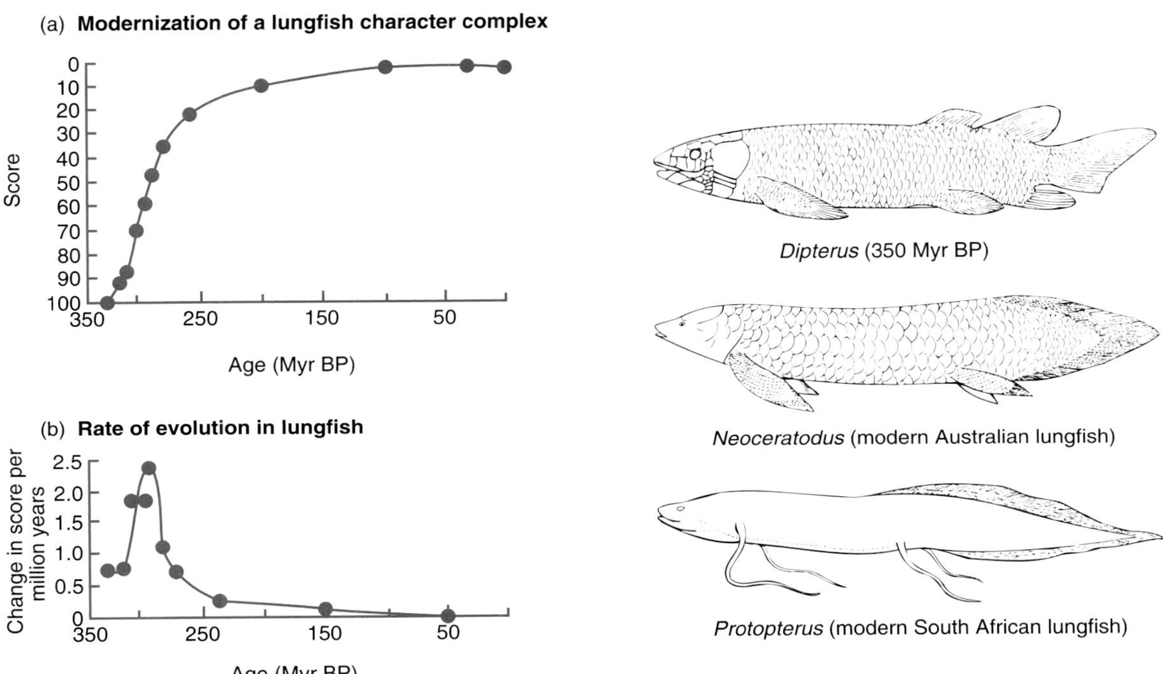

Figure 20.11 The modern horseshoe "crab" (in fact, a chelicerate, not a crustacean) *Limulus polyphemus*, which lives along the east coast of the United States, is a living fossil. It is morphologically very similar to forms that lived about 200 million years ago, and not all that different from Cambrian species. Reprinted, by permission of the publisher, from Newell (1959).

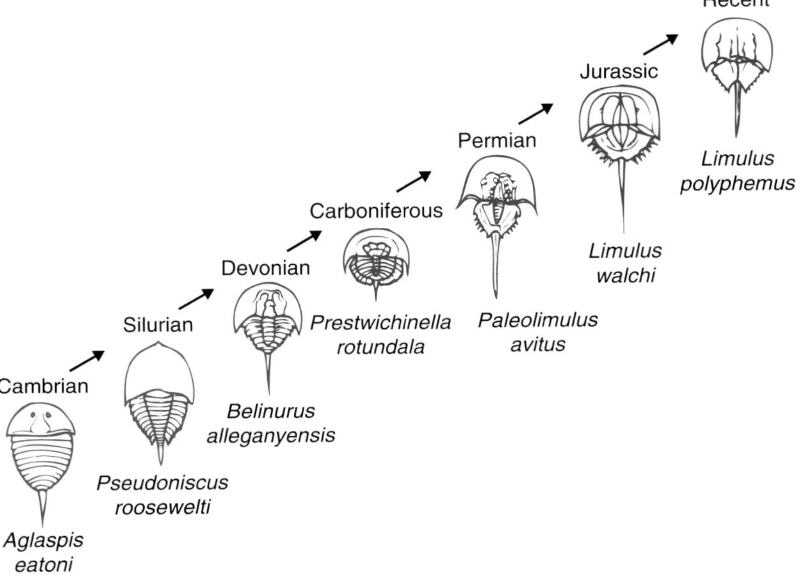

The quantification is mainly useful for purposes of illustration. Figure 20.10 neatly shows how rates of evolution have varied through time in a way that a table of raw character data could not. The division of characters into states, and the assignment of scores to states, is arbitrary, however. The five states for supraorbital canal bone fusion could just as well have been scored 40, 16, 15, 14, 0 or 2, 8, 17, 39, 40 as 4, 3, 2, 1, 0. It would, therefore, be meaningless to compare exact numerical rates of change between characters, or between taxa. The scores are incommensurable. The approximate shape of a graph like Figure 20.10 could be compared with another such graph for another group, but it would be pointless to ask why one group changed at a rate of, for example, 2.1 units per million years and another at 1.3 units per million years. The scores are not intended for that kind of analysis. On the other hand, as an illustration of how rates of change in lungfish have risen and fallen and declined to a virtual standstill, Westoll's analysis is a classic.

20.10 *Taxonomic data can be used to describe the rate of evolution of higher taxonomic groups*

For a question such as "Do mammals evolve faster than bivalve molluscs?", several kinds of measurement of evolutionary rate can be used. We could calculate the evolutionary rates for individual characters, measured either metrically or in discrete states, in mammals and bivalves, and then compare them. With such a method, a danger exists in that the comparisons might be meaningless. Any differences in evolutionary rate may reflect only the way we measure teeth in mammals or shell shape in bivalves, rather than something real. For some purposes—comparisons with rates of nucleotide change, perhaps—such measurements might be useful, however.

We could also use taxonomic evidence to ask this question. When taxonomists divide a set of organisms into a number of species, they make their judgment according to the degree of phenetic differences among the forms. Their judgment will not usually be based on a single character, but on several characters, integrated in the taxonomist's mind into a single dimension of taxonomic similarity. Thus, if two separate but comparable evolutionary lineages were identified, and in the same time interval a taxonomist divided one lineage into two species and the other lineage into three species, it would suggest that the latter lineage had evolved at a more rapid rate (Figure 20.12). This comparison uses a *taxonomic* rate of evolution.

A taxonomic rate of evolution offers an abstract measure of how rapidly change takes place in a group of species. The exact meaning of a taxonomic rate is less easy to specify than for an evolutionary rate of a single character; the taxonomic rate has a relatively imprecise meaning. Such rates do summarize evidence from more than one character, and have greater generality. The reliability of a taxonomic rate depends on the reliability of the judgment of the taxonomist who divided the lineage up into species and genera.

Taxonomic rates of evolution are expressed in two main ways. One is the number of species, or genera (or whatever taxon), per million years. Table 20.3 gives some examples from the two taxa that we have discussed in this chapter:

Figure 20.12 Taxonomic measurement of evolutionary rate. If a taxonomist has divided one group into two species, and another group into three species, in the same time interval, the latter group shows a 50% higher taxonomic rate of evolutionary change.The diagram illustrates only the logic of the argument. In real data, gaps would appear in the lineages, and the real pattern of evolution could have been smooth or jerky, with any number of branches in addition to the lineages shown here.

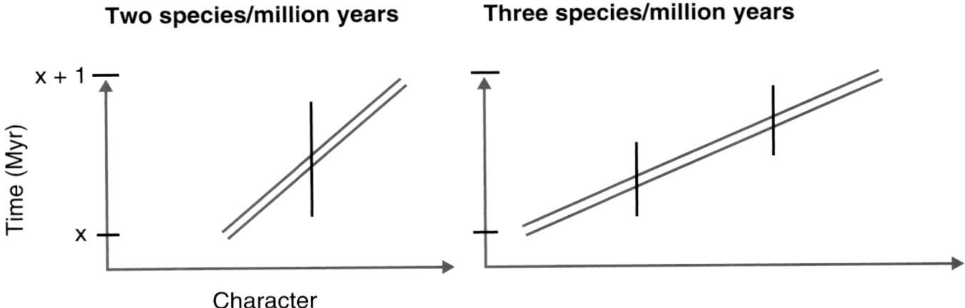

horses and lungfish. The rates for lungfish show how that group initially had a high rate of evolution, which then slowed down so much that they became living fossils.

The same data for a group made up of a larger number of lineages can also be expressed as a *survivorship curve* (Figure 20.13). Survivorship curves are constructed by taking a sample of a number (such as 100) of mammalian genera and measuring how long each one lasts in the fossil record. (Any taxonomic level within the mammals can be used; genera are an example.) The survivorship curve plots the number of genera surviving for different times. Most of the genera are still surviving after a short time, but as time passes the members of the original sample drop out. The slope of a survivorship curve measures the rate of evolution of the group. If the group is evolving rapidly, the survivorship curve falls dramatically as a taxonomist does not recognize a species as surviving for long; the curve is more gradual for a slowly changing group. (Survivorship

TABLE 20.3

Taxonomic rates of evolution in mammals and lungfish. Early in their evolution, lungfish evolved as rapidly as mammals, but they have subsequently slowed down. *Hyracotherium-Equus* is the horse lineage discussed in section 20.1. From Simpson (1953). Copyright © 1953 by Columbia University Press. Reprinted with permission of the publisher.

Group or Line	Average Duration of Genus (Myr)
Hyracotherium-Equus	7.7
Lungfish	
Devonian	7
Permo-Carboniferous	34
Mesozoic	115

Figure 20.13 Survivorship curves for (a) bivalve molluscs and (b) carnivores (Mammalia). The curves express the numbers of genera surviving for different amounts of time. Note bivalve genera tend to last longer than carnivore genera, as is clear from the different scales of the *x*-axes in the two figures. The average duration of a bivalve genus is 78 million years, contrasted with 8.1 million years for a carnivore. From Simpson (1953). Copyright © 1953 by Columbia University Press. Reprinted with permission of the publisher.

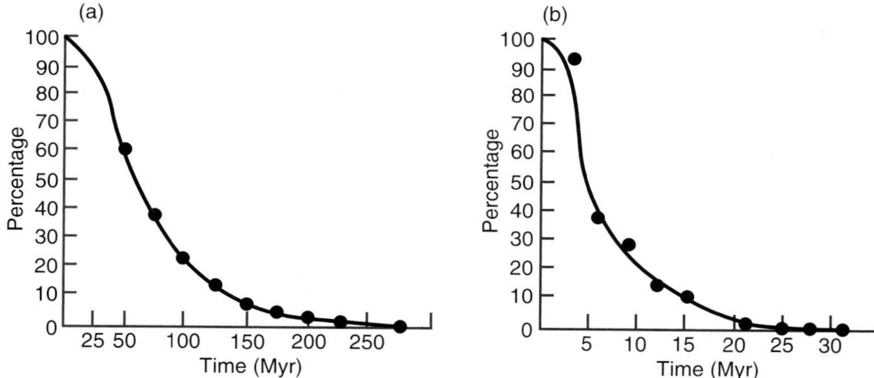

curves are more familiar for populations of individuals. See section 4.1, p. 69. An actuarial survivorship curve plots the survival of a sample of individuals through time, but the same type of graph can be plotted for species and other taxonomic groups.)

A similar problem arises in comparing taxonomic evolutionary rates between groups as is seen with single characters. A bivalve taxonomist may be a good judge of a bivalve genus, and a mammal taxonomist of a mammal genus, but it is still difficult to interpret any differences in the rates of turnover of the two sorts of genera. The variations could simply reflect some difference in the way the two taxonomists work. Within a group, survivorship curves can have revealing features, as we shall see in chapter 22. For now, however, we need only know that taxonomic evidence can provide another sort of measurement of evolutionary rates, along with measurements of single characters.

20.11 *Conclusion*

This chapter has introduced three methods of measuring rates of evolution. For single characters that show continuous variation, we can measure the metrical rate of change and express it in darwins. For characters with discrete states, we can give each state an arbitrary score and measure the rate of change of the score. A cruder measure is provided by the duration of species in the fossil record, because taxonomists will recognize a higher turnover of species in groups that evolve rapidly. All three methods have their particular uses and applications.

Paleobiologists have used measures of evolutionary rates to study a number of general questions. This chapter has concentrated on the relative rates of evolution during, and between, speciation events. The controversy between punctuated equilibrium and phyletic gradualism as a whole has involved several large issues in evolutionary biology, but the empirical controversy, concerning the characteristic pattern of evolution in fossils, remains open for further work.

SUMMARY

1. Evolutionary rates of fossil characters can be measured as simple rates of change through time. Logarithms are often taken of the character measurements and the rate expressed in "darwins." Evolution in horse teeth is a classic example of this technique.
2. Rates of evolution measured in the fossil record are slower than those produced by artificial selection in the laboratory.
3. Evolutionary rates vary between different geological times, taxa, and types of taxa. The science of evolutionary rates is mainly concerned with explaining the pattern of evolutionary rates.
4. Among published measurements of evolutionary rates, the rate and the time interval over which it was measured are inversely related: faster evolution is seen in shorter intervals. The reason is probably that the direction of evolution fluctuates through time.
5. Eldredge and Gould stimulated a controversy about evolutionary rates by their suggestion that rates have a strict pattern (called punctuated equilibrium). According to their theory, evolution occurs rapidly at times of splitting (speciation) and comes to a halt between splits. The opposite pattern, in which evolution has a constant tempo, they called phyletic gradualism.
6. It is difficult to discover the pattern of evolutionary rates at, and between, speciation events because the fossil record is incomplete.
7. There is some evidence for punctuated equilibrium, such as Cheetham's study of Caribbean bryozoans from the Miocene and Pliocene, and some evidence for phyletic gradualism, such as Sheldon's study of Ordovician trilobites in Wales. No general empirical conclusion is yet possible.
8. The theory of punctuated equilibrium has been developed in radical, if speculative, directions. It has been suggested that stasis results from constraints on genetic variability and speciation to the break-down of these constraints, or even macromutation. Evidence exists for stabilizing and directional selection, but little or none is known for constraint and revolutionary speciation.
9. Neither Darwin's theory nor neo-Darwinism necessarily predict pure phyletic gradualism. They were gradualist about the evolution of adaptation. It is historically controversial whether Darwin and the neo-Darwinians mainly held gradualist, or pluralist, views about the rates of evolution during, and between, speciation events.
10. Evolutionary processes and rates can be examined at all taxonomic levels from evolution within populations, through speciation, to the origin of the higher groups. Evolution may have characteristic mechanisms and rates at different levels, or the same set of rates and processes may operate equally at all levels.

11. For large changes, such as from a limb into a wing, evolutionary rates cannot be measured as a continuous variable. The character can, instead, be divided into states. The evolutionary rate can then be studied in the rate of change between states. Westoll studied the evolution of lungfish by this method.
12. The number of species in a lineage per million years is a complex measure of evolutionary rate, called a taxonomic rate of evolution. Taxonomic rates can be expressed as survivorship curves.

FURTHER READING

Simpson (1953) remains a good introduction to the study of evolutionary rates; Fernholm and Sorhannus (1991) is a more recent review. MacFadden (1992) is an excellent book about evolution in fossil horses; Gingerich (1993) updates his work discussed in sections 20.2 and 20.3 and suggests some more sophisticated measures of evolutionary rate than the darwin.

The literature on punctuated equilibrium is now vast. Eldredge and Gould (1972) and Gould and Eldredge (1977) are the key original works; see Brown (1987) on their evidence. Eldredge (1995) contains a popular introduction to this and related topics; Levinton (1988) contains a critique. Gould and Eldredge (1993) is their most recent statement. Erwin and Anstey (eds) (1995) contains papers on fossil speciation. Jackson and Cheetham (1994) is a popular paper about the Caribbean bryozoans, and Cheetham *et al.* (1994) a more recent paper that reconstructs quantitiative genetic variables and infers the roles of selection and drift in the observations. Williamson's (1981a) study of the snails of Lake Turkana is also important, though it is controversial whether it shows real evolution or ecophenotypic changes (Fryer *et al.* 1985). On developmental constraints to explain stasis, see Williamson (1981b) and the comment in *Nature*, 296 (1982), pp. 608–612. Sheldon (1987) is the reference for trilobites; for the controversy, see Maynard Smith (1987c), Eldredge and Gould (1988), and Maynard Smith (1988). Many more case studies are discussed in the multiauthor works cited above. For later developments of the theory: Gould (1980b, 1982, 1983b) and Williamson (1981b). For rapid change without speciation, see MacLeod (1991). For what the population geneticists think, see Charlesworth *et al.* (1982), Turner (1986), and Maynard Smith (1983); Hoffmann and Parsons (1991) discuss environmental stress. See also Dennett (1995, ch. 10) and Williams (1992), particularly on stasis. See Grant (1986, 1991) on the finches, and Christie *et al.* (1992) for evidence of earlier islands in the Galápagos. On history, see (in addition to Gould) Rhodes (1983) and Dawkins (1986). Eldredge and Stanley (eds, 1984) is about living fossils. On taxonomic rates, see (for example) Simpson (1953), Gilinsky (1988), and Archibald (1993).

STUDY AND REVIEW QUESTIONS

1. A character (such as tooth size) has been measured in two populations, at two times (t_1 and t_2). x_1 and x_2 are the average sizes of the character at the two times. Calculate the evolutionary rate in darwins. You may need a calculator that works out natural logarithms.

Character values (in size units) and times (in Myr B.P.)				
x_1	t_1	x_2	t_2	Rate (r)
2	11	4	1	
2	11	20	1	
20	11	40	1	
20	6	40	1	

2. Gingerich (1983) plotted evolutionary rates against the time interval used to measure the rate for over 500 evolutionary lineages. (a) What did he find? (b) How would you interpret his finding?

3. Review the main predictions of the punctuated equilibrium and phyletic gradualist models of speciation in fossils.

4. Describe two or more evolutionary mechanisms that could generate punctuated equilibrium. Assess how well supported they are.

5. Describe two interpretations of what the modern synthesis suggested about the pattern of evolutionary rates during speciation. Do they support, contradict, or have some other relation with, the theory of punctuated equilibrium?

6. How can it be quantitatively shown whether a taxonomic group is a living fossil?

Macroevolutionary Change

In this chapter, we begin with an example of a large-scale evolutionary transition—the origin of mammals from reptiles—that is well documented in the fossil record. It illustrates the neo-Darwinian theory of how major groups originate, in gradual adaptive stages. Most morphological change (large and small) is produced by changes in developmental processes. We move on to look at three themes in the relation between developmental and evolutionary change: evolutionary changes in the relative rates of somatic and reproductive development; the relative ease of evolutionary changes at different development stages; and the idea that macroevolution may proceed by developmental macromutations as well as by gradual adaptation. We conclude by examining how to investigate the reasons why one major group, once it has originated, replaces another in the fossil record.

21.1 The mammals evolved from the reptiles in a long series of small changes

The mammals are a distinct group of vertebrates in many respects. They have warm blood and a constant body temperature, and the high metabolic rate and homeostatic mechanisms that go with it; they have a characteristic mode of locomotion, or gait, in which the body is held upright with the legs underneath (in contrast to the "sprawling" gait of reptiles, such as lizards, in which the legs stick out sideways); mammalian brains are large; the mammalian method of reproduction, including lactation, is also distinctive; the active metabolism of mammals demands efficient feeding; and mammals have powerful jaws and a set of relatively durable teeth, differentiated into a number of tooth types. Therefore, when the mammals evolved from the reptiles, changes on a large scale had to occur in many characters. Thus, the process involved a *macroevolutionary* change. How did this transition take place?

Not all of the distinctively mammalian characters are preserved in the fossil record. The earliest mammalian fossils are a group called the morganucodontids, such as *Kuehneotherium*, from the Triassic-Jurassic border almost 200 million years ago. Whether *Kuehneotherium* was viviparous and lactated is not known directly. It is possible to see that it had a mammalian jaw, gait, and tooth structure, and therefore probably also had warm-blooded physiology. The origin of the mammals can be traced back before the morganucodontids, through a series of reptilian groups called the *mammal-like reptiles*. Mammal-like reptiles is an informal name for the reptilian group Synapsida. These animals evolved over an approximately 100 million year period from the Pennsylvanian to the end of the Triassic, when the first true mammals appear. Some members of Synapsida persisted into the Jurassic, but by then the dinosaurs had proliferated and no other terrestrial tetrapods thrived until after the Cretaceous catastrophe. (Figure 21.13, p. 604, shows the distribution in time of these groups.) The interesting feature of the evolution of the mammal-like reptiles, for our purposes, is what they suggest about how a major new group originates.

The characters that can most clearly be reconstructed in fossils are those concerned with locomotion and feeding, because these are simply related to the form of preserved bones and teeth. The reptilian jaw contrasts in many respects with the mammalian jaw (Figure 21.1). Mammalian teeth have a complex, multicusped structure and are differentiated down the jaw into canines, molars, and so on. In contrast, reptilian teeth form a relatively undifferentiated row and have a simpler structure. The top and bottom of the reptilian jaw articulates (that is, hinges) at the back, where it has muscles that simply snap it shut. The mammalian jaw has cheek muscles that surround the cheek teeth and enable the jaw to close more powerfully and accurately than the reptile's. As the point of jaw articulation moved forward in mammalian evolution, the bones at the rear of the jaw became evolutionarily liberated, and went on to evolve into the ear bones—but we shall not follow that fascinating story. Instead, we will concentrate on how the jaw, and gait, changed in the evolution of the mammal-like reptiles.

We can distinguish three main phases in mammal-like reptilian evolution. The first phase corresponds to one of the two major divisions of the group, the *pelycosaurs*. Pelycosaur fossils are preserved from the Pennsylvanian and Permian, particularly in the U.S. southwest where rocks of this age are located (Color Plate 9). *Archaeothyris* lived in this area about 300 million years ago, and is the oldest known pelycosaur (Figure 21.2). It was a lizard-like animal, about 50 cm long. Its distinctive difference from other reptile groups is an opening in the

Figure 21.1 Articulation of jaw in mammal-like reptiles. In the early form *Biarmosuchus* (a), the jaw muscle is at the basal articulation. In the evolution of mammals, the muscles move forward, and by *Probainognathus* (c), an advanced mammal-like reptile, the masseter has split into two. The superficial masseter joins a characteristic region of the upper jaw, and the presence of the advanced jaw condition can be recognized from the jaw alone (see Figure 21.2, where these three forms are illustrated without the muscles drawn in). *Thrinaxodon* (b) was probably like *Probainognathus*. The positions of muscles in fossils can be inferred from the bone shapes, which imply attachment sites for muscles. Note also the increasing complexity and serial differentiation of the teeth. For the different gaits of reptiles and mammals, see Figure 14.3, p. 378. From Carroll (1988). Copyright © 1988 by W.H. Freeman and Company. Used with permission.

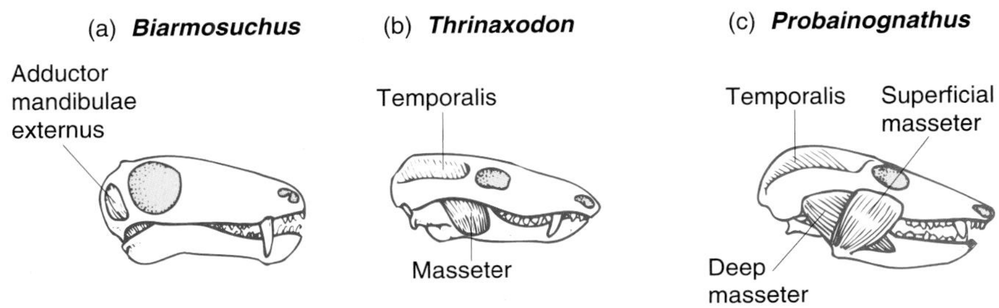

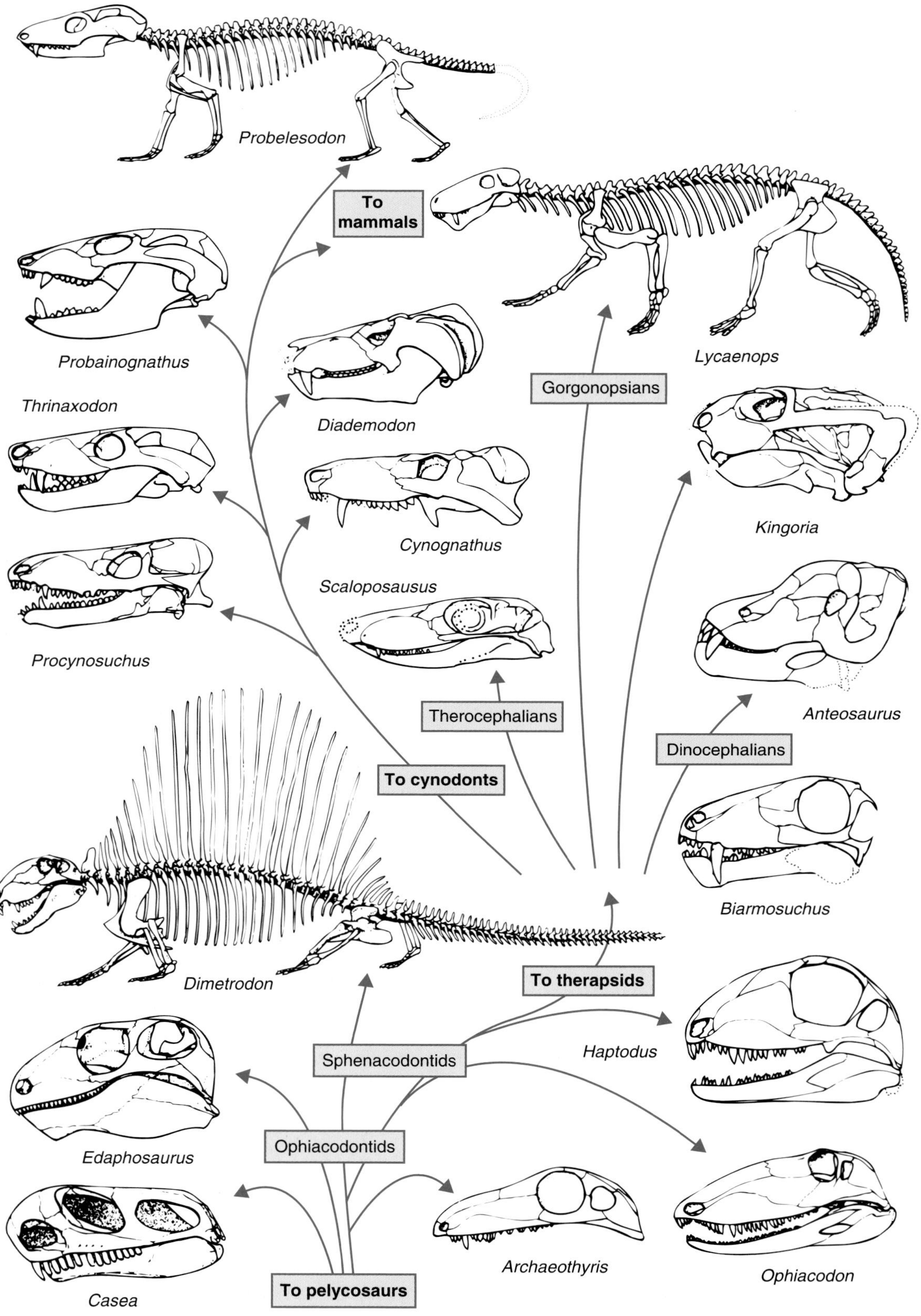
Probelesodon
To mammals
Lycaenops
Probainognathus
Diademodon
Gorgonopsians
Thrinaxodon
Kingoria
Cynognathus
Scaloposausus
Procynosuchus
Anteosaurus
Therocephalians
Dinocephalians
To cynodonts
Biarmosuchus
Dimetrodon
To therapsids
Haptodus
Sphenacodontids
Edaphosaurus
Ophiacodontids
Casea
Archaeothyris
Ophiacodon
To pelycosaurs

Figure 21.2 The evolutionary radiation of the mammal-like reptiles occurred in three main phases: (a) pelycosaurs (sphenacodontids and ophiacodontids in this picture), (b) therapsids, and (c) cynodonts. Each phase contained many smaller evolutionary lineages. Some fossil forms are illustrated. Note again the evolution of the more powerful and precision-action mammalian jaw, and the change from a sprawling gait in *Dimetrodon* to an upright gait in *Probelesodon*. From Kemp (1982b).

bones behind the eye, called a temporal fenestra. In the living animal, a muscle passed through this opening. The muscle acted to close the jaw, and the opening up of the temporal fenestrae is the first sign of the more powerful jaw mechanism of the mammals. (The temporal fenestra, by the way, is the defining character of the Synapsida.) A better known pelycosaur was *Dimetrodon*, with its enigmatic back-sails. Pelycosaurs showed little or no tooth differentiation, and had the reptilian sprawling gait (see Figure 14.3, p. 378). They evolved into three main groups during their 50 million year history, and most of them went extinct quite suddenly about 260 million years ago.

A few of the sphenacodontids survived, however, and it was from an unknown line within the sphenacodontids that the second main group of mammal-like reptiles evolved. The evolution of this group, known as the *therapsids*, makes up the second main phase of the mammal-like reptiles, in the Permian and Triassic. Therapsid fossils are found in many regions of the world, with a site in South Africa having the best deposits. The therapsids underwent a remarkably similar pattern of evolution to the pelycosaurs. Their temporal fenestrae are generally larger and more mammal-like than pelycosaurs, however, and their teeth in some cases show more serial differentiation. Later forms had evolved a secondary palate, which enables the animal to eat and breathe at the same time and is a sign of a more active, perhaps warm-blooded, way of life (see section 13.11, p. 363).

One subgroup of therapsids, the *cynodonts*, are of particular importance in tracing the origin of mammals, and they make up the third phase of mammal-like reptilian evolution. The jaws of cynodonts resemble modern mammal jaws more closely, and their teeth are multicusped and differentiated down the jaw. Some cynodonts show a particularly interesting intermediate stage in jaw evolution. Recall that the reptilian jaw articulates in a different place from the mammalian jaw, a change associated with the evolution both of more precise chewing and of hearing in mammals. Some cynodonts seem to have had a double jaw articulation—that is, their jaws articulated in both the mammalian and the reptilian positions. This development suggests one way in which evolution can proceed from one structure to another without a non-functional intermediate stage. In this case, the structure evolved from state A to state $A + B$, then A was lost, leaving state B alone. The jaw remained a functional structure throughout.

The cynodonts complete the story of the mammal-like reptiles, because it was from a line of cynodonts that the ancestors of the modern mammals such as *Kuehneotherium* evolved. The identity of the exact cynodont line is uncertain, but *Probainognathus* (Figure 21.2) is probably close to it. (Figure 17.18, p. 495, illustrates how some mammal-like reptiles can be arranged in a morphoseries, with one end related to the mammals.)

21.2 *The mammal-like reptiles illustrate the neo-Darwinian theory of macroevolution*

The fossil record for the origin of mammals is superior to that for the origin of any other major group. It is, therefore, an important test case for general theories about how major evolutionary transitions take place. We should notice two important conclusions from the historical narrative. First, the changes from reptilian to mammalian characters evolved in gradual stages. Second, the large-scale differences between mammals and reptiles concern adaptations. The mammals have a high-energy, high metabolic rate kind of physiology, with locomotory adaptations for rapid movements (upright rather than sprawling gait) and adaptations for powerful and efficient feeding (the mammalian teeth and jaw articulation). These changes are surely adaptive, and would have been brought about by natural selection. Thus, the general evolutionary model suggested by the mammal-like reptiles is one of the cumulative action of natural selection over a long period (40 million years); the accumulation of many small-scale changes resulted in the large-scale change from reptile to mammal.

From the previous chapter, this idea can be recognized as an extrapolative theory. Macroevolution is taking place by the same process—natural selection and adaptive improvement—as has been observed within species and at speciation, but the process is operating over a much longer period. While the extrapolative model is not the only model for the evolution of major groups, it is the most important and the only one that can be illustrated with detailed fossil evidence. It can also, in a sense, be thought of as the "neo-Darwinian" theory of macroevolution. In population genetics, neo-Darwinism explains microevolution by changes in the frequencies of preexisting variants. According to this theory, most characters show variation, and the character evolves as its frequency distribution is altered by selection. No extraordinary kinds of variation are needed. Likewise, in a neo-Darwinian account, macroevolution can be explained without recourse to extraordinary kinds of variation. Natural selection, on ordinary variation and mutation, is adequate. That process is how the mammals evolved from the reptiles.

The term "neo-Darwinian" may be potentially confusing in these circumstances. The pattern for the mammal-like reptiles may not represent the universal mode of evolution. Other large groups may have evolved by non-adaptive, or sudden, processes. In that case, their evolution would, in a sense, not have been neo-Darwinian. It is important to understand that the "neo-Darwinian" extrapolative theory of macroevolution is not the same as neo-Darwinism as a whole. Neo-Darwinism refers to various sets of ideas, some of which are more fundamental than others. As a theory of adaptation, neo-Darwinism asserts that adaptations evolve only by natural selection. This assertion is a most fundamental claim, because we have no theory other than natural selection to explain adaptation; without it, we must fall back on miracles as an explanation. The extrapolative theory of macroevolution is a weaker neo-Darwinian claim. If it turns out to be wrong, plenty of scientific alternatives—all compatible with the neo-Darwinian theory of natural selection—still remain. The extrapolative theory of macroevolution can, therefore, easily be jettisoned if it ever turns out to be wrong in fact. If a major evolutionary transition is shown to be sudden, or non-adaptive, neo-

Darwinism in the fundamental sense certainly has not been refuted. Provided that we do not confuse the neo-Darwinian theory of adaptation with the neo-Darwinian theory of macroevolution, no problem arises.

Notice that the extrapolative model does not imply that a single line of evolution stretching from reptile to mammal, with each new step being added on to the one before. Each phase of mammal-like reptilian evolution contains a radiation into many evolutionary lines (Figure 21.2), most of which went extinct without issue. If we retrospectively trace back the line from the mammals to the reptiles of 300 million years ago, however, we pass through a series of stages in which the mammalian adaptations emerge. Within that line, evolution would have made steps both toward and away from the mammalian condition. For the extrapolative model, the most important point is simply that, whatever else was going on, the many mammalian adaptations evolved in stages.

21.3 *Morphological transformations are generally accomplished by developmental changes*

Macroevolution has mainly been studied morphologically, because more taxonomic and fossil evidence exist for morphological characters than for other kinds of characters, such as physiology or chromosomes. Morphological structures are produced by growth, and their form emerges from the process of development. Thus, evolutionary changes in the form of an organ are frequently developmental—they are produced by changes in the rate or timing of developmental events. An organ may evolve to be larger if its growth speeds up, and it may change shape if the growth rate of one of its parts speeds up relative to other parts. In this section, we will encounter several types of evolutionary change in development; we can begin by looking at *recapitulation.* Recapitulation, or the "Biogenetic law," is a bold and influential idea. It is often associated with Haeckel, but had many other adherents in the nineteenth century. Many ideas about the relation of development and evolution have since grown up in opposition to the theory of recapitulation.

According to the theory of recapitulation, the stages of an organism's development correspond to the species' phylogenetic history; this idea might also be stated as "ontogeny recapitulates phylogeny." Each stage in development corresponds to (i.e., recapitulates) an ancestral stage in the evolutionary history of the species. The transitory appearance of structures resembling gill pouches in the development of humans, and other mammals, is a striking example of this concept. Mammals evolved from an ancestral fish stage, and their embryonic gill slits recapitulate the piscine ancestry.

Another example, often quoted in the nineteenth century, is seen in the tail shapes of fish (Figure 21.3). During the development of an individual evolutionarily advanced fish species, such as the flatfish *Pleuronectes*, the tail has a diphycercal stage in the larva, and develops through a heterocercal stage, to the homocercal form of the adult. It is also possible to find different species of fish possessing these three tail types in the adult. Lungfish, sturgeon, and salmon (on the right of Figure 21.3) are examples of current species with these characters. The lungfish is thought to resemble most closely an early fish, the sturgeon to be a later stage,

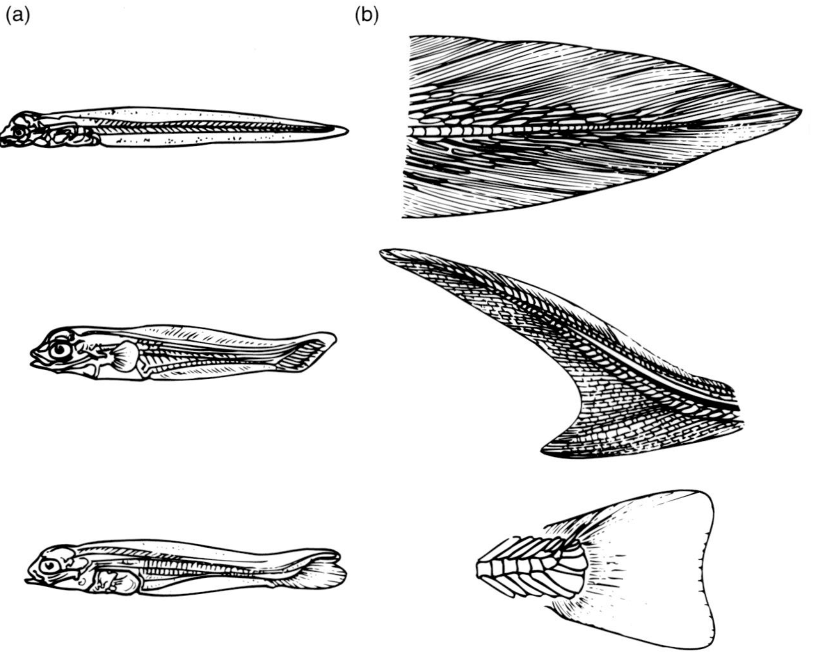

Figure 21.3
Recapitulation, illustrated by fish tails. (a) The development of a modern teleost, the flatfish *Pleuronectes*, passes through (starting at the top) a diphycercal stage, to a stage in which the upper lobe of the tail is larger (heterocercal), and the adult has a tail with equal sized lobes (homocercal). (b) Adult forms in order of evolution of tail form, from top to bottom: lungfish (diphycercal), sturgeon (heterocercal), and salmon (homocercal). Reprinted, by permission of the publisher, from Gould (1977a).

and the salmon to be the most recently evolved form. Thus, evolution has proceeded by adding on successive new stages to the end of development. Let us symbolize the diphycercal, heterocercal, and homocercal tails by A, B, and C, respectively. The development of the early fish advanced to stage A and then stopped. Then, in evolution, a new stage was added on to the end. The development of the fish at this second stage was A B. The final type of development was A B C. Gould has called this mode of evolution *terminal addition* (Figure 21.4a).

Evolution does not inevitably proceed by terminal addition. New, or modified, characters logically could be intruded at earlier developmental stages (Figure 21.4b). Indeed, many examples of specialized larval forms do not involve recapitulated ancestral stages (e.g., the zoea of crabs, the Müller's larva of echinoderms, and the caterpillar of Lepidoptera); they probably evolved by modification of the larva. These exceptions notwithstanding, recapitulation is noticeably common, and evolution has often proceeded by terminal addition.

Two types of reasons explain why recapitulation breaks down. One, as was just mentioned, is the evolution of new early developmental stages. The other is a kind of *heterochrony*. To understand heterochrony, we must first consider how an organism grows. Development proceeds because various cell lines proliferate at distinct times, and places within the embryo, at distinct rates. Some cell lines grow into skin, other cell lines grow into muscles, and so on. A heterochronic change is, in general, a change in the rate or timing of development of some cell lines in the body relative to others. A mutation that alters the rate at which a cell line develops relative to other cell lines is a heterochronic mutation. The kind of heterochrony that has been most investigated, and matters for

Figure 21.4 (a) Evolution by terminal addition. The stages in an individual's development are symbolized by alphabetic letters. Under terminal addition, new stages are added only to the end of the life cycle. (b) Evolution by nonterminal addition. A new evolutionary stage has been added in early development, not to the end of the life cycle in the adult.

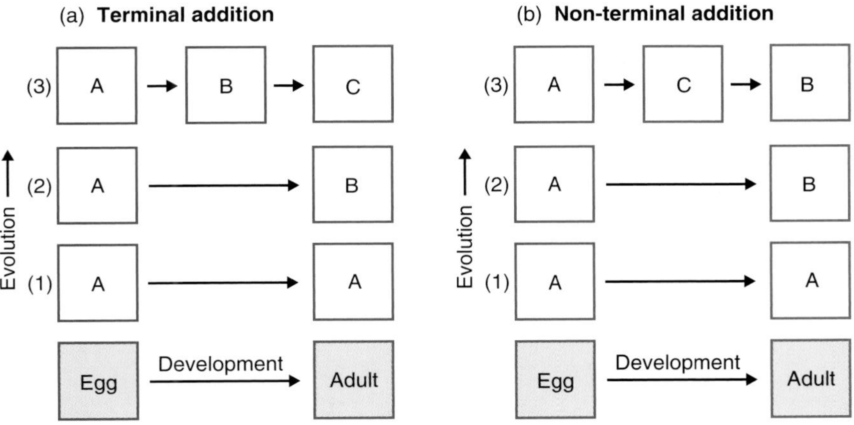

understanding recapitulation, concerns the rates of development of the reproductive (germ) cells on the one hand, and all the rest of the body (the soma) on the other. Any particular body becomes reproductively mature at a certain stage of somatic development; a heterochronic change in this case is one in which the organism comes to reproduce at an earlier (or later) stage of somatic development. Recapitulation is violated when a heterochronic shift allows reproduction earlier in somatic development. The new adult form is not added on terminally, after the ancestral adult form, and the development of the descendant form will not recapitulate the ancestral adult form (which has been lost).

Two processes can lead to this result. As noted above, reproduction is occurring at an earlier stage of somatic development. In such a case, either somatic development is proceeding at the same rate in absolute time as before, and germ line development has speeded up, or reproduction happens at the same absolute age, and somatic development has slowed down (Figure 21.5). The morphological result is the same with either process: reproduction is seen in what was ancestrally a juvenile morphological stage. The morphological result (reproduction in ancestral juvenile form) is called *paedomorphosis*; the two ways of generating this result are called *progenesis* (speeding up the germ line) and *neoteny* (slowing down the soma). We distinguish two morphological results, each with two possible causes (Table 21.1) (some modern analyses use more than the four terms in the table).

21.4 *Examples exist of the different types of heterochronic evolution*

Examples exist of all types of developmental change listed in Table 21.1. Let us look at two cases of neoteny—one in modern species and the other in fossils—to see how the inference can be made.

Figure 21.5 Paedomorphosis, in which a descendant species reproduces at a morphological stage that was juvenile in its ancestors, can be caused by (a) progenesis, in which reproduction is earlier in absolute time, or (b) neoteny, in which reproduction is at the same age but somatic development has slowed down.

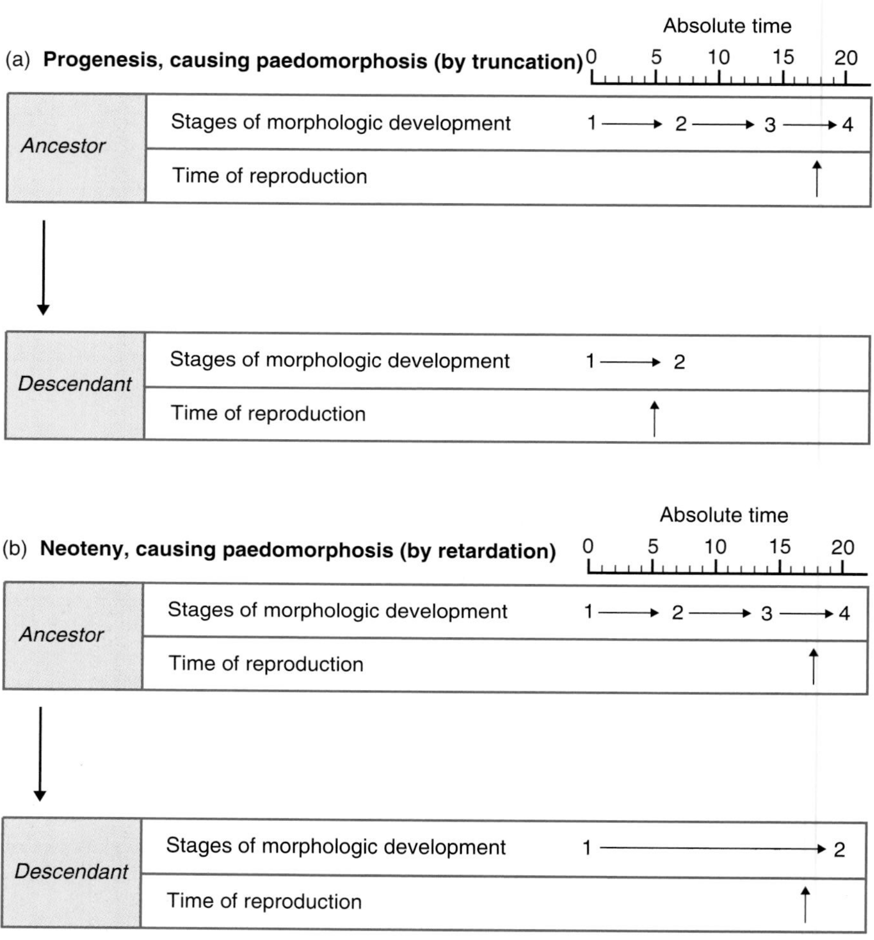

Among modern species, the classic example of neoteny is the Mexican axolotl, *Ambystoma mexicanum*. The axolotl is an aquatic salamander. (Actually, we should say "axolotls" because a number of types are known, and Schaffer's fine-scale genetic work has shown that the kind of larval reproduction described below has evolved many times independently, even within what appears to be a single species.) Most salamanders have an aquatic larval stage, which breathes through gills; the larva later emerges from the water as a metamorphosed terrestrial adult forms, with lungs instead of gills. The Mexican axolotl, however, remains in the water during its entire life and retains its external gills for respiration. It reproduces while it has this juvenile morphology. It takes only a quite straightforward experiment to make a Mexican axolotl grow up into a conventional adult salamander (it can be done, for instance, by injection of thyroid

TABLE 21.1

Categories of heterochrony. In modern work, the term "peramorphosis" is sometimes substituted for "recapitulation." From Gould (1977a), Cambridge, Mass.: Harvard University Press. Copyright © 1977 by the President and Fellows of Harvard College.

Developmental timing		Name of Evolutionary Process	Morphological Result
Somatic Features	Reproductive Organs		
Accelerated	Unchanged	Acceleration	Recapitulation (by acceleration)
Unchanged	Accelerated	Progenesis	Paedomorphosis (by truncation)
Retarded	Unchanged	Neoteny	Paedomorphosis (by retardation)
Unchanged	Retarded	Hypermorphosis	Recapitulation (by prolongation)

extract). These experiments practically confirm that the timing of reproduction has moved earlier in development during the axolotl's evolution. Otherwise, the Mexican axolotl there would have no reason to possess the unexpressed adaptive information of the terrestrial adult.

So the Mexican axolotl is paedomorphic. But is it neotenous or progenetic? Its age of breeding (and the body size at which it breeds) is not abnormally early (or small) for a salamander. It is, therefore, a reasonable inference that the time of reproduction has stayed roughly constant, while somatic development has slowed down. Thus, the axolotl is neotenous. Humans have also been argued to be neotenous. As adults, we are morphologically similar to the juvenile forms of great apes. This paedomorphosis, if it is real (and serious argument claims that it is not), would be neotenous rather than progenetic because our age of breeding has not shifted earlier relative to other apes; our age of first breeding is actually later than other apes. Human somatic development has not simply slowed down while reproductive development has stayed the same. Instead, somatic development has slowed down even more than human reproductive development.

In fossils, we do not know the time of breeding, but an inference can be made from the size of the forms. Consider three fossil cockles (Figure 21.6). The juvenile *Cardium plicatum* (Figure 21.6a) has fewer ribs than the adult (Figure 21.6b). A later descendant, *Cardium fittoni* (Figure 21.6c), has a rib pattern like the juvenile of *C. plicatum*. *C. fittoni* is, therefore, paedomorphic. Is it neotenous or progenetic? An answer is suggested by the sizes. The juvenile *C. plicatum* is about 5 mm, the adult about 17 mm. *C. fittoni* is about 35 mm long. If both species grow at about the same rate in absolute time, then the juvenile morphology of *C. fittoni* must be neotenous. That is, the development of the ribbing pattern has slowed down in absolute time.

If the adult *C. fittoni* had been about 5 mm long, the stronger inference would be that it was progenetic. The cockle would then have both the form of the ancestral juvenile and its size—suggesting it bred at an earlier age than its ancestor. The rule of inference with fossils is that if the adult descendant looks

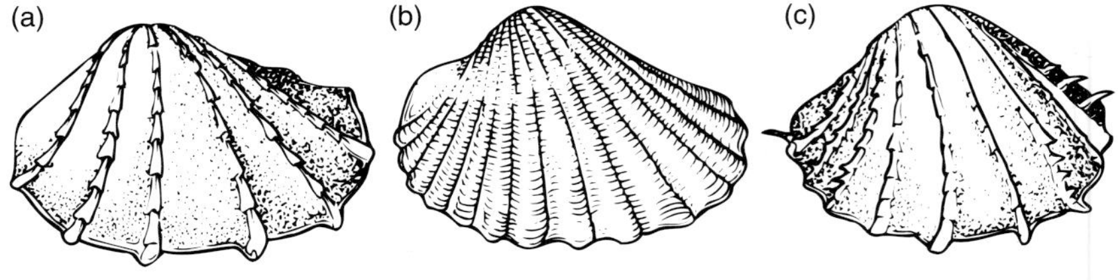

Figure 21.6 Neoteny in fossil cockles. (a) *Cardium plicatum* juvenile, 5 mm in length. (b) Same species, adult, 17 mm long. (c) *C. fittoni*, 35 mm long, and descended from *C. plicatum*. Reprinted, by permission of the publisher, from Gould (1977a), Cambridge, Mass.: Harvard University Press. Copyright © 1977 by the President and Fellows of Harvard College.

like the juvenile ancestor (i.e., it is paedomorphic), then if the descendant is as large as (or larger than) the ancestor it is probably neotenous; if it is the size of a juvenile ancestor, on the other hand, it is probably progenetic.

Clearly, the inference could sometimes be wrong. It makes two main assumptions. First, it assumes that size is proportional to age of breeding. Second, it assumes that we can tell that the juvenile-formed descendant species actually is an adult (our evidence is that no larger specimens have been found—evidence that will be stronger for a richer fossil record). The assumptions could be in error, which would mean that the inference is uncertain, not unreasonable.

21.5 *The question of the relative frequencies of the different types of developmental change is interesting but not yet answered*

We have considered a number of different types of developmental change in evolution. What are the relative frequencies of these processes? Are the four kinds of heterochrony equally frequent? Or is one more common than the others? Perhaps the different types of evolution are correlated with different ecological conditions, with different types of development, or with different kinds of evolutionary change.

One study by McNamara illustrates the kind of comparison that can be made. For 44 trilobites, he divided the species into those that were (inferred to show) progenesis, neoteny, hypermorphosis, and acceleration, and displayed the frequency of the different types of change. McNamara's main result was a difference in the frequency distribution between the Cambrian and post-Cambrian. Progenesis seems to have occurred more frequently in the Cambrian and neoteny afterward (Figure 21.7). No explanation has been suggested. Work like McNamara's will continue to emerge in the future, but at present we have only limited factual knowledge. In addition, evolutionary biologists have occasionally discussed the possibility that one or other mode of evolution is particularly important. Let us look at some of these ideas, beginning with recapitulation.

Many explanations for recapitulation were offered in the nineteenth century. Only a few of them remain interesting today, and they apply equally well to a related but crucially different phenomenon called *von Baer's law* (section 17.8.2, p. 478), in which the early developmental stages of a group of related species are more similar than are the later stages. In different mammalian species, for example, the early embryonic stages are more similar than the adult forms.

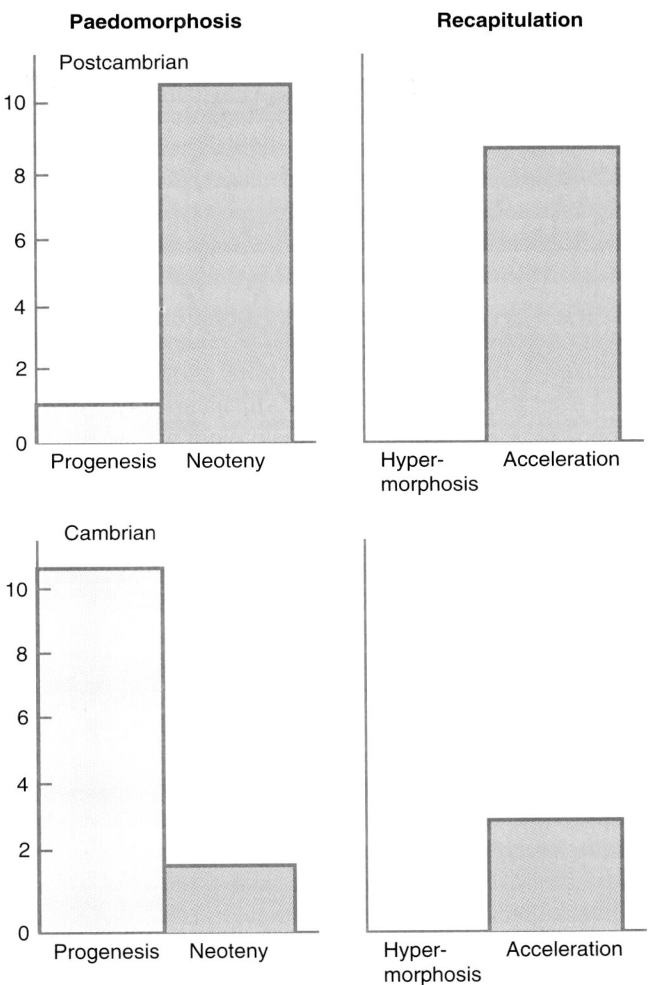

Figure 21.7 Frequency of different types of heterochrony in trilobite evolution. Simplified from McNamara, in McKinney (ed, 1988) by permission of Oxford University Press.

What is the relation between recapitulation and von Baer's law? In a case in which recapitulation occurs, related evolutionary lineages must then have accumulated their new characteristics in the adult, and von Baer's law will also be satisfied (Figure 21.8a). If recapitulation is true, von Baer's law will also hold true. What if recapitulation does not occur? Is von Baer's law then violated? We have seen two reasons why recapitulation may not happen: the modification of early stages (Figure 21.8b) and paedomorphic heterochrony (Figure 21.8c). In either case, the ontogenetic stages do not appear in phylogenetic order and the principle of recapitulation is wrong. The ancestor of the Mexican axolotl, for example, was a terrestrial form, which the axolotl does not recapitulate. Likewise, a modified early stage, such as the aquatic larva of a crab, is not a recapitulated ancestral form.

Von Baer's law is violated by one of these processes but not the other. It requires the relative rarity of modifications to early stages, but it does not rule

Figure 21.8 (a) Evolution by terminal addition (as in Figure 21.4). Both von Baer's law and the principle of recapitulation are correct: species A and B are more similar at earlier than at later developmental stages and the ontogeny of each recapitulates its phylogeny. (b) Nonterminal addition of new characters: both von Baer's law and recapitulation are violated. (c) Heterochronic evolution (paedomorphosis): von Baer's law is correct, but the principle of recapitulation is violated in species B.

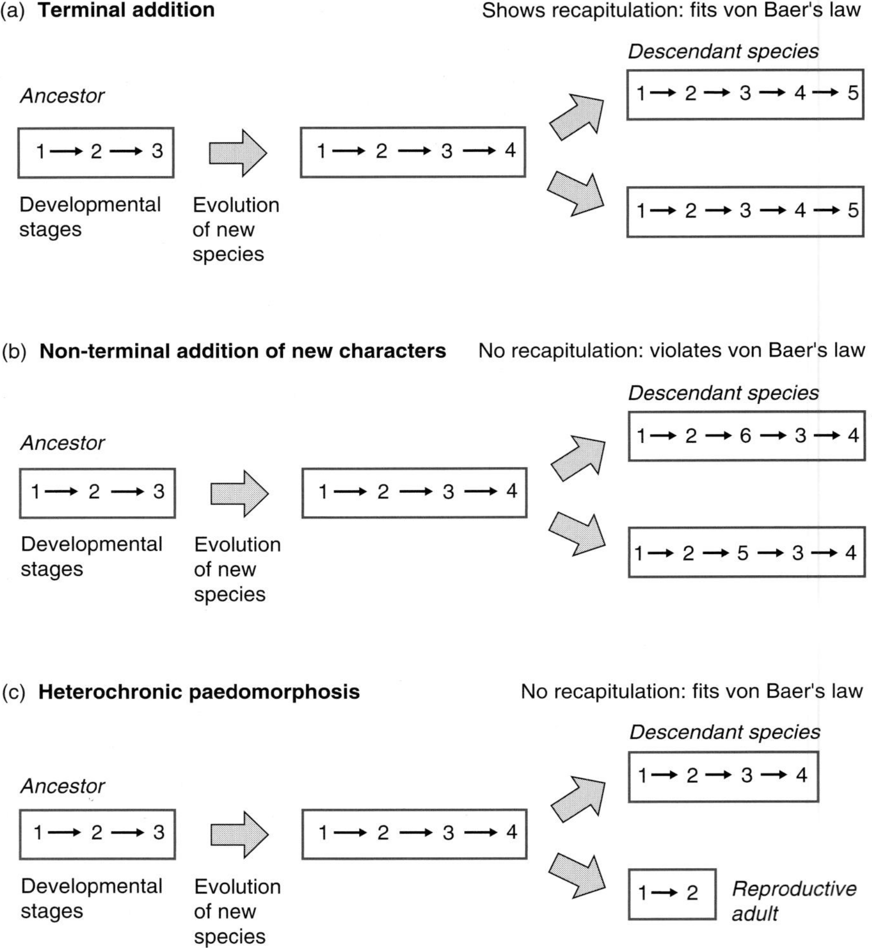

out heterochrony. It is still true of the axolotl and a salamander, for example, that their earlier developmental stages are more similar; the truncation of adult stages does not alter the similarity of earlier stages. In summary, recapitulatory evolution leads to von Baer's law, but non-recapitulatory evolution may or may not. Paedomorphosis is consistent with the law, but the modification of larval stages violates it.

Von Baer's law is a weaker claim than the theory of recapitulation, and stands a better chance of being correct. Indeed, many biologists believe that the law is a sufficiently good generalization for it to need an explanation. If the law is correct, it must be because evolutionary change happens most often in the later stages of development; early stages must be evolutionarily more conservative. Why should this distinction arise? One possibility is a special case of the general argument against the evolutionary importance of macromutations (Box 7.3, p. 182; section 13.3, p. 341). A mutation that influences early development is more likely to have a large phenotypic effect than one that influences a later stage. Development (excepting metamorphoses) is continuous and cumulative, and changes early on will have ever magnifying and ramifying consequences later in the adult form. Some of these consequences may well be deleterious. A late-acting mutation has a higher probability of being a fine-tuning change, with small phenotypic effect. Fine-tuning, in turn, has a higher probability of being selectively advantageous. Changes late in development should, therefore, be more common than early changes. Hence, we have the relative frequencies of changes at different stages and their manifestation as von Baer's law. The argument does not require that macromutations never be selectively advantageous, only that small mutations have a greater likelihood of achieving this state.

A second (and not incompatible) theory reasons from external, ecological selective forces rather than the magnitude of mutations. Early developmental stages are shielded from the environment. In many species, the mother broods her offspring, but even in species in which the eggs are abandoned, the embryo is provided with yolk to support its early growth. As the embryo grows up, it increasingly must fend for itself. Selection is relatively unlikely to demand changes in an embryo being brooded inside its mother, and such an embryo will tend to retain its ancestral form. Once it is exposed to the external environment, it will be selected to adapt to any change in conditions. Early stages will (as von Baer's law describes) tend to be more similar, and later stages will accumulate their own adaptations in different species.

The same reason may explain why some groups fit von Baer's law better than do others. In the nineteenth century, Balfour argued that species with brooding will tend to retain their ancestral forms in the embryo. Species with less yolk and independent development, on the other hand, will tend to evolve individual embryonic adaptations. Species that brood their young should then fit the law better than species that discharge eggs. Many examples of specialized larval forms come from groups in which the larva must fend for itself, but Balfour's idea must still be systematically tested.

In summary, two theories have been suggested to explain why evolutionary changes should be more common in the adult than earlier in development. One derives from the magnitude of mutations acting at different developmental stages, while the other comes from external selective circumstances.

Another idea concerns the evolutionary importance of paedomorphosis. It has been argued that paedomorphosis is especially important in the origin of higher taxa—in the kind of evolutionary breakthroughs that generate new classes or phyla. This argument, if correct, would "uncouple" macro- and micro-evolution. In that case, paedomorphosis might be relatively unimportant in ordinary microevolution, and even in speciation. Instead, its importance would lie in the origin of groups at higher levels. Macroevolution would then have a different characteristic process from microevolution. Doubts have been cast on this argument by observations, such as the one mentioned earlier for the axolotl, of microevolution by paedomorphosis. Here, however, we will look at the case for this theory.

The first component of the argument is empirical. For many large groups of animals, the adults appear to resemble an early developmental stage of a possible ancestor. The chordates provide the classic example. They must have evolved from one of the invertebrate groups, but the distinctive chordate characters—a backbone or notochord, a dorsal nerve chord, and segmented muscles—are not found in any of the invertebrates. They are found in the larvae of a group called the tunicates, or sea-squirts. Although tunicates are included in the chordates because of their chordate larva, adult forms lack any chordate characters. Instead, adult tunicates look like perfectly good invertebrates. It has, therefore, been argued that the ancestral chordate resembled a larval tunicate, and the chordates originated by paedomorphosis. Paedomorphosis has subsequently been suggested in the origin of many animal groups. De Beer, for example, in his important work on the subject (*Embryology and Evolution*, 1931; renamed *Embryos and Ancestors*, 1940) suggested that the following groups had originated from paedomorphic ancestors: ctenophores, siphonophores, hexacorals, proparian trilobites, cladocerans, copepods, insects, graptolites, pteropods, appendicularians, chordates, ratite birds, hominids.

If each case in this long list was correct, paedomorphosis would definitely have been a major—perhaps the major—process in the origin of higher taxa. We will consider in a moment how convincing the evidence is, but let us first ask why the trend might exist. The classic explanation was that paedomorphosis enables an "escape from specialization" (in Hardy's phrase). Within a higher taxon, the argument said, most changes occur in the adult form, and the species become increasingly specialized to narrow niches by these adult adaptations. A return to the juvenile stage means that the slate of evolutionary specialization can be wiped clean, and new large niches may then be invaded. This idea is debatable at best. We will not discuss it in detail here because the phenomenon it is intended to explain is too uncertain. None of the cases in de Beer's list—not even the origin of the chordates—is well confirmed.

Even if the chordates did evolve from a form like the modern tunicate larva, it does not prove it was a larval form at the time of the origin. As Gould and others have pointed out, 500 million years ago the tunicate larva could have been an adult form. During the subsequent millions of years the adult could have evolved a new sessile (and invertebrated) adult stage by terminal addition. The chordates would then have originated from an ancestral tunicate, but not by paedomorphosis. With unique events, happening millions of years ago, it is difficult to rule out such a possibility. The importance of paedomorphosis in the origin of higher taxa remains only a conjecture.

In summary, theoretical work has outstripped the empirical in the study of the relative frequency of different types of developmental change. Apart from McNamara's compilation, no quantitative work has been done on the frequency of the different types of heterochrony. Moreover, no systematic study has been made of the relation between heterochrony and macroevolution. That lack of information has not prevented work on provisional theories of why evolutionary change should be concentrated in the adult, and of why paedomorphosis should be associated with macroevolution. How well these theories will be factually supported remains to be determined.

21.6 Heterochronic change can occur between different somatic cell lines as well as between the timing of reproductive and somatic development

The examples of heterochrony we have discussed so far have concerned the relative timing of reproduction and somatic development as a whole. Within the soma, many different cell lines develop and the form of the adult can be altered by heterochronic changes between them. This kind of heterochrony can be studied by two of the classic methods of studying morphological evolution. The simpler of the two was largely invented by Huxley and is called *allometry*. A typical allometric graph plots body size on the *x*-axis and some other character, such as brain or eye-stalk size, on the *y*-axis (Figure 13.9, p. 359). The points on these graphs can represent data for the same individual measured at different ages, for different individuals of a species (in which case the scatter will mainly be due to variation in age), or for different species in a higher taxon.

For the study of heterochrony, the most interesting allometric graphs show the intraspecific variation (or individual growth) for two or more species. If brain size increases more rapidly relative to body size in one species than in another, the growth of the brain cell lines has probably been speeded up relative to the other cell lines in the body (Figure 21.9). In other words, a heterochronic change has occurred between the species. Heterochrony was introduced earlier as potential change in the somatic and germ lines of the body; but it can also operate between any two cell lines, or sets of cell lines. The term allometry, on the other hand, is customarily applied to the relations between the sizes of two organs. Nevertheless, we can see that allometry and heterochrony are, in an

Figure 21.9 Imaginary graph of allometric relation between brain and body size in two species. The points on the graph for each species could either be for many different individuals, or for one individual as it grows up. Species 2 has a larger brain at each body size, which could occur because the growth of the brain cell lines have been heterochronically speeded up relative to the growth of other cell lines.

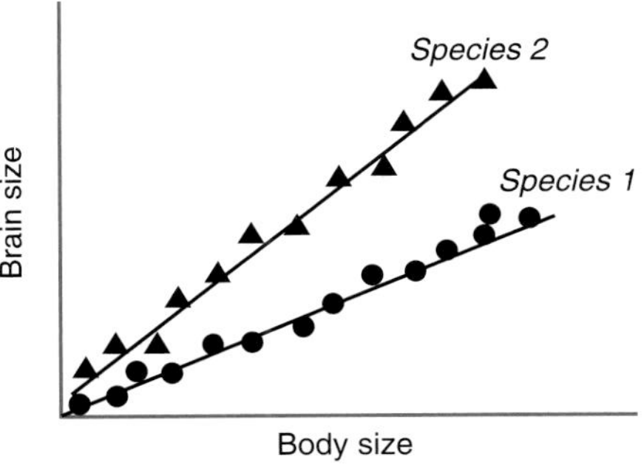

abstract sense, two ways of talking about the same thing. If the *y*-axis of Figure 21.9 were relabelled to express some measure of the maturity of the reproductive cells, then a shift from species 1 to 2 would be paedomorphic and a shift from species 2 to species 1 would be recapitulatory. In general, we can think of the two axes on the allometric graph as representing two cell lines, or sets of cell lines. When a heterochronic change affects the rate at which the two cell lines develop, the allometric graph will change shape.

Allometry is an important method for describing morphological evolution. The body shape of a species can usually be partly represented in an allometric graph, and allometry provides a method of representing formally the shape differences between species. The axis of an allometric graph may or may not represent a true growth gradient in individual development. If it does, then the shifts in the graph slope between species will portray the changes in the development regulatory system that produced the evolution. Even if the variable on an axis does not correspond to an actual developmental mechanism, it will to some extent be correlated with one, and the graph for two species gives some idea of the underlying developmental changes.

Huxley's method works easily for two characters, and its many-dimensioned extension is obvious (just add a third axis for the third and subsequent characters). Multidimensional measurement, however, is clumsy for complex shapes. For complex shapes, *D'Arcy Thompson's transformations* can be more illuminating (Figure 21.10). D'Arcy Thompson found that related species that superficially look very different could, in some cases, be represented as simple Cartesian transformations of one another. We saw the most thoroughly worked-out modern example in an earlier chapter (Raup's analysis of snail shell shapes; see Figure 13.8, p. 357). With some simplification, we can conceive of the axes on the fish grids in Figure 21.10 or the snails of Figure 13.9 as growth gradients. The evolutionary change between the species would then have been produced by a genetic change in the regulatory mechanisms controlling those gradients. If we examined, for example, *Scarus* and *Pomacanthus* without the grids of Figure 21.10, we might think that an evolutionary change from one into the other would be at least moderately complicated. D'Arcy Thompson's diagrams enable us to show that shape changes could have been produced by simple regulatory changes in growth gradients.

Regulatory changes in growth gradients, including heterochronic changes, probably provide the means of many evolutionary changes in shape. Some people would extend the argument and say that almost all morphological evolution employs this process. The main difficulty is accounting for evolutionary innovations, such as the origin of a new organ like a liver or a heart in a lineage in which the early species lack that character. The organ will have evolved from some precursor stage, and the changes may be interpretable as heterochronic. Others would argue that the concept of heterochrony can, in the attempt to make it account for everything, be made so diffuse as to lose its meaning. Heterochrony is uncontroversially a good theory to explain some shape changes, and changes in the relative time of reproductive and somatic development. On the other hand, it is a matter of controversy how general an evolutionary mechanism heterochrony is.

Figure 21.10 A D'Arcy Thompson transformational diagram. The shapes of two species of fish have been plotted on Cartesian grids. *Argyropelecus olfersi* could have evolved from *Sternoptyx diaphana* by changes in growth patterns corresponding to the distortions of the axes, evolution could have occurred in the other direction, or the two species could have evolved from common ancestral species. Reprinted, by permission of the publisher, from D'Arcy Thompson (1942).

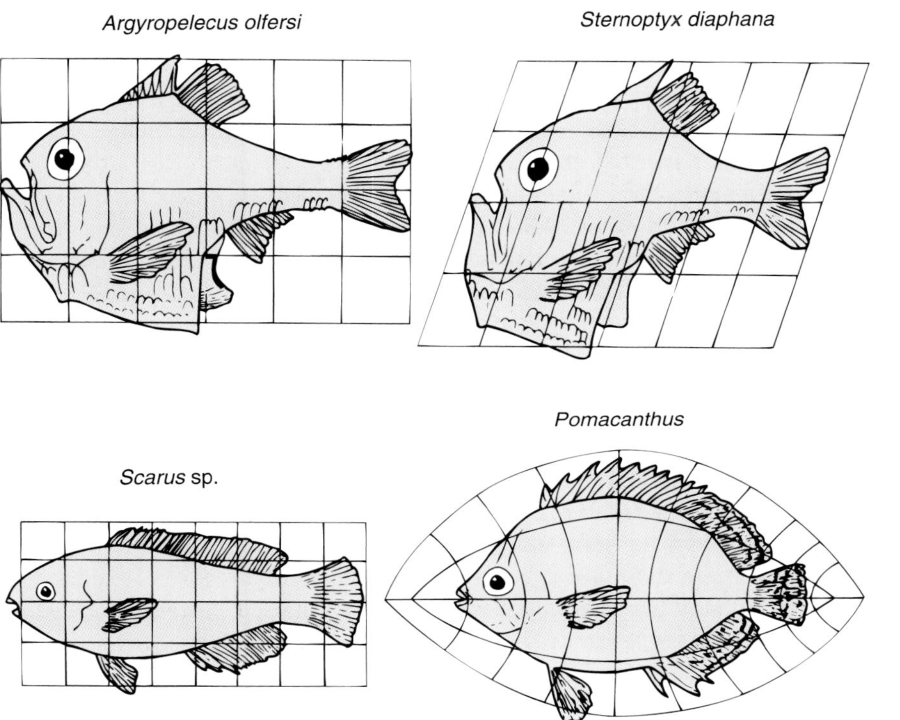

21.7 *Changes in the genes controlling development can produce macromutational monsters*

If a developmental change is to have any evolutionary importance, it must have a genetic basis. The genetics of development is a fast-moving area of biology, but most statements made about evolutionary changes in developmental genetic systems rely on deduction or extrapolation from other systems rather than on straightforward facts. The main, if obvious, deduction is that the general theory of evolution should apply to evolutionary changes in development in much the same way as it applies to any other character, such as structural molecules, enzymes, or behavior. Therefore, evolutionary changes in development probably proceed by small genetic changes. Growth systems have proved amenable to quantitative genetic study (chapter 9) just like any other character. Small genetic changes in growth patterns can be produced, and presumably serve as the source of most developmental change in evolution.

As we have seen, mutations with effects in early stages of development can have large phenotypic consequences. The *homeotic mutations* of fruitflies

(*Drosophila*) are clear examples. Homeotic mutations typically transpose part of the body from one region to another. The mutant *bithorax* is a well-known case. It has two rear-thoraxes instead of one front and one rear thorax, giving it two pairs of wings instead of a pair of wings and a pair of halteres. *Antennapaedia* is another example, in which legs grow out of the antennal sockets instead of antennae. At an abstract level, it is easy to imagine how homeotic mutations work. A set of genes presumably encodes for the growth of a leg and another set specifies where these leg-genes are switched on. Mutations in the position-specifying genes could result in the genes encoding for leg growth being switched on in the wrong place.

Homeotic mutations are examples of macromutations, or "monsters." The mere existence of these mutations of large phenotypic effect demonstrates that macromutations can arise. We can imagine that even more monstrous phenotypes might be generated by mutations in genes controlling early development, although mutations with too radical effects would be lethal and never emerge from the egg stage. How important are these developmental macromutations in evolution?

Let us consider two types of anwsers to this question. The standard neo-Darwinian argument, due to Fisher, is that developmental macromutations are evolutionarily inconsequential. They may arise from time to time, but will always be selectively disadvantageous because they introduce a gross change into a fairly well-adjusted machine. As a result, they will soon disappear from the population.

It is necessary to make a distinction about whether macromutations can happen. The homeotic mutations of fruitflies illustrate that macromutations can happen, but they produce their large effect by reshuffling existing information. They produce new forms showing large phenotypic changes, but they do not produce large adaptive innovations. No one can deny that macromutations happen, but we have good reason to deny that macromutations generating novel adaptations ever arise. The many adaptations (e.g., in tooth shape, jaw, locomotion, and circulatory physiology) by which a mammal differs from a reptile could have evolved in small stages, but it is improbable to the point of impossibility that they all appeared in one single mutation in an ordinary reptile. A macromutation producing a new complex set of adaptations would require a miracle. This argument only applies against *adaptive* macromutations, not phenotypic macromutations in general. A phenotypic macromutation that shuffles the parts of a reptile to produce any monster might happen; an adaptive macromutation that shuffled the parts of a reptile to produce a mammal all at once would not.

On the other hand, Goldschmidt and his followers have suggested that macromutations in the genes controlling development are the source of many evolutionary innovations. These mutations are often known as *hopeful monsters*. The idea is that radically new phenotypes are produced by mutations in developmental genes, and these mutations are the mechanism by which new major groups evolve. The origin of major groups, according to Goldschmidt, takes place by a different process from microevolution. Thus, this argument provides another example of the (hypothetical) uncoupling of micro- and macro-evolution. Microevolution may take place by natural selection of small genetic variants, but major groups (Goldschmidt said) evolve by a different process—via hopeful monsters.

This idea has some modern supporters and, as a rare theoretical possibility, it cannot be ruled out. Perhaps selectively advantageous macromutations do sometimes arise. No evidence exists to prove that they do, however, and theory suggests they would not.

At the beginning of this chapter, we saw how the mammals evolved by gradual and adaptive evolution. They provide a counter-example to Goldschmidt's theory—and to any other theory suggesting major groups evolve by a sudden saltation from their ancestors. The origin of no other major group is understood so well, although we also have evidence of fossil intermediate stages and gradual evolution for birds. The transition from reptiles to birds was adaptive, because most of the characters of birds are adaptations for flying; it must, therefore, have proceeded in many stages. It is more difficult to rule out macromutations in the origin of groups, such as many invertebrates, for which no appropriate fossil evidence is available. Saying that a theory cannot be ruled out by negative evidence is very different from saying that it is correct, however. The theory of evolution by hopeful monsters should be regarded as no more than a hypothesis. It is implausible in theory and unsupported in fact.

21.8 *Conclusions: developmental change and evolution*

The relation between development and evolution is one of the classic topics in evolutionary biology. The nineteenth century generalization, that new evolutionary stages are added on to the end of the life cycle, was no longer believed when the modern synthesis was being put together in the 1920s. For roughly 50 years, many evolutionary biologists showed less interest in the subject. Today, this topic is again an active field of research. Work is being done on the frequency, and circumstances, of the different types of developmental change; on many individual case studies of heterochronic evolution in modern and fossil species; and on the genetics of developmental change. The best understood macroevolutionary transitions suggest that macroevolution occurs by extrapolated microevolution. The evidence does not suggest that any kinds of developmental change characteristically happen any more frequently in the origin of higher groups than in smaller-scale evolution.

21.9 *Higher taxa rise, fall, and replace one another*

After the origin of a higher taxon, the number of species in it usually increase; so, too, do the number of genera, families, and other taxonomic levels. This pattern of increase is called *radiation,* and its exact form can differ between groups. In the mammals, for example, the major subgroups—dogs, cats, primates, ungulates—all originated relatively suddenly in a burst of evolution during the early Paleocene; once established, the number of groups remained constant to the present. In bivalves, by contrast, the number of families has steadily increased (Figure 21.11). Eventually, after a phase of expansion, a higher taxon goes into decline and ends with extinction.

What factors explain the timing and pattern of a group's rise and fall? Why, for example, did the mammals remain a minor group for about 150 million

Figure 21.11 (a) The number of families of mammals increased abruptly in the Paleocene and Eocene, after which it has remained constant. (b) The number of families of bivalves has increased steadily through time. From Stanley (1979). Copyright © 1979 by W.H. Freeman and Company. Used with permission.

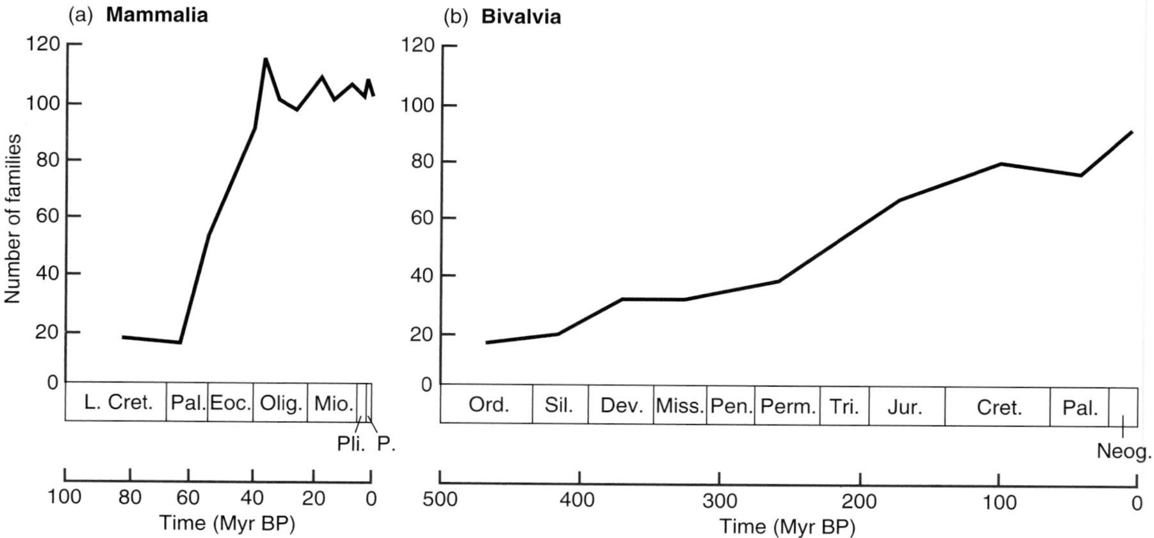

years after their origin and then radiate rapidly 50-60 million years ago? The obvious answer is that until the Paleocene the ecological niches later to be occupied by mammals were already filled by dinosaurs. Thus, the mammals radiated as the ecological replacements for the dinosaurs. The rise and fall of a higher taxon has often been connected with the fall of one set of ecologically equivalent species and the later rise of another set.

Why should one higher taxon replace another higher taxon of ecologically similar species? Two theories can be tested. Competitive displacement says that the later group outcompeted the first, and drove it extinct. Independent replacement, on the other hand, says that the first group declined and went extinct for some reason—environmental change, perhaps—unrelated to the presence of the second group, and the second group radiated only after the first had been cleared away. The pattern of change in diversity of the two groups provides the best evidence to test between the two theories (Figure 21.12). If the first group declines before the second expands, it suggests competition was not influential; whereas if the first group declines in proportion to the increase in the second, it suggests competition.

The tests for these theories are not fool-proof. In practice, environmental change could produce the pattern of competitive replacement (Figure 21.12b) if the conditions favoring the second group gradually appear while those favoring the first group disappear. Alternatively, the environmental conditions could themselves influence the relative competitive power of the two groups. The environment might change from an earlier condition under which the first group outcompeted the second to a later condition under which the second outcompeted the first. A pattern of gradual replacement, does not therefore, rule out factors

Figure 21.12 The exact pattern of replacement of one group by another suggests whether competition was at work. (a) If the initially dominant group declines before the second group expands, it suggests that the replacement was not caused by competitive displacement. The dominant group may decline either gradually or catastrophically. (b) If the dominant group declines as the other group gains at its expense, competition and relative adaptation are more likely to have influenced the replacement.

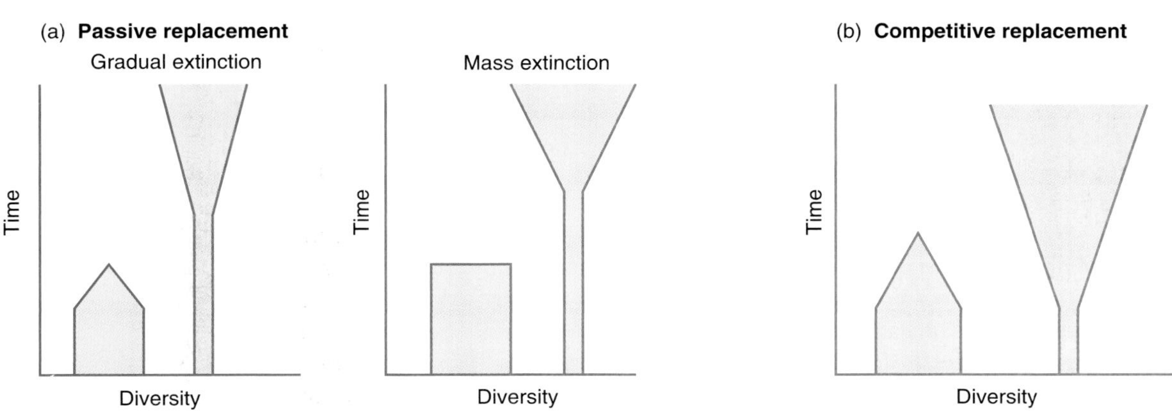

other than ecological competition. The test (Figure 21.12) still holds some interest. If the earlier group does go extinct clearly before the rise of the later group, it is difficult to explain the replacement as being due to competition. If the takeover is correlated in time and the two groups are ecological analogous, competition is at least suggested.

Let us consider some examples. In the Permo-Triassic period, the mammal-like reptiles were the dominant group of large terrestrial vertebrates; herbivorous, carnivorous, and smaller insectivorous taxa existed for this group. By the end of the Triassic, these mammal-like reptiles had been replaced in most of these niches by the dinosaurs. It had been argued that the dinosaurs were competitively superior (in Mesozoic conditions) to the mammalian lineage, which brought about the replacement. The most detailed study, by Benton, of the diversity changes of the two groups does not strongly support that idea, however, as the evidence is ambiguous (Figure 21.13). The principal Triassic herbivores comprised a group of mammal-like reptiles (Diademodontoidea, a group of cynodonts) and another unrelated group of reptiles called rhyncosaurs. The herbivorous prosauropod dinosaurs radiated at about the time the two groups went extinct, but little evidence suggests a parallel rise and fall of the groups. The decline of the Diademodontoidea was almost complete by the time the dinosaurs expanded significantly. A more suggestive example of competitive displacement can be found lower in Figure 21.13. The dicynodonts did decline more or less as the diademodontids expanded; this parallel action could be due to competition. The main conclusion we reach from this first example, however, is that the evidence for the competitive displacement of the mammal-like reptiles by the dinosaurs is unconvincing.

A second example is the replacement through the Cenozoic of the perissodactyl mammals (horses and their relatives) by the artiodactyls (cattle, deer, and

Figure 21.13 Patterns of diversity among the main terrestrial reptile groups of the Triassic. The width of the stippled area for each group corresponds to the abundance of the group. Dinosaurs are the saurischians and ornithischians. Note the replacements of the diademodontoids and rhynchosaurs by the dinosaurs in the Norian, and of dicynodonts by diademodontoids in the Scythian. See Figure 21.2 for some mammal-like reptile forms. Reprinted, by permission of the publisher, from Benton (1983).

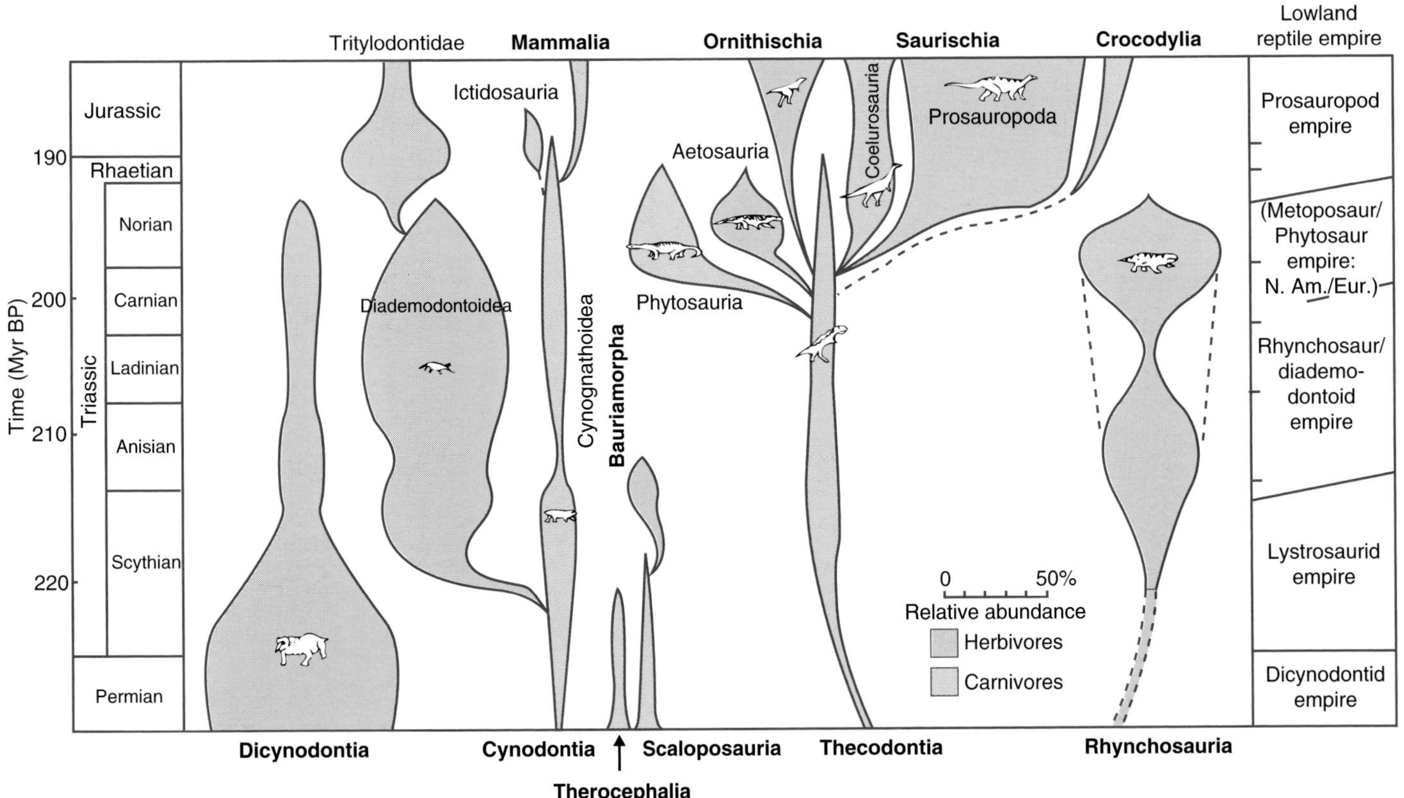

their relatives). Perissodactyls such as horses and zebra still exist, but members of the group were relatively much more abundant in the Eocene. The artiodactyls and perissodactyls are ecologically similar groups of large-bodied herbivorous grazers, and the rise of the artiodactyls has been explained by their competitive superiority, due to the evolution of such factors as their special kind of teeth (called selenodont molars) and ruminant digestion. Again, the facts are not clear-cut. Cifelli analyzed the data for the diversity of the two groups for four separate continents. He found that the classic pattern of a correlated rise and fall existed only in North America. In Europe, Africa, and Asia, the artiodactyls do not appear to have proliferated at the expense of the perissodactyls (Figure 21.14). It is possible, therefore, that a competitive displacement occurred in North America, but the evidence for a consistent competitive superiority of artiodactyls to perissodactyls is poor.

The angiosperms (flowering plants) have replaced the conifers and ferns as the dominant terrestrial plants (Figure 21.15). The rise of the angiosperms and their replacement of the gymnosperms has been attributed to a number of factors. Angiosperms can grow more rapidly than gymnosperms under many conditions, which gives them a competitive advantage. Pollination and seed dispersal by insects is mainly confined to flowering plants, and it may have enabled them to colonize new habitats more rapidly. Specialized pollinator-relationships in angiosperms may also have increased their speciation rate (see section 22.2). Figure 21.15 suggests that the increase of the flowering plants may well have been at the expense of the gymnosperms. The two groups show a parallel increase and decline, rather than a decline of one group preceding the increase of the other.

A final example of a faunal replacement shows how both factors may operate in the same case. In the Palaeozoic, the dominant sessile filter-feeding invertebrate group was the brachiopods. They still survive today, but have largely been replaced

Figure 21.14 The diversities of perissodactyl (P) and artiodactyl (A) mammals in four continents. Diversities are numbers of genera. Reprinted, by permission of the publisher, from Cifelli (1981).

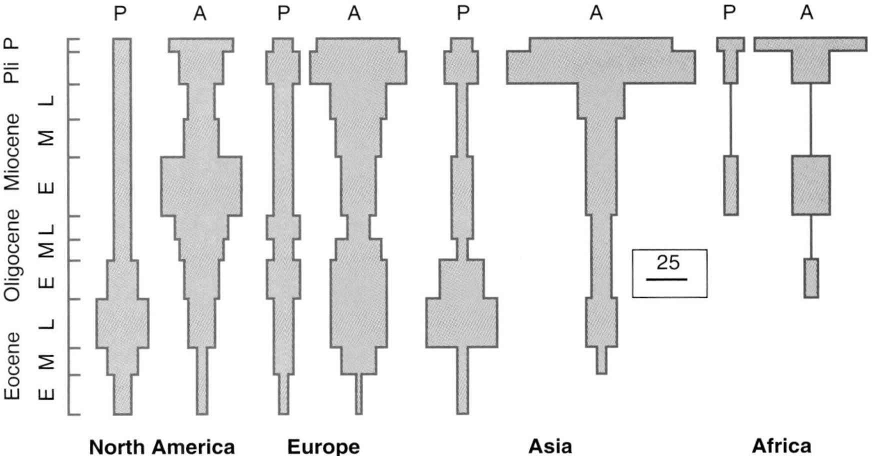

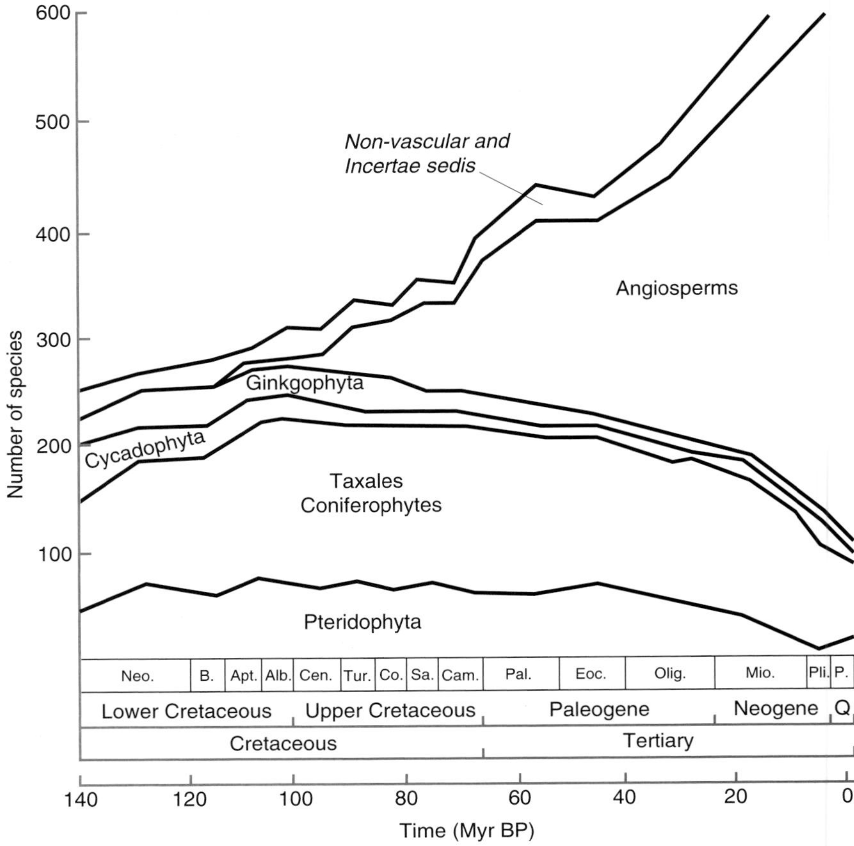

Figure 21.15 The rise of the angiosperms. Angiosperms (flowering plants) have gradually expanded in diversity since the Cretaceous. Gymnosperms (conifers, cycads, and ginkgos) have declined, as have pteridophytes. For some species, it is not known to which group they belong; these species are called "incertae sedis." Reprinted, by permission of the publisher, from Niklas (1986).

by another taxonomic group, the bivalve molluscs. Gould and Calloway plotted the diversity changes of the two groups during the crucial period around the Permo-Triassic (Figure 21.16). The bivalves increased relative to the brachiopods during the Permian. They then suffered less than the brachiopods in the mass extinction (see chapter 23) at the end of the Permian. Since that time, the bivalves have expanded at a more rapid rate. It is possible, therefore, that the replacement was due both to a competitive superiority of bivalves and to the better fortunes of the bivalves in the extraordinary conditions of the Permian mass extinction. (It is also possible that the bivalves' better survival in the mass extinction relates to their competitive superiority. That relation, however, would depend on what events took place at that time, and we have no evidence either way.) Mass extinctions, by causing a temporary reduction in the level of competition, may lead to phases in which ecological replacements are particularly likely to occur.

In summary, we have considered two reasons why one higher taxon may replace another, and how to test which theory has operated. First, competitive displacement has often been put forward as a hypothesis, but is difficult to support with evidence. Second, the rises and falls of higher taxa suggest that replacements often happen after the earlier group has gone extinct; the second group then radiates into the empty ecological space.

Figure 21.16 Replacement of brachiopods by bivalves through the Permo-Triassic mass extinction. The bivalves were gaining before the mass extinction, and survived through it relatively better. This example suggests (see Figure 21.13) that both relative adaptation and the fortunes of survival in the mass extinction may have influenced the replacement. Time is for the 25 million years preceding, and the 40 million years following, the Permo-Triassic boundary. Reprinted, by permission of the publisher, from Gould and Calloway (1980).

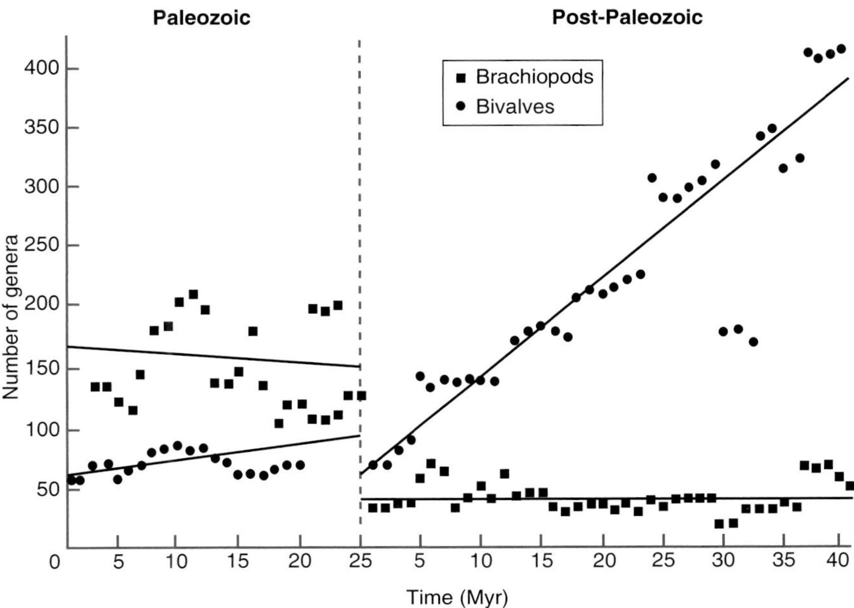

SUMMARY

1. The evolution of mammals from reptiles provides an example of adaptive evolution, and the fossil record reveals that it proceeded in a series of stages, through various groups of mammal-like reptiles.
2. The neo-Darwinian theory of the origin of higher taxa suggests that they evolve in many small adaptive stages, by natural selection. Macroevolution represents microevolution extrapolated over long periods.
3. Morphological transitions often take place by changes in the rate, and timing, of developmental processes.
4. A formerly popular idea—the principle of recapitulation—held that successive new evolutionary stages can only be added on to the adult stage; an individual then develops by "climbing up its family tree."
5. New evolutionary stages may be added at any stage of an individual's development. The relative timing and rate of different developmental processes can also shift, in a process called heterochrony.
6. The time of reproduction may be shifted earlier relative to somatic development (paedomorphosis) either because somatic development is slowed down (neoteny, as seen in the axolotl) or reproduction is accelerated (progenesis). Heterochrony can also shift reproduction to a relatively later stage of somatic development (by "acceleration" or "hypermorphosis").

7. Evolutionary biologists are interested in the relative frequencies of evolutionary changes early and late in development, and the frequencies of the different types of heterochrony.
8. Two theories have been put forward to explain a possible tendency for new evolutionary stages to be added more often in the adult than earlier in development: (1) early-acting mutations will be more likely to be macromutations; and (2) the forces of natural selection from the external environment are more intense on the independent adult than the protected early embryo.
9. It has been argued that new higher taxa often evolve by paedomorphosis from the larval stage of ancestors. The possible origin of the chordates from an animal like the tunicate larva is an example. The reason could be that larval stages are evolutionarily more flexible. The argument as a whole is questionable and its factual support uncertain.
10. Changes in the relative sizes of two organs can be studied by allometric graphs.
11. Changes in complex shapes can be studied in D'Arcy Thompson's transformational diagrams.
12. Mutations influencing developmental processes can produce macromutational phenotypic effects. Homeotic mutations in fruitflies, such as *antennapaedia* and *bithorax*, are examples. Developmental macromutations are usually (or even always) deleterious and evolutionarily inconsequential.
13. Goldschmidt and his followers argued that developmental macromutations serve as the source of new large evolutionary changes. No evidence supports this idea, and it is theoretically implausible, if not impossible.
14. Large-scale evolutionary replacements of one higher taxon by another may be due to competitive ecological displacement, or to the extinction of the earlier group for an unrelated reason followed by the radiation of the second group. The two ideas can be tested by calculating the exact timing of the rise and fall of the two groups.

FURTHER READING

Simpson (1953) and Rensch (1959), for example, give classic neo-Darwinian accounts of macroevolutionary morphological changes. On the mammal-like reptiles see Kemp (1982a,b, 1985), Hotton *et al.* (eds, 1986), Bramble and Jenkins (1989), Benton (1990), and Carroll (1988). See Wellnhofer (1990) on *Archaeopteryx*. There is a large body of modern literature on the relation between developmental change and macroevolution. Gould's (1977a) book discusses the history of recapitulatory ideas and modern work on heterochrony. Hall (1992) and Raff and Kaufman (1983) are more and less recent books. On heterochrony, see the book by McKinney and McNamara (1991) and a conference edited by McKinney (1988) that contains general chapters by Gould (on its significance), McKinney (on analysis), and McNamara (on frequency). McNamara (1988) discusses further fossil examples of heterochrony. McNamara (1989) is a popular article, and Slatkin (1987) discusses its genetics. For neoteny in human evolution, Gould (1977a) gives the case for and Shea (1989) the case against, and McKinney and McNamara (1991) include a chapter. Godfrey and Sutherland (1995) criticize the inference of heterochrony from size in fossils. The chapters in Spalding (1993) discuss the conservation of developmental mechanisms. See Kenyon (1994) and

Carroll (1995) on homeotic genes. On allometry see chapter 13. D'Arcy Thompson's book (1942) is one of the great books of biology; see also Medawar (1958), Gould (1971), and Bookstein (1977). Heterochrony is not the only idea in macroevolutionary morphology; for others, see Nitecki (ed, 1990). On replacements, see the general article by Benton (1987), the particular studies in the text and picture captions, and Crane *et al.* (1995) for more on angiosperms.

STUDY AND REVIEW QUESTIONS

1. What are the main differences between mammals and reptiles in (a) the skeletons of fossil species, and (b) modern living species?
2. What does the fossil record suggest about the mode of evolution of the mammalian characteristics?
3. What are two kinds of evolutionary change in development that cause descendant species not to recapitulate the forms of their ancestors? Do they invalidate von Baer's law?
4. If a descendant species, in its reproductive (adult) form, morphologically resembles a juvenile ancestral stage, (a) what is the descriptive term for this morphological pattern, and (b) what are two possible heterochronic processes that could produce it? How would you test which process identified in (b) applies in (c) modern species and (d) fossil species?
5. Describe two ideas about developmental change in evolution in which microevolution is uncoupled from macroevolution.
6. Why are the earlier stages of development evolutionarily constrained? In what kinds of species would you predict the earlier stages to be more constrained than in what other kinds?

Coevolution

COEVOLUTION happens when two or more species influence one another's evolution. It is most often invoked to explain coadaptations between species, and we begin this chapter by considering whether coadaptation provides good evidence for coevolution. Strictly speaking, coevolution requires reciprocal influences between species, but in a related phenomenon, called sequential evolution, changes in one species influence the other but not the reverse. We then consider parasites and hosts, concentrating on two topics, the evolution of virulence in parasites and whether the phylogenies of parasites and their hosts form mirror-images. Parasites and hosts provide an example of antagonistic coevolution, and antagonistic coevolution in general can lead to evolutionary "arms races." We look at some possible examples of escalating coevolution in the fossil record. The Red Queen hypothesis suggests that species continually evolve to maintain a level of adaptation against competing species, and Van Valen invented the hypothesis to explain a general result he discovered in the fossil record—the chance of extinction of a species in a taxonomic group is independent of the age of the species.

22.1 Coadaptation alone suggests, but is not conclusive evidence for, coevolution

Figure 22.1a shows an ant (*Formica fusca*) feeding on the caterpillar of the lycaenid butterfly *Glaucopsyche lygdamus.* The ant is not eating the caterpillar; it is drinking "honeydew" from a special organ (Newcomer's organ), the sole purpose of which seems to be to provide food for ants. The reason why the caterpillars feed the ants has been the subject of several hypotheses, and in 1981 Pierce and Mead carried out an experiment that suggested that the caterpillars, at least in *Glaucopsyche lygdamus*, feed ants in return for protection from parasites.

The caterpillars are parasitized by braconid wasps and tachinid flies. Alone, they are almost defenseless against these lethal parasites. The tending ants will fight off parasites from their caterpillars (Figure 22.1b). Pierce and Mead experimentally prevented ants from tending caterpillars, and measured the rates of parasitism in the experimentally unprotected and in normally protected (control) caterpillars. Their results show that ants reduce the rate of parasitism in *Glaucopsyche lygdamus* (Table 22.1). The ants and caterpillars are, therefore, closely adapted to one another; the ants gain food, and the caterpillars gain protection. Thus, the two form a kind of interspecific *coadaptation.* (Here the term refers to the mutual adaptation of two species; it has also been used to describe the

Figure 22.1 (a) An ant (*Formica fusca*) tends a caterpillar of the lycaenid butterfly species *Glaucopsyche lygdamus*. The ant is drinking honeydew, secreted from a special organ. (b) *Formica fusca* defending a caterpillar of *Glaucopsyche lygdamus* against a parasitic braconid wasp. The ant has seized the wasp in its mandibles. Bars indicate 1 mm (photos: Naomi Pierce).

mutual adaptation of genotypes (section 8.2, p. 197) and of parts (section 13.3, p. 342) within an organism.) Relationships like that between ants and lycaenids are called *mutualism*. The many examples of mutualism known provide some of the most charming stories in natural history.

How could the coadaptation between ant and lycaenid have evolved? The morphological structure and the behavior patterns of both ant and caterpillar appear to have evolved in relation to one another. It seems likely that, after the ancestors of the two species had become associated, natural selection would have favored mutually adapted changes in each species. Changes in one species, such as to increase honeydew production, would favor changes in the other (to increase protection) as the caterpillars became more beneficial to the ants. This kind of reciprocal influence is described as *coevolution*: each species exerts selection pressures on, and evolves in response to, the other species. The two lineages evolve together (Figure 22.2). We have previously encountered the idea that species evolve in relation to changes in the environment; in coevolution, the species' environment is itself evolving.

TABLE 22.1

Caterpillars of the lycaenid butterfly *Glaucopsyche lygdamus* are more likely to be parasitized if they are not tended by ants. Ants were experimentally excluded from some caterpillars, and the rate of parasitism on these and untreated control caterpillars was measured. The two sites are in Gunnison County, Colorado. Parasites were wasps and flies. *N* is the sample size. Reprinted, with permission, from Pierce and Mead (1981). Copyright 1981 American Association for the Advancement of Science.

	Caterpillars without Ants		Caterpillars with Ants	
Site	% Parasitized	*N*	% Parasitized	*N*
Gold Basin	42	38	18	57
Naked Hills	48	27	23	39

Figure 22.2 Coevolution means that two separate lineages mutually influence one another's evolution. The two lineages tend to (a) change together, and (b) speciate together. Lineages 1 and 2 could be, for example, an ant lineage and a lycaenid butterfly lineage.

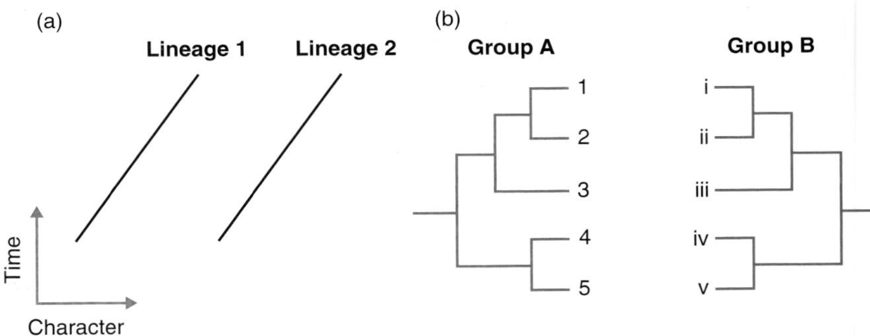

Logically, the observation of coadaptation between two species, such as an ant and a caterpillar, is not enough to confirm that the two have coevolved together. As Janzen pointed out, the two lineages might have been evolving independently, and it just turned out at some stage that the two forms were mutually adapted to each other. The ancestors of *Glaucopsyche lygdamus* might have evolved their Newcomer's organs for some other reason than feeding *Formica,* and the ants might have evolved anti-wasp behavior patterns for some other reason than defending caterpillars. When the two came together, they were coadapted. To demonstrate coevolution requires showing not only that two forms are coadapted now but also that their ancestors evolved together, exerting selective forces on one another.

That is a tall order. In practice, biologists tend to assume that interspecific coadaptations result from a long history of coevolution unless a convincing alternative hypothesis can be put forward. Janzen's stricture is logically correct, but difficult to carry out in practical biology. Further evidence that a coadapted system arose by coevolution can come from comparison with related species. The relation between *Glaucopsyche lygdamus* and *Formica* is not unique. Lycaenids and ants have evolved a large number of relationships in different species, which suggests the two groups have been evolving together for some time.

22.2 *Coevolution should be distinguished from sequential evolution*

In a paper that is perhaps the most influential modern discussion of coevolution, Ehrlich and Raven (1964) listed the food plants of the main butterfly taxa. Each family of butterflies feeds on a restricted range of plants, but these plants are in many cases not phylogenetically closely related. Ehrlich and Raven explained the diet patterns mainly in terms of plant biochemistry. Plants produce natural insecticides—chemicals like alkaloids that can poison herbivorous (phytophagous) insects. Insects, in the manner of pest species evolving resistance to artificial pesticides (section 5.8, p. 109), may evolve resistance to these chemicals, perhaps by means of detoxifying mechanisms. When a new detoxifying mechanism arises, it will open up a new array of food supplies, consisting of all plants that produce the now harmless chemical. The insects can feed on those plants, and will diversify to exploit the resource. The result will be that each insect group can feed on a range of food plants, the range being set by the capabilities of the insect's detoxifying mechanisms. The range of food plants will form a biochemical group, but need not form a phylogenetic group because unrelated plants could use the same defensive chemicals. Ehrlich and Raven's pattern of butterfly-plant relationships could arise as a result.

In turn, selection will act on the plants to evolve improved insecticides. Plant-insect coevolution should, therefore, consist of cycles, as plant groups are drawn into, and removed from, the diets of insect groups, and the insects "move" between plant types according to their biochemical abilities. The biochemical arms race between plants and insects should persistently favor new mechanisms on both sides, and might have promoted the diversification of insects and of angiosperms (section 16.6.3, p. 445; section 21.9, p. 601). In Ehrlich and Raven's words "the fantastic diversification of modern insects has developed in large measure as the result of a stepwise pattern of coevolutionary stages superimposed on the changing pattern of angiosperm variation."

The defining property of coevolution is the reciprocal selective influence between the two parties, over a prolonged period. For true coevolution between insects and plants, insects must influence plant evolution, and plants must influence insect evolution. Another possibility would be for one party to influence the evolution of the other, but not the reverse. Plants and insects appear to influence one another's fitnesses strongly, and it is widely believed that they have coevolved. Some biologists, however, have doubted the strength of these reciprocal influences. Jermy, for instance, suggests that plants and insects show

Figure 22.3 Sequential evolution means that change in one lineage selects for change in the other lineage, but not the reverse. Jermy hypothesizes that sequential evolution operates in insects and plants. Plant evolution then influences insects, but a change in an insect lineage does not select for a change in its foodplants. The pattern of change in (a) lineages and (b) phylogenies differs from strict coevolution (compare with Figure 22.2b). (a) Change in plants coevolves with changes in insects, but changes (for some reason other than changes in plants) in insects do not cause changes in plants. (b) When plants speciate, so do insects; when insects speciate, it has no effect on plants.

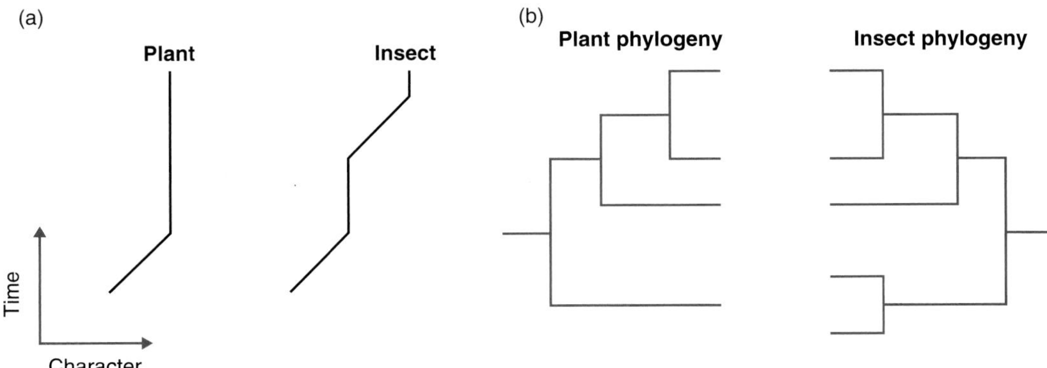

sequential evolution rather than coevolution—that is, plants influence the evolution of insects, but insects have less effect on evolution in plants (Figure 22.3). He offers a number of reasons. One is that many insects eat only one type of plant, whereas plants are eaten by many insects. When a plant changes, its insects will all have to change to keep up; when one insect species changes, it alone will exert only a small selective pressure on its food plant.

It is not yet known whether coevolution, sequential evolution, or even independent evolution is the better description of insect-plant relations. The challenge for research is to test between these possibilities by combining ecological studies of the reciprocal influences of plants and insects with phylogenetic evidence for the species involved.

22.3 *Coevolutionary relations will often be diffuse*

The clearest examples of coevolution come from ecologically coupled pairs of species. In practice, each species will experience, and exert, selective pressures on many other species. The evolution of a species will consist of an aggregate response to all of its mutualists and competitors, and any evolutionary change may not be easy to explain in terms of any single competitor. This process, which is called *diffuse coevolution*, undoubtedly operates in nature. Indeed, diffuse coevolution may be the main force shaping the evolution of communities of species. This concept is difficult to study, however, and its importance is consequently controversial.

22.4 *Parasites and hosts coevolve*

Step-by-step coevolution is particularly likely to take place between parasites and their hosts. They can have specific and close relations, and it is easy to imagine how a change in a parasite that improves its ability to penetrate its hosts will reciprocally set up selection for a change in the host. If the range of genetic variants in parasite and host is limited, coevolution can be cyclic (section 11.3.6, p. 294); but if new mutants continually arise the parasite and host may undergo unending coupled changes that may or may not be directional according to the type of mutations that arise. Coevolution in parasites and hosts is *antagonistic*, unlike the mutualistic coevolution of ants and caterpillars. Coevolution can be antagonistic or mutualistic according to the circumstances of the species.

Many properties of the biology of parasites and hosts have been attributed to coevolution. In this chapter we will concentrate on two such properties. The first is parasitic virulence, which indicates how destructive the parasite is of its host. This property can be measured precisely as the reduction in fitness of a parasitized host relative to an unparasitized host. A highly virulent parasite is one that kills its host quickly, reducing the host's fitness to zero. The virulence of a parasite is normally thought to be a side effect of the manner in which the parasite lives off its host. If, for instance, a parasite consumes a large proportion of its host's cells, it will be more likely to kill its host and will, therefore, be more virulent than a parasite that consumes less host cells. The second topic we shall examine is whether the phylogenies of parasites and their hosts have the same shape, implying that the two speciate at the same time.

22.5 *The evolution of parasitic virulence*

22.5.1 *Parasitic virulence and host resistance showed evolutionary changes in Australian myxoma virus and rabbits*

The myxoma virus (which causes myxomatosis) in Australian rabbits provides the classic illustration that the virulence of a parasite can change evolutionarily. The rabbits in question belong to a species (*Oryctolagus cuniculus*) that is native to Europe but was introduced to Australia, where it thrived and became a pest. The natural host of the myxoma virus is another kind of rabbit, *Sylvilagus brasiliensis*, of South America; the myxoma virus probably has low virulence in *S. brasiliensis*. In 1950, the virus was deliberately introduced into Australia in an attempt to control the pestiferous rabbits. It was, initially, a deadly success. It spreads (in Australia, at least) from rabbit to rabbit by means of mosquitoes. The large population of those biting insects enabled the myxoma virus to sweep through the southeast Australian rabbit population, and it had moved around the south coast as far as Perth in the west by 1953. Myxomatosis initially almost annihilated the rabbit population—it declined by 99% in some hard-hit areas. The virus was introduced into France in 1952, and began to spread through Europe; it was surreptitiously introduced into the United Kingdom in 1953.

The myxoma virus was highly virulent when it first hit the Australian (and the European) rabbit population. In fact, it killed 100% of infected hosts. Soon, however, the kill rate declined. This decline could have resulted from any combination of increasing host resistance and decreasing viral virulence, and normally we should not know which force was operating. In this case, a carefully controlled set of experiments allowed the two factors to be teased apart.

The decline in virulence of the myxoma virus was demonstrated by infecting standard rabbit strains in the lab with the viruses taken from the wild in successive years. Because the rabbit strain was controlled and constant, any decline in the kill rate must be due to a decline in virulence in the virus. Table 22.2 shows the results for Australia and Europe. In both places the virus started off maximally virulent (killing 100% of infected rabbits), until a rapid increase in the less virulent strains in the viral population occurred. The less virulent strains kill a lower proportion of infected rabbits and take longer to kill them when they do.

Meanwhile, the rabbits were also evolving resistance. This process could be shown by challenging wild rabbits from a series of times with standard strains of the virus. In that case, the virus is held constant and any decline in kill rate must be due to changes in the rabbits. Table 22.3 shows the results of a series of such experiments through the 1960s and 1970s, in which resistance manifestly increased.

Therefore, parasitic virulence and host resistance can evolve. Natural selection will always favor increased resistance in hosts, but how will it operate on virulence in parasites?

TABLE 22.2

Myxoma virus evolved lower virulence over time after its introduction into Australia, France, and Great Britain. Strains of the virus are classified into five virulence grades: I is the most virulent, V is the least. The table shows the percentages of the different strains in rabbits in the wild through time. Modified from Ross (1982), who compiled the results from a number of sources.

	Virulence Grade					
	I	II	IIIA	IIIB	IV	V
Australia						
1950–1951	100	0	0	0	0	0
1958–1959	0	25	29	27	14	5
1963–1964	0	0.3	26	34	31.3	8.3
France						
1953	100	0	0	0	0	0
1962	11	19.3	34.6	20.8	13.5	0.8
1968	2	4.1	14.4	20.7	58.8	4.3
Great Britain						
1953	100	0	0	0	0	0
1962–1967	3	15.1	48.4	22.7	10.3	0.7
1968–1970	0	0	78	22	0	0
1971–1973	0	3.3	36.7	56.7	3.3	0
1974–1976	1.3	23.3	55	11.8	8.6	0
1977–1980	0	30.4	56.5	8.7	4.3	0

22.5.2 *Natural selection can favor higher or lower virulence according to the transmission mode of the parasite, and other factors*

One idea about how natural selection will work on virulence is that it will usually act to reduce it. Parasites depend on their hosts, and if they kill their hosts they will soon be dead as well. It has been argued, therefore, that parasites will evolve

TABLE 22.3

Rabbits evolved resistance over time after the introduction of myxoma virus. These results are for wild rabbits (*Oryctolagus cuniculus*) caught at different times in two regions (called Mallee and Gippsland) of Victoria, Australis. The rabbits were then challenged with a highly virulent standard laboratory strain (SLS) of the myxoma virus. The strain caused 100% mortality in unselected rabbits. From Fenner and Myers (1978), data from Douglas *et al.*

	Mallee		Gippsland	
	Number Tested	Mortality (%)	Number Tested	Mortality (%)
Unselected rabbits		100		100
1961–1966	241	68	169	94
1967–1971	119	66	55	90
1972–1975	73	67	482	85

to keep their hosts alive. The objection to this argument, and the reason why it is almost universally rejected by evolutionary biologists, is that it is group selectionist (section 12.2.5, p. 325). While the long-term interest of a parasite species dictates that it not destroy the resource it lives off, natural selection will favor individual parasites that reproduce themselves in the greatest numbers over those that restrain themselves in the interest of preserving their hosts. The short-term individual advantage of greater reproduction will usually outweigh any long-term group or species advantage of reproductive restraint.

A number of other factors may operate in this case, including the number of parasites that infect a host. If the host is infected by one parasite, all parasitic individuals will be the offspring of the original colonizer and they will all be genetically related brothers and sisters. Kin selection (section 12.2.4, p. 321) will then operate to reduce any selfish proliferation within the host. If the host has suffered multiple infections, by contrast, the parasites will be unrelated. Natural selection will favor individual parasites that can consume as much of the host as possible, as fast as possible, before any of the other parasites take advantage of the resource. Thus, virulence will increase. If an individual restrains itself to preserve the host, other parasites will step in to take it over. We therefore predict lower virulence in diseases with single than with multiple infections.

A second factor is whether *vertical* or *horizontal transmission* of the parasites between hosts occurs. In an external parasite, transmission may mean the movement of an adult parasite that has been living off one individual host to another host. In internal parasites, it typically means the movement of the offspring of parasites living inside one host to another host. In vertical transmission, a parasite transfers from its host to the offspring of that host. This transition can take place by a variety of mechanisms—by the mother's milk, by jumping from host parent to host offspring when the two are near one another, or inside the gamete. In horizontal transmission, the parasite transfers between unrelated hosts, and not necessarily from parent to offspring. Transmission may occur through breathing, by a vector such as a biting insect, or by copulation of one host with another. What consequence does the method of transmission have for the evolution of virulence? A vertically transmitted parasite requires its host to reproduce to provide resources for itself or its immediate offspring; horizontally transmitted hosts have no such requirement.

Consider the success of a more and a less virulent strain of parasite in the two cases. A vertically transmitted parasite experiences a trade-off between making more offspring and the success of those offspring. A parasite that reproduces more will have greater virulence as it uses up a higher proportion of the host, but it will reduce the host's reproduction and the resources available for its own offspring. This trade-off will place an upper limit on virulence. A horizontally transmitted parasite experiences no such trade-off—the success of its offspring is independent of the reproduction of its host. Virulence is, therefore, much less constrained.

In nature, single infections and vertical transmission often occur together, and both factors may work side by side to reduce parasitic virulence. A comparative study, by Herre, of 11 species of nematode worms that parasitize fig wasps in Panama illustrates the idea. The fig wasp life cycle is as follows. The adult female fig wasp, which carries pollen from the fig from which she emerged, enters one of the structures that eventually ripen into a fig; there she pollinates the fig, lays her eggs, and dies. The eggs grow up and emerge within the fig. They then mate, and the females pick up pollen and exit the fig in search of another fig in which to lay their eggs. An important point is that fig wasp species vary in the number of fig wasps that enter a fig. In some species, only one colonizes each fig; in other species, a number of females typically enter and lay their eggs in the same fig.

Nematode worms live off fig wasps. In Panama, a different species of nematode lives off each species of fig wasp. The immature nematodes crawl on to a fig wasp after she emerges in the fig. "At some point [in Herre's words], the nematodes enter the body cavity of the wasp and begin to consume it." The nematodes emerge as adults from the body of the dead wasp and mate and lay their eggs in the same fig as the wasp did; the cycle can then repeat itself. Nematodes that live off the kind of fig wasps in which only one wasp enters a fig will tend to be vertically transmitted, and the nematodes on any one host will be genetically related. Those living off the kind of wasp in which several females may enter the same fig can be horizontally transmitted; nematodes from a number of different parents may crawl on to the same wasp and they will be genetically unrelated. We can, therefore, predict that the parasitic nematodes of fig wasps in which only one female typically enters a fig will have evolved lower virulence than those of fig wasps in which more than one female typically enters a fig. Herre's results appear in Figure 22.4 and show the predicted relation. The virulence of

Figure 22.4 Virulence is higher in nematode species that parasitize fig wasps in which more individual wasps lay their eggs in a fig. The results are for 11 species of fig wasps and the 11 species of nematodes that parasitize them (one nematode species per fig wasp species). Virulence is measured by the average number of offspring produced by parasitized relative to unparasitized wasps in each species; virulence is higher down the *y*-axis. A virulence of 1 means parasitized and unparasitized wasps leave the same number of offspring. These vertically transmitted parasites are so mild as practically to be commensals. The offspring leave a record inside the fig and can be counted accurately. The proportion of figs entered by one or more wasps can also be measured. Reprinted with permission from Herre (1993). Copyright 1993 American Association for the Advancement of Science.

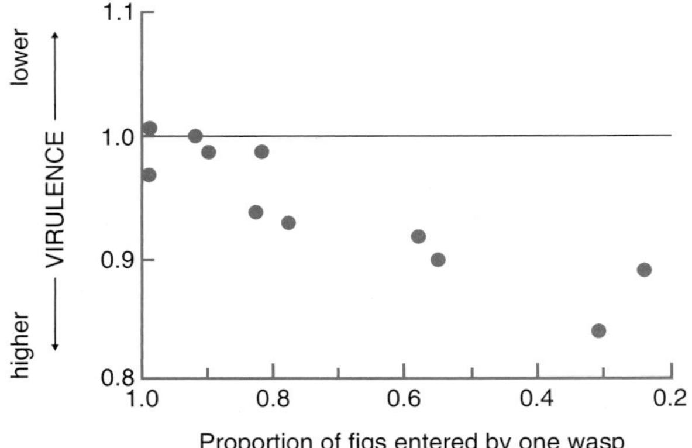

the parasite appears to have been tuned by natural selection to the habits of the host.

Kin selection, and vertical as opposed to horizontal transmission, are just two of the evolutionary factors that have been hypothesized to influence virulence. Most of the other factors have not been so well supported empirically as these two are in Herre's study, in which lower virulence had evolved when the parasites on a host were related and vertically transmitted than when they were less closely related and more horizontally transmitted.

22.6 *The phylogenetic branching of parasites and hosts may be simultaneous*

A further coevolutionary result can be seen in the phylogenies, or classifications that more or less reflect those phylogenies, of parasites and their hosts. They often form the kind of mirror-images we expect for coevolving taxa (Figure 22.2). The correspondence suggests that speciation in parasites and their hosts is approximately simultaneous, a relation called Fahrenholz's rule.

Recent work by Hafner *et al.* (1994) on the rodent family Geomyidae (the pocket gophers) and their ectoparasitic lice (Mallophaga) provides a good example of Fahrenholz's rule. Hafner and his colleagues sequenced a mitochondrial gene in 14 species of pocket gophers and their parasitic lice, and used the data to construct the phylogenies of the two groups (Figure 22.5a). The phylogenies are nearly mirror images; with some deviations. The mirror pattern is attributed to *cospeciation*, and the deviations to *host switching*. In the figure, for example, the top of the figure on both sides includes a split that produced the branches labeled E and F; the branch E on the host side probably represents an ancestral pocket gopher that was parasitized by the lousy ancestor of the four species in the branch E of the parasites. That ancestral louse species then split twice. The events down the E to C to A + B branch (moving up the figure) look very like cospeciation. Other events, such as the parasitism of *Thomomys bottae* at the bottom of the figure by *Geomydoecus actuosi*, appear more like host switching.

It is possible to count the frequencies of cospeciation and host switching by inspecting phylogenies like those in Figure 22.5a, in a statistically approriate way. This case demonstrates a significant tendency toward cospeciation. Figure 22.5b provides a stronger test of cospeciation. Hafner *et al.* used the estimated number of changes in each branch as a molecular clock to estimate the time when the branch originated. The clock runs faster in the lice, probably because their generation times are shorter than those of their hosts (section 7.10.2, p. 178). If real cospeciation occurred, the speciation events in host and parasite should have been simultaneous. Hafner *et al.* used two molecular clocks—one for all nucleotide substitutions and another for the synonymous nucleotide changes. As we saw in chapter 7, synonymous changes are more likely to be neutral and probably provide a more accurate clock. In both cases (Figure 22.5b), the points for the branch lengths cluster around the line for simultaneity, but the fit is better for the synonymous change clock. Figure 22.5b provides good evidence that the host and parasite species tended to speciate at the same time.

Figure 22.5 Mirror-image phylogenies in parasites and hosts. (a) The phylogenies of 14 species of pocket gophers (Geomyidae) and 17 of their mallophagan parasites. The phylogenies were reconstructed from the sequence of a mitochondrial gene (cytochrome oxidase subunit 1) using the parsimony principle. The phylogenies mainly form mirror images, but several cases of probable host switching occur as well. The pattern illustrates Fahrenholz's rule. A pocket gopher (*Geomys bursarius*) and louse (*Geomydoecus geomydis*) are also illustrated. (b) Test of simultaneity of speciation in parasites and hosts. The estimated number of substitutions in various branches in the host phylogeny are plotted against the numbers in the mirror-image branches in the parasite phylogeny. Letters on the graph refer to lettered branches in (a). The clocks in the two taxa probably run at different rates because of difference in generation time. If the speciation events were really simultaneous, the points should fall on the line. The fit is better when only synonymous changes are counted. Reprinted with permission from Hafner *et al.* (1994).

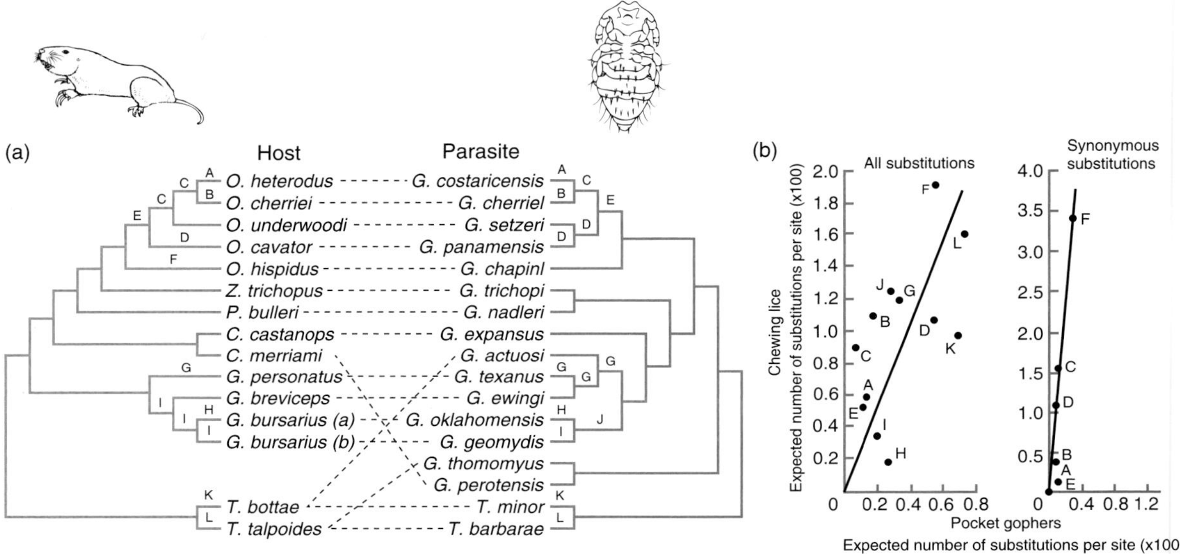

Why should host and parasite speciate synchronously? Probably because the same circumstances favor speciation in both groups. Suppose, for example, that pocket gophers speciate geographically. When the host population becomes geographically subdivided, their parasites' populations will automatically be subdivided as well. (Mallophaga have limited independent powers of dispersal.) The conditions for speciation in the parasite are then set up whenever the host is speciating. The process could be reinforced if each parasite needs a distinct set of adaptations to live on a particular host species.

The process requires that the parasite cannot move independently of its host. Timm has found evidence that a comparable group of lice that can move independently do not show mirror-image phylogenies. The dispersal abilities of parasites should generally influence whether they speciate simultaneously with their hosts. The parasitic lice in Figure 22.5 move with their hosts; for them, therefore, the observed mirror-image phylogenies of host and parasite make sense.

22.7 *Coevolution can proceed in an "arms race"*

A graph of brain size against body size for many vertebrate species reveals that larger vertebrates have larger brains (Figure 22.6). The idea is that brain size is usually measured as the relative *encephalization quotient*—that is, the deviation from the line for all species in Figure 22.6. A "large" brain is one above the line; it is larger than would be expected for an animal of that body size. Brain size is determined by two factors: the main one is body size, with intelligence constituting a secondary factor. The more intelligent animals, in a loose sense, are those that deviate further above the line; they have greater relative encephalization.

Let us provisionally accept here that the encephalization quotient offers an index of intelligence. We can then consider, from the work of Jerison, a possible example of coevolution between prey and predator. (We met a related part of Jerison's general study when we discussed the Great American Interchange in section 18.6, p. 525.)

In Cenozoic mammals, predators typically possess relatively larger brains than their prey. This relation can be seen in Figure 22.7. The figure also shows another, more interesting fact. The relative brain sizes of both predators and prey have increased through time. To estimate encephalization in a fossil, it is necessary to estimate body size, and the method used by Jerison has been criticized. The result is therefore uncertain, but Jerison's explanation is nevertheless interesting. He suggests that natural selection has favored higher intelligence both in the prey, to escape predators, and in predators, to catch prey. There has been a coevolutionary arms race between predator and prey, leading to ever larger brain sizes in both. The selective forces might well be truly coevolutionary, with each party exerting a reciprocal selective pressure on the other. That is, as predators become more clever at catching prey, the prey are selected to avoid them more intelligently, and vice versa. In this case, the evidence remains inconclusive, as it is difficult to demonstrate coevolutionary relations in fossils and Jerison's work is well worth noticing as a rare example in which coevolution is a plausible explanation.

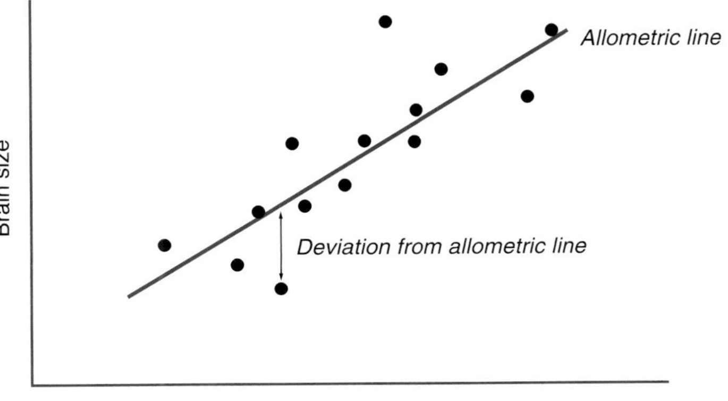

Figure 22.6 Relative brain size may be measured as relative encephalization by the deviation of a species' brain size from the allometric line for many species. Relative encephalization measures whether a species has a brain larger or smaller than would be expected for an animal with its body size. The species illustrated in the figure has a relatively small brain.

Figure 22.7 The distribution of relative brain sizes for carnivores (predators) and ungulates (prey) through the Cenozoic. Brain size increased through time, and at any one time carnivores had bigger brains than ungulates. Reprinted, by permission of the publisher, from Jerison (1973).

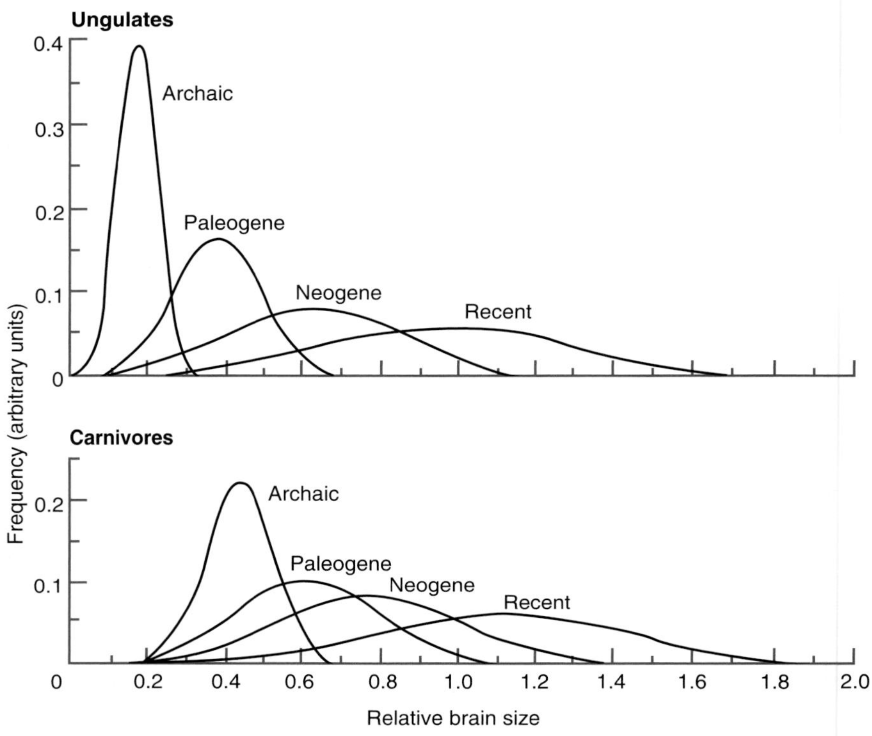

22.8 *Coevolutionary arms races can result in evolutionary escalation*

Vermeij has applied the same argument as was used by Jerison, but much more generally. Vermeij suggests that predators and prey typically show an evolutionary pattern that he calls *escalation*. By escalation, he means that life has become more dangerous over evolutionary time—predators have evolved more powerful weapons and prey have evolved more powerful defenses against them. Vermeij distinguishes escalation from evolutionary progress. If evolution is progressive, organisms will become better adapted to their surroundings through evolutionary time; if it is escalatory, the improvement in predatory adaptations may be matched by improvements in prey defenses, and neither gains any advantage over the other. If evolution is progressive in predators, for example, then later predators would be better at catching their prey than were earlier predators. If, however, evolution is escalatory, later predators will be no better than their ancestors at catching their contemporary prey types—but if those later predators were transported in a time machine and set loose on the prey hunted by ancestral predators, later predators should cut through ancestral prey like a modern jet fighter in a dogfight with an early biplane.

In his book *Evolution and Escalation*, Vermeij discusses both biogeographical and paleontological evidence for escalation. His evidence comes mainly from molluscs in shallow-water marine environments. The molluscs are abundant as fossils, and the nature of a shell itself can reveal how strongly adapted a species was, as either predator or prey. More strongly defended shells have properties such as general thickening, or thickening concentrated around their apertures. Burrowing species, or those that cement themselves down, are better defended that those that lie loose on the bottom surface. The degree of escalation in a habitat can be estimated by the proportions of more and less powerfully defended prey, and equivalent measures of the power of predators. Vermeij's biogeographical evidence suggests that some places are currently more dangerous than others for molluscs. The faunas of freshwater habitats, for instance, show fewer signs of escalation than shallow marine ones, and within shallow marine environments, escalation is highest in the western Pacific and Indian Oceans, intermediate in the East Pacific, and lowest in the Atlantic.

We can look at Vermeij's fossil evidence for escalatory coevolution in both predatory and prey species. The evidence for predators is, as we shall see, less convincing than for prey, but both classes of evidence are worth considering. The earliest predatory "molluscivores" in the fossil record were what Vermeij calls "whole body ingestors" that simply swallowed molluscan prey whole—an unspecialized method of eating a mollusc and for that reason perhaps not a particularly menacing one. By the Devonian, specialized shell breakers have appeared—in the case of fish, this development can be inferred from their teeth—and by the late Devonian 10 fish families of specialized shell breakers existed. Decapods, which crush shells in their claws, proliferated in the Triassic; gastropods that drill holes in other shelled species appeared in the late Triassic; and starfishes, with their specialized suckers and pullers that enable them to pry bivalved shells apart, first appeared in the Jurassic. By the late Mesozoic, many more shell-breaking fish forms had appeared.

This narrative is consistent with an increase in the number, and specialization, of molluscivores through time, but it does not provide persuasive evidence for escalation. The total number of forms recorded as fossils has increased through time, and trends through time inevitably lead toward increased numbers of many things. It is not possible to show that life had become more dangerous for molluscan prey by showing that larger numbers of specialist predators on molluscs existed later in the fossil record. There may also be more non-molluscivorous predators and weakly molluscivorous predators as well as more specialized molluscivores. To show escalation, we must show the number of escalated predatory forms has proportionally increased. For the prey species, Vermeij has found some evidence of that kind.

The most obvious measure for escalation in prey defences would perhaps be shell thickness, but Vermeij's best evidence is for other characters. Consider first the evidence on shell repair. When a mollusc is attacked but survives, it repairs the damage to its shell and the repair pattern can be observed in the shell. Proportions of shells showing signs of repair have been measured in several fossil faunas, and the trend appears to be toward increasing amounts of repair over time (Figure 22.8). Vermeij interprets this evidence as meaning that the prey have been suffering higher frequencies of predatory attacks over evolutionary

Figure 22.8 Incidence of repair at five successive time periods, for shells divided into three size classes. Note that the incidence of repair is higher in more recent times. Larger shells show a relatively high incidence of repair compared with small shells in more recent times, relative to earlier times. Large shells tend to be more resistant to breakage than small ones. One interpretation of this trend is that more recent predators have become stronger, and therefore able to injure large-shelled animals. From Vermeij (1987).

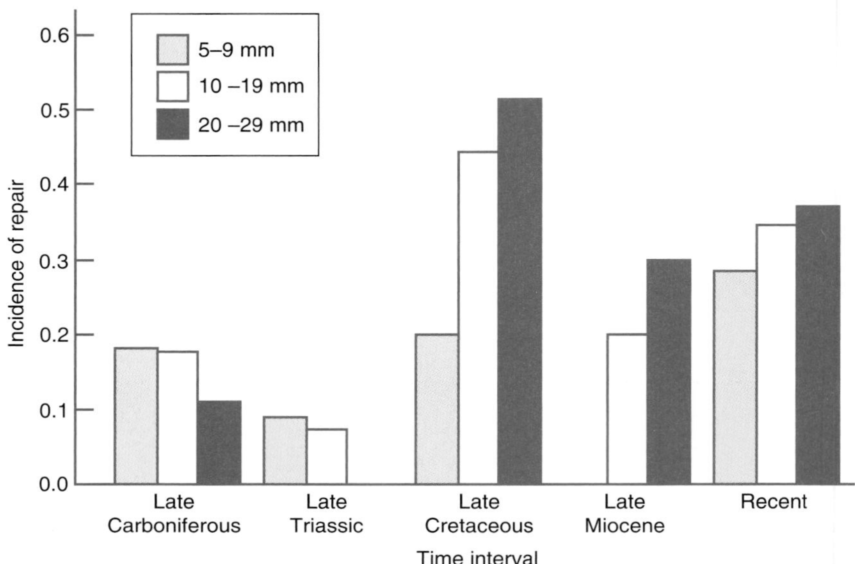

time. (It could also mean—though this is unlikely—that the predators have de-escalated from forms that destroyed their prey to forms that sometimes merely injured them!) The escalation of molluscan prey defenses is also suggested by a trend in the proportion of different types of gastropod shells through time. Shells that lie loose are less well defended than burrowers and attached shells, and if we look at the proportions of the different types through time, we see that the loosely attached forms have decreased (Figure 22.9). Vermeij also found limited evidence that the better defended burrowers increased from being about 5–10% of genera in the Late Carboniferous-Late Triassic to about 37% in the Late Cretaceous and 62–75% in modern formations. Internal thickening or narrowing of the aperture is another form of escalated defense; these types have also proportionally increased through time (Figure 22.10). Thus, we can conclude that more recent molluscs are more strongly defended than were their earlier ancestors.

Vermeij's evidence would not be persuasive enough to convince a skeptic. The evidence is not abundant; it is noisy; the patterns are not all consistent; and it is difficult to prove that the fossil record is unbiased. For example, could some unidentified reason explain why shell forms with narrower apertures became relatively more likely to be preserved as time passed? Biases of this general sort (in which some forms are more likely than others to be preserved at certain times) do exist in the fossil record. In the absence of any particular argument that sampling biases caused these trends, we can give the evidence the benefit

Figure 22.9 The incidence of sessile or sedentary uncemented gastropods through time. Note that the proportion decreases. Each point is for one fossil assemblage. For each assemblage, Vermeij divided the gastropods into different types (burrowers, sessile, attached forms, and so on), adding up to 100%. This graph gives the proportion of gastropods that lie unattached on the bottom surface. Neogene = Miocene + Pliocene (see Figure 19.1, p. 539). Reprinted, by permission of the publisher, from Vermeij (1987).

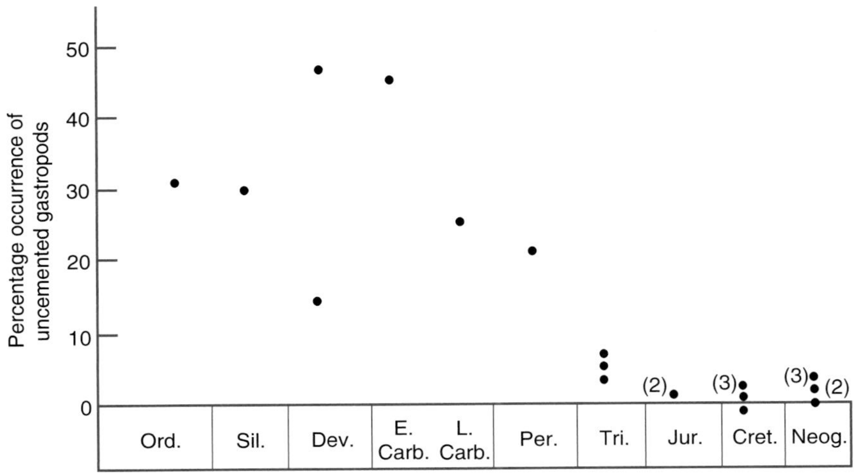

Figure 22.10 (a) Total number of gastropod subfamilies through time. (b) Proportion of subfamilies with members that have evolved internally thickened or narrowed apertures, a character that probably evolved as a defense against predators. Because the total number of subfamilies has increased, it is necessary to plot the *proportion*, not the total *number* of subfamilies, to show a trend. Reprinted, by permission of the publisher, from Vermeij (1987).

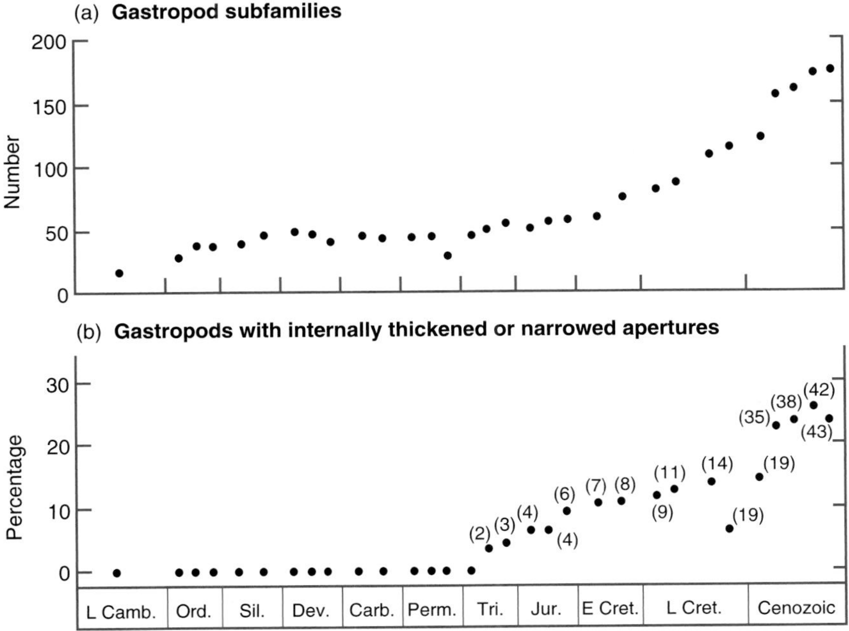

of the doubt. It does then suggest that some escalation of prey defenses and predatory weapons has taken place in evolution.

22.9 *The probability that a species will go extinct is approximately independent of how long it has existed*

We shall now stay with the question of how influential coevolution has been in the history of life, but shift the scale of the evidence to a more abstract, and general, level. We shall look at Van Valen's inference, from the shape of taxonomic survivorship curves, that macroevolution is shaped not only by coevolution, but also by a particular mode of coevolution called the "Red Queen" mode.

We encountered taxonomic survivorship curves in chapter 20 as a means of studying evolutionary rates (section 20.10, p. 576). Here we will use them to demonstrate a property of extinction rates. To plot a taxonomic survivorship curve, take a higher taxonomic group, such as a family or order, measure the duration in the fossil record of its member species, and plot the number (or percent) of species that survive for each duration (Figure 20.13, p. 578). In 1973, Van Valen published a large study, based on measurements of the durations of 24,000 taxa, but with one crucial difference from earlier studies—he plotted the graphs after taking logarithms of the numbers surviving. He found that survivorship tends to be approximately linear on the log scale, so that survivorship is log linear (Figure 22.11).

Van Valen's result created new interest in using survivorship curves in evolutionary theory. With the survivorship curves like Figure 20.13 on an arithmetic scale, it could be seen that different taxa evolved (taxonomically) at different rates, and it was possible to argue about why that trend occurred. When it was noticed that the curves were linear on a log scale, more interesting questions could be asked. The degree of linearity of Van Valen's results has been questioned, and many of the curves do not apply to species, but to genera or even families within a higher taxon. Moreover, some of the extinctions will almost certainly have been pseudo-extinctions, due to taxonomic division of a continuous lineage rather than the true extinction of lineage (section 23.2, p. 640). Van Valen's work does show a strong enough suggestion of logarithmic linearity in taxonomic survivorship curves for us to question what it means.

The answer is that species do not evolve to become any better (or worse) at avoiding extinction. The straight line on the logarithmic survivorship curve means that the chance a species will go extinct is independent of its age. Of the species that survived to a time t million years after their origins, a certain proportion go extinct by time $t + 1$ million years; of the survivors to time $t + 1$, the same proportion go extinct by time $t + 2$; and so on. An exponential decay of species occurs, with a constant proportion of the survivors going extinct in the next age unit. This situation need not have occurred. If, for example, evolution is progressive, the probability of extinction might decrease with time, as the level of adaptation improves; later species would then last longer. For this reason, Van Valen's result provides an important piece of evidence that evolution is not generally progressive. Alternatively, if evolution is escalatory (section 22.8),

Figure 22.11 Three taxonomic survivorship curves plotting the number of genera (for mammals and Osteichthyes) or families (for reptiles) surviving for various durations in the fossil record. Note that the lines, with this logarithmic *y*-axis, are approximately straight. Osteichthyes are bony fish, and the two groups drawn on that graph are subgroups of bony fish. Reprinted, by permission of the publisher, from Van Valen (1973).

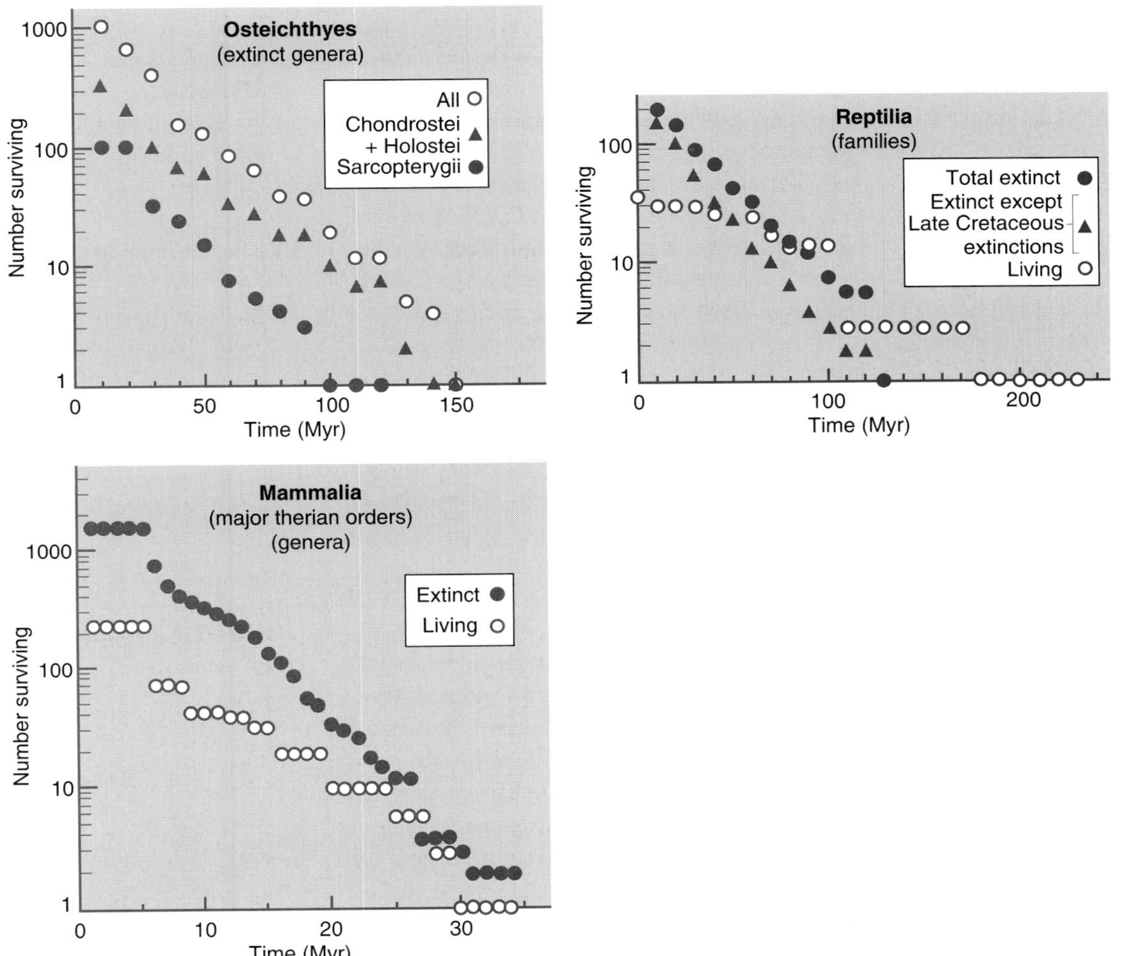

the level of individual adaptation under normal conditions relative to competitors may not change in any particular direction, but as a species such as the molluscs discussed earlier come to invest in heavier armaments, it might become more vulnerable to environmental stresses and extinction rate might increase with age. As a purely theoretical possibility, every species might also have a fixed lifespan and go extinct a certain time t after its origins. Van Valen's result suggests that no such process leading to an increase or decrease in the chance of extinction generally operates.

We can distinguish two extreme interpretations of the log linear survivorship curves, though in any real taxonomic group both could operate in any mixture (Figure 22.12). Interpretation (1) is that each species has a constant chance of going extinct each year. We can call this constant extinction. Alternatively, (2) the chance a species goes extinct could vary between years, but be the same for all species, independently of how long they have existed. The two interpretations can be illustrated by analogy with a human actuarial survivorship curve. A log linear survivorship curve for individuals means that the chance that an individual dies at any age is independent of age. Thus, the chance of dying at age 20 equals the chance of dying at age 21, 22, and so on. The two interpretations are then as follows.

1. Each individual has a constant chance of dying each year—a chance *d* in 1996, the same chance *d* in 1997, in 1998, and so on (we also assume no "baby boom" effect is present, and that people are born at random times).
2. The chance of dying varies between years. The chance might be very low, perhaps 0.001, every year through the 1990s, but then shoot up to perhaps 0.8 in the year 2000, if a plague or some other disaster strikes. If the disaster hits young and old alike, the survivorship curve will again be log linear. To obtain a nonlinear curve, the chance of death must vary with age. For example, if the plague carries off more old people than young, the survivorship would slope downward.

A log linear survivorship in a human population would mean that people do not become better (or worse) at avoiding death as they grow older. Similarly, a log linear taxonomic survivorship curve means that species do not become better or worse at avoiding extinction as they persist through time. Extinction factors—whatever they are—hit species independently of their length of existence.

Figure 22.12 A linear survivorship curve need not imply that the probability of species extinction is constant through the lifetime of the higher taxon. The two graph have crucially different *x*-axes. (a) Extinction probabilities in absolute time: constant and non-constant. (These probabilities correspond to interpretations (1) and (2) respectively in the text.) Both will produce (b) a log linear taxonomic survivorship curve. The *x*-axis has the origin of every species at time 0 and plots the numbers surviving for various durations.

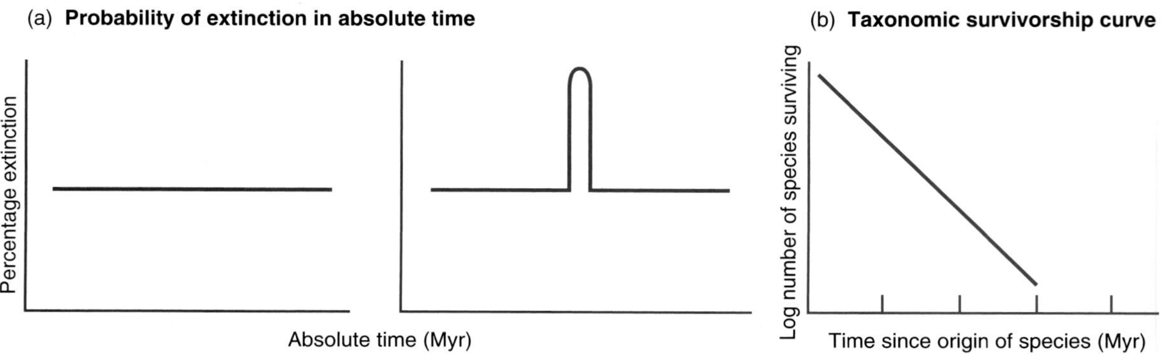

22.10 *Antagonistic coevolution can have various forms*

What can we deduce from the taxonomic survivorship curves about the factors that cause extinction? A first division of extinction factors classifies them as either physical or biological. Physical factors could be catastrophic or more gradual in action. A geological catastrophe might, for instance, hit an area once every million years and wipe out a certain proportion of the species. Alternatively, a steadier elimination of species might be caused by a more constant physical, perhaps climatic, change. There is no puzzle about which pattern of extinction rates results in either case. In the first case, they are low with sudden increases at the times of the catastrophes; in the second case, they are intermediate in level and less variable. Either could lead to log linear survivorship curves under interpretation (2) above.

The biological factor most likely to be causing extinction is antagonistic coevolution. As ecological competitors, or parasites and hosts, evolve against one another, if one competitor fails to evolve an adaptive improvement to keep up with its antagonists, it may go extinct. If a host species evolves a new kind of immunity, then if the parasite does not soon evolve a way of penetrating the defense, it will go extinct. What pattern of extinction rates is likely to apply under this process?

Stenseth and Maynard Smith have discussed the question theoretically in terms of what they call *lag load*. The lag load has the standard form for loads (section 7.3, p. 157). We imagine that at any time the members of a species could have an optimum state; a genotype with that optimum has fitness w_{opt}; it is defined such that $w_{opt} = 1$. The genotype that is needed to have that optimum fitness changes continually because of the changes in competing species. Each species will evolve toward its optimum (i.e., to have all of its members with the optimum genotype), but will lag a certain distance behind it. The actual mean fitness of the population is $\bar{w}$, a number usually less than 1. If all members of the population have the optimum genotype, then $\bar{w} = 1$. The lag load L is then defined as

$$L = \frac{w_{opt} - \bar{w}}{w_{opt}}$$

As the lag load L increases, the rate of evolution of the species will increase, because it is subject to stronger selection pressure. The lag load also controls the chance that the species will go extinct. As L increases, the species lags increasingly further behind its competitors and its chance of extinction goes up. As L approaches zero, its chance of extinction also decreases.

Coevolution, in this model, can have four forms. The first two are unstable. First, if a species lags behind its competitors and does not evolve fast enough to keep up with them, it will fall further behind until it goes extinct; this process is called the "contractionary" mode of evolution. Second, a species could be ahead of its competitors and outevolving them; it could theoretically expand until it had an infinite number of descendants. This process is called the "expansionary" mode. Clearly, the expansionary mode could not continue for long. In real systems, the main possibility is that a species could alternate between periods of expansionary and contractionary evolution as the fortunes of natural selection favored one species and then another among a group of competitors.

In the third evolutionary mode, which is called "stationary," the competing species evolve to a set of optimal states and then stay there. The lag load reduces to zero, no species changes, and no species goes extinct. In practice, biological coevolution would always take the species to a stable equilibrium, from which it could be perturbed by physical events. After a perturbation, the species would evolve back to a stationary equilibrium. At the stationary equilibrium, coevolution is not driving species extinct; any extinctions would be due to physical processes. Stationary evolution may be a common kind of coevolution. If an organism has one optimum form to compete with members of other species, its species will evolve to that form and stay there.

The fourth result is also an equilibrium, but it is dynamic. It is the *Red Queen* equilibrium. Instead of evolving to an optimal state and then staying there, this result arises when possibilities for adaptive improvement constantly emerge, and the species continually evolves toward them. All species have constant positive values for their lag loads, and the system neither expands nor contracts. Van Valen originally suggested the idea to explain the log linear survivorship curves he had documented. His name for it—the Red Queen hypothesis—alludes to the Red Queen's remark in Lewis Carroll's *Alice Through the Looking Glass:* "Here, you see, it takes all the running you can do, to keep in the same place." The analogy for running is coevolutionary change. In the Red Queen mode of coevolution, natural selection continually operates on each species to keep up with improvements made by competing species; each species environment deteriorates as its competitors evolve new adaptations. This deterioration causes extinctions in the model. On average, competing species have balanced levels of adaptation, and they all lag behind their best possible states. At any one time, a species may experience some random run of bad reproductive luck, and go extinct. Coevolution will result in a log linear survivorship curve if the rate of environmental deterioration is roughly constant through time. If the species' competitive environments deteriorate in fits and starts, the survivorship curve will be non-linear.

Why should the rate of environmental deterioration be approximately constant? Van Valen reasoned that it would follow from the zero-sum nature of competitive ecological interactions. The total resources available, he thought, will stay approximately constant. If one species adaptively improves, it will temporarily at least be able to take more of the resources. Consequently, its population will expand. This increase will be experienced by its competitors as an equivalent decrease in the resources available to them. The selection pressure on them to improve will increase, by an amount proportional to the loss in resources caused by the competitor's improvement. They will then tend to improve their competitive abilities, and make up the ground lost to the competitor. On average, the lag load of each species will, therefore, be constant. The justification may be correct, but it is debatable. Resource levels, for example, have probably changed through evolutionary time, and competitors may not always compete for a constant-size pie. If resource levels increase, Van Valen's argument might predict that extinction rates would decrease, and the change in the resource level would cause a change in the extinction rate. The point here is only that

22.13 *Competition may have driven some grand scale trends in the history of life*

The history of life, from its origin to the present, probably shows an increase in average body size. Life most likely originated as a replicating molecular system. Since then the range of the sizes of living things has increased while the smallest size has remained about constant—replicating molecules still exist. As the range has expanded with a constant lower limit, the average size of a living thing must have increased through time. Even if we take as our starting point a single-celled prokaryote, there will have been a trend to increase average size in the history of life. The trend in average size cannot be plotted exactly, because we do not know the average size of a living organism in the world today; one estimate, by Vogel, puts the average size somewhere between 1 mm and 1 cm. The estimate depends crucially on the abundance of very small bacteria. In any case, we cannot plot a graph of average size against time because no equivalent estimates have been made for other stages in the history of life. Bonner has plotted the size of the largest known species against time, and the average size has probably mirrored the same trend, though at a much lower scale (Figure 21.14a,).

Figure 21.14a illustrates an evolutionary trend on the largest possible scale. It is a trend in the state of all life through time. Bonner has also tried to quantify a number of similar macroevolutionary trends. Figure 21.14b suggests a trend toward increased complexity in living things. Organismic complexity is notoriously difficult to measure, and Bonner used the number of cell types in an organism as an approximate criterion. Rough estimates of the number of cell types in different sized modern species suggests that larger species have more cell types. It is a reasonable inference that complexity (by this measure) will have shown a comparable increase in the history of life to that shown by size. Whatever the quantitative pattern, it is almost necessarily true that complexity must have on average increased, because the first living things would have been at the simple extreme of the range of complexity seen in modern species.

What is the explanation for these very large-scale trends? Maynard Smith has pointed out that the increase in size and complexity is almost inevitable, because life started small and simple, and had nowhere to go except up. Bonner favors a similar theory; he calls it "pioneering" (it is essentially the same as Darwin's principle of divergence, section 14.8, p. 391). As ecological resources become fully exploited, selection favors forms that evolve to make new ways of living; ecological competition is a recurrent squeeze on living species. Because earlier living things were small and simple, species could always escape from competition by evolving to become larger and more complex. The ecological niche that could be occupied by organisms that were larger and more complex than any other organisms existing at the time was always vacant, and selection could have favored "pioneers" that evolved to occupy it. The net effect of the process will be for ever larger and more complex organisms to arise. The principle of divergence forces species to explore extreme niches. If Bonner is correct, biological competition, and the divergence it produces, has acted as the main force behind the pattern of increasing size and complexity—of "progress," some would say—in the history of life.

Figure 22.14 Graphs of rough estimates of (a) the maximum size of living thing at different times in the history of life (both axes are logarithmic), and (b) number of cell types in modern species of different sizes. Larger individuals tend to be more complex, as measured by number of cell types. From Bonner (1988).

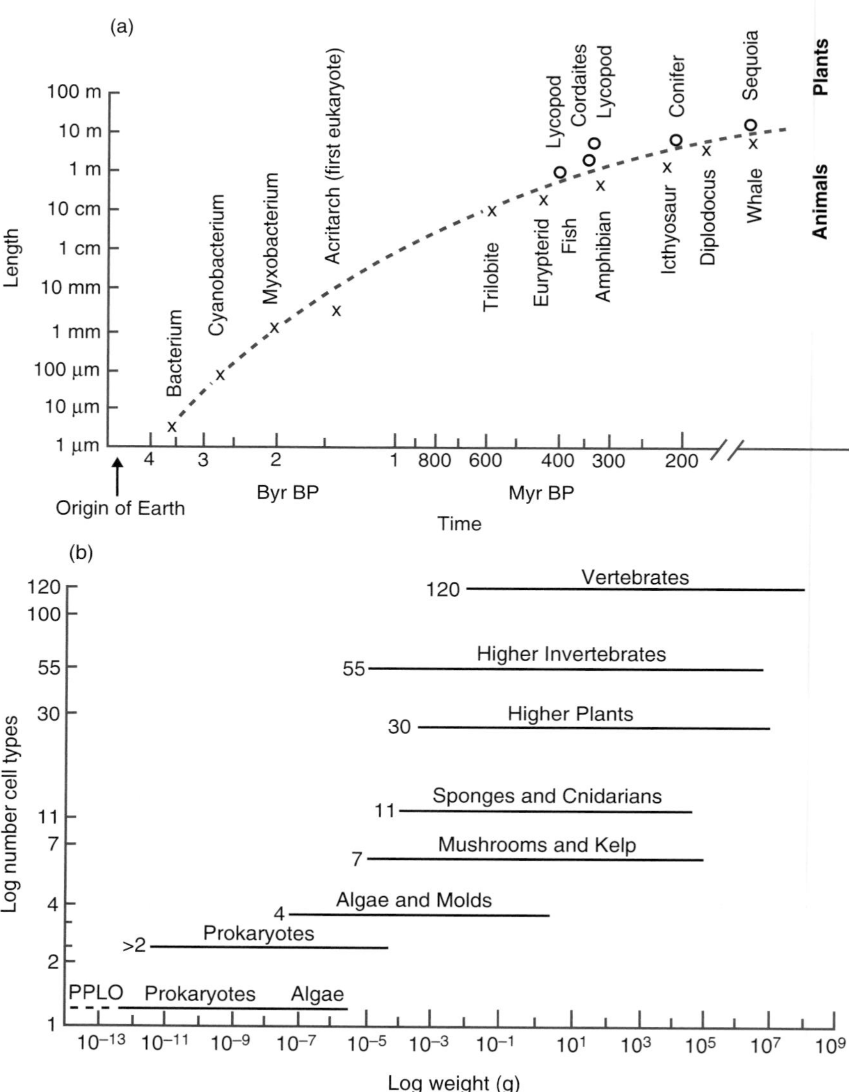

SUMMARY

1. Coevolution occurs when two or more lineages reciprocally influence one another's evolution. Coadaptation between species, such as in any example of mutualism, is probably but not necessarily the result of coevolution.
2. Insects and plants influence one another's evolution, by the evolution of insecticides in plants and of detoxification and avoidance mechanisms in insects. The evolutionary relations between insects and plants may more often take the form of sequential evolution, in which plant evolution influences insect evolution but not vice versa, than of fully reciprocal coevolution.

3. The level of virulence of parasites can evolutionarily decrease or increase. It can be understood in terms of the parasite-host relationship. Two factors that influence it are kin selection and the mode of transmission of the parasite between hosts.
4. Speciation in parasites and hosts is likely to occur simultaneously if the parasites have limited independent powers of dispersal. The phylogeny of the two form mirror-images.
5. Coevolutionary "arms races" between predators and prey produce escalatory long-term evolutionary trends. These trends can be seen in the evolution of brain sizes in mammals and of armor and weapons in molluscs and their predators.
6. The extinction rates of species are independent of how long the species has existed: a species does not become more likely to go extinct as time passes. Taxonomic survivorship curves are logarithmically linear.
7. Van Valen explained the log linearity of survivorship curves by his Red Queen hypothesis. It suggests that (1) each species' environment deteriorates as competing species evolve new, superior adaptations; (2) the competing species improve at a constant rate relative to one another; and (3) the constant deterioration in the environment causes the chance of extinction of any species to be probabilistically constant.
8. Coevolution within groups of competing species need not have the Red Queen mode. Rates of change, and relative competitive abilities, need not be constant.
9. The Red Queen hypothesis could be tested by seeing whether extinction rates within taxa in absolute time are constant. Little conclusive work of this kind has been done.
10. Trends in the whole history of life can be measured by working out the average state for a character such as body size for all species alive at a series of times. Body size, and complexity, have probably increased from the origin of life to the present.

FURTHER READING

On coevolution in general, see Futuyma and Slatkin (1983), Kim (1986), Howe and Westley (1988), Thompson (1989, 1994), and the special issue on coevolution in animal-plant relations, in *Bioscience*, vol. 42 (1992), pp. 12–57. See Pierce (1987) on the symbiosis between lycaenid butterflies and ants, and Janzen (1980) on coevolution and coadaptation. Singer and Parmesan (1993) is a recent study of non-1:1 relations between insects and plants. Bronstein (1994) is a recent review of mutualism.

Ehrlich and Raven (1964) is the classic paper on plant-insect coevolution. Farrell and Mitter (1994) and Pellmyr (1992) are recent accounts; see also part of Crane *et al.*'s (1995) review about the rise of the angiosperms. Labandeira *et al.* (1994) cast doubt on the idea that butterflies radiated in relation to angiosperm evolution; they offer fossil evidence that the Lepidoptera were already diversified before the angiosperm radiation.

On the coevolution of parasites and hosts, see May and Anderson (1983) and Toft *et al.* (eds) (1991) in general; Bull (1994), Ewald (1993, 1994), Frank (1996), and Roughgarden (1975) for the evolution of virulence, and the influence of transmission mode and multiple infections in particular; and Fenner and Ratcliffe (1965) on myxomatosis. On cospeciation, see also Hafner and Page (1995), Page (1994) on the methodology, and Moran and Baumann (1994) for another example. Dawkins and Krebs (1979) and Abrams (1986a) argue for different views of "arms race" coevolution. Vermeij (1987, 1994) discusses evolutionary escalation; see also several of the papers in a special issue of *Paleobiology*, vol. 19, pp. 287–397 (1993). See Jerison (1973) on brain evolution; Gould (1977b, ch. 23) popularizes Jerison's work. Bakker (1983) studied coevolution of the same ungulates and carnivores as Jerison, but in their morphological adaptations for running.

References for Van Valen's law of constant extinction and Red Queen hypothesis are Van Valen (1973, 1985), Archibald (1993), Gilinsky and Bambach (1987), Hoffman (1991), Levinton (1988), McCune (1982), and Stenseth and Maynard Smith (1984). Some of them also discuss the role of physical factors; on this issue see also Vrba (1993). On grand-scale trends, see Bonner (1988) and Valentine *et al.* (1994).

STUDY AND REVIEW QUESTIONS

1. (a) How can the phylogenies of two taxonomic groups be used to find out whether they show coevolution, sequential evolution, or neither? (b) Some groups of parasites and hosts show cospeciation, others do not. What property (or properties) of their biology explain the difference?
2. It has been argued that the transmission mode of parasites influences the evolution of virulence. From this general idea, what rank order of virulence would you predict for otherwise similar parasites that are transmitted by the following means: (a) by the breathing of the host, (b) by the water supply, (c) by insect vectors, and (d) by copulation of the host.
3. Summarize the hypothesis of evolutionary escalation. What kind of fossil evidence can be used to study it?
4. What does the log linear survivorship curve of fossil taxa tell us about their rates of extinction in (a) absolute time, and (b) relative to the age of the taxon?
5. (a) Define lag load. (b) In terms of the lag loads of a group of interacting species, what are the theoretically possible coevolutionary modes? (c) What evolutionary, or ecological, process might generate the Red Queen mode?

CHAPTER **23**

Extinction

THIS chapter looks at the two factors that govern diversity in the fossil record: extinction and radiation. In particular, it concentrates on extinction. We start by looking at extinction and radiation within one group (snails) over a short geological time period, and ask whether some kinds of species have higher extinction or speciation rates than others. The study illustrates the theory of species selection and leads us into a discussion of how microevolution can be uncoupled from macroevolution. We then shift attention to the whole of life and the full fossil record since the Cambrian. We examine the evidence for mass extinctions, the possibility that they follow a 26 million year cycle, and some of the theories to explain these events. We finish with two questions related to mass extinctions: Is survival through them a matter of luck, or are some kinds of species better equipped to survive them than others? Does the fossil record contain a continuum of extinction rates, with mass extinctions at one extreme, or do two clearcut extinction regimes exist, with normal and mass extinction rates?

23.1 *The diversity of life in the fossil record is due to a balance of extinction and speciation*

The diversity of life through time reflects the rates of loss and gain of new life forms. The loss of species occurs by extinction, the gain of species by speciation.

At some times in the fossil record it can be seen that a taxonomic group proliferates, as the rate at which new species arise in that group far exceeds the rate at which species go extinct. These rounds of proliferation are called *radiation.* The increase in the number of mammal families 60–40 million years ago is an example of a radiation (Figure 21.11, p. 602).

The discovery that species go extinct was made relatively recently in human history, and dates from only the late eighteenth and early nineteenth centuries. Fossils had been known about long before that time, but when a fossil was found that differed from any known species, it could still have been alive in some unexplored region of the globe. As the global flora and fauna became better known through the eighteenth century, it became increasingly likely that some fossil forms were no longer alive. By the end of the century, several naturalists accepted that some marine invertebrate groups, such as the ammonites, were extinct. The best known taxa—the vertebrates, and mammals in particular—posed a special problem, however. Fossil bones are preserved as isolated, disarticulated fragments, and it is even more difficult to show that a single bone does not belong to any modern species than it is for a complete specimen (such as a shell). The decisive work is usually credited to Cuvier. Cuvier reconstructed, with new standards of rigor, whole skeletons from bone fragments. It is easier to see whether the whole skeleton, rather than just the disarticulated bones, of a vertebrate belong to any living species. The most convincing cases of extinction were for gigantic forms like mastodons—it was hardly plausible that the explorers would have overlooked them.

Why do species go extinct? Some modern extinctions have been witnessed closely enough for the cause to be known with certainty. The enormous Steller's sea cow was discovered by a shipwrecked German naturalist called Georg Steller in 1742, but he was the only naturalist ever to see it alive. The animals were completely tame—Steller records how he could stroke them—and by 1769 they had been hunted to extinction.

The extinction of modern species by analogous human means is all too easy to observe, but for fossil species we do not have evidence as direct as sailors shooting sea cows. The quality of the evidence depends on the age of the fossils, and for very recent fossils we can have quite convincing evidence about the cause of extinctions. The most recent ice age, for example, which peaked about 18,000 years ago, almost certainly caused many local extinctions. If a species disappears before the advancing ice cap and does not return, little doubt exists about the cause of the extinction. The tulip tree and hemlock are only two of the species lost from the European flora at that time, although both survived in North America. As we move further back in time, however, the uncertainty mounts rapidly. We saw in the well-studied case of the Great American Interchange how uncertain the evidence is about the causes of the many extinctions, and the Interchange took place only 2 million years ago and has left a good fossil record (section 18.6, p. 525).

The difficulty of studying the causes of individual extinctions in the fossil record does not mean that fossils cannot be used to study extinction. It means instead that we should shift our attention to larger-scale patterns. In this chapter we shall concentrate on three particular topics. We will first look at how different kinds of species may vary in their rates of extinction and speciation; some kinds may proliferate relative to others in consequence. This example will illustrate how microevolutionary natural selection is not a sufficient theory to explain macroevolutionary patterns of the survival and proliferation of large taxa. The second topic is mass extinction. We then take the first two topics together and ask whether some kinds of taxa survive mass extinctions better than do others. Third, we notice that mass extinctions may clear space for the radiation of a new taxon.

23.2 *Real extinction should be distinguished from pseudoextinction*

Species (or higher taxa) may go extinct for two reasons. One is "real" extinction in the sense that the lineage has died out and left no descendants. For modern species, the meaning is unambiguous, but for fossils real extinction must be distinguished from *pseudoextinction* (Figure 23.1). As a lineage evolves, later forms may look sufficiently different from earlier ones that a taxonomist who

Figure 23.1 If a continuous phylogenetic lineage is taxonomically subdivided, the earlier species will go "extinct" at the dividing line, even though the lineage persists just as it did before. The "extinction" of species A at time *t* is called pseudoextinction.

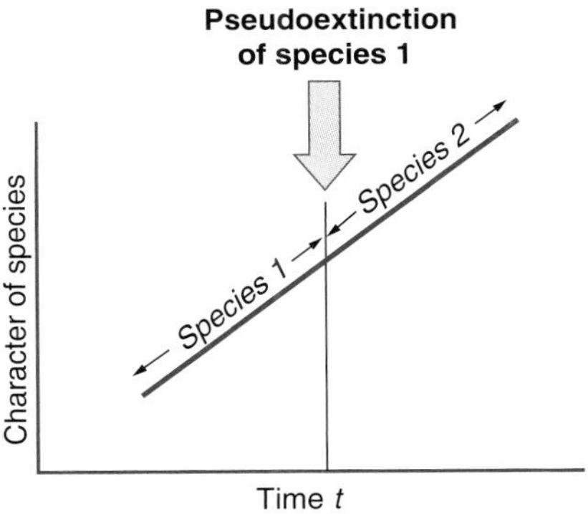

only has a few specimens will classify them as different species, even though a continuous breeding lineage exists. We have discussed before (section 15.3, p. 418) the purely taxonomic question of whether lineages that undergo evolutionary change ought to be split into more than one species. Both viewpoints have their supporters, but either way there can be no doubt that this kind of taxonomic extinction is conceptually different from the literal death of a reproducing lineage. The taxonomic survivorship curves that we considered in sections 20.10, p. 576, and 22.9, p. 626, contain some usually unknown mix of real extinction and pseudoextinction.

The distinction is worth keeping in mind when considering theories to explain extinction and diversity patterns. Most of the theories we shall address in this chapter make sense only for real extinctions. For example, the theory that mass extinctions are caused by asteroid impacts makes sense if the mass extinctions are real, but not if they are largely composed of pseudoextinctions.

23.3 *The characters that evolve within taxa may influence extinction and speciation rates, as is illustrated by snails with planktonic and direct development*

What factors determine the patterns of speciation and radiation? The question has been studied in various ways, and in this and the next section we shall concentrate on two ideas: one in which the attributes of the organisms, and the other in which external ecological factors, may influence a taxon's probabilities of survival and of speciation.

Different molluscs grow up in different ways. In gastropod snails, planktonic and direct development are two of the main types of development. With planktonic development, the egg is released into the surface waters of the ocean and develops into a larval form that disperses among, and feeds on, the microscopic organisms (called "plankton") that float near the ocean surface. The larva eventually settles and metamorphoses into an adult snail. With direct development, the eggs and young grow up near or (initially) inside the parental snail. Various ecological trends are known among modern forms, such as that planktonic development is more common among shallow-water than deep-water species, and among tropical than among polar species.

The relation between larval type and speciation and extinction rates can be studied in fossil gastropods. Larval types in fossils are inferred by analogy with modern species. These kinds of inference were pioneered in the work of Thorson, and several criteria have now been used. Figure 23.2 shows one such method that uses the size of regions in the larval shell. Modern species with planktonic development typically have small, yolk-poor eggs; in the larval shell, a region called prodissoconch I tends to be small and another region called prodissoconch

Figure 23.2 Larval shell form is correlated with type of development in molluscs. The species in (a) and (b) are modern gastropods; (c) and (d) are Late Cretaceous fossil bivalves. Note the relative sizes of the regions labeled PdI and PdII. (a) *Rissoa guerini*, known to have a planktonic larva. (b) *Barleeia rubra*, known to develop directly without a stage in the plankton. (c) *Uddenia texana*, which has small PdI and large PdII like (a) and is inferred to have had planktonic development. (d) *Vetericardiella crenalirata*, which had large PdI and small PdII like (b) and is inferred to have had direct development. Reprinted, by permission of the publisher, from Jablonski and Lutz (1983).

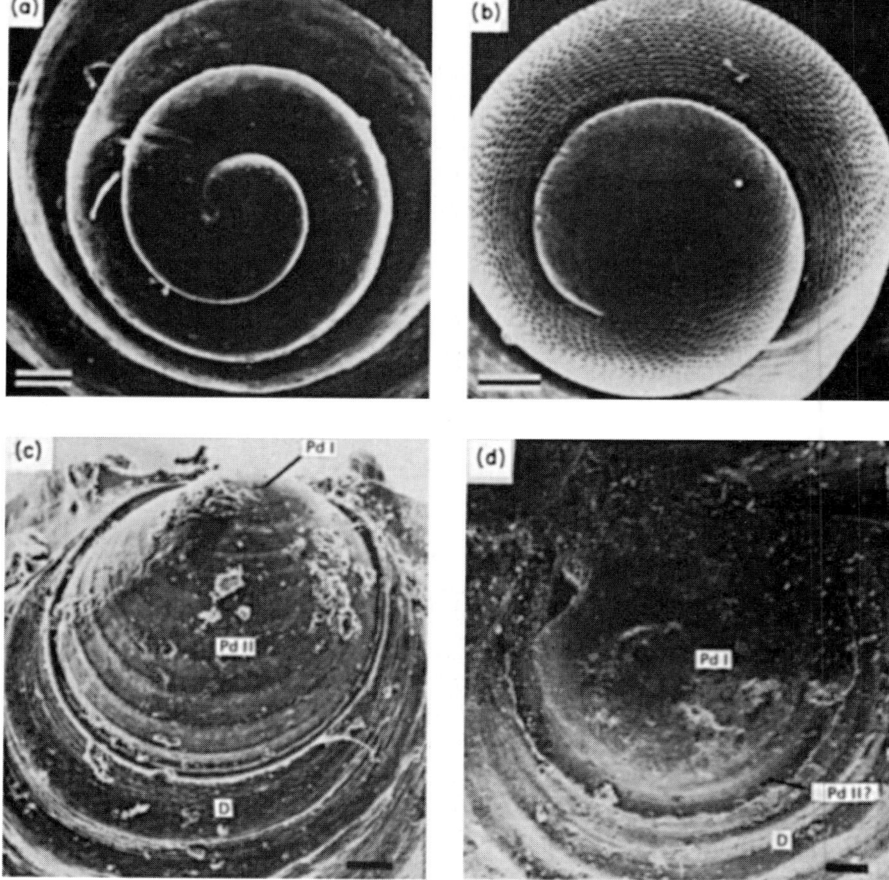

II is larger (Figure 23.2a). Species with direct development show the reversed condition (Figure 23.2b). These morphological regions can be distinguished in fossil larval shells, by scanning electron microscope, and it is a reasonable inference that shell form is correlated with development type in the same way as modern forms.

What is the relation between larval type and extinction rate? Several studies have found that species with planktonic larva have lower extinction rates (Figure 23.3). As the figure shows, these species also have wider geographic ranges. This fact may account for their lower extinction rates, because a species with a wider range is less vulnerable to local circumstances. Alternatively, it may mean merely that planktonic forms have a higher chance of being preserved as fossils than have directly developing forms, because their wider distribution increases the chance that the conditions in one site will allow fossilization; the difference could then be just a bias in the fossil record.

The relation with speciation rate is more interesting. Hansen predicted that snails with direct development will speciate more rapidly than species with planktonic larvae, because the species with non-planktonic development will be more likely to be geographically localized and isolated, which makes allopatric speciation easier. Planktonic development increases gene flow and makes allopatric speciation less probable. Hansen used this idea to explain an observed trend

Figure 23.3 Duration in the fossil record and geographic ranges for Late Cretaceous gastropods from North America. Species with planktonic larvae last longer in the fossil record (i.e., they have lower extinction rates) and also have wider geographic ranges. Extinction rate is given as the chance that a species-lineage will go extinct per million years. See Table 23.3 for related results. Reprinted, by permission of the publisher, from Jablonski and Lutz (1983).

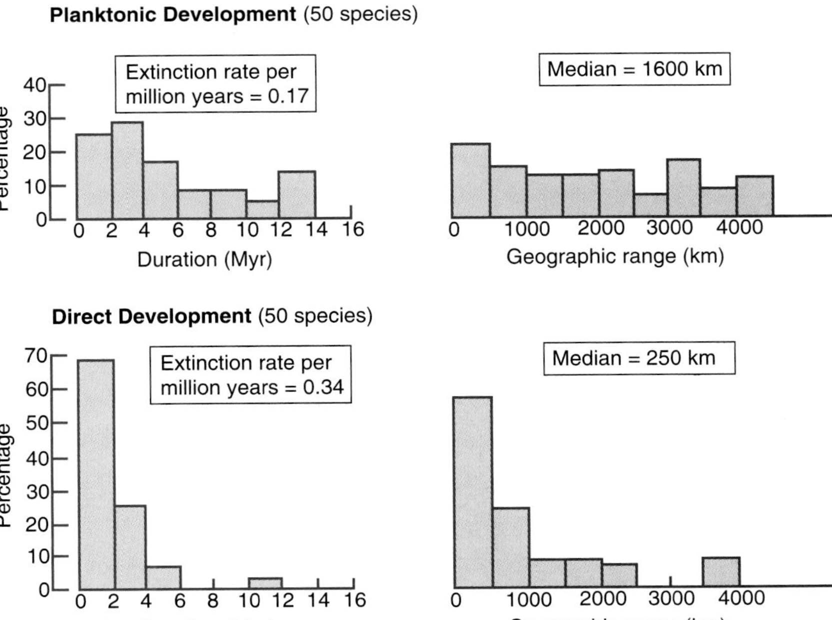

in snails of the early Tertiary (Figure 23.4). The proportion of planktonically developing species declined through the Paleocene and Eocene. The trend was not the result of a difference in extinction rates. As usual, the planktonically developing species had lower extinction rates (Figure 23.4b), which would tend to produce the opposite trend from what was observed.

Two alternatives are left. First, natural selection could have been favoring direct development within the majority of lineages. Hansen "suggested" this possibility was not true, although he gave no evidence to prove this point. The period was a time of global cooling (see also section 23.7.3), which might favor direct development, given the latitudinal trend mentioned earlier. Hansen states that the decline in planktonically developing forms preceded the global cooling; concrete evidence, however, rather than a vague statement, would be needed to persuade a skeptic. Second, the increase might have been due to a higher speciation rate of the directly developing forms, simply because forms with lower dispersal rates are more likely to speciate. This alternative was favored by Hansen. The results published so far are not conclusive, but Hansen's theory is consistent with them. The decline in planktonically developing forms in the early Tertiary may have happened because the speciation rate of these organisms was so low that it did not make up for their longer average survival.

23.4 *Differences in the persistence of ecological niches will influence macroevolutionary patterns*

A species must inevitably go extinct if its ecological niche disappears. A parasite, for example, will go extinct if its host (or hosts) become extinct. The duration over geological time of ecological niches is, therefore, another factor to investigate.

Williams (1992) discussed the three-spined stickleback (*Gasterosteus aculeatus*) as a concrete example to illustrate the general idea. This fish is widespread in coastal waters in the northern hemisphere, on both sides of the North Atlantic and Pacific Oceans. From these coastal waters (it appears), many populations have separately colonized the local freshwater rivers and their tributaries inland. Some of these populations show various local adaptations to the rivers they occupy; they have formed local races or subspecies in a complex pattern. Thus, at the microevolutionary level the fish have evolved, by natural selection, adaptations to the local conditions.

The evolution of *G. aculeatus* over longer time periods, however, is dominated by another factor. The populations that colonize the freshwater rivers are probably evolutionarily short-lived. This could be for either of two reasons. They may be poor competitors against the established freshwater forms, or their niches may be short-lived, as ecological change removes the freshwater habitats to which they have adapted. We concentrate here on the second possibility, for purposes of argument. The coastal niche, however, probably lasts longer and acts repeatedly as the source for new colonists and the ancestor of descendant freshwater populations. The pattern over geological time is set by the durations of the niches, with a persistent coastal, and ephemeral inland freshwater, niches.

Under this view, all subspecies, races, and varieties of sticklebacks around the northern hemisphere are not all equally likely to be the evolutionary source

Figure 23.4 (a) The proportion of gastropod species with direct, rather than planktonic, development increases through the Paleocene and Eocene. The effect exists within several of the six families, and as an average for all six families. (b) Detailed observations for Volutidae, from the Gulf Coast of the United States. (Volutidae are furthest left of the six families in (a).) Note the proliferation of directly developing species. The extinction rate of species with planktonic development appears to be lower—which counters the trend toward more directly developing species. Small discrepancies between the numbers in (a) and the number of lineages in (b) may be due to extra data in the later paper (a). Reprinted with permission from Hansen (1978, 1983). Copyright 1983 American Association for the Advancement of Science.

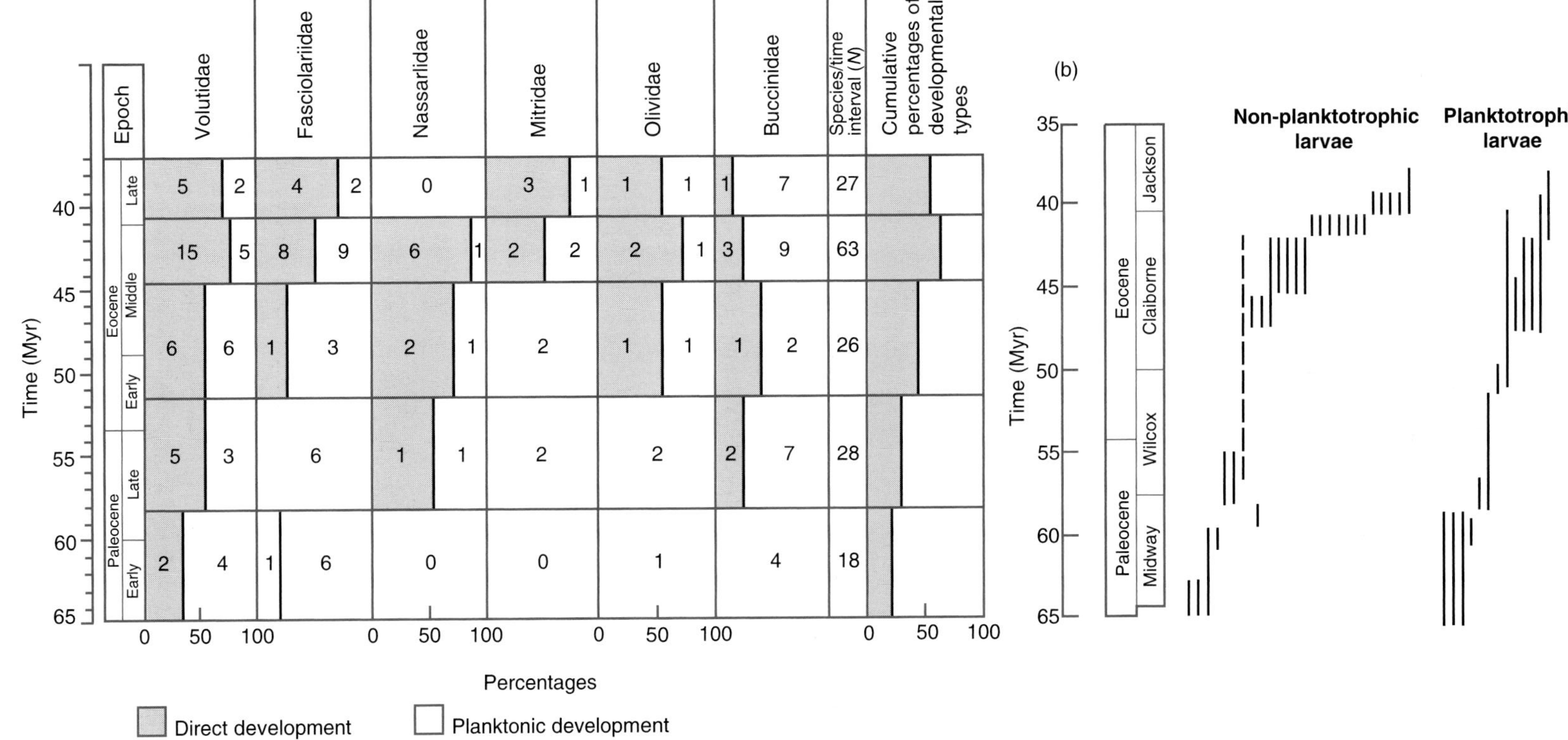

of new species, or higher taxa. The forms that occupy long-lasting niches are more likely to be the source of a large-scale evolutionary event than are the many short-lived forms. The extinction rates of the taxa through time, and their contribution to new radiations, can be understood in terms of the differential persistency of their ecological niches. Although this interpretation of stickleback evolution is uncertain, the general idea that macroevolution will be influenced by the nature of the ecological niches occupied by different taxa has logical force and is likely to operate.

23.5 *The factors that control macroevolution may differ from those that control microevolution*

The trend toward increasing numbers of snail species with direct development is an example of what is sometimes called *species selection.* We can expect trends—all other things being equal—toward increasing numbers of the kinds of species that have lower extinction and higher speciation rates. When a trend arises for this reason, species selection is the process at work.

Species selection should not be confused with group selection (section 12.2.5, p. 325). Group selection aims to explain why individuals sacrifice themselves for the good of the group (or species) to which they belong; we saw that it is difficult for adaptations of this sort to arise. In species selection, individuals do not use a disadvantageous developmental mode in order to boost the speciation rate of its taxonomic group. Instead, direct and planktonic development are favored by natural selection in different taxonomic groups for good ecological reasons within each species, but they can then have different long-term consequences for radiation and extinction. We have no reason to suppose that the characters favored by the short-term process of natural selection will always be those that allow a species to last a long time or split at a high rate. Natural selection may favor adaptations within some species that result in reduced long-term survival and adaptations that increase it in others.

Species selection provides another reason why macroevolution and microevolution may be uncoupled (sections 20.8, p. 570, 21.5, p. 595, and 21.7, p. 599). Within a species, natural selection favors one character in one species and another character in a second species. Species selection over long periods may cause the species with one of the characters to proliferate because of the character's consequences for speciation or extinction rates. This does not mean that the long-term process contradicts, or is incompatible with, the short-term process; it simply means that we cannot understand the long-term evolutionary pattern by studying natural selection in the short-term alone, and extrapolating it.

A similar moral can be drawn from the argument about niches: microevolution and macroevolution are again uncoupled. A microevolutionary study would reveal how natural selection favored various characters in the stickleback populations, according to the aquatic environments they were occupying. The key to macroevolution, however, is the persistency of the niches over time, which is irrelevant to the short-term process of natural selection. (Natural selection does not favor one adaptation over another because it allows the organisms to occupy a longer-lasting niche.) Thus, additional factors beside those studied in the short-term matter when we try to understand evolutionary phenomena on the grand scale.

23.6 *Mass extinctions*

23.6.1 *At certain moments in the history of life, there have been mass extinctions of many species*

The geologists of the nineteenth century who worked out the main eras of the Earth's history did so by looking for characteristic fossil faunas that lasted for a noticeable time (or rather, depth) in the sediments. They used different characteristic faunas to identify different time periods. They recognized three large-scale faunal types and named them Paleozoic, Mesozoic, and Cenozoic; shorter-term characteristic faunas were identified within each of these three divisions. Two major faunal transitions divide the three Eras: the Permo-Triassic boundary, between the Paleozoic and Mesozoic, and Cretaceous-Tertiary boundary, between the Mesozoic and Cenozoic. In the nineteenth century, it was not known whether these two transitions corresponded to long gaps in the sedimentary record, or to the massive and rapid extinction of one fauna and its replacement by another. We now know, from radioisotope dating, that the second interpretation is correct. At the end of the Permian, about 225 million years ago, and the end of the Cretaceous, 65 million years ago, *mass extinctions* occurred.

And they were massive. Raup (1979) estimated that at the end of the Permian as many as 96% of marine species went extinct. Stanley and Yang recently revised the estimate downward, to about 80% of marine species—which still represents a massive extinction. In the Cretaceous extinction at least half, and perhaps 60–75%, of all marine species went extinct. These mass extinctions pose immediate questions of description and cause. How long did they last? Were they global, or local? Did they affect all groups equally? We shall begin with the Cretaceous-Tertiary mass extinction, which has been most thoroughly studied. The Cretaceous extinction may be but one instance of a more general phenomenon, however, and we will therefore move on to consider its explanation in a broader context.

23.6.2 *The best studied mass extinction occurred at the Cretaceous-Tertiary boundary*

The mass extinction at the end of the Cretaceous has been found in all regions of the globe, and affected more or less every group of plants and animals (Figure 23.5). The fossil record of small, abundant, microfossil groups such as Foraminifera provides the best evidence for the fine-scale pattern of the extinction, but the demise of larger groups provides its drama. Some groups, such as the dinosaurs and ammonites, were driven finally extinct; most large groups were drastically reduced in diversity, although some odd groups, such as crocodiles, were not noticeably affected at all. The obvious question is: why did it happen?

In 1980, Alvarez *et al.* published an influential observation. They sampled the rocks of the Cretaceous-Tertiary border from Gubbio in Italy, and found exceptionally large concentrations of rare earth elements, particularly iridium (Figure 23.6). These elements also have high concentrations in extraterrestrial

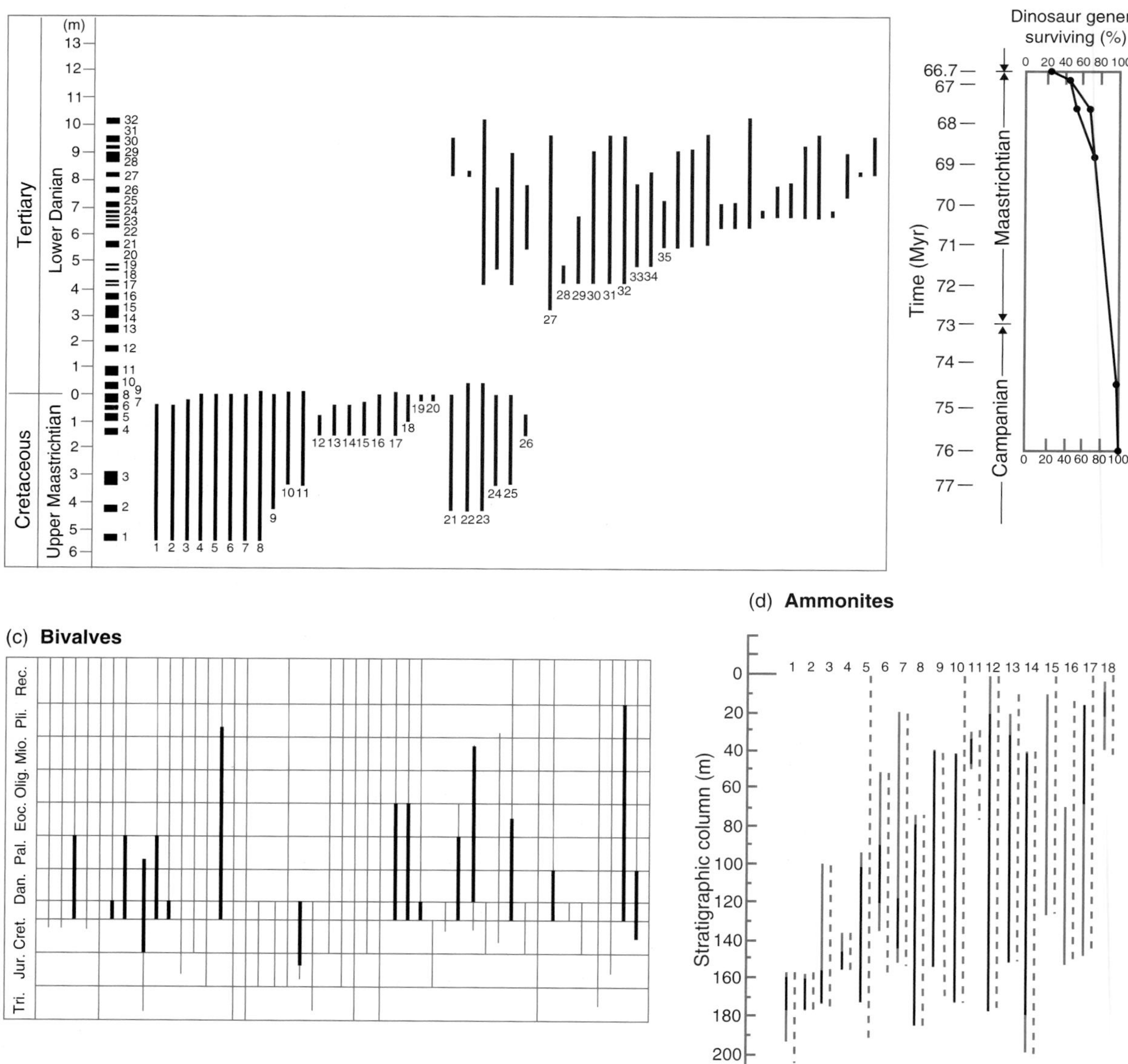

Figure 23.5 The mass extinction at the end of the Cretaceous affected all main taxa, but the evidence about whether they all went extinct suddenly at the same time is controversial. (a) Brachiopods, from Nye Kløv, Denmark. These extinctions look synchronous. Some argue that a sedimentary hiatus occurred in this example; others disagree: but note the gap at the base of the Tertiary. (b) Dinosaurs, from Hells Creek, Montana. These extinctions look gradual. It has been argued that the gradual extinctions, and persistence of dinosaurs into the Tertiary, is due to secondary reworking of the fossils, and the real extinction pattern is sudden and synchronous at the end of the Cretaceous (see Smit and Kaars 1984, and Sheehan *et al.* 1991). (c) Bivalves, from Stevens Klint in Denmark. These extinctions appear synchronous. It has been argued that the sudden extinctions are only apparent, being due to a gap in the sedimentary record; most students of the site accept that the sedimentary record is continuous, and the extinctions real (see Stanley 1984). (d) Ammonites, from the Zumaya section in northern Spain. The results are shown for two seasons of collecting. Note improved evidence in the larger data set for synchronous extinctions at the Cretaceous-Tertiary border. Perhaps the non-synchronous pattern in the earlier data was due to incomplete evidence and the real pattern is synchronous. Reprinted, by permission of the publisher, from (a) Surlyk and Johansen (1984) Copyright 1984 American Association for the Advancement of Science; (b) Sloan *et al.* (1986) Copyright 1986 American Association for the Advancement of Science; (c) Alvarez *et al.* (1984) Copyright 1984 American Association for the Advancement of Science; and (d) Ward (1990).

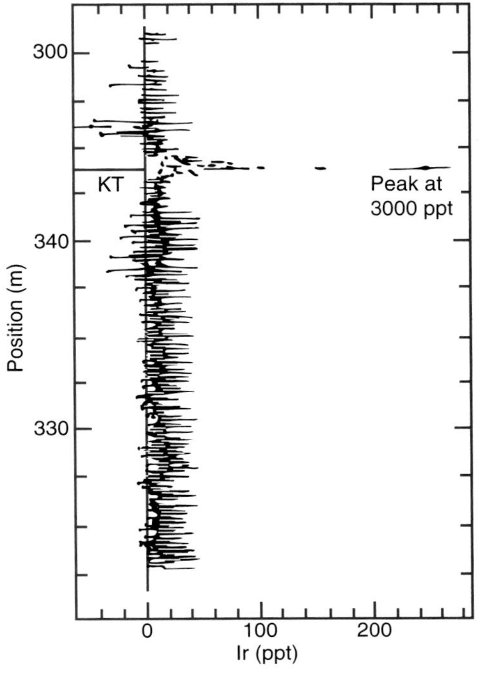

Figure 23.6 The iridium concentration increases suddenly by 2–3 orders of magnitude at the Cretaceous-Tertiary (marked KT) boundary rocks at Gubbio, Italy. Reprinted with permission from Alvarez *et al.* (1990). Copyright 1990 American Association for the Advancement of Science.

objects, which led Alvarez and his colleagues to explain the biological mass extinction, and the geochemical *iridium anomaly*, by the collision of a large asteroid with the Earth. Since then, similar iridium anomalies have been found in Cretaceous-Tertiary boundary rocks at several other sites. Some geologists have argued that the iridium anomaly could have had a terrestrial cause, by volcanic eruptions, but asteroids remain the most widely accepted explanation.

The exact means by which such an impact could have precipitated the mass extinction has been considered in detail by Alvarez, and other authors. Alvarez *et al.* originally suggested that the impact would have thrown up a global dust cloud, which would have blocked out sunlight for several years until it settled again. When Krakatoa erupted in 1883, it ejected an estimated 18 km^3 of matter into the atmosphere, which took 2.5 years to fall down again. From the iridium concentrations at Gubbio, and in known asteroids, Alvarez *et al.* estimated the size of their hypothesized asteroid at a diameter of about 10 km. Such an asteroid, whose kinetic energy they described as "approximately equivalent to that of 10^8 megatons of TNT," would have produced an explosion about 1000 times as large as the eruption of Krakatoa. The loss of sunlight alone would have been enough to cause the extinctions, but the impact could have had other destructive side-effects as well. Global warming, acid rain, extreme vulcanism, and perhaps an associated global fire are some of the possibilities. We need not enter into the geological and climatic ideas here; it is enough to note that an impact on the scale suggested by Alvarez *et al.* would surely have been capable of causing the mass extinction at the end of the Cretaceous.

Since Alvarez *et al.*'s original publication, geologists have found an increasing quantity of evidence that supports this idea. The evidence is of four main kinds.

First, the geochemical evidence, of which the iridium anomaly was the first example, has broadened in space, as similar anomalies have been found in Cretaceous-Tertiary boundary rocks at other sites. It has also expanded in kind, as other chemical signatures of an asteroid impact have been detected.

Second, evidence of the impact crater itself has probably been located. About 120 impact craters are known from around the globe today; most of them are too small to have been produced by one asteroid of the size calculated by Alvarez and most (possibly all) of them date from times other than the Cretaceous-Tertiary boundary. Through the 1980s, this missing crater presented one of the main difficulties in accepting Alvarez's theory. It is now becoming widely accepted that a geological structure (called the Chicxulub crater), buried beneath sediments on the Yucatan coast of Mexico, was the site of the impact. The structure is large enough, with a diameter of probably about 180 km, and it dates to the Cretaceous-Tertiary border.

The third kind of evidence concerns physical structures that would have been generated by the impact. Rocks, such as shocked tektites and quartzes that are suggestive of a high velocity collision, have been found from several Cretaceous-Tertiary sites, including Chicxulub.

The fourth kind of evidence comes from the pattern of extinctions in the fossil record and is of most interest in the theory of evolution. If Alvarez's theory is correct, the extinctions at the Cretaceous-Tertiary border should have been sudden, concentrated in a short interval of time, and not preceded by any decline through the Cretaceous; they should be synchronous in different taxa and geographical localities; and they should coincide with the iridium anomaly. This set of predictions is both highly testable and stimulating.

As the evidence has fallen into place, many geologists have come to accept that the mass extinction was caused by an asteroid, but not everyone is convinced.The evidence presents several problems. You might think you could simply peer into the fossil record and observe whether extinctions were sudden or gradual, synchronous or spread out in time. In reality, this type of study is not so easy. How, for instance, do we observe the exact time of an extinction? The last appearance of a species in the fossil record will usually precede its final, true extinction (and a species certainly cannot appear *after* its true extinction). The species' population may decline before it finally disappears, which would reduce the chance of leaving fossils. Even if the population is constant, its chance of fossilization will still be much less than one. Species therefore appear to go extinct in the fossil record before they actually did, and this "push backward" is greater for forms that are less likely to leave fossils.

It can also be difficult to correlate events at different geographic localities, because absolute dates are often unavailable. The incompleteness of the fossil record also introduces uncertainty—a species may appear to go extinct suddenly at what is really a gap in the sedimentary record (look at Figure 23.5a and 23.5c for what may be controversial examples). On the other hand, a sudden extinction can be smoothed out by the secondary reworking of the fossils—fossils may be moved up and down after their original deposition. For all these reasons, evidence from the fossil record is controversial when used to show either sudden or gradual, or synchronous or asynchronous, extinction patterns.

Despite these problems, the evidence can still be used (Figure 23.5). An increasing amount of evidence supports the idea of a sudden and synchronous mass extinction. Let us look at one such study, by Ward. He first collected ammonites from around the time of the Cretaceous-Tertiary border, at a site in Spain, in 1986; further collections were made later and a larger study was possible in 1989. If the real extinctions were synchronous, the 1989 evidence should show more synchronous extinctions than the 1986 evidence; the reverse would be true if the real pattern was non-synchronous. Ward's result tends to support the idea of an exactly synchronous extinction at the Cretaceous-Tertiary boundary (Figure 23.5d), although it is not enough to convince a skeptic. It concerns only one taxon in one region, and the extinctions there could easily have been synchronous without the same being true of the rest of the world. Thus, the supporters of Alvarez's theory accept evidence such as Figures 23.5a,c as showing synchroneity and attribute evidence like Figure 23.5b to the imperfections of the fossil record; critics make the reverse argument.

In summary, good evidence exists for both suddenness and synchroneity of extinctions at the end of the Cretaceous, and the evidence appears to improve in more thorough fossil samples. On the other hand, the evidence is not complete enough to have persuaded everyone.

23.6.3 *It has been suggested that mass extinctions occur at regular intervals of 26 million years in the history of life*

Sepkoski has compiled, from the published literature, the distributions in the fossil record of all families of marine organisms. Sepkoski's compilation was not the first of its kind, but it is the most up-to-date and is now the most widely used. The data set is continually expanding, and in 1984, Sepkoski had evidence for 3500 families. Along with Raup, Sepkoski used this evidence to examine the grand pattern of extinctions through the past 250 million years. The data set has more and less inclusive versions according to which relatively poor parts are excluded; for the initial study the 3500 families were reduced to a sample of 567 families—still a large number. The striking result was that the probability of extinction appeared to vary cyclically, with a wavelength of about 26 million years (Figure 23.7a). The initial study suggested as many as 12 peaks of extinction; but (as the figure shows) some peaks are clearer than others. Raup and Sepkoski have repeated their analysis as new data have come in; their later results mirror those found in 1984. The two have also documented extinction patterns for all life taken together and for individual taxa (see Figure 22.13, p. 632).

Raup and Sepkoski's factual claim has proved highly controversial. The time scale they used, the taxonomic data, and the statistical methods are all open to criticism. More than one geological time scale is known (the Harland time scale in Figure 19.1, p. 539, is only one among several), and Hoffman and others have reexamined Raup and Sepkoski's evidence and found that the cyclical pattern is peculiar to some of the time scales and is not found in others. The finding is not all that surprising, but the evidence for a cyclical pattern would be more convincing if the result did not depend on picking the right time scale.

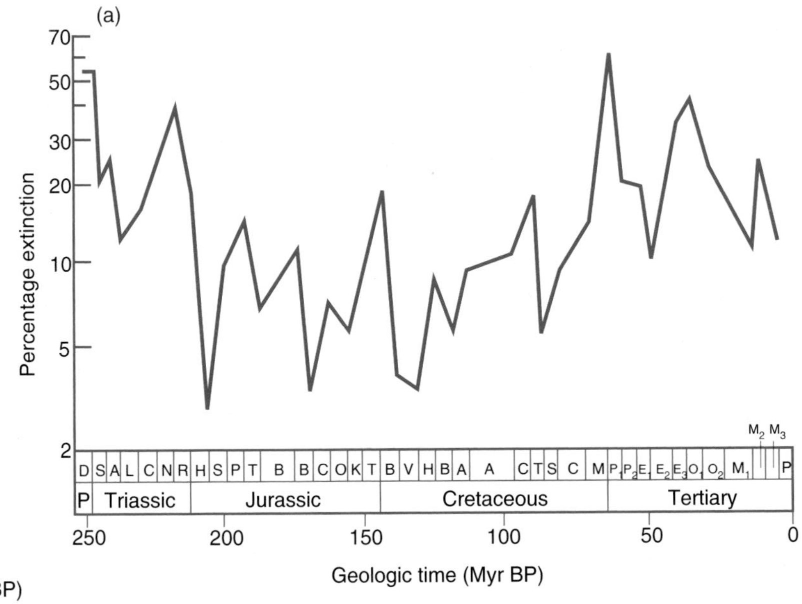

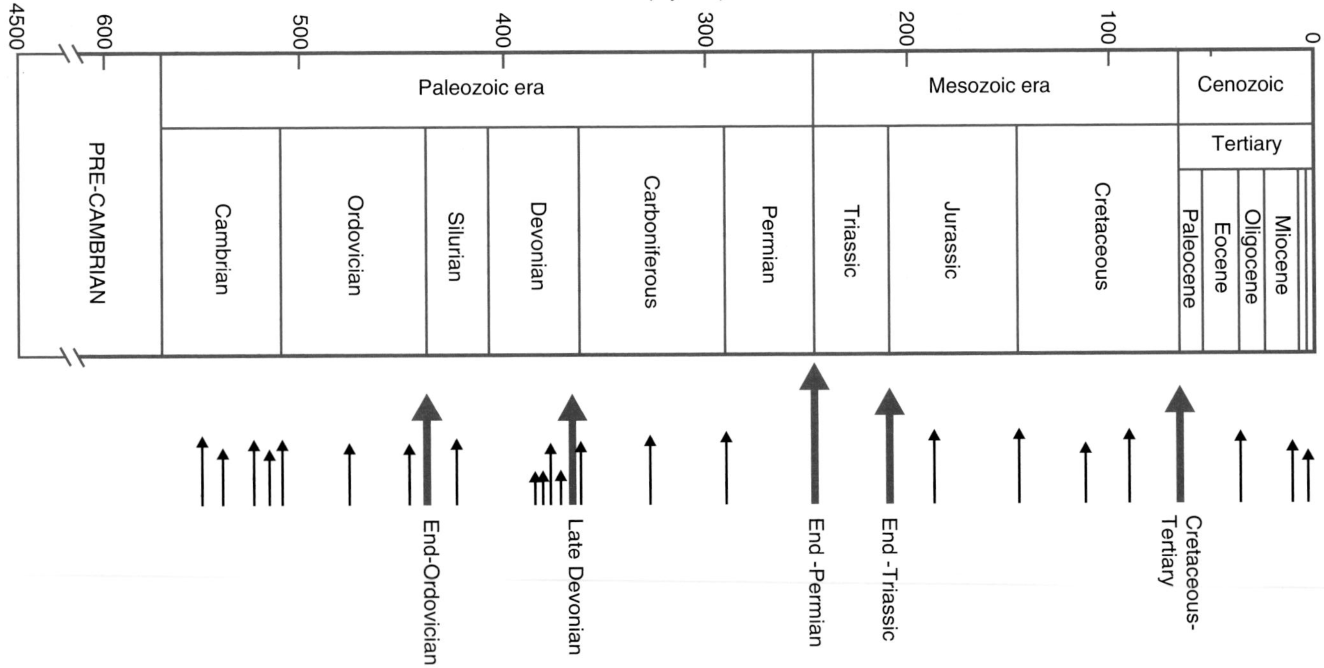

Figure 23.7 The pattern of extinction rates through time. (a) Cyclical changes in extinction probability are suggested by this figure, which shows the initial results of Raup and Sepkoski in 1984 for 567 marine families. Subsequent work has found much the same pattern, but has larger sample sizes. (b) Another illustration, to emphasize the strength of the evidence for various mass extinctions; arrows represent the intensity of extinctions. Five mass extinctions are particularly clear: Cretaceous, Triassic, Permian, Devonian, and Ordovician. Reprinted, by permission of the publisher, from Raup and Sepkoski (1984) and Raup (1991).

The taxonomic problem is that of pseudoextinctions. It is clearly important to know whether the 26 million year cycle is for real extinctions or pseudo-extinctions. If the extinctions were real, then some physical cause or other must have been operating, and we should be able to look for it; but if they were pseudoextinctions, the cycles may simply be some artifact of the habits of taxonomists. A number of taxa in the data have been examined in detail and a high proportion of the extinctions suggested to be pseudoextinctions (the danger particularly arises with paraphyletic taxa, because they are phenetically defined).

As a consequence of these problems, the 26 million year cycle is not widely accepted. The pattern, however, has not been shown to be false, although severe problems have been identified in the evidence. When we consider the causes of mass extinctions later in the following section, we should keep cyclicity in mind as a possibility. If a theory of mass extinctions can account for a cyclic pattern, it may turn out to be a virtue. For now, the evidence for cycles is not strong enough to rule out a theory that lacks the property.

23.7 *Several theories have been suggested to explain why mass extinctions happen*

23.7.1 *It is difficult to test the theories because the facts are not settled*

We can proceed in either (or both)of two ways to study mass extinctions. One is to take Raup and Sepkoski's result at face value—we then provisionally accept that the 26 million year cycle of mass extinctions, and look for an explanation. The other is to be more skeptical and doubt whether documentation for the cyclical pattern is adequate to merit our consideration. Even those who prefer the skeptical position cannot doubt that at least *some* mass extinctions occurred. About five such events are widely accepted (Figure 23.7b): three of them are the tallest peaks in Raup and Sepkoski's time series (Cretaceous, Triassic, and Permian) and two others precede the Permian (Devonian and Ordovician). Thus, even if cyclicity is not accepted, we need an explanation for a recurrent phenomenon. Mass extinctions, we have seen, may either be regular and cyclic, or rare and erratic features in the history of life.

Which position we take on the time distribution of mass extinctions will influence the kind of explanation we seek. Physical processes, in particular, are more likely to have 26 million year periodicities than are biological processes. Some large-scale biological convulsion could perhaps have caused one, or a few scattered, mass extinctions. It is difficult to imagine, however, biological cycles on the time scale of 26 million years that could generate mass extinctions. Here, we shall consider several of the more important theories. Some are more plausible for cyclic patterns of mass extinction, others for erratic patterns of mass extinction.

23.7.2 *The evidence does not suggest that mass extinctions other than the Cretaceous-Tertiary event are driven by periodic asteroid impacts*

Three kinds of evidence for an asteroidal collision with the Earth have been offered: the "geochemical signature" (that is, the iridium anomaly), the characteristic signs of an impact (shocked quartz), and impact craters. Only the Cretaceous-

Tertiary boundary has been extensively investigated in the first two ways. Iridium measurements have been made for other mass extinctions (Table 23.1); some have revealed small increases; but most do not. Small increases, of about one order of magnitude, may be better explained by terrestrial processes that concentrate iridium, rather than by an asteroid collision. The iridium spike at Gubbio is much larger, by three to four orders of magnitude (Figure 23.6). No good evidence exists for an extraterrestrial collision for any mass extinction except the one at the end of the Cretaceous. The largest mass extinction, at the end of the Permian, has no striking iridium anomaly and almost certainly had a different cause from the Cretaceous-Tertiary event.

The fossil record provides further reason to doubt that the 26 million year pattern (if it is real) is driven by periodic collisions with large asteroids. The record for the Cretaceous-Tertiary event (Figure 23.5) is consistent with a global and synchronous extinction. Other mass extinctions have been less thoroughly studied, but there are strong hints that some do not show the same pattern. Hallam, for example, analyzed two of Raup and Sepkoski's mass extinctions from the Jurassic, and found that they were regional, rather than global, in extent. The Plienisbachian extinction took place in North American bivalves, but not in their South American counterparts. In addition, the extinction at the Jurassic-Cretaceous boundary can be seen clearly in samples from all over Europe, but not from Argentina and Chile. In both cases, small and local causes were probably operating.

23.7.3 *Mass extinctions may have been caused by changes of climate and sea level*

Some mass extinctions have coincided with times of climatic change, especially cooling; and climatic change is the sort of physical factor that might be expected

TABLE 23.1

Iridium anomalies at times other than the Cretaceous-Tertiary boundary. Most times were for mass extinctions, although one may not be. Magnitudes for iridium spikes are expressed relative to the average for the type of rock before and afterward. Compare Figure 23.6 for the iridium anomaly at the end of the Cretaceous. It is doubted whether any of the iridium anomalies in the table have extraterrestrial causes. From Orth (1989), who gives the references for all sources except Holser *et al.* (1989), who are the third Permo-Triassic source and Donovan (1989), who are the Precambrian-Cambrian source.

Period	Extinction	Iridium
Eocene-Oligocene	Mass	Small spike
Eocene-Oligocene	Mass	Small spike
Eocene-Oligocene	Mass	Small spike
Permo-Triassic	Mass	Spike
Permo-Triassic	Mass	Normal
Permo-Triassic	Mass	10 × spike
Late Devonian	Mass	3-7 × spike
Late Devonian	Mass	Small spike
Ordovician-Silurian	Mass	Normal
Ordovician-Silurian	Mass	Normal
Precambrian-Cambrian	Mass?	Small spike
Precambrian-Cambrian	Normal?	Small spike

to cause extinctions. The evidence for the association remains imperfect. The five principal mass extinctions illustrate this point. The mass extinction in the late Ordovician coincides with a relatively sudden cooling. The one in the late Devonian coincides with a sudden cooling according to some authors, but not according to others. For the Permian-Triassic extinction, some authors have found an association with a bout of cooling, or with salinity changes in the ocean that would probably indicate a change in climate. No evidence of cooling has been found associated with the late Triassic extinction. The Cretaceous-Tertiary extinction has been argued both ways. We should consider not only whether extinctions are associated with cooling, but also whether cooling is associated with extinction. Periods of cooling, such as a bout of glaciation at the end of the Carboniferous, may not also be accompanied by an exceptional number of extinctions. The theory, therefore, is not supported by the evidence; it convincingly fits only one of five mass extinctions and makes erroneous predictions elsewhere.

Even if climatic change is not the general cause of mass extinctions, it may operate during smaller extinction events. This explanation has been suggested for some relatively recent extinctions. A round of extinctions in the late Eocene 36 million years ago coincides with a time of climatic cooling, when average temperatures may have decreased by about 4°C. Stanley, who has particularly argued that climatic cooling has caused extinctions, has documented an extinction event 2.4–3 million years ago in molluscs that took place when the climate cooled down. The hypothesis still needs to be tested systematically with all the evidence, as most of the evidence consists of pointing to cases in which the climatic change and extinction happen to be associated. Further evidence for the influence of climate on extinctions comes from the selective pattern of extinctions. In some extinction events, species that are more tolerant of cool climates survived better than did warm-adapted species (see section 23.9.2).

Asteroidal collisions would cause large short-term climatic changes, as would intense global volcanic eruptions. The climatic theory is not necessarily an alternative to these two ideas. Climatic change may also be connected with another theory of mass extinction—changes in sea level. The times of lower sea level (called marine regressions) in the Mesozoic appear to have coincided with times of mass extinction (Figure 23.8). The correlation is better than for the climatic theory, but again some regressions do not coincide with mass extinctions. In recent times, particularly in the Pleistocene, fluctuations of sea level have also not coincided with extinctions, though it is possible to argue around the exceptions. Changes in sea level should be taken into account when studying extinctions.

In summary, evidence shows that climatic and sea level changes are associated with extinctions. Both factors could, in principle, cause extinctions. They are likely to be connected because of the way the Earth's oceans and climate interact. However,the evidence for an association with mass extinctions is unpersuasive. Proper statistical testing is needed, to see whether the majority of the evidence supports the idea, or whether a few confirmatory instances have been selected from a wide range of data.

23.7.4 *Extinctions may be connected with changes in the shape of continents*

Valentine and Moores have suggested that changing land and shore areas, due to plate tectonic wandering, should be associated with global species diversity

Figure 23.8 Three successive rounds of ammonite extinction were correlated with declines in the sea level (expressed here as area of continents covered by sea; when the sea is higher, it covers more of continental land). Reprinted, by permission of the publisher, from Hallam (1983).

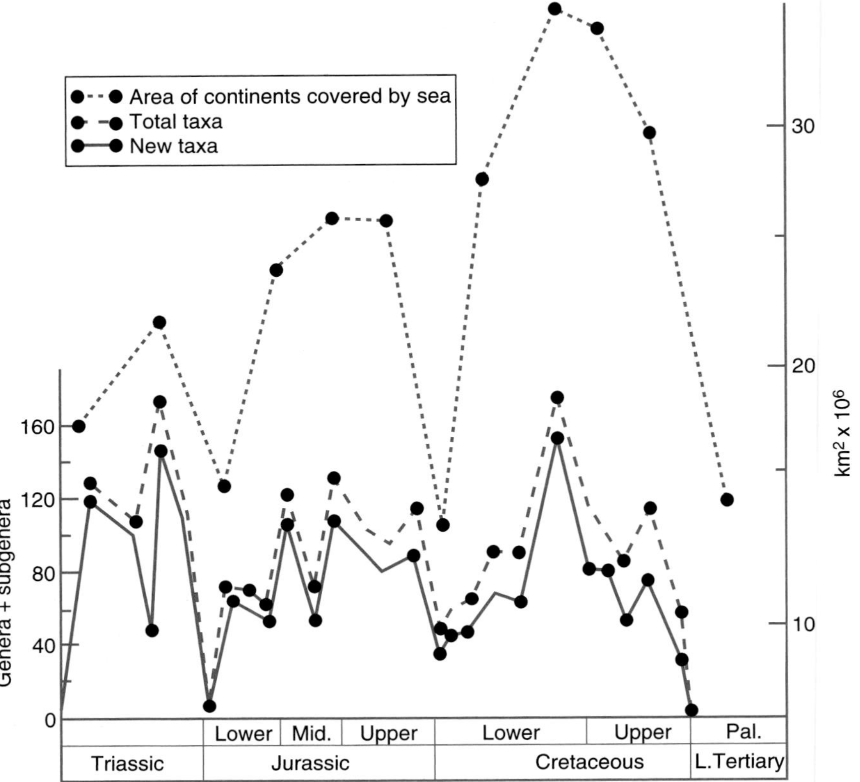

and extinction rates. The number of species in a locality is usually proportional to its geographic area, for good ecological reasons, and a reduction in area should produce extinctions. The best fit with the pattern of mass extinctions is found at the end of the Permian. If, as has been suggested, the continents coalesced at the end of the Permian to produce the supercontinent Pangaea (Figure 18.5, p. 517), the length of continental perimeter would have been reduced for simple geometric reasons. Many fossils are marine invertebrates that inhabited the shoreline habitats. As the area of their habitat decreased, the diversity of these species would have decreased as well. Hence, in Valentine and Moores' theory, the Permian mass extinction would have occurred.

This idea is difficult to test because of uncertainty about the exact continental positions at times so distant in the past. The theory therefore remains poorly supported by evidence. Changes in habitat area are less clear for other mass extinctions, but the theory should operate generally. As continents wander, habitat areas will change, and species diversity can be expected to alter accordingly. Thus, Valentine and Moores have identified a potentially general influence on extinction rates, and its most plausible application so far involves the extinction at the end of the Permian.

23.8 *Summary for mass extinctions*

Mass extinctions are real events in the history of life. The exact number of mass extinctions that have occurred since the Cambrian probably lies somewhere between the five clear events and the 23 or so implied by a 26 million year cycle and a 600 million year span. Whatever the actual number, we are clearly dealing with a recurrent phenomenon. Mass extinctions are more likely to have a physical than a biological cause. For none of the mass extinctions has universal agreement been reached as to its cause. The best studied mass extinction happened at the Cretaceous/Tertiary boundary, for which a growing body of evidence suggests that it was caused by a collision with an asteroid. This idea has clear implications for the pattern of fossil extinctions, and those implications are the subject of active research.

The causes of other mass extinctions are under research as well, but they are even less certain. If the 26 million year cycle of mass extinctions is not an artifact, then a good theory of mass extinctions should be able to account for a regular, and not just a recurrent and erratic, event. Given the present quality of the evidence for cycles, identifying such a theory is not a pressing demand. Several geological processes—including volcanism, changes in sea level, climate, and continental geometry—may contribute to mass extinctions. Table 23.2 summarizes the explanations for various mass extinctions that were favored at an expert symposium in 1989. The summary is not definitive, however! Knowledge of this subject moves quickly, and it is not appropriate to hold firm opinions.

Other theories have been put forward occasionally, and the large number of hypotheses—many of them frivolous—for the extinction of the dinosaurs is notorious. A good hypothesis for the extinction of the dinosaurs must be general enough to explain the entire mass extinction at the end of the Cretaceous, because

TABLE 23.2

Mass extinctions and their possible causes, suggested by various authorities. Simplified from Donovan (1989).

Eocene to Oligocene	Stepwise extinction associated with severe cooling, glaciation, and changes of oceanographic circulation, driven by the development of the circum-Antarctic current
End Cretaceous	Bolide impact producing catastrophic environmental disturbance
Late Triassic	Possibly related to increased rainfall with implied regression
End Permian	Gradual reduction in diversity produced by sustained period of refrigeration, associated with widespread regression and reduction in area of warm, shallow seas
Late Devonian	Global cooling associated with (causing?) widespread anoxia of epeiric seas
Late Ordovician	Controlled by the growth and decay of the Gondwanan ice sheet following a sustained period of environmental stability associated with high sea level

it is implausible that one process was driving the dinosaurs extinct while other processes were independently disposing of the many other groups that suffered extinction at exactly the same time. Any causal hypothesis must be tuned to the generality of the facts it is meant to explain, and thus further work on the fossil record for mass extinctions should continue to influence thinking about the cause, or causes, of these extraordinary catastrophes.

23.9 *Some kinds of species may survive mass extinctions more effectively than other species*

23.9.1 *Natural selection will not have equipped species with adaptations to survive the sort of extraordinary conditions that probably prevail in mass extinctions*

The traumatic environmental circumstances—whatever they were—of mass extinctions must have imposed exceptional selective forces, in kind as well as degree. Living things cannot become adapted by natural selection to conditions that recur on the time scale of at least 26 million years, and we should not expect organisms to possess adaptations for survival in mass extinctions. It could almost be a matter of luck whether a taxon survives through a mass extinction. It is also possible that some taxa will incidentally evolve characters that enable them to survive better than others. Raup (1991) has expressed this distinction by asking whether taxa go extinct because of bad luck or bad genes. We looked earlier (section 21.3) at species selection, which suggests that some kinds of species might have higher extinction rates than others. We now turn to look at species selection in the exceptional conditions of mass extinctions.

23.9.2 *Species with different temperature adaptations may have survived mass extinctions differentially*

Stanley, as we have seen, has argued that mass extinctions are caused by climatic cooling. A major part of his evidence is the selective pattern of extinctions at those times. In the late Eocene extinction, for instance, warm-water molluscan species suffered more extinctions than polar species. Likewise, many foraminiferid groups went extinct, but the globigerines (a cool-water group) survived and became the dominant Oligocene foraminifers. Other anecdotal evidence suggests that tropical species suffer more in mass extinctions, but it is always important to check that a few confirmatory instances have not been selected from evidence that points in all directions. The evidence also needs to be corrected for the greater species diversity, and the narrower geographical range of individual species, at the tropics. Tropical species will have a higher total extinction rate than polar species even if extinction hits species at random, because more tropical species exist. The narrower geographic range of tropical species increases their apparent extinction rate, because they are less likely to be preserved in the fossil record. Thus, the evidence that temperature adaptations influence survival through mass extinctions is suggestive, but not yet rigorous enough to be persuasive.

23.9.3 *The selectivity of extinction at the Cretaceous-Tertiary boundary differed from the pattern in the late Cretaceous*

Jablonski has carried out a more systematic study to find out whether extinctions were selective at the end of the Cretaceous. He looked at the relation between extinction rates and three properties of various gastropods and bivalves: the species richness of the taxon, the geographic range of the constituent species, and direct versus planktonic larval development.

We saw in section 21.3 that, in the normal times of the late Cretaceous before the mass extinction, the extinction rate was higher in species with direct development than in those with planktonic development. Jablonski found similar relations for the other two variables: taxa that contained more species and had broader geographic ranges had lower extinction rates than taxa with less species and smaller ranges. He compared these results with those from the Cretaceous-Tertiary mass extinction and found that at that time two of the three correlations disappeared (Table 23.3). Species-rich taxa had the same chance of extinction as species-poor ones, and planktonic species had the same chance of extinction as directly developing ones. Only broad geographic range continued to be associ-

TABLE 23.3

Survival of kinds of snail taxa through the Cretaceous-Tertiary mass extinction and at other times (showing a "background" extinction pattern). Background extinction rates vary with snail type, whereas survival through the mass extinction may have been a matter of luck. (a) Relation between chance of extinction and developmental mode. The background rate of extinction is lower for snails with planktonic, than with direct, development; this evidence is the same as in Figure 23.3. In mass extinctions, however, snail genera of the two types have the same chance of surviving. (b) Relation between chance of generic extinction and number of species in the genus. The background extinction rate is lower for genera that are species-rich (contain three or more species) than for genera that are species-poor (contain one or two species). In mass extinctions, however, a species-rich genus had about the same chance of going extinct as a species-poor genus; in both cases, about 40% of genera went extinct and about 60% survived. In the table, *n* is number of genera. The genera studied at the two times are not all the same. Results of Jablonski (1986).

(a) Extinction rate and mode of development

	Background Extinctions		Mass Extinction		
				Percent:	
Mode of Development	*n*	Median Geological Longevity (Myr)	*n*	Surviving	Extinct
Planktonic development	50	6	28	60	40
Direct development	50	2	21	60	40

(b) Extinction rate and species-richness of genus

	Background Extinctions		Mass Extinction	
			Number of Genera	
Species Richness	*n*	Median Geological Longevity (Myr)	Surviving	Extinct
Species poor	145	32	31	38
Species rich	114	49	22	25

ated with lower extinction rate. The extinction seems to have been so massive as to have taken out groups almost at random.

At any rate, the relations between the characters of a taxon and its extinction probability were significantly altered. In normal times, planktonically developing and species-rich taxa have lower probabilities of extinction than directly developing and species poor taxa. In contrast, in the Cretaceous-Tertiary mass extinction, any such differences disappeared.

23.10 *A two-geared engine of macroevolution?*

The different relations between a taxon's extinction rate and its characters has led to the suggestion that there are two *macroevolutionary regimes.* Evolution may alternate between "normal" periods and mass extinctions, and the kinds of taxa favored in the two periods differ. Mass extinctions are extraordinary and intense and remove whole taxa in a predominantly random way. In the normal periods between such events, planktonically developing forms (for example) will proliferate relative to directly developing forms by species selection, as planktonic development reduces extinction rate in those quiet times. It does, indeed, seem likely that the relation between the characters possessed by taxa and their extinction rates will differ between mass and normal extinctions, because the circumstances are probably so different that the same character would not assist in avoiding extinction in both.

One question we can ask is whether two distinct macroevolutionary regimes exist, or if there is a continuum with mass extinctions at one of the extremes. A bimodal pattern might be observed if, for example, mass extinctions have a different cause, such as asteroidal collision, from the extinctions at other times. This question has been tackled by looking at the frequency distribution of extinction rates for the whole history of life (Figure 23.9a,b). If there are two regimes, then the frequency distribution should be bimodal, with one peak for the normal low extinction rate and another higher peak for the mass extinctions. If extinction rates have one long-term average for the history of life, and the actual rate changes continuously between low and high values, the frequency distribution would probably be a Poisson distribution (the Poisson distribution is the frequency distribution for the number of events per time unit for a random process). The real pattern shows little deviation from the random Poisson pattern (Figure 23.9c); it contains a small secondary peak, but it is entirely due to the Cretaceous-Tertiary mass extinction and is statistically insignificant. Thus, the evidence counts against the strong prediction of bimodality.

As ecological and geological conditions change through time, they place varying demands on living species. Sometimes the environment will be harsh and extinction rates will rise; sometimes there will be a period of calm and extinction rates will decline. The species that survive best in the different environment will vary, both because of their adaptations to the environment and because of chance relations between adaptations and the probability of extinction (by the mechanism of species selection). Mass extinctions, therefore, may have been important influences in the replacements of one taxon by another in the history of life (section 21.9, p. 601). After the late Triassic extinction, the dinosaurs rose

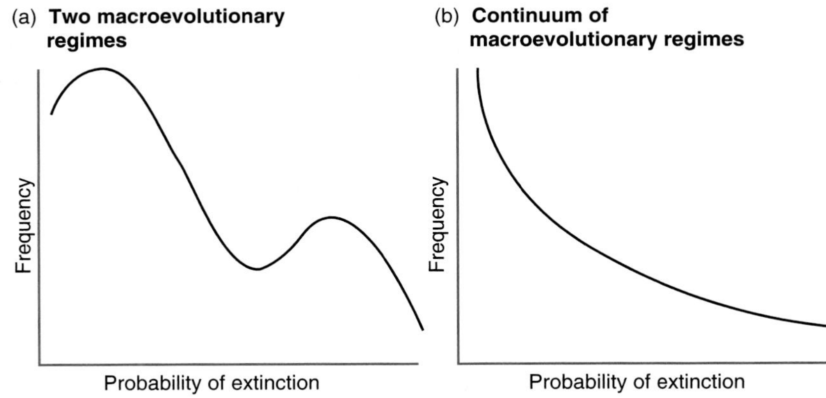

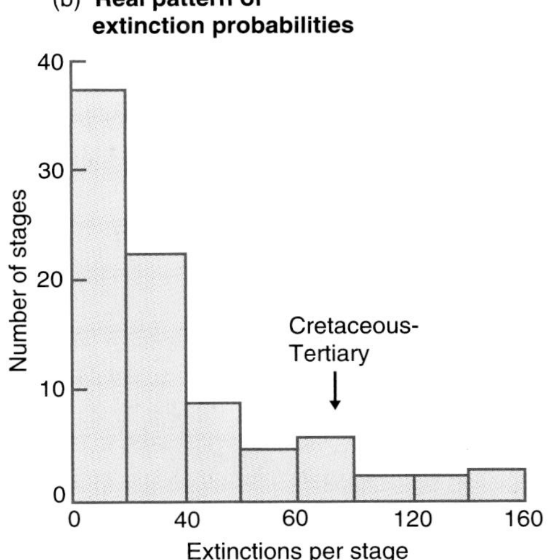

Figure 23.9 (a) If two macroevolutionary regimes occurred, the frequency distribution of extinction probabilities in the fossil record should be bimodal. (b) If there is a continuum of macroevolutionary regimes, it will be a continuous Poisson distribution. (c) The actual distribution for 2316 marine animal families in the 79 generally recognized divisions of geological time since the Cambrian is continuous. The extinction intensity at the Cretaceous-Tertiary boundary is indicated for comparison. Reprinted with permission from Raup (1986). Copyright 1986 American Association for the Advancement of Science.

to become the dominant large terrestrial vertebrates, replacing the formerly dominant mammal-like reptiles (Figure 21.13, p. 604). Then, between the Cretaceous and the Tertiary, the tables were turned—the dinosaurs were extinguished and the mammals took over.

The replacement may have been competitive, taking place over a long period; the evidence cannot rule out such a possibility. However, it is also possible that the replacement was determined by the relative survival of the two groups in the Cretaceous-Tertiary mass extinction. There could have been two causes. If the mass extinction was indiscriminate, and survival a matter of luck, then the mammals survived to radiate into the empty ecological space after the mass extinction simply because they were luckier than the dinosaurs. Alternatively, the characters of the two groups might have influenced their chance of survival. Species selection in a mass extinction might have favored a group of small-bodied animals, which were adapted to survive in marginal habitats. While the dinosaurs were dominant on land, mammals persisted as small, mainly nocturnal, species. Perhaps mammals were better equipped than dinosaurs to survive the extraordinary selective conditions of the Cretaceous catastrophe.

Through the history of life on Earth, the changing geological conditions have probably acted to sift among the groups of plants and animals and to generate the grand patterns of taxonomic rises, falls, and replacements—patterns that Medawar has called "evolutionary dynastics." The continuous variation of extinction rates (Figure 23.9) suggests that the conditions change continuously, moving among an infinite variety of states, rather than banging back and forth between only two kinds of macroevolutionary regime.

SUMMARY

1. The ecological causes of particular species extinctions are best studied in modern, not fossil, forms.
2. The real extinction of lineages should be distinguished from pseudoextinction, which is due to the taxonomic subdivision of a continuous lineage.
3. Factors influencing extinction and speciation rates in different taxa can be investigated. These factors may be internal properties of the organisms, or external ecological factors.
4. If natural selection favors one form of a character in one species and another form in another species, and if the different forms of the character cause species to have differing speciation or extinction rates, then a trend may operate to favor the kind of species with higher speciation, or lower extinction, rates. In Tertiary snails, species with planktonic development have lower speciation rates than do species with direct development; this difference may have caused a trend to an increasing proportion of directly developing forms. The process is called species selection.
5. If the niches of some species last longer than others, those species will have lower extinction rates. If some niches are positioned so that new species can easily evolve from them, then the species occupying them will be more likely than average to give rise to new taxa. Modern sticklebacks may illustrate this process.
6. Long-term evolutionary patterns will often be determined by a particular non-random subsample of all short-term evolutionary events. Macroevolution will then be "uncoupled" from microevolution.
7. At various times in the history of life, an exceptionally large number of species have gone extinct together; these events are called mass extinctions. The best studied mass extinction occurred at the boundary of the Cretaceous and Tertiary, 65 million years ago.
8. Alvarez *et al.*'s discovery of anomalously large concentrations of the rare earth element iridium at the Cretaceous-Tertiary boundary rocks at Gubbio, in Italy, suggested that the mass extinction may have been caused by the collision of an asteroid, about 10 km in diameter, with the Earth. Much evidence now supports their idea.
9. The impact theory of mass extinctions predicts that extinctions in different taxa should be sudden, synchronous, and global. The evidence for the Cretaceous-Tertiary extinction mainly fits the prediction, although other interpretations of the pattern are possible.

10. Raup and Sepkoski's survey of extinction rates in the history of life suggests that extinction rates vary cyclically, with mass extinctions recurring about every 26 million years. Their evidence has been challenged taxonomically, statistically, and geochronologically. The result is controversial and uncertain.
11. Asteroidal collisions, volcanic eruption, climatic cooling, sea level changes, and changes in habitat area caused by continental drift are the five potentially general theories of mass extinction. The evidence does not suggest mass extinctions are generally caused by asteroidal collisions; the effects of climatic sea level changes need to be tested systematically; and the effect of continental drift is difficult to test at present.
12. Different kinds of species may suffer differentially in mass extinctions. The relation between the characters of taxa and their extinction rates changed between the late Cretaceous and the Cretaceous-Tertiary mass extinction.
13. Mass extinctions may represent an end point of a continuum of macroevolutionary extinction regimes. The kind of taxa favored in macroevolution will depend on the prevailing conditions.

FURTHER READING

Recent popular books about, or partly about, extinction include Eldredge (1995), Raup (1991), and Ward (1995). Gould's popular monthly column in *Natural History* (collected in Gould 1977b, 1980a, 1983a, 1985, 1991, 1993) contains a number of excellent essays on extinction. At a more advanced level, symposia include Larwood (1988), Donovan (1989), *Philosophical Transactions of the Royal Society of London*, vol. B 325 (1989), pp. 239–488, and Novacek and Wheeler (1992). Raup (1986), Marshall (1988), Jablonski (1994), and Benton (1995) are general review papers; Hoffman (1989) discusses the subject too.

On species selection, Eldredge and Gould (1972) is the influential source. Further discussion is found in, among many others, Williams (1966, 1982—who prefers the term "clade selection"), Gould and Eldredge (1977, 1993), Dawkins (1982), Levinton (1988), the exchange between Maynard Smith (1987c, 1988) and Eldredge and Gould (1988), and Hoffman (1989). On the effects of niche persistency, see Williams (1992), and Bell (1988) for background on sticklebacks. Likewise, Jablonski and Bottler (1991) show that 20/26 well-preserved higher taxa originated in onshore habitats.

On the Cretaceous-Tertiary mass extinction, see the pair of papers in *Scientific American* by Alvarez and Asaro (1990) and Courtillot (1990). The key original Alvarez papers are Alvarez *et al.* (1980, 1984); see Grieve (1990) on impact craters. Officer *et al.* (1987) give a different perspective. Benton (1994) discusses how to weight the evidence. Blum (1993) is a short guide to some of the recent literature on the Chicxulub structure; Melosh (1995) is an update on its size. Meyerhoff *et al.* (1994) give a dissenting opinion; they offer evidence against the asteroidal theory and attribute the structure to volcanism. The literature moves fast; non-experts can follow most of it in the journal *Geology Today*. The end-Permian extinction is discussed by Erwin (1993, 1994). On cycles, the original paper is Raup and Sepkoski (1984). Raup and Boyajian (1988) and Sepkoski (1992) are more recent updates; the controversy can be traced through these papers and the general references given above, particularly Hoffman (1989). See

Pease (1992) on the trend toward declining extinction rates with time. On the effect of climate, in addition to the general works cited above, see Crowley and North (1988) and Stanley (1984). Hallam (1983) discusses the effect of sea level; see also Officer *et al.* (1987).

On replacement, see the general references in chapter 21, and for dinosaurs and mammals see Gould (1983a, ch. 30) and Van Valen and Sloan (1977). Feduccia (1995) describes a similar pattern of radiations in birds as in mammals. For further topics in radiation, see the authors in Ross and Allman (1990) on many topics, Losos (1994) on radiation rules in lizards, and Martin (1992) on the tendency for taxa with smaller body size to speciate at a higher rate; Levinton (1988) and Gould (1989) discuss the number of higher groups through geological time. See also chapter 14 on divergence and chapter 15 on character displacement.

STUDY AND REVIEW QUESTIONS

1. Why could it not be known until relatively recently in human history that a species had gone extinct?
2. (a) What is the distinction between a real extinction and a pseudoextinction? (b) How does the distinction matter for theories of species selection and mass extinctions?
3. Give two reasons why the macroevolutionary pattern of extinction rates in different taxa may not simply be extrapolated from microevolution in the taxa.
4. (a) What is the best evidence that the mass extinction at the Cretaceous-Tertiary boundary was caused by an asteroidal impact? (b) What predictions about the pattern of extinctions in the fossil record can be made if it was indeed caused by an asteroidal impact? What difficulties arise in testing them?
5. When did the five best documented mass extinctions occur in the history of life?
6. What is the relation between the developmental mode of snail taxa and the chance of extinction at the time of mass extinctions and at times of normal (or background) extinction rates? How do you explain the trends?
7. How has it been tested whether mass extinctions are a distinct macroevolutionary regime, or whether they are one end of a continuum of such regimes? What did the test find?

GLOSSARY

Words in italics refer to separate entries found elsewhere in the glossary.

adaptation Feature of an organism enabling it to survive and reproduce in its natural environment better than if it lacked the feature.

adaptive landscape A graph of the average *fitness* of a population in relation to the frequencies of genotypes in it; peaks on the landscape correspond to genotypic frequencies at which the average fitness is high, valleys to genotypic frequencies at which the average fitness is low. Also called a fitness surface.

allele A variant of a single gene, inherited at a particular genetic locus. It consists of a particular sequence of *nucleotides*, coding for *messenger RNA*.

allometry The relation between the size of an organism and the size of any of its parts. For example, an allometric relation exists between brain size and body size, such that (in this case) animals with bigger bodies have bigger brains. Allometric relations can be studied during the growth of a single organism, between different organisms within a species, or between organisms in different species.

allopatric speciation Speciation via geographically separated populations.

allopatry Living in separate places. Compare with *sympatry*.

amino acid Unit molecular building block of *proteins*; a protein is a chain of amino acids in a certain sequence. There are 20 main amino acids in the proteins of living things, and the properties of a protein are determined by its particular amino acid sequence.

amniotes The group of reptiles, birds, and mammals. They all develop through an embryo that is enclosed within a membrane called an amnion. The amnion surrounds the embryo with a watery substance, and is probably an adaptation for breeding on land.

analogy Character shared by a set of species but not present in their common ancestor. A convergently evolved character. Compare with *homology*.

anatomy (1) The structure of an organism, or one of its parts. (2) The science that studies those structures.

ancestral homology *Homology* that evolved before the common ancestor of a set of species, and is present in other species outside that set of species. Compare with *derived homology*.

area cladogram Branching diagram (or *phylogeny*) of a set of species (or other taxa) showing the geographic areas they occupy. According to the theory of vicariance biogeography, the branching diagram represents the history of range splits (probably driven by geological processes such as continental drift) in the ancestry of the species.

artificial selection Selective breeding, carried out by humans, to alter a population. The forms of most domesticated and agricultural species have been produced by artificial selection; it is also an important experimental technique for studying evolution.

asexual reproduction Production of offspring by virgin birth or by vegetative reproduction; that is, reproduction without sexual fertilization of eggs.

assortative mating Tendency of like to mate with like. It can be for a certain genotype (e.g., individuals with genotype *AA* tend to mate with other individuals of genotype *AA*) or phenotype (e.g., tall individuals mate with other tall individuals).

atomistic (as applied to theory of inheritance) Inheritance in which the entities controlling heredity are relatively distinct, permanent, and capable of independent action. *Mendelian inheritance* is an atomistic theory because, in it, inheritance is controlled by distinct *genes*.

autosome Any *chromosome* other than a sex chromosome.

base The DNA is a chain of nucleotide units; each unit consists of a backbone made of a sugar and a phosphate group, with a nitrogenous base attached. The base in a unit is one of adenine (A), guanine (G), cytosine (C), or thymine (T). In RNA, uracil (U) is used instead of thymine. A and G belong to the chemical class called purines; C, T, and U are pyrimidines.

Batesian mimicry A kind of *mimicry* in which one non-poisonous species (the Batesian mimic) mimics another poisonous species.

Biogenetic law Name given by Haeckel to *recapitulation.*

biological species concept Concept of species, according to which a species is a set of organisms that can interbreed among each other. Compare with *cladistic species concept, ecological species concept, phenetic species concept,* and *recognition species concept.*

biometrics Quantitative study of characters of organisms.

blending inheritance The historically influential but factually erroneous theory that organisms contain a blend of their parents' hereditary factors and pass that blend on to their offspring. Compare with *Mendelian inheritance.* See chapter 2, section 2.8.

character Any recognizable trait, feature, or property of an organism.

character displacement Increased difference between two closely related species where they live in the same geographic region (sympatry) as compared with where they live in different geographic regions (allopatry). Explained by the relative influences of intra- and inter-specific competition in sympatry and allopatry.

chloroplast Structure (or *organelle*) found in some cells of plants; its function is photosynthesis.

chromosomal inversion See *inversion.*

chromosome Structure in the cell nucleus that carries DNA. At certain times in the cell cycle, chromosomes are visible as string-like entities. Chromosomes consist of the DNA with various proteins, particularly histones, bound to it.

clade Set of species descended from a common ancestral species. Synonym of *monophyletic group.*

cladism Phylogenetic classification. The members of a group in a cladistic classification share a more recent common ancestor with one another than with the members of any other group. A group at any level in the classificatory hierarchy, such as a "family," is formed by combining a subgroup (at the next lowest level, perhaps the genus in this case) with the subgroup with which it shares its most recent common ancestor. Compare with *evolutionary classification* and *phenetic classification.*

cladistic species concept Concept of species, according to which a species is a *lineage* of populations between two phylogenetic branch points (or speciation events). Compare with *biological species concept, ecological species concept, phenetic species concept,* and *recognition species concept.*

classification Arrangement of organisms into hierarchical groups. Modern biological classifications are *Linnaean* and classify organisms into species, genus, family, order, class, phylum, kingdom, and certain intermediate categoric levels. *Cladism, evolutionary classification,* and *phenetic classification* are three methods of classification.

cline A geographic gradient in the frequency of a gene, or in the average value of a character.

clock See *molecular clock.*

clone A set of genetically identical organisms asexually reproduced from one ancestral organism.

coadaptation Beneficial interaction between a number of (1) genes at different loci within an organism, (2) different parts of an organism, or (3) organisms belonging to different species.

codon Triplet of *bases* (or nucleotides) in the *DNA* coding for one amino acid. The relation between codons and amino acids is given by the *genetic code.* The triplet of bases that is complementary to a condon is called an anti-codon; conventionally, the triplet in the *mRNA* is called the codon and the triplet in the *tRNA* is called the anticodon.

coevolution Evolution in two or more species in which the evolutionary changes of each species influence the evolution of the other species.

comparative biology Study of patterns among more than one species.

comparative method Study of adaptation by comparing many species.

concerted evolution Tendency of the different genes in a *gene family* to evolve in concert; that is, each gene locus in the family comes to have the same genetic variant.

convergence Process by which a similar character evolves independently in two species. Also, a synonym for *analogy*; that is, an instance of a convergently evolved character, or a similar character in two species that was not present in their common ancestor.

Cope's rule Evolutionary increase in body size over geological time in a lineage of populations.

creationism See *separate creation.*

crossing-over The process during meiosis in which the chromosomes of a *diploid* pair exchange genetic material. It is visible in the light microscope. At a genetic level, it produces *recombination.*

cytoplasm Region of *eukaryotic cell* outside the nucleus.

Darwinism Darwin's theory that species originated by evolution from other species and that evolution is mainly driven by natural selection. Differs from *neo-Darwinism* mainly in that Darwin did not know about *Mendelian inheritance.*

derived homology *Homology* that first evolved in the common ancestor of a set of species and is unique to those species. Compare with *ancestral homology.*

diploid Having two sets of genes and two sets of chromosomes (one from the mother, one from the father). Many common species, including humans, are diploid. Compare with *haploid* and *polyploid.*

directional selection Selection causing a consistent directional change in the form of a population through time (e.g., selection for larger body size).

disruptive selection Selection favoring forms that deviate in either direction from the population average. Selection favors forms that are larger or smaller than average, but works against the average forms between the extremes.

distance In *taxonomy,* refers to the quantitatively measured difference between the phenetic appearance of two groups of individuals, such as populations or species (phenetic distance), or the difference in their gene frequencies (genetic distance).

DNA Deoxyribose nucleic acid; the molecule that controls inheritance.

dominance (genetic) An allele (*A*) is dominant if the phenotype of the heterozygote *Aa* is the same as the homozygote *AA*. The allele *a* does not influence the heterozygote's phenotype and is called *recessive.* An allele may be partly, rather than fully, dominant; in that case, the heterozygous phenotype is nearer to, rather than identical with, the homozygote of the dominant allele.

drift Synonym of *genetic drift.*

duplication The occurrence of a second copy of a particular sequence of DNA. The duplicate sequence may appear next to the original, or be copied elsewhere into the *genome.* When the duplicated sequence is a *gene,* the event is called gene duplication.

ecological genetics Study of evolution in action in nature, by a combination of field work and laboratory genetics.

ecological species concept Concept of species, according to which a species is a set of organisms adapted to a particular, discrete set of resources (or "niche") in the environment. Compare with *biological species concept, cladistic species concept, phenetic species concept,* and *recognition species concept.*

electrophoresis Method of distinguishing entities according to their motility in an electric field. In evolutionary biology, it has been mainly used to distinguish different forms of proteins. The electrophoretic motility of a molecule is influenced by its size and electric charge.

epistasis An interaction between the genes at two or more loci, such that the phenotype differs from what would be expected if the loci were expressed independently.

eukaryote Made up of eukaryotic cells. A eukaryotic cell is a cell with a distinct nucleus. Almost all multicellular organisms are eukaryotic. Compare with *prokaryote.*

evolution Darwin defined this term as "descent with modification." It is the change in a lineage of populations between generations.

evolutionary classification Method of classification using both *cladistic* and *phenetic* classificatory principles. To be exact, it permits *paraphyletic* groups (which are allowed in phenetic but not in cladistic classification) and *monophyletic* groups (which are allowed in both cladistic and phenetic classification) but excludes *polyphyletic* groups (which are banned from cladistic classification but permitted in phenetic classification).

exon The nucleotide sequences of some genes consist of parts that code for amino acids, and other parts that do not code for amino acids interspersed among them. The coding parts, which are translated, are called exons; the interspersed non-coding parts are called introns.

fitness The average number of offspring produced by individuals with a certain genotype, relative to the number produced by individuals with other genotypes. When genotypes differ in fitness because of their effects on survival, fitness can be measured as the ratio of a genotype's frequency among the adults divided by its frequency among individuals at birth.

fixation A gene has achieved fixation when its frequency has reached 100% in the population.

fixed (1) In *population genetics*, a gene is "fixed" when it has a frequency of 100%. (2) In *creationism*, species are described as "fixed" in the sense that they are believed not to change their form, or appearance, through time.

founder effect The loss of genetic variation when a new colony is formed by a very small number of individuals from a larger population.

frequency-dependent selection Selection in which the *fitness* of a genotype (or phenotype) depends on its frequency in the population.

gamete Haploid reproductive cells that combine at fertilization to form the zygote; sperm (or pollen) in the male and eggs in females.

gene Sequence of nucleotides coding for a protein (or, in some case, part of a protein).

gene duplication See *duplication.*

gene family Set of related genes occupying various loci in the DNA, almost certainly formed by *duplication* of an ancestral gene, and having a recognizably similar sequence. The globin gene family is an example.

gene flow The movement of genes into, or through, a population by interbreeding or by migration and interbreeding.

gene frequency The frequency in the population of a particular gene relative to other genes at its *locus.* Expressed as a proportion (between 0 and 1) or percentage (between 0 and 100%).

gene pool All the genes in a population at a particular time.

genetic code The code relating nucleotide triplets in the mRNA (or DNA) to amino acids in the proteins. It has been decoded (see chapter 2, Figure 2.2).

genetic distance See *distance.*

genetic drift Random changes in gene frequencies in a population.

genetic load A reduction in the average fitness of the members of a population because of the deleterious genes, or gene combinations, in the population. It has many particular forms, such as "mutational load," "segregational load," "recombinational load."

genetic locus See *locus.*

genome The full set of DNA in a cell or organism.

genotype The set of two genes at a locus possessed by an individual.

geographic isolation See *reproductive isolation.*

geographic speciation See *allopatric speciation.*

germ plasm The reproductive cells in an organism; the cells that produce the *gametes.* All cells in an organism can be divided into the *soma* (the cells that ultimately die) and the germ cells (that are perpetuated by reproduction).

group selection Selection operating between groups of individuals rather than between individuals. It would produce attributes beneficial to a group in competition with other groups, rather than attributes beneficial to individuals.

haploid Condition of having only one set of genes or chromosomes. In normally *diploid* organisms such as humans, only the *gametes* are haploid.

haplotype Set of genes at more than one *locus* inherited by an individual from one of its parents. It is the multi-locus analog of an *allele.*

Hardy–Weinberg ratio Ratio of genotype frequencies that evolve when mating is random and neither selection nor drift are operating. For two alleles (*A* and *a*) with frequencies p and q, there are three genotypes *AA*, *Aa*, and *aa*; the Hardy–Weinberg ratio for the three is $p^2\ AA : 2pq\ Aa : q^2\ aa$. It is the starting point for much of the theory of population genetics.

heritability Broadly, the proportion of variation (more strictly *variance*) in a phenotypic character in a population that is due to individual differences in genotypes. Narrowly, the proportion of variation (more strictly *variance*) in a phenotypic character in a population that is due to individual genetic differences that will be inherited in the offspring.

heterogametic Sex with two different *sex chromosomes* (males in mammals, because they are XY). Compare with *homogametic.*

heterozygosity (for most purposes) Proportion of individuals in a population that are heterozygotes.

heterozygote Individual having two different *alleles* at a genetic *locus*. Compare with *homozygote*.

heterozygote advantage Condition in which the *fitness* of a heterozygote is higher than the fitness of either homozygote.

homeostasis (developmental) Self-regulating process in development, such that the organism grows up to have much the same form independently of the external influences it experiences while growing up.

homeotic mutation *Mutation* causing one structure of an organism to grow in the place appropriate to another. For example, in the mutation called "antennapedia" in the fruitfly, a foot grows in the antennal socket.

homogametic Sex with two of the same kind of *sex chromosomes* (females in mammals, because they are XX). Compare with *heterogametic*.

homology Character shared by a set of species and present in their common ancestor. Compare with *analogy*. (Some molecular biologists, when comparing two sequences, call the corresponding sites "homologous" if they have the same nucleotide—regardless of whether the similarity is evolutionarily shared from a common ancestor or convergent; they likewise talk about percent homology between the two sequences. Homology then simply means similarity. This usage is frowned upon by many evolutionary biologists, but is established in much of the molecular literature.)

homozygote Individual having two copies of the same *allele* at a genetic *locus*. Also sometimes applied to larger genetic entities, such as a whole chromosome; a homozygote is then an individual having two copies of the same chromosome.

hybrid Offspring of a cross between two species.

idealism Philosophical theory that there are fundamental non-material "ideas," "plans," or "forms" underlying the phenomena we observe in nature. It has been historically influential in classification.

inheritance of acquired characters Historically influential but factually erroneous theory that an individual inherits characters that its parents acquired during their lifetimes.

intron The nucleotide sequences of some genes consist of parts that code for amino acids, and other parts that do not code for amino acids interspersed among them. The interspersed non-coding parts, which are not translated, are called introns; the coding parts are called exons.

inversion An event (or the product of the event) in which a sequence of nucleotides in the DNA is reversed, or inverted. Sometimes inversions are visible in the structure of the chromosomes.

isolating mechanism Any mechanism, such as a difference between species in courtship behavior or breeding season, that results in *reproductive isolation* between the species.

isolation Synonym for *reproductive isolation*.

Lamarckian inheritance Historically misleading synonym for *inheritance of acquired characters*.

larva (and **larval stage**) Prereproductive stage of many animals; the term is particularly apt when the immature stage has a different form from the adult.

lineage An ancestor-descendant sequence of (1) populations, (2) cells, or (3) genes.

linkage disequilibrium Condition in which the *haplotype* frequencies in a population deviate from the values they would have if the genes at each locus were combined at random. (When no deviation exists, the population is said to be in linkage equilibrium.)

linked Of genes, present on the same chromosome.

Linnaean classification Hierarchical method of naming classificatory groups, invented by the eighteenth century Swedish naturalist Carl von Linné, or Linnaeus. Each individual is assigned to a species, genus, family, order, class, phylum, and kingdom, and some intermediate classificatory levels. Species are referred to by a Linnaean binomial of its genus and species, such as *Magnolia grandiflora*. Universally used by educated persons.

locus The location in the DNA occupied by a particular *gene*.

macroevolution Evolution on the grand scale. The term refers to events above the species level. The origin of a new higher group, such as the vertebrates, would be an example of a macroevolutionary event.

macromutation *Mutation* of large phenotypic effect; one that produces a phenotype well outside the range of variation previously existing in the population.

mean The average of a set of numbers. For example, the mean of 6, 4, and 8 is $(6 + 4 + 8)/3 = 6$.

meiosis Special kind of cell division that occurs during the reproduction of *diploid* organisms to produce the *gametes*. The double set of genes and chromosomes of the normal diploid cells is reduced during meiosis to a single *haploid* set. *Crossing-over* and therefore *recombination* occur during a phase of meiosis.

Mendelian inheritance The mode of inheritance of all *diploid* species, and therefore of nearly all multicellular organisms. Inheritance is controled by *genes*, which are passed on to the offspring in the same form as they were inherited from the previous generation. At each *locus*, an individual has two genes—one inherited from its father and the other from its mother. The two genes are represented in equal proportions in its *gametes*.

messenger RNA (mRNA) Kind of *RNA* produced by *transcription* from the DNA and which acts as the message that is decoded to form *proteins*.

microevolution Evolutionary changes on the small scale, such as changes in gene frequencies within a population.

mimicry A case in which one species looks more or less similar to another species. See *Batesian mimicry* and *Müllerian mimicry*.

mitochondrion A kind of organelle in eukaryotic cells. Mitochondria burn the digested products of food to produce energy. They contain DNA coding for some mitochondrial proteins.

mitosis Cell division. All cell division in multicellular organisms occurs by mitosis except for the special division called *meiosis* that generates the gametes.

modern synthesis Synthesis of natural selection and Mendelian inheritance. Also called *neo-Darwinism*.

molecular clock Theory that molecules evolve at an approximately constant rate. The difference between the form of a molecule in two species is then proportional to the time since the species diverged from a common ancestor, and molecules become of great value in the inference of *phylogeny*.

monophyletic group Set of species containing a common ancestor and all of its descendants.

morphology The study of the form, shape, and structure of organisms.

Müllerian mimicry A kind of *mimicry* in which two poisonous species evolve to look like one another.

natural selection Process by which the forms of organisms in a population that are best adapted to the environment increase in frequency relative to less well-adapted forms over a number of generations.

neo-Darwinism (1) Darwin's theory of natural selection plus Mendelian inheritance. (2) The larger body of evolutionary thought that was inspired by the unification of natural selection and Mendelism. A synonym of the *modern synthesis*.

neutral drift Synonym of *genetic drift*.

neutral mutation *Mutation* with the same *fitness* as the other *allele* or alleles) at its *locus*.

neutral theory (and neutralism) Theory that most evolution at the molecular level occurs by *genetic drift*.

niche The ecological role of a species; the set of resources it consumes, and habitats it occupies.

nucleotide Unit building block of DNA and RNA; a nucleotide consists of a sugar and phosphate backbone with a *base* attached.

nucleus Region of *eukaryotic cells* containing the DNA.

numerical taxonomy In general, any method of *taxonomy* using numerical measurements; in particular, it often refers to *phenetic classification* using large numbers of quantitatively measured *characters*.

organelle Any of a number of distinct small structures found in the cytoplasm (and therefore outside the nucleus) of eukaryotic cells (e.g. *mitochondrion, chloroplast*).

orthogenesis The erroneous idea that species tend to evolve in a fixed direction because of some inherent force driving them to do so.

paleobiology Biological study of fossils.

paleontology Scientific study of fossils.

panmixis Random mating throughout a population.

parapatric speciation Speciation in which the new species forms from a population contiguous with the ancestral species' geographic range.

paraphyletic group Set of species containing an ancestral species together with some, but not all, of its descendants. The species included in the group are those that have continued to resemble the ancestor; the excluded species have evolved rapidly and no longer resemble their ancestor.

parsimony Principle of phylogenetic reconstruction in which the phylogeny of a group of species is inferred to be the branching pattern requiring the smallest number of evolutionary changes.

parthenogenesis Reproduction by virgin birth; a term for *asexual reproduction.*

particulate (as property of theory of inheritance) Synonym of *atomistic.*

peripatric speciation Synonym of *peripheral isolate speciation.*

peripheral isolate speciation A form of *allopatric* speciation in which the new species is formed from a small population isolated at the edge of the ancestral population's geographic range. Also called peripatric speciation.

phenetic classification Method of classification in which species are grouped together with other species that they most closely resemble phenotypically.

phenetic species concept Concept of species, according to which a species is a set of organisms that are phenetically similar to one another. Compare with *biological species concept, cladistic species concept, ecological species concept,* and *recognition species concept.*

phenotype The characters of an organism, whether due to the genotype or environment.

phylogeny "Tree of life"; branching diagram showing the ancestral relations among species, or other taxa. It shows, for each species, with which other species it shares its most recent common ancestor.

plan of nature Philosophical theory that nature is organized according to a plan. It has been influential in classification, and is a kind of *idealism.*

plankton Refers to the microscopic animals and plants that float in the water near the surface. In the top meter or two of water, both in the sea and in freshwater, small plants can photosynthesize, and abundant microscopic life can be observed. Many organisms that are sessile as adults disperse by means of a planktonic larval stage.

plasmid A genetic element that exists (or can exist) independently of the main DNA in the cell. In bacteria, plasmids can exist as small loops of DNA and be passed between cells independently.

Poisson distribution Frequency distribution for number of events per unit time, when the number of events is determined randomly and the probability of each event is low.

polymorphism Condition in which a population possesses more than one allele at a locus. Sometimes it is defined as the condition of having more than one allele with a frequency of over 5% in the population.

polyphyletic group Set of species descended from more than one common ancestor. The ultimate common ancestor of all species in the group is not a member of the polyphyletic group.

polyploid An individual containing more than two sets of genes and chromosomes.

population A group of organisms, usually a group of sexual organisms that interbreed and share a *gene pool.*

population genetics Study of processes influencing gene frequencies.

postzygotic isolation *Reproductive isolation* in which a *zygote* is successfully formed but then either fails to develop or develops into a sterile adult. Donkeys and horses are postzygotically isolated from one another; a male donkey and a female horse can mate to produce a mule, but the mule is sterile.

prezygotic isolation *Reproductive isolation* in which the two species never reach the stage of successful mating, and thus no *zygote* is formed. Examples would be species with different breeding seasons or courtship displays, and which therefore never recognize one another as potential mates.

prokaryote Cell without a distinct nucleus. Bacteria and some other simple organisms are prokaryotic. Compare with *eukaryote.* In classificatory terms, the group of all prokaryotes is *paraphyletic.*

protein A molecule made up of a sequence of amino acids. Many of the important molecules in a living thing are proteins. All enzymes, for example, are proteins.

pseudogene A sequence of nucleotides in the DNA that resembles a *gene* but is non-functional for some reason.

purine A kind of *base*; in the DNA, adenine (A) and guanine (G) are purines.

pyrimidine A kind of *base*; in DNA, cytosine (C) and thymine (T), and in RNA, cytosine (C) and uracil (U) are pyrimidines.

quantitative character A *character* showing continuous variation in a population.

random drift Synonym of *genetic drift.*

random mating Mating pattern in which the probability of mating with another individual of a particular genotype (or phenotype) equals the frequency of that genotype (or phenotype) in the population.

recapitulation Partly or wholly erroneous theory that an individual, during its development, passes through a series of stages corresponding to its successive evolutionary ancestors. An individual thus develops by "climbing up its family tree."

recessive An allele (*A*) is recessive if the phenotype of the heterozygote *Aa* is the same as the homozygote (*aa*) for the alternative allele *a* and different from the homozygote for the recessive (*AA*). The allele *a* controls the heterozygote's phenotype and is called *dominant.* An allele may be partly, rather than fully, recessive; in that case, the heterozygous phenotype is nearer to, rather than identical with, the homozygote for the dominant allele.

recognition species concept Concept of species, according to which a species is a set of organisms that recognize one another as potential mates; they have a shared mate recognition system. Compare with *biological species concept, cladistic species concept, ecological species concept,* and *phenetic species concept.*

recombination Event, occurring by the *crossing-over* of *chromosomes* during *meiosis,* in which DNA is exchanged between a pair of chromosomesof a pair. Thus, two genes that were previously unlinked, being on different chromosomes, can become *linked* because of recombination, and linked genes may become unlinked.

reinforcement Increase in *reproductive isolation* between incipient species by natural selection. Natural selection can directly favor only an increase in prezygotic isolation; reinforcement therefore amounts to selection for *assortative mating* between the incipiently speciating forms.

reproductive character displacement Increased reproductive isolation between two closely related species where they live in the same geographic region (sympatry) as compared with where they live in separate geographic regions. A kind of *character displacement,* in which the character concerned influences reproductive isolation, not ecological competition.

reproductive isolation Two populations, or individuals of opposite sex, are reproductively isolated from one another if they cannot together produce fertile offspring.

ribosomal RNA (rRNA) Kind of RNA that constitutes the ribosomes and provides the site for translation.

ribosome Site of protein synthesis (or translation) in the cell, and mainly consisting of *ribosomal RNA.*

ring species Situation in which two reproductively isolated populations (see *reproductive isolation*), living in the same region, are connected by a geographic ring of populations that can interbreed.

RNA Ribonucleic acid. *Messenger RNA, ribosomal RNA,* and *transfer RNA* are its three main forms. They act as the intermediaries by which the hereditary code of DNA is converted into proteins. In some viruses, RNA is itself the hereditary molecule.

selection Synonym of *natural selection.*

selectionism Theory that some class of evolutionary events, such as molecular or phenotypic changes, have mainly been caused by natural selection.

separate creation The theory that species have separate origins and never change after their origin. Most versions of the theory of separate creation are religiously inspired and suggest that the origin of species occurs by supernatural action.

sex chromosome *Chromosome* that influences sex determination. In mammals, including humans, the X and Y chromosomes are the sex chromosomes (females are XX, males XY). Compare with *autosome.*

sexual selection Selection on mating behavior, either through competition among members of one sex (usually males) for access to members of the other sex or through choice by members of one sex (usually females) of certain members of the other sex. In sexual selection, individuals are favored by their fitness relative to other members of the same sex, whereas natural selection works on the fitness of a genotype relative to the whole population.

spacer region Sequence of nucleotides in the DNA between coding *genes.*

species An important classificatory category, which can be variously defined by the *biological species concept, cladistic species concept, ecological species concept, phenetic species concept,* and *recognition species concept.* The biological species concept, according to which a species is a set of interbreeding organisms, is the most widely used definition, at least by biologists who study vertebrates. A particular species is referred to by a *Linnaean* binomial, such as *Homo sapiens* for human beings.

stabilizing selection Selection tending to keep the form of a population constant. Individuals with the mean value for a character have high *fitness*; those with extreme values have low fitness.

stepped cline *Cline* with a sudden change in gene (or character) frequency.

substitution The evolutionary replacement of one allele by another in a population.

sympatric speciation Speciation via populations with overlapping geographic ranges.

sympatry Living in the same geographic region. Compare with *allopatry.*

systematics A near synonym of *taxonomy.*

taxon (plural **taxa**) Any named taxonomic group, such as the family Felidae, or the genus *Homo*, or species *Homo sapiens.* A formally recognized group, as distinct from any other group (such as the group of herbivores, or tree-climbers).

taxonomy Theory and practice of biological *classification.*

tetrapod A member of the group made up of amphibians, reptiles, birds, and mammals.

transcription Process by which *messenger RNA* is read off the DNA forming a gene.

transfer RNA (tRNA) Kind of *RNA* that brings the amino acids to the *ribosomes* to make proteins. There are 20 kinds of transfer RNA molecules, one for each of the 20 main amino acids. A transfer RNA molecule has an amino acid attached to it, and contains the anti-*codon* corresponding to that amino acid in another part of its structure. In protein synthesis, each codon in the *messenger RNA* combines with the appropriate tRNA's anti-codon, and the amino acids are arranged in order to make the protein.

transformism Evolutionary theory of Lamarck, in which changes occur within a *lineage* of populations, but in which lineages do not split (i.e., no speciation occurs, at least not in the sense of the *cladistic species concept*) and do not go extinct.

transition *Mutation* changing one *purine* into the other purine, or one *pyrimidine* into the other pyrimidine (i.e., changes from A to G, or vice versa, and changes from C to T, or vice versa).

translation The process by which a protein is manufactured at a *ribosome*, using *messenger RNA* code and *transfer RNA* to supply the *amino acids.*

transversion *Mutation* changing a *purine* into a *pyrimidine*, or vice versa (i.e., changes from A or G to C or T and changes from C or T to A or G).

typology (1) Definition of classificatory groups by *phenetic* similarity to a "type" specimen. A species, for example, might be defined as all individuals less than *x* phenetic units from the species' type. (2) Theory that distinct "types" exist in nature, perhaps because they are part of some *plan of nature.* (See also *idealism.*) The type of the species is then the most important form of it, and variants around that type are "noise" or "mistakes." Neo-Darwinism opposes typology because in a *gene pool* no one variant is any more important than any others.

unequal crossing-over *Crossing-over* in which the two chromosomes do not exchange equal lengths of DNA; one receives more than the other.

variance A measure of how variable a set of numbers are. Technically, it is the sum of squared deviations from the *mean* divided by (n-1) (the number of numbers in the sample minus one). Thus, to find the variance of the set of numbers, 4, 6, and 8, we first calculate the mean, which is 6. We then sum the squared deviations from the mean $(4 - 6)^2 + (6 - 6)^2 + (8 - 6)^2$, which comes to 8, and divide by (n-1) (which is 2 in this case). The variance of the three numbers is $8/2 = 4$. The more variable the set of numbers, the higher the variance. The variance of a set of identical numbers (such as 6, 6, and 6) is zero.

virus A kind of intracellular parasite that can replicate only inside a living cell. In its dispersal stage between host cells, a virus consists of nucleic acid that codes for a small number of genes, surrounded by a

virus (***continued***)
protein coat. (Less formally, according to Medawar's definition, a virus is "a piece of bad news wrapped in a protein.")

vitamin A member of a chemically heterogeneous class of organic compounds that are essential, in small quantities, for life.

wild type The genotype, or phenotype, out of a set of genotypes, or phenotypes, of a species that is found in nature. The expression is mainly used in lab genetics, to distinguish rare mutant forms of a species from the lab stock of normal individuals.

wobble The ability of the third base in some anticodons of tRNA to bond with more than one kind of base in the complementary position in the mRNA codon.

zygote The cell formed by the fertilization of male and female gametes.

REFERENCES

Abbott, R. J. (1992) Plant invasions, interspecific hybridization, and the evolution of new plant taxa. *Trends Ecol. Evol.* **7:** 401–405.

Abrams, P. A. (1986a) Adaptive responses of predators to prey and prey to predators: the failure of the arms race analogy. *Evolution* **40:** 1229–1247.

Abrams, P. A. (1986b) Character displacement and niche shift analyzed using consumer-resource models of competition. *Theoretical Population Biology* **29:** 107–160.

Adams, M. B. (ed.) (1994) *The Evolution of Theodosius Dobzhansky.* Princeton University Press, Princeton, New Jersey.

Ahlberg, P. E. and Milner, A. R. (1994) The origins and early diversification of tetrapods. *Nature* **368:** 507–514.

Alberts, B., Bray, D., Lewis, J., Raff, M., Roberts, K., and Watson, J. D. (1994) *Molecular Biology of the Cell,* 3rd edn. Garland, New York.

Alexander, R. McN. (1985) The ideal and the feasible: physical constraints on evolution. *Biol. J. Linn. Soc.* **26:** 345–58.

Alvarez, L. W., Alvarez, W., Asaro, F., and Michel, H. V. (1980) Extraterrestrial cause for the Cretaceous-Tertiary extinction. *Science* **208:** 1095–1108.

Alvarez, W. and Asaro, F. (1990) An extraterrestrial impact. *Sci. Am.* **263** (October): 78–84.

Alvarez, W., Asaro, F., and Montanari, A. (1990) Iridium profile for 10 million years across the Cretaceous-Tertiary boundary at Gubbio (Italy). *Science* **250:** 1700–1702.

Alvarez, W., Kauffman, E. G., Surlyk, F., Alvarez, L. W., Asaro, F., and Michel, H. V. (1984) Impact theory of mass extinctions and the invertebrate fossil record. *Science* **223:** 1135–1141

Andersson, M. (1994) *Sexual Selection.* Princeton University Press, Princeton, New Jersey.

Andrews, P. (1995) Ecological apes and ancestors. *Nature* **376:** 555–556.

Antonovics, J., Bradshaw, A. D., and Turner, J. R. G. (1971) Heavy metal tolerance in plants. *Adv. Ecol. Research* **7:** 1–85.

Antonovics, J., Ellstrand, N. C., and Brandon, R. N. (1988) Genetic variation and environmental variation: expectations and experiments. In: Gottlieb, L. D. and Jain, S. K. (eds) *Plant Evolutionary Biology,* pp. 275–303. Chapman & Hall, London.

Archibald, J. D. (1993) The importance of phylogenetic analysis for the assessment of species turnover: a case history of Paleocene mammals in North America. *Paleobiology* **19:** 1–27.

Arnheim, N. (1983) Concerted evolution of multigene families. In: Nei, M. and Koehn, R. K. (eds) *Evolution of Genes and Proteins,* pp. 38–61. Sinauer, Sunderland, Massachusetts.

Arnold, M. L. (1994) Natural hybridization and Louisiana irises. *Bioscience* **44:** 141–147.

Arnold, M. L. and Bennett, B. D. (1993) Natural hybridization in Louisiana irises: genetic variation and ecological determinants. In: Harrison, R. G. (ed.) *Hybrid Zones and the Evolutionary Process,* pp. 115–139. Oxford University Press, New York.

Arnold, M. L. and Hodges, S. A. (1995) Are natural hybrids fit or unfit relative to their parents? *Trends Ecol. Evol.* **10:** 67–71.

Atchley, W. R. and Woodruff, D. S. (eds) (1981) *Evolution and Speciation.* Cambridge University Press, New York.

Avise, J. C. (1994) *Molecular Markers, Natural History, and Evolution.* Chapman & Hall, New York.

Bakker, R. T. (1983) The deer flees, the wolf pursues: incongruencies in predator-prey coevolution. In: Futuyma, D. J. and Slatkin, M. (eds) *Coevolution,* pp. 350–382. Sinauer, Sunderland, Massachusetts.

Barigozzi, C. (ed.) (1982) *Mechanisms of Speciation.* A. R. Liss, New York.

Barthélemy-Madaule, M. (1982) *Lamarck the Mythical Precursor.* MIT Press, Cambridge, Massachusetts.

Barton, N. H. (1988) Speciation. In: Myers, A. A. and Giller, P. S. (eds) *Analytical Biogeography,* pp. 185–218. Chapman & Hall, New York.

Barton, N. H. (1989) Founder effect speciation. In: Otte, D. and Endler, J. A. (eds) *Speciation and Its Consequences,* pp. 229–256. Sinauer, Sunderland, Massachusetts.

Barton, N. H. and Hewitt, G. M. (1985) Analysis of hybrid zones. *Ann. Rev. Ecol. System.* **16:** 113–148.

Barton, N. H. and Turelli, M. (1989) Evolutionary quantitative genetics: how much do we know? *Ann. Rev. Genetics* **23:** 337–370.

Beatty, J. (1994) Theoretical pluralism in biology, including systematics. In: Grande, L. and Rieppel, O. (eds) *Interpreting The Hierarchy of Nature: From Systematic Patterns to Evolutionary Process Theories,* pp. 33–60. Academic Press, New York.

Beer, G. R. de (1931) *Embryology and Evolution.* Oxford University Press, Oxford, UK.

Beer, G. R. de (1940) *Embryos and Ancestors.* Oxford University Press, Oxford, UK. (2nd edn, 1958.)

Begon, M., Harper, J. L., and Townsend, C. R. (1990) *Ecology: Individuals, Populations, and Communities.* 2nd edn. Blackwell Science, Boston, Massachusetts.

Bell, G. (1982) *The Masterpiece of Nature.* University of California Press, Berkeley, and Croom Helm, London.

Bell, M. A. (1988) Stickleback fishes: bridging the gap between population biology and paleobiology. *Trends Ecol. Evol.* **3:** 320–325.

Bentley, D. and Hoy, R. R. (1974) The neurobiology of cricket song. *Sci. Am.* **231** (August): 34–44.

Benton, M. J. (1983) Dinosaur success in the Triassic: a noncompetitive ecological model. *Quart. Rev. Biol.* **58:** 29–55.

Benton, M. J. (1987) Progress and competition in macroevolution. *Biol. Rev.* **62:** 305–338

Benton, M. J. (1990) *Vertebrate Paleontology: Biology and Evolution.* Unwin Hyman, Boston.

Benton, M. J. (ed.) (1993) *The Fossil Record 2.* Chapman & Hall, New York.

Benton, M. J. (1994) Palaeontological data and identifying mass extinctions. *Trends Ecol. Evol.* **9:** 181–185.

Benton, M. J. (1995) Diversification and extinction in the history of life. *Science* **268:** 52–58.

Berlin, B. (1992) *Ethnobiological Classification: Principles of Categorization of Plants and Animals in Traditional Societies.* Princeton University Press, Princeton, New Jersey.

Bertram, B. (1978) *Pride of Lions.* Scribner, New York.

Bishop, M. J. and Friday, A. E. (1988) Estimating the interrelationships of tetrapod groups on the basis of molecular sequence data. In: Benton, M. J. (ed.) *The Phylogeny and Classification of the Tetrapods,* vol. 1, p. 33–58. Oxford University Press, Oxford, UK.

Blair, W. F. (1964) Isolating mechanisms and interspecies interactions in anuran amphibians. *Quart. Rev. Biol.* **39:** 334–344.

Blum, J. D. (1993) Zircon takes the heat. *Nature* **366:** 718.

Bodmer, W. F. (1983) Gene clusters and genome evolution. In: Bendall, D. S. (ed.) *Evolution from Molecules to Men,* pp. 197–208. Cambridge University Press, Cambridge, UK.

Bodmer, W. F. and Cavalli-Sforza, L. L. (1976) *Genetics, Evolution, and Man.* W. H. Freeman, San Francisco.

Bonner, J. T. (1988) *The Evolution of Complexity.* Princeton University Press, Princeton, New Jersey.

Bookstein, F. L. (1977) The study of shape transformation after D'Arcy Thompson. *Math. Biosciences* **34:** 177–219.

Bowcock, A. M., Ruiz-Linares, A., Tomfohrde, J., Minch, E., Kidd, J. R., and Cavalli-Sforza, L. L. (1994) High resolution of human evolutionary trees with polymorphic microsatellites. *Nature* **368:** 455–457.

Bowler, P. J. (1989) *Evolution: The History of an Idea.* Revised edn. University of California Press, Berkeley, California.

Box, J. F. (1978) *R. A. Fisher. The Life of a Scientist.* John Wiley, New York.

Bradshaw, A. D. (1971) Plant evolution in extreme environments. In: Creed, E. R. (ed.) *Ecological Genetics and Evolution,* pp. 20–50. Appleton-Century-Croft, New York.

Brakefield, P. M. (1987) Industrial melanism: do we have the answers? *Trends Ecol. Evol.* **2:** 117–122.

Bramble, D. M. and Jenkins, F. A. (1989) Structural and functional integration across the reptile-mammal boundary: the locomotor system. In: Wake, D. B. and Roth, G. (eds) *Complex Organismal Functions,* pp. 133–146. Wiley, Chichester, UK.

Briggs, S. P. and Johal, G. S. (1994) Genetic patterns of plant host-parasite interactions. *Trends in Genetics* **10:** 12–16.

Britten, R. J. (1994) Evolutionary selection against change in many Alu repeat sequences interspersed through primate genomes. *Proc. Nat. Acad. Sci. USA* **91:** 5992–5996.

Bronstein, J. L. (1994) Our current understanding of mutualism. *Quart. Rev. Biol.* **69:** 31–51.

Brookfield, J. F. Y. and Sharp, P. M. (1994) Neutralism and selectionism face up to DNA data. *Trends in Genetics* **10:** 109–111.

Brown, D. D., Wensink, P. C., and Jordan, E. (1972) A comparison of the ribosomal DNAs of *Xenopus laevis* and *Xenopus mulleri*: the evolution of tandem genes. *J. Mol. Biol.* **63:** 57–73.

Brown, J. H. (1971) The desert pupfish. *Sci. Am.* **225** (November): 104–110.

Brown, J. H. and Gibson, A. C. (1983) *Biogeography.* Mosby, St. Louis, Missouri.

Brown, W. L. (1987) Punctuated equilibrium excused: the original examples fail to support it. *Biol. J. Linn. Soc.* **31:** 383–404

Brown, W. L. and Wilson, E. O. (1958) Character displacement. *System. Zoo.* **5:** 49–64.

Browne, J. (1995) *Charles Darwin Voyaging.* Vol. 1 of a biography. Simon & Schuster, New York, and Jonathan Cape, London.

Brundin, L. (1988) Phylogenetic biogeography. In: Myers, A. A. and Giller, P. S. (eds) *Analytical Biogeography*, pp. 343–369. Chapman & Hall, New York.

Bull, J. J. (1994) Virulence. *Evolution* **48:** 1423–1437.

Bull, J. J. and Charnov, E. L. (1988) How fundamental are Fisherian sex ratios? *Oxford Surv. Evol. Biol.* **5:** 96–135.

Bulmer, M. (1988a) Evolutionary aspects of protein synthesis. *Oxford Surv. Evol. Biol.* **5:** 1–40.

Bulmer, M. (1988b) Estimating the variability of substitution rates. *Genetics* **123:** 615–619.

Bulmer, M. G. (1989) Maintenance of genetic variability by mutation-selection balance: a child's guide through the jungle. *Genome* **31:** 761–767.

Bulmer, M. (1994) *Theoretical Evolutionary Ecology.* Sinauer, Sunderland, Massachusetts.

Bulmer, M., Wolfe, K. H., and Sharp, P. M. (1991). Synonymous nucleotide substitution rates in mammalian genes: implications for the molecular clock and the relationships of mammalian orders. *Proc. Nat. Acad. Sci. USA* **88:** 5974–5978.

Burkhardt, F. and Smith, S. (eds) (1985–) *The Correspondence of Charles Darwin.* vol. 1 –. Cambridge University Press, Cambridge, UK.

Burkhardt, R. W. (1977) *The Spirit of System: Lamarck and Evolutionary Biology.* Harvard University Press, Cambridge, Massachusetts.

Bush, G. L. (1975) Modes of animal speciation. *Ann. Rev. Ecol. System.* **6:** 339–364.

Bush, G. L. (1994) Sympatric speciation in animals: new wine in old bottles. *Trends Ecol. Evol.* **9:** 285–288.

Bush, G. L., Case, S. M., Wilson, A. C., and Patton, J. L. (1977) Rapid speciation and chromosomal evolution in mammals. *Proc. Nat. Acad. Sci. USA* **74:** 3942–3946.

Bush, M. B. and Whittaker, R. J. (1991) Krakatau: colonization patterns and hierarchies. *J. Biogeography* **18:** 341–356.

Buss, L. W. (1987) *The Evolution of Individuality.* Princeton University Press, Princeton, New Jersey.

Butlin, R. K. (1987) Speciation by reinforcement. *Trends Ecol. Evol.* **2:** 8–13.

Butlin, R. K. (1989) Reinforcement of premating isolation. In: Otte, D. and Endler, J. A. (eds) *Speciation and Its Consequences*, pp. 158–179. Sinauer, Sunderland, Massachusetts.

Cain, A. J. (1954) *Animal Species and Their Evolution.* Hutchinson, London. (Reprinted 1993 by Princeton University Press, Princeton, New Jersey.)

Cain, A. J. (1964) The perfection of animals. In: Carthy, J. D. and Duddington, C. L. (eds) *Viewpoints in Biology*, vol. 3, p. 36–63. Butterworths, London. (Reprinted in *Biol. J. Linn. Soc.* **36** (1989): 3–29.)

Cain, A. J. and Sheppard, P. M. (1954) Natural selection in *Cepaea. Genetics* **39:** 89–116

Cairns, J., Overbaugh, J., and Miller, S. (1988). The origin of mutants. *Nature* **335:** 142–145

Calhoun, J. B. (1947) The role of temperature and natural selection in the variations of the English sparrow in the United States. *Am. Naturalist* **81:** 203–228

Calvo, R. N. (1990) Inflorescence size and fruit distribution among individuals in three orchid species. *Am. J. Botany* **77:** 1378–1381.

Camin, J. H. and Ehrlich, P. R. (1958) Natural selection in water snakes (*Natrix sipedon* L.) on islands in Lake Erie. *Evolution* **12:** 504–511

Carlquist, S. (1970) *Hawaii, a Natural History.* Natural History Press, New York.

Carr, G. D., Robichaux, R. H., Witter, M. S., and

Kyhos, D. W. (1989) Adaptive radiation of the Hawaiian silversword alliance (Compositae—Madiinae): a comparison with Hawaiian picture-winged *Drosophila.* In: Giddings, L. V., Kaneshiro, K. Y., and Anderson, W. W. (eds) *Genetics, Speciation, and the Founder Principle*, pp. 79–95. Oxford University Press, New York.

Carroll, R. L. (1988) *Vertebrate Paleontology and Evolution.* W. H. Freeman, New York.

Carroll, S. B. (1995) Homeotic genes and the evolution of arthropods and chordates. *Nature* **376:** 479–485.

Carson, H. L. (1983) Chromosomal sequences and inter-island colonizations in Hawaiian *Drosophila. Genetics* **103:** 465–482.

Carson, H. L. (1990) Evolutionary process as studied in population genetics: clues from phylogeny. *Oxford Surv. Evol. Biol.* **7:** 129–156.

Carson, H. L. and Kaneshiro, K. Y. (1976) *Drosophila* of Hawaii: systematics and ecological genetics. *Ann. Rev. Ecol. System.* **7:** 311–345.

Cavalli-Sforza, L. L. and Bodmer, W. F. (1971) *Genetics of Human Populations.* W. H. Freeman, San Francisco.

Chao, L. and Carr, D. E. (1993) The molecular clock and the relationship between population size and generation time. *Evolution* **47:** 688–690

Charlesworth, B. (1990) Mutation-selection balance and the evolutionary advantage of sex and recombination. *Genetical Research* **55:** 199–221.

Charlesworth, B. (1994) *Evolution in Age-Structured Populations.* 2nd edn. Cambridge University Press, Cambridge, UK.

Charlesworth, B., Lande, R., and Slatkin, M. (1982) A neo-Darwinian commentary on macroevolution. *Evolution* **36:** 474–498.

Charlesworth, B., Sniegowski, P., and Stephan, W. (1994) The evolutionary dynamics of repetitive DNA in eukaryotes. *Nature* **371:** 215–220.

Cheetham, A. (1986) Tempo of evolution in a Neogene bryozoan: rates of morphologic change within and across species boundaries. *Paleobiology* **12:** 199–202.

Cheetham, A. H., Jackson, J. B. C., and Hayek, L. C. (1994) Quantitative analysis of bryozoan phenotypic evolution. II. Analysis of selection and random change in fossil species using reconstructed genetic parameters. *Evolution* **48:** 360–375.

Chesser, R. T. and Zink, R. M. (1994) Modes of speciation in birds: a test of Lynch's method. *Evolution* **48:** 490–497.

Christie, D. M., Duncan, R. A., McBirney, A. R., Richards, M. A., White, W. M., Harpp, K. S., and Fox, C. G. (1992) Drowned islands downstream from the Galapagos hotspot imply extended speciation times. *Nature* **355:** 246–248. (And comment: *Nature* **355** [1992]: 202–203.)

Cifelli, R. L. (1981) Patterns of evolution among the Artiodactyla and Perissodactyla (Mammalia). *Evolution* **35:** 433–440.

Clark, A. G. (1994). Invasion and maintenance of a gene duplication. *Proc. Nat. Acad. Sci. USA* **91:** 2950–2954.

Clark, R. W. (1969) *JBS: The Life and Work of J. B. S. Haldane.* Coward-McCann, New York.

Clarke, B. C. (1979) The evolution of genetic diversity. *Proc. Royal Soc. London B* **205:** 453–474.

Clarke, C. A. and Sheppard, P. M. (1969) Further studies on the genetics of the mimetic butterfly *Papilio memnon. Phil. Trans. Royal Soc. London B* **263:** 35–70.

Clarke, C. A., Sheppard, P. M., and Thornton, I. W. B. (1968) The genetics of the mimetic butterfly Papilio memnon. *Phil. Trans. Royal Soc. London B* **254:** 37–89.

Clarkson, E. N. K. (1992) *Invertebrate Palaeontology and Evolution.* 3rd edn. Chapman & Hall, New York.

Clayton, G. A. and Robertson, A. (1955) Mutation and quantitative variation. *Am. Naturalist* **89:** 151–158.

Clutton-Brock, T. H. (ed.) (1988) *Reproductive Success.* University of Chicago Press, Chicago.

Clutton-Brock, T. H., Albon, S. D., and Guinness, F. E. (1984) Maternal dominance, breeding success, and birth sex ratios in red deer. *Nature* **308:** 358–360.

Colinvaux, P. A. (1993) Pleistocene biogeography and diversity in tropical forests of South America. In: Goldblatt, P. (ed.), *Biological Relationships Between Africa and South America*, pp. 473–499. Yale University Press, New Haven, Connecticut.

Conway Morris, S. (1993) The fossil record and the early evolution of Metazoa. *Nature* **361:** 219–225.

Cook, A. (1975) Changes in the carrion/hooded crow hybrid zone and the possible importance of climate. *Bird Study* **22:** 165–168.

Coope, G. R. (1979) Late Cenozoic fossil Coleoptera:

evolution, biogeography, and ecology. *Ann. Rev. Ecol. System.* **10:** 247–267.

Cooper, A., Atkinson, I. A. E., Lee, W. G., and Worthy, T. H. (1993). Evolution of the moa and their effect on the New Zealand flora. *Trends Ecol. Evol.* **8:** 433–437.

Corsi, P. (1988) *The Age of Lamarck.* University of California Press, Berkeley, California.

Cosmides, L. M. and Tooby, J. (1981). Cytoplasmic inheritance and intragenomic conflict. *J. Theor. Biol.* **89:** 83–129.

Courtillot, V. E. (1990) A volcanic eruption. *Sci. Am.* **263** (October): 85–92.

Cox, C. B. and Moore, P. D. (1993) *Biogeography.* 5th edn. Blackwell Science, Boston, Massachusetts.

Coyne, J. A. (1992) Genetics and speciation. *Nature* **355:** 511–515.

Coyne, J. A. (1994a) Ernst Mayr and the origin of species. *Evolution* **48:** 19–30.

Coyne, J. A. (1994b) Rules for Haldane's rule. *Nature* **369:** 189–190.

Coyne, J. A. and Orr, H. A. (1989a) Patterns of speciation in *Drosophila. Evolution* **43:** 362–381.

Coyne, J. A. and Orr, H. A. (1989b) Two rules of speciation. In: Otte, D. and Endler, J. A. (eds) *Speciation and Its Consequences*, pp. 180–207. Sinauer, Sunderland, Massachusetts.

Coyne, J. A., Orr, H. A., and Futuyma, D. J. (1989) Do we need a new species concept? *System. Zoo.* **37:** 190–200.

Crane, P. R., Friis, E. M., and Pedersen, K. R. (1995). The origin and early diversification of angiosperms. *Nature* **374:** 27–33.

Creed, E. R., Lees, D. R., and Bulmer, M. G. (1980) Pre-adult viability differences of melanic *Biston betularia* (L.) (Lepidoptera). *Biol. J. Linn. Soc.* **13:** 25–62.

Croizat, L., Nelson, G., and Rosen, D. E. (1974) Centers of origin and related concepts. *System. Zoo.* **23:** 265–287.

Cronin, H. (1991) *The Ant and the Peacock.* Cambridge University Press, Cambridge, UK.

Cronin, T. M. and Schneider, C. E. (1990) Climatic influences on species: evidence from the fossil record. *Trends Ecol. Evol.* **5:** 275–279.

Crow, J. F. (1979) Genes that violate Mendel's rules. *Sci. Am.* **240** (February): 134–146.

Crow, J. F. (1986) *Basic Concepts in Population, Quantitative, and Evolutionary Genetics.* W. H. Freeman, New York.

Crow, J. F. (1993) Mutation, mean fitness, and genetic load. *Oxford Surv. Evol. Biol.* **9:** 3–42.

Crow, J. F. and Kimura, M. (1970) *An Introduction to Population Genetics Theory.* Harper & Row, New York.

Crowley, T. J. and North, G. R. (1988) Abrupt climate change and extinction events in Earth history. *Science* **240:** 996–1002

Crowson, R. A. (1970) *Classification and Biology.* Atherton Press, New York.

Curio, E. (1973) Towards a methodology for teleonomy. *Experientia* **29:** 1045–1058.

Curtis, C. F., Cook, L. M., and Wood, R. J. (1978) Selection for and against insecticide resistance and possible methods of inhibiting the evolution of resistance in mosquitoes. *Ecol. Ent.* **3:** 273–287.

Cushing, D. H. (1975) *Marine Ecology and Fisheries.* Cambridge University Press, Cambridge, UK.

Cvancara, A. M. (1990) *Sleuthing Fossils.* John Wiley, New York.

Darwin, C. R. (1859) *On the Origin of Species.* John Murray, London.

Darwin, C. R. (1871) *The Descent of Man, and Selection in Relation to Sex.* John Murray, London. (Paperback edition, Princeton University Press, Princeton, New Jersey.)

Darwin, C. (1872) *The Expression of the Emotions in Man and Animals.* John Murray, London. (Paperback edition, University of Chicago Press, Chicago.)

Darwin, F. (ed.) (1887) *The Life and Letters of Charles Darwin.* 3 vols. John Murray, London.

Darwin, F. and A. C. Seward (eds) (1903) *More Letters of Charles Darwin.* 2 vols. John Murray, London.

Dawkins, R. (1982) *The Extended Phenotype.* W. H. Freeman, Oxford, UK. (Paperback edition, Oxford University Press, Oxford, UK.)

Dawkins, R. (1986) *The Blind Watchmaker.* W. W. Norton, New York and Longman, London.

Dawkins, R. (1989) *The Selfish Gene.* 2nd edn. Oxford University Press, Oxford, UK.

Dawkins, R. (1996) *Climbing Mount Improbable.* W. W. Norton, New York, and Viking Penguin, London.

Dawkins, R. and Krebs, J. R. (1979) Arms races between and within species. *Proc. Royal Soc. London B* **205:** 489–511.

Dean, G. (1972) *The Porphyrias.* Pitman, London.

de Queiroz, K. and Gauthier, J. (1994) Toward a phylogenetic system of biological nomenclature. *Trends Ecol. Evol.* **9:** 27–31.

Devonshire, A. L. and Field, L. M. (1991) Gene amplification and insecticide resistance. *Ann. Rev. Ent.* **36:** 1–23.

Diamond, J. (1990a) Alone in a crowded universe. *Nat. Hist.* June 1990: 30–34.

Diamond, J. (1990b) Biological effects of ghosts. *Nature* **345:** 769–770.

Diamond, J. (1991) *The Third Chimpanzee.* Harper Collins, New York, and Hutchinson Radius, London.

Dobzhansky, T. (1951) *Genetics and the Origin of Species.* 3rd edn. Columbia University Press, New York.

Dobzhansky, T. (1970) *Genetics of the Evolutionary Process.* Columbia University Press, New York.

Dobzhansky, T. (1973) Nothing in biology makes sense except in the light of evolution. *Am. Biol. Teacher* **35:** 125–129.

Dobzhansky, T. and Pavlovsky, O. (1957) An experimental study of interaction between genetic drift and natural selection. *Evolution* **11:** 311–319.

Dodd, D. M. B. (1989) Reproductive isolation as a consequence of adaptive divergence in *Drosophila pseudoobscura. Evolution* **43:** 1308–1311.

Donoghue, M. J., Doyle, J. A., Gauthier, J., Kluge, A. G., and Rowe, T. (1989) The importance of fossils in phylogeny reconstruction. *Ann. Rev. Ecol. System.* **20:** 431–460.

Donovan, S. K. (ed.) (1989) *Mass Extinctions: Processes and Evidence.* Columbia University Press, New York.

Donovan, S. K. (ed.) (1991) *The Processes of Fossilization.* Columbia University Press, New York.

Donovan, S. K. (ed.) (1994) *The Paleobiology of Trace Fossils.* Wiley, New York.

Doolittle, R. F. (1994) Convergent evolution: the need to be explicit. *Trends Biochem. Sci.* **19:** 15–18.

Doolittle, W. F. and Brown, J. R. (1994) Tempo, mode, the progenote, and the universal root. *Proc. Nat. Acad. Sci. USA* **91:** 6721–6728.

Doolittle, W. F. and Sapienza, C. (1980) Selfish genes, the phenotype paradigm, and genome evolution. *Nature* **284:** 601–603.

Dudley, J. W. (1977) 76 generations of selection for oil and protein percentage in maize. In: Pollack, E., Kempthorne, O., and Bailey, T. B. (eds), *Proceedings of the International Conference on Quanititative Genetics,* pp. 459–473. Iowa State University Press, Ames, Iowa.

Dybas, H. S. and Lloyd, M. (1962) Isolation by habitat in two synchronized species of periodical cicadas (Homoptera: Cicadidae: *Magicicada*). *Ecology* **43:** 444–459.

Dykhuizen, D. E. and Dean, A. M. (1990) Enzyme activity and fitness: evolution in solution. *Trends Ecol. Evol.* **5:** 257–262.

Dykhuizen, D. E. and Green, L. (1991) Recombination in *Escheriichia coli* and the definition of biological species. *J. Bacteriology* **173:** 7257–7268.

Eberhard, W. G. (1980) Evolutionary consequences of intracellular competition. *Quart. Rev. Biol.* **55:** 231–249.

Eberhard, W. G. (1990) Evolution in bacterial plasmids and levels of selection. *Quart. Rev. Biol.* **65:** 3–22.

Edwards, A. W. F. (1977) *Foundations of Mathematical Genetics.* Cambridge University Press, Cambridge, UK.

Eernisse, D. J. and Kluge, A. G. (1993) Taxonomic convergence versus total evidence, and amniote phylogeny inferred from fossils, molecules, and morphology. *Mol. Biol. Evol.* **10:** 1170–1195.

Ehrlich, P. R. and Raven, P. H. (1964) Butterflies and plants: a study in coevolution. *Evolution* **18:** 586–608.

Ehrlich, P. R. and Raven, P. H. (1969) Differentiation of populations. *Science* **165:** 1228–1232.

Elder, J. F. and Turner, B. J. (1995). Concerted evolution of repetitive DNA sequences in Eukaryotes. *Quart. Rev. Biology* **70:** 297–320.

Eldredge, N. (1982) *The Monkey Business: A Scientist Looks at Creationism.* Washington Square Press, New York.

Eldredge, N. (1995) *Reinventing Darwin.* John Wiley, New York, and Weidenfeld & Nicolson, London.

Eldredge, N. and Gould, S. J. (1972) Punctuated equilibria: an alternative to phyletic gradualism. In: T. J. M. Schopf (ed.) *Models in Paleobiology,* pp. 82–115. Freeman, Cooper, & Co., San Francisco.

Eldredge, N. and Gould, S. J. (1988) Punctuated equilibrium prevails. *Nature* **332:** 211–212.

Eldredge, N. and Stanley, S. M. (eds) (1984) *Living Fossils.* Springer-Verlag, New York.

Elias, S. A. (1994) *Quaternary Insects and Their Environments.* Smithsonian Institution Press, Washington, D.C..

Endler, J. A. (1977) *Geographic Variation, Speciation, and Clines.* Princeton University Press, Princeton, New Jersey.

Endler, J. A. (1986) *Natural Selection in the Wild.* Princeton University Press, Princeton, New Jersey.

Ereshefsky, M. (ed.) (1992) *The Units of Evolution: Essays on the Nature of Species.* MIT Press, Cambridge, Massachusetts.

Erwin, D. H. (1993) *The Great Paleozoic Crisis.* Columbia University Press, New York.

Erwin, D. H. (1994) The Permo-Triassic extinction. *Nature* **367:** 231–236.

Erwin, D. H. and Anstey, R. L. (eds.) (1995) *New Approaches to Speciation in the Fossil Record.* Columbia University Press, New York.

Ewald, P. W. (1993) The evolution of virulence. *Sci. Am.* **268** (April): 56–62.

Ewald, P. W. (1994) *Evolution of Infectious Disease.* Oxford University Press, New York.

Ewens, W. J. (1979) *Mathematical Population Genetics.* Springer-Verlag, Berlin.

Falconer, D. S. (1989) *Introduction to Quantitative Genetics.* 3rd edn. Longman, London.

Farrell, B. D. and Mitter, C. (1994) Adaptive radiation in insects and plants: time and opportunity. *Am. Zoo.* **34:** 57–69.

Federoff, N. and Botstein, D. (eds) (1993) *The Dynamic Genome: Barbara Mcclintock's Ideas in the Century of Genetics.* Cold Spring Harbor Laboratory Press, New York.

Feduccia, A. (1995) Explosive evolution in Tertiary birds and mammals. *Science* **267:** 637–638.

Felsenstein, J. (1993) *PHYLIP: Phylogeny inference package. Version 3.5.* (Computer software package distributed by the author, Dept. Genetics, Univ. Washington, Seattle, WA 98195.)

Fenner, F. and Myers, K. (1978) Myxoma virus and myxomatosis in retrospect: the first quarter century of a new disease. In: Kurstak, E. and Maramorosch, K. (eds) *Viruses and Environment*, pp. 539–570. Academic Press, New York.

Fenner, F. and Ratcliffe, R. N. (1965) *Myxomatosis.* Cambridge University Press, London, UK.

Fenster, E. J. and Sorhannus, U. (1991) On the measurement of morphological rates of evolution: a review. *Evol. Biol.* **25:** 375–410.

Fernholm, B., Bremer, K., and Jörnvall, H. (eds) (1989) *The Hierarchy of Life.* Excerpta Medica, Amsterdam.

Fisher, R. A. (1918) The correlation between relatives under the supposition of Mendelian inheritance. *Trans. Royal Soc. Edinburgh* **52:** 399–433.

Fisher, R. A. (1930) *The Genetical Theory of Natural Selection.* Oxford University Press, Oxford, UK. (2nd edn, 1958, published by Dover Books, New York.)

Fisher, R. A. and Ford, E. B. (1947) The spread of a gene in natural conditions in a colony of the moth *Panaxia dominula* L. *Heredity* **1:** 143–174.

Fitch, W. M. (1976) Molecular evolutionary clocks. In Ayala, F. J. (ed.) *Molecular Evolution*, pp. 160–178. Sinauer, Sunderland, Massachusetts.

Flessa, K. W., Barnett, S. G., Cornue, D. B., Lomaga, M. A., Lombardi, N., Miyazaki, J. M., and Murer, A. S. (1979) Geologic implications of the relationship between mammalian faunal similarity and geographic distance. *Geology* **7:** 15–18.

Ford, E. B. (1975) *Ecological Genetics.* 3rd edn. Chapman & Hall, London.

Forey, P. and Janvier, P. (1994). Evolution of the early vertebrates. *Am. Scientist* **82:** 554–565.

Fortey, R. (1991) *Fossils: The Key to the Past.* 2nd edn. Van Nostrand Reinhold, New York; and Natural History Museum, London.

Fortey, R. (1993) *The Hidden Landscape.* Jonathon Cape, London.

Frank, S. A. (1991) Divergence of meiotic drive suppression systems as an explanation for sex-biased hybrid sterility and inviability. *Evolution* **45:** 262–267.

Frank, S. A. (1996) Models of parasite virulence. *Quarterly Review of Biology* 71:37–78.

Fryer, G., Greenwood, P. H., and Peake, J. F. (1985) The demonstration of speciation in fossil molluscs and living fishes. *Biol. J. Linn. Soc. London* **26:** 325–336.

Futuyma, D. J. (1983) *Science on Trial: The Case for Evolution.* Pantheon Books, New York.

Futuyma, D. J. and Slatkin, M. (eds) (1983) *Coevolution.* Sinauer, Sunderland, Massachusetts.

Gale, J. S. (1990) *Theoretical Population Genetics.* Unwin Hyman, Boston.

Gardiner, B. G. (1982) Tetrapod classification. *Zoo. J. Linn. Soc. London* **74:** 207–232.

Gardiner, B. G. (1993) Haematothermia: warm-blooded amniotes. *Cladistics* **9:** 369–395.

Gauthier, J., Kluge, A. G., and Rowe, T. (1988) Amniote phylogeny and the importance of fossils. *Cladistics* **4:** 105–209.

Ghiselin, M. T. (1969) *The Triumph of the Darwinian Method.* University of California Press, Berkeley, California.

Ghiselin, M. T. (1984) "Definition," "character," and other equivocal terms. *System. Zoo.* **33:** 104–110.

Gibbs, H. L. and Grant, P. R. (1987) Oscillating selection on Darwin's finches. *Nature* **327:** 511–513.

Giddings, L. V., Kaneshiro, K. Y., and Anderson, W. W. (eds) (1989) *Genetics, Speciation, and the Founder Principle.* Oxford University Press, New York.

Gilinsky, N. L. (1988) Survivorship in the Bivalvia: comparing living and extinct genera and families. *Paleobiology* **14:** 370–386.

Gilinsky, N. L. and Bambach, R. K. (1987) Asymmetrical patterns of origination and extinction in higher taxa. *Paleobiology* **13:** 427–445

Gill, D. E. (1989) Fruiting failure, pollinator inefficiency, and speciation in orchids. In: Otte, D. and Endler, J. (eds) *Speciation and Its Consequences*, pp. 458–481. Sinauer, Sunderland, Massachusetts.

Gillespie, J. H. (1991) *The Causes of Molecular Evolution.* Oxford University Press, New York.

Gillespie, J. H. (1993) Episodic evolution of RNA viruses. *Proc. Nat. Acad. Sci. USA* **90:** 10411–10422.

Gillespie, R. G., Croom, H. B., and Palumbi, S. R. (1994) Multiple origins of a spider radiation in Hawaii. *Proc. Nat. Acad. Sci. USA* **91:** 2290–2294.

Gingerich, P. D. (1983) Rates of evolution: effects of time and temporal scaling. *Science* **222:** 159–161.

Gingerich, P. D. (1993) Quantification and comparison of evolutionary rates. *Am. J. Sci.* **293–A:** 453–478.

Gingerich, P. D., Smith, B. H., and Simons, E. L. (1990) Hind limbs of Eocene *Basilosaurus*: evidence of feet in whales. *Science* **249:** 154–157.

Givnish, T. J., Sytsma, K. J., Smith, J. F., and Hahn, W. J. (1994). Thorn-like prickles and heterophylly in *Cyanea*: adaptation to extinct avian browsers in Hawaii? *Proc. Nat. Acad. Sci. USA* **91:** 2810–2814.

Glaessner, M. F. (1984) *The Dawn of Animal Life.* Cambridge University Press, Cambridge, UK.

Godfrey, L. R. and Sutherland, M. R. (1995) Flawed inference: why size-based tests of heterochronic processes do not work. *J. Theor. Biol.* **172:** 43–61.

Goldblatt, P. (ed.) (1993) *Biological Relationships Between Africa and South America.* Yale University Press, New Haven, Connecticut.

Golding, G. B. (1987) Nonrandom patterns of mutation are reflected in evolutionary divergence and may cause some of the unusual patterns observed in sequences. In: Loeschcke, V. (ed.) *Genetic Constraints on Adaptive Evolution,* pp. 151–172. Springer-Verlag, Berlin

Golding, G. B. (ed.) (1994) *Non-neutral Evolution: Theories and Molecular Data.* Chapman & Hall, New York.

Goldschmidt, R. B. (1940) *The Material Basis of Evolution.* Yale University Press, New Haven, Connecticut.

Gomendio, M., Clutton-Brock, T. H., Albon, S. D., Guinness, F. E., and Simpson, M. J. (1990). Mammalian sex ratios and variation in costs of rearing sons and daughters. *Nature* **343:** 261–263.

Goodman, M. (1963) Man's place in the phylogeny of the primates as reflected in serum proteins. In: Washburn, S. L. (ed.) *Classification and Human Evolution*, pp. 204–234. Aldine, Chicago.

Gould, S. J. (1971) D'Arcy Thompson and the science of form. *New Literary History* **2:** 229–258.

Gould, S. J. (1977a) *Ontogeny and Phylogeny.* Harvard University Press, Cambridge, Massachusetts.

Gould, S. J. (1977b) *Ever since Darwin.* W.W. Norton, New York

Gould, S. J. (1980) *The Panda's Thumb.* W. W. Norton, New York.

Gould, S. J. (1982) Darwinism and the expansion of evolutionary theory. *Science* **216:** 380–387.

Gould, S. J. (1983a) *Hen's Teeth and Horse's Toes.* W.W. Norton, New York.

Gould, S. J. (1983b) Irrelevance, submission, and partnership: the changing role of palaeontology in Darwin's three centennials, and a modest proposal for macroevolution. In: Bendall, D. S. (ed.) *Evolution from Molecules to Men,* pp. 347–366. Cambridge University Press, Cambridge, UK.

Gould, S. J. (1985) *The Flamingo's Smile*. W. W. Norton, New York.

Gould, S. J. (1989) *Wonderful Life*. W. W. Norton, New York.

Gould, S. J. (1991) *Bully for Brontosaurus*. W. W. Norton, New York.

Gould, S. J. (1993) *Eight Little Piggies*. W. W. Norton, New York.

Gould, S.J. (1996) *Dinosaur in a Haystack*. W.W. Norton, New York.

Gould, S. J. and Calloway, C. B. (1980) Clams and brachiopods—ships that pass in the night. *Paleobiology* **6:** 383–396.

Gould, S. J. and Eldredge, N. (1977) Punctuated equilibria: the tempo and mode of evolution reconsidered. *Paleobiology* **3:** 115–151.

Gould, S. J. and Eldredge, N. (1993) Punctuated equilibrium comes of age. *Nature* **366:** 223–227.

Gould, S. J. and Johnston, R. F. (1972) Geographic variation. *Ann. Rev. Ecol. System.* **3:** 457–498

Gould, S. J. and Lewontin, R. C. (1979) The spandrels of San Marco and the panglossian paradigm: a critique of the adaptationist program. *Proc. Royal Soc. London B* **205:** 581–598.

Gould, S. J. and Vrba, E. S. (1982) Exaptation—a missing term in the science of form. *Paleobiology* **8:** 4–15.

Grafen, A. (1984) Natural selection, kin selection, and group selection. In: Krebs, J. R. and Davies, N. B. (eds) *Behavioural Ecology: An Evolutionary Approach*, 2nd edn., pp. 62–84. Blackwell Scientific, Boston, Massachusetts.

Grafen, A. (1991) Modelling in behavioural ecology. In: Krebs, J. R. and Davies, N. B. (eds) *Behavioural Ecology: An Evolutionary Approach*, 3rd edn., pp. 1–31. Blackwell Scientific, Boston, Massachusetts.

Graham, L. (1993) *Origin of Land Plants*. John Wiley, New York.

Graham, R. W. and Grimm, E. C. (1990) Effects of global climatic change on the patterns of terrestrial biological communities. *Trends Ecol. Evol.* **5:** 289–292.

Grant, P. R. (1975) The classic case of character displacement. *Evol. Biol.* **8:** 237–337.

Grant, P. R. (1986) *Ecology and Evolution of Darwin's Finches*. Princeton University Press, Princeton, New Jersey.

Grant, P. R. (1991) Natural selection and Darwin's finches. *Sci. Am.* **265** (October): 82–87.

Grant, P. R. and Grant, B. R. (1995) Predicting microevolutionary responses to directional selection on heritable variation. *Evolution* **49:** 241–251.

Grant, V. (1981) *Plant Speciation*. 2nd edn. Columbia University Press, New York.

Grantham, R., Perrin, P., and Mouchiroud, D. (1986) Patterns in codon usage of different species. *Oxford Surv. Evol. Biol.* **3:** 48–81.

Grieve, R. A. F. (1990) Impact cratering on the Earth. *Sci. Am.* **262** (April): 66–73.

Griffiths, A. J. F., Miller, J. F., Suzuki, D. T., Lewontin, R. C., and Gelbart, W. M. (1993) *An Introduction to Genetic Analysis*. 5th edn. W. H. Freeman, New York.

Haffer, J. (1969) Speciation in Amazonian forest birds. *Science* **165:** 131–137.

Haffer, J. (1974) Avian speciation in tropical South America. *Pub. Nuttall Orn. Club* **14:** 1–390.

Hafner, M. S. and Page, R. D. M. (1995) Molecular phylogenies and host-parasite cospeciation: gophers and lice as a model system. *Phil. Trans. Royal Soc. London B* **349:** 77–83.

Hafner, M. S., Sudman, P. D., Villablanca, F. X., Spradling, T. A., Demastes, J. W., and Nadler, S. A. (1994) Disparate rates of molecular evolution in cospeciating hosts and parasites. *Science* **265:** 1087–1090.

Haig, D. (1993) Alternatives to meiosis: the unusual genetics of red algae, microsporidia, and others. *J. Theor. Biol.* **163:** 15–31.

Haig, D. and Grafen, A. (1991) Genetic scrambling as a defence against meiotic drive. *J. Theor. Biol.* **153:** 531–558.

Haldane, J. B. S. (1922) Sex ratio and unisexual sterility in animals. *J. Genetics* **12:** 101–109.

Haldane, J. B. S. (1924) A mathematical theory of natural and artificial selection. Part I. *Trans. Cambridge Phil. Soc.* **23:** 19–41.

Haldane, J. B. S. (1932) *The Causes of Evolution*. Longman, London. (Reprints by Cornell University Press, New York, 1966, and by Princeton University Press, Princeton, New Jersey, 1990.)

Haldane, J. B. S. (1949a) Disease and evolution. *La Ricercha Scientifica* **19** (supplement): 68–76.

Haldane, J. B. S. (1949b) Suggestions as to quantitative measurement of rates of evolution. *Evolution* **3:** 51–56.

Haldane, J. B. S. (1957) The cost of natural selection. *J. Genetics* **55:** 511–524.

Haldane, J. B. S. (1958) The theory of selection for melanism in Lepidoptera. *Proc. Royal Soc. London* **145:** 303–306.

Haldane, J. B. S. (1963) A defense of beanbag genetics. *Persp. Biol. Med.* **7:** 343–359.

Hall, B. K. (1992) *Evolutionary Developmental Biology.* Chapman & Hall, London.

Hall, B. K. (ed.) (1994) *Homology.* Academic Press, San Diego, California.

Hallam, A. (1983). Plate tectonics and evolution. In: Bendall, D. S. (ed.) *Evolution from Molecules to Men,* pp. 367–386. Cambridge University Press, Cambridge, UK.

Hallam, A. (1986) The Pliensbachian and Tithonian extinction events. *Nature* **319:** 765–768.

Hamilton, W. D. (1964) The genetical evolution of social behaviour. [Parts I and II] *J. Theor. Biol.* **6:** 1–52.

Hamilton, W. D. (1967) Extraordinary sex ratios. *Science* **156:** 477–488.

Hamilton, W. D. (1972) Altruism and related phenomena, mainly in the social insects. *Ann. Rev. Ecol. System.* **3:** 193–232.

Hamilton, W. D. (1991) The seething genetics of health and the evolution of sex. In: Osawa, S. and Honjo, T. (eds.) *Evolution of Life: Fossils, Molecules, and Culture,* pp. 229–252. Springer-Verlag, Berlin.

Hamilton, W. D. (1993) Haploid dynamic polymorphism in a host with matching parasites: effects of mutation/subdivision, linkage, and pattern of selection. *J. Heredity* **84:** 328–338.

Hamilton, W. D. and Zuk, M. (1982) Heritable true fitness and bright birds: a role for parasites. *Science* **218:** s384–387.

Hamilton, W. D., Axelrod, R., and Tanese, R. (1990) Sexual selection as an adaptation to resist parasites (a review). *Proc. Nat. Acad. Sci. USA* **87:** 3566–3573.

Hansen, T. A. (1978) Larval dispersal and species longevity in Lower Tertiary neogastropods. *Science* **199:** 885–887.

Hansen, T. A. (1983) Modes of larval development and rates of speciation in early Tertiary neogastropods. *Science* **220:** 501–502.

Hardison, R. C. (1991) Evolution of globin gene families. In: Selander, R. K., Clark, A. G., & Whittam, T. S. (eds), *Evolution at the Molecular Level,* pp. 272–289. Sinauer, Sunderland, Massachusetts.

Hardy, A. C. (1954) Escape from specialisation. In: Huxley, J. S., Hardy, A. C., and Ford, E. B. (eds) *Evolution as a Process,* pp. 122–142. Allen & Unwin, London.

Harland, W. B., Armstrong, R. L., Cox, A. V., Craig, L. E., Smith, A. G., and Smith, D. G. (1990) *A Geologic Time Scale 1989.* Cambridge University Press, Cambridge, UK.

Harris, H. (1966) Enzyme polymorphisms in man. *Proc. Royal Soc. London B* **164:** 298–310.

Harrison, R. G. (1990) Hybrid zones: windows on evolutionary processes. *Oxford Surv. Evol. Biol.* **7:** 129–156.

Harrison, R. G. (1991) Molecular changes at speciation. *Ann. Rev. Ecol. System.* **22:** 281–308.

Harrison, R. G. (ed.) (1993) *Hybrid Zones and the Evolutionary Process.* Oxford University Press, New York.

Harrison, R. G. and Rand, D. M. (1989) Mosaic hybrid zones and the nature of species boundaries. In: Otte, D. and Endler, J. (eds), *Speciation and Its Consequences,* pp. 111–133. Sinauer, Sunderland, Massachusetts.

Hartl, D. L. (1988) *A Primer of Population Genetics.* 2nd edn. Sinauer, Sunderland, Massachusetts.

Hartl, D. L. and Clark, A. G. (1989). *Principles of Population Genetics.* 2nd edn. Sinauer, Sunderland, Massachusetts.

Hartl, D. L., Dykhuizen, D. E., and Dean, A. M. (1985) Limits of adaptation: the evolution of selective neutrality. *Genetics* **111:** 655–674.

Harvey, P. H. and Pagel, M. D. (1991) *The Comparative Method in Evolutionary Biology.* Oxford University Press, Oxford, UK.

Hayden, M. (1981) *Huntington's Chorea.* Springer-Verlag, Berlin.

Hedges, S. B., Moberg, K. D., and Maxson, L. R. (1990) Tetrapod phylogeny inferred from 18S and 28S ribosomal RNA sequences and a review of the evidence for amniote relationships. *Mol. Biol. Evol.* **7:** 607–633.

Hedrick, P. W. (1983) *Genetics of Populations.* Science Books International, Boston.

Hedrick, P. W. (1986) Genetic polymorphism in heterogeneous environments: a decade later. *Ann. Rev. Ecol. System.* **17:** 535–566.

Hedrick, P. W. (1994) Evolutionary genetics of the major histocompatibility locus. *Am. Naturalist* **143:** 945–964.

Hedrick, P., Jain, S., and Holden, L. (1978) Multilocus systems in evolution. *Evol. Biol.* **11:** 101–184.

Hedrick, P. W., Klitz, W., Robinson, W. P., Kuhner, M. K., and Thomson, G. (1991) Evolutionary genetics of HLA. In: Selander, R. K., Clark, A. G., and Whittam, T. S. (eds), *Evolution at the Molecular Level*, pp. 248–271. Sinauer, Sunderland, Massachusetts.

Hennig, W. (1966) *Phylogenetic Systematics.* University of Illinois Press, Urbana.

Hennig, W. (1981) *Insect Phylogeny.* John Wiley, Chichester, UK.

Henry, C. S. (1985) Sibling species, call differences, and speciation in green lacewings (Neuroptera: Chrysopidae: *Chrysoperla*). *Evolution* **39:** 965–984.

Herre, E. A. (1993) Population structure and the evolution of virulence in nematode parasites of fig wasps. *Science* **259:** 1442–1446.

Hewitt, G. M. (1988) Hybrid zones—natural laboratories for evolutionary studies. *Trends Ecol. Evol.* **3:** 158–167.

Hill, W. G. (ed.) (1984) *Quantitative Genetics.* 2 vols. (Benchmark Papers in Genetics series.) Van Nostrand Reinhold, New York.

Hillis, D. M., Huelsenbeck, J. P. and Cunningham, C. W. (1994) Application and accuracy of molecular phylogenies. *Science* **264:** 671–676.

Hillis, D. M., B. K. Moritz, and Mable, C. (eds) (1996) *Molecular Systematics,* 2nd edn. Sinauer, Sunderland, Massachusetts.

Hillis, D. M., Moritz, C., Porter, C. A., and Baker, R. J. (1991) Evidence for biased gene conversion in concerted evolution of ribosomal DNA. *Science* **251:** 308–310.

Hoffman, A. (1989) *Arguments on Evolution: A Paleontologist's Perspective.* Oxford University Press, New York.

Hoffman, A. (1991) Testing the Red Queen hypothesis. *J. Evol. Biol.* **4:** 1–7.

Hoffmann, A. A. and Parsons, P. A. (1991) *Evolutionary Genetics and Environmental Stress.* Oxford University Press, New York.

Holman, E. W. (1987) Recognizability of sexual and asexual species of rotifers. *System. Zoo.* **36:** 381–386.

Holmes, E. C. and Harvey, P. H. (1994) Spinning the web of life. *Current Biol.* **4:** 841–843.

Holser, W. T. and 14 co-authors. (1989) A unique geochemical record at the Permo-Triassic boundary. *Nature* **337:** 39–44.

Hotton, N. H., MacLean, P. D., Roth, J. J., and Roth, E. C. (eds) (1986) *Ecology and Biology of Mammal-like Reptiles.* Smithsonian Institution, Washington, D.C.

Houle, D., Hughes, K. A., Hoffmaster, D. K., Ihara, J., Assimacopoulos, S., Canada, D., and Charlesworth, B. (1994) The effects of spontaneous mutation on quantitative traits. I. Variances and covariances of life history traits. *Genetics* **138:** 773–785.

Howard, D. J. (1993a) Reinforcement: origins, dynamics, and fate of an evolutionary hypothesis. In: Harrison, R. G. (ed.) *Hybrid Zones and the Evolutionary Process*, pp. 46–69. Oxford University Press, New York.

Howard, D. J. (1993b) Small populations, inbreeding, and speciation. In: Thornhill, N. W. (ed.) *The Natural History of Inbreeding and Outbreeding*, pp. 118–142. University of Chicago Press, Chicago.

Howard, R. S. and Lively, C. M. (1994) Parasitism, mutation accumulation, and the maintenance of sex. *Nature* **367:** 554–557. (And erratum: *Nature* **368, 358;** 1994.)

Howe, H. F. and Westley, L. C. (1988) *Ecological Relationships of Plants and Animals.* Oxford University Press, New York.

Hudson, R. R. (1994) How can the low levels of DNA sequence variation in regions of the *Drosophila* genome with low recombination rates be explained? *Proc. Nat. Acad. Sci. USA* **96:** 6815–6818.

Hudson, R. R., Kreitman, M., and Aguadè, M. (1987) A test of neutral molecular evolution based on nucleotide data. *Genetics* **116:** 153–159.

Hughes, A. H. and Nei, M. (1988) Pattern of nucleotide substitution at major histocompatibility complex class I loci reveals overdominant selection. *Nature* **335:** 167–170.

Hughes, A. H. and Nei, M. (1989) Nucleotide substitution at major histocompatibility complex class II loci: evidence for overdominant selection. *Proc. Nat. Acad. Sci. USA* **86:** 958–962.

Hull, D. L. (1965) The effect of essentialism on taxonomy. *Brit. J. Phil. Sci.* **15:** 314–326; **16:** 1–18.

Hull, D. L. (1967) Certainty and circularity in evolutionary taxonomy. *Evolution* **21:** 174–189.

Hull, D. L. (1988) *Science as a Process.* University of Chicago Press, Chicago.

Hume, D. (1779) *Dialogues Concerning Natural Reli-*

gion. Edinburgh. (There are various modern editions.)

Hunt, H. R., Hoppert, C. A., and Rosen, S. (1955) Genetic factors in experimental rat caries. In: Sognnaes, R. F. (ed.), *Advances in Experimental Caries Research*, pp. 66–81. American Association for the Advancement of Science, Washington, D.C.

Huntley, B. and Webb, T. (1989) Migration: species response to climatic variations caused by changes in the earth's orbit. *J. Biogeography* **16:** 5–19.

Hurst, L. (1993) The incidences, mechanisms, and evolution of cytoplasmic sex ratio distorters. *Biol. Rev.* **68:** 121–193.

Huxley, J. S. (1932) *Problems of Relative Growth*. Methuen, London, UK.

Huxley, J. S. (1942) *Evolution: The Modern Synthesis*. Allen & Unwin, London, UK.

Huxley, J. S. (1970–1973) *Memories*. 2 vols. Allen & Unwin, London, UK.

Huxley, J. S. (ed.) (1940) *The New Systematics*. Oxford University Press, Oxford, UK.

Jablonski, D. (1986) Background and mass extinctions: the alternation of macroevolutionary regimes. *Science* **231:** 129–133

Jablonski, D. (1994) Extinction in the fossil record. *Phil. Trans. Royal Soc. London B* **344:** 11–17.

Jablonski, D. and Bottler, J. B. C. (1991) Environmental patterns in the origin of higher taxa: the post-Paleozoic record. *Science* **252:** 1831–1833.

Jablonski, D. and Lutz, R. A. (1983) Larval ecology of marine benthic invertebrates: paleobiological implications. *Biol. Rev.* **58:** 21–89.

Jackman, T. R. and Wake, D. B. (1994) Evolutionary and historical analysis of protein variation in the blotched forms of salamanders of the *Ensatina* complex (Amphibia: Plethodontidae). *Evolution* **48:** 876–897.

Jackson, J. and Cheetham, A. (1994) On the importance of doing nothing. *Nat. Hist.* June: 56–59.

Jackson, J. F. and Pounds, J. A. (1979) Comments on assessing the dedifferentiating effects of gene flow. *System. Zoo.* **28:** 78–85.

Janzen, D. H. (1980) When is it coevolution? *Evolution* **34:** 611–612.

Janzen, D. H. (1983) Dispersal of seeds by vertebrate guts. In: Futuyma, D. J. and Slatkin, M. (eds) *Coevolution*, pp. 232–262. Sinauer, Sunderland, Massachusetts.

Janzen, D. H. and Martin, P. S. (1982) Neotropical anachronisms: the fruits the gomphotheres ate. *Science* **215:** 19–27.

Jeffreys, A. J., Harris, S., Barrie, P. A., Wood, D., Blanchetot, A., and Adams, S. M. (1983) Evolution of gene families: the globin genes. In: Bendall, D. S. (ed.) *Evolution from Molecules to Men*, pp. 175–195. Cambridge University Press, Cambridge, UK.

Jeffreys, A. J., MacLeod, A., Tamaki, K., Neil, D. L., and Monkton, D. G. (1991) Minisatellite repeat coding as a digital approach to DNA typing. *Nature* **354:** 204–209.

Jeffreys, A. J., Royle, N. J., Wilson, V., and Wong, Z. (1988) Spontaneous mutation rate to new length alleles at tandem-repetitive hypervariable loci in human DNA. *Nature* **332:** 278–281.

Jeffreys, A. J. , Wilson, V., and Thein, S. L. (1985) Hypervariable minisatellite regions in human DNA. *Nature* **314:** 67–73.

Jepsen, G. L., Mayr, E., and Simpson, G. G. (eds) (1949) *Genetics, Paleontology, and Evolution*. Princeton University Press, Princeton, New Jersey.

Jerison, H. J. (1973) *Evolution of the Brain and Intelligence*. Academic Press, New York.

Jermy, T. (1984) Evolution of insect/host relationships. *Am. Naturalist* **124:** 609–630.

John, B. and Miklos, G. L. G. (1988) *The Eukaryote Genome in Development and Evolution*. Allen & Unwin, London, UK.

Johnson, C. (1976) *Introduction to Natural Selection*. University Park Press, Baltimore, Maryland.

Johnson, C. (1979) An overview of selection theory and analysis. In: Thompson, J. N. and Thoday, J. M. (eds) *Quantitative Genetic Variation*, pp. 111–119. Academic Press, New York.

Johnson, L. A. S. (1970) Rainbow's end. *System. Zoo.* **19:** 203–239.

Johnston, M. O. and Schoen, D. J. (1995) Mutation rates and dominance levels of genes affecting total fitness in two angiosperm species. *Science* **267:** 226–229.

Johnston, R. F. and Selander, R. K. (1964) House sparrow: rapid evolution of races in North America. *Science* **144:** 548–550.

Johnston, R. F. and Selander, R. K. (1971) Evolution in the house sparrow. II. Adaptive differentiation in North American populations. *Evolution* **25:** 1–28.

Jones, J. S. (1993) *The Language of the Genes.* Harper Collins, London.

Jones, J. S., Leith, B., and Rawlings, P. (1977) Polymorphism in *Cepaea*: a problem with too many solutions?. *Ann. Rev. Ecol. System.* **8:** 109–143.

Jones, J. S. and Probert, R. F. (1980) Habitat selection maintains a deleterious allele in a heterogeneous environment. *Nature* **287:** 632–633.

Kaneshiro, K. (1988) Speciation in Hawaiian *Drosophila. Bioscience* **38:** 258–263.

Kaplan, N. L., Hudson, R. R., and Langley, C. H. (1989) The "hitchhiking effect" revisited. *Genetics* **123:** 887–899.

Karn, M. N. and Penrose, L. S. (1951) Birth weight and gestation time in relation to maternal age, parity, and infant survival. *Ann. Eugenics* **16:** 147–164.

Kay, R. F. and Simons, E. L. (1983) A reassessment of the relationship between later Miocene and subsequent Hominoidea. In: Ciochin, R. L. and Corruccini, R. S. (eds) *New Interpretations of Ape and Human Ancestry*, pp. 577–624. Plenum, New York.

Keller, E. F. (1983) *A Feeling for the Organism: The Life and Work of Barbara McClintock.* W. H. Freeman, San Francisco.

Keller, E. F. and Lloyd, E. A. (eds) (1992) *Keywords in Evolutionary Biology.* Harvard University Press, Cambridge, Massachusetts.

Kelley, S. E. (1994) Viral pathogens and the advantage of sex in the perennial grass *Anthoxanthum odoratum. Phil. Trans. Royal Soc. London B* **346:** 295–302.

Kemp, T. S. (1982a) *Mammal-like Reptiles and the Origin of Mammals.* Academic Press, London, UK.

Kemp, T. S. (1982b) The reptiles that became mammals. *New Scientist* **93:** 581–584. (Reprinted in Cherfas, J.J. [ed.] (1982) *Darwin Up to Date*, pp. 31–34. New Scientist, London, UK.)

Kemp, T. S. (1985) Synapsid reptiles and the origin of higher taxa. *Special Papers in Paleontology* **33:** 175–184.

Kemp, T. S. (1988). Haemothermia or Archosauria? The interrelationships of mammals, birds, and crocodiles. *Zoo. J. Linn. Soc.* **92:** 67–104.

Kenyon, C. (1994) If birds can fly, why can't we? Homeotic genes and evolution. *Cell* **78:** 175–180.

Kessler, S. (1966) Selection for and against ethological isolation between *Drosophila pseudoobscura* and *Drosophila persimilis. Evolution* **20:** 634–645.

Kettlewell, H. B. D. (1973) *The Evolution of Melanism.* Oxford University Press, Oxford, UK.

Kidwell, M. G. (1972) Genetic change of recombination value in *Drosophila melanogaster.* I. Artificial selection for high and low recombination and some properties of recombination-modifying genes. *Genetics* **70:** 419–432.

Kim, K. C. (ed.) (1986) *Coevolution of Parasitic Arthropods and Mammals.* Academic Press, New York.

Kimura, M. (1965) A stochastic model concerning the maintenance of genetic variability in quantitative characters. *Proc. Nat. Acad. Sci. USA* **68:** 984–986.

Kimura, M. (1968) Evolutionary rate at the molecular level. *Nature* **217:** 624–626.

Kimura, M. (1979) The neutral theory of molecular evolution. *Sci. Am.* **241** (November): 98–126.

Kimura, M. (1983) *The Neutral Theory of Molecular Evolution.* Cambridge University Press, Cambridge, UK.

Kimura, M. (1991) Recent developments of the neutral theory viewed from the Wrightian tradition of theoretical population genetics. *Proc. Nat. Acad. Sci. USA* **88:** 5969–5973.

Kimura, M. and Maruyama, T. (1966) The mutation load with epistatic gene interactions in fitness. *Genetics* **54:** 1337–1351.

King, L. and Jukes, T. (1969) Non-darwinian evolution. *Science* **164:** 788–789.

King, M. (1993) *Species Evolution: The Role of Chromosomal Change.* Cambridge University Press, New York.

King, R. B. (1993) Color-pattern variation in Lake Erie water snakes: prediction and measurement of natural selection. *Evolution* **47:** 1819–1833.

Kirkpatrick, M. (ed.) (1994) *The Evolution of Haploid-Diploid Life Cycles.* Lectures on Mathematics in the Life Sciences 25, American Mathematical Society, Providence, Rhode Island.

Kirkpatrick, M. and Jenkins, C. D. (1989) Genetic segregation and the maintenance of sexual reproduction. *Nature* **339:** 300–301.

Klein, R. G. (1989) *The Human Career.* University of Chicago Press, Chicago.

Koehn, R. K. and Hiblich, T. J. (1987) The adaptive importance of genetic variation. *Am. Sci.* **75:** 134–141.

Koepfer, H. R. (1987) Selection for isolation between geographic forms of *Drosophila mojavensis*. I. Interactions between the selected forms. *Evolution* **41:** 37–48.

Kondrashov, A. S. (1988) Deleterious mutations and the evolution of sexual reproduction. *Nature* **336:** 435–440.

Kondrashov, A. (1992) The third phase of Wright's shifting-balance: a simple analysis of the extreme case. *Evolution* **46:** 1972–1975.

Kondrashov, A. S. (1993) Classification of hypotheses on the advantage of amphimixis. *J. Heredity* **84:** 372–387.

Kondrashov, A. S. and Turelli, M. (1992) Deleterious mutations, apparent stabilizing selection, and the maintenance of quantitative variation. *Genetics* **132:** 603–618.

Krebs, J. R. and Davies, N. B. (1993) *An Introduction to Behavioural Ecology*, 3rd edn. Blackwell Science, Boston, Massachusetts.

Kreitman, M. (1983) Nucleotide polymorphism at the alcohol dehydrogenase locus of *Drosophila melanogaster. Nature* **304:** 412–417.

Kreitman, M. and Akashi, H. (1995) Molecular evidence for natural selection. *Annual Review of Ecology and Systematics*, 26, 403–422.

Labandeira, C. C., Dilcher, D. L., Davis, D. R., and Wagner, D. L. (1994) Ninety-seven million years of angiosperm-insect associations: paleobiological insights into the meaning of coevolution. *Proc. Nat. Acad. Sci. USA* **91:** 12278–12282.

Lack, D. (1947) *Darwin's Finches.* Cambridge University Press, Cambridge, UK.

Lake, J. A. (1990) Origin of the Metazoa. *Proc. Nat. Acad. Sci. USA* **87:** 763–766.

Lake, J. A. (1994) Reconstructing evolutionary trees from DNA and protein sequences: paralinear distances. *Proc. Nat. Acad. Sci. USA* **91:** 1455–1459.

Lake, J. A. and Rivera, M. C. (1994) Was the nucleus the first endosymbiont? *Proc. Nat. Acad. Sci. USA* **91:** 2880–2881.

Lamarck, J-B. (1809) *Philosophie Zoologique.* Paris.

Lambert, D. M. and Spencer, H. E. (eds) (1994) *Speciation and the Recognition concept: Theory and Application.* Johns Hopkins University Press, Baltimore, Maryland.

Lande, R. (1976) The maintenance of genetic variability by mutation in a polygenic character with linked loci. *Genetical Research* **26:** 221–235.

Lande, R. (1979) Effective deme size during long-term evolution estimated from rates of chromosomal rearrangement. *Evolution* **33:** 234–251.

Lande, R. (1980) Genetic variation and phenotypic evolution during allopatric speciation. *Am. Naturalist* **116:** 463–479.

Lande, R. (1985) The fixation of chromosomal rearrangements in a subdivided population with local extinction and recolonization. *Heredity* **54:** 323–332.

Lande, R. (1986) The dynamics of peak shifts and the pattern of morphological evolution. *Paleobiology* **12:** 343–354.

Lane, N. G. (1992) *Life of the Past.* 3rd edn. Macmillan, New York.

Langley, C. H. (1977) Nonrandom associations between allozymes in natural poulations of *Drosophila melanogaster.* In: Christiansen, F. B. and Fenchel, T. M. (eds) *Measuring Selection in Natural Populations*, pp. 265–273. Springer-Verlag, Berlin.

Langley, C. H., Voeolker, R. A., Leigh Brown, A. J., Ohnishi, S., Dickson, B., and Montgomery, E. (1981) Null allele frequencies at allozyme loci in natural populations of *Drosophila melanogaster. Genetics* **99:** 151–156.

Larson, E. J. (1989) *Trial and Error. The American Controversy over Creation and Evolution.* Updated edn. Oxford University Press, New York.

Larwood, G. P. (1988) *Extinction and Survival in the Fossil Record.* Systematics Association Special Volume, no. 34. Oxford University Press, Oxford, UK.

Latter, B. D. H. (1970) Selection in finite populations with multiple alleles. II. Centripetal selection, mutation, and isoallelic variation. *Genetics* **66:** 165–86.

Law, R. (1991) Fishing in evolutionary waters. *New Scientist* 2 March: 35–37.

Leary, R. F. and Allendorf, F. W. (1989) Fluctuating asymmetry as an indicator of stress: implications for conservation biology. *Trends Ecol. Evol.* **4:** 214–217.

Lee, B. T. O. and Parsons, P. A. (1968) Selection, prediction, and response. *Biol. Rev.* **43:** 139–174.

Lees, D. R. (1971) Industrial melanism: genetic adaptation of animals to air pollution. In: Bishop, J. A. and Cook, L. M. (eds) *Genetic Consequences of Man Made Change*, pp. 129–176. Academic Press, London.

Lenski, R. (1993) A lack of interest in evolution. *Current Biol.* **3:** 121–123.

Levene, H. (1953) Genetic equilibrium when more than one niche is available. *Am. Naturalist* **87:** 331–333.

Leverich, W. J. and Levin, D. A. (1979) Age-specific survivorship and reproduction in *Phlox drummondii. Am. Naturalist* **113:** 881–903.

Levin, D. A. (1978) The origin of isolating mechanisms in flowering plants. *Evol. Biol.* **11:** 185–317.

Levin, D. A. (1979) The nature of plant species. *Science* **204:** 381–384.

Levinton, J. (1988) *Genetics, Paleontology, and Macroevolution.* Cambridge University Press, Cambridge, UK.

Lewin, B. (1994) *Genes V.* Oxford University Press, New York.

Lewis, W. H. (ed.) (1980) *Polyploidy: Biological Relevance.* Plenum Press, New York.

Lewontin, R. C. (1974) *The Genetic Basis of Evolutionary Change.* Columbia University Press, New York.

Lewontin, R. C. (1985a) Population genetics. In: Greenwood, P. J., Harvey, P. H., and Slatkin, M. (eds) *Evolution: Essays in Honour of John Maynard Smith*, pp. 3–18. Cambridge Unversity Press, Cambridge, UK.

Lewontin, R. C. (1985b) Population genetics. *Ann. Rev. Genetics* **19:** 81–102.

Lewontin, R. C. (1986) How important is population genetics for an understanding of evolution? *Am. Zoo.* **26:** 811–820.

Lewontin, R. C. and Hubby, J. L. (1966) A molecular approach to the study of genic heterozygosity in natural populations. II. Amount of variation and degree of heterozygosity in natural populations of *Drosophila pseudoobscura. Genetics* **54:** 595–605.

Lewontin, R. C, Moore, J. A., Provine, W. B., and Wallace, B. (eds) (1981) *Dobzhansky's Genetics of Natural Populations I-XLIII.* Columbia University Press, New York.

Li, C. C. (1976) *First Course in Population Genetics.* Boxwood Press, Pacific Grove, California.

Li, W-H. (ed.) (1977) *Stochastic Models in Population Genetics.* Benchmark Papers in Genetics, vol. 7. Dowden, Hutchinson, & Ross, Stroudsburg, Pennsylvania.

Li, W-H. and Graur, D. (1991) *Fundamentals of Molecular Evolution.* Sinauer, Sunderland, Massachusetts.

Li, W-H., Wu, C-I., and Luo, C-C. (1985) A new method for estimating synonymous and nonsynonymous rates of nucleotide substitution considering the relative likelihood of nucleotide and codon changes. *Mol. Biol. Evol.* **2:** 150–174.

Li, W-H., Tanimura, M., and Sharp, P. M. (1987) An evaluation of the molecular clock hypothesis using mammalian DNA sequences. *J. Mol. Evol.* **25:** 330–342.

Lively, C. M. (1987) Evidence from a New Zealand snail for the maintenance of sex by parasitism. *Nature* **328:** 519–521.

Lively, C. M., Craddock, C., and Vrijenhoek, R. C. (1990) Red Queen hypothesis supported by parasitism in sexual and clonal fish. *Nature* **344:** 864–866.

Lockhart, P. J., Steel, M. A., Hendy, M. D., and Penny, D. (1994) Recovering evolutionary trees under a more realistic model of sequence evolution. *Mol. Biol. Evol.* **11:** 605–612.

Loeschcke, V. (ed.) (1987) *Genetic Constraints on Adaptive Evolution.* Springer-Verlag, Berlin.

Long, J. A. (1995) *The Rise of Fishes: 500 Million Years of Evolution.* Johns Hopkins University Press, Baltimore, Maryland.

Losos, J. (1994) Integrative approaches to evolutionary ecology: *Anolis* lizards as model systems. *Ann. Rev. Ecol. System.* **24:** 467–493.

Luria, S. E. and Delbruck, M. (1943) Mutations of bacteria from virus sensitivity to virus resistance. *Genetics* **28:** 491–511.

Lynch, J. D. (1988) Refugia. In: Myers, A. A. and Giller, P. S. (eds) *Analytical Biogeography*, pp. 311–342. Chapman & Hall, New York.

Lynch, J. D. (1989) The gauge of speciation: on the frequencies of modes of speciation. In: Otte, D. and Endler, J. A. (eds) *Speciation and Its Consequences*, pp. 527–553. Sinauer, Sunderland, Massachusetts.

Lynch, M. (1988) The rate of polygenic mutation. *Genetical Research* **51:** 137–148.

Lynch, M. D., Burger, R., Butcher, D., and Gabriel, W. (1993). The mutational meltdown in asexual populations. *J Hered* **84:** 339–344.

Lynch, M. and Jarrell, P. E. (1993) A method for calibrating molecular clocks and its application to animal mitochondrial DNA. *Genetics* **135:** 1197–1208.

MacArthur, R. H. (1958) Population ecology of some warblers of northeastern coniferous forests. *Ecology* **39:** 599–619.

MacFadden, B. J. (1992) *Fossil Horses. Systematics, Paleobiology, and Evolution of the Family Equidae.* Cambridge University Press, New York.

Macgregor, H. C. (1991) Chromosomal heteromorphism in newts (*Triturus*) and its significance in relation to evolution and development. In: Green, D. M. and Sessions, S. K. (eds), *Amphibian Cytogenetics and Evolution*, pp. 175–196. Academic Press, San Diego, California.

Macgregor, H. C. and Horner, H. A. (1980) Heteromorphism for chromosome 1, a requirement for normal development in crested newts. *Chromosoma* **76:** 111–122.

Macleod, N. (1991) Punctuated anagenesis and the importance of stratigraphy to paleobiology. *Paleobiology* **17:** 167–188.

Macnair, M. R. (1991) Why the evolution of resistance to anthropogenic toxins normally involves major gene changes: the limits to natural selection. *Genetica* **84:** 213–219.

Macnair, M. R. and Christie, P. (1983) Reproductive isolation as a pleiotropic effect of copper tolerance in *Mimulus guttatus? Heredity* **50:** 295–302.

Maddison, W. P. and Maddison, D. R. (1992) *Macclade: Analysis of Phylogeny and Character Evolution.* Sinauer, Sunderland, Massachusetts.

Mallett, J. (1989) The evolution of insecticide resistance: have the insects won? *Trends Ecol. Evol.* **4:** 336–340.

Mani, G. S. (1982) A theoretical analysis of the morph frequency variation in the peppered moth over England and Wales. *Biol. J. Linn. Soc.* **17:** 259–267.

Marshall, L. G. (1988) Extinction. In: Myers, A. A. and Giller, P. S. (eds) *Analytic Biogeography* pp. 219–254. Chapman & Hall, New York.

Marshall, L. G., Webb, S. D., Sepkoski, J. J., and Raup, D. M. (1982) Mammalian evolution and the Great American Interchange. *Science* **215:** 1351–1357.

Martin, L. (1985) Significance of enamel thickness in hominoid evolution. *Nature* **314:** 260–263.

Martin, R. A. (1992) Generic species richness and body mass in North American mammals: support for the inverse relationship of body size and speciation rate. *Hist. Biol.* **6:** 73–90.

Marx, J. (1994) Chromosome ends catch fire. *Science* **265:** 1656–1658.

Mather, K. (1943) Polygenic inheritance and natural selection. *Biol. Rev.* **18:** 32–64.

May, A. W. (1967) Fecundity of Atlantic cod. *J. Fisheries Research Board of Canada* **24:** 1531–1551.

May, R. M. (1993) Resisting resistance. *Nature* **361:** 593–594.

May, R. M. and Anderson, R. M. (1983) Parasite-host coevolution. In: Futuyma, D. J. and Slatkin, M. (eds) *Coevolution*, pp. 186–206. Sinauer, Sunderland, Massachusetts.

Maynard Smith, J. (1968) *Mathematical Ideas in Biology.* Cambridge University Press, Cambridge, UK.

Maynard Smith, J. (1976) Group selection. *Quart. Rev. Biol.* **51:** 277–283.

Maynard Smith, J. (1978a) *The Evolution of Sex.* Cambridge University Press, Cambridge, UK.

Maynard Smith, J. (1978b) Optimization theory in evolution. *Ann. Rev. Ecol. System.* **9:** 31–56.

Maynard Smith, J. (1983) The genetics of stasis and punctuation. *Ann. Rev. Genetics* **17:** 11–25.

Maynard Smith, J. (1986) *The Problems of Biology.* Oxford University Press, Oxford, UK.

Maynard Smith, J. (1987a) J. B. S. Haldane. *Oxford Surv. Evol. Biol.* **4:** 1–9.

Maynard Smith, J. (1987b) How to model evolution. In: Dupré, J. (ed.), *The Latest on the Best*, pp. 119–31. MIT Press, Cambridge, Massachusetts.

Maynard Smith, J. (1987c) Darwin stays unpunctured. *Nature* **330:** 516.

Maynard Smith, J. (1988) Punctuation in perspective. *Nature* **332:** 311–312.

Maynard Smith, J. (1989) *Evolutionary Genetics.* Oxford University Press, Oxford, UK.

Maynard Smith, J. (1994) Estimating selection by comparing synonymous and substitutional changes. *J. Mol. Evol.* **39:** 123–128.

Maynard Smith, J., Burian, R., Kauffman, S., Alberch, P., Campbell, J., Goodwin, B., Lande, R., Raup, D., and Wolpert, L. (1985) Developmental constraints and evolution. *Quart. Rev. Biol.* **60:** 265–287.

Maynard Smith, J. and Haigh, J. (1974) The hitch-hiking effect of a favourable gene. *Genetical Research* **23:** 23–35.

Maynard Smith, J., Smith, N.H., O'Rourke, M., and Spratt, B.G. (1993) How clonal are bacteria? *Proc. Nat. Acad. Sci. USA* **90:** 4384–4388.

Maynard Smith, J. and Szathmáry, E. (1995) *The Major*

Transitions in Evolution. W.H.Freeman/Spektrum, Oxford, UK, and New York.

Mayr, E. (1942) *Systematics and the Origin of Species*. Columbia University Press, New York.

Mayr, E. (1954) Change of genetic environment and evolution. In: Huxley, J. S., Hardy, A. C., and Ford, E. B. (eds) *Evolution as a Process*, pp. 157–180. Allen & Unwin, London.

Mayr, E. (1963) *Animal Species and Evolution*. Harvard University Press, Cambridge, Massachusetts.

Mayr, E. (1976) *Evolution and the Diversity of Life*. Harvard University Press, Cambridge, Massachusetts.

Mayr, E. (1981) Biological classification: toward a synthesis of opposing methodologies. *Science* **214:** 510–516.

Mayr, E. (1982a) Processes of speciation in animals. In: Barigozzi, C. (ed.) *Mechanisms of Speciation*, pp. 1–19. Alan Liss, New York. (Reprinted in Mayr, E. [1987]. *Toward a New Philosophy of Biology*. Harvard University Press, Cambridge, Massachusetts.)

Mayr, E. (1982b) *The Growth of Biological Thought: Diversity, Evolution, and Inheritance*. Harvard University Press, Cambridge, Massachusetts.

Mayr, E. and Ashlock, P. D. (1991) *Principles of Systematic Zoology*. 2nd edn. McGraw-Hill, New York.

Mayr, E. and Provine, W. B. (eds) (1980) *The Evolutionary Synthesis*. Harvard University Press, Cambridge, Massachusetts.

McCune, A. R. (1982) On the fallacy of constant extinction rates. *Evolution* **36:** 610–614.

McDonald, J. H. and Kreitman, M. (1991) Adaptive evolution at the *Adh* locus in *Drosophila*. *Nature* **351:** 652–654.

McKenzie, J. A. and O'Farrell, K. (1993) Modification of developmental instability and fitness: malathion-resistance in the Australian sheep blowfly. *Genetica* **89:** 67–76.

McKenzie, J. A. and Batterham, P. (1994) The genetic, molecular, and phenotypic consequences of selection for insecticide resistance. *Trends Ecol. Evol.* **9:** 166–169.

McKinney, M. L. (1987) Taxonomic selectivity and continuous variation in mass and background extinctions of marine taxa. *Nature* **325:** 143–145

McKinney, M. L. (ed.) (1988) *Heterochrony in Evolution: An Interdisciplinary Approach*. Plenum Press, New York.

McKinney, M. L. (1991) Completeness of the fossil record: an overview. In: Donovan, S. K. (ed.) *The Processes of Fossilization*, pp. 66–83. Columbia University Press, New York.

McKinney, M. L. and K. McNamara (1991) *Heterochrony: The Evolution of Ontogeny*. Plenum Press, New York.

McNamara, K. J. (1989) The great evolutionary handicap. *New Scientist* **123** (10 September): 47–51.

Medawar, P. B. (1958) Postscript: D'Arcy Thompson and *Growth and Form*. In: R. D'Arcy Thompson, *D'Arcy Thompson: The Scholar Naturalist*. Oxford University Press, London, UK. (Reprinted in P. B. Medawar [1982]. *Pluto's Republic*. Oxford University Press, Oxford, UK.)

Melosh, H. J. (1995) Around and around we go. *Nature* **376:** 386–387.

Meyer, A. (1993) Phylogenetic relationships and evolutionary processes in East African cichlid fishes. *Trends Ecol. Evol.* **8:** 279–284.

Meyerhoff, A. A., Lyons, J. B., and Officer, C. B. (1994) Chicxulub structure: a volcanic sequence of Late Cretaceous age. *Geology* **22:** 3–4.

Michod, R. E. (1995) *Eros and Evolution*. Addison-Wesley, Reading, Massachusetts.

Milinkovitch, M. C., Ortí, G., and Meyer, A. (1993) Revised phylogeny of whales suggested by mitochondrial ribosomal DNA sequences. *Nature* **361:** 346–348.

Mindell, D. P. and Honeycutt, R. L. (1990) Ribosomal RNA in vertebrates and phylogenetic applications. *Ann. Rev. Ecol. System.* **21:** 541–566.

Mishler, B. D. and Brandon, R. N. (1987) Individuality, pluralism, and the phylogenetic species concept. *Biol. Phil.* **2:** 397–414.

Mitchell-Olds, T. and Rutledge, J. J. (1986) Quantitative genetics in natural plant populations: a review of the theory. *Am. Naturalist* **127:** 379–402.

Miyata, T., Hayashida, H., Kikuno, R., *et al.* (1982) Molecular clock of silent substitution: at least six-fold preponderance of silent changes in mitochondrial genes over those in nuclear genes. *J. Mol. Evol.* **19:** 28–35.

Møller, A. P. (1994) *Sexual Selection and the Barn Swallow*. Oxford University Press, Oxford, UK.

Moore, J. A. (1949) Patterns of evolution in the genus *Rana*. In: Jepsen, G. L., Mayr, E., and Simpson, G.

G. (eds) *Genetics, Paleontology, and Evolution*, pp. 315–338. Princeton University Press, Princeton, New Jersey.

Moore, W. S. (1977) An evaluation of narrow hybrid zones in vertebrates. *Quart. Rev. Biol.* **53:** 263–277.

Moran, N. and Baumann, P. (1994) Phylogenetics of cytoplasmically inherited microorganisms of arthropods. *Trends Ecol. Evol.* **9:** 15–20.

Mukai, T. (1964) Spontaneous mutation rate of polygenes controlling viability. *Genetics* **50:** 1–19.

Mukai, T., Chigusa, S. I., Mettler, L. E., and Crow, J. F. (1972) Mutation rate and dominance of genes affecting viability in *Drosophila melanogaster. Genetics* **72:** 335–355.

Muller, H. J. (1959) One hundred years without Darwinism are enough. *School Sci. Math.* **49:** 314–318.

Mumme, R. L. (1992) Do helpers increase reproductive success: an experimental analysis in the Florida scrub jay. *Behavioral Ecol. Sociobiol.* **31:** 319–328.

Murray, J. and Clarke, B. (1980) The genus *Partula* on Moorea: speciation in progress. *Proc. Royal Soc. London B* **211:** 83–117.

Myers, A. A. and Giller, P. S. (eds) (1988) *Analytical Biogeography.* Chapman & Hall, London.

Nei, M. (1987) *Molecular Evolutionary Genetics.* Columbia University Press, New York.

Nelkin, D. (1982) *The Creation Controversy: Science or Scripture in the Schools.* W. W. Norton, New York.

Nelson, G. and Rosen, D. E. (eds) (1981) *Vicariance Biogeography: A Critique.* Columbia University Press, New York.

Neufeld, P. J. and Colman, N. (1990) Genetic fingerprinting. *Sci. Am.* **262** (May): 46–53.

Nevo, E. (1988) Genetic diversity in nature. *Evol. Biol.* **23:** 217–246.

Newell, N. D. (1959) The nature of the fossil record. *Proc. Am. Phil. Soc.* **103:** 264–285.

Nichol, S. T., Rowe, J. E., and Fitch, W. M. (1993) Punctuated equilibrium and positive Darwinian evolution in vesicular stomatitis virus. *Proc. Nat. Acad. Sci. USA* **90:** 10424–10428.

Nielsen, C. (1994) *Animal Evolution: Interrelationships of the Living Phyla.* Oxford University Press, Oxford, UK.

Niklas, K. J. (1986) Large-scale changes in animal and plant terrestrial communities. In: Raup, D. M. and Jablonski, D. (eds) *Patterns and Processes in the History of Life.* Dahlem Workshop. John Wiley, Chichester, UK.

Nilsson, A. L. (1992) Orchid pollination biology. *Trends Ecol. Evol.* **7:** 255–259.

Nilsson, D-E. (1989) Vision optics and evolution. *Bioscience* **39:** 298–307.

Nilsson, D-E. and Pelger, S. (1994) A pessimistic estimate of the time required for an eye to evolove. *Proc. Royal Soc. London B* **256:** 53–58.

Nitecki, M. H. (ed.) (1990) *Evolutionary Innovations.* University of Chicago Press, Chicago.

Nordenskiöld, E. (1929) *The History of Biology.* Knopf, New York.

Nordström, K. and Austin, S. J. (1989) Mechanisms that contribute to the stable segregation of plasmids. *Ann. Rev. Genetics* **23:** 37–69.

Norell, M. A. and Novacek, M. J. (1992) The fossil record and evolution: comparing cladistic and paleontological evidence for vertebrate history. *Science* **255:** 1690–1693.

Novacek, M. J. (1992) Mammalian phylogeny: shaking the tree. *Nature* **356:** 121–125.

Novacek, M. J. and Wheeler, Q. D. (eds) (1992) *Extinction and Phylogeny.* Columbia University Press, New York.

Novak, S. J., Soltis, D. E., and Soltis, P. S. (1991) Ownbey's Tragopogons: 40 years later. *Am. J. Botany* **78:** 1586–1600.

Numbers, R. L. (1992) *The Creationists: The Evolution of Scientific Creationism.* Knopf, New York (paperback edn. 1993, University of California Press, Berkeley, California).

Ochman, H., Jones, J. S., and Selander, R. K. (1983) Molecular area effects in *Cepaea. Proc. Nat. Acad. Sci. USA* **80:** 4189–4193.

Officer, C. B., Hallam, A., Drake, C. L., and Devine, J. D. (1987) Late Cretaceous and paroxysmal Cretaceous/Tertiary extinctions. *Nature* **326:** 143–149.

Ohta, T. (1988) Multigene and supergene families. *Oxford Surv. Evol. Biol.* **5:** 41–65.

Ohta, T. (1992) The nearly neutral theory of molecular evolution. *Ann. Rev. Ecol. System.* **23:** 263–286.

Ohta, T. (1993) An examination of the generations-time effect on molecular evolution. *Proc. Nat. Acad. Sci. USA* **90:** 10676–10680.

Orgel, L. E. and Crick, F. H. C. (1980) Selfish DNA: the ultimate parasite. *Nature* **284:** 604–607.

Orr, H. A. and Coyne, J. A. (1992) The genetics of adaptation: a reassessment. *Am. Naturalist* **140:** 725–742.

Orth, C. J. (1989) Geochemistry of the bio-event horizons. In: Donovan, S. K. (ed.) *Mass Extinctions*, pp. 37–72. Columbia University Press, New York.

Ospovat, D. (1981) *The Development of Darwin's Theory.* Cambridge University Press, Cambridge, UK.

Otte, D. and Endler, J. A. (eds) (1989) *Speciation and Its Consequences.* Sinauer, Sunderland, Massachusetts.

Ownbey, M. (1950) Natural hybridization and amphiploidy in the genus *Tragopogon. Am. J. Botany* **27:** 487–499.

Page, R. E. M. (1994) Maps between trees and cladistic analysis of historical associations among genes, organisms, and areas. *System. Biol.* **43:** 58–77.

Parker, G. A. and Maynard Smith, J. (1990) Optimality theory in evolutionary biology. *Nature* **349:** 27–33.

Parker, M. A. (1994) Pathogens and sex in plants. *Evol. Ecol.* **8:** 560–584.

Paterson, H. E. H. (1993) *Evolution and the Recognition Concept of Species: Collected Writings.* Johns Hopkins University Press, Baltimore, Maryland.

Patterson, C. (1981) Methods of paleobiogeography. In: Nelson, G. and Rosen, D. E. (eds) *Vicariance Biogeography: A Critique*, pp. 446–489. Columbia University Press, New York.

Patterson, C., Williams, D. M., and Humphries, C. J. (1993) Congruence between molecular and morphological phylogenies. *Ann. Rev. Ecol. System.* **24:** 153–188.

Pease, C. (1992) On the declining extinction and origination rates of fossil taxa. *Paleobiology* **18:** 89–92.

Pease, C., Lande, R., and Bull, J. J. (1989) A model of population growth, dispersal, and evolution in a changing environment. *Ecology* **70:** 1657–1664.

Pellmyr, O. (1992) Evolution of insect pollination and angiosperm diversification. *Trends Ecol. Evol.* **7:** 46–49.

Penny, D., Foulds, L. R., and Hendy, M. D. (1982) Testing the theory of evolution by comparing the phylogenetic trees constructed from five different protein sequences. *Nature* **297:** 197–200.

Penny, D., Hendy, M. D., and Steel, M. A. (1992) Progress with methods for constructing evolutionary trees. *Trends Ecol. Evol.* **7:** 73–79.

Pielou, E. C. (1991) *After the Ice Age: The Return of Life to Glaciated North America.* University of Chicago Press, Chicago.

Pierce, N. E. (1987) The evolution and biogeography of associations between lycaenid butterflies and ants. *Oxford Surv. Evol. Biol.* **4:** 89–116.

Pierce, N. E. and Mead, P. S. (1981) Parasitoids as selective agents in the symbiosis between lycaenid butterfly larvae and ants. *Science* **211:** 1185–1187.

Pilbeam, D. R. (1986) Hominoid evolution and hominoid origins. *Am. Anthropol.* **88:** 295–312.

Pinker, S. and Bloom, P. (1990) Natural language and natural selection. *Behavioral and Brain Sciences* **13:** 707–784. (Reprinted in Barkow, J., Cosmides, L., and Tooby, J. (eds) (1992). *The Adapted Mind*, pp. 451–493. Oxford University Press, New York.)

Potts, W. K. and Wakeland, E. K. (1990) Evolution of diversity at the major histocompatibility complex. *Trends Ecol. Evol.* **5:** 181–187.

Price, T. and Langen, T. (1992) Evolution of correlated characters. *Trends Ecol. Evol.* **7:** 307–310.

Primack, R. B. and Kang, H. (1989) Measuring fitness and natural selection in wild plant populations. *Ann. Rev. Ecol. System.* **20:** 367–396.

Provine, W. B. (1971) *The Origins of Theoretical Population Genetics.* University of Chicago Press, Chicago.

Provine, W. B. (1986) *Sewall Wright and Evolutionary Biology.* University of Chicago Press, Chicago.

Queller, D. C., Strassmann, J. E., and Hughes, C. R. (1993) Microsatellites and kinship. *Trends Ecol. Evol.* **8:** 285–288.

Raff, R. A. and Kaufman, T. C. (1983) *Embryos, Genes, and Evolution.* Macmillan, New York.

Raup, D. M. (1966) Geometric analysis of shell coiling. *J. Paleontology* **40:** 1178–1190.

Raup, D. M. (1979) Size of the Permo-Triassic bottleneck and its evolutionary implications. *Science* **206:** 217–218.

Raup, D. M. (1986) Biological extinction in earth history. *Science* **231:** 1528–1533.

Raup, D. M. (1991) *Extinction: Bad Genes or Bad Luck?* W. W. Norton, New York.

Raup, D. M. and Boyajian, G. E. (1988) Patterns of generic extinction in the fossil record. *Paleobiology* **14:** 109–125.

Raup, D. M. and Sepkoski, J. J. (1984) Periodicity of extinctions in the geologic past. *Proc. Nat. Acad. Sci. USA* **81:** 801–805.

Raven, P. H. (1976) Systematics and plant population biology. *Systematic Botany* **1:** 284–316.

Read, A. F. (1988) Sexual selection and the role of parasites. *Trends Ecol. Evol.* **3:** 97–102.

Redfield, R. J. (1994) Male mutation rates and the cost of sex for females. *Nature* **369:** 145–147.

Reeve, H. K. and Sherman, P. W. (1993) Adaptation and the goals of evolutionary research. *Quart. Rev. Biol.* **68:** 1–32.

Rendel, J. M. (1967) *Canalisation and Gene Control.* Logos, London.

Rensch, B. (1959) *Evolution Above the Species Level.* Methuen, London.

Rhodes, F. H. T. (1983) Gradualism, punctuated equilibrium, and the *Origin of Species. Nature* **305:** 269–272.

Rice, W. R. (1994) Degeneration of a nonrecombining chromosome. *Science* **263:** 230–232.

Rice, W. R. and Hostert, E. E. (1993) Laboratory experiments on speciation: what have we learned in 40 years? *Evolution* **47:** 1637–1653.

Ricker, W. E. (1981) Changes in the average size and average age of Pacific salmon. *Canad. J. Fisheries and Aquatic Sci.* **38:** 1636–1656.

Ridley, M. (1986) *Evolution and Classification: The Reformation of Cladism.* Longman, London.

Ridley, M. (1989) The cladistic solution to the species problem. *Biol. Phil.* **4:** 1–16.

Rieseberg, L. H. and Wendel, J. F. (1993) Introgression and its consequences in plants. In: R. G. Harrison (ed.) *Hybrid Zones and the Evolutionary Process*, pp. 70–109. Oxford University Press, New York.

Robson, G. C. and Richards, O. W. (1936) *The Variations of Animals in Nature.* Longman, London.

Roff, A. and Mousseau, T. A. (1987) Quantitative genetics and fitness. *Heredity* **58:** 103–118.

Rogers, R. A., Rogers, L. A., Hoffmann, R. S., and Martin, L. D. (1991) Native American biological diversity and the biogeographic influence of Ice Age refugia. *J. Biogeography* **18:** 623–630.

Roose, M. L. and Gottlieb, L. D. (1976) Genetic and biochemical consequences of polyploidy in *Tragopogon. Evolution* **30:** 818–830.

Ross, J. (1982) Myxomatosis: the natural evolution of the disease. In: Edwards, M. A. and McDonnell, U. (eds) *Animal Disease in Relation to Animal Conservation.* Symposia of the Zoological Society of London, 50, pp. 77–95. Academic Press, London.

Ross, R. M. and Allman, W. D. (eds) (1990) *Causes of Evolution.* University of Chicago Press, Chicago.

Roth, V. L. (1992) Quantitative variation in elephant dentitions: implications for the delimitation of fossil species. *Paleobiology* **18:** 184–202.

Roughgarden, J. (1975) Evolution of marine symbiosis—a simple cost-benefit model. *Ecology* **56:** 1201–1208.

Rudwick, M. J. S. (1964) The inference of function from structure in fossils. *Brit. J. Phil. Sci.* **15:** 27–40.

Ruse, M. (1982) *Darwinism Defended: A Guide to the Evolution Controversies.* Addison-Wesley, Reading, Massachusetts.

Sarich, V. and Wilson, A. C. (1967) Immunological time scale for hominid evolution. *Science* **158:** 1200–1203.

Scharloo, W. (1987) Constraints in selective response. In: Loeschcke, V. (ed.) *Genetic Constraints on Adaptive Evolution*, pp. 125–149. Springer-Verlag, Berlin.

Scharloo, W. (1991) Canalization: genetic and developmental aspects. *Ann. Rev. Ecol. System.* **22:** 65–94.

Scherer, S. (1990) The protein molecular clock. *Evol. Biol.* **24:** 83–106.

Schimenti, J. C. and Duncan, C. H. (1984) Ruminant globin gene structures suggest an evolutionary role for Alu-type repeats. *Nucleic Acid Research* **12:** 1641–1655.

Schimenti, J. C. and Duncan, C. H. (1985) Concerted evolution of the cow ϵ^2 and ϵ^4 β-globin genes. *Mol. Biol. Evol.* **2:** 505–513.

Schindel, D. E. (1982) The gaps in the fossil record. *Nature* **297:** 282–284.

Schluter, D. (1994) Experimental evidence that competition promotes divergence in adaptive radiation. *Science* **266:** 798–801.

Schluter, D., Price, T. D., and Grant, P. R. (1985) Ecological character displacement in Darwin's finches. *Science* **227:** 1056–1059.

Schopf, J. W. (1993) Microfossils of the Early Archean Apex Chert: new evidence of the antiquity of life. *Science* **260:** 640–645.

Schopf, J. W. (1994) The early evolution of life: solution to Darwin's dilemma. *Trends Ecol. Evol.* **9:** 375–377.

Schopf, J. W. and Klein, C. (eds) (1992) *The Proterozoic Biosphere: A Multidisciplinary Study.* Cambridge University Press, New York.

Schram, F. R. (1986) *Crustacea.* Oxford University Press, New York.

Scotland, R. W., Siebert, D. J., and Williams, D. M. (eds) (1994) *Models in Phylogeny Reconstruction.* Oxford University Press, Oxford, UK.

Seger, J. (1985) Intraspecific resource competition as a cause of sympatric speciation. In: Greenwood, P. J., Harvey, P. H., and Slatkin, M. (eds) *Evolution,* pp. 43–53. Cambridge University Press, Cambridge, UK.

Selander, R. K., Clark, A. G., and Whittam, T. S. (eds) (1991) *Evolution at the Molecular Level.* Sinauer, Sunderland, Massachusetts.

Sepkoski, J. J. (1992) Ten years in the library: new data confirm paleontological patterns. *Paleobiology* **19:** 43–51.

Shaffer, H. B. (1984) Evolution in a paedomorphic lineage. I. An electrophoretic analysis of the Mexican ambystomatid salamanders. *Evolution* **38:** 1194–1126.

Shapiro, J. A. (1995) Adaptive mutation: who's really in the garden? *Science* **268:** 373–374.

Sharp, P. M., Averoff, M., Lloyd, A. T., Matassi, G., and Peden, J. F. (1995). DNA sequence evolution: the sounds of silence. *Philosophical transactions of the Royal Society of London* B 349, 241–247.

Shaw, D. D. (1994) Centromeres: moving chromosomes through space and time. *Trends Ecol. Evol.* **9:** 170–175.

Shea, B. T. (1989) Heterochrony in human evolution: the case for neoteny reconsidered. *Yearbook of Phys. Anthro.* **32:** 69–101.

Shear, W. A. (1991) The early development of terrestrial ecosystems. *Nature* **351:** 283–289.

Sheehan, P. M., Fastovsky, D. E., Hoffmann, R. G., Berghaus, C. B., and Gabriel, D. L. (1991) Sudden extinction of the dinosaurs: latest Cretaceous, Upper Great Plains, USA. *Science* **254:** 835–839.

Sheldon, P. R. (1987) Parallel gradualistic evolution of Ordovician trilobites. *Nature* **330:** 561–563.

Sheppard, P. M. (1975) *Natural Selection and Heredity.* 5th edn. Hutchinson, London.

Sibley, C. G. and Ahlquist, J. E. (1987) DNA hybridization evidence of hominoid phylogeny: results from an expanded data set. *J. Mol. Evol.* **26:** 99–121.

Sibley, C. G. and Ahlquist, J. E. (1990) *Phylogeny and Classification of Birds: A Study in Molecular Evolution.* Yale University Press, New Haven, Connecticut.

Silberglied, R. E., Ainello, A., and Windsor, D. M. (1980) Disruptive coloration in butterflies: lack of support in *Anartia fatima. Science* **209:** 617–619.

Simpson, G. G. (1944) *Tempo and Mode in Evolution.* Columbia University Press, New York.

Simpson, G. G. (1949) *The Meaning of Evolution.* Yale University Press, New Haven, Connecticut.

Simpson, G. G. (1953) *The Major Features of Evolution.* Columbia University Press, New York.

Simpson, G. G. (1961a) One hundred years without Darwin are enough. *Teachers College Record* **60:** 617–626. (Reprinted in Simpson 1964, q.v.)

Simpson, G. G. (1961b) *Principles of Animal Taxonomy.* Columbia University Press, New York.

Simpson, G. G. (1964) *This View of Life.* Harcourt, Brace & World, New York.

Simpson, G. G. (1978) *Concession to the Improbable.* Yale University Press, New Haven, Connecticut.

Simpson, G. G. (1980) *Splendid Isolation.* Yale University Press, New Haven, Connecticut.

Simpson, G. G. (1983) *Fossils and the History of Life.* Scientific American Library, New York.

Singer, M. C. and Parmesan, C. (1993) Sources of variation in patterns of plant-insect association. *Nature* **361:** 251–253.

Slatkin, M. (1985) Gene flow in natural populations. *Ann. Rev. Ecol. System.* **16:** 393–430.

Slatkin, M. (1987) Quantitative genetics of heterochrony. *Evolution* **41:** 799–811.

Sloan, R. E., Rigby, J. K., Van Valen, L. M., and Gabriel, D. (1986). Gradual dinosaur extinction and simultaneous ungulate radiation in the Hell Creek Formation. *Science* **232:** 629–633.

Smit, J. and van der Kaars, S. (1984) Terminal Cretaceous extinctions in the Hell Creek area, Montana: compatible with catastrophic extinction. *Science* **223:** 1177–1179.

Smith, A. W. (1994) *Systematics and the Fossil Record: Documenting Evolutionary Patterns.* Blackwell Science, Boston, Massachusetts.

Smith, H. G. and Montgomerie, R. (1991) Sexual selection and the tail ornaments of North American barn swallows. *Behavioral Ecol. Sociobiol.* **28:** 195–201.

Smith, T. B. (1993) Disruptive selection and the genetic basis of bill size polymorphism in the African finch *Pyrenestes*. *Nature* **363:** 618–620.

Sneath, P. H. A. and Sokal, R. R. (1973) *Numerical Taxonomy*. 2nd edn. W. H. Freeman, San Francisco.

Sniegowski, P.D. and Lenski, R.E. (1995). Mutation and adaptation: the directed mutation controversy in evolutionary perspective. *Annual review of ecology and systematics*, 26, 553–578.

Sober, E. (1984) *The Nature of Selection*. MIT Press, Cambridge, Massachusetts.

Sober, E. (1989) *Reconstructing the Past*. MIT Press, Cambridge, Massachusetts.

Sober, E. (ed.) (1994) *Conceptual Issues in Evolutionary Biology*. 2nd edn. MIT Press, Cambridge, Massachusetts.

Sokal, R. R. (1966) Numerical taxonomy. *Sci. Am.* **215** (December): 106–116.

Soltis, D. E. and Soltis, P. S. (1990) Chloroplast DNA and nuclear rDNA variation: insights into autopolyploid and allopolyploid evolution. In: Kawano, S. (ed.), *Biological Approaches and Evolutionary Trends in Plants*, pp. 97–117. Academic Press, London, UK, and San Diego, California.

Spalding, A. C. (ed.) (1993) *Evolutionary Conservation of Developmental Mechanisms*. Wiley-Liss, New York.

Spiess, E. B. (1989) *Genes in Populations*. 2nd edn. John Wiley, New York.

Stanley, S. M. (1984) Mass extinctions in the ocean. *Sci. Am.* **250** (June): 46–54.

Stanley, S. M. and Yang, X. (1994) A double mass extinction at the end of the Paleozoic era. *Science* **266:** 1340–1344.

Stebbins, R. 1994 colorplate #1 bk ref *Life on the Edge*. Heyday Books, San Francisco.

Stebbins, G. L. (1950) *Plant Variation and Evolution*. Columbia University Press, New York.

Stehli, F. G. and Webb, S. D. (eds) (1985) *The Great American Biotic Interchange*. Plenum Press, New York.

Stenseth, N. C. and Maynard Smith, J. (1984) Coevolution in ecosystems: Red Queen evolution or stasis? *Evolution* **38:** 870–880.

Stephan, W. and Cho, S. (1994) Possible role of natural selection in the formation of tandem repetitive non-coding DNA. *Genetics* **136:** 333–341.

Stewart, C-B. (1993) The powers and pitfalls of parsimony. *Nature* **361:** 603–607.

Strickberger, M. (1990) *Evolution*. Jones & Bartlett, Boston, Massachusetts.

Strobeck, C., Maynard Smith, J., and Charlesworth, B. (1976) The effects of hitch-hiking on a gene for recombination. *Genetics* **82:** 547–558.

Sundström, L. (1994) Sex ratio bias, relatedness asymmetry, and queen mating frequency in ants. *Nature* **367:** 266–268.

Surlyk, F. and Johansen, M. B. (1984) End-Cretaceous brachiopod extinctions in the chalk of Denmark. *Science* **223:** 1174–1177.

Swofford, D. L. (1992) PAUP: *Phylogenetic Analysis Using Parsimony. Version 3.0 s.* (Computer program and manual distributed by the Center for Biodiversity, Illinois Natural History Survey, Champaign, IL 61802.)

Swofford, D. L., Olsen, G. J., and Waddell, P. (1996) Phylogeny reconstruction. In: Hillis, D. M. & Moritz, C., (eds) *Molecular Systematics II*, Sinauer, Sunderland, Massachusetts.

Sytsma, K. (1990) DNA and morphology: inference of plant phylogeny. *Trends Ecol. Evol.* **5:** 104–110.

Taper, M. L. and Case, T. J. (1992) Coevolution among competitors. *Oxford Surv. Evol. Biol.* **8:** 63–109.

Tauber, C. A. and Tauber, M. J. (1989) Sympatric speciation in insects: perception and perspective. In: Otte, D. and Endler, J. A. (eds) *Speciation and Its Consequences*, pp. 307–344. Sinauer, Sunderland, Massachusetts.

Taylor, C. E. (1986) Genetics and evolution of resistance to insecticides. *Biol. J. Linn. Soc.* **27:** 103–112.

Templeton, A. R. (1981) Mechanisms of speciation—a population genetic approach. *Ann. Rev. Ecol. System.* **12:** 23–48.

Templeton, A. R. (1993) The "Eve" hypothesis: a genetic critique and reanalysis. *Am. Anthro.* **95:** 51–72.

Thaler, D. S. (1994) The evolution of genetic intelligence. *Science* **264:** 224–225.

Thoday, J. M. and Gibson, J. B. (1962) Isolation by disruptive selection. *Nature* **193:** 1164–1166.

Thompson, D'A. W. (1942) *On Growth and Form*. 2nd edn. Cambridge University Press, Cambridge, UK.

Thompson, J. D. and Lumaret, R. (1992) The evolutionary dynamics of polyploid plants: origins, establishment, and persistence. *Trends Ecol. Evol.* **7:** 302–307.

Thompson, J. N. (1989) Concepts of coevolution. *Trends Ecol. Evol.* **4:** 179–183.

Thompson, J. N. (1994) *The Coevolutionary Process.* University of Chicago Press, Chicago.

Thompson, J. N. and Burdon, J. J. (1982) Gene-for-gene coevolution between plants and parasites. *Nature* **360:** 121–125.

Timm, R. M. (1983) Fahrenholz's rule and resource tracking: a study of host-parasite coevolution. In: Nitecki, R. M. (ed.) *Coevolution*, pp. 225–265. University of Chicago Press, Chicago.

Toft, C. A., Aeschlimann, A., and Bolis, L. (eds) (1991) *Parasite-Host Interactions: Coexistence or Conflict?* Oxford University Press, Oxford, UK.

Travis, J. (1989) The role of optimizing selection in natural populations. *Ann. Rev. Ecol. System.* **20:** 279–296.

Trivers, R. L. (1985) *Social Evolution.* Benjamin/Cummings, Palo Alto, California.

Trivers, R. L. and Willard, D. E. (1973) Natural selection of parental ability to vary the sex ratio of offspring. *Science* **179:** 90–92.

Tucker, P. K. and Lundriga, B. L. (1993) Rapid evolution of the sex determining locus in Old World mice and rats. *Nature* **364:** 715–717.

Tudge, C. (1992) Last stand for Society snails. *New Scientist* **135** (11 July): 25–29.

Turelli, M. (1984) Heritable genetic variation via mutation-selection balance: Lerch's zeta meets the abdominal bristle. *Theor. Pop. Biol.* **25:** 138–193.

Turelli, M. (1985) Effects of pleiotropy on predictions concerning mutation selection balance for polygenic traits. *Genetics* **111:** 165–195.

Turelli, M. (1986) Gaussian versus non-Gaussian genetic analyses of polygenic mutation-selection balance. In: Karlin, S. and Nevo, E. (eds) *Evolutionary Processes and Theory*, pp. 607–628. Academic Press, New York.

Turelli, M. (1988) Population genetic models for polygenic variation and evolution. In: Weir, B. S. *et al.* (eds) *Proceedings of the Second International Conference on Quantitative Genetics*, pp. 601–618. Sinauer, Sunderland, Massachusetts.

Turelli, M. and Hoffmann, A. A. (1995) Cytoplasmic incompatibility factor in *Drosophila Simulans*: Dynamics and parameter estimates from natural populations. *Genetics*, 140: 1319–1338. California *Drosophila. Nature* **353:** 440–442.

Turelli, M. and Orr, H. A. (1995). The dominance theory of Haldane's rule. *Genetics*, **140:** 389–402.

Turner, J. R. G. (1967) Why does the genotype not congeal? *Evolution* **21:** 645–656.

Turner, J. R. G. (1976) Muellerian mimicry: classical "beanbag" evolution, and the role of ecological islands in race formation. In: Karlin, S. and Nevo, E. (eds) *Population Genetics and Ecology*, pp. 185–218. Academic Press, New York.

Turner, J. R. G. (1977) Butterfly mimicry: the genetical evolution of an adaptation. *Evol. Biol.* **11:** 163–206.

Turner, J. R. G. (1984) Mimicry: the palatability spectrum and its consequences. In: Vane-Wright, R. I. and Ackery, P. R. (eds) *The Biology of Butterflies*, pp. 141–161. Academic Press, London.

Turner, J. R. G. (1986) The evolution of mimicry: a solution to the problem of punctuated equilibrium. In: Raup, D. M. and Jablonski, D. (eds) *Patterns and Processes in the History of Life*, pp. 42–66. John Wiley, Chichester, UK.

Turner, J. R. G. (1987) Random genetic drift, R. A. Fisher, and the Oxford School of Ecological Genetics. In: Gigerenzer, G., Krüger, L., and Morgan, M. (eds) *The Probabilistic Revolution. Vol. 2. Ideas in the Sciences*, pp. 313–354. MIT Press, Cambridge, Massachusetts.

Ulizzi, L. and Manzotti, C. (1988) Birth weight and natural selection: an example of selection relaxation in man. *Human Heredity* **38:** 129–135.

Ulizzi, L. and Terrenato, L. (1992) Natural selection associated with birth weight. VI. Towards the end of the stabilizing component. *Annals of Human Genetics* **56:** 113–118.

Valentine, J. W. (1989) Phanerozoic marine faunas and the stability of the Earth system. *Palaeogeography, Palaeoclimatology, Palaepecology* **75:** 137–155.

Valentine, J. W., Collins, A. G., and Meyer, C. P. (1994) Morphological complexity increase in metazoans. *Paleobiology* **20:** 131–142.

Valentine, J. W. and Moores, E. M. (1972) Global tectonics and the fossil record. *J. Geology* **80:** 167–184.

Van Valen, L. M. (1971) Adaptive zones and the orders of mammals. *Evolution* **25:** 420–428.

Van Valen, L. M. (1973) A new evolutionary law. *Evolutionary Theory* **1:** 1–30.

Van Valen, L. M. (1976) Ecological species, multispecies, and oaks. *Taxon* **25:** 233–239.

Van Valen, L. M. (1985) How constant is extinction? *Evolutionary Theory* **7:** 93–106.

Van Valen, L. M. and Sloan, R. E. (1977) Ecology and the extinction of the dinosaurs. *Evolutionary Theory* **2:** 37–64.

Vermeij, G. J. (1987) *Evolution and Escalation.* Princeton University Press, Princeton, New Jersey.

Vermeij, G. J. (1991) When biotas meet: understanding biotic interchange. *Science* **253:** 1099–1104.

Vermeij, G. J. (1994) The evolutionary interaction among species: selection, escalation, and coevolution. *Ann. Rev. Ecol. System.* **25:** 219–236.

Vigilant, L., Stoneking, M., Harpending, H., Hawkes, K., and Wilson, A. C. (1991) African populations and the evolution of human mitochondrial DNA. *Science* **25:** 1503–1507.

Vogel, S. (1988) *Life's Devices.* Princeton University Press, Princeton, New Jersey.

Vrba, E. S. (1992) Mammals as a key to evolutionary theory. *J. Mammalogy* **73:** 1–28.

Vrba, E. S. (1993) Turnover-pulses, the Red Queen, and related topics. *Am. J. Sci.* **293–A:** 418–452.

Waage, J. K. (1979) Dual function of the damselfly penis: sperm removal and transfer. *Science* **203:** 916–918.

Waddington, C. H. (1957) *The Strategy of the Genes.* Allen & Unwin, London.

Wade, M. J. (1976) Group selection among laboratory populations of *Tribolium. Proc. Nat. Acad. Sci. USA* **73:** 4604–4607.

Wade, M. J. (1992) Sewall Wright: genetic interaction and the shifting balance theory. *Oxford Surv. Evol. Biol.* **8:** 35–62.

Wake, D. B. (1991) Homoplasy: the result of natural selection, or evidence of design limitation? *Am. Naturalist* **138:** 543–567.

Wake, D. B. and Roth, G. (eds) (1989) *Complex Organismal Functions: Integration and Evolution in Vertebrates.* Dahlem Workshop. John Wiley, Chichester, UK.

Wake, D. B., and Yanev, K. P. (1986) Geographic variation in allozymes in a "ring species," the plethodontid salamander *Ensatina eschscholtzii* of western North America. *Evolution* **40:** 702–715.

Wake, D. B., Yanev, K. P., and Brown, C. W. (1986) Intraspecific sympatry in allozymes in a "ring species," the plethodontid salamander *Ensatina eschscholtzii,* in southern California. *Evolution* **40:** 866–868.

Wake, D. B., Yanev, K. P., and Frelow, M. M. (1989) Sympatry and hybridization in a "ring species": the plethodontid salamander *Ensatina eschscholtzii.* In: Otte, D. and Endler, J. A. (eds) *Speciation and Its Consequences,* pp. 134–157. Sinauer, Sunderland, Massachusetts.

Wallace, B. (1981) *Basic Population Genetics.* Columbia University Press, New York.

Walsh, J. B. (1987) Sequence-dependent gene conversion: can duplicated genes diverge fast enough to escape conversion? *Genetics* **117:** 543–557.

Ward, P. D. (1990) The Cretaceous/Tertiary extinctions in the marine realm; a 1990 perspective. *Special Paper. Geological Society of America* **247:** 425–432.

Ward, P. (1995) *The End of Evolution.* Bantam Books, New York, and Weidenfeld and Nicolson, London.

Wasserman, M. and Koepfer, H. R. (1977) Character displacement for sexual isolation between *Drosophila mojavensis* and *Drosophila arizonensis. Evolution* **31:** 812–823.

Watts, W. B. (1991) Biochemistry, physiological ecology, and population genetics—the mechanistic tools of evolutionary biology. *Functional Ecol.* **5:** 145–154.

Weaver, R. F. and Hedrick, P. W. (1991) *Genetics.* 2nd edn. Wm. C. Brown, Dubuque, Iowa.

Weiner, J. (1994) *The Beak of the Finch.* Knopf, New York, and Cape, London.

Weir, B. S., Eisen, E. J., Goodman, M. M., and Namkoong, G. (eds) (1988) *Proceedings of the Second International Conference on Quantitative Genetics.* Sinauer, Sunderland, Massachusetts.

Weismann, A. (1891–1892) *Essays Upon Heredity and Kindred Biological Problems.* 2 vols. Oxford University Press, London.

Wellnhofer, P. (1990) Archaeopteryx. *Sci. Am.* **262** (May): 70–77.

Werman, S. D., Springer, M. S., and Britten, R. J. (1996) Nucleic acids I: DNA–DNA hybridization. In : Hillis, D. M., Moritz, C., and Mable, B. K. (eds), P. 169–203. Sinauer, Sunderland, Massachusetts

Werren, J. H. (1980) Sex ratio adaptations to local mate competition in a parasitic wasp. *Science* **208:** 1157–1159.

Westoll, T. S. (1949) On the evolution of the Dipnoi.

In: Jepsen, G. L., Mayr, E., and Simpson, G. G., (eds) *Genetics, Paleontology, and Evolution*, pp. 121–184. Princeton University Press, Princeton, New Jersey.

Wheeler, W. C., Cartwright, P., and Hayashi, C. Y. (1993) Arthropod phylogeny: a combined approach. *Cladistics* **9:** 1–39.

White, M. J. D. (1973) *Animal Cytology and Evolution*. 3rd edn. Cambridge University Press, Cambridge, UK.

White, M. J. D. (1978) *Modes of Speciation*. W. H. Freeman, San Francisco.

Whitfield, L. S., Lovell-Badge, R., and Goodfellow, P. N. (1993) Rapid sequence evolution of the mammalian sex determining gene SRY. *Nature* **364:** 713–715.

Whitham, T. G. and Slobodchikoff, C. N. (1981) Evolution by individuals, plant-herbivore interactions, and mosaics of genetic variability: the adaptive significance of somatic mutations in plants. *Oecologia* **49:** 287–292.

Wiley, E. O. (1988) Vicariance biogeography. *Ann. Rev. Ecol. System.* **19:** 513–542.

Wiley, E. O., Siegel-Causey, D., Brooks, D. R., and Funk, V. A. (1991). *The Compleat Cladist.* Museum of Natural History, University of Kansas, Lawrence, Kansas.

Wilkinson, G. S. (1993) Artificial sexual selection alters allometry in the stalk-eyed fly *Cyrtodiopsis dalmanni* (Diptera: Diopsidae). *Genetical Research* **62:** 213–222.

Williams, G. C. (1966) *Adaptation and Natural Selection.* Princeton University Press, Princeton, New Jersey.

Williams, G. C. (1975) *Sex and Evolution.* Princeton University Press, Princeton, New Jersey.

Williams, G. C. (1985) A defense of reductionism in evolutionary biology. *Oxford Surv. Evol. Biol.* **2:** 1–27.

Williams, G. C. (1992) *Natural Selection: Domains, Levels, and Challenges.* Oxford University Press, New York.

Williamson, M. (1981) *Island Populations.* Oxford University Press, Oxford, UK.

Williamson, P. G. (1981a) Paleontological documentation of speciation in Cainozoic molluscs from Turkana Basin. *Nature* **293:** 437–443.

Williamson, P. G. (1981b) Morphological stasis and developmental constraint: real problems for neo-Darwinism. *Nature* **294:** 214–215.

Wilson, A. C., Carlson, S. S., and White, T. J. (1977) Biochemical evolution. *Ann. Rev. Biochem.* **46:** 573–639.

Wilson, D. S. and Sober, E. (1994) Reintroducing group selection to the human behavioral sciences. *Behavioral and Brain Sciences* **17:** 585–654.

Wilson, E. O. (1992) *The Diversity of Life.* Harvard University Press, Cambridge, Massachusetts.

Woolfenden, G. E. and Fitzpatrick, J. W. (1984) *The Florida Scrub Jay.* Princeton University Press, Princeton, New Jersey.

Wright, S. (1931) Evolution in Mendelian populations. *Genetics* **16:** 97–159.

Wright, S. (1932) The roles of mutation, inbreeding, crossbreeding, and selection in evolution. *Proc. VI International Congress of Genetics* **1:** 356–366.

Wright, S. (1968–1978) *Evolution and Genetics of Populations.* 4 vols. University of Chicago Press, Chicago.

Wright, S. (1986) *Evolution: Selected Papers.* (Provine, W. B., ed.). University of Chicago Press, Chicago.

Wynne-Edwards, V. C. (1962) *Animal Dispersion in Relation to Social Behaviour.* Oliver & Boyd, Edinburgh, UK.

Zahavi, A. (1975) Mate selection—a selection for a handicap. *J. Theor. Biol.* **53:** 205–214.

Answers to Study and Review Questions

For each end-of-chapter question, we have included either an answer (for calculations, problems, and short answers), a reference to the relevant sections of the chapter (for longer answers and definitions or explanations of technical terms), or a reference to further reading (for topics not explicitly discussed in the text).

Chapter 1

1. see section 1.1
2. adaptation
3. The popular concept had evolution as progressive, with species ascending a one-dimensional line from lower forms to higher. Evolution in Darwin's theory is tree-like and branching, and no species is any "higher" than any other—forms are adapted only to the environments in which they live.
4. Darwin's theory of natural selection and Mendel's theory of heredity

Chapter 2

1. The terms are explained in the chapter.
2. (i) 100% *AA;* (ii) 1 *AA* : 1 *Aa;* (iii) 1 *AA* : 2 *Aa* : 1 *aa;* (iv) 100% *AB*/*AB;* (v) 100% *AB*/*AB;* (vi) $\frac{1}{16}$ *AB*/*AB,* $\frac{1}{8}$ *AB*/*Ab,* $\frac{1}{8}$ *AB*/*aB,* $\frac{1}{8}$ *AB*/*ab,* $\frac{1}{16}$ *Ab*/*Ab,* $\frac{1}{8}$ *Ab*/*aB,* $\frac{1}{8}$ *Ab*/*ab,* $\frac{1}{16}$ *aB*/*aB,* $\frac{1}{8}$ *aB*/*ab,* $\frac{1}{16}$ *ab*/*ab;* (vii) $\frac{1}{4}\, r(1 - r)$ *AB*/*AB,* $\frac{1}{4}\, (r^2 + (1 - r)^2)$ *AB*/*Ab,* $\frac{1}{4}\, (r^2 + (1 - r)^2)$ *AB*/*aB,* $\frac{1}{2}\, r(1 - r)$ *AB*/*ab,* $\frac{1}{4}\, r(1 - r)$ *Ab*/*Ab,* $\frac{1}{2}\, r(1 - r)$ *Ab*/*aB,* $\frac{1}{4}\, (r^2 + (1 - r)^2)$ *Ab*/*ab,* $\frac{1}{4}\, r(1 - r)$ *aB*/*aB,* $\frac{1}{4}\, (r^2 + (1 - r)^2)$ *aB*/*ab,* $\frac{1}{4}\, r(1 - r)$ *ab*/*ab*

Chapter 3

1. Approximately the species level; the pigeon example might stretch it to genera: but higher categories do not evolve in human lifetime.
2. (a) If you look at any one time and place, living things usually fall into distinct recognizable groups that could be called "kinds." (b) If you look over a range of space (if the "kinds" in question are species) the kinds break down; if you look through a range of times (if the kinds are species or any higher category) the kinds break down. The differences between higher categories can also be broken down by studying the full range of diversity on Earth. You might think that plants and animals are less clear categories after studying the range of unicellular organisms.
3. (a), (b), (d) homologies; (c) analogy
4. An accident in the sense that other codes with the same four letters could work equally well. Frozen in the sense that changes in it are selected against. See section 3.8.
5. This question is intended as a discussion topic; for the idea see section 3.9.
6. Because a form would have existed in time before the series of fossils (of vertebrates from fish to mammals) that can be strongly argued to be its ancestors.

Chapter 4

1.

Age interval (days)	Number (as density/m^2) surviving to day x	Proportion of original cohort surviving to day x	Proportion of original cohort dying during interval
0–250	100	100	0.9989
251–500	0.11	0.11	0.273
501–750	0.08	0.08	0.75
751–1000	0.06	0.06	—

2. See section 4.2. (b) In technical terms, drift (see chapter 6). The gene frequencies would change between generations because heritability operates (condition 2) and some individuals produce more offspring than others (condition 3). If the differences in reproduction are not systematically associated with some character or other, the changes in gene frequency between generations will be random or directionless. (c) No evolution at all. If the character conferring higher than average fitness is not inherited by the individual's offspring, natural selection cannot increase its frequency in the population.
3. The requirements of inheritance and association between high reproductive success and some character have also to be met.
4. (a) directional selection (for smaller brains); (b) stabilizing selection; (c) no selection
5. The mechanism has to (i) perceive the change in environment, (ii) work out what adaptation is appropriate to the new environment, and (iii) alter the genes in the germ line to code for the new adaptation. (i) is possible, (ii) could vary from possible in a case such as simple camouflage to impossible in a case requiring a new complex adaptation such as the adaptations for living on land of the first terrestrial tetrapods, and (iii) would contradict what is known about genetics and it is difficult to see how it could be accomplished. The mechanism would have to work backward from the new phenotype (something like a long neck in a giraffe) to deduce the needed genetic changes, even though the phenotype was produced by multiple interacting genetic and environmental effects.
6. (a) Here are two arguments. (i) If every pair produced two offspring, natural selection would favor new genetic variants that produced three, or more, offspring. After the more fecund form had spread through the population,

the average would still be two but the greater competition among individuals to survive would lead to variation in the success of the broods of different parents. (ii) Random accidents alone will guarantee that some individuals fail to breed. For the average to be two offspring, as it must be for any population that is reasonably stable in the long term, all successfully reproducing individuals will produce more than two offspring. (b) Ecologists discuss this in terms of r and K selection, or life history theory. Thus, in some environments little competition takes place and selection favors producing large numbers of small offspring; in other environments massive competition takes place and selection favors producing fewer offspring and investing a large amount in each. Many other factors can also operate.

Chapter 5

1. (a) and (e) are in Hardy–Weinberg equilibrium, (b), (c), and (d) are not. As to why they are not, in (d) it looks like *AA* is lethal, (b) may include heterozygote advantage, and (c) may include heterozygote disadvantage. (b) could also be produced by disassortative mating, and (c) by assortative mating. In (c) there could also be a Wahlund effect. All deviations are so large that random sampling is unlikely to be the whole explanation.
2. (a) $p/(1 - sq^2)$; (b) $1 - sq^2$
3. $(1/3)(3 - s) = 1 - (s/3)$
4. If you do it in your head, $s \approx 0.1$. To be exact, $s = 0.095181429619$.
5. The fitness differences are in survivorship (not fertility), and in particular in survivorship during the life stage investigated in the mark-recapture experiment. (Note that mark-recapture experiments are also used by ecologists to estimate absolute survival rates. They require the additional assumptions that the animals do not become "trap shy" or "trap happy," and that the mark and release treatment does not reduce survival; these assumptions are not needed when estimating relative survival. We do need the second-order assumption that these factors are the same for all genotypes.)
6. *AA* 1/2 ; *aa* 1/2. The gene frequencies are 0.5 and the Hardy–Weinberg ratio 1/4 : 1/2 : 1/4. The observed/expected ratios are 2/3 : 4/3 : 2/3, which when scaled to a maximum of 1 give the fitnesses 1/2 : 1 : 1/2.
7. (a) 0.5. (b) 0.5625. You need equation 5.13. $t = 1$. For (a) the equation is $0.625 = 0.5 + (0.75 - 0.5)(1 - m)$. For (b) the equation is $? = 0.5 + (0.625 - 0.5)(1 - 0.5)$.
8. (a) The *aa* genotype is likely to be fixed. If *aa* mate only among themselves and *AA* and *Aa* mate only with *AA* and *Aa*, whenever an *Aa* × *Aa* mating occurs, some *aa* progeny are produced, which will subsequently mate only with other *aa* individuals. (b) *AA* mate only with *AA*, *Aa* with *Aa*, and *aa* with *aa*. The homozygous matings preserve their genotypes, but when *Aa* mate together they produce 1/4 *aa* and 1/4 *AA* progeny. In the extreme case, the population diverges into two species (one *AA* and one *aa*) and the heterozygotes are lost. (c) (i) the dominant allele will be fixed, (ii) the recessive allele will be fixed.
9. $p' = \dfrac{p(1 - s)}{1 - p^2s - 2pqs}$.
   The denominator can be variously rearranged.
10. $p^* \approx \sqrt{\dfrac{m}{s}}$
    The derivation starts with the equilibrium condition, $p^2s = qm$. We then note that $q \approx 1$, and $p^2s \approx m$; divide both sides by s and take square roots.

Chapter 6

1. Either 100% *A* or 100% *a* (there is an equal chance of each).
2. See (a) section 6.1, and (b) section 6.4.
3. (a) 0.5; (b) 0.5; (c) 0.625; (d) 0. See section 6.6.
4. (a) and (b) 10^{-8}. Population size cancels out in the formula for the rate of neutral evolution.
5. (a) $1/(2N)$; (b) $(1 - (1/(2N))$.
6. Both manipulations requires substituting $1 - H$ for f and then some canceling and multiplying though by -1 to make the sign positive.

Chapter 7

1. See Figure 7.1a,b.
2. The main observations suggesting neutral molecular evolution are not also seen in morphology. This concept was discussed for the constancy of evolutionary rates in section 7.9. The other original observations (for absolute rates and heterozygosities, and for the relation between rate and constraint) either have not been made, or such observations do not suggest the problems found in selective explanations for molecular evolution also apply to morphology.
3. $L = \dfrac{1 - 0.92}{1} = 0.08$
4. For: the implied cost of selection exceeds Haldane's upper bound. Against: the cost of selection is relaxed if selection is soft, and evolution at different loci is not independent.
5. The key variable in the neutral explanation is the chance that a mutation is neutral; it is arguably higher for regions with less functional constraint (section 7.11.2). The key variable in the selective explanation is the chance that a mutation has small rather than large effect, and so may cause a fine-tuning improvement (section 7.11.2 and Box 7.3).
6. No: the main evidence is from codon usage biases (section 7.11.4).

7. Under neutral evolution, both ratios are simply the ratio of the neutral mutation rates. The ratio of heterozygosity (or evolutionary rate) for silent sites : heterozygosity (or evolutionary rate) for replacement sites equals the ratio of neutral silent mutation rate : neutral replacement mutation rate.

Chapter 8

1.

	Frequency of:			Value of:
	A_1B_1	A_1	B_1	D
(a)	7/16	1/2	1/2	+3/16
(b)	1/4	1/2	1/2	0
(c)	1/9	1/3	1/3	0
(d)	11/162	1/3	1/3	−7/162

Note: The haplotype frequency is found by the sum of homozygotes plus 1/2 the heterozygotes (as for a gene frequency; see section 5.1). If you have figures in the A_1B_1 frequency column like 11/16 or 14/16 for (a) you may have not divided the frequency of A_1B_1/A_1B_2 by 2.

2. (a) and (d) may show fitness epistasis, in which in (a) A_1 has higher fitness in combination with B_1 than with B_2, and vice versa in (b). Fitnesses are independent (maybe multiplicative or additive) in (b) and (c).
3. (a) and (b) should equilibrate at haplotype frequencies of 1/4, 1/4, 1/4, and 1/4 for the four haplotypes; (c) and (d) should equilibrate at 1/9, 2/9, 2/9, and 4/9. (b) and (c) are already at equilibrium and should not change through time; (a) and (d) will evolve toward the equilibrium frequencies at a rate determined by the recombination rate between the two loci.
4. Observed heterozygosities are arguably a little on the low side for the neutral theory (section 7.6); the effect of section at one locus is on average to reduce heterozygosities at linked loci, producing a net reduction in average heterozygosity through the genome.
5. See Figure 8.8(b). The equilibrium is at the top of the hill.

Chapter 9

1. See section 9.2, and particularly Figures 9.3 and 9.4. In statistical theory, the argument is formalized as the central limit theorem.
2. +1. The answer is incomplete without the sign.
3. (a) $V_P = 300/8 = 37.5$, $V_A = 48/8 = 6$, $h^2 = 6/37.5 = 6.25$; (b) +3: you add the additive effects inherited from each parent.
4. 106
5. Not explicitly discussed in the chapter, but see section 9.13 (p. 252). You might predict it will evolve toward a canalizing type relation, as in Figure 9.11b, because then an individual is most likely to have the optimal phenotype.
6. It will go through an intermediate phase with many recombinant genotypes produced by crossing-over between the three initial chromosomes; it should end up with only the chromosomal type that yields the optimal character by means of a homozygote: all + + + ———.
7. Superficially they do not add up. If the second chromosome has 3000 loci each with a mutation rate of , the mutation rate of the whole chromosome will be 0.003. They are out by a factor of 50. Their reconciliation has not been discussed in the text. Possible answers include the following. (a) The classic mutation rate figure is too low, because the figure derives from gross visible mutations and these instances may be a minority—there will be others as well. Kondrashov (1988) makes this argument clearly. (b) There are more than 10,000 loci in a fruitfly. It is not known for certain how many gene loci are present in a fruitfly, or a human; 10,000 is a guess, and aims to be a guess on the low side. The figure could be 10 times larger. (c) Logically, the Mukai-type figure could be an overestimate, but this is unlikely; it is more likely to be an underestimate. The 0.15 figure was the lowest that as compatible with Mukai's observations, and the real figure may be 2× higher. (d) Other possibilities also exist (though I hope you were not one of those readers who suggested authorial blunder).

Chapter 10

1. (b) only
2. They should be about the same. (Whereas in fact they often are not.)
3. (a) See Figures 10.2 and 10.6. (b) natural selection or drift
4. I. Tandem repeats I.i Microsatellites, I.ii Minisatellites, I.iii Satellite DNA. II. Scattered repeats. Genetic mechanisms: I.i slippage, I.ii slippage and unequal crossing-over, I.iii slippage (for the short units), unequal crossing-over (and maybe rolling circle replication), II. Transposition.
5. (a) a positive relation; (b) None of them explains it, at least in a straightforward manner. The average mutation rate may well vary among sequences, but there is no reason why it should tend to be higher in the more heterozygous ones—it will just be more variable.

Chapter 11

1. (a) 33%; (b) 67%
2. (a) Crudely, it has to be high; exactly, a total deleterious mutation rate of more than one per organism per generation is needed. On its realism, see the end of section 11.3.5; the evidence is inconclusive and neither rules it out or in. (b) Relation (i). Relation (ii) corresponds to independent fitness effects, in which sex (before the 50%

cost) is indifferent; (iii) is the diminishing returns type of epistasis, in which sex is positively daft, even before the 50% cost. Again, you can argue reality either way; see the end of section 11.3.5.

3. The material in the text (section 11.3.6) would suggest looking at the relation between the frequency of sex and parasitism in taxa that can reproduce both ways; alternatively, you might look into the genetics of host-parasite relations and measure the frequency of resistance genes in hosts or penetration genes in parasites. Other answers would be possible as well, going beyond the textual materials.
4. See section 11.4.4. If the character were cheap to produce, males of all genetic qualities would evolve to produce it.
5. See section 11.4.3. A female who did not choose extreme males would on average mate with a less extreme male than would other females in the population; she would produce less extreme than average sons; and they would grow up into a population in which most females prefer extreme males. Thus, these sons would have low reproductive success and their mother's lack of preference would be selected against.
6. See section 11.5. If more daughters than sons were produced by most members of the population, the fitness of a male would be higher than that of a female. Individuals who produced more sons than daughters would be favored by selection. A sex ratio of 1:1 is a stable point at which no advantage is gained by producing more offspring of either sex.
7. (a) positively, and (b) negatively frequency-dependent selection

Chapter 12

1. Many answers are possible, but the main examples in this chapter were (a) adaptations for finding food; (b) eating as much food as possible to maximize reproductive rate, cannibalism, destructive fighting, or producing a 50:50 sex ratio in a polygynous species; (c) restraining reproduction, to preserve the local food supply; (d) segregation distortion in which the total fertility of the organism is reduced.
2. You might explain it in terms of two factors, the relative rates of extinction of altruistic and selfish groups and the rate of migration. Alternatively, you might reduce them to the one variable m, which is the average number of successful emigrants produced by a selfish group during the time the group exists (before it goes extinct). The fate of the model is then determined by whether m is greater than or less than 1 (section 12.2.5).
3. B can be estimated as the number of extra offspring produced by the nests with helpers: $2.2 - 1.24 \approx 1$. That benefit is produced by 1.7 helpers, giving $b \approx 1/1.7 \approx 0.6$. c can be estimated either as zero (if the helper has no other option) or as the number of offspring produced by an unhelped pair (if it could breed alone), in which case $c = 1.24$. With $r = 1/2$, the bird should help if it cannot breed alone but should breed if it can. On the difficulties in estimating b and c this way, see Grafen (1984).
4. Kin selection applies to a family group, or more generally a group of kin (indeed it is not theoretically necessary that the kin live in groups, though they do have to be able to influence one another's fitness); group selection, at least in the pure sense, applies to groups of unrelated individuals. Kin selection is a plausible process, because the conditions for an individual to produce more copies of a gene in itself by helping relatives than by breeding more itself will often apply. Group selection requires more awkward conditions (see question 2!).
5. (a) the whole genome; (b) the chromosome.

Chapter 13

1. They cannot explain adaptation. There is no reason except chance why a new genetic variant should arise in the direction of improved adaptation, and random chance change will not produce adaptation. If (as in the "Lamarckian" theory) the new genetic variants arise in the direction of adaptation, it implies the presence of some adaptive mechanism behind the production of new variants. Natural selection is the only known theory that could explain such a mechanism.
2. See Figures 13.2 and 13.3.
3. It depends on the definition. If adaptation is defined including the criterion of historical constancy of function, then the answer is no—the character is an exaptation in Gould and Vrba's term. If historical constancy of function is irrelevant in the definition of adaptation, then the answer is yes.
4. (a) (1) Natural selection, in the form of negative selection. The absent regions represent maladaptive forms that, when they arise as mutations, are selected out. (2) Developmental constraint. Something about the way the organisms develop embryonically makes it impossible, or at least difficult, for these forms to arise. (b) Four kinds of evidence were mentioned in section 13.9. The kind discussed in the greatest depth was the use of artificial selection. If the character can be altered, its form is unlikely to be due to constraint.
5. (a) yes, (b) no. When different levels of selection conflict, adaptation cannot be perfect at all levels.

Chapter 14

1. Evolutionary: paraphyletic, monophyletic.
   Cladistic: monophyletic
   Phenetic: polyphyletic, paraphyletic, monophyletic.
   See Table 14.1.

2. (a) (cow (lungfish, salmon));
   (b) (cow (lungfish, salmon));
   (c) ((cow, lungfish), salmon).
   See end of section 14.6.2.
3. The Euclidean distances are obtained by Pythagoras's theorem, and the numbers were picked to give a 3, 4, 5 triangle: the three species can be drawn on a graph with one character per axis. MCD is the average of the distances for the two characters. See section 14.5.

	Species 1	Species 2	Species 3
Species 1		3	4
Species 2	1.5		5
Species 3	2	3.5	

   Here, Euclidian distances are given above the diagonal and MCD below.

   The two distance measures imply contradictory groupings of the three species. In this case, the classification chosen by the numerical phenetic method would be ambiguous, and therefore arguably subjective.
4. With a nearest neighbor cluster statistic the grouping is (1, 2)(3(4, 5)); with a nearest average neighbor cluster statistic it is ((1, 2)3)(4, 5). See section 14.5 for the moral.
5. For the species in Figure 14.9:

1.1	*ornata*
1.2.1.1	*hamifera*
1.2.1.2	*paenehamifera*
1.2.1.3	*truncipennis*
1.2.1.4	*varipennis*
1.2.2.1.1	*neogrimshawi*
1.2.2.1.2	*clavisetae*
1.2.2.2.1.1	*touchardiae*
1.2.2.2.1.2	*cilifera*
1.2.2.2.1.3	*adiostola*
1.2.2.2.1.4	*peniculipedis*
1.2.2.2.1.5	*spectabilis*
1.2.2.2.2.1	*setosimentum*
1.2.2.2.2.2	*ochrobasis*

6. A critic would reply that the same problem would resurface in another form. Perhaps the average and nearest neighbor statistics in the case of Figure 14.5 could be made to agree by adding five further characters. These two are just two possibilities out of many cluster statistics, however, and the result would almost certainly still be ambiguous with respect to some other cluster statistic. The ambiguity could be removed only if nature contains one non-ambiguous phenetic hierarchy; there is no reason to suppose such a hierarchy exists.
7. (a) The difference reflects evolutionary theory's scientific peculiarity as a historical theory. The hierarchy in a phylogenetic classification is historical, and is used for the reasons discussed in the chapter. The periodic table is non-historical and non-hierarchical. Its structure represents two of the fundamental properties that determine the nature of an element, and the position of an element in the table can be used to predict what the element will be like. The position of an organism in a phylogenetic classification cannot be used to predict much about what the organism will be like. Much more could be said about the nature of different theories in science, and the way different theories imply different kinds of classifications. (b) See section 14.8.

Chapter 15

1. See sections 15.2 and 15.3.
2. (a) 1; (b) 2; (iii) 1
3. They are in section 15.2.3. See particularly Table 15.1.
4. The main types are mentioned in section 15.2.3; they involve coadaptation between either the two sex chromosomes or the sex chromosomes and autosomes. However, no existing hypothesis fits all the facts; look at Coyne (1992).
5. It can be argued both ways; see the end of section 15.2.6. If asexual species are discrete in the same way that sexual species seem to be, it suggests that the force maintaining species as discrete clusters is ecological rather than interbreeding.
6. (a) 3; (b) 3; (c) 1
7. (i) (a) and (b) (probably) yes; (ii) and (iii) (a) yes, (b) no; (iv) (a) and (b) yes

Chapter 16

1. Adaptation to a range of climates is one reason, discussed in section 16.3. Adaptation to other environmental factors could also be the reason, or genetic drift.
2. Populations that are isolated at the periphery of a species range are often particularly distinct in form. See section 16.4.3.
3. (a) Secondary contact: the species (or near-species) diverged in allopatry, and changes in their ranges have brought them into contact. Based on whether reinforcement operates, the two forms that meet at a hybrid zone could evolve into two full species, or merge into one (or one could drive the other extinct. (b) Divergence in space, along an environmental gradient.
4. Two main ones: the intensity of selection against hybrids (heterozygotes in a simple model) and the dispersal distance of members of the species. Look at Figure 16.10 and associated text.
5. Crude evidence is that the species is morphologically intermediate between the hypothetical parental species. Finer evidence uses progressively more exact genetic markers. If the hypothetical hybrid has twice as many chromosomes as the hypothetical parents, then the form of the chromosomes can be examined to see whether the hybrid has a mix of the parental forms. Other genetic

markers that have been used are gel electrophoretic (to show that the hybrid has a combination of the protein forms in the parents) and DNA sequences identified by restriction enzymes (to show that the hybrid has the same DNA sequence at a certain site, of both its parents). See section 16.7, and also section 3.6.

6. See section 16.4.2. The main snags are extinction of the minority form and fusion of the two forms by interbreeding and gene flow before they have had time to evolve full isolation.
7. If the I of Table 16.1 and Figure 16.2 is used, I = (a) 1, (b) 0.5, (c) 0; if the I of Box 16.1 is used, I = (a) 1, (b) 2/3, (c) 0. If you invent some other index, it ought to vary between 1 (or 100%) for (a) and zero for (c), with (b) somewhere in between.
8. Reproductive character displacement, or character displacement for prezygotic isolation. Two main explanations: (1) *Reinforcement.* Females in allopatry have not been selected to discriminate against heterospecific males, because in evolutionary history the ancestors of modern females have never met those males; females in sympatry are descended from females that have been exposed to both kinds of male. Females who mated heterospecific males produced hybrid offspring of low fitness, so selection favored discrimination. (2) *Without reinforcement.* There are various versions of the alternative explanation; the one most explained in the text (section 16.8) is as follows. Different individuals of the two species in the past may have shown various degrees of isolation from the other species. In areas where they now coexist in sympatry, if reproductive isolation was low, the two species would fuse and probably now look more like one of the species (and so be classified as a member of it). If reproductive isolation was high in those areas, the two would coexist and remain distinct. Thus, only where isolation was high do we now see the two species in sympatry. In areas where the species are now allopatric, they continue to exist, whatever the reproductive isolation. Thus, the average isolation will be lower than for sympatry.
9. The answer is not known for sure. Two favored hypotheses are (i) the more rapid rate of evolution of recessive genes on sex chromosomes than autosomes (see section 16.9), and (ii) the greater susceptibility of sex chromosomes than autosomes to the evolution of segregation distorting "drivers." If different speciating populations evolve different drivers, each population may have genes at other loci to neutralize their own drivers, but not the drivers of the other population (because natural selection will have favored neutralizing genes within each population, but not the neutralizing genes needed against a driver that does not exist in that population). See section 15.2.3.

Chapter 17

1.

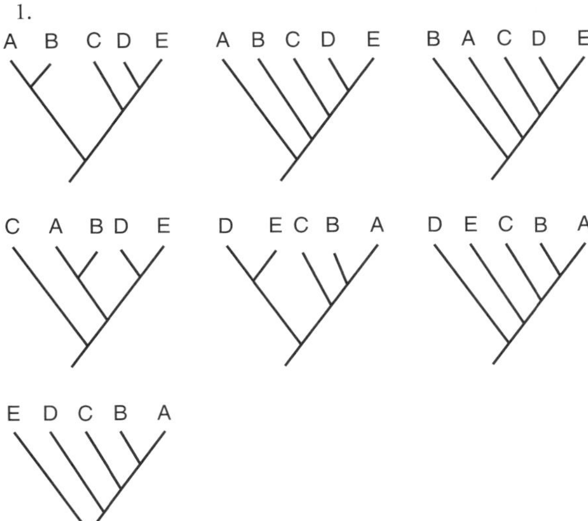

2.

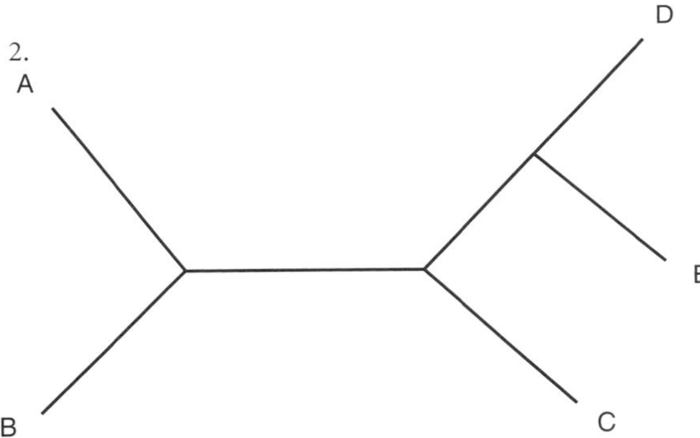

3. ((A, B)(C, D)): 160 events. ((A, C)(B, D)): 175 events. ((A, D)(B, C)): 175 events.
4. Evolutionary rates approximately equal in all lineages.
5. Read section 17.10, p. 481.
6. A is ancestral and A' derived in all three, but the inference is most certain in (i) and least certain in (iii). (If A is ancestral in the group of species 1 + 2, then the minimum number of events in (i), (ii), and (iii), respectively, are 2, 3, and 3, whereas if A' were ancestral, the minimums would be 3, 4, and 4.)
7. Read section 17.10.
8. globin—vertebrate classes; micro and minisatellites—populations within species; histones—kingdoms. (See Table 7.1, p. 156, and Figure 17.14.)
9. Read Box 17.1 and section 17.6.2 for one particular possibility. You should be able to work out a general principle behind it, however. Try reading Bull *et al.* (1993).
10. 2 ↔ 3 ↔ 1

Chapter 18

1. See section 18.1 and Figure 18.2.
2. 1/2, 1/2, 1/2, 1
3. My first three counts gave 26 to 13, 27 to 12, and 26 to 12 from older to younger, and from younger to older respectively! The correct answer is something close to those numbers, but you have the idea if your figures are in this region. Compare Figure 18.11. No reason exists as to why there should be so many more from older to younger if speciation was by splitting a larger range, whereas it makes sense if it occurred by dispersal because the older island would have been occupied first.
4.

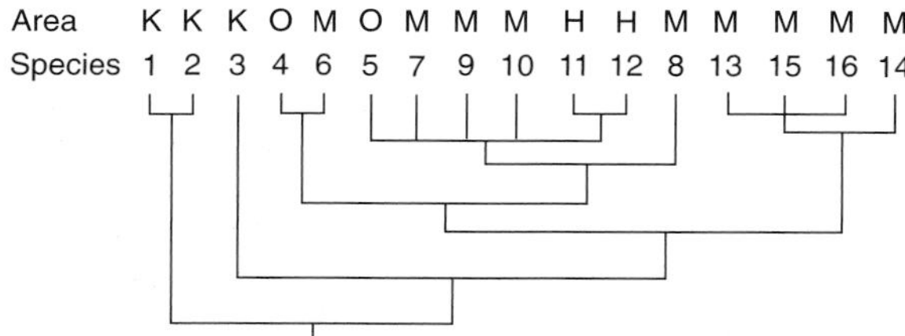

Some minor variants would also be possible, depending on how ancestral species like 8 and 14 are represented.

5. (a)–(b), (a)–(c), and (e)–(f) are congruent. (a)–(d), (a)–(f), (c)–(d), (c)–(f), (e)–(b), and (e)–(d) are incongruent.
6. (a) Competitive superiority of North American mammals, perhaps due to a history of more intense competition, and reflected in their relative encephalization. (b) Environmental change, such that the North American mammals were competitively superior in South American environments after the Interchange.

Chapter 19

1. See section 19.1 and Figure 19.1.
2. With the rounded figure of 1.2×10^{-4} for the decay constant:

14C : 14N	Age	
(a)	1 : 1	5776
(b)	2 : 1	3379
(c)	1 : 2	9155

3.

Time unit (yr)	(a) Completeness (%)	(b) Completeness (%)
100,000	100	100
10,000	100	100
1000	10	2
100	10	2

Chapter 20

1. Character values (in size units) and times (in Myr B.P.)

x_1	t_1	x_2	t_2	Rate
2	11	4	1	0.0693
2	11	20	1	0.2303
20	11	40	1	0.0693
20	6	40	1	0.1386

If you have answers like 0.2, 1.8, 2, and 4, you forgot to take logs. If you have negative numbers, you have reversed x_1 and x_2, or t_1 and t_2.

2. (a) An inverse relation; see Figure 20.2. (b) One possibility is that long periods with rapid change and short periods with slow change have been excluded from study, perhaps because the former would transform the character beyond commensurability and the latter seemed unworthy of notice.
3. See section 20.4, which contains two lists of four points each.
4. (a) Allopatric speciation, particularly by the peripheral isolate model. Plausible in theory and supported at least in the sense that much evidence is consistent with it (see section 16.4.4, p. 434). (b) Saltational macromutations. Possible but implausible in theory (for reasons explained in Box 7.3, p. 182) and unsupported by evidence. (a) and (b) are not mutually exclusive, or mutually dependent.
5. See Table 20.2 and section 20.8. Either (i) the modern synthesis favored a phyletic gradualist interpretation of speciation, and is thus damaged by any evidence of punctuated equilibrium, or (ii) it pluralistically acknowledged many patterns of rates of evolution, and punctuated equilibrium is consistent with it (though it did not predict that all speciation occurred by punctuated equilibrium).
6. The text contains two types of evidence: (i) rates of change in arbitrarily coded characters (see Figure 20.10) and (ii) taxonomic rates, in which the longevity of living fossil genera is longer than average (Table 20.3).

Chapter 21

1. (a) Jaw shape and articulation; degree of differentiation of teeth down the jaw; upright or sprawling gait; secondary palate. (b) Lactation and viviparity (some reptiles are viviparous, but then the details of it differ from mammals); warm- or cold-blooded, or to be exact homeothermy or poikilothermy; other characters are known as well, such as fur or scales.
2. It evolved in many stages over a period of about 100 million years, with the mammalian characters appearing gradually; each step was probably adaptive and evolved by natural selection. The macroevolutionary change appears to have arisen by extrapolated microevolution.
3. (a) non-terminal addition; (b) heterochrony leading to paedomorphosis. (ai) spoils von Baer's law; (b) does not.

4. (a) paedomorphosis; (b) neoteny and progenesis (see Figure 20.4); (c) look at the time of reproductive maturity; (d) look at body size, on the assumption size is proportional to age in years.
5. (a) Macroevolution is by macromutational changes in regulatory genes, whereas microevolution is by changes in other genes (or by smaller changes in regulatory genes). (b) Macroevolution is by a paedomorphosis, enabling an "escape from specialization," whereas microevolution is by other kinds of genetic change, perhaps more often including terminal addition.
6. (a) Changes in early stages are likely to be disruptive, deleterious macromutations because of ramifying effects in later development. (b) Species with extensive parental investment in the early stages of development of the young may be less likely to evolve at least morphological changes in the young than species in which the young are cast into the environment to fend for themselves. (This idea is uncertain but is an example of the kind of argument that might be made; perhaps you can think of other such arguments.)

Chapter 22

1. (a) See Figures 22.2 and 22.3. (b) The factor mentioned in the text was whether the parasite disperses independently of the host, other possibilities exist, such as whether a 1:1 relation is found between parasite and host species.
2. In order of increasing virulence: (iv) < (i) < (ii) < (iii). (ii) and (iii) might be about the same. This question was not specifically discussed in the text. See Ewald (1993, 1994).
3. (a) See section 22.8. Antagonistic biological interactions, such as between predator and prey, have evolved to become more dangerous over time. Predators have become more dangerously armed, prey more powerfully defended. (b) The level of defensive adaptation in prey can be measured in such features as the thickness of molluscan shells and the habitats they occupy. Predatory adaptations have primarily been studied by the numbers of specialist as opposed to generalist predators; the presence of specialists suggests a more dangerous condition. It is important to test escalation by the proportion of species types through time, because more recent fossil records contain more of everything. See Figures 22.8–22.10.
4. (a) Nothing for sure; it is compatible with a constant rate in absolute time, but it can also be explained by non-constant rates. See Figure 22.12. (b) It is constant; at any moment, old and young taxa are equally likely to be struck down by extinction.
5. (a) It has the standard form of a genetic load. $L = \frac{w_{opt} - \bar{w}}{w_{opt}}$, where the deviation of $\bar{w}$ from w_{opt} is due to the lag of the population behind environmental change. (b) Contractionary (L decreasing), expansionary (L increasing), or stable (L constant). If it is stable, it can be either because evolution in all species has come to a stop or because all species are improving relative to one another at exactly the same rate (Red Queen evolutionary mode). (c) Van Valen suggested that the total ecological resources may be constant through time, and the selective pressure on a species is proportional to the loss of resources it suffers due to lagging behind competing species.

Chapter 23

1. A lack of knowledge of the global distribution of species and (for large-bodied animals, whose geographic distributions were best known) the difficulty of assigning disarticulated fossil bone fragments to species (section 23.1).
2. (a) In a real extinction, all members of a lineage die without leaving descendants; in a pseudoextinction, the lineage continues to reproduce but its taxonomic name changes in mid-lineage (section 23.2). (b) It snarls up tests of both. If the extinctions of species with differing developmental modes in Figure 23.3, or in the test of synchroneity in Figure 23.5, were pseudoextinctions, the explanation for the trend, or synchronous pattern, would simply relate to the habits of taxonomists. (The test of the Red Queen hypothesis by survivorship curves in chapter 22 may be less damaged, however: the hypothesis might be recast in terms of rate of change rather than chance of extinction.)
3. Species selection and differential persistency of niches (section 23.5)
4. (a) the iridium anomaly (Figure 23.6); (b) They should be sudden and synchronous in all taxa, rather than gradual; they should not in general be preceded by reductions in population size. For the difficulties of testing these predictions, look at section 23.6.2.
5. At the end of the Cretaceous, Triassic, Permian, Devonian, and Ordovician.
6. At mass extinctions, no relation. In background extinctions, taxa with planktonic development have one-third [or half?] the extinction rate of taxa with direct development. The background extinction difference can be explained either by bias in the fossil record (if planktonically developing species are more likely to be preserved) or by their being more likely to survive local difficulties by their dispersing larval stage.
7. Plot out a frequency distribution of extinction rates. If they are randomly distributed, the rates will have a continuous (and Poisson, to be precise) distribution; if mass extinctions are a distinct macroevolutionary regime, they should make the distribution bimodal. In fact, the graph looks continuous (see Figure 23.9).

INDEX